Oxford Handbook of Nucleic Acid Structure

Oxford Handbook of Nucleic Acid Structure

Edited by

Stephen Neidle

The CRC Biomolecular Structure Unit,
The Institute of Cancer Research, Sutton, Surrey, UK

OXFORD
UNIVERSITY PRESS

OXFORD

UNIVERSITY PRESS

Great Clarendon Street, Oxford OX2 6DP

Oxford University Press is a department of the University of Oxford
and furthers the University's aim of excellence in research, scholarship,
and education by publishing worldwide in

Oxford New York

Athens Auckland Bangkok Bogotá Buenos Aires Calcutta
Cape Town Chennai Dar es Salaam Delhi Florence Hong Kong Istanbul
Karachi Kuala Lumpur Madrid Melbourne Mexico City Mumbai
Nairobi Paris São Paulo Singapore Taipei Tokyo Toronto Warsaw

and associated companies in Berlin Ibadan

Oxford is a registered trade mark of Oxford University Press

Published in the United States
by Oxford University Press Inc., New York

A catalogue record for this book is available from the British Library

Library of Congress Cataloging in Publication Data
Oxford handbook of nucleic acid structure / edited by S. Neidle.
Includes bibliographical references and index.
1. Nucleic acids–Structure. 2. X-ray crystallography.
3. Nuclear magnetic resonance spectroscopy. I. Neidle, Stephen.
QD433.5.S77094 1998 547′.790442–dc21 98–34431

ISBN 0 19 850038 6 (Hbk)

Typeset by EXPO Holdings, Malaysia

Printed in Great Britain by
Bookcraft (Bath) Ltd,
Midsomer Norton, Avon

Preface

The study of nucleic acid structure is now some 45 years old. It has grown into a vast multifaceted field, which continues to play a key role in furthering our understanding of gene regulation and expression, and of ways for intervening with these processes. It has become a fertile meeting-ground for crystallographers, NMR spectroscopists, and theoreticians, and now even has its own database for structure deposition and study.

It is a truism that nucleic acids are conformationally more complex than proteins. This complexity, which to some extent has been masked by the simplicity of the classic DNA double helix, is shown by the readiness of both oligonucleotides and polynucleotides to be structurally responsive to changes in local environment. Such conformational plurality may be caused by water molecules and counterions, or by ligand (drug, protein) binding, and can be highly sequence dependent: as shown by the ability of particular sequences to undergo bending and deformation. The extent of local and global alterations in nucleic acid conformation is constrained by base pairing; once significant stretches of non-helical regions are present, then nucleic acid sequences are capable of folding into altogether more complex, non-linear structures, which typically involve extensive non-Watson–Crick base–base interactions. Our knowledge of these structures is still rather rudimentary.

The chapters of this book describe in detail the variety of DNA and RNA nucleic acid structural types discovered to date, all of which ultimately depend on the conformational plurality of individual nucleotide repeating units. Their underlying conformational and structural properties were extensively studied in the two decades following the elucidation of the structure of DNA itself. NMR and crystallographic structural studies were almost entirely confined to mononucleosides and nucleotides up to the mid-1970s. A number of studies at that time focused on the backbone conformations and sugar puckers evident from these monomers, which provided valuable information on the range of conformations likely to be accessible to oligo- and poly-nucleotides, and on possible correlations between them.

This, the early phase of nucleic acid structural studies, produced atomic-resolution ($c.0.7$ Å) single-crystal analyses of a large number of nucleobases, mononucleosides, and mononucleotides. These have provided highly accurate geometric data for the five standard DNA/RNA bases, as well as for the rare bases occurring in some RNAs and for several protonated bases. This body of data is available for individual structures in the small molecule Cambridge Crystallographic Database, and has recently been collated and statistically analysed by the Nucleic Acid Database in order to produce standardized sets of values. The availability of this data is of particular importance for fibre diffraction, single-crystal, and NMR refinements of poly- and oligo-nucleotide structures and their complexes, all of which rely on accurate geometries for the definition of reliable constraints and restraints. The parameterization of force fields to be used in molecular dynamics simulation studies similarly requires the inclusion of high quality geometric data.

The development of automated chemical synthesis of defined sequence DNA (and, more recently, RNA) oligonucleotides has undoubtedly made a key contribution to the many major advances in nucleic acid structure since the early 1970s. At the same time, advances in both crystallographic and NMR methodology, together with computing and visualization developments, have enabled increasingly complex structures to be analysed effectively. It is perhaps invidious to select highlights of the past 25 years, but structure determinations of tRNA, the Dickerson–Drew dodecamer, Z-DNA, ribozymes, and telomeric DNAs, all represent significant landmarks. What about the next two decades? History tells us that in this field, as in many others, prediction is foolhardy. However, some trends are already apparent. Thus, it is clear that the patterns of folding in complex RNA structures represent a major challenge. DNA itself still has much to reveal. As crystallographic and NMR data become more accurate, features such as hydration and mobility (including sequence dependency), will become better defined. DNA folding, including that of catalytic and aptamer DNA, has yet to be explored at a molecular level.

This handbook has its origins in an earlier short introductory monograph on DNA structure. Feedback from numerous colleagues suggested that there is a need for a comprehensive survey and work of reference for both DNA and RNA structure, at an advanced level. It is no longer possible for one person to emulate the excellent (1984) text by Wolfram Saenger, such has been the growth in this field since then. I have been fortunate in being able to persuade so many of my colleagues of this need, and to contribute to this volume. All the major topics concerned with 'native' structures are represented. There is no explicit discussion on either protein- or drug-nucleic acid complexes; these, if covered at the same level, would require separate volumes, such is the quantity of information on them.

The book is set out in a systematic manner, progressing through the polymorphs of double helical DNA through to the higher order organizations of triplexes, quadruplexes, and junctions, then on to RNA structures in their various degrees of complexity. The two principal tools of molecular structure determination, X-ray crystallography and nuclear magnetic resonance, have been given equal weight in the book. Authors have been encouraged to be comprehensive, but not encyclopaedic, and not to shy away from controversy. It is to be hoped that the reader will arrive at a balanced view of the complementarity of these two approaches, as well as of their current scope and limitations.

I am very grateful to a number of friends and colleagues for their wisdom and helpful advice during this project, especially Helen Berman and Dick Dickerson. I have also been fortunate in a remarkable set of contributors, who have not only put much effort into their individual chapters, but worked together to provide coherence and minimal overlap between chapters. Thanks are due to my editors at Oxford University Press, who have been instrumental in guiding the contributors (and me) through the many minefields of a multi-author volume.

Surrey   S. N.
September 1998

Contents

Contributors

John G. Arnez: Laboratoire de Biologie Structurale, Institut de Génétique et de Biologie Moléculaire et Cellulaire, CNRS/INSERM/ULP, 1, rue L. Fries–BP 163, F-67404 Illkirch, France

Struther Arnott: The University, St. Andrews, Fife KY16 9AR, Scotland

Beth Basham: Department of Biochemistry and Biophysics, ALS 2011, Oregon State University, Corvallis, OR 97331, USA

Helen M. Berman: Department of Chemistry, Rutgers University, Piscataway, NJ 08854–8087, USA

Forrest J.H. Blocker: Department of Pharmaceutical Chemistry, University of California, San Francisco, San Francisco, CA 94143-446, USA

Tom Brown: Department of Chemistry, University of Southampton, Southampton, SO17 1BJ, UK

Serge Bouaziz: Cellular Biochemistry and Biophysics Program, Memorial Sloan-Kettering Cancer Center, New York, NY 10021, USA

Thomas E. Cheatham III: Laboratory for Structural Biology, MGSL/DCRT/12A-2041, National Institutes of Health, Bethesda, MD 20814, USA

Shan-Ho Chou: Institute of Biochemistry, National Chung-Hsing University, Taichung 40227, Taiwan

Donald M. Crothers: Department of Chemistry, Yale University, New Haven, CT 06520, USA

Richard E. Dickerson: Molecular Biology Institute, University of California at Los Angeles, Los Angeles, CA 90025-1570, USA

Brandt F. Eichman: Department of Biochemistry and Biophysics, ALS 2011, Oregon State University, Corvallis, OR 97331, USA

Juli Feigon: Department of Chemistry and Biochemistry, University of California, Los Angeles, CA 90095, USA

P. Shing Ho: Department of Biochemistry and Biophysics, ALS 2011, Oregon State University, Corvallis, OR 97331, USA

William N. Hunter: Department of Biochemistry, University of Dundee, Dundee, DD1 5EH, UK

Thomas L. James: Department of Pharmaceutical Chemistry, University of California, San Francisco, San Francisco, CA 94143-446, USA

Abdelali Kettani: Cellular Biochemistry and Biophysics Program, Memorial Sloan-Kettering Cancer Center, New York, NY 10021, USA

Peter A. Kollman: Department of Pharmaceutical Chemistry, Box 0446, University of California, San Francisco, San Francisco, CA 94143, USA

Richard Lavery: Laboratoire de Biochimie Théorique, CNRS UPR 9080, Institut de Biologie Physico-Chimique, 13, Rue Pierre et Marie Curie, Paris 75005, France

David M.J. Lilley: CRC Nucleic Acid Structure Research Group, Department of Biochemistry, The University, Dundee DD1 4HN, UK

Benoît Masquida: Institut de Biologie Moléculaire et Cellulaire, Centre National de la Recherche Scientifique, UPR 9002, 15, rue R. Descartes, F-67084 Strasbourg, France

Jennifer L. Miller: Department of Pharmaceutical Chemistry, Box 0446, University of California, San Francisco, San Francisco, CA 94143, USA

Dino Moras: Laboratoire de Biologie Structurale, Institut de Génétique et de Biologie Moléculaire et Cellulaire, CNRS/INSERM/ULP, 1, rue L. Fries–BP 163, F-67404 Illkirch, France

Jacek Nowakowski: Department of Chemistry and Molecular Biology, Scripps Research Institute, La Jolla, CA 92037, USA

Wilma K. Olson: Department of Chemistry, Rutgers, State University of New Jersey, New Brunswick, NJ 08903, USA

Dinshaw J. Patel: Cellular Biochemistry and Biophysics Program, Memorial Sloan-Kettering Cancer Center, New York, NY 10021, USA

Brian R. Reid: Departments of Chemistry and Biochemistry, University of Washington, Seattle, WA 98195, USA

Uli Schmitz: Department of Pharmaceutical Chemistry, University of California, San Francisco, San Francisco, CA 94143-446, USA

Bohdan Schneider: Heyrovsky Institute of Physical Chemistry, Academy of Sciences of the Czech Republic, 18223 Prague, Czech Republic

Zippora Shakked: Department of Structural Biology, Weizmann Institute of Science, Rehovot, Israel

Muttaiya Sundaralingam: Ohio State University, Biological Macromolecular Structure Center, Departments of Chemistry and Biochemistry and The Ohio State Biochemistry Program, 012 Rughtmore Hall, 1060 Carmack Road, Columbus, OH 43210, USA

Ignacio Tinoco, Jr: Department of Chemistry, University of California, Berkeley and Structural Biology Division, Lawrence Berkeley National Laboratory, Berkeley, CA 94720-1460, USA

Markus C. Wahl: Ohio State University, Biological Macromolecular Structure Center, Departments of Chemistry and Biochemistry and The Ohio State Biochemistry Program, 012 Rughtmire Hall, 1060 Carmack Road, Columbus, OH 43210, USA

Edmond Wang: Department of Chemistry and Biochemistry, University of California, Los Angeles, CA 90095, USA

Yong Wang: Cellular Biochemistry and Biophysics Program, Memorial Sloan-Kettering Cancer Center, New York, NY 10021, USA

John Westbrook: Department of Chemistry, Rutgers University, Piscataway, NJ 08854–8087, USA

Eric Westhof: Institut de Biologie Moléculaire et Cellulaire, Centre National de la Recherche Scientifique, UPR 9002, 15, rue R. Descartes, F-67084 Strasbourg, France

Christine Zardecki: Department of Chemistry, Rutgers University, Piscataway, NJ 08854–8087, USA

Krystyna Zakrzewska: Laboratoire de Biochimie Théorique, CNRS UPR 9080, Institut de Biologie Physico-Chimique, 13, Rue Pierre et Marie Curie, Paris 75005, France

Abbreviations

aa-tRNA	aminoacylated tRNA
aaRS	aminoacyl-tRNA synthetase
AMP	adenosine monophosphate
APP	alternating pyrimidine-purine
APT	antiparallel triplex
ATP	adenosine triphosphate
bHLH	basic helix-loop-helix
bZIP	basic leucine zipper
CAP	catabolite activator protein
COSY	correlated spectroscopy
CS	cationic strength
CSD	Cambridge Structural Database
DIF	dimeric irregularity function
dn	dinucleotide
dzaX	7-deaza-2'-deoxyxanthosine
edA	1,N6-ethenoadenosine
EF	elongation factor
FMN	flavin mononucleotide
g	*gauche*
GDP	guanosine 5'-diphosphate
GTP	guanosine 5'-triphosphate
HETCOR	heteronuclear correlated spectroscopy
HPLC	high performance liquid chromatography
HTH	helix-turn-helix
IHF	integration host factor
IR	infrared
ISPA	isolated spin-pair approximation
MD	molecular dynamics
MG	magnesium only form of d(CGCGCG)
MGSD	magnesium and spermidine form of d(CGCGCG)
MGSP	magnesium and spermine form of d(CGCGCG)
mmCIF	macromolecular crystallographic information file
MMD	multiple molecular dynamics
mRNA	messenger RNA
NDB	Nucleic Acid Database
NMR	nuclear magnetic resonance
NOE	nuclear Overhauser effect
NOESY	NOE spectroscopy
nt	nucleotide
O6MeG	06-methylguanine
O8A	8-oxoadenine
O8G	8-oxoguanine
PAGE	polyacrylamide gel electrophoresis
PDB	Protein Data Bank
PME	particle mesh Ewald

PNA	peptide nucleic acid
ppm	parts per million
PT	parallel triplex
r	rotation
RESP	restrained electrostatic potential
rMC	restrained Monte Carlo
rMD	restrained molecular dynamic
rms	root mean square
rmsd	root mean square difference
RNAase	ribonuclease
RNP	ribonucleoprotein
RRE	Rev response element
rRNA	ribosomal RNA
SAS	solvent-accessible surfaces
SFE	solvent free energy
SP	spermine only form of d(CGCGCG)
SQL	structure query language
t	*trans*
t	translation
TAR	*trans*-activation response
TBP	TATA-binding protein
tRNA	transfer RNA
UV	ultraviolet
WWW	world-wide web

1

Polynucleotide secondary structures: an historical perspective

Struther Arnott

The University, St. Andrews, Fife KY16 9AR, Scotland

1. Introduction

In this chapter I shall describe the fibre-derived X-ray analyses upon which studies of polynucleotide helical conformations mainly depended from 1950 to 1980. The first of these three decades started off with the dramatic events that showed that DNA, the large complex polymer within whose primary structure genetic information was stored, had an unexpectedly simple secondary structure. Soon it became clear that it could have two secondary structures and for much of the 1950s the efforts of molecular biophysicists were concentrated on putting the details of these two allomorphs beyond cavil. In the 1960s, when as much effort was put into RNA structures as DNA structures, it became evident that polynucleotide double helices belonged to two *sets* of secondary structures related to the original two eponymous DNA allomorphs, A and B. In the same decade the technology of X-ray diffraction analysis of fibres became more sophisticated so that by the 1970s the fine details of synthetic polynucleotide duplexes of defined sequence could begin to be explored routinely. This exploration, and the emerging parallel studies of oligonucleotides in single crystals, uncovered a third set of helical allomorphs, Z, of opposite hand to the two original sets that had become familiar during the previous 20 years or so. These two sets of investigations also promoted speculations that the base sequences within helices might be emphasized by characteristic conformations and morphological wrinkles on the surfaces of helices. There are, indeed, wrinkles on the surfaces both of polymer duplex helices and of quasi-helical oligomer duplexes, but whether they are of much significance biologically in DNA remains to be established. DNA is obviously very plastic and this is important for its role as the substrate in many interactions.

Fibres, metaphorically and literally, are the continuous thread in the story of DNA (1) and related polynucleotides, from before 1950 right up to the present day. The important polynucleotide secondary structures are all helical whether they are single-, double-, triple-, or quadruple-stranded. Long helices are more likely to be ordered in oriented fibres than in large single crystals. (Who has yet crystallized a quasi-helical oligonucleotide with 20 or 30 residues?) Helices imply a motif contained within one pitch length, which is repeated linearly along one polymer molecule. The process of spinning a fibre orients such polymer molecules with their repeated motifs at least parallel to one another. These ordered arrays make X-ray diffraction analyses possible, mainly because the X-rays scattered by them are greatly amplified versions of the

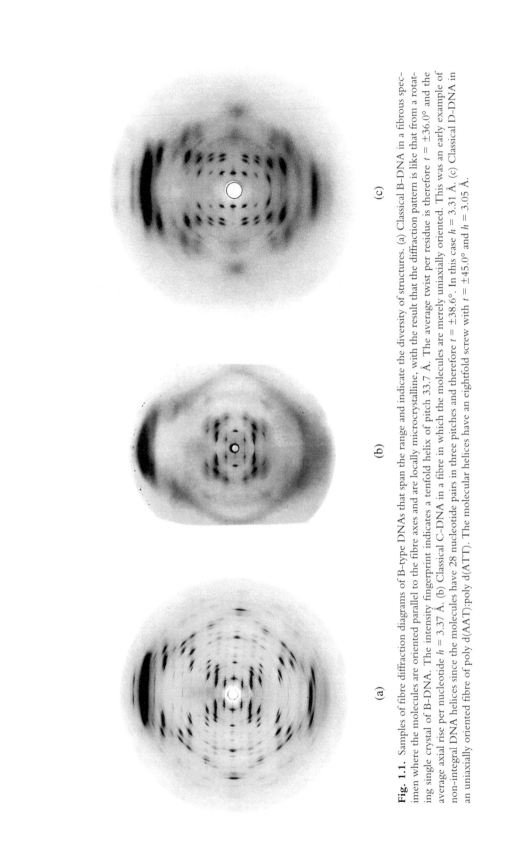

Fig. 1.1. Samples of fibre diffraction diagrams of B-type DNAs that span the range and indicate the diversity of structures. (a) Classical B-DNA in a fibrous specimen where the molecules are oriented parallel to the fibre axes and are locally microcrystalline, with the result that the diffraction pattern is like that from a rotating single crystal of B-DNA. The intensity fingerprint indicates a tenfold helix of pitch 33.7 Å. The average twist per residue is therefore $t = \pm 36.0°$ and the average axial rise per nucleotide $h = 3.37$ Å. (b) Classical C-DNA in a fibre in which the molecules are merely uniaxially oriented. This was an early example of non-integral DNA helices since the molecules have 28 nucleotide pairs in three pitches and therefore $t = \pm 38.6°$. In this case $h = 3.31$ Å. (c) Classical D-DNA in an uniaxially oriented fibre of poly d(AAT)·poly d(ATT). The molecular helices have an eightfold screw with $t = \pm 45.0°$ and $h = 3.05$ Å.

(a)   (b)   (c)

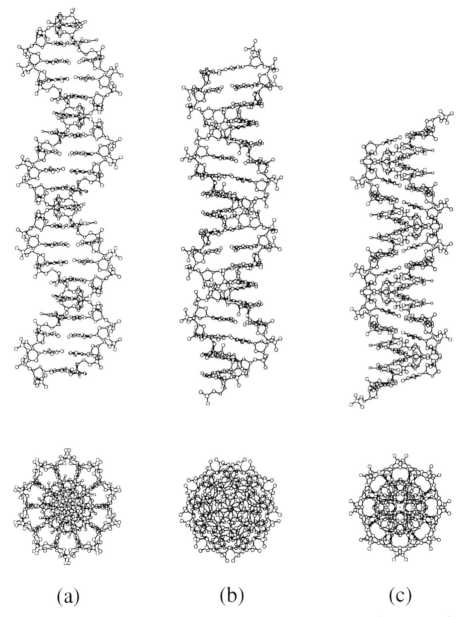

Fig. 1.2. Mutually perpendicular projections of segments of B-type polynucleotide duplexes corresponding to the diffraction patterns of Fig. 1.1. All the helices are right-handed, the chains antiparallel, and in each duplex all the nucleotides conformations are identical. Thus the molecular symmetries are: (a) B-DNA, 10_122; (b) C-DNA, 28_32; (c) D-DNA, 8_122. Morphologically an open and deep major groove is the persistent property of these allomorphs, but as t increases from 36.0 to 45.0° and h declines from 3.37 to 3.05 Å, the almost as deep minor grooves close. At the same time the inclination of the base pairs becomes more negative.

scattering from a single motif. The diffraction patterns from uniaxially oriented fibres give mainly non-Bragg distributions of continuous intensity along layer lines (2). Good examples are shown in Figs 1.1c and 1.5a,b. In this respect they are different from the spotty, Bragg patterns given by crystals where a motif is repeated in a regular three-dimensional array. The diffraction consequences of such three-dimensional regularity is a three-powered amplification of the repeated motif's scattering pattern in specific directions. This amplification is a benefit that usually outweighs the corresponding extinction of the scattering pattern in the many directions that do not obey the Bragg conditions.

X-ray diffraction analyses of merely oriented systems can be just as illuminating as analyses of fully crystalline systems: the structural studies of tobacco mosaic virus (3) and of bacteriophage Pf1 (4) have demonstrated this amply, as have the analyses of fibres of the synthetic DNA:RNA hybrids (5) that provide more non-Bragg X-ray diffraction (Fig. 1.5a,b) than Bragg diffraction. With nucleic acids there are often even more favourable situations when the uniaxially oriented systems are, in addition, microcrystalline and therefore provide only Bragg-type data (e.g. Figs 1.1a and 1.3a,b). Using contemporary methods of measuring intensities, current structure determinations of repeated oligonucleotide sequences in fibres that are both uniaxially oriented and polycrystalline can compete with single-crystal analyses of oligonucleotides, except in the few cases of the latter where exceptional crystal perfection (6) provides an unusually rich set of high resolution data.

To study oligonucleotide systems only in crystals is needlessly remote from polymer structures when the object of the study is to determine the effect of sequence on local conformations on a naked polymer. Certainly, in terms of the secondary structures of Watson–Crick base-paired duplexes, there have been no discoveries with oligonucleotides that have overturned previous, fibre-derived insights with respect to the prevalent right-handed helical conformations. The one true novelty to emerge from oligonucleotide crystallography was the exotic left-handed helical conformations (Z-DNA) available to oligo(dGC):oligo(dGC) (7,8) and later recognized in certain polymers also with alternating purine–pyrimidine (9) base sequences. High resolution, single-crystal analyses are also essential when visualizations of the precise interactions between specific oligonucleotide sequences and adducts are needed (10), or when the subtle adjustments in local structure required to accommodate a mismatched base pair have to be scrutinized (e.g. ref. 11).

2. The DNA duplex: discovery and definition

It may be too procrustean to squeeze the progression of knowledge about polynucleotide secondary structures into exact decades, but there is a certain convenience in doing so. In the 'dark age' before 1950, diffraction patterns of oriented DNA existed. These were confusing because, as we can now see, they were of poorly ordered mixtures of the A and B allomorphs of DNA. Nevertheless, their very existence for a polymer containing complex base sequences encouraged the hope that these diverse sequences might be accommodated in a very simple framework. Maurice Wilkins' first achievement (12) was a clean pattern of the commonest allomorph of DNA, later called B (Fig. 1.1a). Rosalind Franklin's main contribution (13) was the discovery that

DNA was dimorphic (Fig. 1.3c). Interestingly, she named her later discovered form A and the prior Wilkins' form B, perhaps because, in her hands, the uniaxially oriented fibres of the former were always of the 'superior' polycrystalline type whereas those of the latter were polycrystalline only accidentally. The precise experimental circumstances that would provide, routinely, oriented and polycrystalline B patterns (Fig. 1.1a) had to await Wilkins' meticulous further experiments. Meanwhile, both A and B patterns helped Watson and Crick (14) to the conclusion that DNA had helical secondary structures and provided the dimensions and symmetries that were imposed upon their first DNA models, which incorporated antiparallel, duplex, right-handed helices (e.g. Figs 1.2 and 1.4) and isomorphous A:T and G:C pairs. However, it was these isomorphous, complementary pairs that were the key revelation that was immediately exploited in order to understand the molecular biology of genes. To begin with the helical frameworks were incidental and even an embarrassment: the fact that the two helical chains were intertwined posed the difficult problem of visualizing unwinding during replication or transcription; also, the coordinates of all the atoms in the helical models (15) allowed diffraction intensity distributions to be calculated and these were found at once to be seriously different from those observed. This provoked a series of challenges to the Watson and Crick conjecture by (notably) Donahue (e.g. ref. 16). The response by Wilkins and his group (17,18) was a decade of painstaking refinements of the original model, which contrived to preserve the original base-pairing hypothesis while remedying the initial very poor fit with the diffraction data.

The fit of the original Crick and Watson model, incidentally, was so poor that the residual error, as measured by the crystallographers' R-factor, was about 0.80, a value so large as to indicate to conventional chemical crystallographers a structure so erroneous as to be beyond rescue. Ironically, Wilkins' rescue was possible because of the polymorphism of DNA. The original Crick and Watson model for B-DNA (15) was, unwittingly, what we would now call an A structure. It had reasonable stereochemistry but incorporated, not C2'-*endo*-puckered, but C3'-*endo*-puckered furanose rings. Such duplexes have base pairs 4 Å nearer the helix periphery than in B-DNA—a major difference in the distribution of electron density that led to the incompatibility of the calculated with the observed diffraction pattern.

In the 1950s there were no well-developed protocols for melding low resolution diffraction data with stereochemical restraints and constraints. Consequently, the refinement of models to prove the Watson–Crick conjecture was a labour-intensive, manual process that persisted until 1960, accompanied as it was by the equally slow processes of obtaining purer DNA specimens, better methods of spinning DNA fibres and of collecting higher resolution X-ray data. Nevertheless, by the end of this 'decade of discovery and definition' two distinct allomorphs for DNA duplexes had been defined (17,18), the B- and A-forms, which were most obviously distinguished by the position of the base pairs: astride the helix axis in the former but noticeably displaced ($dx = -4$ Å) in the latter. An immediate consequence of this are the equally distinctive groove structures: in B the major and minor grooves are equally deep (Fig. 1.2a), whereas in A the major groove is a relatively deep chasm, contrasting with the minor groove, which is merely a shallow depression (Fig. 1.4c). Other features of the base pairs in both structures were their mild propeller distortion from complete

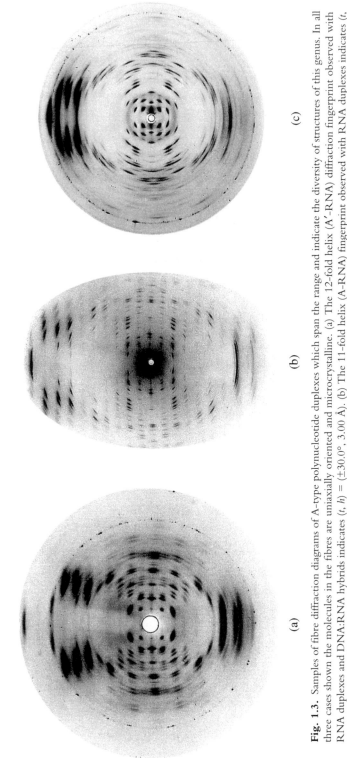

(a)

(b)

(c)

Fig. 1.3. Samples of fibre diffraction diagrams of A-type polynucleotide duplexes which span the range and indicate the diversity of structures of this genus. In all three cases shown the molecules in the fibres are uniaxially oriented and microcrystalline. (a) The 12-fold helix (A'-RNA) diffraction fingerprint observed with RNA duplexes and DNA:RNA hybrids indicates $(t, h) = (\pm 30.0°, 3.00$ Å). (b) The 11-fold helix (A-RNA) fingerprint observed with RNA duplexes indicates $(t, h) = \pm (32.7°, 2.81$ Å). (c) The fingerprint of classical A-DNA also indicates an 11-fold helix with $(t, h) = (\pm 32.7°, 2.56$ Å).

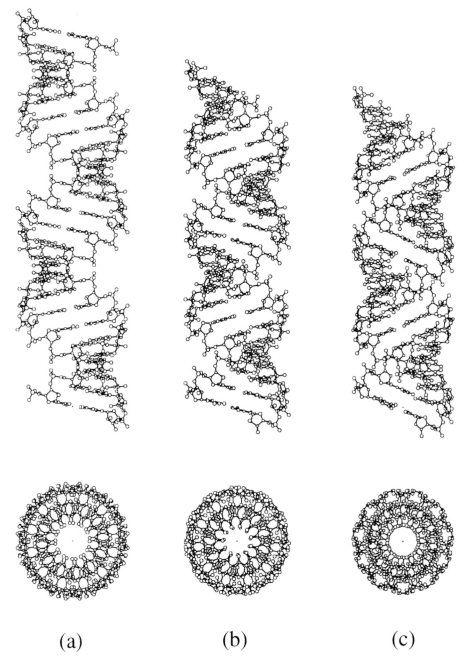

(a)   (b)   (c)

Fig. 1.4. Mutually perpendicular projections of the range of A-type duplex helices corresponding to Fig. 1.3. All are regular and right-handed and have identical antiparallel chains and therefore their molecular symmetries are: (a) A′-RNA, $12_1 22$; (b) A-RNA, $11_1 2$; (c) A-DNA, $11_1 2$. The common molecular feature of these double helices is their shallow minor grooves and very deep major grooves. In (a), where h is maximum, the major groove is also wide open, but in (c), axially the most compact conformational variant, the major groove is essentially closed.

coplanarity and the large inclination of about 20° in A, associated with the shorter (2.56 Å) rise per residue, compared with the essentially 0° inclination in B, which has a longer (3.37 Å) rise per residue. The helical twist in A (32.7°) is also lower than that in B (36.0°).

Towards the end of the1950s a third (19) and a fourth (20) allomorph, C and D, were also discovered, both B-like in structure (Fig. 1.2) but with reduced rises per residue (3.31 and 3.05 Å, respectively) and increased helical twists (38.6–40.0 and 45.0°, respectively). These discoveries heralded the next decade (1960s) which may be thought of as the 'decade of expansion and exploration'.

3. Expansion

By the 1960s it was evident that there might be many polynucleotide structures to determine and that, therefore, computerized model building (21) had to take over from manual procedures and valuable analytical methods, such as least-squares (21,22) refinements and Fourier syntheses of electron density (23,24), that were commonplace in orthodox X-ray diffraction analyses of crystal structures had to be adapted for further studies. While this was in train an important event occurred in the discovery and determination of the structures of two allomorphs of duplex RNA (24,25), both A-type (Figs 1.3a,b and 1.4a,b), which immediately extended the range of polymorphism of this set of right-handed polynucleotide helices and showed that the range of helical twists available to A structures was only 30.0–32.7° (cf. 36.0–45.0° available to B structures), but that rises per residue might be just as variable for A structures (2.56–3.00 Å) as for B (3.05–3.37 Å). It was also realized explicitly that the distinctive morphologies of the A and B structures correlated with C3′-*endo* furanose rings in the former versus C2′-*endo* rings in the latter (21), and that these conformations were the origin of the very negative dx displacements of the base pairs.

A quarter of a century later, and after more than a hundred very expensive oligonu-cleotide crystal structure determinations, it has had to be concluded, reluctantly (26), that: (i) B-like structures have a mean helical twist (and standard deviation) of 36.1° (5.9°) and a mean axial rise per base pair (and standard deviation) of 3.37 Å (0.46 Å); (ii) A-like structures have mean values for helical twist and rise per residue of 31.1° and 2.90 Å, respectively; and (3) the most persistent morphological feature differenti-ating the two families is the 4–5 Å relative base pair displacement that gives rise to their distinctive groove structures!

Rarely in the history of scientific endeavour has so much effort by so many investi-gators provided so few new insights of significance.

4. Discrimination and exploration

The introduction of automatic least-squares refinement to X-ray diffraction analysis of polymers in fibres (21) in the 1960s not only allowed easier and faster production of the polynucleotide models with the best coordinates, but also provided a means of discriminating between alternative structural hypotheses. Suppose Watson and Crick had been aware that for their first model of B-DNA they would have to consider left-handed as well as right-handed duplex helices, and that furanose rings could be C2'-*endo* as well as C3'-*endo*. They should have found it necessary and possible to cobble together four versions of a DNA model each with isomorphous A:T and G:C pairs. The rise per residue of 3.37 Å would not have been very discriminating, nor would a helical twist of ±36°. These generous dimensions result in a fairly open structure for B-DNA and, therefore, none of the initial models would have been embarrassing stereochemically. Since they would also be building isolated molecules that did not have to fit into a tight unit cell, another source of discrimination would have been absent. Only when they had to fit the X-ray intensities optimally, while maintaining viable stereochemistry, would it have been found that the two right- and left-handed models with C2'-*endo* rings were noticably superior to the right- and left-handed models with C3'-*endo* rings. The best right-handed double helix with C2'-*endo* rings might have been somewhat superior to the best left-handed structure, but could only have been judged to be significantly superior by applying statistical tests, such as those that were only later introduced by Walter Hamilton (27), to the best least-squares models of each kind.

During the 'decade of discrimination' (1970s) the possibility of least-squares-optimized models of polynucleotides, and the existence of Hamilton's tests, removed much of the uncertainty that had come to be associated with the fibre diffraction analysis of polynucleotides.

This uncertainty would not have arisen so acutely if meticulous experimental studies of fibrous polynucleotide systems had been commonplace in laboratories other than that of Maurice Wilkins. Unfortunately, they were not. Encouraged by the Watson and Crick model-building coup, which owed little to local experimental effort, many other analyses of fibrous polynucleotide systems were undertaken with just as little experimental investment, but with much less insight. Deservedly, most of the conclusions from these forays were wrong, but from these failures grew an understandable lack of confidence in fibre studies of polynucleotides, which, during the 1960s accumulated an appalling negative record: no fibrous nucleic acid structure produced by a laboratory not of Maurice Wilkins' school survived critical re-examination: the model for B-DNA by Crick and Watson (15) turned out to be a model for a member of the A-family; Rich's three-stranded model (28) for polyinosinic acid should have been four-stranded (29); the double-stranded model of Langridge and Rich (30) for polycytidylic acid should have been single-stranded (31); and Mitsui *et al.* produced a left-handed model for D-DNA (32), which is, in fact, right-handed (33). The point is not that one can easily be wrong in modelling a fibrous structure, but that with today's technology scrupulously applied, most gross errors are detectable if enough effort is invested in alternative structures.

5. Polymorphism

Polymorphism in polynucleotide helices has a number of aspects: How polymorphous are duplexes containing isomorphous Watson–Crick A:T and G:C base pairs, no matter what the base sequence is? How polymorphous are they when a particular base sequence is monotonously repeated along the polymer? Further questions arise when one chain is RNA but the other is DNA; when triplex helices occur in which a

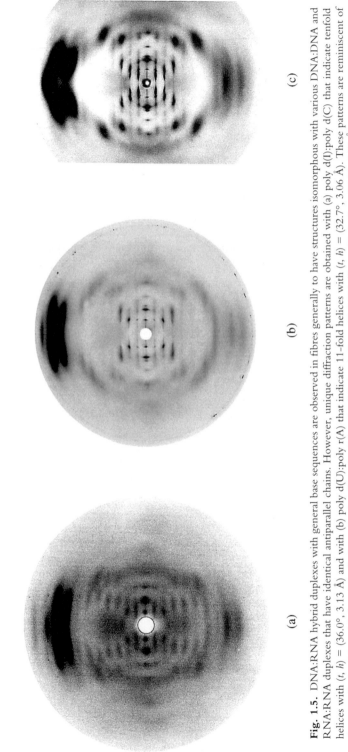

(a)   (b)   (c)

Fig. 1.5. DNA:RNA hybrid duplexes with general base sequences are observed in fibres generally to have structures isomorphous with various DNA:DNA and RNA:RNA duplexes that have identical antiparallel chains. However, unique diffraction patterns are obtained with (a) poly d(I):poly d(C) that indicate tenfold helices with $(t, h) = (36.0°, 3.13$ Å) and with (b) poly d(U):poly r(A) that indicate 11-fold helices with $(t, h) = (32.7°, 3.06$ Å). These patterns are reminiscent of the ones obtained from (c) the triplex helices of poly r(A):poly r(U):poly r(U) that indicate 11-fold helices with $(t, h) = (32.7°, 3.05$ Å).

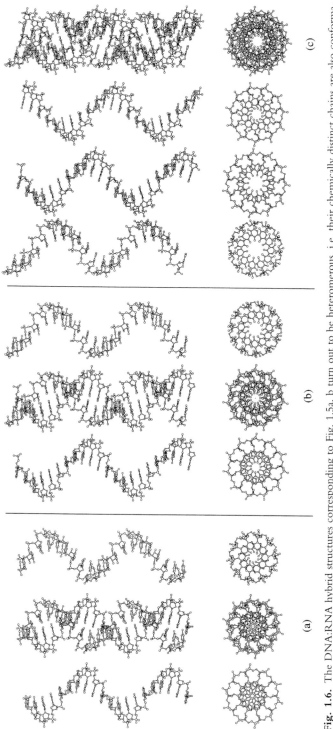

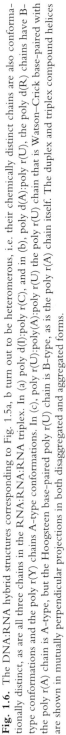

(a)   (b)   (c)

Fig. 1.6. The DNA:RNA hybrid structures corresponding to Fig. 1.5a, b turn out to be heteromerous, i.e. their chemically distinct chains are also conformationally distinct, as are all three chains in the RNA:RNA:RNA triplex. In (a) poly d(I):poly r(C), and in (b), poly d(A):poly r(U), the poly d(R) chains have B-type conformations and the poly r(Y) chains A-type conformations. In (c), poly r(U):poly(A):poly r(U) the poly r(U) chain that is Watson–Crick base-paired with the poly r(A) chain is A-type, but the Hoogsteen base-paired poly r(U) chain is B-type, as is the poly r(A) chain itself. The duplex and triplex compound helices are shown in mutually perpendicular projections in both disaggregated and aggregated forms.

Watson–Crick duplex of special sequence has a third strand attached that involves non-Watson–Crick base–base interactions; and when duplexes, triplexes, and quadruplexes are studied in which none of the base–base interactions can be isomorphous with the classical Watson–Crick base pairs. All these situations began to be explored before the 1970s but it was only when the technology of fibre diffraction analysis had been systematized that they could be explored scrupulously and reasonably rapidly. An additional non-trivial requirement was better data from better fibres, which could be contrived only after there was ready availability, and in quantity, of truly polymeric homopolynucleotides and polyoligonucleotides of well-defined sequence.

Discrimination is a persistent feature of polynucleotide structure analyses in fibres and of oligonucleotide analyses in single crystals. As the precision of analyses becomes finer, the issues move on from questions of the handedness of helices, and from questions of one ring pucker or another, to whether a conformational wrinkle on the surface of a helix is real, and, if real, is its existence predetermined by primary structure or merely an accident of local crystal interactions or the effect of an odd cation or two? How many blobs of electron density represent real water molecules and, if real, are they important and, if important, are they truly important to molecular biologists rather than merely comforting to crystallographers worried by less-than-atom resolution data?

To anticipate the detailed conclusions of the 'decade of discrimination and polymorphism' (1970s) it should be said that polynucleotide helices have turned out to be much less polymorphic than a polymer chemist might have supposed. Any nucleotide residue has six variable conformation angles in its phosphate diester backbone and each of these angles has two or three regions of variation. In addition, there are two regions of variation available to bases at their glycosylic attachments. The naïve expectation has to be that polynucleotide helices should be *very* polymorphic. Even if it is insisted that bases are 'stacked', it is not obvious that the expected polymorphism should be reduced to merely three classes; namely, the original right-handed A and B chains that incorporate either C3′ *endo* or C2′-*endo* furanose rings, and the unique, left-handed Z chains that incorporate the two kinds of rings alternately! Nor is it obvious that requiring a few hydrogen bonds in Watson–Crick or any other base pairing would seriously limit further macropolymorphism. Yet, this does seem to be the case.

This is not to say that micropolymorphism does not exist. It does: not all chains of the A-, B-, or Z-types are identical to one another; nor need the two chains in any particular A-, B-, or Z-duplex be identical to one another, nor even similar, since duplexes with A and B chains exist, as do triplexes that incorporate mixtures of A and B chains (Fig. 1.6). It is also the case that local nucleotide conformations in oligonucleotides sometimes vary, apparently in a sequence-dependent way. Much of the extent and limits of these polymorphisms have been revealed in polynucleotide fibres. These conclusions have been confirmed and a few of them have been extended by detailed analyses of oligonucleotides in single crystals.

6. Homopolymers

Uniaxially oriented fibres of poly A, poly U, poly (thioU), poly C, poly G, poly I, poly X have all been fabricated. The diffraction patterns of the first three polymers have all been interpreted as deriving from double-stranded molecules and that of poly I from a triple-stranded molecule (28). This pattern and that of the essential fibres of poly G have since been shown to arise from quadruplexes (29) with A-like polynucleotide chains. Oriented fibres of poly C fibres can also be polycrystalline and are now firmly established as containing single, not double, strands of A-like poly C helices (31). No satisfactory analyses of poly A or poly U fibres have been completed. Poly (thioU) and poly X give surprising similar diffraction patterns that are even more surprisingly similar to A-DNA (34,35) and must therefore correspond to duplex arrangements of identical right-handed, antiparallel chains with conformations in right-handed, 11-fold helices with a 2.56 Å rise per residue! Apparently, such identical, antiparallel, sugar–phosphate chains can, by mutual rotation about their common helix axis, contrive duplex structures that can accommodate purine–purine (X:X), purine–pyrimidine (A:T or G:C), or pyrimidine–pyrimidine ($s^2U:s^2U$) base pairs without any significant conformational readjustment. This truly remarkable result has important implications for the lack of genetic specificity implicit in polynucleotide secondary structures by themselves.

To emphasize how adept polynucleotide helices of conventional conformation are at accommodating exotic base sequences with complementary (but non-Watson–Crick) base pairs, one only has to consider the structures of duplexes and triplexes containing mixtures of homopolynucleotides such as poly I:poly A:poly I, where there are two kinds of purine–purine pairs and yet the polynucleotide strands are conformationally conventionally A-type (36), albeit not conformationally identical. Other triple-stranded homopolymer systems, such as poly U:poly A:poly U (37) (Figs 1.7 and 1.8) and poly dT:poly dA:poly T (38), have also been investigated. These contain both Watson–Crick and Hoogsteen base pairs. Originally it was assumed (36–38) that all the chains would be A-type, i.e. the structures would be merely an A-type Watson–Crick duplex with the third strand, also A-type, filling the wide, deep major groove. Comprehensive review (39) of alternative models with the best least-squares results, Hamilton-tested, has shown that poly dT:poly dA:poly dT in fact has a structure with all chains B, but poly U:poly A:poly U has an A:B:B structure. The original conjecture that poly I:poly A:poly I has an A:A:A triplex has, however, survived rescrutiny.

It always had to be thinkable that DNA:RNA hybrids might have a heteromerous duplex structure with two conformationally non-identical strands. In fact, DNA:RNA hybrids most often have fibrous structures isomorphous with A-DNA or A′-RNA (5,40) (Figs 1.3 and 1.4) and must, therefore, form duplexes with polynucleotide chains that are conformationally identical despite their chemical difference. That heteromerous structures indeed exist has been demonstrated with synthetic DNA:RNA hybrids where the chains are homopolymers, like poly dA:poly rU and the related, but not isomorphous, poly dI:poly rC (Figs 1.7 and 1.8). In each of these duplexes the DNA strand is B-type and the RNA strand A-type (40). It was originally thought that the unique (B′) diffraction pattern of poly dA:poly dT (38) (Fig. 1.7c) was also the consequence of just such an heteromerous structure (41), but more intensive analyses of a variety of crystal forms of poly dA:poly dT (42,43), poly dA:poly dU (44), and poly d(AI):poly d(CT) (45) have shown that all these structures, although heteromerous with two non-identical polynucleotide strands, contain two B-type strands (Fig. 1.8c).

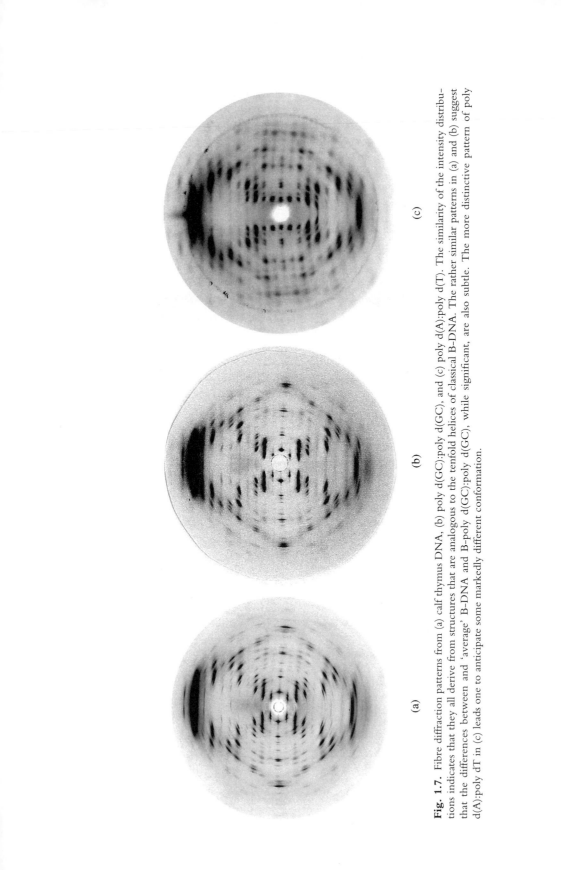

(a)

(b)

(c)

Fig. 1.7. Fibre diffraction patterns from (a) calf thymus DNA, (b) poly d(GC):poly d(GC), and (c) poly d(A):poly d(T). The similarity of the intensity distributions indicates that they all derive from structures that are analogous to the tenfold helices of classical B-DNA. The rather similar patterns in (a) and (b) suggest that the differences between and 'average' B-DNA and B-poly d(GC):poly d(GC), while significant, are also subtle. The more distinctive pattern of poly d(A):poly dT in (c) leads one to anticipate some markedly different conformation.

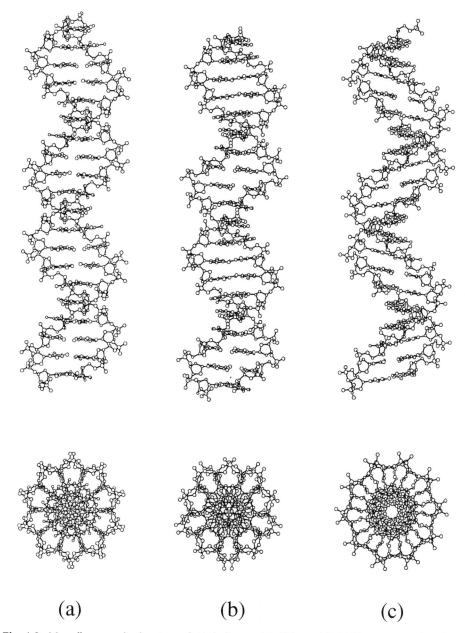

Fig. 1.8. Mutually perpendicular views of: (a) the 'average' (calf thymus) B-DNA structure with molecular symmetry $10_1 22$; (b) the B-form of poly d(GC):poly d(GC) which has $5_1 22$ symmetry, i.e. a right-handed helical duplex with identical antiparallel chains, each of which is a fivefold helix of dinucleotides with GpC conformationally distinct from CpG; (c) the so-called B'-form of poly d(A):poly dT where the molecular symmetry is 10_1, and there is no dyad axis of symmetry relating the two chains, i.e. the poly d(A) and poly d(T) chains have the same pitch and symmetry but the nucleotides in the different chains do not have the same conformations.

7. Polyoligonucleotide duplexes

Following on from the polymononucleotides, chemically, the simplest synthetic polynucleotides are the polydinucleotides with alternating, self-complementary base sequences, poly d(GC):poly d(GC) and poly d(AT):poly d(AT), both of which, in different ways, turned out to be very important in extending the range of DNA polymorphism (Figs 1.7–1.12). As mentioned before, poly d(AT):poly d(AT) was important for its B-like, D structure, which strictly is a fourfold helix of dinucleotides (46), but to a good approximation is an eightfold helix, with twist = 45.0°, and with a reduced rise per residue (3.02 Å) compared with tenfold helical B (3.37 Å). Unlike the classical B structure, the base pairs are inclined, but in the opposite sense to A. This D structure, with C, broke the classical B monopoly and indicated that the twists per residue in B structures could vary markedly, and that the variation could be expected to be upwards from the classical value of 36.0°. Poly d(AT):poly d(AT) also forms orthodox B helices (47) and, reluctantly, classical A helices. The rarity of A helices for this polymer and their complete absence in poly d(A):poly d(T) reemphasizes an older discovery that (AT)-rich DNAs find the B→A transition more difficult than (GC)-rich DNAs.

Poly d(GC):poly d(GC) can be obtained (and in fibres of well-washed DNA, always is) in the A or B forms (47): the A form is classical, a regular 11-fold helix with no conformational evidence of the underlying polydinucleotide sequence; not so the B form which has both a crystal structure that is different from native B-DNA (Fig. 1.7b) and contains fivefold helices of dinucleotides, despite the generally close resemblance of its diffraction pattern to the classical B form of DNA. The root of the difference lies in the different local conformations in GpC and CpG where the con-

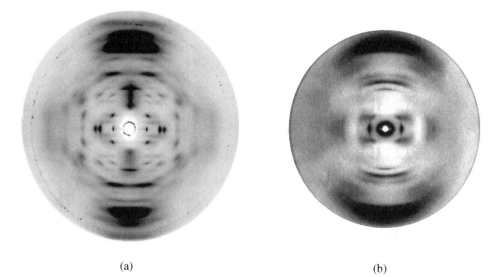

(a)   (b)

Fig. 1.9. Fibre diffraction patterns from two forms of polymeric Z-DNA: (a) from poly d(GC):poly d(GC) a sixfold helix of pitch 43.5 Å, (t, h = ±60.0°, 7.25 Å); (b) from poly d(A^{s4}T):poly d (A^{s4}T) a sevenfold helix of pitch 53.2 Å, (t, h = ±51.4°, 7.60 Å).

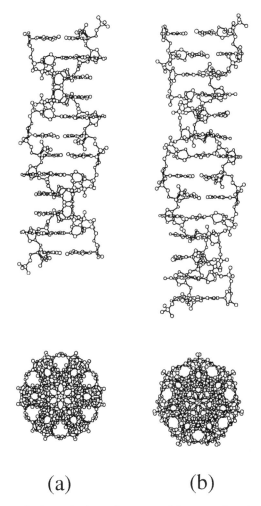

(a)   (b)

Fig. 1.10. Mutually perpendicular projections of segments of the two polynucleotide duplexes that cor-respond to the diffraction patterns of Fig. 1.9. Both are left-handed helices with antiparallel chains in which the unit of structure is a dinucleotide: (a) has molecular symmetry $6_5 22$; (b) has $7_6 2$. The mor-phologies of both are compact and quasi-cylindrical.

formations (ε, ζ) are (g^-, t) and (t, t), rather than both (t, t) as they are, on average, in native B-DNA (46). This apparently sequence-related wrinkle (Fig. 1.14b) was the first detected in a polymeric DNA. A more modest version of the same wrinkle is present in the D forms of poly d(AT):poly d(AT) and its isomorph, poly d(IC):poly d(IC) (Figs 1.11 and 1.12).

There is also an interesting variant of the D form of poly d(AT):poly d(AT) which has a hexanucleotide structural repeat (40) (Figs 1.11 and 1.12) because successive A:T nucleotides have all their (ε, ζ) conformations successively, but not identically (t, g^-), but successive TA nucleotides are (g^-, t), (g^-, t), and (t, t). In other words the nondescript conformation, (t, t), is intruded every sixth nucleotide in place of the

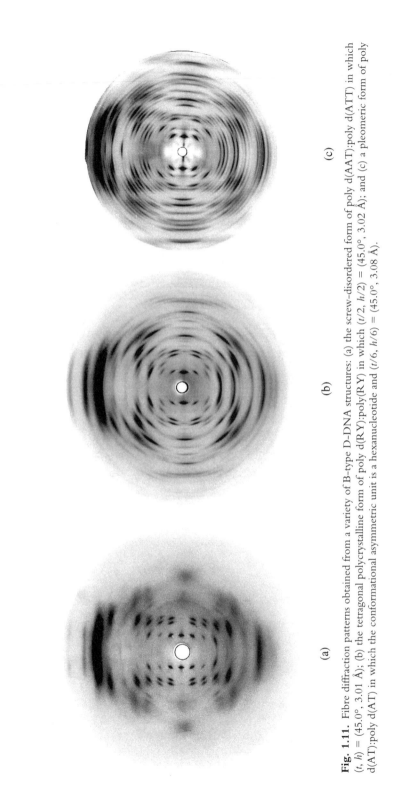

(a)   (b)   (c)

Fig. 1.11. Fibre diffraction patterns obtained from a variety of B-type D-DNA structures: (a) the screw-disordered form of poly d(AAT):poly d(ATT) in which $(t, h) = (45.0°, 3.01\ \text{Å})$; (b) the tetragonal polycrystalline form of poly d(RY):poly(RY) in which $(t/2, h/2) = (45.0°, 3.02\ \text{Å})$; and (c) a pleomeric form of poly d(AT):poly d(AT) in which the conformational asymmetric unit is a hexanucleotide and $(t/6, h/6) = (45.0°, 3.08\ \text{Å})$.

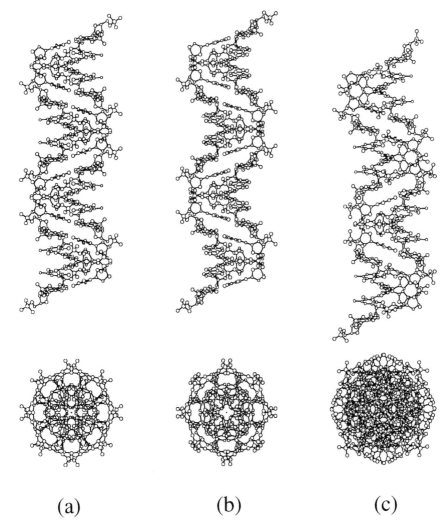

Fig. 1.12. Mutually perpendicular projections of segments of the D helices that furnished the diffraction patterns in Fig. 1.11. The regular $8_1 22$ helix of average mononucleotides in (a) is fairly closely mimicked by the $4_1 22$ helix of dinucleotides in (b), but less so in the 4_3 helix of hexanucleotides in (c), as is evident when one views the overall morphologies perpendicular to the helix axes. Then, the distinctive surfaces are more apparent than when one contemplates the projections parallel to the helix axes.

discriminating conformations (t, g^-) for (purine, pyrimidine) steps and (g^-, t) for (pyrimidine, purine) steps. The important messages to be taken from this structure are that not every variation of sequence produces a wrinkle and that only some wrinkles may be diagnostic of sequences. Thus, when one comes to examine detailed conformations in various B-type polymer structures, such as poly d(GGT):poly d(ACC) (48, 46) (Figs 1.13 and 1.14), poly d(AG):poly d(CT) (46), poly d(AI):poly d(CT) (45), and poly d(AATT):poly d(AATT) (49), one does indeed find that the nondescript (t,

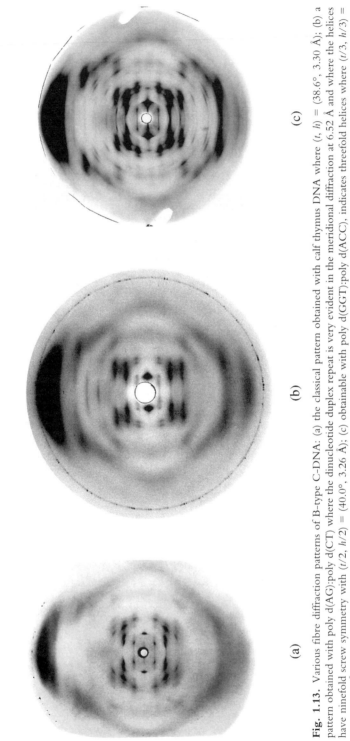

Fig. 1.13. Various fibre diffraction patterns of B-type C-DNA: (a) the classical pattern obtained with calf thymus DNA where $(t, h) = (38.6°, 3.30 \text{ Å})$; (b) a pattern obtained with poly d(AG):poly d(CT) where the dinucleotide duplex repeat is very evident in the meridional diffraction at 6.52 Å and where the helices have ninefold screw symmetry with $(t/2, h/2) = (40.0°, 3.26 \text{ Å})$; (c) obtainable with poly d(GGT):poly d(ACC), indicates threefold helices where $(t/3, h/3) = (40.0°, 3.31 \text{ Å})$.

(a)   (b)   (c)

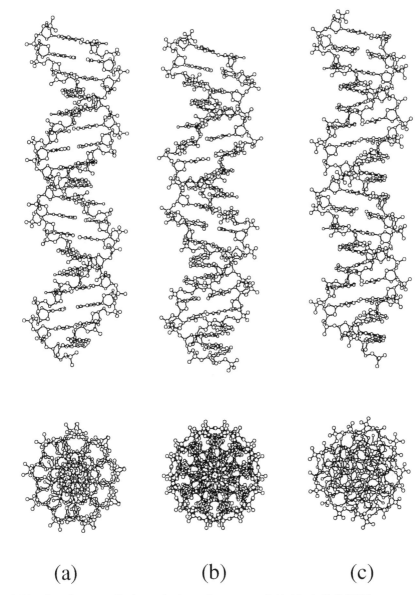

Fig. 1.14. Mutually perpendicular projections of segments of: (a) (classical) C-DNA, symmetry 28_32; (b) poly d(AG):poly d(CT), symmetry 9_2; and (c) poly d(GGT):poly d(ACC), symmetry 3_1. The views down the helix axes emphasize best how much the surfaces of these helices would 'feel' different to exploring interactants.

t) conformations are quite common. The discriminating (g^-, *t*) and (*t*, g^-) conformations for (ε, ζ) also occur, and may indeed represent a conformational language of likely nucleotide sequences. The morphological consequences of this language may be braille-like wrinkles on the surface of DNA, but so far all the evidence indicates that this language has a sloppy vocabulary and that it is impressionistic rather than precise, just as one would expect from a potentially rather polymorphic polymer that is most often merely a substrate.

The most dramatically new allomorphs of DNA, the left-handed forms, called trivially Z, were discovered during the 1970s, also with alternating purine–pyrimidine base sequences. The first allomorph was detected in an exotic variant of poly d(AT):poly d(AT), namely poly d(^{s4}TA):poly d(^{s4}TA), by Saenger *et al.* (9), (Fig. 1.9b). It has a structure (Fig. 1.10b) which is a sevenfold helix of dinucleotides (i.e. the helix twist is ±51.4°) with an axial rise per dinucleotide that is 7.60 Å. Unfortunately, Saenger *et al.* did not even contemplate seriously a Watson–Crick base-paired structure for their exotic new complex, far less a left-handed duplex, and so a great opportunity went unrecognized until pointed out by Arnott *et al.* (8) when they discovered a similar novel diffraction pattern (Fig. 1.9a) for poly d(GC):poly d(GC) in an old fibre that earlier had been shown to contain B-DNA duplex helices. Their new allomorph was a sixfold helix of dinucleotides, with, therefore a helix twist of ±60.0°. Its axial rise per dinucleotide was 7.25 Å. Unfortunately for these researchers too, the new allomorph had already been visualized from a single-crystal analysis of oligo d(GC):oligo d(GC) (7) and shown to be, unprecedentedly, left-handed. Even so, the fibre structures (Fig. 1.10a,b) attest to two important conclusions: first, Z-DNAs are also polymorphic; and secondly, the B to Z transition can take place in a not very wet or plastic fibre, suggesting that inversion of helix sense involves a mechanism with limited local melting, base unstacking, and rotation, followed by total rotations of individual quasi-cyclindical molecules. All of this could conceivably take place in the hydrated solid state.

8. Envoi

In the 1980s and 1990s, fibre diffraction analyses of polymers have largely given way to single-crystal analyses of oligonucleotides. It would be a pity if the former were extinguished altogether. The structures of polymer molecules are not subject to end-effects, nor are they terrorized by lattices; the sizes, shapes, and space groups of their lattices are more likely to reflect their intrinsic dimensions and symmetries rather than the reverse. Certainly, interactions of polynucleotides with drugs and the like may be visualized more precisely in high resolution single-crystal analyses, but it could be that subsequent direct measurements in a polymeric system of the effects of the interactions would be more convincing than extrapolatory modelling. Even if such collaborations do not evolve, it would be a denial of an important pioneering era in the history of molecular biology to disguise or diminish how much information about nucleic acid secondary structures was distilled from X-ray studies of fibres in the third quarter of this century.

9. Appendix: further details of fibrous polynucleotide structures together with some comments

A comprehensive survey of fibrous polynucleotide studies was prepared and published by Chandrasekaran and Arnott in the mid-1980s and published (50) in 1989. Some of these results are reproduced here but with a different emphasis and with revisions of certain complex structures that have been reviewed since, such as the double- and tripled-stranded helices where each strand in the complex has a different conformation from the other(s).

9.1 Fibre diffraction analysis

The number, quality, and resolving power of the X-ray diffraction intensities from fibrous specimens are rarely sufficient for the relative atomic positions in the diffracting molecules to be established independently with useful accuracy. However, as with crystallography of oligonucleotides, there are systematic schemes for augmenting these data with non-controversial stereochemical information, which certainly includes the primary structure of the polymer and the most probable values of its bond lengths and angles. Further metrical constraints may be provided by the dimensions and symmetry of the unit cell, by the requirements that non-bonded atoms should never be less than certain distances apart, and by the requirement that hydrogen-bonded and polar interactions should be characterized by a narrow range of distances. The meeting together of these rather different kinds of data can lead to very detailed structures in which most of the atomic positions are defined to within a few hundredths of a nanometer, which is a precision adequate for identifying the critical interactions within and between molecules.

How far one proceeds varies from case to case, since there are a great many kinds of partially ordered systems of helical molecules, each giving rise to different types of fibre diffraction patterns in which both continuous intensity and Bragg maxima occur. If we wish to analyse quantitatively a diffraction pattern, we of course must succeed in modelling not only the molecular structure, but also the molecular packing. This is true for any diffraction pattern, but for fibre diffraction patterns there is additional complexity because the modes of packing are more varied and complex than in single crystals. With fibrous structures, solving the X-ray phase problem, and arbitration between plausible alternative models devised to provide the initial solution of this problem, is more of an issue than with crystallographic analyses, where multiple isomorphous replacement can lead to an unbiased experimental solution. Although a direct or experimental solution of the X-ray phase problem is not usually possible for fibrous structures, the extensive symmetry of helical molecules means that the molecular asymmetric unit is commonly a relatively small chemical unit such as a few nucleotides. It is therefore not difficult to fabricate a preliminary model that provides an approximate solution to the phase problem and then to refine this model to provide a 'best' solution. This process, however, provides no assurance that the solution is unique. Other stereochemically plausible models may have to be considered. Fortunately, the linked-atom least-squares approach (21,22) provides a very good framework for objective arbitration; independent refinements of competing models provide the best model of each kind; the final values of the residuals provide measures of the acceptability of various models; and these measures of relative acceptability can be compared using standard statistical tests (27) and the decision made whether or not a particular model is significantly superior to any other. This approach has been consistently applied to the structures detailed in this Appendix.

Table 1.1. List of nucleic acid structures

	Structure	References
1	A-DNA (calf thymus)	18,50
2	A-DNA poly d(ABr⁵U) : polyd(ABr⁵U)	50
3	A-DNA (calf thymus) : poly d(A1T2C3G4G5A6A7T8G9G10T11) : poly d(A1C2C3A4T5T6C7C8G9A10T11)	50
4	B-DNA (calf thymus)	51,50
5	B-DNA poly d(GC) : poly d(GC)	52
6	B-DNA (calf thymus) Poly d(C1C2C3C4C5) : poly d(G6G7G8G9G10)	53
7	C-DNA (calf thymus)	54
8	C-DNA poly d(GGT) : poly d(ACC)	54
9	C-DNA poly d(G1G2T3) : poly d(A4C5C6)	54
10	C-DNA poly d(AG) : poly d(CT)	50
11	C-DNA poly d(A1G2) : poly d(C3T4)	50
12	D-DNA poly d(AAT) : poly d(ATT)	55,50
13	D-DNA poly d(IC) : poly d(IC) or poly d(AT) : poly d(AT)	52
14	D-DNA poly d(A1T2A3T4A5T6) : poly d(A1T2A3T4A5T6)	47
15	Z-DNA poly d(GC) : poly d(GC)	8,50
16	Z-DNA poly d(As4T) : poly d(As⁴T)	8,50
17	L-DNA (calf thymus) poly d(RY) : poly d(RY)	56,50
18	B′-DNA α poly d(A) : poly d(T)	57
19	B′-DNA β_2 Poly d(A) : poly d(T)	58
20	A-RNA poly(A) : poly (U)	59,50
21	A′-RNA poly(I) : poly(C)	59,50
22	Poly(A) : poly d(T)	60,50
23	Poly d(G) : poly (C)	60,50
24	Poly d(I) : poly (C)	61
25	Poly d(A) : poly (U)	61
26	Poly(X) : poly(X) (10-fold)	35
27	Poly(X) : poly(X) (11-fold)	35
28	Poly(s²U) : poly(s²U) (symmetric base pair)	34
29	Poly(s²U) : poly(s²U) (asymmetric base pair)	34
30	Poly d(C) : poly d(I) : poly d(C)	62
31	Poly d(T) : poly d(A) : poly d(T)	63,62
32	Poly(U) : poly (A) : Poly (U) (11-fold)	62
33	Poly(U) : poly (A) : Poly (U) (12-fold)	63
34	Poly(I) : poly(A) : poly (I)	63,62
35	Poly(I) : poly(I) : poly (I) : poly (I)	64,29
36	Poly(C) or poly(mC) or Poly (eC)	50,31,65
37	B′-DNA β_2 poly d(A) : poly d(U)	44
38	B′-DNA β_1 poly d(A) : poly d(T)	43
39	B′-DNA β_2 poly d(AI) : poly d(CT)	45
40	B′-DNA β_1 poly d(AI) : poly d(CT)	45
41	B′-DNA poly d(AATT) : poly d(AATT)	49

9.2 The structures and tables

The development of the methodologies for analysing fibre diffraction patterns proceeded concurrently with the discovery of new patterns and with the availability of more powerful computers. Consequently, some structures in the earlier literature are flawed in having no hydrogen atoms and in retaining more steric compression than need be tolerated now. Among the 41 structures listed in Table 1.1, with the exception of a few (7,30,33,35, and 36), this has been remedied in that the models presented here come either from recent analyses of new structures or modern re-refinements of older models.

For each structure, the helix symmetry (P_Q) and the unit-cell dimensions are given in Table 1.2; under repeating unit, n, is listed the number of nucleotides in one, two, or three chains that constitute the molecular asymmetric unit. In some duplexes the two chains are (or are assumed to be) antiparallel and identical. This implies that there is a diad axis perpendicular to the screw axis. Formally, this is indicated as $2P_Q$. When P is an even integer (as in B-DNA), there is necessarily another diad perpendicular to the first at half a pitch along the helix axis. This situation is indicated formally by $22P_Q$.

The conformation angles are listed in Table 1.3. If more than one chain is involved in the molecular asymmetric unit of a structure, it is indicated by chain 1, 2, etc., immediately after the structure number. The angles α, β, γ, δ, ε, and ξ are the backbone conformation angles at bonds P–O5′, O5′–C5′, C5′–C4′, C4′–C3′, C3′–O3′, and O3′–P, respectively; the glycosidic conformation, χ, is the conformation at the C1′–N bond; the endocyclic conformation angles of the sugar rings are $\nu_0,\ldots,\nu_4$.

The dispositions and shapes of base pairs (Table 1.4) are of some interest and in this presentation the older descriptions are provided to allow comparison with reference 50. The radial shift d and the lateral shear s are the orthogonal components of the displacement of a base pair from the helix axis in the xy-plane that is perpendicular to it. The propeller twist, θ_P, of the two bases in a pair is defined like a conformation angle. The angles between base normals and the helix axis, γ_1 and γ_2, are equal or similar in most structures. The tilt of the whole base pair is θ_T, while θ_R is the roll angle of the ith base pair. The relative roll $\Delta\theta_R = \theta_R(i-1)-\theta_R(i)$ is also of interest, as, of course, is t, the local helical twist. All these parameters are defined in Millane *et al.* (40).

The dimensions of grooves in Watson–Crick base-paired 'smooth' duplexes, wherein only one nucleotide per chain constitutes the molecular asymmetric unit, are given in Table 1.5. These have been calculated following Arnott (66).

The orientation of the phosphate group relative to the helix axis in each of the structures is provided in Table 1.6. θ_1 and θ_2 are, respectively, the angles that the P–O1 and P–O2 bonds make with the helix axis. Similarly, θ_3 and θ_4 are, respectively, the angles that the line O1···O2 and the bisector of the O1–P–O2 plane make with the helix axis.

Finally, Table 1.7 shows the mean values for many morphological and conformational features of polynucleotide helices derived from single-crystal diffraction analyses of oligonucleotides (26).

Table 1.2. Molecular and crystal structures. Number of nucleotides in the molecular asymmetric unit n, helix symmetry P_Q, unit cell dimensions a, b, c, α, β, γ. For structure description and references see Table 2.1

Structure	n	P	Q	a (nm)	b (nm)	c (nm)	α (°)	β (°)	γ (°)
1	1	11	1	2.17	3.99	2.80	90.0	96.8	90.0
2	2	11	2	2.23	4.14	5.60	90.0	90.0	90.0
3	11	1	1	2.17	3.99	2.80	90.0	96.8	90.0
4	1	10	1	3.08	2.24	3.37	90.0	90.0	90.0
5	2	5	1	3.79	3.61	3.36	90.0	90.0	90.0
6	5+5	2	1	3.08	2.24	3.37	90.0	90.0	90.0
7	1	28	3	3.50	3.50	9.24	90.0	90.0	120.0
				3.22	2.02	9.33	90.0	90.0	90.0
8	1	9	1	3.34	3.34	2.98	90.0	90.0	120.0
9	3+3	3	1	3.34	3.34	2.98	90.0	90.0	120.0
10	2	9	2	2.20	2.20	5.87	90.0	90.0	120.0
11	2+2	9	2	2.20	2.20	5.87	90.0	90.0	120.0
12	1	8	1	1.95	1.95	2.41	90.0	90.0	120.0
13	2	4	1	1.69	1.69	2.42	90.0	90.0	90.0
		4	1	1.70	1.70	2.43	90.0	90.0	90.0
14	6	4	3	1.72	1.72	7.40	90.0	90.0	90.0
15	2	6	−1	1.91	1.91	4.35	90.0	90.0	120.0
16	2	7	−1	1.77	1.77	5.32	90.0	90.0	120.0
17	2	1	1	2.00	1.15	1.02	90.0	90.0	90.0
18	1+1	10	1	2.32	2.32	3.32	90.0	90.0	120.0
19	1+1	10	1	1.87	3.55	3.23	90.0	90.0	90.0
20	1+1	11	1	3.97	3.97	3.09	90.0	90.0	120.0
21	1	12	1	3.94	3.94	3.60	90.0	90.0	120.0
22	1	11	1	2.36	2.36	2.81	90.0	90.0	120.0
23	1	45	4	2.32	2.32	11.32	90.0	90.0	120.0
24	1+1	10	1	2.32	2.32	3.13	90.0	90.0	120.0
25	1+1	11	1	2.48	2.48	3.37	90.0	90.0	120.0
26	1	10	1	2.11	2.11	3.01	90.0	90.0	120.0
27	1	11	1	2.35	2.35	2.77	90.0	90.0	120.0
28	1+1	11	1	2.15	3.73	2.86	90.0	90.0	90.0
29	1+1	11	1	2.15	3.73	2.86	90.0	90.0	90.0
30	1+1+1	11	1	–	–	3.48	–	–	–
31	1+1+1	12	1	4.95	4.95	3.84	90.0	90.0	120.0
32	1+1+1	11	1	4.58	4.58	3.35	90.0	90.0	120.0
33	1+1+1	12	1	2.71	2.71	3.65	90.0	90.0	120.0
34	1+1+1	12	1	4.03	4.03	3.97	90.0	90.0	120.0
35	1	23	2	2.79	2.79	7.84	90.0	90.0	120.0
36	1	6	1	2.32	2.32	1.86	90.0	90.0	120.0
		6	1	1.58	2.16	1.89	90.0	90.0	90.0
		6	1	1.65	2.19	1.89	90.0	90.0	90.0
37	1+1	10	1	1.84	3.49	3.20	90.0	90.0	90.0
38	1+1	10	1	1.86	2.27	3.24	90.0	90.0	99.9
39	2+2	5	1	1.93	2.32	3.21	90.0	90.0	98.7
40	2+2	5	1	1.93	2.32	3.21	90.0	90.0	98.7
41	4	5	2	3.11	2.26	3.39	90.0	90.0	90.0

The two entries for structure 13 are successively for poly d(IC) : poly d(IC) and poly d(AT) : poly d(AT). The three entries for structure 37 are successively for poly (C), poly (mC), and poly (eC).

Table 1.3. Conformation angles. Conformation angles (°) in the nucleotide backbone (α to ζ), about the glycosyl bond (χ), and the endocyclic conformation angles in the sugar ring (ν_0 to ν_4). For the definition of the angles see text. For the references see Table 2.1

Structure	Chain	Nucleotide	α	β	γ	δ	ε	ζ	χ	ν_0	ν_1	ν_2	ν_3	ν_4
1		N	−52	175	42	79	−148	−75	−157	8	−34	44	−40	21
2		A	−58	176	47	81	−149	−75	−157	2	−26	39	−38	23
		U	−58	173	48	77	−150	−72	−156	4	−30	43	−41	23
3		A1	−71	−177	−50	−79	−154	−64	−152	−1	−24	−38	−40	26
		T2	−67	176	56	76	−161	−68	−153	−5	−22	−38	−42	30
		C3	−66	−174	46	80	−151	−70	−150	3	−27	39	−39	23
		G4	−68	174	58	78	−156	−73	−156	−1	−25	39	−40	26
		G5	−66	178	52	83	−152	−65	−157	5	−28	38	−36	20
		A6	−69	−175	50	81	−153	−67	−152	0	−24	37	−38	24
		A7	−68	180	52	78	−155	−69	−150	0	−25	39	−40	25
		T8	−65	180	51	80	−154	−66	−151	0	−24	38	−39	24
		G9	−68	173	59	79	−160	−69	−156	−5	−20	35	−39	28
		G10	−68	−178	52	82	−146	−70	−155	−1	−22	36	−37	24
		T11	−68	175	55	80	−160	−67	−153	−6	−19	35	−39	29
4		N	−30	136	31	143	−141	−161	−98	−33	45	−40	23	6
5		C	−30	126	47	143	−85	−169	−97	−35	53	−48	30	3
		G	−66	145	22	147	−156	−158	−74	−16	37	−42	34	−12
6	1	C1	−59	173	51	137	−92	172	−105	−40	47	−36	15	16
		C2	−64	128	41	125	−163	−102	−110	−38	37	−23	2	22
		C3	−50	173	39	143	−150	171	−90	−33	44	−38	20	8
		C4	−46	120	68	127	−151	−106	−134	−36	37	−24	4	20
		C5	−75	−173	52	140	−173	−102	−101	−29	39	−33	17	8
	2	G6	−32	174	39	153	−175	−95	−108	−6	25	−34	31	−16
		G7	−61	180	43	144	−145	−136	−103	−30	42	−37	20	6
		G8	−44	154	32	146	−72	158	−98	−31	44	−39	23	5
		G9	−84	158	27	153	−161	−129	−95	−15	33	−38	30	−10
		G10	−47	152	47	141	−176	−147	−109	−33	42	−36	18	9
7		N	−37	−160	37	157	161	−106	−97	−4	25	−35	33	−19
8		N	−65	134	43	145	−100	−179	−84	−24	39	−38	26	−2
9	1	G1	−57	149	55	147	−148	−151	−96	−11	28	−33	27	−11
		G2	−62	143	67	144	−166	−108	−104	−9	−30	25	−10	
		T3	−63	168	57	138	−141	−158	−100	−25	36	−32	18	4
	2	A4	−64	124	44	141	−95	170	−95	−32	45	−39	22	6
		C5	−69	125	51	133	−96	−152	−113	−34	41	−32	14	13
		C6	−61	128	45	139	−96	174	−81	−33	44	−38	20	8
10		A	−63	110	98	146	−148	−89	−137	−23	37	−36	23	0
		G	−83	−173	39	146	−166	−155	−88	−23	37	−36	23	0
11	1	A1	−90	147	88	147	−153	−164	−108	−31	45	−41	24	4
		G2	−50	153	48	147	−141	−113	−122	−13	29	−33	26	−8
	2	C3	−82	131	62	147	−162	−87	−113	−23	38	−38	25	−2
		T4	−64	173	36	131	−95	145	−91	−44	48	−35	11	21
12		N	−59	156	64	145	−163	−131	−102	−13	36	−42	36	−15
13		C	−51	140	61	146	−128	−141	−115	−28	41	−40	25	2

Table 1.3. *Continued*

Structure	Chain	Nucleotide	α	β	γ	δ	ε	ζ	χ	ν_0	ν_1	ν_2	ν_3	ν_4
		I	−76	142	68	148	−152	−154	−105	−28	43	−42	27	0
14		A1	−64	138	78	141	−169	−104	−118	−10	30	−37	32	−15
		T2	−84	178	74	140	−96	−160	−103	−23	42	−42	30	−5
		A3	−103	124	85	143	172	−101	−101	−10	31	−38	33	−15
		T4	−78	−176	74	138	−114	−160	−105	−20	38	−38	25	−5
		A5	−38	126	81	141	−179	−96	−99	−11	30	−36	31	−13
		T6	−72	171	68	130	−152	−149	−111	−31	43	−37	20	7
15		G	52	179	−174	95	−104	−65	59	−4	−11	21	−24	17
		C	−140	−137	51	138	−97	82	−154	−28	36	−31	16	7
16		A	58	−175	−179	93	−107	−61	61	−8	−9	21	−25	21
		T	−139	−137	49	133	−98	79	−149	−32	37	−29	11	13
17		R	82	−162	180	76	171	125	26	−3	−24	39	−42	29
		Y	−60	−133	−139	147	−84	98	167	−37	53	−48	29	5
18		A	−36	127	35	137	−127	−166	−107	−39	49	−42	21	11
		T	−40	138	46	133	−144	−148	−111	−44	50	−37	13	19
19	1	A	−42	135	43	136	−135	−156	−113	−49	56	−42	16	20
	2	T	−43	147	40	143	−146	−147	−116	−38	49	−40	20	11
20		A	−69	179	55	82	−154	−71	−161	2	−25	37	−37	22
		U	−64	178	51	83	−152	−173	−161	2	−25	37	−37	22
21		N	−70	177	61	77	−153	−70	−163	−3	−23	38	−42	28
22		N	−85	−153	48	83	180	−50	−155	3	−26	37	−36	21
23		G	−58	176	46	83	−148	−78	−167	3	−26	37	−36	21
		C	−60	178	47	83	−148	−78	−167	3	−26	37	−36	21
24		I	−81	180	63	134	−169	−106	−119	−32	37	10	14	
		C	−82	169	72	86	−146	−75	−155	4	−25	35	−34	19
25		A	−69	−176	51	130	−174	−101	−121	−36	38	−27	7	18
		U	−74	180	60	84	−153	−72	−160	8	−30	39	−36	17
26		X	−75	171	63	87	−142	−80	−156	7	−27	36	−33	17
27		X	−66	−179	51	80	−153	−70	−163	3	−27	39	−38	22
28	1	U	−46	169	41	77	−147	−76	−157	0	−26	40	−41	26
2		U	−45	172	37	80	−148	−77	−156	3	−28	41	−40	24
29	1	U	−43	172	37	77	−146	−77	−162	−3	−24	39	−41	28
	2	U	−43	163	41	80	−148	−80	−149	4	−29	42	−40	23
30	1	C	−61	176	51	83	−155	−71	−158	3	−26	37	−36	21
	2	I	−82	172	72	83	−151	−73	−157	3	−26	37	−36	21
	3	C	−65	178	54	83	−153	−72	−156	3	−26	37	−36	21
31	1	T	−48	131	28	135	−114	−162	−117	−46	53	−42	17	18
	2	A	−40	155	41	127	−158	−128	−113	−49	51	−34	6	27
	3	T	−38	154	28	135	−149	−135	−111	−43	49	−36	12	19
32	1	U	−74	177	62	83	−146	−78	−166	5	−30	41	−37	21
	2	A	−99	−167	74	138	−174	−110	−123	−42	50	−39	14	19
	3	U	−44	−178	26	132	−170	−101	−131	−39	44	−31	7	21
33	1	U	−28	171	23	83	−156	−75	−154	3	−26	37	−36	21
	2	A	−66	−179	53	83	−163	−67	−149	3	−26	37	−36	21
	3	U	−40	167	37	83	−149	−83	−156	3	−26	37	−36	21
34	1	I	−40	131	52	80	−120	−114	−173	4	−28	40	−38	22

Table 1.3. *Continued*

Structure	Chain	Nucleotide	α	β	γ	δ	ε	ζ	χ	ν_0	ν_1	ν_2	ν_3	ν_4
	2	A	−74	179	63	82	−160	−68	−168	1	−23	36	−35	22
	3	I	−75	178	64	82	−155	−72	−163	3	−26	38	−36	21
35		I	−103	176	92	83	−156	−69	−169	3	−26	37	−36	21
36		C	−78	173	64	83	−125	−67	−161	3	−26	37	−36	21
37	1	A	−53	137	49	136	−133	−150	−117	−51	56	−40	13	23
	2	U	−58	146	66	122	−157	−120	−127	−45	46	−28	3	26
38	1	A	−45	128	37	139	−119	−170	−109	−48	56	−43	19	17
	2	T	−41	136	38	141	−133	−160	−115	−42	52	−42	19	14
39	1	A1	−74	177	70	139	−174	−124	−126	−28	36	−31	16	8
		I2	−53	165	53	127	−169	−105	−123	−44	45	−29	4	25
	2	C3	−47	161	58	138	−162	−97	−134	−44	46	−29	5	22
		T4	−43	164	28	131	−175	−139	−124	−29	36	−30	17	9
40	1	A1	−44	161	25	144	−150	−137	−111	−40	52	−44	22	11
		I2	−78	132	58	149	−106	−173	−109	−39	54	−47	27	7
	2	C3	−46	168	32	138	−160	−128	−108	−46	57	−45	21	15
		T4	−71	135	75	143	−134	−148	−101	−46	54	−41	17	17
41		A1	−44	125	24	129	176	−132	−101	−38	40	−28	7	19
		A2	−44	163	59	−140	−163	−98	−115	−17	28	−28	18	−1
		T3	−45	172	27	143	−166	−135	−98	−37	48	−41	21	10
		T4	−56	172	50	159	−96	168	−96	−22	44	−47	36	−9

9.3 Commentary

What is evident from the 41 structures listed in Table 1.1 is the wide coverage of polynucleotide helices that is provided by fibre diffraction analysis. Most of them are Watson–Crick paired duplexes, but not all; some base sequences from native material are, in effect, random, but some are special in the extreme—homopolymers, for example.

The diverse crystal structures in Table 1.2 attest to another important consideration and that is the range of environments inhabited by the different molecular helices. Fibres of polynucleotides, like single crystals of oligonucleotides, are awash with water, some of it firmly bound and contributing to the diffraction in a cooperative, crystal-like fashion, but a great deal of it more indifferently distributed from cell to cell in a more liquid-like fashion. The point is that the polynucleotide helices examined in fibres are not only unperturbed by the end-effects that have to be suspected in crystals of polynucleotide fragments, but are unlikely to be perturbed by lateral packing effects in their spacious fibrous environments. In these environments microcrystallinity is an option for molecular packing but is not obligatory—in many of the fibres the constituent molecules are merely uniaxially oriented. In studies of oligonucleotide crystals there has been selection for only those conformations that have ended up in crystals.

Beneath the diversity of structure apparent in Tables 1.1 and 1.2 there is the much more conservative framework indicating that all the nucleotides belong to one of two major genera, A or B, and a very few belong to a third genus Z. The very common A

Table 1.4. Base pair orientations and helical twists. Base pair positions, orientations and helical twists in the Watson–Crick base-paired duplexes. For definition of symbols see text. For the references see Table 1.1

Structure	Nucleotide	d (nm)	s (nm)	θ_P (°)	γ_1 (°)	γ_2 (°)	θ_T (°)	θ_R (°)	$\Delta\theta_R$ (°)	t (°)
1	N	0.48	0.00	−10.5	23.2	23.2	22.6	0.0	0.0	32.7
2	A	0.46	0.00	−11.9	22.4	23.2	22.0	−1.6	3.1	32.7
	U	0.46	0.00	−11.9	23.2	22.4	22.0	1.6	−3.1	31.8
3	A1	0.48	0.03	−13.0	22.5	24.1	22.2	−2.8	2.8	33.8
	T2	0.43	−0.01	−11.0	25.2	25.4	24.7	−0.4	−2.4	31.3
	C3	0.47	−0.03	−10.6	25.4	25.6	25.0	−0.4	0.0	32.4
	G4	0.47	−0.04	−10.2	25..4	26.1	25.2	−1.7	1.3	31.6
	G5	0.51	−0.01	−13.2	25.0	25.0	24.1	0.1	−1.8	33.7
	A6	0.51	0.01	−13.2	25.0	24.1	25.0	−0.1	0.2	34.5
	A7	0.47	0.04	−10.2	26.1	25.4	25.2	1.7	−1.8	33.7
	T8	0.47	0.03	−10.6	25.6	25.4	25.0	0.4	1.3	31.6
	A9	0.43	0.01	−11.0	25.4	25.2	24.7	0.4	0.0	32.6
	A10	0.48	−0.03	−13.0	24.0	22.5	22.2	2.8	−2.4	31.3
	T11	0.48	0.00	−9.0	22.8	22.8	22.4	0.0	2.8	33.8
4	N	−0.02	0.00	−15.1	8.1	8.1	2.8	0.0	0.0	36.0
5	C	−0.05	−0.01	−14.2	9.1	9.2	5.7	0.0	0.0	29.5
	G	−0.05	0.01	−14.2	9.2	9.1	5.7	0.0	0.0	42.5
6	C1	−0.06	−0.03	−1.8	1.3	2.3	1.3	−1.0	3.5	35.7
	C2	0.02	−0.01	−23.9	18.4	5.8	1.6	6.4	−7.3	37.1
	C3	−0.09	0.00	5.6	6.1	11.0	4.1	7.4	−1.0	36.5
	C4	−0.02	0.00	−18.6	17.5	1.1	−0.4	8.2	−0.8	34.2
	C5	0.07	0.00	−17.3	11.3	6.3	−1.5	2.5	5.7	36.5
7	N	−0.05	0.00	−1.8	8.2	8.2	−8.2	0.0	0.0	38.6
8	N	−0.29	0.00	−18.5	11.2	11.2	−6.4	0.0	0.0	40.0
9	G1	−0.21	0.02	−19.8	12.2	12.1	7.1	−0.1	3.4	39.2
	G2	−0.21	0.08	−28.6	16.4	18.6	10.2	1.4	−1.5	42.9
	T3	−0.27	0.00	−12.1	9.1	12.7	8.7	3.3	−1.9	37.9
10	A	−0.18	0.01	−17.3	20.9	11.0	−11.0	9.1	−18.2	42.4
	G	−0.18	−0.01	−17.3	11.0	20.9	−11.0	−9.1	18.2	37.6
11	A1	−0.25	0.09	−5.3	11.0	10.1	−10.1	1.8	−9.3	46.0
	G2	−0.14	0.04	−23.7	16.8	25.4	−16.3	−7.5	9.3	34.0
12	N	−0.17	0.00	−21.0	16.6	16.6	−13.0	0.0	0.0	45.0
13	C	−0.19	−0.02	−16.9	17.8	16.3	−14.8	1.5	−3.0	44.8
	I	−0.19	0.02	−16.9	16.3	17.8	−14.8	−1.5	3.0	45.2
14	A1	−0.39	−0.09	−14.5	20.1	18.0	−17.5	2.8	−5.6	49.6
	T2	−0.23	−0.22	−27.3	19.1	22.7	−15.8	−2.8	5.6	36.5
	A3	−0.09	−0.01	−21.4	23.3	20.0	−18.7	3.3	−6.1	54.4
	T4	−0.09	0.01	−21.4	20.0	23.3	−18.7	−3.3	6.6	38.7
	A5	−0.23	0.23	−27.3	22.7	19.1	−15.8	2.8	−6.1	54.4
	T6	−0.39	0.09	−14.5	18.0	20.1	−17.5	−2.8	5.6	36.5
15	G	−0.30	−0.26	8.3	6.9	1.4	0.1	2.8	−5.6	−10.7
	C	−0.30	0.26	8.3	1.4	6.9	0.1	−2.8	5.6	−49.3

Table 1.4. *Continued*

Structure	Nucleotide	d (nm)	s (nm)	θ_P (°)	γ_1 (°)	γ_2 (°)	θ_T (°)	θ_R (°)	$\Delta\theta_R$ (°)	t (°)
16	A	−0.25	−0.25	−8.6	1.7	7.8	−1.5	3.4	−6.8	−8.1
	T	−0.25	0.25	−8.6	7.8	1.7	−1.5	−3.4	6.8	−43.4
17	R	0.12	−0.18	−21.1	6.2	16.9	−4.0	5.8	−11.7	12.0
	Y	0.12	0.18	−21.1	16.9	6.2	−4.0	−5.8	11.7	−12.0
18	A	0.03	−0.03	−22.0	8.0	16.1	−4.7	−4.5	0.0	36.0
19	A	0.08	−0.01	−15.1	12.9	6.9	−5.9	4.0	0.0	36.0
20	A	0.44	0.01	−2.1	15.6	15.5	−15.5	−0.8	0.0	32.7
21	I	0.51	0.00	2.3	10.6	10.6	−10.6	0.0	0.0	30.0
22	A	0.50	0.00	10.5	20.2	20.2	19.5	0.0	0.0	32.7
23	G	0.51	0.00	16.1	18.0	18.0	16.1	0.0	0.0	32.0
24	I	0.25	0.00	−14.3	14.9	19.5	14.9	−5.4	0.0	36.0
25	A	0.38	−0.04	−4.3	12.7	13.3	12.7	−2.0	0.0	32.7
30	I	0.33	0.04	−13.3	12.9	8.8	8.2	4.0	0.0	32.7
31	A	0.31	0.02	−5.5	17.9	13.1	7.3	13.5	9.7	30.0
32	A	0.48	0.00	−1.0	17.2	17.2	17.2	−0.1	6.0	32.0
33	A	0.39	0.07	−8.8	13.4	10.6	10.6	5.1	0.0	30.0
34	A	0.25	0.00	−10.1	6.9	5.1	3.1	0.4	0.0	30.0
37	A	0.09	0.01	−21.5	11.2	12.1	−4.4	−0.5	0.0	36.0
38	A	0.03	0.01	−12.0	10.6	8.3	7.2	−1.8	0.0	36.0
39	A	0.15	0.04	−12.3	12.1	9.6	−8.8	2.2	4.6	35.0
	I	0.07	0.04	−5.4	12.8	9.5	−8.5	6.8	−4.6	37.0
40	A	0.11	−0.07	−19.8	2.0	21.7	0.5	−11.8	5.3	31.0
	I	0.11	0.04	−22.3	4.6	17.7	0.7	−6.5	−5.3	40.0
41	A1	−0.03	0.00	−8.9	6.7	15.4	−1.8	10.9	−21.7	37.0
	A2	0.02	0.01	−20.4	6.3	15.4	3.4	4.8	6.0	39.0
	T3	0.02	−0.01	−20.4	15.4	6.3	3.4	−4.8	9.7	27.0
	T4	−0.03	0.00	−8.9	15.4	6.7	−1.8	−10.9	6.0	39.0

and B conformations each aggregate into right-handed helices; only the Z conformations aggregate into left-handed helices.

The diagnostic conformational difference between A and B was long ago identified as the sugar ring pucker which is C3′-*endo* in A structures and C2′-*endo* in B. This translates into there being a very different set of endocyclic conformation angles ($\nu_0,\ldots,\nu_4$) for the furanose rings [cf. (A-DNA) structure 1 and (B-DNA) structure 4, in Table 1.3]. More simply, one can use δ (which is equivalent to ν_3), and of the order of 80° in A and 140° in B. Associated with the different furanose conformations are different values for χ, the glycosidic conformation which has a (60°) greater magnitude in A than in B. Other local nucleotide conformational differences between A and B are evident in ζ (O3′–P) which in A is invariably g^- with a mean value of −80° but in B can be g^- or t with a mean value of −120° but a wide range. The neighbouring conformation angle, ε (C3′–O3′), is t in A (mean value −160°) but can be t or

Table 1.5. Groove dimensions. Dimensions of major and minor grooves in 'smooth' Watson–Crick base-paired duplexes, i.e. those in which all the nucleotides are assumed to have the same conformations

Structure	Major		Minor	
	Width (nm)	Depth (nm)	Width (nm)	Depth (nm)
1	0.22	1.30	1.11	0.26
4	1.16	0.85	0.60	0.82
7	1.05	0.76	0.48	0.79
8	1.10	0.50	0.35	0.72
12	0.96	0.62	0.08	0.74
18	1.38	0.90	0.29	0.71
19	1.41	0.96	0.28	0.70
20	0.47	1.29	1.08	0.33
21	0.89	1.44	1.05	0.34
22	0.27	1.36	1.09	0.29
23	0.46	1.36	1.11	0.25
24	0.65	1.17	0.95	0.53
25	0.87	1.31	0.93	0.45
26	0.07	1.14	1.38	0.39
27	0.08	1.36	1.24	0.28
28	0.36	1.27	1.00	0.24
30	0.79	1.22	1.02	0.44
31	1.57	1.39	0.62	0.67
32	0.85	1.43	0.95	0.39
33	0.85	1.27	1.08	0.41
34	0.87	1.28	1.26	0.55
37	1.38	0.96	0.30	0.72
38	1.40	0.89	0.27	0.71

Table 1.6. Phosphate group orientations. Phosphate group orientations relative to helix-axis in fibrous polynucleotide structures

Structure	Chain	Nucleotide	θ_1 (°)	θ_2 (°)	θ_3 (°)	θ_4 (°)
1		N	145	89	61	140
2		A	140	95	66	144
		U	139	96	67	143
3		A1	132	105	76	151
		T2	136	99	71	147
		C3	137	101	71	151
		G4	140	94	66	142
		G5	142	96	66	147
		A6	136	103	73	154
		A7	135	101	72	148
		T8	138	98	69	147
		G9	137	99	71	147

Table 1.6. *Continued*

Structure	Chain	Nucleotide	θ_1 (°)	θ_2 (°)	θ_3 (°)	θ_4 (°)
		G10	134	106	76	156
		T11	130	104	76	147
4		N	59	139	138	103
5		C	67	128	126	102
		G	18	124	153	68
6	1	C1	82	115	109	105
		C2	25	136	163	80
		C3	73	117	116	98
		C4	67	135	130	108
		C5	59	133	135	100
	2	G6	95	128	108	131
		G7	75	125	119	107
		G8	66	118	121	93
		G9	18	131	162	74
		G10	70	125	123	102
7		N	93	123	107	124
8		N	18	118	147	63
9	1	G1	51	133	141	93
		G2	53	136	141	96
		T3	64	121	125	94
	2	A4	19	115	143	60
		C5	21	128	156	73
		C6	20	108	138	54
10		A	66	121	123	96
		G	47	117	132	78
11	1	A1	52	112	126	77
		G2	65	143	136	111
	2	C3	23	137	167	79
		T4	64	101	112	76
12		N	62	133	132	102
13		C	56	135	138	98
		I	36	122	143	75
14		A1	62	131	132	100
		T2	64	129	129	101
		A3	14	109	140	52
		T4	73	126	121	106
		A5	22	122	149	68
		T6	69	117	118	96
15		G	28	95	125	41
		C	109	36	48	63
16		A	27	99	128	46
		T	118	38	42	73
17		R	42	140	153	91
		Y	90	101	96	101
18		A	48	140	148	95
		T	54	135	140	97
19	1	A	50	136	143	94
	2	T	62	135	133	104

Table 1.6. *Continued*

Structure	Chain	Nucleotide	θ_1 (°)	θ_2 (°)	θ_3 (°)	θ_4 (°)
20		A	137	99	70	146
		U	137	97	69	144
21		N	135	99	71	144
22		N	135	108	76	162
23		G	144	89	61	139
		C	144	89	61	139
24		I	92	130	111	129
		C	124	106	80	142
25		A	95	128	108	132
		U	136	99	70	145
26		X	135	97	70	141
27		X	141	96	67	147
28	1	U	140	92	65	139
	2	U	141	91	63	139
29	1	U	147	85	57	135
	2	U	130	100	74	140
30	1	C	130	101	75	142
	2	I	123	108	82	144
	3	C	135	97	70	142
31	1	T	57	136	138	100
	2	A	73	131	124	110
	3	T	83	128	116	118
32	1	U	143	92	63	141
	2	A	102	130	105	143
	3	U	100	123	103	133
33	1	U	133	94	69	135
	2	A	125	108	81	146
	3	U	134	92	67	134
34	1	I	117	91	75	117
	2	A	124	107	81	144
	3	I	131	101	74	143
35		I	117	116	89	141
36		C	141	91	63	138
37	1	A	54	131	137	94
	2	U	69	131	127	106
38	1	A	43	136	149	89
	2	T	59	138	138	102
39	1	A1	81	133	120	119
		I2	73	127	122	107
		C3	93	124	108	125
		T4	72	127	123	106
40	1	A1	68	132	128	106
		I2	37	128	147	80
	2	C3	47	127	139	86
		T4	64	133	131	103
41		A1	30	142	168	86
		A2	90	127	111	125
		T3	66	124	125	99
		T4	80	119	113	106

Table 1.7. Comparison of helix parameters for A-, B-, and Z-DNA derived from crystal structures (from ref. 26)

1. Base step parameters

Helix	Step	Roll $\Delta\theta_R$ (°)	Tilt (°)	Cup (°)	Slide (nm)	Twist t (°)	Rise h (nm)	Rad (P) (nm)
B	NpN	0.6	0.0	0.0	0.04	36.1	0.34	0.94
A	NpN	6.3	–	–	−0.16	31.1	0.29	0.95
Z	RpY	5.8	0.0	−12.5	−0.11	−50.6	0.35	0.73
	YpR	−5.8	0.0	12.5	0.54	−9.4	0.39	0.63

2. Base pair parameters

Helix	Base	Tip θ_R (°)	Incl θ_T (°)	Prop θ_P (°)	Buck (°)	X d (nm)	Y s (nm)	P–P (nm)
B	N	0.0	2.4	−11.1	−0.2	0.08	0.01	0.88–1.40
A	N	11.0	12.0	−8.3	−2.4	0.41	–	0.12–1.59
Z	R	−2.9	−6.2	−1.3	6.2	0.30	0.23	0.77
	Y	2.9	−6.2	−1.3	−6.2	0.30	−0.23	1.37

3. Main chain conformation angles (°)

Helix	pN	α	β	γ	δ	ε	ζ	χ	Sugar
B	N	−65	167	51	129	−157	−120	−103	2′*en*
A	N	−73	173	64	78	−151	−77	−165	3′*en*
Z	R	48	179	−170	100	−104	−69	67	3′*en*
	Y	−137	−139	55	138	−94	80	−159	2′*en*

g^- in B, producing, therefore, a mean value lower in magnitude. The effect of changes in ε and ζ is to alter greatly the orientation of the phosphate groups. The 'wrinkles' on many B helices usually take the form of altered phosphate orientations and therefore distinctly different values for ε and or ζ (cf. structures 4 and 5, and structures 12, 13, and 14).

The two Z structures (15 and 16) emerge from alternating purine–pyrimidine (RY) poly(dinucleotides). The R nucleotide is not only C3′-*endo* but the glycosidic angle is *syn* (+60°) rather than the usual *anti* (–0 to −160°). The Y nucleotide is C2′-*endo* with an *anti* glycocidic conformation. All the other conformational angles in the two nucleotides are different also.

When one examines the summary of oligonucleotide structures in Table 1.7 what is depressing or reassuring—depending on one's vantage point—is that nothing dramatically new has been revealed about the common conformations of nucleotides in helices. Much has been made of the fine morphological differences from one structure to another. Older terms such as *tilt* and *twist* have been given new meanings and now *incline* and *propeller*, and bases and base pairs also *roll*, *cup*, *slide*, *tip*, and *buckle* as, up to a point, they must in lower symmetry arrangements. Nevertheless, it is hard not to conclude that only one revelation of significance has emerged from oligonucleotide crystal studies, i.e. the possibility of the existence of Z structures in polynucleotides containing some alternating purine and pyrimidine nucleotides.

Acknowledgements

The presidency of Scotland's oldest university, St Andrews, founded 1410–1413 AD, is not the best vantage point from which to write a review of polynucleotide structure, even an historical one. I am, therefore, very indebted to my long-time friend and former colleague Professor R. Chandrasekaran of Purdue University for keeping me aware of recent developments. His own contributions to polynucleotide structure determinations are substantial and the extent to which he has inherited the Wilkins' tradition of meticulous fibre diffraction studies is amply illustrated by the many useful fibre structures of polysaccharides as well as polynucleotides that are emerging from his laboratory.

References

1. Olby, R. (1974) *The Path to the Double Helix*, p. 65. University of Washington Press, Seattle.
2. Arnott, S. (1973) *Trans. Am. Cryst. Assoc.* **9**, 93.
3. Stubbs, G., Warren, Se. and Holmes, K. (1977) *Nature* **267**, 216.
4. Makowski, L. (1978) *J. Appl. Cryst.* **11**, 273.
5. Arnott, S., Chandrasekaran R., Millane, R.P. and Park, H.-S. (1986) *J. Mol. Biol.* **188**, 631.
6. Privé, G.G., Yanagi, K. and Dickerson, R.E. (1991) *J. Mol. Biol.* **217**, 177.
7. Wang, A.H-J., Quigley, G.J., Kolpak, F.J., Crawford, J.L., van Broom, J.H., van der Marel, G. and Rich, A. (1979) *Nature* **281**, 680.
8. Arnott, S., Chandrasekaran, R., Birdsall, D.L., Leslie, A.G.W. and Ratliff, R.L. (1980) *Nature* **283**, 743.
9. Saenger, W., Landmann, H. and Lazius, A.G. (1973) in *Jerusalem Symposium on Quantum Chemistry V*, p. 457. The Israeli Academy of Sciences and Humanities, Jerusalem.
10. Coll, M., Frederick, C.A., Wang, A. H.-J. and Rich, A. (1987) *Proc. Natl. Acad. Sci. USA* **84**, 8385.
11. Brown, T., Leonard, G.A., Booth, E.D. and Chambers, J. (1989) *J. Mol. Biol.* **207**, 455.
12. Wilkins, M.H.F., Stokes, A.R. and Wilson, H.R. (1953) *Nature* **171**, 738.
13. Franklin, R.E. and Gosling, R.G. (1953) *Acta Cryst.* **6**, 673.
14. Watson, J.D. and Crick, F.H.C. (1953) *Nature* **171**, 737.
15. Crick, F.H.C. and Watson, J.D. (1954) *Proc. R. Soc. (London) Ser. A.* **223**, 80.
16. Donahue, J. (1956) *Proc. Natl. Acad. Sci. USA* **42**, 60.
17. Langridge, R., Marvin, D.A., Seeds, W.E., Wilson, H.R., Hooper, C.W., Wilkins, M.H.F. and Hamilton, L.D. (1960) *J. Mol. Biol.* **2**, 28.
18. Fuller, W., Wilkins, M.H.F., Wilson, H.R., Hamilton, L.D., and Arnott, S. (1965) *J. Mol. Biol.* **12**, 60.
19. Marvin, D.A., Spencer, M., Wilkins, M.H.F. and Hamilton, L.D. (1961) *J. Mol. Biol.* **3**, 547.
20. Davies, D.R. and Baldwin, R.L. (1963) *J. Mol. Biol.* **6**, 251.
21. Arnott, S., Dover, S.D. and Wonacott, A.J. (1969) *Acta Cryst.* B **25**, 2142.

22. Smith, P.J.C. and Arnott, S. (1978) *Acta Cryst. A* **34**, 3.
23. Arnott, S., Wilkins, M.H.F., Hamilton, L.D. and Langridge, R. (1965) *J. Mol. Biol.* **27**, 391.
24. Arnott, S., Wilkins, M.H.F., Fuller, W., Venable, J. and Langridge, R. (1967) *J. Mol. Biol.* **11**, 391.
25. Arnott, S., Fuller, W., Hodgson, A. and Prutton, I. (1968) *Nature* **220**, 561.
26. Dickerson, R.E. (1992) *Meth. Enzymol.* **211**, 67.
27. Hamilton, W.D. (1965) *Acta Cryst.* **18**, 502.
28. Rich, A. (1958) *Biochim. Biophys. Acta* **29**, 502.
29. Arnott, S., Chandrasekaran, R. and Martilla, C. (1974) *Biochem. J.* **141**, 537.
30. Langridge, R. and Rich, A. (1963) *Nature* **198**, 725.
31. Arnott, S., Chandrasekaran, R. and Leslie, A.G.W. (1976) *J. Mol. Biol.* **106**, 735.
32. Mitsui, Y., Langridge, R., Shortle, B.E., Cantor, C.R., Grant, R.C., Kodama, M. and Wells, R.D. (1970) *Nature* **228**, 1166.
33. Arnott, S., Chandrasekaran, R., Hukins, D.W.L., Smith, P.J.C. and Watts, L. (1974) *J. Mol. Biol.* **88**, 523.
34. Arnott, S., Chandrasekaran, R., Leslie, A.G.W., Puigjaner, L.C. and Saenger, W. (1981) *J. Mol. Biol.* **149**, 507.
35. Arnott, S., Chandrasekaran, R., Day, W.A., Puigjaner, L.C. and Watts, L. (1981) *J. Mol. Biol.* **149**, 489.
36. Arnott, S. and Bond, P.J. (1973) *Science* **181**, 68.
37. Arnott, S. and Bond, P.J. (1973) *Nature New Biology* **244**, 99.
38. Arnott, S. and Selsing, E. (1974) *J. Mol. Biol.* **88**, 509.
39. Giacometti, A. and Chandrasekaran, R. (1998) (in preparation).
40. Millane, R.P., Walker, J.K., Arnott, S., Chandrasekaran, R. and Birdsall, D.L. (1984) *Nucl. Acids Res.* **12**, 5475.
41. Arnott, S., Chandrasekaran, R., Hall, I.H. and Puigjaner, L.C. (1983) *Nucl. Acids Res.* **11**, 4141.
42. Chandrasekaran, R. and Radha, A. (1992) *J. Biomol. Struct. Dynamics* **10**, 153.
43. Chandrasekaran, R., Radha, A. and Park, H.-S. (1995) *Acta Cryst.* **D51**, 1024.
44. Chandrasekaran, R., Radha, A., Park, H.-S. and Arnott, S. (1989) *J. Biomol. Struct. Dynamics* **6**, 1203.
45. Chandrasekaran, R., Radha, A. and Park, H.-S. (1997) *J. Biomol. Struct. Dynamics* **15**, 285.
46. Arnott, S., Chandrasekaran, R., Puigjaner, L.C., Walker, J.K., Hall, I.H., Birdsall, D.L. and Ratliff, R.L. (1983) *Nucl. Acids Res* **11**, 1457.
47. Leslie, A.G.W., Arnott, S., Chandrasekaran, R. and Ratliff, R.L. (1980) *J. Mol. Biol.* **143**, 49.
48. Arnott, S. and Selsing, E. (1975) *J. Mol. Biol.* **98**, 265.
49. Chandrasekaran, R., Radha, A. and Ratliff R.L. (1994) *J. Biomol. Struct. Dynamics* **11**, 741.
50. Chandrasekaran, R. and Arnott, S. (1989) Landolt–Börnstein Numerical Data and Functional Relationships in Science and Technology (Group VII, Biophysics), Subvolume VII 1b, p. 31. Springer-Verlag, Berlin, Heidelberg.
51. Arnott, S. and Hukins, D.W.L. (1973) *J. Mol. Biol.* **81**, 93.
52. Arnott, S., Chandrasekaran, R., Puigjaner, L.C., Walker, J.K., Hall, I.H. and Birdsall, D.L (1983) *Nucl. Acids Res.* **11**, 1457.
53. Chandrasekaran, R., Arnott, S., He, R.-G., Millane, R.P., Park, H.-S., Puigjaner, L.C. and Walker, J.K. (1985) *J. Macromol. Sci. Phys.* **24**, 1.
54. Arnott, S. and Selsing, E. (1975) *J. Mol. Biol.* **98**, 243.
55. Selsing, E., Arnott, S. and Ratliff, R.L. (1975) *J. Mol. Biol.* **98**, 243.

56. Arnott, S. and Chandrasekaran, R. (1980) *Nature* **287**, 561
57. Park, H.-S., Arnott, S., Chandrasekaran, R., Millane, R.P. and Campagnari, F. (1987) *J. Mol. Biol.* **197**, 513.
58. Chandrasekaran, R. and Radha, A. (1992) *J. Biomol. Struct. Dynamics* **10**, 153.
59. Arnott, S., Hukins, D.W.L., Dover, S.S., Fuller, W. and Hodgson., A.R. (1973) *J. Mol. Biol.* **81**, 107.
60. Arnott, S., Chandrasekaran, R., Puigjaner, L.C., Walker, J.K., Hall, I.H. and Birdsall, D.L (1983) *Nucl. Acids Res.* **11**, 1457.
61. Arnott, S., Chandrasekaran, R., Millane, R.P., and Park, H.-S (1986) *J. Mol. Biol.* **188**, 631.
62. Arnott, S., Bond, P.J., Selsing, E. and Smith, P.J.C. (1976) *Nucl. Acids Res.* **3**, 2459.
63. Giacometti, A. and Chandrasekaran, R. (1997) in preparation.
64. Chou, C.H., Thomas, Jr, G.J., Arnott, S. and Smith, P.J.C. (1977) *Nucl. Acids Res.* **4**, 2407.
65. Leslie, A.G.W. and Arnott, S. (1978) *J. Mol. Biol.* **119**, 399.
66. Arnott, S (1976) in *Organisation and Expression of Chromosomes*, (Bautz, E.K.F., McCarthy, B.J., Schimke, R.T. and Tissieres, A. eds), p. 209. Dahlem Konferenzen, Berlin.

2

Base and base pair morphologies, helical parameters, and definitions

Richard Lavery and Krystyna Zakrzewska

Laboratoire de Biochimie Théorique, CNRS UPR 9080, Institut de Biologie Physico-Chimique, 13, Rue Pierre et Marie Curie, Paris 75005, France

1. Introduction

As time passes, the complexity of nucleic acid structure and conformation continues to increase rapidly. The beautifully regular double helix of Watson and Crick has lost its symmetry with the appearance of major base sequence effects and local perturbations caused by base modifications, mispairing, bulges, and abasic sites. In addition, larger scale deformations such as curvature and groove width variations have come to light and are particularly important for understanding drug and protein binding. The standard duplex has also been joined by an ever-growing collection of new structures, including triple and quadruple helices, parallel duplexes, mutually intercalated duplexes, stem loops and three- and four-branch junctions. This growth has also been fuelled by rapid progress involving RNAs, which has revealed a host of complex tertiary conformations, and also by the creation of novel oligonucleotides destined to bind to specific DNA or RNA targets as part of the antigene and antisense strategies for artificial genetic control.

This increase in complexity requires a parallel effort in developing the means for describing and analysing the new structures. This need exists on several different levels: in classifying the basic elements of the structures (strand direction, pairing schemes, etc.), in describing the detail of conformation (notably, to enable structures to be compared in a quantitative way), and in dealing with data including conformational dynamics (such as the trajectories generated in increasingly realistic MD simulations). This chapter attempts to summarize the present state of affairs in each of these areas and to point out the difficulties that still exist.

2. Nucleic acid bases

The standard nucleic acid bases are illustrated in Fig. 2.1. In the case of DNA, they comprise two purines (abbreviated, Pur or R), adenine (Ade or A), and guanine (Gua or G), each containing two fused rings with five and six atoms, respectively, and two pyrimidines (Pyr or Y), thymine (Thy or T), and cytosine (Cyt or C), each containing a single six-atom ring. Within RNA, thymine is replaced by uracil (Ura or U) which differs only in the lack of a methyl group at position 5. The figure shows the standard notation of the base atoms and their geometries are listed in Table 2.1 (which, for

Table 2.1. Standard base geometries. [Taken from the most recent fibre coordinates for canonical B-DNA (53)]. For reference, backbone bond lengths have also been included

(a) Bonds lengths (Å)

Adenine

N1–C2	1.332	N1–C6	1.346	C2–N3	1.315	C2–H2	1.000
N3–C4	1.349	C4–C5	1.365	C4–N9	1.370	C5–C6	1.404
C5–N7	1.388	C6–N6	1.341	N6–H61	1.000	N6–H62	1.000
N7–C8	1.297	C8–N9	1.366	C8–H8	1.000		

Guanine

N1–C2	1.381	N1–C6	1.402	N1–H1	1.000	C2–N2	1.335
C2–N3	1.331	N2–H21	1.000	N2–H22	1.000	N3–C4	1.359
C4–C5	1.375	C4–N9	1.378	C5–C6	1.419	C5–N7	1.394
C6–O6	1.228	N7–C8	1.311	C8–N9	1.378	C8–H8	1.000

Thymine

N1–C2	1.374	N1–C6	1.370	C2–O2	1.219	C2–N3	1.381
N3–C4	1.380	N3–H3	1.000	C4–O4	1.233	C4–C5	1.444
C5–C6	1.343	C5–C7	1.500	C6–H6	1.000	C7–H71	1.090
C7–H72	1.090	C7–H73	1.090				

Cytosine

N1–C2	1.392	N1–C6	1.360	C2–O2	1.237	C2–N3	1.358
N3–C4	1.339	C4–N4	1.324	C4–C5	1.433	N4–H41	1.000
N4–H42	1.000	C5–C6	1.357	C5–H5	1.000	C6–H6	1.000

Backbone

P–O1′	1.480	P–O2′	1.480	P–O3′	1.600	P–O5′	1.600
O3′–C3′	1.422	O5′–C5′	1.440	C1′–C2′	1.525	C1′–O4′	1.419
C1′–H1′	1.090	C1′–N9	1.490	C2′–C3′	1.529	C2′–H2′	1.090
C2′–H2′	1.090	C3′–C4′	1.529	C3′–H3	1.090	C4′–C5′	1.516
C4′–O4′	1.457	C4′–H4	1.091	C5′–H5′	1.090	C5′–H5′	1.090

(b) Bond angles (°)

Adenine

C2′–C1′–N9	113.71	O4′–C1′–N9	108.11	H1′–C1′–N9	109.46
C2–N1–C6	118.83	N1–C2–N3	129.18	N1–C2–H2	115.41
N3–C2–H2	115.42	C2–N3–C4	110.82	N3–C4–C5	126.69
N3–C4–N9	127.20	C5–C4–N9	106.11	C4–C5–C6	117.11
C4–C5–N7	110.48	C6–C5–N7	132.41	N1–C6–C5	117.38
N1–C6–N6	119.12	C5–C6–N6	123.50	C6–N6–H61	119.99
C6–N6–H62	120.02	H61–N6–H62	119.99	C5–N7–C8	103.97
N7–C8–N9	113.83	N7–C8–H8	123.06	N9–C8–H8	123.11
C1′–N9–C4	126.00	C1′–N9–C8	128.39	C4–N9–C8	105.60

Guanine

C2′–C1′–N9	113.71	O4′–C1′–N9	108.11	H1′–C1′–N9	109.46
C2–N1–C6	125.23	C2–N1–H1	117.37	C6–N1–H1	117.39
N1–C2–N2	116.05	N1–C2–N3	123.30	N2–C2–N3	120.65
C2–N2–H21	119.99	C2–N2–H22	120.01	H21–N2–H22	120.00
C2–N3–C4	112.25	N3–C4–C5	128.51	N3–C4–N9	125.07
C5–C4–N9	106.43	C4–C5–C6	119.31	C4–C5–N7	110.61
C6–C5–N7	130.07	N1–C6–C5	111.40	N1–C6–O6	119.80

Table 2.1. *Continued*

C5–C6–O6	128.80	C5–N7–C8	103.75	N7–C8–N9	113.99
N7–C8–H8	123.01	N9–C8–H8	122.99	C1′–N9–C4	125.60
C1′–N9–C8	129.18	C4–N9–C8	105.22		
Thymine					
C2′–C1′–N1	113.71	O4′–C1′–N1	108.11	H1′–C1′–N1	109.46
C1′–N1–C2	117.09	C1′–N1–C6	120.84	C2–N1–C6	122.07
N1–C2–O2	122.93	N1–C2–N3	115.43	O2–C2–N3	121.64
C2–N3–C4	126.40	C2–N3–H3	116.78	C4–N3–H3	116.82
N3–C4–O4	120.55	N3–C4–C5	114.09	O4–C4–C5	125.36
C4–C5–C6	120.75	C4–C5–C7	117.53	C6–C5–C7	121.72
N1–C6–C5	121.26	N1–C6–H6	119.39	C5–C6–H6	119.35
C5–C7–H71	109.49	C5–C7–H72	109.45	C5–C7–H73	109.45
H71–C7–H72	109.49	H71–C7–H73	109.50	H72–C7–H73	109.45
Cytosine					
C2′–C1′–N1	113.71	O4′–C1′–N1	108.10	H1′–C1′–N1	109.46
C1′–N1–C2	117.80	C1′–N1–C6	121.05	C2–N1–C6	121.15
N1–C2–O2	118.85	N1–C2–N3	118.70	O2–C2–N3	122.45
C2–N3–C4	120.63	N3–C4–N4	118.32	N3–C4–C5	121.55
N4–C4–C5	120.13	C4–N4–H41	120.01	C4–N4–H42	119.99
H41–N4–H42	120.00	C4–C5–C6	116.89	C4–C5–H5	121.56
C6–C5–H5	121.55	N1–C6–C5	121.08	N1–C6–H6	119.48
C5–C6–H6	119.44				

Amino and methyl hydrogens are named by adding 1, 2, or 3 to the parent atom number, thus, G(N2) carries the hydrogens H21 and H22. The methyl group of thymine is numbered C7.

completeness, also provides the bond lengths within the phosphodiester backbone). Since all the bases contain conjugated rings, their most stable conformations are planar. They can, nevertheless, undergo non-planar deformations as a result of thermal agitation, steric strain, or the presence of other species.

In addition to these standard bases many others are found within nucleic acids. These may occur naturally, as in the case of RNAs that contain both modified bases (m^2G, m^7G, m^1A, m^5C, wybutine, etc.) and unconventional linkages (e.g. pseudouracil) (1). Other unusual bases are the result of chemical modifications (see below), while still others are voluntarily introduced into oligonucleotides with specific goals in mind. This is the case for efforts aimed at generating so-called 'universal' bases which could recognize more than one paired partner and thus be very useful in designing antisense or antigene oligonucleotides (2). The reader is referred to Wolfram Saenger's book on nucleic acids for an overview of modified base structures (1).

Amongst the various chemical modifications that the bases can undergo, protonation and methylation merit consideration. Protonation occurs most readily at C(N3), A(N1), G(N7), and T(O4). Such changes considerably modify the charge distribution within the conjugated bases and also modify the pairing schemes they can adopt. A well-known example of such modifications involves cytosine, for which protonation

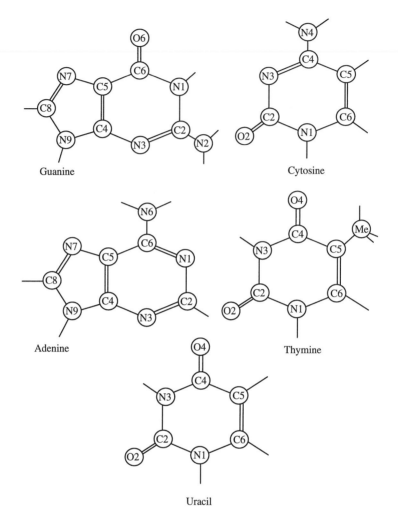

Fig. 2.1. Standard nucleic acid bases.

at N3 is a necessary step in the formation of G:C$^+$ Hoogsteen pairing within triple helices (3) or the formation of the novel i motif, which is comprised of two mutually intercalated C:C$^+$ parallel duplexes (4). The influence of the protonation on the base geometry is generally limited and local in nature, as shown by crystallographic (5) and quantum chemical studies (6).

Base methylation plays an important biological role, since it functions as a genetic control mechanism (7). The most prominent reaction occurs at C(C5), mainly in CpG sequences. The next most important site involves the external proton of A(N6). Methylation has an important effect on interactions with proteins and, for example, generally protects from endonucleases (although some enzymes of this class actually require methylated bases to function).

3. Base pairing

Standard Watson–Crick base pairs are formed by specific recognition between a purine and a pyrimidine base: adenine with thymine (or uracil) and guanine with cytosine (Fig. 2.2). These combinations lead to virtually identical base geometries as illustrated in Fig. 2.3. This identity was the basis of the realization that it is possible to build a regular helical double helix with an arbitrary base sequence and it was also the basis for understanding the replication of the genetic code. A:T pairs are maintained by two hydrogen bonds, while G:C pairs have three bonds. For isolated pairs this leads to stronger binding in the latter case (G:C −21 kcal/mol versus A:T −13 kcal/mol, measured in vacuum, ref. 8), and, in general, G:C pairs are less easily deformed or broken within DNA than A:T pairs. It should, however, be recalled that base pairing is much

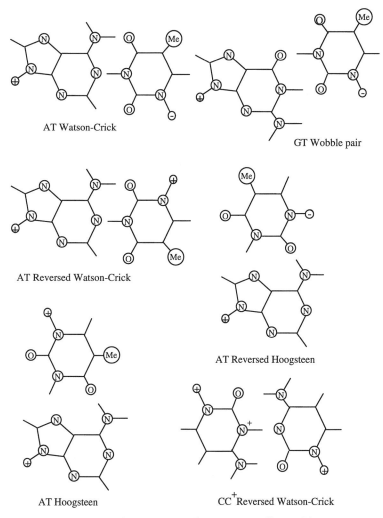

Fig. 2.2. Schematic views of various types of base pairing.

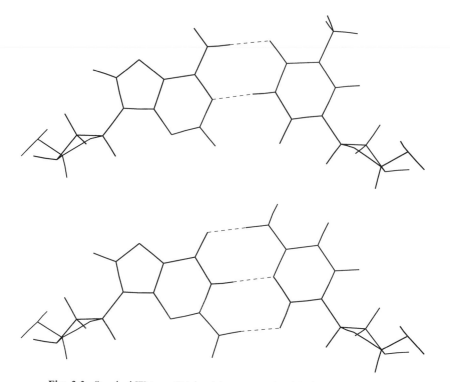

Fig. 2.3. Standard Watson–Crick pairing geometries: A:T (top), G:C (bottom).

weaker in water than in vacuum [the values are unknown in water, since isolated bases prefer to stack rather than to pair, but G:C pairing in chloroform (9) is reduced to −5.8 kcal/mol].

It should be noted that canonical Watson–Crick base pairs involve bases in the keto and amino forms. In parallel with the appearance of the double helical model for DNA, Watson and Crick proposed that tautomerism from keto to enol and from amino to imino forms could be at the origin of the point mutations necessary to power evolution (10). As shown in Fig. 2.4, such tautomerism permits the formation of A:C and G:T pairs with overall geometries very close to those of the canonical base pairs. Much effort has since been put into attempts to demonstrate the importance of such mechanisms for mutagenesis (11). The present state of knowledge, however, suggests that point mutations occur most frequently as a result of the formation of G:T (Fig. 2.2) and A⁺:C wobble pairs rather than of tautomeric forms. This is supported by an increasing number of crystallographic structures containing mispairs (12; see also Chapter 10). It is also clear, today, that the flexibility of the double helix also allows it to accommodate R:R and Y:Y pairs whose C1′–C1′ separations (when the interaction involves the Watson–Crick faces) are, respectively, much wider (12.5 Å) and much narrower (8.4 Å) than those of the canonical base pairs (11 Å). In the case of R:R mispairing it is, however, also possible to diminish these steric constraints by changing to *syn* conformations (13,14), while NMR data show that Y:Y pairs can be extended by water bridging (15).

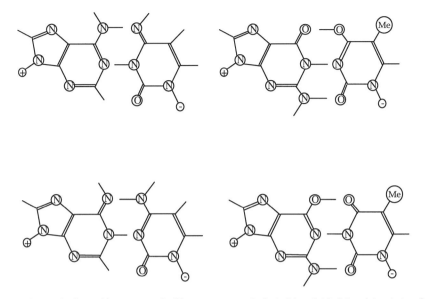

Fig. 2.4. Base pairs formed by non-standard base tautomers. Left: A:C⋆ and A⋆:C involving imino forms. Right: G:T⋆ and G⋆:T involving enol forms.

Despite the primordial importance of the standard Watson–Crick pairs, the ways that bases can be assembled by hydrogen bonding are remarkably varied. As the range of nucleic acid conformations has progressed, more and more structures containing non-canonical base interactions have been generated (see examples in Fig. 2.2). The most important alternative pairings are probably the Hoogsteen and reversed Hoogsteen schemes, which occur notably within triplet helices (see Chapter 12). These pairings involve either purines or pyrimidines interacting with the sites on purine bases that are not involved in Watson–Crick hydrogen bonding (N7 and O6 in the case of guanine, N7 and N6 in the case of adenine). This explains why a combination of Watson–Crick and Hoogsteen (or reversed Hoogsteen) pairing can coexist within a triplex (Fig. 2.5). It should be noted that Watson–Crick pairing can also be 'reversed' in the case of certain bases. This occurs notably in parallel-stranded DNA (16). Most unusual pairs are less stable than their canonical cousins, but they are often sterically advantageous, by being adapted to narrow or wide strand separations or to particular backbone orientations. Examples of unusual pairings occur commonly within the complex folded structures of RNAs, within loops, within mispairs, and within chemically modified nucleic acids.

Lastly, with certain bases, it is also possible to form four-stranded (or quadruplex) structures as shown in Fig. 2.6 involving G tetrads (17–19; see also Chapter 13). It is also interesting to note that two identical base pairs can also form favourable interactions between their major groove faces (20). This type of interaction is the basis of a four-stranded structure that could play a role in homologous recombination (21).

It is possible to describe these multiple pairing schemes in an ordered way. The first important step in this direction came from the work of Rose *et al.* (22) who remarked that the nucleic acid bases have two distinct faces (because they have no twofold

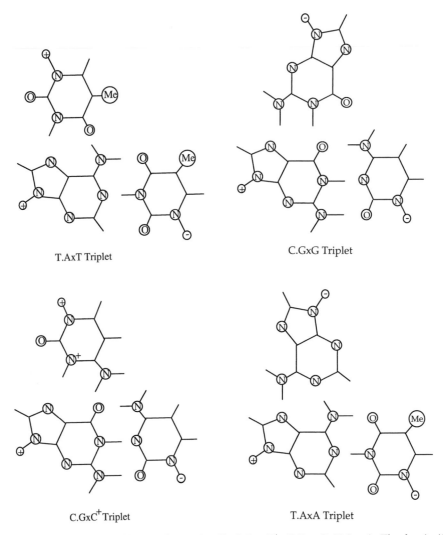

Fig. 2.5. Schematic views of base triplets: T.A × T, C.G × C⁺, C.G × G, T.A × A. The dot signifies Watson–Crick pairing between the first two strands and the cross either Hoogsteen (TAT, CGC⁺) or reverse Hoogsteen pairing (CGG, TAA) between the first and third strands.

symmetry axis in their π-plane). This point, which became important when discussing the difference between B- and Z-DNA (23), led to the idea that these faces should be distinguished when describing pairing interactions. A unique definition can be made using the right-hand rule, with the fingers of the right hand pointing around the shortest distance from the glycosidic bond to the Watson–Crick pairing edge of the base. The direction of the thumb then indicates a unique face which we will conventionally colour white, the opposing face being black. To make a simplified diagram that shows not only base orientation, but also strand orientation, we draw a rectangle for the base (longer for purines than for pyrimidines), add a line representing the

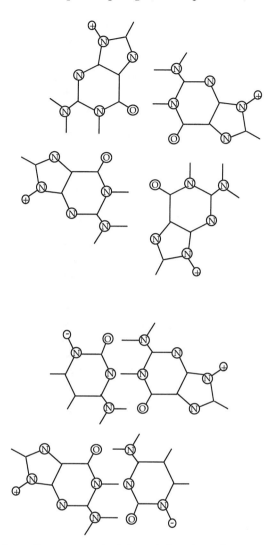

Fig. 2.6. Schematic views of base quartets: (top) G_4, with four identical reverse Hoogsteen pairings, and (bottom) $(AT)_2$, where two Watson–Crick A:T pairs interact through their major groove faces.

glycosidic bond, and add a small circle representing the strand: a white circle means that the $5' \rightarrow 3'$ direction points upwards, while a black circle means that it points downwards. For canonical nucleotides, with an *anti* conformation around the glyco-sidic bond, a white circle will always accompany a white base face, whereas, for *syn* nucleotides, the circle and the base face will have opposite colours.

We can use this scheme to classify duplex structures fully (Fig. 2.7). Taking into account the Watson–Crick (W), reversed Watson–Crick (C), Hoogsteen (H), and reversed Hoogsteen (R) pairing schemes, two nucleotides may be paired in four dif-ferent ways. Since, in addition, each nucleotide can be in one of two possible states, there are a total of 16 distinct combinations. In fact, as the figure shows, there are

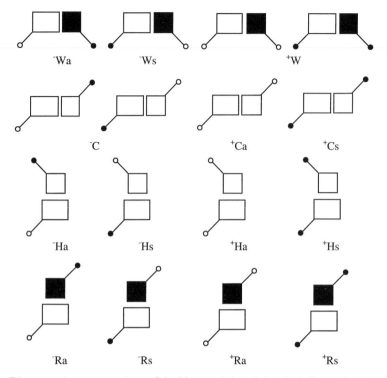

Fig. 2.7. Diagrammatic representations of double-stranded nucleic acid helices: W, Watson–Crick; C, reversed Watson–Crick; R, reversed Hoogsteen; H, Hoogsteen.

actually only 14 unique classes because of two degeneracies created by the pseudo-dyad symmetry of W and C pairing (note each base pair in Fig. 2.7 is oriented so that the left-hand base, or lower base, shows its white face). Each structural family can be defined by a notation consisting of a letter to specify the base pairing (W, C, H, or R), a prefix indicating whether the strand directions are parallel (+) or antiparallel (–) and a suffix specifying whether the left-hand (or lower) nucleotide is of type 'a' or type 's' (this index can be dropped in the case of the degenerate pairs $^+W_a/^+W_s$ and $^-C_a/^-C_s$).

If we consider the classical B-DNA duplex (corresponding to the diagram in the top left-hand corner of Fig. 2.7), the combination of Watson–Crick pairing and *anti* nucleotides automatically leads to antiparallel strands. The same result can be obtained with reverse Hoogsteen pairing, which also has one white and one black base face exposed, but not with reversed Watson–Crick or Hoogsteen pairing. This result points to the utility of such a scheme. It links together three factors: the type of pairing, the *anti/syn* conformation of the nucleotides, and the strand directions. This can be very useful in building nucleic acid structures since if any one of these data is absent, its nature can be deduced from the other two (24,25): when the base pair faces are of a common colour, parallel strands imply nucleotides of the same type and antiparallel strands imply nucleotides of different types. The opposite is true when two colours appear on the base pair faces. One must, however, be cautious concerning one point—the strand direction referred to here is local. If we attempt to apply these rules to Z-DNA, Watson–Crick pairing combined with a *syn* purine base and an *anti* pyrimidine base would imply parallel strands. In fact, this is true on a local level, despite the fact that, macroscopically, Z-DNA is an antiparallel duplex. This apparent conflict is created by the strong zigzag in the phosphodiester backbone which gave this conformation its name (26).

A number of the conformational families shown in Fig. 2.7 are already known. The family ^-W_a corresponds to B-DNA (or A-DNA) with antiparallel strands, Watson–Crick hydrogen bonding, and *anti* nucleotides. The family ^-W_s corresponds to Z-DNA. The representation of ^-W_s makes it clear that base pairs have to be turned over in passing from the B to the Z conformation, since to align the strand directions between the first two families in the figure it is necessary to invert the ^-W_s diagram around a horizontal axis leading to a base pair with a black face on the left and a white face on the right. Changing the nucleotide stereochemistry at C1′ by using α-nucleotides is one route to new Watson–Crick families. This change diminishes the steric hindrance associated with *syn* conformation pyrimidines. It is thus not surprising that an all *a*-nucleotide duplex belonging to the family ^-W_s can be formed (27). The final family that can be made with Watson–Crick base pairs, ^+W, has also been observed in parallel-stranded duplexes where one strand is again composed of α-nucleotides in the *syn* conformation (28). For the reversed Watson–Crick duplexes only the ^+C_a family is currently known. It is found in parallel-stranded DNA (16) and in the unusual four-stranded i motif structure (4).

Hoogsteen and reversed Hoogsteen pairings are seen within triple helices and, indeed, it is possible to extend this classification scheme to both triplex and quadruplex structures (24). Figure 2.8 shows an example of this for the 16 triplex families that can be built from Watson–Crick duplexes. Each of these triplexes is named on the basis of its two constituent duplexes. The first family, built from a ^-W_a Watson–Crick duplex and a ^-H_a Hoogsteen duplex, thus becomes $^-W_a^-H_a$. In fact, the nucleotide type indicated for the second base pair can be dropped since it must be identical to that of the first pair. The nucleotide type refers to the left-hand, or lower, nucleotides for the constituent duplexes and is necessarily common to any pair of duplexes forming a triplex.

The best known triple helix made by adding a Hoogsteen-bonded thymidine strand to a poly (dA):poly (dT) double helix (29) corresponds to the family $^-W_a^+H$ since all nucleotides are *anti* and the Hoogsteen-bound poly (dT) strand is parallel to the adenosine strand of the duplex. An identical triple helix family CGC$^+$ can also be formed under acidic conditions by adding a protonated cytosine strand to a poly (dG):poly (dC) duplex, again using Hoogsteen hydrogen bonding (see also Fig. 2.5). The only other way to form an all *anti* triple helix starting from a Watson–Crick duplex is the family $^-W_a^-R$, which has indeed been experimentally observed for T.A×A and C.G×G triple helices (30,31) where the two purine strands form an antiparallel, reversed Hoogsteen duplex (see Fig. 2.5). A related family containing *syn* nucleotides in the third strand, $^-W_a^+R$, is also known to exist when α-thymidine nucleotides, which can easily adopt the *syn* conformation, are built into the third strand of a T.A×T triplex (32).

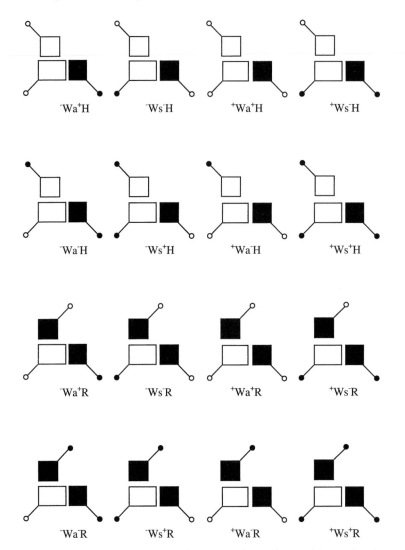

Fig. 2.8. Diagrammatic representations of triple-stranded nucleic acid helices based on Watson–Crick duplexes.

Forming the triple helices shown in the second column of Fig. 2.8 seems unlikely since the only known form of the ^-W_s duplex is Z-DNA, in which the major groove face of the base pairs is sterically hindered. In contrast, starting from a parallel-stranded Watson–Crick duplex ^+W (formed using an α-nucleotide pyrimidine strand) one could form triple helices belonging to the families $^+W_a{}^+H$ or $^+W_a{}^-R$, which only require *syn* conformations in the Watson–Crick-bound pyrimidine strand. It is possible to continue this classification to deal with other triplexes, for example based on reversed Watson–Crick duplexes, and also with quadruplexes. The reader may refer to an earlier publication for details (24).

4. Helical parameter definitions

Although a completely detailed description of the conformation of a nucleic acid fragment requires $3N$ Cartesian coordinates for N atoms, it is possible to reduce significantly this number of variables. Since bond lengths and valence angles vary only slightly, one can describe a conformation successfully using only torsion angles. Since the intracyclic torsion angles of the sugar rings are dependent on one another, it is actually easier to describe ring puckering using the well-known pseudo-rotational variables (1) phase (P) and amplitude (A). In this case, a nucleotide can be described by a total of eight variables: the backbone torsions α (P–O5′), β (O5′–C5′), γ (C5′–C4′), ε(C3′–O3′), and ζ(O3′–P); the sugar conformation given by P and A (which also fix the δ torsion around C4′–C3′); and the glycosidic bond χ (C1′–N9 for purines or C1′–N1 for pyrimidines). However, these parameters are not very helpful for judging the overall shape of the molecule. Since nucleic acids often form helical structures it is clearly useful to be able to describe their helical geometry in a more direct way. Such parameters have been employed since the very first nucleic acid conformations were obtained, but because these conformations resulted from fibre diffraction experiments, which average out base sequence information, only perfectly helical conformations were considered.

For regular helices, the helical axis can be located rather easily. When the tails of difference vectors joining symmetrically related atoms (for example, successive C1′ sugar atoms within an oligonucleotide strand) are brought together, their heads generate a plane and the helical axis is defined by the perpendicular to this plane. A point on the helical axis may be located by joining the heads of successive vectors (which lie on a circle around the axis) and projecting perpendiculars to these lines into the plane described. The helical axis must lie at the intersection of these perpendiculars (33). Once the helical axis is known, the position of the base pairs can be described in terms of the rise and twist between successive pairs, leading also to the pitch of the helix and the number of base pairs per turn. If a reference axis system is defined for each base pair [this is often taken as the line joining the R(C8)–Y(C6) atoms], it is possible to fix the distance of the base pair from the axis and the inclination of the base pair with respect to the axis. The calculation of such parameters has been described in detail by Struther Arnott (34).

When the first single-crystal nucleic acid conformations appeared, it was clear that such descriptions were insufficient. The conformation of the famous oligomer d(CGCGAATTCGCG)$_2$ (35) showed that base sequence effects led to a deformed double helix with non-planar base pairs, fluctuating rise and twist values, and a kinked helical axis. If such deformations remain small it is still reasonable to look for an optimal linear axis to describe the structure. This can be done with the method described above, but using an eigenvalue approach to find the shortest principal axis of the ellipsoidal cloud now formed by the heads of the vectors. A point on the axis is similarly found by looking for the barycentre of the dispersed intersection points of the perpendiculars to the projected vectors (33).

Observed deformations, however, led to the need for an increased number of parameters for describing the base pairs, such as the propeller twist angle, formed by the contra-rotation of the bases around their long axes, or slide, characterizing the lateral displacement of successive base pairs. Such parameters were still generally calculated with respect to the R(C8)–Y(C6) axis, or, in some cases, with respect to the line joining the glycosidic C1′–C1′ atoms. In the period following the appearance of high resolution oligomer conformations, new parameters were added to the existing set as the need was felt, without much attention being paid to coherence in definitions, names, or sign conventions. Since different groups used different parameters and calculation techniques it became extremely difficult to compare existing structures and it was clearly necessary to review the situation. This review was carried out at a meeting in Cambridge in 1988, where an effort was made to define, and name, a complete set of parameters for describing helical nucleic acids (36). Figure 2.9 shows these parameters, which were divided into three families: base pair-axis parameters, intra-base pair parameters, and inter-base pair parameters. Each family contained the geometrically required combination of rotations (r) and translations (t), 2t+2r in the first case, to position a body with respect to a vector, and 3t+3r in the other cases, which positions two bodies with respect to one another. Both base pair-axis and inter-base pair parameters can be further broken down into parameters describing individual base positions. Although it is important to have complete families of parameters for mathematical reasons, it is clear that they are not all equally interesting. Certain parameters show only small variability within standard nucleic acid conformations (notably, shear, stretch, stagger, and opening), but even these parameters can become important in describing the growing number of deformed nucleic acid conformations (see Section 6).

The Cambridge meeting fixed the names (and abbreviations) for all parameters and defined a right-handed axis reference system. The orientations of these axes, which set the positive direction for translational variables, are shown in Fig. 2.9 and notably have the pseudo-dyad axis pointing towards the major groove. The sign of all rotational parameters was chosen to correspond to right-handed rotation around the associated axes. Finally, rules were invented for building up compound parameters from the underlying parameters referring to individual bases. Thus, base pair parameters such as propeller are obtained by adding the base tip of the left-hand strand to that of the right-hand strand (left and right refer to the orientation shown in Fig. 2.9, with the viewer looking into the minor groove). Other parameters, such as buckle, are obtained by subtracting the inclination of the right-hand strand from that of the left-hand strand. These definitions are given in Table 2.2. Note that it is necessary to take into account the fact that the parameters Ydisp and tip refer to an axis that points towards the backbone of each strand (36,37). Rules also define the derivation of base pair step parameters, which are obtained by subtracting the value for the lower base pair from that of the upper base pair (again with the nucleic acid oriented as shown). With these rules, all the parameters in Fig. 2.9 (and in the various publications resulting from the Cambridge meeting) are positive. Since 1988 these parameters have been almost completely respected, although a disagreement has arisen concerning buckle, which is defined by Dickerson with a reverse sign (Fig. 2.9 would show a negative buckle in this case). A new parameter, termed cup, has also been introduced to characterize the space created when two successive base pairs buckle away from one another.

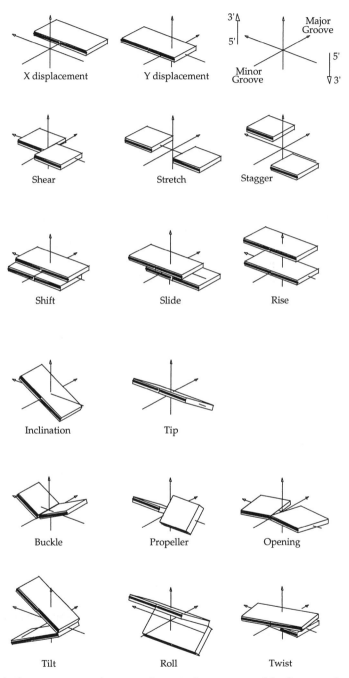

Fig. 2.9. Helical parameters. Translations are shown in the upper part of the diagram and rotations in the lower part. Each section contains base pair-axis, intrabase pair, and interbase pair parameters, respectively.

Table 2.2. Helicoidal parameter names and definitions for base pair values

Name	Family	Code	Symbol	Base pair value
X-displacement	Base–axis	XDP	dx	$(dx_L + dx_R)/2$
Y-displacement		YDP	dy	$(dy_L - dy_R)/2$
Inclination		INC	η	$(\eta_L + \eta_R)/2$
Tip		TIP	θ	$[\theta_L - \theta_R]/2$
Shear	Intra-base	SHR	S_x	$dx_L - dx_R$
Stretch		STR	S_y	$dy_L + dy_R$
Stagger		STG	S_z	$\sum_{m=1}^{i}(Dz_L - Dz_R)$
Buckle		BKL	κ	$\eta_L - \eta_R$
Propeller		PRP	ω	$\theta_L + \theta_R$
Opening		OPN	σ	$\sum_{m=1}^{i}[\Omega_L - \Omega_R]$
Shift	Inter-base	SHF	D_x	$dx(i) + A_x - dx(i-1)$
Slide		SLD	D_y	$dy(i) + A_y - dy(i-1)$
Rise		RIS	D_z	
Tilt		TLT	τ	$\eta(i) + \eta A - \eta(i-1)$
Roll		ROL	ρ	$\theta(i) + \theta_{A} - \theta(i-1)$
Twist		TWT	Ω	
Axis X-disp.	Axis	AXD	A_x	
Axis Y-disp.		AYD	A_y	
Axis inclination		AIN	η_A	
Axis tip		ATP	θ_A	

The definition column above indicates how compound parameters (base pair, interbase) are built up. A full geometrical definition of the parameters can be found in an earlier reference (42). Stagger and opening for the *i*th base pair are, respectively, sums of differences in rise and twist.

5. Helical parameter calculations

Despite the importance of defining names and sign conventions for helical parameters, it is still necessary to define how they are to be calculated. This was not determined by the Cambridge meeting, which invited those interested to compare the behaviour of the various existing methods (38–42). Since then the situation has evolved to some extent. First, most programs have been revised to respect the Cambridge recommendations and basic criteria such as the independence of parameters with respect to the direction in which an oligomer is analysed. Secondly, while certain programs have continued to be used relatively frequently, others have more or less disappeared. However, the overall choice of methods has hardly diminished since new programs have also been proposed (43–45). In addition, an important question has arisen concerning the need for defining a helical axis in the case of irregular conformations. This question is linked to two fundamentally different ways of describing nucleic acid conformations, which are termed global and local. The global approach is an extension of the analysis of regular helices. It maintains the notion of a helical axis and parameters that position the bases or base pairs with respect to this axis. The difficulty with this approach arises from the fact that a linear helical axis is no longer appropriate for many conformations. A possible solution is to calculate individual linear axes for helical segments that are more or less straight, but this choice is necessarily subjective.

The alternative local approach to helical parameters abandons the notion of a helical axis and calculates parameters that describe the junctions linking one base or base pair to the next along the nucleic acid fragment. The difference between the local and global approaches has been nicely illustrated by Calladine and Drew (46) in their discussion of the transition between the B- and A-forms of DNA. From the global viewpoint, this transition consists of moving the base pairs away from the axis towards the minor groove (negative displacement), inclining them with respect to the helical axis (positive inclination), and reducing the helical rise and twist. In contrast, from the local viewpoint, the same transition is described as sliding successive base pairs over one another in the direction of their long axes and creating a roll wedge between them. Using either local or global parameters, A- and B-DNA are distinguished from one another, but by different means. In the global case, the distinction is mainly on the level of the base pair–axis parameters, while in the local case, inter-base pair parameters must be used.

Note that while both the local approach and the global approach calculate inter-base pair parameters, these parameters can differ quite dramatically if the conformations analysed do not have base pairs centred on the axis, as in canonical B-DNA. This is owing to the fact that global inter-base pair parameters are calculated after deconvolution of base pair–axis parameters (37). It should be added, in this connection, that the global parameters shift, slide, tilt, and roll are all indicative of dissimilarity between the two strands of the double helix and these parameters are zero by definition for duplexes where dyadic symmetry exists between the two strands. These problems are reconsidered in Section 6, where we present several examples of helical analysis for regular and irregular conformations.

There has been no final decision regarding local versus global parameters and, indeed, both can be useful. Local parameter algorithms have the advantage that they avoid the difficulty of defining axes, they also reflect more closely the models developed for describing intrinsically curved DNA, which only attempt to define the twist, tilt, and roll of successive inter-base pair steps. They also yield parameters that only depend on the geometry of the given dinucleotide step. In contrast, global parameters depend on the conformation of the whole nucleic acid fragment analysed, but they have the major advantage of enabling the extent and the location of curvature to be calculated. (It must be stressed that non-zero local roll or tilt angles certainly do not imply curvature, as shown by local description of regular A-DNA discussed above.) Since axis bending is a common feature of nucleic acid conformations, notably within protein–nucleic acid complexes, the availability of a defined axis has often become a determining argument for the use of global parameters. Global parameters finally have the advantage of distinguishing more easily between the different families of helical conformation.

We present below a brief summary of the various analysis techniques that have been developed to date.

5.1 'Newhelix' (38)

'Newhelix' remains a popular DNA analysis program. It originated from the 'Modhelix' code written by Rabinovich, Reich, and Shakked at the Weizmann Institute of Science on the basis of routines coming from the Helib library of Rosenberg and Dickerson. It is basically a global parameter approach, but it is restricted to calculating an optimal linear helical axis using the technique described in Section 4. Base pair parameters are defined with respect to an R(C8)–Y(C6) reference vector, and thus helical twist becomes the angle between successive reference vectors projected on to the plane perpendicular to the helical axis, while rise is the distance between successive R(C8)–Y(C6) vectors, projected onto the helical axis. Similarly, slide is the relative movement of the base pairs along the direction defined by the averaged R(C8)–Y(C6) vectors for the step in question.

Roll and tilt have two definitions in 'Newhelix'. The original definition involves calculating the angles between the successive base pair normals, which are then resolved into the pseudo-dyad and perpendicular directions based on the helical axis reference system. This technique is appropriate only for small angles. A more recent definition involves a preliminary removal of helical twist for the base pair step in question, to avoid parameter dependencies. The two sets of values for roll and tilt are related by formulae that involve tip and inclination (45). Lastly, propeller is calculated as the angle between the two base normals projected into the plane normal to the R(C8)–Y(C6) vector, and buckle is obtained by projection into a plane defined by the R(C8)–Y(C6) vector and the bisector of the two base normals.

When local kinking is suspected, the molecule can be divided into segments for which individual straight-line axes are calculated. This simple technique nevertheless introduces a degree of subjectivity. The authors of 'Newhelix' have approached the analysis of axis deformation with a supplementary program named 'Bend' (47) which calculates the bending at a given base pair as the angle formed between the mean normal vectors belonging to the base pairs '*i*' steps before and after the test pair (*i* = 1, 2,...). Global curvature is measured as the angle between normal vectors averaged over 10 successive base pairs in order to attenuate local conformational irregularities.

5.2 'von Kitzing/Diekmann' (39)

This program is a local parameter approach. Base plane normals are defined by the atoms of the six-membered rings of the purine and pyrimidine bases. Base pair normals are taken as perpendicular to the least-squares plane of the two bases, which can be weighted to take into account the difference between purines and pyrimidines. The base pair reference vector is then defined as the projection of the R(C8)–Y(C6) vector or of the C1'–C1' vector on to the base pair plane. Decomposing the angle between the base normals along the reference vector and perpendicular to this direction then leads to the buckle and propeller angles.

The relative positions of successive base pairs are defined in a local sense using wedge angles between the base pair normals, which are decomposed along, or perpendicularly to, an axis obtained by averaging the reference vectors of the base pairs involved. This leads to roll and tilt values. The translational parameters rise, 'long axis slide' (slide), and 'short axis slide' (shift) are also calculated with respect to an axis system averaged over the two base pairs involved. In a second, so-called cylinder, approach (termed 'screw axis' by other authors), a local helical axis is defined so that the passage from one base pair axis system to the next can be obtained by rotation around and translation along this axis. The orientation of the base pair reference axes with respect to the cylinder axis is then used to define 'cylinder' roll, tilt, and twist angles, as well as the related translational parameters. These six cylindrical parameters are identical to those used by Arnott *et al.* for regular helices (34). Axis curvature is estimated by bringing the successive base pair normals, or cylinder twist axes, to a common origin. The heads of these vectors then lie on a unit sphere, whose surface can be mapped on to a plane using a Mercator projection. A bent helix will appear as a pathway across this plane.

5.3 'Tung/Soumpasis' (40)

This method uses inertial axes to create the base (or base pair) reference systems. These axes are obtained by diagonalization of the moment of inertia tensor, with the origin being taken as the centre of mass of the base (or base pair). This approach has the advantage of being applicable to both unusual bases and unusual pairing schemes, but requires detailed corrections to avoid apparent irregularity within regular helices resulting simply from variations in the chemical structure of successive base pairs.

The position of successive base pairs is defined by a translation along the difference vector joining the origins of the two reference axis systems and by a 3×3 rotation matrix. Translational parameters are obtained by projecting the difference vector on to a mean axis system between the base pairs, and rotational parameters are obtained from the rotation matrix, which is decomposed into three Euler angles. Parameters for the bases within a base pair are obtained in a similar way.

A straight global axis can also be obtained in this method by diagonalizing the moment of inertia tensor for the entire molecule or for a subset of selected atoms, such as C1' or P, to avoid artefacts related to the chemical structures of the bases. Curvature is again approached by looking at the angles formed between successive base pair normals or by calculating straight-line axes for segments of the molecule, as in the case of 'Newhelix'.

5.4 'Bansal' (41)

This method is similar to that of von Kitzing and Diekmann. The base pair reference vector is again chosen as R(C8)–Y(C6), whose midpoint determines the base pair origin. A mean plane perpendicular to the averaged base normal is then calculated. Wedge parameters are used to describe the relative orientation of two base pairs in terms of an axis system calculated by averaging the base pair reference planes and mid-points. Propeller and buckle angles are decomposed using the mean base pair normal and the base pair reference vector.

The method equally determines local helical axes for each base pair step within the molecule and related 'helical' (otherwise termed cylinder or screw axis) parameters.

A straight global axis can be calculated as a least-squares fit to the successive local axis reference points and an idea of curvature is obtained by plotting the path of the local helical axes in a plane perpendicular to the global axis.

5.5 'Babcock/Olson' (43, 44)

This method, which also calculates local helical parameters, employs a full three-dimensional rotation matrix for relating the positions of bases and base pairs. Bases are considered to rotate around a chosen pivot point and the authors have carefully considered the effect of the choice of this point on the dependence between helical parameters. The principal axis of each base passes through R(C8) or Y(C6) and is parallel to the C1'–C1' direction in the corresponding, ideal Watson–Crick pair. A perpendicular vector lies in the base plane, pointing towards the major groove, and passes through the midpoint of the ideal C1'–C1' vector. As in 'Curves' (see Section 5.7), a set of reference bases are available for fitting to experimental coordinates to avoid the effects of base deformation. Unusual geometries, such as *syn* bases, are dealt with via specially adapted reference systems.

Interbase pair parameters are calculated with respect to a coordinate frame, which is defined as the half-way rotated and translated system between the two base pair references. Parameters correspond to simultaneous rotations around the three axes in the coordinate frame and are decomposed by a formalism adapted from rigid body dynamics. In addition, unfortunately termed 'local helical parameters' (again related to earlier 'cylinder' or 'screw axis' approaches) are calculated from the screw axis linking successive base pairs.

Intrabase pair parameters are derived from a half-way-rotated reference frame derived from the two base reference systems. Both translational (shear, stretch, stagger) and rotational (buckle, propeller, opening) parameters are therefore made up from two half movements on either side of the reference system.

5.6 'El Hassan/Calladine' (45)

This recent local parameter approach (named CEHS: Cambridge University Engineering Department Helix Computation Scheme) shares many common features with the algorithms already described and, for many parameters, gives results similar to 'Newhelix'. Its authors feel strongly that only local parameters are useful for understanding nucleic acid structure. Making a new analysis approach at this late stage is justified by criticisms of a subset of earlier methods ('Newhelix', 'Babcock/Olson', 'von Kitzing/Diekmann'). At a base pair level, the standard R(C8)–Y(C6) vector and the mean base pair normal are used to define the reference axis system. For individual bases, axes parallel to R(N1–C4) or Y(N3–C6) are chosen. Parameters are based on Euler angles with one approximation: twist is treated normally, but roll and tilt are grouped into a single rotation about a 'RollTilt' axis in the *xy*-plane of the reference system between successive base pairs. In a similar way, base pair parameters are separated into a principal propeller twist angle and a grouped opening buckle rotation around a common axis.

5.7 'Curves' (42)

'Curves' was created with the aim of obtaining a global description of nucleic acid conformation. Its development was guided by the desire to extend the approach applicable to regular helical geometries to the description of irregular systems, without losing the notion of a helical axis. In the case of a perfect helix, 'Curves' automatically yields a straight-line axis. In this case, every monomer has the same relative position and orientation with respect to the axis, and consecutive monomers are related by a fixed rotation around, and translation along, the axis.

'Curves' extends these notions to irregular conformations by introducing a least-squares optimization procedure based on a function that mathematically describes departures from ideal helical symmetry. First, it is required that the helical axis should be as straight as possible. If the overall axis is broken down into segments, with one segment per nucleotide (or nucleotide pair), then these segments should ideally be aligned and the reference points on each axis should not be laterally displaced from one another. These two criteria, which are, respectively, rotational and translational in nature, can be expressed as sums of squares, with each term referring to a dinucleotide step within the nucleic acid. Next, it is required that successive nucleotides should, as far as possible, have identical orientations with respect to their local helical axis systems. The translational and rotational differences between successive nucleotides (or nucleotide pairs) again leads to two terms which can be squared and summed over the nucleic acid fragment. This procedure leads to a function with four terms that describes the helical irregularity. If we now consider the parameters that position the individual axis segments as variables (two translations and two rotations with respect to a reference nucleotide at each level), it is possible to search for the variables that minimize the irregularity function. This set of variables then defines the optimal axis describing the given nucleic acid conformation.

Several remarks can be made concerning this approach. As already mentioned, the optimization procedure means that the analysis of a helically regular conformation will automatically lead to a straight axis, since all terms of the irregularity function can simultaneously become zero. In the case of irregular conformations, the axis will be chosen so as to minimize both deformation of the axis and irregular positioning of the bases with respect to this axis. This choice will be optimal in a least-squares sense. It reveals the presence of axial deformation and/or base mispositioning, without any subjective decisions having to be made. It also avoids local changes in helical conformation being incorrectly interpreted as axis curvature. It should also be added that, after optimization, the value of the irregularity function is in itself a useful measure. This value can be broken down into contributions from each dinucleotide step (DIF: dimeric irregularity function), yielding a valuable guide to the location of deformation 'hot spots' within the structure (see Section 6).

It is also important to note that 'Curves' is founded on individual bases and not on base pairs. 'Curves' defines each nucleotide by a base-fixed reference axis system whose origin lies beyond the Watson–Crick base pairing face and whose z-axis is perpendicular to the base plane. This choice was made so that the reference axis systems would be centred on the helical axis within the canonical B-DNA conformation (Xdisp = Ydisp = 0). Since each nucleotide has its own reference axis system, a duplex can be treated in two ways: an optimal helical axis can be calculated for the duplex, or axes can be generated independently for each strand, showing up disparities between them. Similarly, it is easy to treat three- or four-stranded systems, without having to change the reference axes, and it is easier to treat modified nucleic acids containing bulges, abasic sites, or mispairing. Thus, helical deformations resulting from the unorthodox orientation of one or more bases (for example, caused by a transition from *anti* to *syn* conformation) can be avoided by excluding the corresponding bases from the helical axis optimization procedure. Once the unperturbed axis is known, the position of these bases can, nevertheless, be calculated.

It should be noted that care must be taken in determining the base-fixed reference axes in cases where the bases themselves may be deformed. This can arise in low resolution X-ray or NMR conformations, but is most common within conformations coming from molecular dynamics trajectories, where the effect of thermal agitation can lead to major out-of-plane deformations. In such cases it is better to fit an ideal base conformation optimally to the given coordinates before calculating the reference axes.

Although the 'Curves' algorithm was specifically made for global helicoidal analysis, it also calculates local helical parameters, maintaining the nucleotide level approach and the base-fixed reference axis systems described above. Having both global and local parameters available from a single analysis allows a deeper understanding of the conformation in hand and allows easier comparison with other methods, which have almost exclusively chosen the local approach. It is also possible to use 'Curves' to determine an optimal linear axis.

As well as providing numerical parameters describing helical conformation, 'Curves' creates a graphical file including a spline-fitted curve describing the optimal helical axis and a simplified ribbon and plate representation of the nucleic acid backbones and bases (see examples in Section 6). The program can finally carry out an analysis of groove geometry based on spline-fitted curves running through chosen backbone atoms. This approach leads to a continuous measurement of groove width and depth, and also of helical diameter (48).

6. Examples of helical analysis

The fundamental differences between the methods described above can be illustrated by looking at possible helical axis definitions within an irregular structure. Figure 2.10 shows a theoretically generated DNA dodecamer with a 50 Å radius of curvature. Using 'Curves' it is possible to mimic the ways the various analysis schemes distribute this irregularity. First, it is possible to insist on a linear axis and to locate its optimal orientation for the molecule. This is the approach adopted by 'Newhelix'. Secondly, one can calculate separate helical axes for each successive base pair step without worrying about longer range continuity. This corresponds to the so-called 'screw axis' or 'cylinder' approaches discussed above. Thirdly, one can look for an optimal curved axis as is normally done with the 'Curves' algorithm. These choices clearly have an effect on the resulting helical parameters, particularly when they are compounded by different base reference systems and translation/rotation definitions.

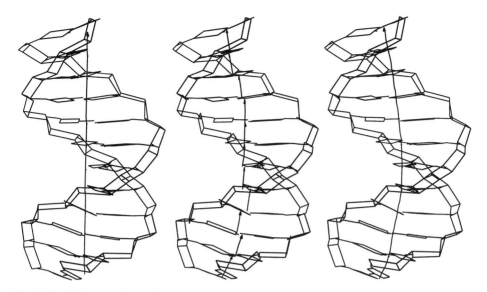

Fig. 2.10. Different analyses of a curved DNA dodecamer, using a global linear axis (left), local helical axes for each base pair step (centre), and a curved global axis (right).

For a comparison of the numerical results obtained by the analysis programs, the reader is referred to a recent study by Elgavish and Harvey (49). These authors have compared the programs discussed above (excepting the most recent program, 'CEHS') for a number of test oligomers, including two B-DNAs, an A-DNA, and two different studies of an Okazaki fragment. The results bring to light several points worthy of note. First, the various programs can disagree dramatically for given helicoidal parameters, including such fundamental values as rise and twist. This is particularly visible when the structure differs significantly from canonical B-DNA, and, in the case of irregular fragments, these disagreements can lead to qualitatively different structural descriptions. Secondly, the programs fall into families based on the differences in base reference system and algorithm described above. Thus, 'Newhelix' and 'Bansal' often agree closely and also show strong correlations with 'von Kitzing/Diekmann' for certain parameters. 'Babcock/Olson' and 'Curves' (local) parameters also agree closely, with exceptions for rise and an offset in slide. The 'Tung/Soumpasis' program stands apart from the others in many cases (a consequence of the authors preference for the use of inertial axes). It is also recalled that 'Newhelix' and 'Bansal' use a different sign convention for buckle compared with the other programs, and tilt also has a inverse sign in 'Tung/Soumpasis'. Lastly, 'Curves' is the only program to propose an optimally curved axis for irregular fragments and coherent sets of local and global parameters.

In order to give a better feeling for the sense of helical parameters and, notably, for the difference between local and global parameters, we present below a number of analyses using 'Curves'.

(a)

i

ii

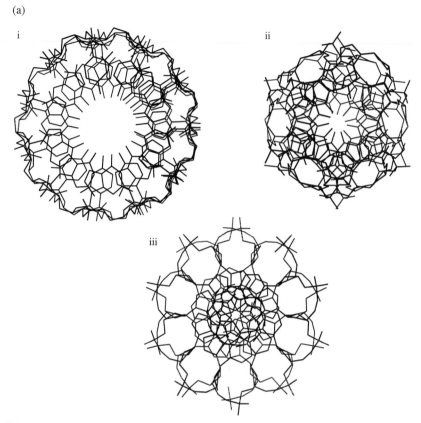

iii

(b)

i

ii

iii

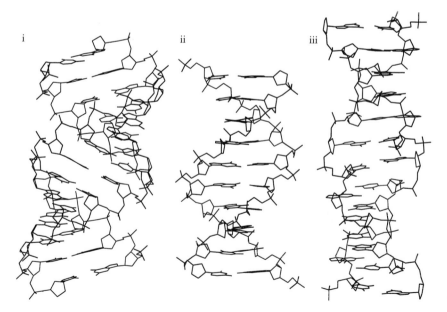

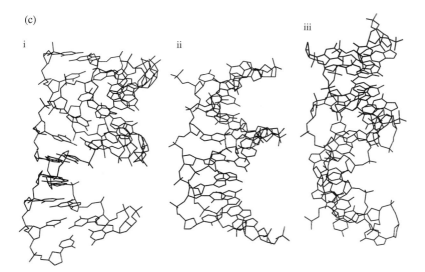

Fig. 2.11. Regular DNA conformations. (i) A-DNA, (ii) B-DNA, (iii) Z-DNA. The three views shown are, respectively, (a) along the axis, (b) perpendicular to the axis, and (c) with an inclined axis to show the groove profiles.

6.1 Regular conformations

We begin with the basic allomorphic forms of DNA (and RNA) which are presented in Table 2.3 (see also Chapters 1 and 5). This table includes both global and local helical parameters (the latter being denoted by the prefix 'L-') and, for completeness, the backbone conformations. (Only parameters referring to base pairs are presented and those that remain close to zero for regular helices, shear, stretch, and stagger, have been excluded.) Different fibre conformations are given for A- (50,51) and B-DNA (51–53) and for A-RNA (54), while the Z conformation is represented by the idealized Z_I- and Z_{II}-forms (55). The overall shape of these helically regular duplexes can be compared in Fig. 2.11, where three different projections illustrate the variations in diameter, base position, helicity, and groove geometry.

Since these conformations are well known, they are useful for a first comparison of global and local parameters. Table 2.3 shows little difference between these two types of parameters for B-DNA, where the bases lie close to the helical axis and are only slightly inclined. The difference, however, becomes much clearer for the A and Z conformations. For A-DNA, as discussed in Section 5, the missing Xdisp and inclination values are replaced in the local parameter description by negative slide and positive roll. One can also note that the shorter and broader nature of the A-DNA helix is visible in the global Xdisp and rise values, whereas the local rise shows, as it should, a value close to that in B-DNA, linked only to local stacking interactions. It is also important to recall that, in the global description, non-zero values of shift, slide, tilt, and roll all signify differences between the two strands of a duplex (see the definitions given in Table 2.2). This is the case for A and B helices that have identical strands (homonomous conformations), but not for Z-DNA owing to its dinucleotide repeat symmetry. Global shift, slide, tilt, and roll (unlike their local equivalents) are thus a

Table 2.3. Helical and backbone parameters for allomorphic conformations of DNA and RNA. (Translations in Å, rotations in degrees. The prefix L- distinguishes local parameters. For Z-DNA, values are given for CG/GC pairs and CpG/GpC steps, respectively)

Parameter	B-DNA (52)	B-DNA (51)	B-DNA (53)	A-DNA (50)	A-DNA (51)	A-RNA (54)	Z_I-DNA (26,55)		Z_{II}-DNA (26,55)	
X–disp	−0.71	0.0	−0.18	−5.43	−5.28	−5.3	−1.16		−2.46	
Y–disp	0.0	0.0	0.0	0.0	0.0	0.0	−1.95	1.95	−2.32	2.32
Inclin	−5.9	1.5	2.7	19.1	20.7	15.8	14.5		4.2	
Tip	0.0	0.0	0.0	0.0	0.0	0.0	−178.5	178.5	−178.2	178.2
Buckle	0.0	0.0	0.1	0.0	0.0	0.0	4.9	−4.9	6.3	−6.3
Prop	3.7	−13.3	−15.1	13.7	−7.5	14.5	1.1		−0.8	
Open	−4.1	0.0	0.4	−4.6	0.0	−4.2	−0.1		5.6	
Shift	0.0	0.0	0.0	0.0	0.0	0.0	0.0		0.0	
Slide	0.0	0.0	0.0	0.0	0.0	0.0	3.90	−3.90	4.63	−4.63
Rise	3.38	3.38	3.38	2.56	2.56	2.81	4.73	2.66	4.35	3.08
Tilt	0.0	0.0	0.0	0.0	0.0	0.0	0.0		0.0	
Roll	0.0	0.0	0.0	0.0	0.0	0.0	−2.9	2.9	−3.6	3.6
Twist	36.0	36.0	36.0	32.7	32.7	32.7	−8.9	−49.2	−3.7	−56.3
L–Shift	0.0	0.0	0.0	0.0	0.0	0.0	0.0		0.0	
L–Slide	−0.76	0.08	0.04	−2.08	−1.92	−2.13	5.11	−1.87	5.13	−1.54
L–Rise	3.32	3.38	3.38	3.42	3.44	3.52	3.58	3.21	4.00	3.21
L–Tilt	0.0	0.0	0.0	0.0	0.0	0.0	0.0		0.0	
L–Roll	−3.6	0.9	1.7	10.7	11.4	8.9	−5.1	−9.2	−4.2	−0.5
L–Twist	35.8	35.6	36.0	30.9	30.7	31.5	−8.7	−47.7	−3.6	−56.2
α	−46.9	−40.7	−29.9	−84.6	−74.8	−62.1	71.7	−137.4	92.4	145.9
β	−146.0	135.6	136.3	−152.1	−179.1	−179.9	−176.0	−169.1	−167.0	163.0
γ	36.4	37.4	31.1	45.5	58.9	47.4	175.5	60.0	156.9	66.4
δ	156.4	139.5	143.4	82.6	78.2	83.5	140.2	103.4	146.9	93.4
ε	155.0	−133.2	−140.8	177.7	−155.0	−151.7	−92.2	−101.8	−100.5	−178.7
ζ	−95.1	−156.9	−160.5	−46.4	−67.1	−73.6	78.3	−53.3	73.6	55.5
χ	−97.9	−101.9	−98.0	−154.3	−158.9	−165.9	−161.2	63.3	−147.4	62.9
Phase	191.6	154.8	154.3	13.1	18.3	13.4	156.8	13.2	163.4	50.4
Amplitude	36.3	39.7	45.9	38.9	41.6	39.0	38.9	17.1	41.0	26.6

guide to strand asymmetry (heteronomous conformations). It is also worth noting that the base inversion involved in the B→Z transition is clearly visible only in the global helical parameter tip, with a value close to 180° for the Z conformation.

6.2 Irregular conformations

We will now look at some more irregular conformations, beginning with the famous dodecamer CGCGAATTCGCG (35, protein data bank 'PDB' entry 1BNA), which clearly revealed base sequence effects within the double helix. Because this oligomer often serves as a reference, we present a rather complete set of helical parameters in Table 2.4. Figure 2.12 shows a molecular graphic and the simplified representation generated by 'Curves', where the axis is shown, the backbone is represented by a ribbon (passing through P, bisecting C3′–C4′ and oriented by the phosphate anionic oxygens), and the bases are replaced by rectangles (completed by a line to their reference points which touch their Watson–Crick partner and lie on the helical axis in a canonical B conformation).

The analysis of the dodecamer nature shows the order of sequence-induced variations within crystallographic B-DNAs, typically of the order of 20° for rotational parameters and 1 Å for translational parameters. The kinking of the axis within the dodecamer is visible in the figure and is characterized by the angle formed between successive helical axis vectors, by the tilt and roll angles, and by the DIF values (see Section 5), which measure the overall irregularity of each dinucleotide step. One can also note large propeller values and strong buckling on either side of the central AATT sequence. The centre of the oligomer also shows positive opening values linked to the

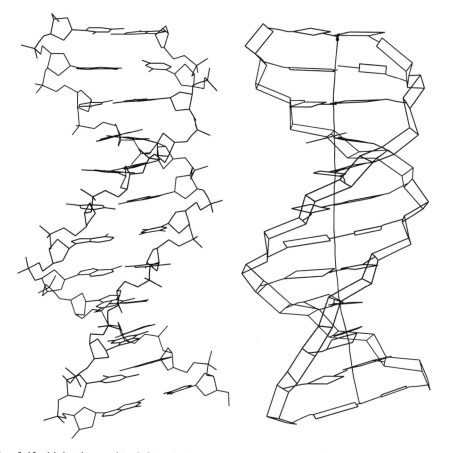

Fig. 2.12. Molecular graphic (left) and 'Curves' schematic view (right) of the B-DNA dodecamer d(CGCGAATTCGCG)$_2$ (35).

Table 2.4. Helical analysis of the dodecamer d(CGCGAATTCGCG)$_2$ (35, PDB entry 1BNA) obtained using Curves

(a) Global base pair-axis and intra-base pair parameters

Base pair	X-disp	Y-disp	Inclin	Tip	Buckle	Propel	Opening
C1	−0.66	0.16	5.9	0.6	3.7	−14.4	−3.0
G2	−0.78	0.09	4.3	0.6	−4.5	−10.6	−3.0
C3	−0.54	0.01	3.9	−4.2	−7.5	−3.9	−1.6
G4	−0.75	0.26	4.3	−1.6	10.1	−11.7	0.4
A5	−0.77	0.22	−0.9	−2.3	4.7	−18.2	3.7
A6	−0.74	0.16	−4.8	−1.7	3.2	−19.6	7.3
T7	−0.55	−0.01	−5.4	−0.6	0.6	−18.4	10.0
T8	−0.59	0.00	−2.5	1.8	−1.7	−19.7	2.8
C9	−0.42	−0.08	0.1	2.3	−10.8	−19.3	0.7
G10	−0.01	0.28	−0.2	5.7	2.4	−6.2	0.4
C11	−0.51	0.26	−1.2	−1.3	−3.9	−19.8	−5.0
G12	−0.12	0.13	−1.1	−4.1	7.2	0.5	−2.5
Average	−0.54	0.12	0.2	−0.4	0.3	−13.4	0.8

(b) Global and local interbase pair parameters

Duplex	Rise	Tilt	Roll	Twist	L-Rise	L-Tilt	L-Roll	L-Twist
C1/G2	3.56	−3.2	2.3	42.7	3.62	−3.4	6.0	42.8
G2/C3	3.49	−0.2	−8.0	36.0	3.48	1.0	−5.3	36.1
C3/G4	3.08	1.8	7.1	27.6	3.14	3.2	9.0	26.5
G4/A5	3.36	−5.0	0.8	39.8	3.40	−3.1	2.1	40.0
A5/A6	3.33	−2.3	2.0	35.2	3.32	−0.7	0.7	35.3
A6/T7	3.39	1.2	−0.5	34.6	3.31	2.1	−3.3	34.6
T7/T8	3.36	3.9	2.3	35.1	3.33	2.9	0.1	35.3
T8/C9	3.39	2.5	−0.1	38.1	3.38	0.8	−0.8	39.0
C9/G10	3.23	−0.7	4.9	32.2	3.28	−2.9	4.8	31.1
G10/C11	3.62	−3.3	−13.1	38.6	3.57	−5.2	−13.2	38.9
C11/G12	3.22	1.2	−2.1	34.8	3.19	3.3	−3.0	34.4
Average	3.37	−0.37	−0.4	35.9	3.37	−0.2	−0.3	35.8

(c) Axis bend and DIF (Dimeric irregularity function)

Duplex	Angle	Diff
C1/G2	2.7	0.55
G2/C3	3.2	0.54
C3/G4	4.8	1.28
G4/A5	1.5	0.48
A5/A6	2.1	0.25
A6/T7	2.4	0.43
T7/T8	1.1	0.22
T8/C9	0.6	0.34
C9/G10	1.5	1.69
G10/C11	6.5	2.06
C11/G12	1.3	1.51

decreased minor groove width. Lastly, note that since this structure is clearly a B-DNA, there are only small differences between local and global parameters.

The same is not true if we move away from the B domain, for example with hybrid decamer d(GGGTATACGC):r(GCG)d(TATACCC) (56, PDB entry 1OFX). The conformation of this oligomer is shown in Fig. 2.13 and analysed in Table 2.5. It is globally closer to an A conformation, with a strong negative Xdisp and strong positive inclination for all but the first two base pairs. Less obvious is the kink within the structure, which is largely concentrated at the T4–A5 step following the 5′-hybrid:DNA-3′ junction. Note that such junctions are now generally thought (57) to be more perturbing than 5′-DNA:hybrid-3′ junctions. The global and local inter-base pair parameters for this structure differ significantly, notably in describing the A-like form, which again leads, in the local analysis, to negative slide, positive roll, and increased rise. One can also note that the T4–A5 kink (towards the major groove) is assimilated into a very strong local roll.

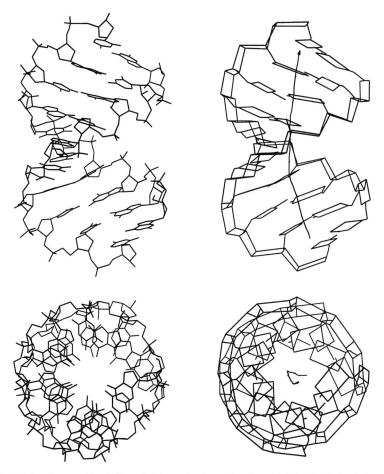

Fig. 2.13. Molecular graphic (left) and 'Curves' schematic view (right) of a hybrid d(GGGTAT-ACGC):r(GCG)d(TATACCC) oligomer (56).

Table 2.5. Helical analysis of the hybrid decamer d(GGGTATACGC).
r(GCG)d(TATACCC) (56, PDB entry 1OFX) obtained using 'Curves'

(a) Selected base pair-axis and intrabase pair parameters

Base pair	X-disp	Inclination	Buckle	Propel	Opening
G1	−4.53	6.2	−2.0	−4.8	−1.7
G2	−4.21	8.0	−9.1	−10.5	3.1
G3	−4.42	11.7	−11.3	−19.3	2.7
T4	−4.54	12.8	8.1	−9.9	1.9
A5	−4.27	14.6	2.1	−16.2	7.3
T6	−3.97	13.6	6.7	−23.8	12.2
A7	−4.02	10.9	13.0	−21.0	3.6
C8	−4.17	12.3	18.2	−14.6	0.6
G9	−4.20	13.3	−5.2	−14.7	−1.8
C10	−3.96	11.9	−4.3	10.6	0.4
Average	−4.23	11.5	1.6	−12.4	2.8

(b) Global and local inter-base pair parameters

Duplex	Slide	Rise	Roll	Twist	L-Slide	L-Rise	L-Roll	L-Twist
G1/G2	−0.25	3.23	1.2	38.5	−2.75	3.56	5.7	38.6
G2/G3	0.01	2.98	1.7	34.2	−2.01	3.39	7.4	34.3
G3/T4	−0.41	2.57	−2.3	22.4	−1.57	2.92	2.7	20.7
T4/A5	0.20	2.11	14.9	33.0	−1.56	3.55	22.3	29.5
A5/T6	0.01	2.94	−1.1	30.8	−1.39	3.41	7.1	29.3
T6/A7	0.17	2.26	8.0	35.3	−1.61	3.11	15.4	32.5
A7/C8	−0.67	3.07	−2.6	27.9	−2.00	3.39	3.1	27.9
C8/G9	−0.07	3.00	8.0	31.6	−1.55	3.91	14.6	31.0
G9/C10	−0.29	3.51	−6.0	32.4	−1.80	3.80	1.0	31.7
Average	−0.15	2.85	2.4	31.8	−1.80	3.45	8.8	30.6

It is also interesting to consider a very perturbed duplex, d(CGCAGAATTCGCG)$_2$, which contains both bulges and opened base pairs (58, PDB entry 1D31). Both A4 adenines in this oligomer are bulged bases, but while A4 in the first strand is excluded from the helix, the equivalent base in the second strand maintains stacking. Two such oligomers interact head-to-tail in the crystal, the excluded adenine filling the space opposite the stacked adenine. In addition, the terminal G:C pair is disrupted and the bases point outwards (Fig. 2.14). Using the options in 'Curves', it is possible to ignore the positions of the excluded A4 and the two opened bases during the global helical analysis, revealing a more or less straight helical axis and typical B-DNA helical parameters (⟨Xdisp⟩ = 0.6 Å, ⟨Ydisp⟩ = 0.8 Å, ⟨Incl⟩ = 0.9°, ⟨Rise⟩ = 3.4 Å, ⟨Twist⟩ = 36.6°). However, it is also possible to characterize the local deformations, by showing, for example, that the stacked and unstacked bulge sites both have similar twist values (40.5 and 39.7°), but clearly differ in rise (3.8 and 6.5 Å). The disrupted base pair can also be fully characterized, notably by a spectacular opening of 208°. Note that a local parameter analysis is not well adapted to dealing with structures containing major local deformations.

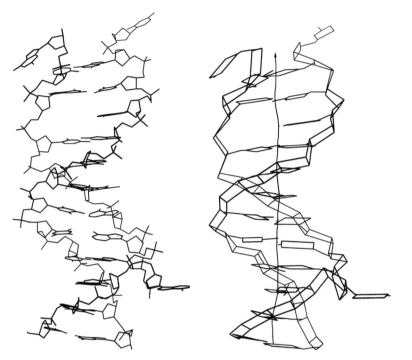

Fig. 2.14. Molecular graphic (left) and 'Curves' schematic view (right) of a B-DNA oligomer d(CGCA-GAATTCGCG)$_2$ containing two adenine bulges and opened bases (58).

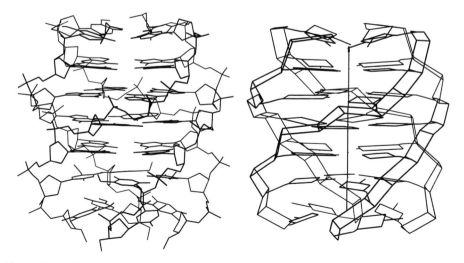

Fig. 2.15. Molecular graphic (left) and 'Curves' schematic view (right) of the tetraplex (TTGGGGT)$_4$ containing both G$_4$ and T$_4$ tetrads (59).

We finally consider a multistranded conformation (Fig. 2.15). A parallel-stranded tetraplex, d(TTGGGGT)$_4$, containing both T and G tetrads, has been chosen for this purpose (59, PDB entry 201D). This structure has a rather regular overall conformation, with bases displaced roughly 2.9 Å towards the minor groove side. Twist angles vary from 24 to 36°, with the largest value between the first and second G tetrads. It should be noted that Hoogsteen base pairing leads to unusual base–base parameters (e.g. G tetrads have ⟨Shear⟩ = −5.8 Å, ⟨Stretch⟩ = 2.8 Å and ⟨Opening⟩ = −90°), but this does not perturb the inter-base pair parameters or the characterization of tetrad deformations, such as the buckling of the terminal T tetrads (roughly −18° for the first tetrad, versus roughly 37° for the last tetrad).

Today, more and more irregular DNA conformations are being analysed within protein:DNA complexes. Since the deformations induced by protein binding can be severe, it is often difficult to understand their nature without the help of a detailed helical analysis. The interested reader might look at the example provided by the complex between DNA and the TATA box-binding protein (60,61). A visual inspection of this complex shows DNA to be very significantly bent away from the protein and also helically unwound, but a global helical parameter analysis goes further in revealing a new, virtually regular, helical conformation in the protein-bound region (62). The new conformation shows a striking resemblance to A-DNA, differing only in strongly, positively inclined base pairs resulting from less negative glycosidic torsions.

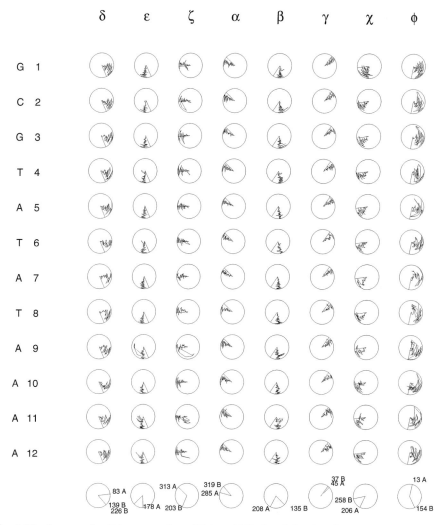

Fig. 2.16. An example of the output from 'Dials and Windows' showing the temporal fluctuations of selected helical and backbone parameters within a B-DNA oligomer. The time axis points upwards within the rectangular diagrams and is radial for the circular diagrams. The data shown cover an 850 ps simulation in water (67). See also p. 70.

7. Analysing nucleic acid dynamics

Dynamic conformational information is available both from experimental and theoretical studies (Chapter 4) and treating this type of information poses a number of
new problems. The principal problem in analysing the trajectories that result from
molecular dynamic simulations is that a very large mass of information must be made
readable. Nucleic acid simulations are today generally carried out in water (typically
in boxes containing roughly 5000 water molecules) and often last for one or more
nanoseconds. Since structures are typically saved about every 0.5 ps, this means that
the complete trajectory is represented by roughly 2000 sets of coordinates and velocities, each of which contains roughly 15 000 atoms. With double-precision data this
represents about 1.5 Gb of information to be processed. Although the mass of data is
not the same, NMR studies can also lead to a large number of structures compatible
with the spectroscopic measurements, whose dispersion contains valuable information on the dynamics of the molecule in solution (Chapter 8). Although not representing a time series, such data also need to be presented in a comprehensible
fashion.

A first step to analysis is to use molecular graphics to generate optimally super-
posed structures or an animation of the time evolution. This gives a good overall
impression of the conformational changes taking place during the trajectory, but is not
adapted to extracting any quantitative information. Such information can be obtained
by plotting the time evolution of individual conformational variables such as backbone
angles, interatomic distances, sugar puckers, or helicoidal parameters, but there are a
very large number of such variables. To overcome this problem, the 'Dials and
Windows' program developed at Wesleyan University (63) provides a compact representation that plots time series using rectangular 'windows' for translational variables
and circular 'dials' for rotational variables, with the time axis running vertically
upwards in the former case and radially outwards in the latter. (This has the advantage
of avoiding problems owing to the cyclic nature of torsion angles, but the disadvantage
that variations in parameters occurring early in the trajectory are somewhat compressed.) Reference values, typically for the A- and B-forms of DNA, enable the
range of variations to be estimated. An example of the output from 'Dials and
Windows' is shown in Fig. 2.16. This output can be adjusted interactively on a graphics workstation and enables many elements of an entire trajectory to be viewed on a
single page. Although 'Dials and Windows' uses 'Curves' to obtain the helicoidal parameters and axis variables, similar techniques can be used with any other analysis
approach and many authors have developed their own programs.

On a more global level, a number of measurements can be useful, such as axis
bending, persistence length (64), or rms differences with known conformations (65).
In the latter case, it is particularly informative to follow the evolution of a trajectory
with respect to two or more reference conformations. This sort of triangulation often
gives very good insight into the basic nature of a complex conformational pathway. A
useful extension of rms calculations consists of building a two-dimensional matrix
where every conformation saved along the trajectory is compared with every other
conformation (66,67). The results can be represented in terms of shading, with darker
squares referring to smaller rms differences (Fig. 2.17). One would expect such a plot

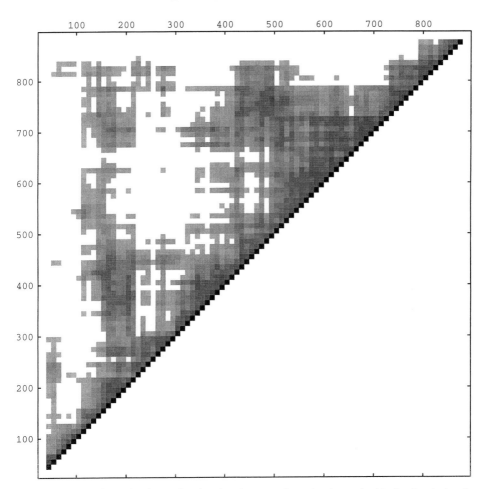

Fig. 2.17. A two-dimensional rms plot showing that a conformational state is visited twice during the 850 ps simulation of a B-DNA oligomer (67). Structures along the trajectory (separated by 10 ps) are compared with one another. Darker shading indicates smaller rms values (with white areas corresponding to all values > 2.2 Å).

to show a dark band close to the diagonal, since neighbouring points along the trajectory will generally be related to one another, but if, in addition, off-diagonal dark zones appear, there is clear evidence for the reoccurrence of a given conformation (which may be taken as evidence for the existence of a conformational substate).

8. Conclusions

This chapter has summarized present approaches to describing the structure of nucleic acids. The enormous growth in the variety of such structures has posed a number of problems that are not yet completely solved. At the simplest level of description (number of strands, strand direction, *syn/anti* conformation, and base pairing) some order has been introduced, but challenges to this order continue to appear, such as the triad DNA proposed by Kuryavyi and Jovin (68) or the related 'adenosine platforms' (69) found in RNA. At the level of helical conformation, although a number of different analysis schemes still coexist, there is a better level of understanding of the meaning of helical parameters and, in particular, of the differences between local and global parameters. Non-helical structures, however, continue to pose problems and more work is needed on loops, multi-arm junctions, and the range of baroque architectures of RNA. Lastly, while some useful steps have been made in the analysis of dynamic data, problems persist and are particularly challenging for the organizers of structural databases.

References

1. Saenger, W. (1984) *Principles of Nucleic Acid Structure*. Springer-Verlag, New-York.
2. Hélène, C. and Toulmé, J.J. (1990) *Biochem. Biophys. Acta* **1049**, 99.
3. Sun, J.S., Garestier, T. and Hélène C. (1996) *Curr. Opin. Struct. Biol.* **6**, 327.
4. Leroy, J.L. and Guéron, M. (1995) *Structure* **3**, 101.
5. Taylor, R. and Kennard, O. (1982) *J. Mol. Struct.* **78**, 1.
6. Jiang, S.-P., Raghunathan, G., Ting, K.-L., Xuan, J.C. and Jernigan, R.L. (1994) *J. Biomol. Struct. Dynamics* **12**, 367.
7. Jost, J.P. and Saluz, H.P. (eds) (1993) *DNA Methylation: Molecular Biology and Biological Significance*. Birkhäuser, Basel.
8. Yanson, I.K., Teplitsky, A.B. and Sukhodub, L.F. (1979) *Biopolymers* **18**, 1149.
9. Williams, L., Chawla, B. and Shaw, B. (1987) *Biopolymers* **26**, 591.
10. Watson, J.D. and Crick, F.H.C. (1953) *Nature* **171**, 964.
11. Morgan, A.R. (1993) *TIBS* **18**, 160.
12. Hartmann, B. and Lavery, R. (1996) *Quart. Rev. Biophys.* **29**, 309.
13. Leonard, G.A., Thomson, J., Watson, W.P. and Brown, T. (1990) *Proc. Natl. Acad. Sci. USA* **87**, 9573.
14. Kennard, O. (1985) *J. Biomol. Struct. Dynamics* **3**,205.
15. Jaishree, T.N. and Wang, A.H.J. (1993) *Nucl. Acids Res.* **16**, 3839.
16. Rippe, K. and Jovin, T.M. (1992) *Meth. Enzymol.* **211**, 199.
17. Aboul-ela, F., Murchie, A.I.H. and Lilley, D.M.J. (1992) *Nature* **360**, 280.
18. Kang, C., Zhang, X., Ratliff, R., Moyzis, R. and Rich, A. (1992) *Nature* **356**, 126.
19. Smith, F.W. and Feigon, J. (1992) *Nature* **356**, 164.
20. Kettani, A., Kumar, R.A. and Patel, D.J. (1995) *J. Mol. Biol.* **254**, 638.
21. Lebrun, A. and Lavery, R. (1996) *J. Biomol. Struct. Dynamics* **13**, 459.
22. Rose, I.A., Hanson, K.R., Wilkinson, K.D. and Wimmer, M.J. (1980) *Proc. Natl. Acad. Sci. USA* **77**, 2439.
23. Harvey, S.C. (1983) *Nucl. Acids Res.* **11**, 4867.
24. Lavery, R., Zakrzewska, K., Sun, J.S. and Harvey, S.C. (1992). *Nucl. Acids Res.* **20**, 5011.
25. Westhof, E. (1992) *Nature* **358**, 459.
26. Wang, A.H.J., Quigley, G.J., Kolpak, F.J., Crawford, J.L., van Boom, J.H., van der Marel, G. and Rich, A. (1979) *Nature* **282**, 680.
27. Morvan, F., Rayner, B., Imbach, J.L., Chang, D.K. and Lown, J.W. (1987) *Nucl. Acids Res.* **15**, 4241.
28. Sun, J.S., François, J-.C., Lavery, R., Saison-Behmoaras, T., Montenay-Garestier, T., Thuong, N.T. and Hélène, C. (1988) *Biochemistry* **27**, 6039.
29. Arnott, S. and Selsing, E. (1974) *J. Mol. Biol.* **88**, 509.

30. Broitman, S.L., Im, D.D. and Fresco, J.R. (1987) *Proc. Natl. Acad. Sci. USA* **84**, 5120.
31. Pilch, D.S., Levensen, C. and Shafer, R.H. (1991) *Biochemistry* **30**, 6081.
32. Sun, J.S., Mergny, J.-L., Lavery, R., Montenay-Garestier, T. and Hélène, C. (1991) *J. Biomol. Struct. Dynamics* **9**, 411.
33. Rosenberg, J.M., Seeman, N.C., Day, R.O. and Rich, A. (1976) *Biochem. Biophys. Res. Commun.* **69**, 979.
34. Arnott, S. (1970) *Progr. Biophys. Mol. Biol.* **21**, 265.
35. Drew, H.R., Wing, R.M., Takano, T., Broka, C., Tanaka, S., Itakura, K. and Dickerson R.E. (1981) *Proc. Natl. Acad. Sci. USA* **78**, 2179.
36. Dickerson, R.E., Bansal, M., Calladine, C.R., Diekmann, S., Hunter, W.N., Kennard, O., Lavery, R., Nelson, H.C.M., Olson, W.K., Saenger, W., Shakked, Z., Sklenar, H., Soumpasis, D.M., Tung, C.-S., Von Kitzing, E., Wang, A.H.-J. and Zhurkin, V.B. (1989) *J. Mol. Biol.* **205**, 787.
37. Lavery, R. and Sklenar, H. (1990) in *Structure and Methods*, Vol.2, *DNA Protein Complexes and Proteins*, (Sarma, R.H. and Sarma, M.H., eds), p. 412. Adenine Press, New York.
38. Fratini A.V., Kopka M.L., Drew H.R. and Dickerson R.E. (1982) *J. Biol. Chem.* **257**, 14686.
39. von Kitzing, E. and Diekman, S. (1987). *Eur. Biophys. J.* **15**, 13.
40. Soumpasis, D.M., Tung, C.-S. and Garcia, A.E. (1991) *J. Biomol. Struct. Dynamics* **8**, 867.
41. Bhattacharyya, D. and Bansal, M. (1989) *J. Biomol. Struct. Dynamics* **6**, 635.
42. Lavery, R. and Sklenar, H. (1989) *J. Biomol. Struct. Dynamics* **6**, 655.
43. Babcock M.S., Pednault E.P.D. and Olson W.K. (1994) *J. Mol. Biol.* **237**, 125.
44. Babcock M.S. and Olson W.K. (1994) *J. Mol. Biol.* **237**, 98.
45. El Hassan M.A. and Calladine C.R. (1995) *J. Mol. Biol.* **251**, 648.
46. Calladine C.R. and Drew H.R. (1984) *J. Mol. Biol.* **178**, 773
47. Goodsell D.S. and Dickerson R.E. (1994) *Nucl. Acids Res.* **22**, 5497
48. Stofer, E. and Lavery, R. (1993) *Biopolymers* **34**, 337.
49. Elgavish, T. and Harvey, S.C. (1998) in preparation.
50. Arnott, S. and Hukins, D.W.L. 1972) *Biochem. Biophys. Res. Commun.* **47**, 1504.
51. Arnott, S., Chandrasekaran, R., Birdsall, D.L., Leslie, A.G.W. and Ratliff, R.L. (1980) *Nature* **283**, 743 (and coordinates communicated to our laboratory by S. Arnott).
52. Arnott, S. and Hukins, D.W.L. (1973) *J. Mol. Biol.* **81**, 93.
53. Chandrasekaran, R. and Arnott, S. (1996) *J. Biomol. Struct. Dynamics* **13**, 1015.
54. Arnott, S., Hukins, D.W.L., Dover, S.D., Fuller, W. and Hodgson, A.R. (1973) *J. Mol. Biol.* **81**, 107.
55. Wang, A.H.-J., Quigley, G.J., Kolpak, F.J., van Der Marel, G., van Boom, J.H. and Rich, A. (1981) *Science* **211**, 171.
56. Egli, M., Usman, N., Zhang, S. and Rich, A. (1992) *Proc. Natl. Acad. Sci. USA* **89**, 534.
57. Nishizaki, T., Iwai, S., Ohkubo, T., Kojima, C., Nakamura, H., Kyogoku, Y. and Ohtsuka, E. (1996) *Biochemistry* **35**, 4016.
58. Joshua-Tor, L., Frolow, F., Appella, E., Hope, H., Rabinovich, D. and Sussman, J.L. (1992) *J. Mol. Biol.* **225**, 397.
59. Wang, Y. and Patel, D.J. (1995) *J. Mol. Biol.* **251**, 76.
60. Kim, Y., Gieger, J.H., Hahn, S. and Sigler, P.B. (1993) *Nature* **365**, 512.
61. Kim, J.L., Nikolov, D.B. and Burley, S.K. (1993) *Nature* **365**, 520.
62. Guzikevich-Guerstein, G. and Shakked, Z. (1996) *Nature Struct. Biol.* **3**, 32.
63. Ravishankar, G., Swaminathan, S., Beveridge, D.L., Lavery, R. and Sklenar, H. (1989) *J. Biomol. Struct. Dynamics* **6**, 669.
64. Prévost, C., Louise-May, S., Ravishankar, G., Lavery, R. and Beveridge, D.L. (1992) *Biopolymers* **33**, 335.

65. Goodfellow, J. M., de Souza, O.N., Parker, K. and Cruzeiro-Hansson, L. (1993) in *Computer Simulation of Biomolecular Systems*, Vol. 2, (van Gunteren, W.F., Weiner, P.K. and Wilkinson, A.J., eds), p. 483. Escom., Leiden.
66. McConnell, K.J., Nirmala, R., Young, M.A., Ravishankar, M.A. and Beveridge, D.L. (1994) *J. Am. Chem. Soc.* **116**, 4461.
67. Flatters, D., Young, M.A., Beverdige, D.L. and Lavery, R. (1997) *J. Biomol. Struct. Dynamics* **14**, 757.
68. Kuryavyi, V.V. and Jovin, T.M. (1995) *Nature Genetics* **9**, 339.
69. Cate, J.H., Gooding, A.R., Podell, E., Zhou, K., Golden, B.L., Szewczak, A.A., Kundrot, C.E., Cech, T.R. and Doudna, J.A. (1996) *Science* **273**, 1696.

3

The Nucleic Acid Database: a research and teaching tool

Helen M. Berman, Christine Zardecki, and John Westbrook
Department of Chemistry, Rutgers University, Piscataway, NJ 08854–8087, USA

1. Introduction

The Nucleic Acid Database (NDB) was established in 1991 in response to the need for a repository of the specialized information contained within nucleic acid crystal structures (1). The vision of the founders was to create a research tool that would allow for data exploration and inspire knowledge discovery about the three dimensional structure of nucleic acids.

Since its inception in 1992, the NDB has grown from a database of 100 structures to well over 800 structures (Fig. 3.1) in 1998. The information content contained

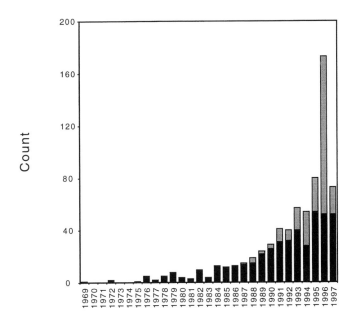

Fig. 3.1. Distribution of the number of structures released by the NDB per year. All structures, including protein:nucleic acid complexes, are shown in grey; all nucleic acid structures are shown in black. It should be noted that in 1996, 87 structures were deposited for a series of DNA Polymerase Beta complexes by the same authors (H. Pelletier and M. R. Sawaya).

within the database has also continued to expand. In 1996, the NDB graduated from being a special interest research database to serving as the direct deposition site for DNA and RNA crystal structures. As the technology for data retrieval continues to improve, full advantage is being taken of the primary NDB WorldWide Web (WWW) site at http://ndbserver.rutgers.edu/, which has more than 250 000 'hits' per month.

In this review, we shall describe the contents and structure of the database, the query capabilities, the determination of standard dictionary values, and the procedures used for processing and distributing data. We shall also describe how the NDB can be used for both research and teaching.

2. The infrastructure of the NDB

2.1 Information: content and format

For each individual crystal structure, the NDB encodes information about experimental features, including the crystallization conditions, the data collection methods, and the refinement procedures. The coordinates of the oligonucleotide are clearly differentiated from those of each ligand type. Quantitative derived features, such as valence geometry and their root mean square (rms) comparisons with standard values, torsion angles, base morphology, groove dimensions, and intermolecular interactions, are calculated and stored. Additionally, qualitative classifications of conformation type and intermolecular interactions are contained in the NDB. The structure factors are also stored in the NDB Archives. Table 3.1 summarizes the information content of the NDB.

Table 3.1. Primary experimental data stored in the NDB

Primary Data
Structure features
Descriptor
NDB, PDB, and CSD IDs
Coordinate availability
Structure description
Sequence
Conformation type
Base, phosphate, and sugar modifier descriptions
Mismatched base pairs
Drug name and binding type
Base pairing, asymmetric unit, and biological unit descriptions
Citation
Authors
Title
Journal
Volume
Pages
Year
Crystal data
Cell dimensions
Space group
Data collection description
Radiation source

Table 3.1. *Continued*

Data collection device
Radiation wavelength
Temperature
Resolution range
Total number of unique reflections
Crystallization description
Method
Temperature
pH value
Composition of solutions
BMCD ID
Refinement information
Method
Program
Number of reflections
Data cut-off
Resolution range
R-factor
Refinement of temperature factors and occupancies
Coordinate information
Atomic coordinates, occupancies, and temperature factors
Structure factors
Availability
Derivative data
Distances
Covalent bond lengths
Non-bonded contacts
Virtual bonds involving phosphorus atoms
Torsions
Backbone and side chain torsion angles
Pseudo-rotation parameters
Angles
Valence bond angles
Virtual angles involving phosphorus atoms
Base morphology
Parameters calculated by a variety of algorithms
Groove dimension
Dimensions of major and minor groove
RMS deviations
Deviations and RMS deviations from small molecule, standard values for covalent bond
   distances and angles
Crystal types
Isomorphous groups of structures
Packing motifs
Classification of intermolecular crystalline interactions
Protein types
Function classification of nucleic acid-binding proteins
Biological units
Symmetry operations to generate the biological unit

Over the years the format of the data files has evolved. Since the intent was to load the data into a searchable database, it has been crucial that the semantics and syntax of the data are strictly defined. The original NDB format resembled the Protein Data Bank (PDB) format (2), but with richer content and greater structure. Currently, the NDB has adopted the macromolecular Crystallographic Information File (mmCIF) (3) as its standard. This format has several advantages from the point of view of building a database: (1) the definitions for the data items are based on a comprehensive dictionary of crystallographic terminology and molecular structure description; (2) it is self-defining; and (3) the syntax contains explicit rules that further define the characteristics of the data items, particularly the relationships between data items. This latter feature is important because it allows for rigorous checking of the data.

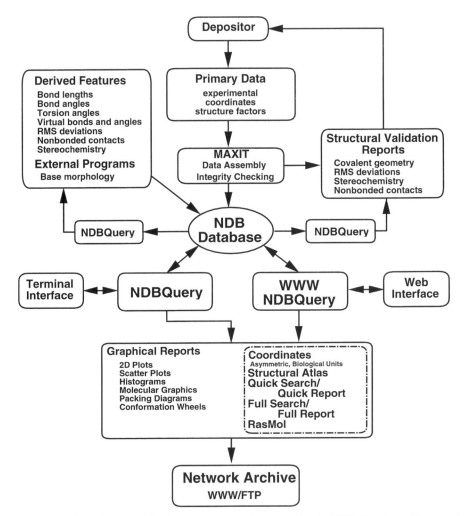

Fig. 3.2. Functional diagram of the data management scheme used by the NDB. The figure illustrates the generation of derived structural features by the NDBQuery program using both internal and encapsulated programs. The graphical and tabular reports created by the NDBQuery application are accessible via the NDB network server.

2.2 The features of the database

The NDB is a relational database that uses SYBASE (4) as the database management system. All of the primary and derived features are stored as tables. The program NDBQuery (5) is used to access the database and is the centre of data processing and distribution (Fig. 3.2).

There are two forms of NDBQuery: a WWW interface and a terminal menu. The WWW interface is publicly available and has both interactive structure selection and report generation. The Quick Search/Report WWW option allows for simple queries and reports while the Full Search/Report option allows the user to build more complex queries and reports. The terminal menu interface is an English language interface to SQL (Structure Query Language). This interface provides a broader range of queries and reports. However, its use requires a terminal session on the computer that supports the SYBASE SQL server, thus limiting its general availability.

By using either interface, reports can be generated for a particular group of structures and coordinate files can be written for structures within the database (Fig. 3.3). Structure selection, which is the first step in the query process, involves creating combinations of the features within the database to constrain the structure search.

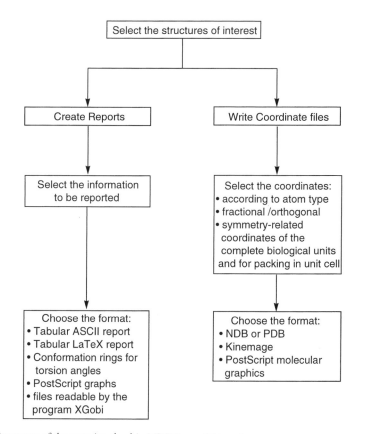

Fig. 3.3. Summary of the steps involved in NDBQuery (5) to select structures, create reports, and write coordinate files.

Table 3.2. Example of the Boolean logic used to construct a query in NDBQuery: Structure selection of B-DNAs containing the residue sequence 'C G C G' without base modifiers, mismatches, or drugs

Table	Attribute (Column)	Operator	Operand	Logical
structure_summary	Conformation_Type	=	B	AND
structure_summary	Classification	=	DNA	AND
sequence	Sequence	like	%CGCG%	AND
structure_summary	Base_Modifier_Y_N	=	N	AND
structure_summary	Mismatch_Y_N	=	N	AND
structure_summary	Drug_Y_N	=	N	

The simplest use of constraints is the selection of one structure by its NDB Structure ID. The power of NDBQuery, however, is the ability to search across the Archive. For example, it is possible to search for all structures of a particular conformational class determined by a particular investigator; one can also select all structures with a particular combination of conformation angles that crystallize in a particular space group. An example of a structure search and the Boolean logic involved is given in Table 3.2, which looks for B-DNA structures with the sequence pattern 'C G C G'.

Once the structure(s) have been selected, a large variety of reports can be written. Tables of information about the experiment or about the structural features can be produced. Additionally, graphical reports, including scattergrams, pie charts, and histograms, can be generated. Molecular graphics of the molecule alone, or of the molecule packed in the unit cell, are another output option. One of the most useful reports created by NDBQuery is the 'Atlas' page (Fig. 3.4) which provides a summary of the experimental and structural features and figures of the biological unit and the crystal packing. Figure 3.5 provides some additional examples of the variety of reports possible using NDBQuery.

Another important feature of NDBQuery is the query-saving option. Once a complex query has been successfully constructed, the constraints used in this particular search can be saved for use at a later time. NDBQuery can also search using batches of these saved queries, such as in the creation of a large number of different reports about one group of structures. Another example of a batch query is the creation of one particular type of report about many different groups of structures. This capability is used routinely by the NDB staff to produce summary reports that are made available on the WWW.

2.3 Standards for the validation of structures

The development of standard dictionaries of geometric values was the first requirement for the evaluation of structures in the NDB. To do so, it was necessary to search the Cambridge Structural Database (CSD) (7) for well-determined structures of the constituents of nucleic acids: the bases (8) and the sugars and the phosphates (9). The results of those analyses produced considerably updated values that take the sugar conformation into account. The values obtained were also used to create new parameter files for use with the refinement program X-PLOR (10,11). This has resulted in the greatly increased use of this program for nucleic acid–containing crystals.

It is useful to determine how much the values obtained for the valence geometry deviate from the standard values (Fig. 3.6). The root mean square differences (rmsds) between the standard values and the refined results have been calculated for all the values in each structure, as well as for each of the component parts. In general, the distributions for the structures are Gaussian and are centred around the correct dictionary value. The values for the bases are more precise than those of the sugars. The distribution of the rmsds for the phosphates is even broader. This is a reflection of the relatively high error in the positions of the phosphates, as well as of the relatively poorly determined dictionary values.

The distributions and correlations between the torsion angles of high resolution structures in the NDB were used to determine the expected values for the common helical types (Fig. 3.7). These values are useful for structure comparison as well as for refinement.

3. Production characteristics of the NDB

3.1 Data deposition and processing

Data for nucleic acid crystal structures are accepted in electronic format and then immediately transferred into macromolecular crystallographic information files (mmCIF). The mmCIF files are then populated with the additional experimental details that are extracted from the manuscript. The data are loaded into the database to begin the checking procedure.

All the relevant geometric features of the structure are calculated. These features include valence geometry, torsion angles, and intermolecular contacts. The values are checked against standard values where appropriate. The Atlas pages produced at this stage provide valuable checks about the content of the data files. An electronic summary is made of the results of structure validation and these are sent to the author for review and possible revision.

The structure is released publicly after checking procedures are completed and the file is approved by the author. In addition, the files are converted to PDB format and sent to the PDB Archive. Overall, it takes less than two weeks from the time the coordinates are received to the time they are released to the public.

3.2 Data distribution

The database itself has several levels of access. The WWW forms interface can search the NDB via a quick query, a full query, and a processing status query. The quick query interface provides a limited menu for the most common queries. The more extensive full query option allows for the construction of more complex queries. Once structures have been selected, there are several output options. For each structure, the coordinates can be retrieved in PDB or mmCIF format, an NDB Atlas entry can be viewed, or a RasMol (12) graphic can be displayed. In addition, tabular reports about a variety of features can be created interactively. Reports created from

(a)

NDB ID: URX035

Features

RNA, RNA hammerhead ribozyme, catalytic RNA, loop

Compound name

RNA hammerhead ribozyme

Sequence in asymmetric unit

- Chain A: GUGGUCUGAUGAGGCC
- Chain B: GGCCGAAACUCGUAAGAGUCACCAC
- Chain C: GUGGUCUGAUGAGGCC
- Chain D: GGCCGAAACUCGUAAGAGUCACCAC

Citation

W.G. Scott, J.T. Finch, A. Klug
The Crystal Structure of an All-RNA Hammerhead Ribozyme: A Proposed Mechanism for RNA
Catalytic Cleavage
Cell, 81, pp. 991–1002, 1995.

Space group

P3₁21

Cell constants

```
A = 64.980    B = 64.980    C = 138.140   (Angstroms)
alpha = 90.00   beta = 90.00   gamma = 120.00   (degrees)
```

Crystallization conditions

- Method: vapor diffusion, sitting drop
- Solution: nucleic acid, water, PEG 6000, glycerol, spermine, magnesium acetate, ammonium acetate, ammonium cacodylate

Refinement

The structure was refined using the X-PLOR 3.1 program. The R value is 25.1 for 6199 reflections in the resolution range 15.0 to 3.1 Å with Fobs > 2.0 sigma(Fobs).

Coordinates

The coordinates for the asymmetric unit of this structure are stored in the NDB archive.

Views of URX035

(b)

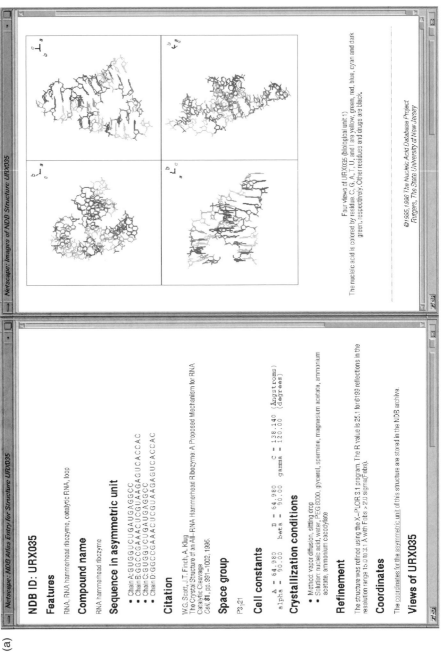

Four views of URX035 (biological unit)

The nucleic acid is colored by residue: C, G, A, T, U, and I are yellow, green, red, blue, cyan and dark green, respectively. Other residues and drugs are black.

©1995, 1996 The Nucleic Acid Database Project
Rutgers, The State University of New Jersey

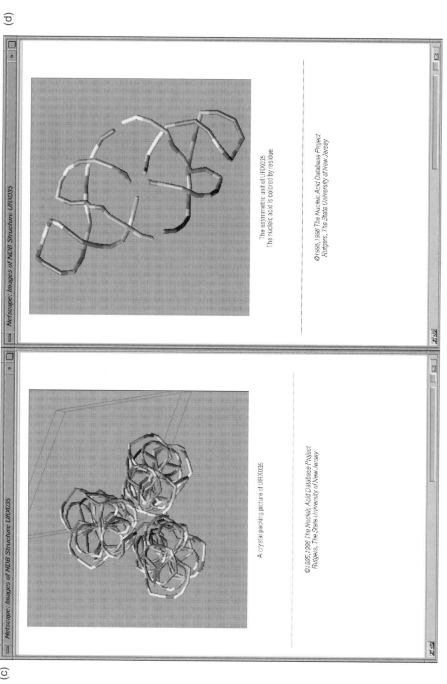

Fig. 3.4. Sample of an atlas page for the structure URX035 (6). (a) The tabular information summarizes some of the contents of the database tables for this structure; (b) four quarter views of one of the biological units are shown; (c) a view of the crystal packing; (d) a view of the asymmetric unit.

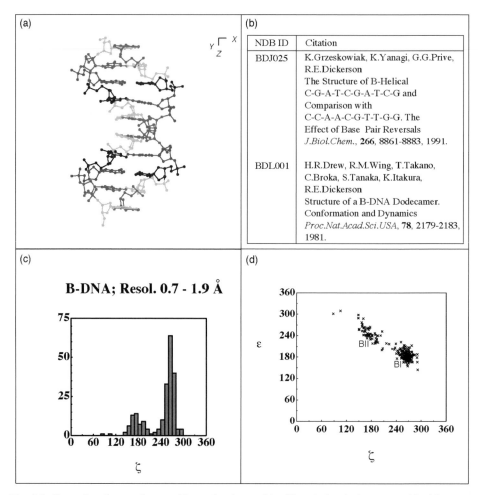

Fig. 3.5. Examples of report formats: (a) a molecular graphic; (b) a tabular citation report; (c) a histogram of the distribution of the values of one torsion angle; (d) a scattergram of torsion angles ε vs. ζ; the main conformational classes are labelled B_I and B_{II}.

this interface are shown in Fig. 3.8. The status query creates a processing status report for any structure in the database.

The WWW site also distributes information about nucleic acid structure through the 'Atlas' and 'Archives' sections. The Atlas section has prepared Atlas pages for all the structures in the NDB (Fig. 3.9). Each entry presents useful summaries of each structure that can aide in the preparation of manuscripts and in teaching about structure. The Archives section contains a large number of summary reports from the NDB, including citation and cell dimension reports for all structures in the database. These reports are prepared on a regular basis using the batch query option of NDBQuery. The Archives section also contains the standard geometry dictionaries and links to the NDB FTP server.

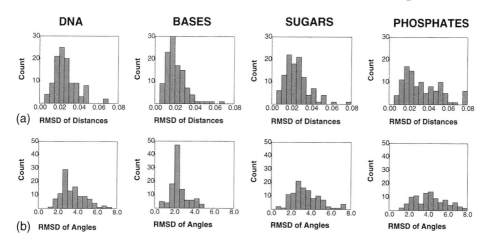

Fig. 3.6. (a) The distributions of the rmsds between bond distances and their target values derived by Gelbin *et al.* and Clowney *et al.* (8,9) for each structure in the NDB. The distributions are shown, from left to right, for all structures in the NDB, in bases, in sugars, and in phosphates. (b) The rmsds between bond angles and the target values as in (a).

Although the WWW is a very powerful distribution mechanism, it popularity has exceeded its ability to provide information over an unlimited geography. For this reason the NDB project has set up mirror sites in Europe (http://www.ebi.ac.uk/NDB/), Japan (http://ndbserver.nibh.go.jp/NDB/) and the US (http://ndb.sdsc.edu/). These sites fully reproduce the content and capability of the database, and are updated regularly.

4. Practical uses of the NDB for research and teaching

The NDB has been designed to be used by people at every level of understanding and interest in biomolecular structure. For those users who wish simply to extract coordinates for further studies, the search capabilities of NDB allow for rapid delineation of the desired structures. One can reliably and rapidly find structures of a particular conformational class, or by a particular author, or within a particular resolution range, etc. Searches such as these have been done by a growing group of researchers who have used their own analysis tools to understand further the features of a particular class of molecules (12–16).

All users may create tables of information using the tools contained within the NDB Quick Search/Report and Full Search/Report options. This capability is very useful for comparative analysis. To create more detailed graphical reports it is necessary to have direct access to the NDBQuery server. For those users, it has been possible to do more extensive analyses of structural features (17). In the case that a user does not have access to the server, the NDB staff provides custom reports for queries sent to ndbadmin@ndbserver.rutgers.edu.

The Archives section of the NDB WWW site contains a wealth of information that can be of use to researchers. There are summary reports of information that have been produced by NDBQuery that can be extracted and used; for example, the citations for all structures, grouped by type, are available. Specific information about structures that

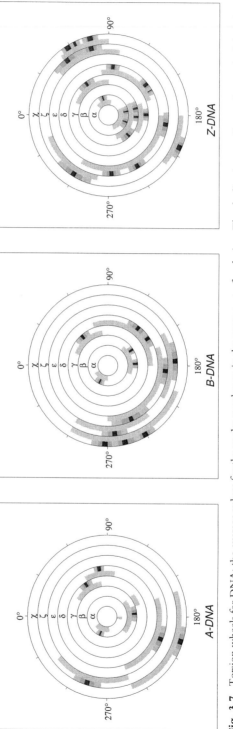

Fig. 3.7. Torsion wheels for DNA; the average values for the angles are shown in the outer part of each ring. The shading is according to the standard deviation of the average. The ranges are shown on the inner part of each ring (17).

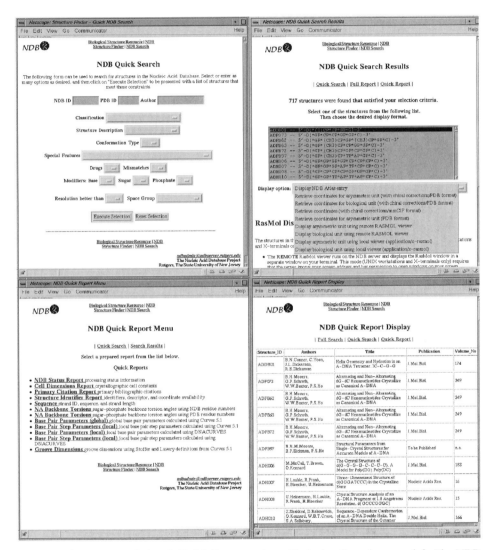

Fig. 3.8. The NDB Quick Search/Quick Report options. Clockwise, from the upper left: The NDB Quick Search page; The NDB Quick Search Results; a sample citation report; the NDB Quick Report Menu.

(a)

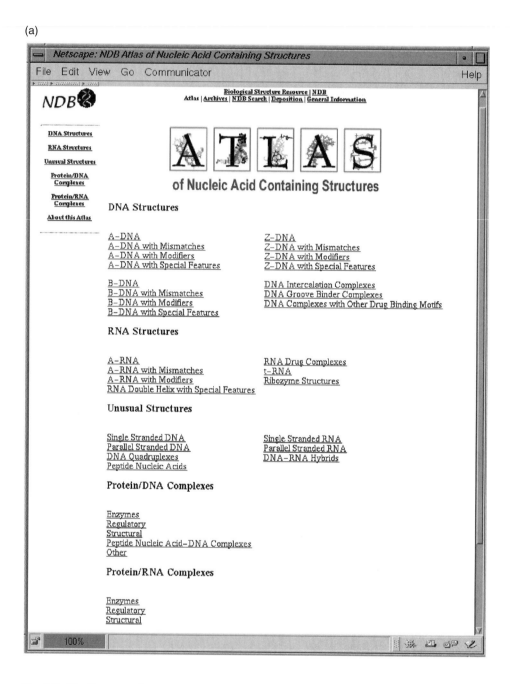

Fig. 3.9. Caption opposite

(b)

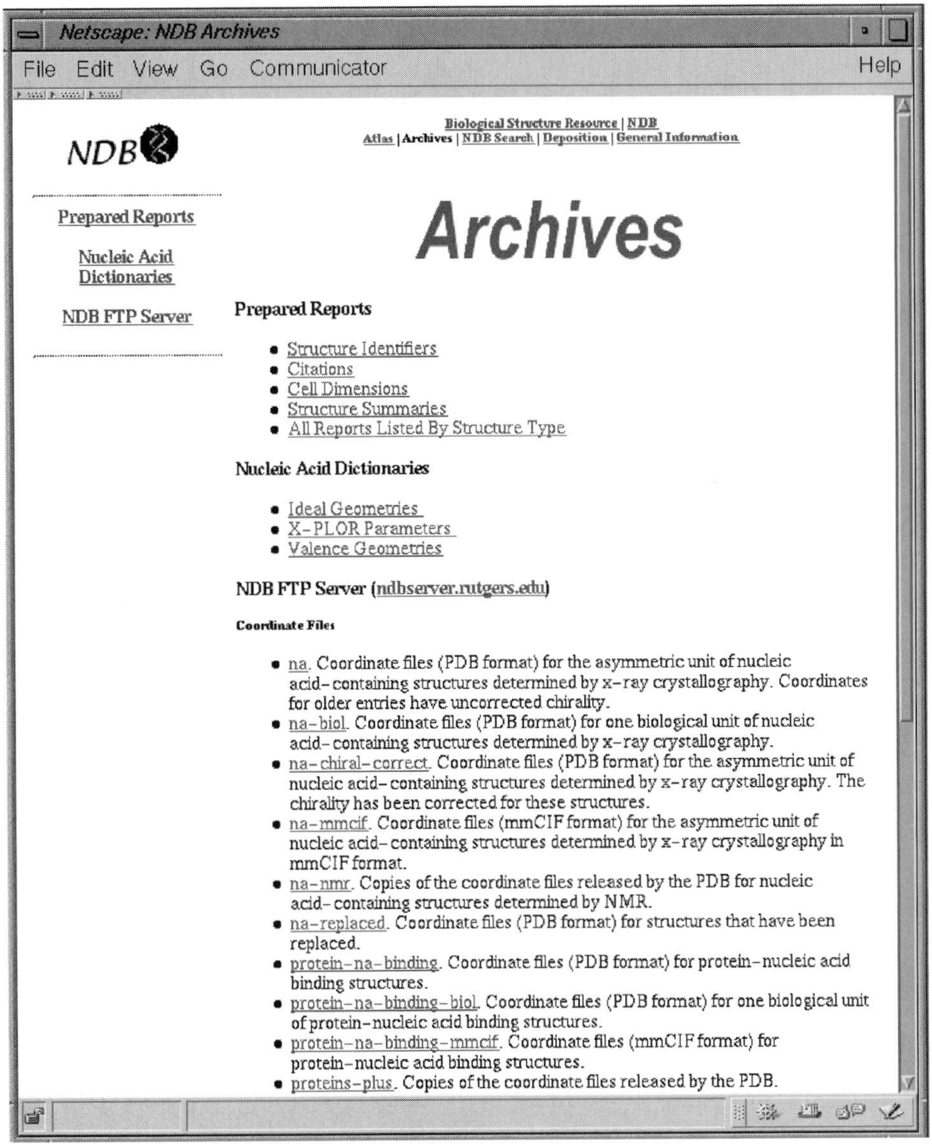

Netscape: NDB Archives

File Edit View Go Communicator Help

NDB

Biological Structure Resource | NDB
Atlas | Archives | NDB Search | Deposition | General Information

Prepared Reports

Nucleic Acid Dictionaries

NDB FTP Server

Archives

Prepared Reports

- Structure Identifiers
- Citations
- Cell Dimensions
- Structure Summaries
- All Reports Listed By Structure Type

Nucleic Acid Dictionaries

- Ideal Geometries
- X-PLOR Parameters
- Valence Geometries

NDB FTP Server (ndbserver.rutgers.edu)

Coordinate Files

- na. Coordinate files (PDB format) for the asymmetric unit of nucleic acid-containing structures determined by x-ray crystallography. Coordinates for older entries have uncorrected chirality.
- na-biol. Coordinate files (PDB format) for one biological unit of nucleic acid-containing structures determined by x-ray crystallography.
- na-chiral-correct. Coordinate files (PDB format) for the asymmetric unit of nucleic acid-containing structures determined by x-ray crystallography. The chirality has been corrected for these structures.
- na-mmcif. Coordinate files (mmCIF format) for the asymmetric unit of nucleic acid-containing structures determined by x-ray crystallography in mmCIF format.
- na-nmr. Copies of the coordinate files released by the PDB for nucleic acid-containing structures determined by NMR.
- na-replaced. Coordinate files (PDB format) for structures that have been replaced.
- protein-na-binding. Coordinate files (PDB format) for protein-nucleic acid binding structures.
- protein-na-binding-biol. Coordinate files (PDB format) for one biological unit of protein-nucleic acid binding structures.
- protein-na-binding-mmcif. Coordinate files (mmCIF format) for protein-nucleic acid binding structures.
- proteins-plus. Copies of the coordinate files released by the PDB.

Fig. 3.9. (a) The main NDB Atlas and (b) Archives index page available on the NDB WWW server (http://ndbserver.rutgers.edu, http://www.ebi.ac.uk/NDB/, http://ndbserver.nibh.go.jp/NDB/, and http://ndb.sdsc.edu/).

contain mispairs is also available. The Archives section also stores the standard geometry information that can be used for modelling and refinement.

The Atlas section is particularly useful for learning about nucleic acid structures by providing an easy and immediate method for browsing the contents of the NDB. The Atlas is sorted by structure type so that it is possible to learn about a particular class of molecule simply by reviewing the contents of those pages. The information contained within the Atlas has been used to create a tutorial entitled 'An Introduction to Nucleic Acids' which is designed to give an overview of nucleic acid structure and is available through the NDB Archives section at http://ndbserver.rutgers.edu/NDB/archives/NAintro/index.html.

5. Prospects

This summary of the NDB describes the state of the project in 1998. Because the project has always attempted to keep up with the latest technology, it is likely that most aspects of the infrastructure will continue to evolve. This will ensure even stronger and more diverse features to enable the research and teaching community.

Acknowledgements

We thank David Beveridge, Stephen Neidle, Wilma Olson, and Bohdan Schneider for their input into the various stages of the development of this project. We are also grateful for the many contributions of Shu-Hsin Hsieh, Zukang Feng, Les Clowney, and Anke Gelbin towards the creation and support of the NDB infrastructure. The NDB Project is funded by the National Science Foundation (BIR 95 10703) and the Department of Energy.

References

1. Berman, H.M., Olson, W.K., Beveridge, D.L., Westbrook, J., Gelbin, A., Demeny, T., Hsieh, S.-H., Srinivasan, A.R. and Schneider, B. (1992) *Biophys. J.* **63**, 751.
2. Bernstein, F.C., Koetzle, T.F., Williams, G.J.B., Meyer, E.F., Brice, M.D., Rodgers, J.R., Kennard, O., Shimanouchi, T. and Tasumi, M. (1977) *J. Mol. Biol.*, **112**, 535.
3. Fitzgerald, P.M.D., Berman, H.M., Bourne, P.E., McMahon, B., Watenpaugh, K. and Westbrook, J. (1996) *Acta Cryst.*, **A52**(Suppl.), MSWK.CF.06.
4. Sybase, Inc. (1995) *SYBASE SQL server release 11.0*. Emeryville, CA, 70202-01-1100-01.
5. Westbrook, J., Demeny, T. and Hsieh, S.-H. (1996) *NDBQUERY, V.4.0, A Simplified User Interface to the Nucleic Acid Database*, NDB-99. Rutgers University, New Brunswick.
6. Scott, W.G., Finch, J.T. and Klug, A. (1995) *Cell* **81**, 991.
7. Allen, F.H., Bellard, S., Brice, M.D., Cartright, B.A., Doubleday, A., Higgs, H., Hummelink, T., Hummelink-Peters, B.G., Kennard, O., Motherwell, W.D.S., Rodgers, J.R. and Watson, D.G. (1979) *Acta Cryst.* **B35**, 2331.
8. Clowney, L., Jain, S.C., Srinivasan, A.R., Westbrook, J., Olson, W.K. and Berman, H.M. (1996) *J. Am. Chem. Soc.* **118**, 509.
9. Gelbin, A., Schneider, B., Clowney, L., Hsieh, S.-H., Olson, W.K. and Berman, H.M. (1996) *J. Am. Chem. Soc.* **118**, 519.
10. Brünger, A.T. (1991) *X-PLOR Manual. v. 3.1.* Yale University Press, New Haven.

11. Parkinson, G., Vojtechovsky, J., Clowney, L., Brünger, A.T. and Berman, H.M. (1996) *Acta Cryst.* **D52**, 57.
12. Sayle, R. and Milner-White, E. J. (1995). *TIBS* **20**, 374.
13. Gorin, A.A., Zhurkin, V.B. and Olson, W.K. (1995) *J. Mol. Biol.* **247**, 34.
14. Young, M.A., Ravishanker, G., Beveridge, D.L. and Berman, H.M. (1995) *Biophys. J.* **68**, 2454.
15. Suzuki, M. and Yagi, N. (1995) *Nucl. Acids Res.* **23**, 2083.
16. Tabernero, L., Bella, J. and Alemán, C. (1996) *Nucl. Acids Res.* **24**, 3458.
17. Schneider, B., Neidle, S. and Berman, H.M. (1997) *Biopolymers* **42**, 113–24.

4
Simulation of nucleic acid structure

Jennifer L. Miller,[1,3] *Thomas E. Cheatham III,*[1,2] *and Peter A. Kollman*[1]

[1]*Department of Pharmaceutical Chemistry, Box 0446, University of California, San Francisco, San Francisco, CA 94143, USA*

[2]*Current address: Laboratory for Structural Biology, MGSL/DCRT/12A-2041, National Institutes of Health, Bethesda, MD 20814, USA*

[3]*Current address: CombiChem, Inc., 1804 Embarcadero Road, Suite 201, Paolo Alto, CA 94303, USA*

In this chapter, we briefly introduce the importance and usefulness of applying molecular mechanical models to studies of nucleic acid systems. Section 1, on force fields for nucleic acids, and Section 2, an introduction to simulation methods, discuss the most important methodological issues encountered in theoretical studies of nucleic acids. Section 3, on applications of molecular mechanics and dynamics to nucleic acid systems, reviews the history of both *in vacuo* and solution phase simulations.

1. Force fields for nucleic acids

1.1 General principles

The development of potential energy functions for nucleic acids has followed similar pathways to those for other organic and biological molecules. Early studies used rigid bond lengths and angles and allowed flexibility only in dihedral angles (1,2). The development of Cartesian coordinate energy gradients by Lifson and Warshel (3) opened the way for more general molecular mechanical models and molecular dynamics; such studies were pioneered by Levitt (4,5).

Key insights into the flexibility of the furanose puckering came from Levitt (4). The C2′ tendency of deoxyribo in contrast to ribo sugars was interpreted as owing to the *gauche* tendency of the OCCO torsion by Olson (6). Newton's quantum mechanical calculations on dimethyl phosphate showed that the *gauche* tendency of OPOC units was an electronic effect and an intrinsic property of this fragment, thus rationalizing why nucleic acids preferred (*g,g*) around the phosphodiester bond, not the sterically least crowded (*t,t*) conformation (7).

In the 1980s, force fields were developed for both proteins and nucleic acids and these were applied using molecular mechanics and dynamics; those by Weiner *et al.* (8), Nilsson and Karplus (9), and van Gunsteren *et al.* (10) were probably the most used. In the 1990s, improvements on these earlier force fields have appeared from Cornell *et al.* (11), MacKerrell *et al.* (12), and van Gunsteren *et al.* (13). These are mainly directed at molecular dynamics simulations in explicit solvent. All use a 'generic' functional form (eqn 4.1). In addition to these, OPLS (optimized potentials for liquid simulations) parameters have been developed for the nucleic acid bases for applications to molecular recognition in host–guest systems (14). Finally, Lavery and coworkers (15) have developed a force field whose focus has been its use of molecular mechanical minimization in internal coordinate space. This force field uses helicoidal parameters, sugar puckers, and dihedral angles as variables in order to minimize the number of variables to be optimized. It differs from the three generic force fields in its use of rigid bond lengths and angles and its inclusion of explicit angular dependence in hydrogen bonds. The model of Lavery *et al.* is aimed at an implicit solvent representation.

When one applies these molecular mechanical models to systems without including explicit solvent, the most successful approaches use a distance-dependent dielectric constant, and/or reduce the phosphate charges. A common approximation (9) is to make the net charge equal to that suggested by Manning's counterion condensation theory (~ −0.2e per phosphate).

Simulations including explicit solvent use a dielectric constant, $\varepsilon = 1$, assuming that solvent orientational polarization is explicitly included in such a microscopic model. When such a model is used with standard atom–atom or residue–residue non-bonded cut-offs (typically at 8–12 Å), the results are very bad—a DNA duplex will tend to break apart in ~100s of picoseconds (16). Only when the electrostatic forces are shifted to approach zero smoothly (17) or when long-range electrostatic effects are explicitly included with Ewald or particle mesh Ewald methods (18), are stable DNA or RNA MD trajectories achieved in explicit solvent (16). This emphasizes that both the force field (functional form and parameters) and a suitable treatment of long-range electrostatic effects are essential for an accurate representation of nucleic acids using molecular dynamics.

1.2 Specific issues

How does one develop a force field for nucleic acids? There is subjectivity in the choice of functional form and parameters, but we will illustrate the issues using eqn 4.1 and the Cornell *et al.* (11) force field as examples. We consider, in turn, the various terms in eqn 4.1:

$$E_{total} = \sum_{bonds} K_r (r - r_{eq})^2 + \sum_{angles} \frac{1}{2} K_\theta (\theta - \theta_{eq})^2 + \sum_{dihedrals} \frac{V_n}{2} [1 + \cos(n\theta - \gamma)] +$$

$$\sum_{i<j} \left[\frac{A_{ij}}{R_{ij}^{12}} - \frac{B_{ij}}{R_{ij}^{6}} + \frac{q_i q_j}{\varepsilon R_{ij}} \right] \tag{4.1}$$

(a) Bond lengths, K_r and r_{eq}. These are typically derived by using experimental (or high level *ab initio* calculated) bond lengths and vibrational frequencies.

(b) Bond angles, K_θ and θ_{eq}. These are typically derived similarly to bond lengths, but are a significantly softer degree of freedom and thus coupling with the torsional parameter must be considered.

(c) Torsional parameters V_n, n for $n = 1, 2, 3$. For a nucleic acid backbone, $-C(sp^3)-C(sp^3)-$, $-C(sp^3)-O(sp^3)-$, $-C(sp^3)-N(sp^2)-$, and $-P-O(sp^3)-$ torsions need to be considered. (See Chapter 1 for definitions of these angles.) The nucleic acid torsions α and ζ correspond to $C(sp^3)-O(sp^3)-P-O(sp^3)$ and, using *ab initio* calculations on dimethyl phosphate as a reference point, Cornell *et al.* (11) showed that V_2 ($n = 0$) and V_3 ($n = 0$) torsions were required to describe this torsional energy well. This need for the V_2 component comes from the electronic back-bonding first characterized by Newton.

(i) In the case of the very important furanose puckering, one develops an initial set of bond angle and torsional parameters, and then may adjust both of them to assure an accurate calculation of the pseudo-rotation profile in furanose sugars. Another example is the phosphodiester bond, where Gorenstein *et al.* (19) have shown considerable coupling between the O(ester)–P–O(ester) angle and C–O(ester)–P–O(ester) torsion, with the internal angle being significantly smaller when the torsion is *trans* than when it is *gauche*.

(ii) $\gamma(C4'-C5')$, $\nu_1(C1'-C2')$, $\nu_2(C2'-C3')$, and $\delta(C3'-C4')$ are all examples of $X-C(sp^3)-C(sp^3)-Y$ angles. Cornell *et al.* (11) note that a single V_3 ($n = 0$) torsion can be employed for all such torsions, unless X and Y are both electronegative, in which case an additional V_2 torsion for $X-C(sp^3)-C(sp^3)-Y$ is needed. They carried out *ab initio* calculations on 1,2-dimethoxyethane to derive the appropriate V_2 torsion to describe the *gauche* tendency of $O(sp^3)-C(sp^3)-C(sp^3)-O(sp^3)$ units.

(iii) $\beta(C5'-O5')$, $\varepsilon(C3'-O3')$ $\nu_4(C4'-O4')$, and $\nu_0(C1'-O4')$ are all examples of $X-C(sp^3)-O(sp^3)-Y$. Cornell *et al.* use the $V_3(\delta = 0)$ derived from dimethyl ether for all such torsions. In order to describe more accurately the furanose pseudo-rotation, a $V_2 = 0.1$ kcal/mole, $\delta = 180°$ is included for $C(sp^3)-C(sp^3)-O(sp^3)-C(sp^3)$ torsions.

(iv) The glycosidic torsion $-C(sp^3)-N(sp^2)-$, as any other sixfold torsion, should not have any significant electronic contribution to its torsional potential. However, comparison of *ab initio* energies and molecular mechanical ones convinced Cornell *et al.* (11) to include V_1 and V_2 torsions in order to enable molecular mechanical energy around the glycosidic bond to reproduce the *ab initio* values.

(d) The electrostatic partial charges, q_i. These are derived either by using quantum mechanical calculations on fragments (11) or empirically [CHARMM (Chemistry at Harvard Molecular Mechanics, 12), OPLS (14)].

(e) van der Waals parameters A_{ij} and B_{ij}. These are derived by ensuring that the densities and enthalpies of vaporization of liquids [OPLS (14), Cornell *et al.* (11)] or sublimation energies and crystal parameters of solids [CVFF (Consistent Valency Force Field, 20)] are well described.

In summary, there are different approaches to developing nucleic acid force fields but no matter how well these reproduce the properties of small fragments/model systems, the crucial test is how well the force field reproduces the properties of DNA and RNA systems.

2. Introduction to simulation methods

The potential energy hypersurface for nucleic acid molecules is a very rough and complex multidimensional surface. In molecular modelling we are interested in exploring the minima of this surface, especially the global minimum. Unfortunately, a complete search of the surface is impossible, and we are restricted to local searches in the vicinity of a particular conformation, which is usually experimentally determined. Using a potential energy function such as eqn 4.1, it is possible to employ computational methods to find the nearest local minimum on an energy hypersurface, investigate the dynamic behaviour of the system of interest with molecular dynamics or normal mode calculations, or search the energy hypersurface with Monte Carlo techniques. This section will briefly cover each of these methods.

2.1 Minimization and normal mode analysis

Minimization is used during the refinement of experimental structures, to compare the energies of different conformations, and in the initial stages of dynamics studies in order to relieve bad steric contacts and/or bond lengths, angles, etc. This technique has been used for many years in molecular mechanical studies of nucleic acids, typically to compare the energies of two different conformations, such as B-DNA and Z-DNA. With the recent improvement in both computational power and treatment of the electrostatic interactions, minimization plays less of a role in modelling of macromolecular systems. However, it is still an important step in performing normal mode calculations.

Because the potential energy is an analytical function of the atomic coordinates, it is relatively easy to find the nearest local minimum. There are a few techniques that are commonly used to perform the minimization. They are generally split into two classes. The first class only requires the calculation of the first derivatives (e.g. steepest descents and conjugate gradients), while the second class requires that the second derivatives also be determined (e.g. Newton–Raphson). In a typical molecular dynamics simulation, the first few steps of minimization are performed using steepest descents, with the rest of the minimization carried out using the more efficient conjugant gradients method. The second class is required for normal mode calculations where it is very important to get as close to the minimum as possible. However, because of the computational demands of this method and the complex nature of the potential energy hypersurface, this method is usually used after one of the first-order methods has brought the system close to the minimum.

Normal mode analysis of nucleic acid structures can provide an estimate of thermodynamic quantities such as the vibrational free energy and entropy and give some insight into the 'motion' of a molecule. These calculations are based on the assumption that the normal vibrational modes of a molecule are harmonic in nature, something that is reasonable as long as the conformation of interest represents a true minimum on the hypersurface. In molecular mechanics studies, this type of analysis is used in both the construction of the force field (11) and, in some cases, in studies of macromolecular properties (21).

The harmonic approximation, which requires the system to be at a local minimum, precludes the use of normal mode analysis in simulations with explicit solvent since it is intractable to 'minimize' the positions of all the water. Moreoever, solvation is most likely an ensemble property and not well represented by considering a small set of minima. In addition to giving an estimate of the vibrational entropy, the calculation of the normal mode frequencies can give insight into the 'motion' of a molecule. Based on equipartition arguments, each normal mode frequency should be equally populated. However, low frequency modes (<100 cm^{-1}) lead to most of the motion (22). Simulations on proteins suggest that the first 3–8 non-zero normal mode frequencies account for 70% of the motion of the molcule (23). Analysis of these low frequency vibrational modes can give insight into which collective motions dominate. Cheatham has used this technique to gain insight into the relative flexibility of A- and B-form geometries of DNA (24). It should be noted that the harmonic approximation to the normal mode frequencies may not be completely valid since proteins and nucleic acids have significant anharmonic character (21). However, there are techniques for extracting information about the anharmonic modes from molecular dynamics trajectories (25).

A drawback of minimization and normal mode calculations is the need for investigating a representative minimum energy conformation of the molecule; in other words, the insight gained is only as good as the choice of the initial conformation. Therefore, calculations of this type are generally limited to cases where the structure chosen is reasonable, such as an experimentally derived structure, and where it is assumed that the force field will properly treat the structure as a minimum. In the case of nucleic acids and minimizations *in vacuo*, the latter assumption is not always valid; without some representation of solvent in the simulations, the structure of a nucleic acid in solution will tend to distort from experimental values. Since minimization gets trapped into local minima, methods that overcome this difficulty and reasonably represent an 'ensemble' of energetically reasonable conformations are needed. The difficulty in finding all the representative structures for a given potential is often termed the 'local minimum' or 'conformational sampling' problem. Given the expense of minimization calculations and the combinatorial explosion resulting from the independence of the degrees of freedom, exhaustive searching of all possible (or even reasonable) conformations is not possible. Therefore, other methods, such as Monte Carlo calculations and molecular dynamics, are typically applied to investigate the dynamic properties of nucleic acids.

2.2 Molecular dynamics (MD) and Monte Carlo (MC)

Molecular dynamics (MD) is the most widely used method for studying the motions of macromolecules, including nucleic acids, under fully solvated simulation conditions. Here, the 'motion' of a molecule is the time-dependent set of configurations generated by integrating Newton's equations of motion. MD studies have been carried out for many years on nucleic acids with differing degrees of success, mostly because of the treatment of the electrostatic interactions. A proper treatment of these interactions is critical in nucleic acid simulations because of the highly charged backbone. However, with the implementation of fast Ewald methods, long and unrestrained simulations are now possible and MD studies of nucleic acids is a rapidly growing field.

Molecular dynamics methods assign random velocities about a mean temperature to each atom in the system and propagate the dynamics of the atoms by integrating the equations of motion. In Cartesian coordinates (r_i), where m_i is the mass of each atom (and t is time), the force is equal to the mass times the acceleration:

$$F_i(t) = m_i \frac{\partial^2 r_i(t)}{\partial t^2} = -\frac{\partial}{\partial r_i} E_{\text{total}} \qquad (4.2)$$

With the Lagrangian defined in terms of the kinetic and potential energies, the force is simply the negative gradient of the potential energy (which is calculated analytically). For example, in AMBER, the propagation of the dynamics (or integration) is performed using the simple first-order leap-frog algorithm which is derived based on Taylor expansions. A typical step size is 1–2 femtoseconds which implies that a nanosecond simulation requires of the order of a million time steps.

In addition to providing an ensemble of structures near the minimum, MD has the added benefit of providing some estimate of the dynamics of the system and effective configurational entropy. However, given that the kinetic energy added to the system is finite (generally the dynamics are simulated at approximately room temperature or 300 K) the probability of overcoming large energetic barriers is very small. Therefore, molecular dynamics calculations can get 'stuck' in the vicinity of a local minimum; this is exemplified nicely by Cheatham and Kollman (24) where it is observed that, depending on where the calculation is started, a double-stranded RNA duplex gets trapped in A-form and B-form structures at 300 K during the course of multinanosecond simulations.

In Monte Carlo calculations, random moves are made to a new conformation and the move is always retained if the new conformation is more favourable energetically. This random move can be made in a number of ways, including a change in atomic coordinates or torsional values. If it is less favourable energetically, the move is retained (probabilistically) some of the time. In macromolecular systems, Monte Carlo methods have some drawbacks. Not only is there no implicit time evolution to the system, and therefore no straightforward estimate of the dynamics (and no implicit entropic effects), it is often difficult to chose a set of moves that avoids trivial sampling (such as small atomic moves that change the energy and conformation only slightly) yet at the same time avoids unacceptable moves to high energy structures. The latter is particularly troublesome in simulations of flexible macromolecules with explicit solvent. In the limit of infinite sampling, both molecular dynamics and Monte Carlo will ultimately converge to a Boltzmann distribution; however given that molecular dynamics simulations are currently restricted to the nanosecond time-scale and Monte Carlo calculations are restricted to the evaluation of millions of moves (both of which correspond to a roughly equivalent amount of sampling), the conformational sampling problem is still a major issue.

A number of methods have emerged that may prove useful for enhancing the effective sampling, such as locally enhanced sampling (LES) (26) and jumping between wells (27) adaptive Monte Carlo methods. However, it is still unclear whether these will prove useful in the simulation of solvated nucleic acids, although recently, it has been shown that LES can find the correct local minimum in an RNA hairpin loop at least 35 times more efficiently than standard MD (123).

3. Applications of molecular mechanics and dynamics to nucleic acid systems

3.1 Applications without explicit solvent

All of the applications of molecular mechanics and molecular dynamics to nucleic acids pre-1985 were without explicit solvent because of computer limitations. However, even today there are still many applications where the inclusion of explicit solvent is impractical or not warranted. We highlight a selection of some of the applications that we (subjectively) consider interesting. In our view, most of the applications with implicit solvation can best be done within Lavery's JUMNA (15), because it allows infinite periodic repeats of helices and is very efficient because of its use of helicoidal parameters.

3.1.1 Sequence-dependent stabilities

Why is poly dA:poly dT more stable than poly d(AT):poly d(AT) and yet poly d(GC):poly d(GC) is more stable than poly dG:poly dC? To address this question, let us consider the thermodynamic cycle where, because free energy is a state function:

$$
\begin{array}{ccc}
& \Delta G_1 & \Delta G_2 \\
\text{SS} \longrightarrow & \text{B} \longrightarrow & \text{Z} \\
\downarrow \Delta G_5 & \downarrow \Delta G_6 & \downarrow \Delta G_7 \qquad (1) \\
\text{SS}' \longrightarrow & \text{B}' \longrightarrow & \text{Z}' \\
& \Delta G_3 & \Delta G_4 \\
\end{array}
$$

$$\Delta\Delta G \text{ (B–SS)} = \Delta G_3 - \Delta G_1 = \Delta G_6 - \Delta G_5 \qquad (4.3)$$

$$\Delta\Delta G \text{ (Z–B)} = \Delta G_4 - \Delta G_2 = \Delta G_7 - \Delta G_6 \qquad (4.4)$$

$\Delta\Delta G$(B–SS) is the relative stability of B-DNA for one sequence (unprimed) compared with another (primed).

Ideally, to calculate the relative stability of $dG_{10}:C_{10}$ and $d(GC)_5:d(GC)_5$, one would mutate, by free energy perturbation, every alternating base pair from G:C to C:G both in a B structure and in a model for the denatured state (SS). Not only is this a large mutation, but one doesn't know the structural model to use for the denatured state. Thus, it is impractical to use the thermodynamic cycle (1) to address the question posed at the beginning of this section. We have found that one can qualitatively reproduce these homo/heteropolymer stabilities in DNA just using minimized molecular mechanical energies for B-DNA and assuming that any differential energies of the different sequences are zero [or at least smaller than found in the duplex form (28)]. Equally interesting is the decomposition of the energies: we have found, in the

B-form, that the van der Waals energy is more favourable for the homopolymer, given the better purine–purine stacking, and this is the key to why poly dA:poly dT is more stable than poly d(AT):poly d(AT). In the case of the G:C polymers, the larger dipole moment (both G and C have μ ~6 D; A and T have μ ~3 D) and the fact that the moments are almost antiparallel in G compared with C stabilize the heteropolymer and make it more stable.

Interestingly, in RNA, with its T→U and an A structure, the stabilities are reversed (poly G:poly C is more stable than poly (GC) and poly (AU):poly (AU) is more stable than poly A:poly U) and molecular mechanical minimization can at least partially rationalize these differences as well (29). These differences between DNA and RNA stabilities originate in the subtle differences in both the base stacking (van der Waals) and interbase electrostatic energies when the bases are canted (A structure) vs. close to parallel (B structure).

One can use thermodynamic cycle (1) usefully to try to address the questions: Why does 5-CH$_3$C stabilize Z-DNA compared with B-DNA more than does C itself (30)? Why do A:T base pairs destabilize Z compared with G:C base pairs? Here, it is of particular interest to carry out free energy calculations on thermodynamic cycle (1) using eqn 4.4, both with implicit and explicit solvent models. The calculations reproduced the experimental 'Z philicity' of 5-CH$_3$C vs. C and the magnitude of the $\Delta\Delta G$ was twice as large in explicit solvent as in implicit solvent and was in good quantitative agreement with experiment (29). This strongly suggests better van der Waals attraction (30) and hydrophobic effects, because the 5-CH$_3$ is more buried in Z than in B, contribute in almost equal proportion to the Z philicity of 5-CH$_3$.

The Z phobicity of A:T vs. G:C was also reproduced in both implicit and explicit models; in this case, the solvent effect was much smaller (31,32). Here, there is no simple explanation for this Z phobicity, which arises because of both van der Waals and electrostatic interactions between the A:T(G:C) and the bases above and below, as well as the greater stability of G than A in a *syn* conformation of the glycosidic bond.

3.1.2 *Qualitative structural insights*

Does the incorporation of a photodamaged thymine dimer in a DNA duplex strongly bend the DNA? Two modelling efforts using implicit solvent molecular mechanics led to opposite conclusions because of the differences in model building strategy (33,34). But until the study by Rao *et al.* (34), it was more or less assumed that large bending was inevitable upon thymine dimer formation. In recent years, molecular dynamics studies in explicit solvent have supported the results of Rao *et al.*

When Singh and coworkers (35) modelled mitomycin covalent adducts with duplex DNA, it had been assumed that these adducts were attached to the 2-NH$_2$ of guanine in the minor groove. Nonetheless, Singh showed that one could form adducts to both G2(NH) and the O6 in the major groove, and this was interesting because experiments subsequently found predominantly O6 adducts to be favoured (36). Modelling without some use of quantum mechanical calculations could not have assessed whether 2-NH$_2$ and O6 adduct formation was more favourable, but it was able to demonstrate that both were sterically and electrostatically acceptable.

Molecular dynamics on DNA duplexes with implicit solvent (37,38) have led to generally reasonable structures in terms of congruence with experiment, albeit the structures tend to be more 'A-like' than DNA actually is in aqueous solution. (This makes sense since, experimentally, DNA undergoes a conformational change to the A-form in dehydrating conditions.) None the less, Auffinger and Westhof have found that such a model does not work very satisfactorily in modelling non-duplex DNA, whereas an explicit solvent representation with accurate treatment of long-range electrostatics does work well (39).

Keepers *et al.* have modelled base pair opening in DNA using implicit solvent models and have shown nicely how exchangeable proton exposure to solvent (i.e. proton exchange) can occur, particularly in A:T base pairs, while maintaining stacking and structural integrity of the rest of the duplex (40). An analogous analysis has also been done by Lavery (41).

It is intriguing that when the bis intercalator triostin A interacts with d(CGTACG)$_2$, it drives the central T:A base pairs to the Hoogsteen rather than the Watson–Crick form. Molecular mechanical energy minimization with implicit solvent was able to reproduce the fact that in the isolated duplex the structure with Watson–Crick hydrogen bonds was more stable than that with Hoogsteen, but with triostin A complexed, the structure with Hoogsteen hydrogen bonds was more stable (42). One could rationalize this beautifully from the energy components, in that the Hoogsteen hydrogen bonded structure has the interstrand phosphates closer and, thus, a higher electrostatic energy, but the 'narrower backbone' allows almost perfect hand-in-glove van der Waals fit with triostin A, whereas with the backbones further apart, characteristic of Watson–Crick hydrogen bonding, the drug does not experience such stabilizing van der Waals interactions with the DNA.

3.2 Simulation of nucleic acids in solution

Water is a basic and integral part of nucleic acid structure (for reviews see refs 43–45 and Chapter 9). Even under extremely dehydrating conditions, nucleic acids still have some tightly associated water (46,47). The bound water is likely to be important not only structurally, but also functionally. Water may be involved in specific recognition, such as at the protein–nucleic acid interface, and may influence the overall dynamics and flexibility of the nucleic acid. In addition to associated water, salt also plays an important role. The highly charged polyionic backbone of the nucleic acid leads to a condensation of ions around the nucleic acid (48,49). Various salts interact with nucleic acids in different ways and can modulate the structure depending on the concentration; moreover, some ions are critical for structure, such as the Mg^{2+} ions observed in higher order RNA structures. Given the structural role of water bound to nucleic acids, implicit solvent models may not represent completely the interaction of solvent. Therefore, inclusion of some representation of explicit solvent and ions in simulations of nucleic acids seems reasonable.

The addition of explicit solvent significantly increases the computational cost owing to the tremendous growth in the number of pairwise interactions that need to be calculated. Various methods are applied to reduce the computational cost, such as minimal nucleic acid representations and the application of spherical cut-offs to ignore the longer ranged pairwise interactions. However, the cut-off of the long-range electrostatic interactions leads to instability in the simulations. Particularly troublesome is simple truncation of the electrostatic interactions which leads to localized heating, instability, and distortion of the nucleic acid structure. This can, in part, be remedied by smoothing the discontinuities in the energy and forces at the cut-off through the application of atom-based force shifting and other methods (50). A true representation of the long-range electrostatic interactions is obtained through the application of Ewald methods (51,52).

In addition to the computational cost, there are issues with respect to the representation of the solvent that must be considered. Most of the published simulations of nucleic acids involve the application of the simple, rigid, three-point water models, such as TIP3P (53), SPC, and SPC/E (54). The solvent is typically placed around the nucleic acid with a spherical shell of 4–10 Å, or the nucleic acid is immersed into a box of water subject to periodic boundary conditions. While the spherical shell may lead to less water overall and therefore might be expected to be more computationally tractable, the water–vacuum interface at the surface leads to an ordering of the waters and a large surface tension which results in large pressures on the nucleic acid. Moreover, more distortion of the nucleic acid structure has appeared in simulations with a spherical solvent shell compared with those applying periodic boundary conditions. Under conditions of 'true' periodicity in PBC (Periodic boundary conditions) simulations, such as in the application of Ewald methods, the molecules may feel the influence of their periodic images. This influence may become a problem with highly charged solutes, such as nucleic acids, since the nearby periodic images may unduly influence the dynamics and structure. This is particularly acute in simulations of nucleic acid duplexes in rectangular boxes. During nanosecond length simulations, the nucleic acid can rotate in the box such that the long ends of the duplex span the short width of the box. This brings the periodic images very close to the nucleic acid, which may perturb the structure. The problem can be remedied either by removing this inherent rotational motion or by running the simulations in larger cubic or truncated octahedral boxes. Beyond this problem and other obvious issues relating to the choice of solvent boundary, simulations by Smith and Pettitt suggest that imposing true periodicity does not appear to lead to large artefacts (55) in solvents with a reasonably high dielectric constant, such as water. In Sections 3.2.1–3.2.4, a review of the use of Monte Carlo simulations with fixed nucleic acids and of molecular dynamics simulations with explicit solvent is presented. The presentation herein is not meant to be exhaustive but to highlight the work done from the early 1980s up to the present day. For additional reviews see the papers by Beveridge (56,57) or Louise-May *et al.* (58).

3.2.1 Rigid nucleic acids: simulation of water and counterions

The earliest inclusion of explicit solvent in simulations of nucleic acids were the Monte Carlo simulations of Clementi and Corongiu (59). The authors provide a very detailed account of the hydration of minimal nucleic acid duplex models from short Monte Carlo simulations. To reduce the computational burden, small repeating units (such as base pairs or dinucleotide base pairs) were simulated with helical periodic boundary conditions. Later, entire duplex structures were solvated and subjected to hexagonal periodic boundary conditions. A series of Monte Carlo simulations by Beveridge and coworkers investigated the crystal structure of d(CGC-GAATTCGCG)$_2$ with 1777 waters in a hexagonal cell under periodic boundary conditions (60–62). These simulations found 10.4 waters per nucleotide comprising the first coordination sphere, which was surrounded by a second hydration sphere of waters primarily in the minor and major groove, which brought the hydration level to 17.4 waters per nucleotide. This is close to the 20 waters per nucleotide found in early experiments (63). Both G:C and A:T regions can support ordered water structure and the 'spine of hydration' seen in crystal structures (64) was reproduced. Additionally, the expected 'cone of hydration', or triad of water around the phosphates (65), was found. Results on other systems also demonstrate reasonable agreement with the crystal structures (62,66,67) and demonstrate sequence-specific hydration patterns (68,69). As such, these models reasonably represent the hydration; however, they completely neglect the dynamic flexibility seen in the nucleic acid structure, which clearly modulates the specific hydration.

3.2.2 Molecular dynamics of nucleic acids in explicit solvent: the early years

Before 1995, the simulations performed were generally short (<200 ps) but, despite exposing deficiencies in the force fields and methodology, these early simulations gave promise to the field. The inclusion of explicit water lead to results that were in qualitative agreement with what had been seen experimentally and suggested the profound flexibility of nucleic acids. The first reported molecular dynamics simulation of a nucleic acid in explicit water was reported in 1985 where a B-DNA nucleic acid duplex, d[CGCGA], was surrounded by eight Na^+ counterions and a spherical shell of 806 TIP3P water molecules (70). This simulation, run for a reported ~100 ps, required approx. 20 hours on a Cray X-MP 4/8; one of the larger supercomputers of the time. Compared with the *in vacuo* simulations done previously, there were not too many significant differences when solvent was added into the simulation, except for the observation of more twist and tilt in the central base pairs and a number of deviations from the classic C2'-*endo* sugar pucker conformations. Although the structure was more 'B-like' in this simulation than was seen in the *in vacuo* simulations of Levitt (71), 100 ps is not sufficiently long for the instabilities in the cut-off of the electrostatic interactions to become apparent. Also, by the end of the simulation, the water droplet was distorted to such a degree that one end of the helix was nearly exposed to the vacuum.

Perhaps a better representation of the solvent was found in simulations applying a periodic boundary treatment by van Gunsteren and coworkers, who investigated an eight base pair fragment of d[CGCAACGC]/d[GCGTTGCG] with 1245 SPC (simple point charge) water molecules and 14 Na^+ ions in MD simulations applying a twin ranged cut-off on the electrostatic interactions. This simulation was run for 114 ps and the nucleic acid duplex retained 80% of a set of 174 experimental NOE (nuclear Overhauser effect) distances during the simulation, with the number of distance violations increasing as the simulation continued. The structure observed was intermediate between A-DNA and B-DNA, with features closer to B-DNA. The observation of a structure intermediate between canonical A-DNA and B-DNA is a fairly general trend seen in the early MD simulations of DNA in solution. Overall, the simulation suggested that the nucleic acid was fairly dynamic, with breaking and reforming of hydrogen bonds between base pairs during the simulation, particularly at the termini. A better characterization of the overall dynamics of the nucleic acid was presented in the 140 ps MD simulation of d(CGCGAATTCGCG)$_2$ by Swaminathan *et al.* (72). In this simulation, employing the GROMOS force field (10) with 1927 waters and 22 Na$^+$ ions in a hexagonal prism cell with periodic boundary conditions and extra restraints added to maintain Watson–Crick base pairing, correlated transitions in the backbone were evident throughout the simulation, including B$_I$ to B$_{II}$ transitions around the ε and ζ backbone angles and also α, γ crankshaft transitions. Over the course of the simulation, the rmsd from the initial structure continued to rise, and the final structures are characterized by structural distortions, despite the use of a switch on the electrostatic interactions. However, the structural distortions are significantly less than those seen in *in vacuo* simulations. These trajectories were later analysed to investigate sequence-specific flexibility and the 'spatial extent of fluctuations' (73), and also to compare with experimental NOE intensities (74,75). The analysis suggests that a dynamic model of duplex DNA better represents the experimental data than do snapshots of the structure. Results similar to those of Swaminathan and coworkers discussed above were seen in the simulation of the same sequence in a droplet of water by Miaskiewicz *et al.* (76) where extensive motion of the DNA was seen. This led the authors to suggest that DNA may not be characterized by a 'unique' structure. As was seen in the earlier simulations, during the 150 ps of dynamics with the Weiner *et al.* (8) force field, the structure moved continuously away from the starting structure and was characterized by structural distortions.

In the work of van Gunsteren *et al.* (77), analysis of the solvent was limited and restricted to calculating coordination numbers (from pair distribution functions) and first shell neighbour residence times. Similar analysis was applied to the trajectories of Miaskiewicz *et al.* who suggested that G:C base pairs (21.38 waters) are more hydrated than A:T base pairs (20.45 waters) with cytosine being the most hydrated base. However, it is difficult to judge the validity of these results based on these short simulations. Despite being only 40 ps, the simulation of d(CCAACGTTGG)$_2$ with a small solvent shell of 4.8 Å (or 471 waters) provides a very nice visualization of hydration of DNA (78). The results suggest that the spine of hydration in the minor groove is largely determined by minor groove width rather than base sequence.

In addition to B-DNA structures, Z-DNA structures were also studied in simulations ranging from the very short (<4 ps) simulations of Swamy and Clementi (79) to longer (~70 ps) simulations on poly(dGdC) sequences by Laaksonen *et al.* (80). These simulations demonstrated differences in hydration in the Z-DNA structures. In the short simulation, the ions did not have time to move. In the longer simulation, it was demonstrated that counterion diffusion is lowered as is the motion of waters near the duplex (or in the first hydration shell). Longer simulations of d(CGCGCGCGCGCG)$_2$ in a Z-DNA conformation showed considerable kinking of the Z-DNA, base pair opening, and specific ion coordination. The overall dynamics suggested to the authors that the beginnings of a Z-DNA to B-DNA transition were being observed; however, the instability could also have resulted from the force field and simulation methodology employed (81). The authors note that application of an Ewald treatment leads to more consistent results (80).

Ewald simulations were also applied by other groups, such as in the work of Forester and McDonald studying B-DNA duplexes of d(CGCGCGCG)$_2$ with various salt concentrations. These simulations suggest that ions move into the second coordination shell but do not seem to interact directly with the DNA. The phosphates were shown to be extensively hydrated, this is similar to the 'cone of hydration' found in early quantum mechanical simulations (65) and seen in other molecular dynamics simulations. The major drawback with the Ewald methods is the tremendous computational cost (which formally scales as $N^{3/2}$ where N is the number of atoms). A major methodological advance was the development of fast Ewald methods, such as the particle mesh Ewald (18,82,83) and particle–particle particle mesh Ewald (84,85) methods, which scale much better [formally as $N\log(N)$].

3.2.3 *MD of nucleic acids in explicit solvent: nanosecond length simulations and beyond*

The years 1994 and 1995 mark the beginning of the era of routine nanosecond length simulation of nucleic acids in explicit solvent, thanks to advances in computer power, force fields, and the development of fast Ewald methods. As the simulations became longer, deficiencies with the methods and force fields became readily apparent. Despite this, the simulations have moved beyond simply evaluating methods and demonstrating reasonable agreement with the starting simulation conditions (i.e. low rmsd values) by actually representing the sequence-specific structure and flexibility of nucleic acids. Even more recently, simulations have moved to another level, that of demonstrating some effect of the environment on nucleic acid structure. Of course there are still a number of issues with force fields and conformational sampling to be resolved; however, it is an exciting time for the all-atom simulation of nucleic acids in explicit water.

Longer simulations clearly point out deficiencies with the methods and force fields. A particular problem is poor representation of the long-range electrostatic interactions. As mentioned previously, the worst behaviour is seen in simulations that simply truncate the electrostatic interactions in a finte range (i.e. 8–20 Å) (50). This was seen in simulations of d(CCAACGTTGG)$_2$ applying a 9 Å group-based truncation (16). Application of switching functions with less than adequate force field representations can also lead to instability. This is shown in the 1 ns simulations of d(CGC-GAATTCGCG)$_2$ with the GROMOS force field (10), a switch on the electrostatic interactions from 7.5 to 11.5 Å, reduced phosphate charges, and Watson–Crick base pair restraints where the structure moves more than 4.5 Å from canonical B-DNA (86). More recently it has been shown that simulations applying an atom-based force shift on the electrostatic interactions, along with reasonable force fields, can lead to stable simulations of nucleic acids (12,87). However, for little additional cost, the fast Ewald methods can also be applied which give a 'true' representation of the long-range electrostatic interactions, subject to the imposition of true periodicity.

It should also be mentioned that interesting results can still be obtained in the absence of periodic boundary conditions, despite the criticism levied earlier in this discussion, in simulations that surround the nucleic acid by a spherical shell of water. This has been shown in simulations of d(CGCGCG)$_2$ immersed in a 23 Å sphere of TIP3P waters applying stochastic boundary conditions (88). A series of simulations were run at different temperatures and the mean square atomic fluctuations of the atoms and number of hydrogen bonds from the DNA to the water monitored, which led to a prediction of the glass transition temperature in the range 223–234 K, in line with experiment. These simulations also demonstrate the following trend in mobilities, with phosphate groups>sugar riboses>bases, as was seen in the early *in vacuo* simulations (29). This is one of the few published reports applying stochastic boundary conditions in the simulation of nucleic acids.

An obvious means to test the validity of the empirical force fields, and a natural test case for the Ewald methods, is the simulation of nucleic acid crystals. Darden and coworkers applied the particle mesh Ewald (PME) method to simulate the B-DNA d(CGCGAATTCGCG)$_2$ (89), a Z-DNA hexamer crystal (90), and RNA dinucleotides (91). The simulation of the B-DNA crystal, including four complete duplexes in the periodic cell, was run for 2.2 ns and the structures remained extremely close to the crystal structure (~1.2 Å rmsd for all heavy atoms). The structure retained the sequence-specific narrowing in the AATT region and transient α, γ crankshaft and B$_I$ to B$_{II}$ transitions in the backbone angles. A possible criticism of these crystal simulations is that the simulations were run at constant volume. Given the tight packing of the duplexes in the crystal, low rmsd values could have been a result of poor sampling rather than excellent agreement with the force field. However, clearly if an unreasonable force field was used or the long-range electrostatic interactions were improperly treated, the structure would show worse behaviour. A better test would be to start a canonical B-DNA structure with the packed unit cell, running a constant pressure calculation, to see if the structure converged to the crystal and displayed the notable sequence-specific features and crystal packing artefacts.

Crystal packing artefacts are clearly an issue. Given that the same sequence can crystallize into slightly different structures depending on the unit cell (92,93), the structure of the nucleic acid is strongly influenced by the environment. Ideally, a method that does not impose crystal periodicity to represent solution conditions is desired. To this end, simulations applying an Ewald method, but without crystal packing of the unit cell, have been applied in the simulation of duplexes and triplexes of DNA. Nanosecond length simulations of d(CCAACGTTGG)$_2$ and RNA hairpin loops demonstrate stability (16). Constant energy simulations on triplex models of d(CG·G)$_7$ in 1 M NaCl with SPC/E water and the CHARMM 22 parameters (94) also displayed reasonable behaviour (95). These simulations suggest that the backbone fluctuations are larger in the third strand and reasonable, albeit lowered owing to association with the DNA, diffusion of the water is observed. By analysing the fluctuations in the dipoles, effective dielectric constants were estimated which show an effective dielectric constant of ~16 for the DNA and ~3 for the bases and sugars (96). When a GCT mismatch is put into the centre of the triplex, overall the structure does not appear to be destabilized (97). The primary differences seem to be enhanced mobility of the thymine and larger differences in the structure and dynamics appear on the 5′ side of the mismatch compared with the 3′ side. Water is clearly important in the structure and appears highly coordinated between the guanine and thymine bases of the mismatch.

Issues with the force field also become readily apparent in nanosecond-scale simulations. The results show a large dependence on the force field applied. In contrast to the force field of Cornell *et al.* (11), which favours B-DNA, earlier versions of the CHARMM 23 all-hydrogen parameter set favour A-DNA in solution. This was demonstrated in simulations where a spontaneous transition from A-DNA to B-DNA was seen in PME simulations with the Cornell *et al.* force field on the d(CCAACGTTGG)$_2$ duplex within ~500 ps (24), in constast with a B-DNA to

A–DNA transition seen over the course of an approximately 3 ns simulation of d(CGCGAATTCGCG)$_2$ applying an Ewald treatment with the CHARMM 23 all-hydrogen parameter set (Version 6.1, November 1993) (98). A B-DNA to A-DNA transition was also seen in atom-based force shifted cut-off (to 11.5 Å) simulations of d(GCGCGCGCGCGC)$_2$ (99) with a slightly newer version of the CHARMM force field (12). The philosophy employed in the design of Cornell *et al.* (11), including restrained electrostatic potential (RESP)-derived charges (100), van der Waals parameters from simulations of neat liquids, and multiconformation charge fitting (101), seems to represent nucleic acid structure well. Recent PME simulations applying this philosophy to the simulation of 3'-phosphoramidates show the expected preference for A-DNA phosphoramidates and a spontaneous B-DNA to A-DNA transition in these models of d(CGCGAATTCGCG)$_2$ in contrast to standard DNA (102). Simulations also suggest, with the Cornell *et al.* force field and an Ewald treatment, that A-RNA duplexes are stable and that DNA:RNA hybrid duplexes adopt a structure that has features of both DNA and RNA duplexes with helicoidal parameters closer to an A-form geometry, intermediate minor groove widths, DNA strands with a mixture of C3'-*endo* and C2'-*endo* structures, and RNA strands with primarily C3'-*endo* sugar puckers (103). These same simulations also demonstrate that B-RNA is stable on a multinanosecond time-scale. This highlights important issues with respect to conformational sampling of RNA; the larger barrier to repuckering may inhibit sampling and stabilize metastable states. Work is currently in progress to determine the relative free energies of A-RNA and B-RNA duplexes in an attempt to characterize the force field better (124).

Simulations with various different ion environments show that ions may also profoundly influence the structure and dynamics. In particular, ions are shown to interact favourably with electronegative pockets in the minor groove of DNA (104) and the major groove of RNA (103). Magnesium (Mg^{2+}) has been shown to affect the backbone dynamics, leading to decreased fluctuations, and to interact primarily with the phosphates (105). With that force field, little salt effect on the structure is seen, such as an expected transition to A-DNA at high salt. Little effect is also seen in simulations with the Cornell *et al.* force field. In contrast, transitions to A-DNA are seen under high salt conditions in hexamer DNA simulations with the BMS (Bristol-Myers Squibb) force field (87).

The new force fields are able to represent the sequence-specific structure of DNA reasonably well. A 5 ns simulation of d(CGCGAATTCGCG)$_2$, applying the PME method with the Cornell *et al.* force field, showed excellent agreement of the helicoidal parameters (106) with those expected based on the distributions seen in the Nucleic Acid Database (107). These simulations are also able to represent reasonably sequence-specific bending and to confirm the wedge model of bending in A-tract DNA (108). Additionally, the same methods and force field were applied to investigate the structures of radiation-damaged DNA, including thymine dimers and other cross-linked DNA (109,110).

Recent results demonstrate the effect of the environment on nucleic acid structure. These include the stabilization of A-DNA in water/ethanol mixtures (87,103) and the specific B-DNA to A-DNA transition observed when cobalt hexammine binds into G-rich pockets in the major groove and additional ions stabilize the interstrand phosphate repulsion of A-DNA structures (111). That nanosecond length simulations are able to represent the effect of the environment on DNA structure is an exciting result and encouraging to the field.

3.2.4 RNA and other structures: a field in its infancy

While the history of MD simulations of non-helical structures reaches back to the *in vacuo* simulations of tRNA in the early 1980s (112), this field has a relatively sparse history when compared with simulations of DNA and RNA double helices. MD simulations on these types of structures—internal loops, hairpins, etc.—have been hampered both by the lack of experimentally determined structures, and by the difficulty of achieving stable simulations. Improved experimental techniques, especially multidimensional NMR, have provided a number of new structures in recent years (see Chapter 19). Even so, the lack of a proper treatment of the long-range electrostatic interactions, while problematic for helix simulations, is catastrophic for non-helical structure simulations (16,113). In the former, artefactual behaviour is somewhat mediated by extra restraints on the Watson–Crick interactions, but similar restraints in non-helical structures do not have the same effect. With the implementation of fast Ewald methods, alongside the increasing number of NMR-determined structures, the number of published MD studies of non-helical systems has virtually exploded. These studies are revealing insights into the nature of the stabilizing interactions in such systems as an anticodon loop in tRNA, and in the tetraloop hairpins belonging to the UNCG and GNRA families.

Westhof and coworkers have studied the anticodon loop of tRNA^{Asp} for many years, with their investigations focusing on both proper simulation protocol (58,114–116), and, more recently, on the existence and importance of C–H⋯O hydrogen bonds (115). Their early studies on RNA fragments showed that, even with the inclusion of specific restraints, the MD protocol was not able to maintain the experimental structure (117). It was thought that the highly charged nature of the nucleic acid backbone required a more accurate treatment in order to obtain stable trajectories. To this end, their more recent work has employed a multiple molecular dynamics (MMD) protocol, a set of uncorrelated trajectories that can be used to probe different simulation conditions. Using the MMD approach, they have shown the importance of including the long-range electrostatic contributions, as they saw an improvement in the trajectories in going from an 8 Å to a 16 Å cut-off. Artefactual behaviour was still observed in these simulations, implying that even this long truncation distance was not enough.

Using the implementation of the particle mesh Ewald method in the AMBER 4.1 suite of programs, Westhof's group has published a number of studies of the anticodon loop that are stable, of nanosecond length, and point out the importance of a number of hydrogen bonds in the structural stability of this system (39,115). They examined the dynamic behaviour of the standard Watson–Crick base pairs, a non-canonical 'wobble' base pair, a pseudo-base pair, as well as two C–H⋯O hydrogen bonds. These studies, which were not possible without the PME method, showed that a remarkable level of atomic and dynamic resolution can be obtained once the electrostatic interactions of the solute and solvent (e.g. water and counterions) is properly accounted for.

The other types of non-helical RNA structures that have been studied in recent years are the RNA tetraloops. These studies have all investigated the nature of the atomic interactions within the short four-base loops in an attempt to understand their remarkable thermodynamic stability. While an early free energy study of the UUCG tetraloop was limited to simulations of only ~100 ps (118), more recent work on a similar system has used fast Ewald implementations to reach simulations of up to 2.5 ns (119,121). In the first study, Zichi used the OPLS force field, along with an Ewald treatment of the electrostatic interactions, to study the GCAA tetraloop structure determined by Heus and Pardi (120). This 300 ps simulation not only maintained the experimental structure quite well, but also showed the existence of a water-mediated hydrogen bond between the bases in the G:A pair at the base of the loop.

In studies of the UUCG hairpin—a representative of the other extra-stable class of RNA tetraloops—Miller and Kollman did not observe a transition from an incorrect structure to the correct one using unrestrained MD simulations (121). This work, which also employed the PME within AMBER 4.1, demonstrated that state-of-the-art simulation protocols could maintain the experimental tetraloop structure in simulations of up to 2.5 ns. Their simulations of both the incorrect and correct NMR structures did not interconvert, demonstrating that the barriers to conformational change in RNA are quite high. More recently, employing LES, the incorrect structure did convert to the correct one, which was stable and did not interconvert to the incorrect one (123). Also, through the use of a chimeric modification, where the four loop sugars were changed from ribose to deoxyribose, the conversion to the correct structure was observed in the simulation that started in the incorrect conformation. The control simulation, deoxyribose in the loop and starting in the correct conformation, showed more flexibility but stayed in the correct conformation. Moreover, in the simulation of the correct NMR structure, they were able to provide some atomic-level insight into the stabilizing interactions of the 2′-OH of the first loop residue, something that is difficult, if not impossible, to determine experimentally.

In order to build on the work of Cheatham and Kollman, where an A-DNA to B-DNA transition was observed in a double helix, Miller and Kollman undertook a subsequent study of this UUCG tetraloop (122). In this work, they studied the DNA analogue of the tetraloop with the simulations beginning in the correct conformation. In their unrestrained MD simulations, they observed an A to B transition in the stem portion of the structure. This converged structure was then well maintained by the simulation protocol for the remainder of the simulation, which was well over a nanosecond in length. Srinivasan *et al.* (125) have shown how one can combine MD in solution with continuum electrostatics to estimate relative free energies of nucleic acid systems in solution.

While MD studies of non-helical RNA and DNA structures is certainly in its infancy, these studies have already provided some unique insights into the atomic-level interactions that are so crucial to RNA structures. It is clear that simulations such as these will become more prevalent as more experimentally determined structures become available.

References

1. Lakshminarayanan, A. and Sasisekharan, V. (1969) *Biopolymers*, **8**, 475.
2. Olson, W. and Flory, P. (1972) *Biopolymers* **11**, 15.
3. Lifson, S. and Warshel, A. (1968). *J. Chem. Phys.* **49**, 5116.

4. Levitt, M. (1978) *Proc. Natl. Acad. Sci. USA* **75**, 640.
5. Levitt, M. (1983) *Cold Spring Harbor Symp. Quant. Biol.* **47**, 251.
6. Olson, W. (1982) *J. Am. Chem. Soc.* **104**, 278.
7. Newton, M. (1973) *J. Am. Chem. Soc.* **95**, 256.
8. Weiner, S.J., Kollman, P.A., Case, D.A., Singh, U.C., Ghio, C., Alagona, G., Profeta, Jr, S. and Weiner, P. (1984) *J. Am. Chem. Soc.* **106**, 765.
9. Nilsson, L.M. and Karplus, M. (1986) *J. Comput. Chem.* **7**, 591.
10. van Gunsteren, W.F. and Berendsen, H.J.C. (1987) *Groningen Molecular Simulation (GROMOS) Library Manual.* BIOMOS, Nijenborgh, Groningen, The Netherlands.
11. Cornell, W.D., Cieplak, P., Bayly, C.I., Gould, I.R., Merz, Jr., K.M., Ferguson, D.M., Spellmeyer, D.C., Fox, T., Caldwell, J.W. and Kollman, P.A. (1995) *J. Am. Chem. Soc.* **117**, 5179.
12. Mackerell, A., Wiorkiewiczkuczera, J. and Karplus, M. (1995) *J. Am. Chem. Soc.* **117**, 11946.
13. van Gunsteren, W.F. (1996) *Biomolecular Simulation: The GROMOS96 Manual and User Guide.* BIOMOS, Zurich.
14. Pranata, J., Wierschke, S.G. and Jorgensen, W.L. (1991) *J. Am. Chem. Soc.* **113**, 2810.
15. Lavery, R., Zakrzewska, K. and Sklenar, H. (1995) *Comp. Phys. Commun.* **91**, 135.
16. Cheatham, III, T.E., Miller, J.L., Fox, T., Darden, T.A. and Kollman, P.A. (1995) *J. Am. Chem. Soc.* **117**, 4193.
17. Daggett, V. and Levitt, M. (1993) *Annu. Rev. Biophys. Biomol. Struct.* **22**, 353.
18. Essmann, U., Perera, L., Berkowitz, M.L., Darden, T., Lee, H. and Pedersen, L.G. (1995) *J. Chem. Phys.* **103**, 8577.
19. Gorenstein, D. and Kar, D. (1977) *J. Am. Chem. Soc.* **99**, 672.
20. Liang, C.X., Yan, L.Q., Hill, J.R., Ewig, C.S., Stouch, T.R. and Hagler, A.T. (1995) *J. Comput. Chem.* **7**, 883.
21. McCammon, J.A. and Harvey, S.C. (1987) *Dynamics of Proteins and Nucleic Acids.* Cambridge University Press, Cambridge.
22. Tidor, B., Irikura, K.K., Brooks, B.R. and Karplus, M. (1983) *J. Biomol. Struct. Dynamics* **1**, 231.
23. Levitt, M., Sander, C. and Stern, P. (1985) *J. Mol. Biol.* **181**, 423.
24. Cheatham, III, T.E. and Kollman, P.A. (1996) *J. Mol. Biol.* **259**, 434.
25. Levy, R.M., Kushick, J., Perahia, D. and Karplus, M. (1984) *Macromolecules* **17**, 1370.
26. Roitberg, A. and Elber, R. (1991) *J. Chem. Phys.* **95**, 9277.
27. Senderowitz, H., Guarnieri, F. and Still, W.C. (1995) *J. Am. Chem. Soc.* **117**, 8211.
28. Kollman, P.A., Weiner, P.K. and Dearing, A. (1981) *Biopolymers* **20**, 2583.
29. Tilton, R., Weiner, P. and Kollman, P.A. (1983) *Biopolymers* **22**, 969.
30. Kollman, P.A., Weiner, P., Quigley, G. and Wang, A. (1982) *Biopolymers* **21**, 1945.
31. Dang, L.X., Pearlman, D.A. and Kollman, P.A. (1990) *Proc. Natl. Acad. Sci. USA* **87**, 4630.
32. Singh, S.B., Pearlman, D.A. and Kollman, P.A. (1993) *J. Biomol. Struct. Dynamics* **11**, 303.
33. Pearlman, D., Holbrook, S.D., Pirkle, D. and Kim, S. (1985) *Science* **227**, 1304.
34. Rao, S., Keepers, J.W. and Kollman, P.A. (1984) *Nucl. Acids Res.* **12**, 4789.
35. Remers, W.A., Rao, S.N., Singh, U.C. and Kollman, P.A. (1986) *J. Med. Chem.* **7**, 1256.
36. Tomacz, M., Chawla, A. and Lipman, A. (1988) *Biochemistry* **27**, 3182.
37. Rao, S. N. and Kollman, P. A. (1990) *Biopolymers* **29**, 517.
38. Kumar, S., Duan, Y., Kollman, P.A. and Rosenberg, J.M. (1994) *J. Biomol. Struct. Dynamics* **12**, 487.
39. Auffinger, P. and Westhof, E. (1996) *Biophys. J.* **71**, 940.

40. Keepers, J.W., Kollman, P.A., Weiner, P.K. and James, T.L. (1982) *Proc. Natl. Acad. Sci. USA* **79**, 5537.
41. Ramstein, J. and Lavery, R. (1990) *J. Biomol. Struct. Dynamics* **7**, 915.
42. Singh, U.C., Pattabiraman, N., Langridge, R. and Kollman, P.A. (1986) *Proc. Natl. Acad. Sci. USA* **83**, 6402.
43. Texter, J. (1978) *Progr. Biophys. Mol. Biol.* **33**, 83.
44. Westhof, E. (1988) *Annu. Rev. Biophys.. Chem.* **17**, 125.
45. Berman, H.M. (1994) *Curr. Opin. Struct. Biol.* **4**, 345.
46. Franklin, R.E. and Gosling, R.G. (1953) *Acta Cryst.* **6**, 673.
47. Wolf, B. and Hanlon, S. (1975) *Biochemistry* **14**, 1661.
48. Manning, G.S. (1997) *Q. Rev. Biophys.* **2**, 159.
49. Jayaram, B. and Beveridge, D.L. (1996) *Annu. Rev. Biophys. Biomol. Struct.* **25**, 367.
50. Steinbach, P.J. and Brooks, B.R. (1994) *J. Comput. Chem.* **15**, 667.
51. Ewald, P. (1921) *Ann. Phys.* **64**, 253.
52. Allen, M.P. and Tildesley, D.J. (1987) *Computer Simulations of Liquids*. Oxford University Press, Oxford.
53. Jorgensen, W.L., Chandrasekhar, J., Madura, J., Impey, R.W. and Klein, M.L. (1983) *J. Chem. Phys.* **79**, 926.
54. Berendsen, H.J. C., Grigera, J.R. and Straatsma, T.P. (1987) *J. Phys. Chem.* **91**, 6269.
55. Smith, P.E. and Pettitt, B.M. (1996) *J. Chem. Phys.* **105**, 4289.
56. Beveridge, D.L., Swaminathan, S., Ravishanker, G., Withka, J.M., Srinivasan, J., Prevost, C., Louise-May, S., Langley, D.R., DiCapua, F.M. and Bolton, P.H. (1993) in *Water and Biological Molecules*, p. 165. Macmillan Press, London.
57. Beveridge, D.L. and Ravishanker, G. (1994) *Curr. Opin. Struct. Biol.* **4**, 246.
58. Louise-May, S., Auffinger, P. and Westhof, E. (1996) *Curr. Opin. Struct. Biol.* **6**, 289.
59. Clementi, E. and Corongiu, G. (1981) in *Biomolecular Stereodynamics*, p. 209. Adenine Press, New York.
60. Subramanian, P.S., Ravishanker, G. and Beveridge, D.L. (1988) *Proc. Natl. Acad. Sci. USA* **85**, 1836.
61. Subramanian, P.S. and Beveridge, D.L. (1989) *J. Biomol. Struct. Dynamics* **6**, 1093.
62. Subramanian, P.S., Swaminathan, S. and Beveridge, D.L. (1990) *J. Biomol. Struct. Dynamics* **7**, 1161.
63. Falk, M., Hartman, K.A. and Lord, R.C. (1963) *J. Am. Chem. Soc.* **85**, 397.
64. Drew, H.R. and Dickerson, R.E. (1981) *J. Mol. Biol.* **151**, 535.
65. Pullman, A. and Pullman, B. (1975) *Annu. Rev. Biophys.* **7**, 505.
66. Eisenhaber, F., Mannik, J.H. and Tumanyan, V.G. (1990) *Biopolymers* **29**, 1453.
67. Subramanian, P.S. and Beveridge, D.L. (1993) *Theor. Chim. Acta* **85**, 3.
68. Eisenhaber, F., Tumanyan, V.G. and Abagyan, R.A. (1990) *Biopolymers* **30**, 563.
69. Eisenhaber, F., Tumanyan, V.G., Eisenmenger, F. and Gunia, W. (1989) *Biopolymers* **28**, 741.
70. Siebel, G.L., Singh, U.C. and Kollman, P.A. (1985) *Proc. Natl. Acad. Sci. USA* **82**, 6537.
71. Levitt, M. (1983) *Cold Spring Harbor Symp. Quant. Biol.* **47**, 251.
72. Swaminathan, S., Ravishankar, G.D. and Beveridge, D. (1991) *J. Am. Chem. Soc.* **113**, 5027.
73. Prevost, C., Louise-May, S., Ravishanker, G., Lavery, R. and Beveridge, D.L. (1993) *Biopolymers* **33**, 335.
74. Withka, J.M., Swaminathan, S., Beveridge, D.L., and Bolton, P.H. (1991) *J. Am. Chem. Soc.* **113**, 5041.
75. Withka, J.M., Swaminathan, S., Srinivasan, J., Beveridge, D.L., and Bolton, P.H. (1992) *Science* **255**, 597.
76. Miaskiewicz, K., Osman, R. and Weinstein, H. (1993) *J. Am. Chem. Soc.* **115**, 1526.

77. van Gunsteren, W.F., Berendsen, H.J.C., Geurtsen, R.G. and Zwinderman, H.R.J. (1986) in *Computer Simulation of Chemical and Biomolecular Systems*, New York Academy of Science, New York.
78. Chuprina, V.P., Heinemann, U., Nurislamov, A.A., Zielenkiewicz, P., Dickerson, R.E. and Saenger, W. (1991) *Proc. Natl. Acad. Sci. USA* **88**, 593.
79. Swamy, K. and Clementi, E. (1987) *Biopolymers* **26**, 1901.
80. Laaksonen, A., Nilsson, L.G., Joensson, B. and Teleman, O. (1989) *Chem. Phys.* **129**, 175.
81. Eriksson, M.A.L. and Laaksonen, A. (1992) *Biopolymers* **32**, 1035.
82. Darden, T., York, D. and Pedersen, L. (1993) *J. Chem. Phys.* **98**, 10089.
83. Petersen, H.G. (1995) *J. Chem. Phys.* **103**, 3668.
84. Luty, B.A., Davis, M.E., Tironi, I.G. and van Gunsteren, W.F. (1994) *Mol. Simul.* **14**, 11.
85. Luty, B.A., Tironi, I.G. and van Gunsteren, W F. (1995) *J. Chem. Phys.* **103**, 3014.
86. McConnell, J.J., Nirmala, R., Young, M.A., Ravishanker, G. and Beveridge, D.L. (1994) *J. Am. Chem. Soc.* **116**, 4461.
87. Langley, D.R. (1998) *J. Biomol. Struct. Dyn.* (in press).
88. Norberg, J. and Nilsson, L. (1996) *Proc. Natl. Acad. Sci. USA* **93**, 10173.
89. York, D.M., Yang, W., Lee, H., Darden, T.A. and Pedersen, L. (1995) *J. Am. Chem. Soc.* **117**, 5001.
90. Lee, H., Darden, T.A. and Pedersen, L.G. (1995) *J. Chem. Phys.* **102**, 3830.
91. Lee, H., Darden, T.A. and Pedersen, L. (1995) *Chem. Phys. Lett.* **243**, 229.
92. Dickerson, R.E., Goodsell, D.S., Kopka, M.L. and Pjura, P.E. (1987) *J. Biomol. Struct. Dynamics* **5**, 557.
93. Shakked, Z., Guerstein-Guzikevich, G., Eisenstein, M., Frolow, F. and Rabinovich, D. (1989) *Nature* **342**, 456.
94. Brooks, B.R., Bruccoleri, R.E., Olafson, B.D., States, D.J., Swaminathan, S. and Karplus, M. (1983) *J. Comput. Chem.* **4**, 187.
95. Weerasinghe, S., Smith, P.E., Mohan, V., Cheng, Y.K. and Pettitt, B.M. (1995) *J. Am. Chem. Soc.* **117**, 2147.
96. Yang, L., Weerasinghe, S., Smith, P.E. and Pettitt, B.M. (1995) *Biophys. J.* **69**, 1519.
97. Weerasinghe, S., Smith, P.E. and Pettitt, B.M. (1995) *Biochemistry* **34**, 16269.
98. Yang, L.Q. and Pettitt, B.M. (1996) *J. Phys. Chem.* **100**, 2564.
99. Norberg, J. and Nilsson, L. (1996) *J. Chem. Phys.* **104**, 6052.
100. Bayly, C.I., Cieplak, P., Cornell, W.D. and Kollman, P.A. (1993) *J. Phys. Chem.* **97**, 10269.
101. Cieplak, P., Cornell, W.D., Bayly, C. and Kollman, P.A. (1995) *J. Comput. Chem.* **16**, 1357.
102. Cieplak, P., Cheatham, III, T.E. and Kollman, P.A. (1997) *J. Am. Chem. Soc.* **119**, 6722.
103. Cheatham, III, T.E, Crowley, M.F., Fox, T. and Kollman, P.A. (1997) *Proc. Natl. Acad. Sci. USA* **94**, 9626.
104. Young, M.A., Jayaram, B. and Beveridge, D.L. (1997). *J. Am. Chem. Soc.* **119**, 59.
105. MacKerell, A. D. (1997) *J. Phys. Chem.* **B101**, 646.
106. Young, M.A., Ravishankar, G. and Beveridge, D.L. (1997) *Biophys. J.* **73**, 2313.
107. Berman, H.M., Olson, W.K., Beveridge, D.L., Westbrook, J., Gelbin, A., Demeny, T., Hsieh, S.H., Srinivasan, A.R. and Schneider, B. (1992) *Biophys. J.* **63**, 751.
108. Beveridge, D.L. (1997) Personal communication.
109. Miaskiewicz, K., Miller, J., Cooney, M. and Osman, R. (1996) *J. Am. Chem. Soc.* **118**, 9156.
110. Spector, T., Cheatham, T.E. and Kollman, P.A. (1997) *J. Am. Chem. Soc.* **119**, 7095.
111. Cheatham, III, T.E. and Kollman, P.A. (1997) *Structure*, **5**, 1297.

112. Prabhakaran, M., Harvey, S.C., Mao, B. and McCammon, J.C. (1983) *J. Biomol. Struct. Dynamics* **1**, 357.
113. Miller, J.L. (1996) *Solvation, Structure and Dynamics of Nucleic Acids in Solution: Insights from Simulations*. University of California, San Francisco.
114. Auffinger, P., Louise-May, S. and Westhof, E. (1995) *J. Am. Chem. Soc.* **117**, 6720.
115. Auffinger, P., Louise-May, S. and Westhof, E. (1996) *J. Am. Chem. Soc.* **118**, 1181.
116. Louise-May, S., Auffinger, P. and Westhof, E. (1995) in *Ninth Conversation in the Discipline of Biomolecular Stereodynamics*. State University of New York, Albany.
117. Westhof, E. (personal communications).
118. Singh, S.B. and Kollman, P.A. (1996) *Biophys. J.* **70**, 1940.
119. Zichi, D.A. (1995) *J. Am. Chem. Soc.* **117**, 2957.
120. Heus, H.A. and Pardi, A. (1991) *Science* **253**, 191.
121. Miller, J.L. and Kollman, P.A. (1997) *J. Mol. Biol.* **270**, 436.
122. Miller, J.L. and Kollman, P.A. (1997) *Biophys. J.* **73**, 2702.
123. Simmerling, C., Miller, J.L. and Kollman, P.A. (1998) *J. Am. Chem. Soc.* **120**, 7149.
124. Srinivasan, J., Cheatham, T.E., Cieplak, P., Kollman, P.A. and Case, D.A., *J. Am. Chem. Soc.* (in press).
125. Srinivasan, J., Miller, J.L., Kollman, P.A., Case, D.A., *J. Biomol. Struct. Dyn.* (in press).
126. Forester, T.R. and McDonald, I. (1991) *Mol. Physics.* **72**, 643.

5

A-DNA duplexes in the crystal

Markus C. Wahl[1] and Muttaiya Sundaralingam[2]

Ohio State University, Biological Macromolecular Structure Center, Departments of Chemistry and Biochemistry and The Ohio State Biochemistry Program, 012 Rightmire Hall, 1060 Carmack Road, Columbus, OH 43210, USA

1. Introduction

More than four decades after the first crystallographic evidence for A-DNA (1), we are now confronted with a large number of single-crystal structures of short A-DNA duplexes, through which we are able to begin to understand some of the principles governing their helical structure and their interaction with ligands. The stunning diversity of fine structures promotes a picture of the DNA double helix in which A- and B-forms denote the hemispheres of the right-handed structural continuum (2,3). The local variability in helical parameters and the never-ending exceptions to existing rules also emphasize the dimensions of the task to predict accurate three-dimensional structures from primary sequence information. The picture is further complicated by the large influence of crystal packing forces, which may override the intrinsic structural propensies of the base sequences. Nevertheless, some general trends and rules have been elaborated and conclusions can be drawn about certain physico-chemical properties of double helices, such as bending and kinking. Furthermore, the crystal structures provide glimpses of the hydration of A-DNA duplexes and their inter-actions with metal ions and polyamines. The structures of mispaired duplexes provide relevant information regarding the mechanisms of mutation and repair.

Notwithstanding all the above accomplishments, the 'A-DNA structure' in many respects is still enigmatic: How significant is this DNA conformation *in vivo*? Are there general rules that govern the relationship between the base sequence and the detailed helical structure? Do the observed ligand complexes comprise representative and functionally relevant binding modes? What is the mechanism of DNA interconversion between different helical forms (A-, B- and Z-form)? In this chapter we will attempt to summarize our present knowledge regarding these and other questions, as obtained from single-crystal structures. Another review on this subject has recently been published (4).

2. The A-DNA conformation

2.1 Overall geometry from fibres

In the early 1950s it was discovered by fibre diffraction that, by lowering the relative humidity, B-DNA duplexes could be transformed into the more paracrystalline

[1]Present address: Max-Planck-Institut für Biochemie, Abteilung Strukturforschung, Am Klopferspitz 18a, D–82152 Planegg-Martinsreid, Germany.
[2]Corresponding author.

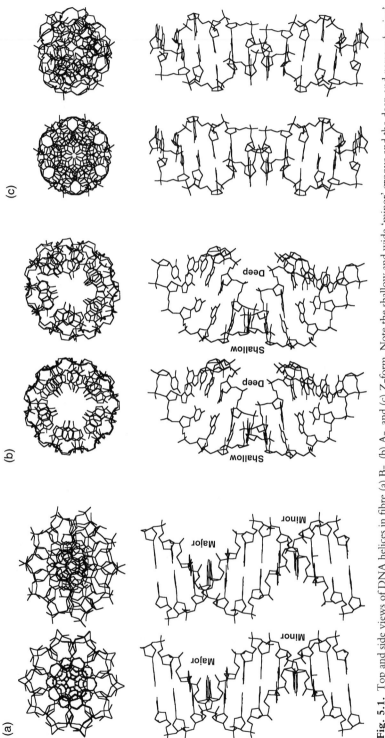

Fig. 5.1. Top and side views of DNA helices in fibre (a) B-, (b) A-, and (c) Z-form. Note the shallow and wide 'minor' groove and the deep and narrow 'major' groove in A-DNA.

A-form (1) (see Chapter 2). Although the technique was unable to determine the structures at atomic resolution, fibre diffraction has defined the global parameters of the A-DNA family (5–7). Accordingly, fibre A-DNA is a right-handed, antiparallel duplex characterized by a C3'-*endo*/N-type sugar pucker (B-DNA: C2'-*endo*/S-type), a strong base pair inclination, ~20° (B-DNA: ~0°), a large x-displacement, ~4 Å, of the base pairs (B-DNA: ~0 Å), and an 11- to 12-fold helix axis (B-DNA: tenfold), corresponding to a 30–32° helical twist (B-DNA: 36°). The rise per residue (~2.6–3.3 Å) is smaller than that of B-DNA (~3.4 Å). The preferred ranges for the backbone torsion angles α (P–O5') to ε (O3'–P) are g^-, t, g^+, g^+, t, g^- in A-DNA. The same ranges are found in B-DNA except for torsion angle δ which adopts the t conformation, corresponding to the change in sugar pucker. The different sugar pucker results in characteristic longitudinal phosphate–phosphate separations, i.e. ~5.9 and ~7.0 Å for A- and B-DNA, respectively. While B-DNA exhibits one wide (major) and one narrow (minor) groove of approximately equal depth, A-DNA entertains a narrow and deep 'major' groove and a shallow and wide 'minor' groove. Hence it is more appropriate to refer to the A-DNA grooves as *deep* and *shallow*, reserving the terms *major* and *minor* for B-DNA. The differential depth of the grooves is a consequence of the x-displacement of the base pairs, which moves the helix axis into the deep groove and produces the typical central void when viewing end-on (Fig. 5.1).

2.2 Structural diversity in crystals and conformational subclasses

While nucleic acid fibres generally exhibit the preferred (major) nucleotide conformations, single-crystal studies detect alternative (minor) conformations in individual nucleotides and oligomers (8). As a consequence, the first single-crystal structures of A-DNA fragments, obtained in the early 1980s, (9–11) were surprisingly non-uniform, and showed clear differences to the fibre structures (5–7). This diversity stems from the conformational spectrum of the individual nucleotide units (see below). Among the unusual findings were a reduced inclination (~10° vs. ~20° in fibres) and a smaller displacement of the base pairs into the shallow groove (~ −3.0 Å vs. ~ −4.0 Å in fibres). A widening of the deep groove was suspected to arise from the unbalanced electrostatic repulsion of phosphate groups across this groove in duplexes shorter than one helical turn (11). Indeed, the groove dimensions of longer fragments are closer to the fibre values (2,12). However, there are now examples of both relatively long (dodecamer) (13) and quite short (hexamer) A-DNA fragments (14), which closely resemble the fibre geometry. It should also be noted that in short A-DNA oligomers (≤ 8 bp) the width of the deep groove cannot be measured exactly since the closest approach of the phosphate groups is not reached in fragments with less than a full turn of the helix. Instead, the groove width has to be estimated by modelling a longer helix or using the next-to-closest approach of the phosphate groups (12).

The local variations in structural parameters observed in single-crystal structures emphasize the A-DNAs' large conformational spectrum (11), which makes it difficult to define precise subclasses. With respect to the twist angle, the A'-form, adopted by fibrous poly (rI):poly (rC) and poly (rI):poly (dC) (15,16) with a 12-fold helical axis, can be distinguished. An additional useful classification with respect to the strongly

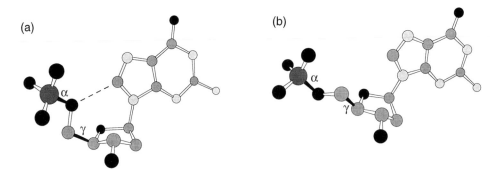

Fig. 5.2. Two alternative nucleotide conformations about the backbone torsion angles α and γ. (a) Major (g^-, g^+) conformation (A_I-form). (b) Minor (t, t) conformation (A_{II}-form). Note the base backbone (C8)–H···O5′ interaction (broken line) in the major conformation which is lost in the minor conformer.

correlated α and γ backbone torsion angles, distinguishes the preferred (g^-,g^+) or A_I-form from the less common (t, t) or A_{II}-form (Fig. 5.2) (17,18). Both A_I- and A_{II}-forms can coexist in the same duplex.

Some special forms of A-DNA are only adopted when nucleic acids interact with other molecules, e.g. a 'tilted A-DNA', or TA-DNA, has been described in the complex of the TATA box-binding protein with the TATA box (19,20) (see also below). The central portion of the complexed DNA can be described as a heavily inclined A-DNA (21).

2.3 *The rigid nucleotide principle*

Despite the variability of the helical parameters and backbone torsion angles, there are clearly preferred domains. The sugar pucker seems to be less flexible in A-DNA (C3′-*endo*) compared with B-DNA (C2′-*endo*) (8), and the glycosidic torsion angle adopts the *anti* conformation in preference. Therefore, the individual nucleotides in A-DNA duplexes seem to adhere closely to their preferred conformations as monomers (8), a theory initially put forward as the *rigid nucleotide principle* (22,23). The diversity in fine structures of oligomers arises from changes in nucleotide conformations, which generally fall within the ranges observed in the structures of the individual subunits, and slight changes in the nucleotide conformations can effect relatively large variations from the regular A-type double helix. However, in polymers, some restraints are imposed on the nucleotide conformation by the continuous sugar–phosphate backbone, so that the whole structural spectrum observed in nucleotides cannot be realized within the helical framework. Distortions from the major backbone conformations in A-DNA occur primarily on the 5′-side of the sugar ring [α/γ in (t, t)] while they are more common on the 3′-side of the sugar ring in B-DNA [ε/δ in (g^-/t)] (8).

3. A-DNA crystal packing

3.1 Crystal systems and arrangement of the duplexes

A-DNA duplexes crystallize in a limited set of space groups, different from the preferred space groups of proteins, representing a large fraction of the crystal systems (Table 5.1). Most prominent among them are the tetragonal (P4$_3$2$_1$2) and hexagonal (P6$_1$ and P6$_1$22) groups for octamers and the orthorhombic (P2$_1$2$_1$2$_1$) and hexagonal (P6$_1$22) groups for decamers. This assembly is further enriched by the structures of DNA:RNA chimeric decamers [P2$_1$2$_1$2$_1$; (24–28)], a DNA:RNA hybrid [P4$_3$22; (29)], and an unusual A-DNA structure with interduplex base pairing [P6$_1$22; (30)]. For the oligomers with a given length and in a given space group the crystalline environment is almost identical, exerting the same lattice forces on these duplexes, and the structure is only slightly influenced by the base sequence. The individual sequences within one group may be regarded as mutants of one another. However, the structures

Table 5.1. A-DNA crystal structures

Length of oligomer	Space group	Number of structures solved	Approximate unit cell	Asymmetric unit
Tetramers	P4$_3$2$_1$2	1	$a = b = 41$, $c = 27$ Å; $\alpha = \beta = \gamma = 90°$	Duplex
Hexamers	C222$_1$	2	$a = 39$, $b = 46$, $c = 40$ Å; $\alpha = \beta = \gamma = 90°$	Duplex
Octamers	P4$_3$2$_1$2	19	$a = b = 43$, $c = 25$ Å; $\alpha = \beta = \gamma = 90°$	Single strand
	P6$_1$	12	$a = b = 46$, $c = 43$ Å; $\alpha = \beta = 90$, $\gamma = 120°$	Duplex
	P6$_1$22	3	$a = b = 32$, $c = 79$Å; $\alpha = \beta = 90$, $\gamma = 120°$	Single strand
	P2$_1$2$_1$2	1	$a = 39$, $b = 51$, $c = 22$ Å; $\alpha = \beta = \gamma = 90°$	Duplex
	R3	1	$a = b = 71$, $c = 53$ Å; $\alpha = \beta = 90$, $\gamma = 120°$	Two duplexes
Nonamers	P4$_3$	1	$a = b = 45$, $c = 25$ Å; $\alpha = \beta = \gamma = 90°$	Duplex
Decamers	P6$_1$22	5	$a = b = 39$, $c = 79$ Å; $\alpha = \beta = 90$, $\gamma = 120°$	Single strand
	P6$_1$22	1	$a = b = 33$, $c = 79$ Å; $\alpha = \beta = 90$, $\gamma = 120°$	Single strand
	P2$_1$2$_1$2$_1$	4	$a = 25$, $b = 45$, $c = 48$ Å; $\alpha = \beta = \gamma = 90°$	Duplex
Dodecamers	P6$_1$22	2	$a = b = 46$, $c = 71$ Å; $\alpha = \beta = 90$, $\gamma = 120°$	Single strand
	P3$_2$21	1	$a = b = 45$, $c = 65$ Å; $\alpha = \beta = 90$, $\gamma = 120°$	Duplex

of the DNA mutants may be dominated by the crystal contacts rather than by intrinsic properties of the molecules (e.g. the base sequence) because, unlike in proteins, where a large portion of the atoms is buried inside the core of the molecules, the rod-shaped DNA fragments expose most of their surface to the environment (31). In order to draw firm conclusions about sequence–structure relationships, it is necessary to analyse the same molecule in various crystal lattices (32,33) and at the same time compare it with different sequences in the same crystal system (31). In this fashion it is sometimes possible to deconvolute the external and internal structural contributions (31,34).

All A-DNA duplexes make use of a universal packing interaction to build up the crystalline lattices (Fig. 5.3). The wide shallow groove exposes hydrophobic carbohydrate portions of the sugar rings and the bases to the environment (M. Sundaralingam, unpublished results), and invites hydrophobic contacts by the aromatic surfaces of the terminal base pairs of symmetry-related duplexes and vice versa. In addition, in longer duplexes, areas of the shallow grooves of neighbouring molecules cross each other at a glancing angle (12). A-DNA duplexes with a larger x-displacement of the base pairs possess a shallower 'minor' groove, pushing the abutting termini of the symmetry-related neighbours further away from their helix axis (and vice versa) and leaving a deeper 'major' groove. We therefore observe an inverse correlation between the crystal volume per duplex and the average base pair x-displacement (35). Conversely, B- and Z-DNA duplexes tend to form pseudo-infinite coaxial helices in the crystals, leaving the grooves largely free of interactions with neighbouring molecules (8). Therefore, single-crystal work on A-DNA is fundamentally different, not only from studies of B- and Z-DNA, but also from fibre diffraction of all three helical forms, because it is the only crystalline state in which DNA does not occur as infinitely stacked helices.

The details of the general scheme of one duplex approaching its neighbour from the shallow groove side vary not only with fragment length but also with crystal system. In the octamer families, those in the tetragonal group, $P4_32_12$, involve interactions at the end of the shallow groove (32), while in the $P6_122$ structures, the contacts

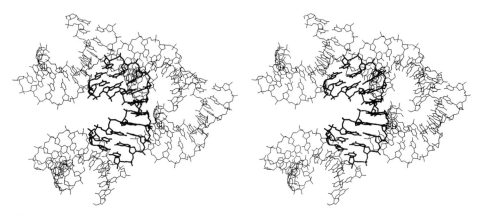

Fig. 5.3. Packing in A-DNA crystal lattices. The reference duplex (thick lines) invades, with its termini, the shallow grooves of two nearest neighbours (at top and bottom). Conversely, the terminal base pairs of two other neighbouring duplexes abut the shallow groove of the reference duplex (centre right). Orthorhombic crystal structure of d(GCGGGCCCGC) (31).

are clustered more in the central region (36). In the hexagonal P6$_1$ structures, only one strand is targeted by symmetry-related molecules, and the other is largely free of interactions (33). The orthorhombic decamer structures form intermolecular, non-planar, shallow groove base multiples (see below) (37), while the hexagonal ones tend to abut the backbone of symmetry-related neighbours (31).

The typical packing of A-DNA duplexes leaves large solvent channels in the crystals and, as a result, A-DNA crystals are often more hydrated than B-DNA. This finding is in contrast to the observation that the A-form is favoured over the B-form in dehydrated environments of DNA fibres at low humidity. The mere presence of water is therefore not sufficient to evoke a conversion of the A-form into the B-form; instead, the water molecules have to be able to contact the DNA directly over its entire length.

As mentioned, the A-form leads to the exposure of more hydrophobic portions of the sugar units of the sugar–phosphate backbone in the shallow groove compared with the B-form helices (M. Sundaralingam, unpublished results). The observed interaction with hydrophobic moieties should therefore stabilize the exposed hydrophobic areas of the shallow groove, and consequently the A-form, which brings them about. Another effect of these interactions is the displacement of water molecules from the shallow groove, creating a local environment of low water content that favours the A-form. It is therefore open to question whether A-DNA is a likely conformation *in vivo* or whether it represents a crystalline artefact. Certain sequences may exhibit an intrinsic preference for this conformation (see below) and simply exploit the shallow groove for further stabilization. It seems that under the crowded conditions in the nucleus similar interactions between nucleic acid portions or between a piece of DNA and some other molecule may prevail and induce or stabilize a local A-conformation. Along these lines it was noted that a family of carcinogens, the benzo-[a]-pyrenes, resembles the shape of a DNA base pair (10). These largely hydrophobic molecules could approach A-DNA in the way suggested by the packing arrangements before covalent modification, and, intriguingly, the functional groups on certain benzo-[a]-pyrenes and their points of attack on the DNA bases could come into close approach in the proposed complexes (10).

The notion of crystal packing stabilizing A-DNA conformation is further undermined by the crystal structure of a DNA:RNA hybrid (29), a decamer consisting of one strand of DNA and a complementary RNA strand, which resembles the A-form overall. While one end of the hybrid helix is involved in A-DNA-like, shallow groove, packing interactions, and is clearly in the A-conformation, the other end displays a stacked, B-DNA-like crystal packing scheme, and the DNA strand of this terminus shows marked B-DNA features, in particular C2'-*endo* sugar puckering.

Interestingly, the A-form is the natural conformation of RNA helices where the sugar units carry a 2'-hydroxyl group that points into the shallow groove. The groove therefore becomes more hydrophilic compared with A-DNA and a stabilization of the A-form in RNA comes from direct and/or water-mediated hydrogen bonds involving the 2'-OHs. At the same time it is observed that RNA duplexes prefer, with the exception of the RNA 14-mer (38), to crystallize with stacked helices, like B- and Z-DNA, a further indication that the hydrophobic shallow groove in A-DNA favours its typical shallow groove lattice contacts and vice versa.

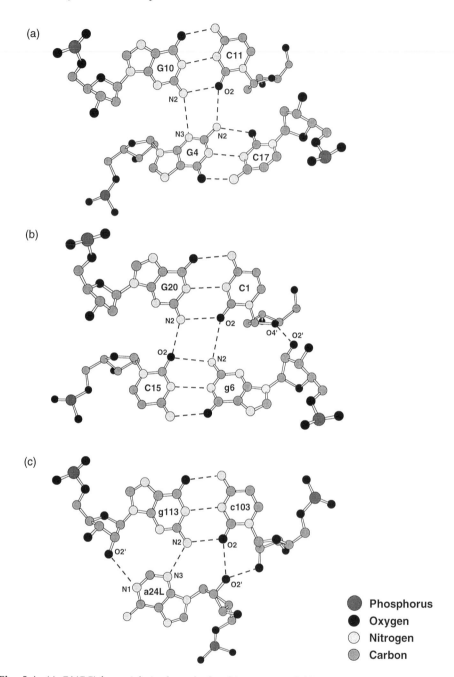

Fig. 5.4. (a) G★(GC) base triple in the orthorhombic structure of d(CCCGGCCGGG) (37). (b) Base quadruple in the orthorhombic structure of d(CCGGC)r(G)d(CCGG) (27). (c) Shallow groove a★(gc) base triple formed between a GNRA tetraloop and a duplex stem, observed in the crystal structure of a hammerhead ribozyme (42). Deoxyribonucleotides are denoted with upper case letters, ribonucleotides with lower case letters. Numbering is according to the published structures. Note the involvement of the 2′-hydroxyl groups in the formation of the multiples whenever present.

3.2 Shallow-groove base multiples

The abutment of the shallow groove by symmetry-related duplexes often results in the formation of base multiples (Fig. 5.4). Such base multiples (27,37) may play a significant role in the compacting of DNA in cell nuclei or virions, e.g. genomic DNA molecules may pack in the condensed state with the termini and internal base pair(s) invading the shallow groove (39). Furthermore, they may be fundamentally related to the shallow groove recognition of DNA by proteins and drugs (39). It has been shown that the geometry of these shallow groove interactions critically depends on the base sequence (37,39). Therefore, while the local geometry of the duplexes may be heavily influenced by the lattice contacts (see below), the base sequence is still the key element in determining these packing contacts.

Interestingly, the base triples observed in A-DNA crystal packing bear some resemblance to the interaction of GNRA-type RNA tetraloops with the shallow grooves of certain complementary duplex sequences in complex RNAs (40–42) (Fig. 5.4c). A-DNA shallow groove packing via formation of base multiples indicate a corresponding folding-potential in the RNAs. The 2′-hydroxyl groups seem to be important players in the stabilization of the shallow groove base multiples when ribonucleotides are present in the structures (Fig. 5.4b, c). It should be noted that the above shallow (minor) groove triples are fundamentally different from the tertiary triple helical interactions in the deep (major) groove, e.g. in tRNA (8).

3.3 The influence of crystal packing vs. sequence dependence

The crystal structures of the octamers d(GTGTACAC) (32,36) and d(GGGCGCCC) (33) have been determined in two different space groups, allowing a detailed comparison. It was consistently found, and later confirmed by a comparative analysis of 14 A-DNA octamers (34), that the molecular structures are dominated by crystal packing forces. A peculiarly small helical twist of 16–25° has been found in the central pyrimidine–purine steps of all the tetragonal $P4_32_12$ octamers, and this conformation of the step is required for the crystallization in this space group. The shallow groove packing contacts are dyad related about the centre and, by pushing the top and bottom halves of the octamers to the deep groove, result in the unwinding of the central step (34). Also, for longer helices, as demonstrated by the analysis of the A-DNA decamer families (31), crystal packing forces exert a larger effect on the conformational parameters than does the base sequence. However, while the influence of the lattice is a curse when trying to extract reliable sequence-dependent data, it is a blessing when one wishes to catch a glimpse of the structural plasticity of DNA (43). The conformational variability under the influence of lattice forces demonstrates the flexibility of the DNA helix.

4. Sequence–structure relationships

4.1 Alternating sequences

One incentive to determine accurate structures of DNA fragments has been to derive rules to predict exact three-dimensional structures from a given sequence (2). Unfortunately, sequence-dependent conformational features seem to be less pronounced in A-DNAs compared with B-DNAs, an observation that is in accord with the notion that A-DNA helices are less malleable than B-DNA helices. Sequences in which purines and pyrimidines alternate along the nucleic acid chain may be expected to exhibit regularly alternating structural features most clearly. Indeed, alternations of the twist, roll, buckle, and rise in phase with the sequence alternation are sometimes seen (38,44,45), even when isomorphous non-alternating sequences do not display such features (14). In general, the local geometry can be rationalized as a consequence of the propeller twisting required to increase the intrastacking interactions and adjustments in the helical parameters to avoid clashes of purines in the shallow groove at YpR steps or the deep groove at RpY steps. In a similar fashion it was possible to explain the observed helical parameters in the B-DNA dodecamer d(CGCGAATTCGCG) by Calladine's rules (46). For B-DNAs, simple sum functions have been used to predict the geometrical parameters with a knowledge of only the base sequence, and similar attempts have had some success in the case of A-DNA structures (47).

The fashion in which the conformational parameters are extracted from a structure requires some consideration. Some computer programs use as a reference a linear global helix axis (48) and this strategy inadequately accounts for local bends and kinks (12), while others (49–51) either employ a local helix axis or a spline-fitted global axis and more accurately determine the local helical parameters. (This is discussed further in Chapter 2.)

4.2 Base pair stacking

One prominent effect of the base sequence on the structures seems to be manifested in the differential stacking of RpY, YpR, and RpR/YpY steps. While in RpY and RpR/YpY steps intrastrand stacking is strong (–17 kcal/mol for R = G and Y = C) and interstrand stacking poor (–5 kcal/mol), the components contribute approximately equally (–11 kcal/mol) to the stabilization of the duplexes at YpR steps (17). The central importance of the base stacking in defining the local DNA geometry is illustrated by the structure of a DNA tetramer that forms a quasi-continuous octamer in the crystal with almost no disruption of the regular helix conformation across the non-covalent step (11). Sequence-dependent variations in helical parameters can also often be explained by considering the optimization of the stacking interactions as a driving force, as pioneered by Calladine (46). (A more extensive discussion of this topic can be found in Chapter 6.)

In A-DNAs the RpY steps usually tend to adopt a larger twist and a less negative roll (45), while the reverse is true for YpR steps. RpY steps exhibit intrastrand stacking of the purine and pyrimidine rings, while YpR steps prefer interstrand stacking between the purine rings (45). It has been shown that even small differences in the rise have a strong influence on the stacking energy (52), suggesting that even minor variations in rise are significant. In some instances the rise shows sequence dependence, but larger values may be associated either with the RpY (14) or YpR steps (45,53). The interstrand stacking of the purines at YpR steps requires opening of the base pair planes towards the shallow groove (negative roll) to avoid Calladine clashes. The roll variation is stronger at YpG than at YpA steps because of the presence of the 2-amino group of guanine in the shallow groove (44). Because A:T base pairs (two Watson–Crick hydrogen bonds) often adopt a larger propeller twist compared with G:C pairs (three Watson–Crick hydrogen bonds), this effect may be somewhat offset owing to the lack of one base–base hydrogen bond aggravating the Calladine-type collisions. Conversely, the decreased roll at RpY steps is a means of increasing intrastrand stacking efficiency (44). Similarly, the smaller twist values for YpR steps seem to indicate the tendency to improve slightly the weak intrastrand stacking for these steps.

4.3 Coupling of structural parameters

Because of restrictions imposed by the continuous sugar–phosphate backbone and the stacking preferences of the base pairs, the structural parameters of the A-DNA duplexes are correlated. Besides the anticorrelated backbone torsion angles α and γ, linked through the crankshaft motion about the intervening O5′–O4′ *trans* bond (54), other such torsion angle pairs exist but with less significant correlation coefficients. Angles χ and δ are linked by the sugar conformation, χ and ε are related by a 'crankshaft' motion involving the entire sugar unit, and ζ and α are joined by the phosphodiester bridge (11,54). Similarly, some of the helical parameters are coupled. It was observed that there is a strong anticorrelation between the rise and inclination of the base pairs (55,56). The rise measures the component of the consecutive base pair separation that is parallel to the helix axis. With increasing inclination the perpendicular component increases, a principle that is clearly seen when comparing A-DNA helices (rise ~2.6 Å, inclination ~20°) with B-DNA (rise ~3.4 Å, inclination ~0°). Furthermore, a correlation between x-displacement and inclination as well as groove width has been described (57). It is also known that the shallow groove width shows some dependence on the roll and propeller twist (12). The above-mentioned motions act hand in hand to avoid the impending clashes at specific base pair steps.

4.4 Individual base pair steps

4.4.1 d(GpG)

One or several GpG steps are generally present in A-DNA but largely absent from B-DNA (58) structures reported to date. Although there are exceptions, it has been suggested that the GpG sequence element could possibly induce or favour the A-form (2,56,59,60). The notion of A-DNA stabilization by runs of Gs is in agreement with the results from fibre diffraction studies of poly (dG):poly (dC) (59,61), which prefers the A-form, with Raman spectroscopic results for GC-rich oligomers (62), and with the recent NMR structural studies of a 34-mer in solution (63).

The source of this stabilization may again be found in the base pair stacking: in the A-DNA geometry the guanines of the GpG step can optimally stack five-ring on to six-ring. Optimization of the stacking of the guanine rings is achieved by adjustments in the slide to relatively high values (60) and occurs with lack of stacking of the pyrimidine rings (56). At the same time GpG steps maintain a negligible roll (60) in order to keep the guanine rings in the preferred parallel stacking orientation.

Interestingly, short stretches of guanines are frequently found at important DNA control regions (30). The binding site for the five-zinc finger motif GLI contains a run of three guanines and adopts an A-like helical conformation in the complex (64), and the 5S DNA-binding site of the zinc finger protein TFIIIA, d(GGATGGGAG):d(CTCC-CATCC), has been crystallized in the A-form (65). In such complexes the 'complementarity' between the deep groove side of the guanine base (O6 and N7) and the guanidinium moiety of arginines is often exploited (66). This interaction is reminiscent of the preference of certain metal cations for the deep groove face of guanine [see below; (30)]. Because both a stretch of guanines and the interaction with metal ions, or, by analogy, with arginine, favour the A-form (see below), it is not unreasonable to assume that these control elements may be recognized and bound in an A-like conformation *in vivo*. An *ideal* A-form would probably resist protein recognition through the narrow deep groove (major groove in B-DNA).

4.4.2 d(CpG)

The central CpG step of all tetragonal $P4_32_12$ octamers has been observed to be underwound by 10–16° and is associated with a fully extended backbone conformation [α/γ in (t, t)]. The extended backbone conformation has sometimes been interpreted as a sequence effect (10,17,55,67) but at times is not found in the chemically equivalent region of the complementary strand (56). While the underwinding at this position is probably induced by the crystal packing (32,34), the CpG step is also found in an unusual conformation elsewhere (2,14) and may constitute a flexible, easily distorted A-DNA element. It seems that in alternating sequences the step is likely to be associated with a low rise, a low twist, and a large roll (14). The low twist, the extended backbone conformation, and the associated slide guarantee the optimal cross-strand stacking of the guanines (55), showing that stacking preferences often dominate the conformation of the nucleotide units (56). The slide preference of the CpG step can be propagated along some length of the helix when the neighbouring steps themselves have no strong preferences for certain stacking modes (55).

5. Interconversions between A-, B-, and Z-forms

5.1 Interconversion of A- and B-DNA

Whether the different helical forms of DNA can coexist *in vivo*, and how a conversion from one form to the other is achieved, constitutes a central question in structural molecular biology. Diffuse meridional reflections at 3.4 Å resolution, matching the helical rise of B-DNA, have at times been observed in A-DNA crystals (12,68) stemming from disordered B-DNA helices in the solvent channels. Although these investigations indicate that A- and B-forms of the same sequence can coexist under similar environmental conditions, it has not yet been possible to crystallize the same molecule in both the A and B conformation. Similarly, no junction between different helical conformations has yet been observed in atomic detail. Model building guided by

NMR (69) and a crystallographic study (70) have shown that an A–B junction, which may occur, in Okazaki fragments for example, introduces a severe bend in the double helix but maintains efficient base stacking. Accordingly, it has been suggested that A–B junctions are not necessarily high-energy conformations since the nucleotides in double helices seem to have quite a wide structural repertoire at their discretion [see above; (57)]. All attempts to visualize an A–B junction with DNA:RNA chimeric molecules have so far failed. Instead, DNA fragments that are found to adopt the B-form, cooperatively convert to the A-form under the influence of single 2′-hydroxyl groups (27,28). Hydroxylation of the C2′ atom seems to confer a strong preference for the A-form on the DNA duplex, the source of which may stem from direct and water-mediated hydrogen bonded interactions within the sugar–phosphate backbone and between sugar and base units, primarily involving nucleotides to the 3′-side of the modified sugar. The different hydration modes may influence the equilibrium between the A- and B-form (see below). It has been suggested that the B-DNA 'spine of hydration' may lock the minor groove in a narrow state (11).

5.2 Interconversion of A- and Z-DNA

An A–Z junction, with a switch of helical handedness, is structurally much more demanding than an A–B junction. It seems that a variety of factors influence the equilibrium between A- and Z-forms, including the nature of the 5′-base of a DNA fragment, the base sequence, and certain chemical modifications of the bases. Until recently, all fully alternating oligomer sequences starting with a 5′-purine were found to be A-DNA (14,57,71–75), while 5′-pyrimidine start fragments show a strong tendency to adopt the Z-form (76). However, it was discovered that the alternating decamer d(GCGCGCGCGC) (73) adopted the Z-form, presumably overcoming the Z destabilizing effect of the 5′-purine start with an extended internal C–G alternation. Interestingly, converting only two G:C pairs to A:T pairs in this sequence, but maintaining the purine–pyrimidine alternation, switched the structure to the A-form (74). It therefore seems that in alternating fragments, the length, the nature of the 5′-nucleotide, and the base composition strongly affect the conformation and the helix handedness. Even more astounding, it was found that simply methylating (m5C) or brominating (Br5C) strategic cytosines in d(GCGCGCGCGC) also causes a switch to the A-form (75). Such findings are not easily rationalized, since m5C has been described as stabilizing both A and Z conformations (77,78). In alternating genomic sequences bearing m5C residues there may therefore be a competition between the A and Z helical forms. Since many A-DNA fragments convert to the B-form in solution (see below) it is also possible that the A-form is only adopted transiently in the interconversion of B- to Z-DNA and vice versa.

6. Chemical modifications of backbone and bases

In order to circumvent the nuclease sensitivity and the difficulties with the cellular uptake of nucleic acid fragments designed for antisense or antigene strategies, it is vital to explore the effects of chemical modifications of the sugar–phosphate backbone. To this end the backbone-modified A-DNA, d(GCCCGpGGC), with a 3′-methylene phosphonate linkage, has been crystallographically examined (79). Shown in comparison to the unmodified isomorphous structure (55), there are some unusual features of this helix that are not a consequence of the backbone alteration but rather result from packing effects. The methylene group seems to exert some structural influence, in particular, effecting a change in the sugar pucker of the associated nucleotide to C2'-exo. However, the hydration and the thermal stability of the duplex are unaltered. These findings demonstrate that modified sugar–phosphate backbones can maintain the overall double helical structure.

The structural influence of base methylation is of interest because of its relevance to the bacterial restriction/modification systems, with possible consequences for the recognition of DNA by endonucleases, and because of the mutagenic effects of alkylating reagents. Several crystal structures of A-DNA fragments with 5-methyl cytosine (m^5C) have been determined (14,75,80–82). In general, it has been found that the 5-methyl group on cytosine has little structural consequence, but some adjustments in the buckle and propeller twist have been observed in response to the crowding of the deep groove at m^5C (80). The methyl groups also disrupt the local water structure in the deep groove (see below) (80). Methylated cytosines lend extra stability to the duplexes, possibly through additional stacking interactions, and aid in crystallization, presumably by rendering the fragments more hydrophobic. A methyl group in the deep groove may block sequence-specific recognition and interactions with restriction endonucleases if the enzymes dock to DNA in the A-form, and, curiously, the 30-mer DNA fragment in the cocrystal structure with the *Eco*RI restriction endonuclease shows some A-DNA-like features (80).

7. Mispairs

Mispairing of bases in duplex DNA may occur during replication, genetic recombination, deamination of m^5C, or by covalent modification of the bases by carcinogens, which is counteracted by the proof-reading activity of DNA polymerases and DNA repair systems [(83) and references therein]. The study of the structures of non-Watson–Crick base pairs at atomic detail shows their effects on the conformation and stability of the DNA helix and may yield insights into their recognition and repair (see Chapters 10 and 11). The thermodynamic destabilization of a DNA helix by incorporation of mispairs depends on the sequence context, and both the unusual structural aspects as well as a reduced stability may be important for the mismatch repair (67).

Base wobbling was proposed in order to explain the observed redundancy of mRNA codons that are recognized by tRNA anticodons (84). While, historically, the 'wobble pairing' refers to this mRNA–tRNA interaction, analogous base pairs can be observed in A-DNA, with thymine (5-methyl 2'-deoxyuracil) replacing uracil. Wobble pairs have so far been the only non-canonical base pairs studied in A-DNA crystal structures. The thermodynamic stability of G:T wobble pairs is comparable with those of G:C or A:T base pairs, which would lead to unacceptable error rates during replication if proof-reading by the DNA polymerases depended solely on differential stability (85). The crystal structures of A-DNA octamers with side-by-side alternating G:T wobble pairs, d(GGGGTCCC) (85) and d(GGGTGCCC) (67), have been determined (Fig. 5.5a) and show that such double mismatches do not perturb the overall conformation of the helix or the backbone. However, to allow wobble formation, the bases are displaced into opposite grooves (T into the deep groove, G into the shallow groove), causing a disparity in the angles subtended at the glycosidic carbons and therefore leading to a disruption of the pseudo-twofold symmetry seen in Watson–Crick base pairs. It seems, therefore, that proof-reading and DNA repair systems must be sensitive to this altered base pair geometry. The thymine O4 and guanine N2 atoms are exposed in the shallow and deep grooves, respectively, since they lack hydrogen bonding partners on the other base, and their hydrogen bonding potential is satisfied by water molecules, which bridge the wobble bases. It has been pointed out that such water bridges may allow the wobble pair to reduce the impending loss of stability, because of the removal of one base–base hydrogen bond compared with the G:C base pair (86,87). The displacement of the bases into the grooves causes a change in the twist angle by about 10° per wobble pair, explaining the unusually high (44°) and low (25°) twists at the central TpG and GpT steps of the above octamers, respectively. Overall, the stacking geometry resembles that of other YpR or RpY steps. Interestingly, crystals of the octamer d(GGGGCTCC), with two isolated G:T mispairs at positions three and six, were found to be more temperature labile than those of the above sequences (83). While this reduced thermal stability may in part be explained by different packing interactions, another reason could be the number of canonical base pairs bordering the wobble pairs: grouping the wobble pairs into adjacent doublets reduces the number of steps with wobble pairs, directly affecting the canonical pairs (three steps) compared with the case with isolated G:Ts (four steps) (67).

Inosine (I), which lacks the guanine 2-amino group, is a non-natural base in DNA but occurs in the wobble position of some tRNA anticodons, where it can pair with C, T, and A (84), and is a valuable tool in nucleic acid structural research (88). Inosine can be seen to resemble guanine in the deep groove and adenine in the shallow groove (89). Not surprisingly, it therefore engages in similar wobble pairing with T, as found in the crystal structure of d(GGIGCTCC) (88) (Fig. 5.5b). Because of the lack of one anchoring point in the shallow groove (2-amino group), no bridging water molecule was found here, rendering the duplex less stable than the corresponding G:T-mismatched molecule. In the lattice, however, the lack of the 2-amino group allows for efficient van der Waals shallow groove packing of the duplexes, yielding crystals with enhanced thermal stability compared with those of the analogous G:T variant (83).

Surprisingly, the A-DNA decamer d(CCIGGCCm⁵CGG) was found to exhibit two I:m⁵C wobble pairs with a protonated N3 on the m⁵C (Fig. 5.5c), in spite of the fact that I and m⁵C could pair with a Watson–Crick geometry (81). The crystal structure of this modified decamer reveals information about the possible mechanisms of mutagenesis: a wobble analogous to that of I:m⁵C may be transiently formed in G:m⁵C pairs, allowing the m⁵C to be deaminated at the groove-exposed 4-amino group, resulting in the formation of a non-canonical G:T pair. Comparison of the I:m⁵C structure with the parent G:C pair suggests that the structural deviation of the wobble arrangement from the Watson–Crick geometry is best described by a rotation of the bases about their centres of mass.

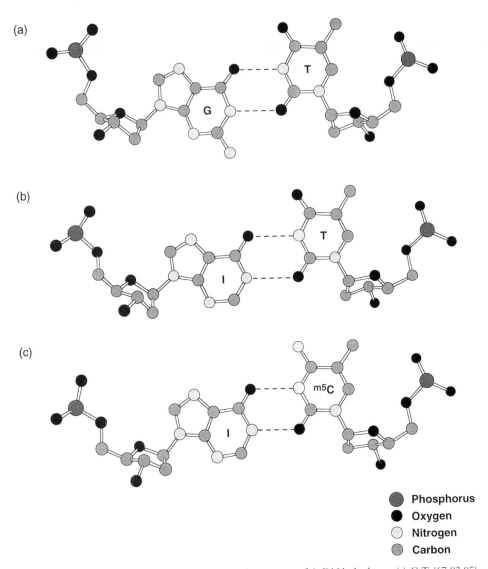

Fig. 5.5. Wobble base pairs observed in the crystal structures of A-DNA duplexes. (a) G:T (67,83,85); (b) I:T (88); (c) I:m⁵C, N3 protonated (81).

8. A-DNA deformability

Various modes of bending have been observed in A-DNA crystal structures. Bending and kinking may often be the result of external forces on the duplexes, stemming, for example, from the crystal lattice or ligands, and the crystal structures reveal how A-DNA can respond to and cope with such conformational stress.

Bending of the duplexes can be achieved by differentially expanding or contracting either of the two strands. Coupled by a crankshaft motion, the backbone torsion angles α and γ may adopt an extended (t, t) conformation, as opposed to the contracted (g^-, g^+) form (17). In the P4$_3$2$_1$2 octamers the extended form occurs in both the dyad-related strands at the central step (17), leaving the structure unaffected. However, in other structures (31,38) the duplexes are unsymmetrically distorted in the above fashion, leading to a kink of about 10° directed towards the deep groove. Similarly, switching the sugar pucker from C3′-*endo* to C2′-*endo* will lead to a lengthening of the corresponding longitudinal phosphate–phosphate distance by about 1 Å. If this stretching of the backbone is not compensated for by a similar sugar pucker in the complementary strand, an abrupt bend results. In d(GTGTACAC) (53) a large negative roll angle is associated with this type of bending, indicating an opening of the base pairs towards the deep groove. The abrupt bend caused by the uncompensated change in sugar pucker is quite different from the smooth bending observed in all tetragonal P4$_3$2$_1$2 octamers. As a result of the crystal packing these sequences are bent towards the shallow groove by about 15° (71,90,91). The hexagonal P6$_1$ octamers exhibit a bend of comparable magnitude which arises from a reduced inclination at the termini of the duplexes compared with the central region, and opens the deep groove (45). Both in the hexagonal P6$_1$22 decamers (31) and dodecamers (12,72) the crystal packing leads to a crowding of the shallow groove in the centre of the duplexes, which is responsible for the observed arching.

Bending of the duplexes towards the deep groove will lead to a closer approach of the negatively charged phosphate groups across this groove, which eventually should resist the deformation of the DNA, through the effect of charge repulsion. Shielding of the negative phosphate charges, e.g. by polyamines, may allow for more severe bends. Three crystal systems have been found for the decamer d(CCGGGCCCGG) and several of its methylated and brominated variants (82), ranging in solvent content from 67 to 40 to 24%. At the same time these structures exhibited an increased number of 0, 1, and 2 well-ordered spermine molecules, respectively, strategically placed in the areas of closest approach of the phosphate groups, allowing for increased bends of 10, 16, and 31°, respectively. These crystallographic results strongly support a role for spermine in DNA condensation and bending (see below).

As mentioned above, it has been shown that a bend of about 25° results at points of transition between A- and B-form helices. This bend is directed towards the deep/major groove, mostly as a result of the large roll angle adopted by flanking base pairs, in order to accommodate the change in inclination between A-DNA and B-DNA. Such a curvature is reminiscent of the deflections observed in the TATA box DNA under the influence of the TATA box-binding protein (19,20), which arises from the abrupt increase in the inclination of the protein-bound stretch of DNA compared with the uncomplexed DNA stretches on either side (21).

9. A-DNA interaction with ligands

9.1 Hydration

DNA crystals harbour a solvent content of about 50% (3). (See Chapter 9 for a full discussion of hydration). About 10 water molecules can be located per base pair in the electron density of a well-resolved A-DNA crystal structure, the vast majority of which belong to the first hydration shell with contacts to the DNA grooves or backbone (Plate I). Crystal structures are time averaged over several hours or days of data collection and only the most tightly bound waters can be seen. A large proportion of the solvent resides in the large channels in A-DNA crystals between the tightly packed duplexes. B-DNA, conversely, is more uniformly enveloped with water molecules in crystals. Several groups have investigated the water structure in the three helical families of oligonucleotides (11,39,86,92–100). While the hydration of A-DNA has often been regarded as more a result of 'opportunity' than of 'order' (11,12), a variety of conserved features have been noted over time.

The anionic phosphate oxygens of the backbone are the preferred hydration sites, with the one pointing to the deep groove (O1P) exhibiting more water contacts in A-DNA (86). Water bridges between anionic phosphate oxygens of consecutive residues are often encountered (Plate Ic). Such bridges are disrupted in B-DNA because of its longer intrastrand P–P distances. There may also be bridges between the O1P and O2P atoms of the same residue (95). After the phosphate groups the sugar O4′, O3′, and O5′ atoms follow, in that order, with a reduced extent of hydration (86). There are a total of six latching points for water molecules, three in each groove for G:C pairs, while there are only five for A:T base pairs, three in the deep and two in the shallow groove (Plate Ia). It may therefore be expected that the deep groove is more extensively hydrated than the shallow groove and this is indeed observed (17,95). Irrespective of the length of the DNA or the crystal system, about 30% of the first shell water molecules are in the deep groove and 20% in the shallow groove, while the remainder associate with the sugar–phosphate backbone, both for sequences containing only G:C base pairs and for mixed GC/AT sequences. However, the narrow deep groove in A-DNA contains better defined hydration sites than the wide shallow groove (100). The available hydrogen bonding sites on the nucleotides seem to be largely hydrated, unless they are obstructed, e.g. by lattice contacts (Plate Ib). The crystal structures show quite intricate deep groove networks of water molecules between donors and acceptors on the bases (31,80).

The water molecules around the bases seem to be more conserved than those around the phosphate groups (101), with distinct patterns seen in A-, B-, and Z-conformations. In A-DNA, guanine N2 and N3 atoms on the shallow groove side are often bridged by a water molecule in the plane of the base. Atoms O6 and N7 in the deep groove are normally separately hydrated. The (C8)–H atoms are usually shielded from the solvent through their interactions with the backbone O5′ but are sometimes found to engage in CH$\cdots$O$_W$ hydrogen bonds to water molecules (102). For cytosine the shallow groove is heavily hydrated at O2, while there is a bimodal hydration in the deep groove with strong water interaction to N4 and weak but significant hydration of C5.

Water bridges are commonly encountered between the backbone and the bases. The N7 of purines and N4/O4 of pyrimidines can be linked through two or three waters to the phosphate O1P (Plate Ib). The O4′ sugar oxygens are frequently linked via a water molecule to the preceding base (O2 of pyrimidines or N3 of purines). Often, the neighbouring or base paired bases are water bridged through their O/N6 and O/N4 sites in the deep groove, or through the O2 atoms in the shallow groove. Such water bridges can, in some cases, lead to elaborate networks. In the structure of the modified octamer, d(GGBr^{5}UABr^{5}UACC) (95), for example, the water molecules in the central deep groove form pentagonal rings with shared edges, resembling clathrate structures (Plate Id). Such water pentagons have been seen before with a DNA:drug complex (92). The water rings are connected via base hetero-atoms and O1P anionic oxygens to the DNA. As part of the network one observes uninterrupted bridging of the consecutive O1P atoms. Interestingly, this intricate pattern seems to be sequence specific, and requires the deep groove to be relatively wide, because it is not observed in other A-DNA octamers. Besides the base sequence, crystal packing also exerts an influence on the hydration (103). Therefore, the water structure adopts to the conformational environment of the nucleic acid, irrespective of whether it is dominated by lattice forces or sequence effects (33,101). It has been shown that the hydration is temperature dependent, and, in A-DNA crystals, the duplexes are more economically hydrated at low temperatures (~100 K) than at room temperature (104,105).

As mentioned, the typical shallow groove packing interactions screen part of crystalline A-DNA from water. Furthermore, A-DNA seems to form preferentially under low humidity/low water activity. As a corollary, the hydrophobic environment created by the neighbouring base pairs abutting the shallow groove plays a role in the stabilization of the A-form (10,45,106). Indeed, it has been argued that loss of the 'spine of hydration', observed on the floor of the minor groove of the B-DNA dodecamer d(CGCGAATTCGCG) (93), may be the reason for the B to A conversion upon dehydration. Also the more economical backbone hydration of A-DNA compared with B-DNA, with single-water-bridged phosphate groups, argues for a role for water structure in the A–B interconversion (107).

The water structure is usually observed to respond to deformations in the local A-DNA structure, suggesting that it, in turn, stabilizes these features (33). While the distortions may be induced by packing effects and may not be guided by the hydration, it was found recently in a comparative study of the *trp* repressor–operator complex and the protein-free DNA that most of the DNA hydration sites are also occupied by solvent molecules in the complex, and that the protein takes advantage of such solvents to achieve indirect protein–DNA interactions (108). The conclusion, that a protein binding to DNA has to cope with the hydration already present in the naked DNA duplex, applies irrespective of which helical form the nucleic acid adopts. The water structure therefore presents itself as an 'integral part of the nucleic acid structure' (86), either providing latching points for the protein, or demanding alternative hydrogen bonding potential from the protein wherever the DNA is dehydrated during complex formation.

9.2 Metal ions

Besides disordered solvent molecules, the channels of A-DNA crystals can harbour various solutes such as metal ions and polyamines, which are usually added for the crystallization. These agents shield the polyanionic DNA duplexes and allow them to build up the lattice, but they are not always located in the electron density maps. Cobalt hexammine is among the compounds that seem to promote the formation of the A-form [(63); B. Ramakrishnan and M. Sundaralingam, unpublished results].

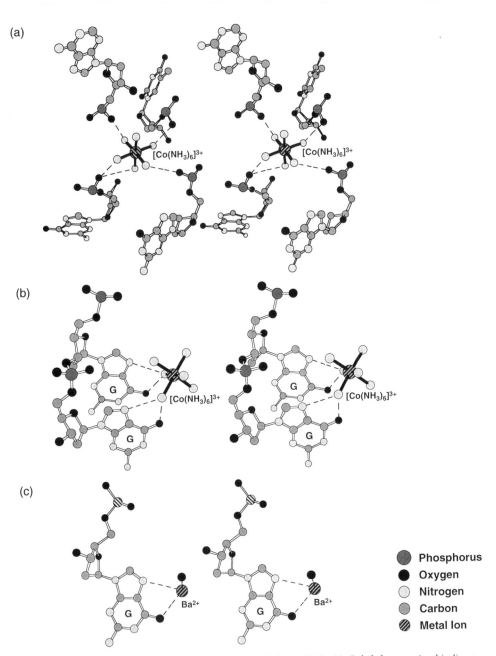

Fig. 5.6. Modes of binding of metal ions to A-DNA duplexes (106). (a) Cobalt hexammine binding to the phosphate groups of four residues from four different duplexes, organizing the crystal packing. (b) Cobalt hexammine complexed to two consecutive guanine bases of a GpG step through their N7 and O6 atoms. (c) Complexation of Ba^{2+} on the deep groove side of guanine.

Several binding modes to crystalline A-DNA were observed (30,106) (Fig. 5.6a, b). In the hexagonal crystal structure of d(ACCGGCCGGT) (106) three independent $[Co(NH_3)_6]^{3+}$ ions were located, one of which makes contacts with the DNA via phosphate groups and seems to organize the crystal packing (Fig. 5.6a), while the other two bind to two consecutive guanines in one strand, hydrogen bonding to the O6 and N7 positions of the bases (Fig. 5.6b). Similar $[Co(NH_3)_6]^{3+}$ binding modes to guanines were seen in the structure of d(AGGCATGCCT), which forms an unusual strand-exchanged duplex with features of both A- and B-DNA (30). The preference of $[Co(NH_3)_6]^{3+}$ for the deep groove face of guanines has been compared to the binding of arginine side chains in some DNA:protein complexes (30). Interestingly, the favoured target of the ions is the GpG step, which seems to induce the A-form in a short DNA fragment (see above). Possibly, the base sequence and the binding of certain ions work hand in hand *in vivo* to stabilize the A conformation in certain DNA stretches. This argument receives further support from the binding mode of a Ba^{2+} ion, found in the structure of d(ACCCGCGGGT) (106), which contacts three consecutive guanines on their deep groove face (Fig. 5.6c). The two independent $[Co(NH_3)_6]^{3+}$ ions observed in the complex with the chimera r(GC)d(GTATACGC) display a phosphate-only binding mode (106). $[Co(NH_3)_6]^{3+}$ has also been found to stabilize Z-DNA (109,110). Binding of certain metal ions may therefore evoke a similar competition between the A- and Z-form, as has been suggested for some base modifications (see above), once the B-form has been destabilized.

9.3 Polyamines

The ubiquitous polyamines are essential mediators of DNA function *in vivo*, affecting condensation, bending, and gene expression [(82) and references therein]. Spermine has been found in several A-DNA crystal structures and one structure of a DNA:RNA chimera, but, on the other hand, has been located in only one structure of a B-DNA (111). It is not clear whether the A-form is conformationally more adept at interacting with spermine or whether the A-type crystal packing, with criss-crossed helices capped by the shallow grooves of neighbouring molecules, is more prone to trap ordered spermine. Some solution studies seem to point to the former possibility (112).

The binding modes of spermine in the A-form structures are quite variable and can be classified into a few types of interaction: backbone/groove, deep groove, shallow groove, and backbone only (Fig. 5.7). In the crystal structure of the dodecamer, d(CCCCCGCGGGGG) (13), two amine functionalities of suspected spermines interact with bases in the deep groove while the remaining two bind to two phosphates of the reference and a neighbouring duplex. The asymmetric binding mode of spermine correlates with the asymmetric geometry of the two CpG steps. In contrast, spermine binds in a perfectly symmetrical fashion in the deep groove of the tetragonal $P4_32_12$ octamer, d(GTGTACAC) (90), but is absent from the structure of the same octamer in a hexagonal space group (36). In the tetragonal isoform, the spermine outlines an S-shaped path in the deep groove, hydrogen bonding to the O4 and N7 atoms of consecutive Ts and Gs, respectively. The complex is further maintained by hydrophobic interactions between the spermine methylene groups and the floor of the deep groove,

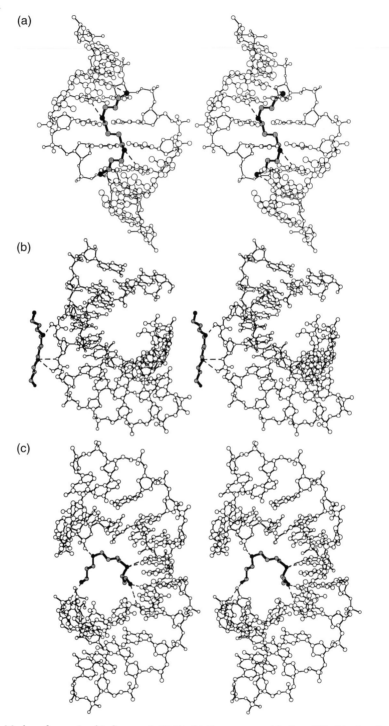

Fig. 5.7. Modes of spermine binding to A-DNA. (a) Deep groove binding (36); (b) phosphate-only binding (28); (c) base-and-phosphate binding (13). After ref. 4.

and by water bridges between the terminal ammonium groups and O6/N7 of the 5'-terminal Gs. A spermine molecule has also been located in the central portion of the shallow groove in the tetragonal crystal structure of d(ATGCGCAT) (113), where it is held by van der Waals contacts between its methylene groups and the DNA sugar rings, as well as by interactions between the terminal ammonium groups and the phosphates of residues 6 and 14. In the crystal structure of the highly distorted chimerical decamer r(C)d(CGGCGCCG)r(G) (28), a spermine molecule stabilizes the crystal lattice by latching on to the phosphates of neighbouring duplexes. This phosphate-only binding mode is accompanied by a vastly reduced unit cell volume, suggesting a role for spermine in the close packing of DNA, as found for example in the cell nucleus. Additionally, the spermine may induce bending of the decamer by allowing a narrowing of the deep groove, as shown for d(CCGGGCCCGG) (see above).

9.4 Protein–nucleic acid complexes

While the majority of protein–DNA complexes show DNA in a B-like conformation, there are several examples in which the DNA adopts a helical structure that resembles the A-form. In the complex of DNase I with the cleavage-resistant octamer d(GGTATACC) (114), the DNA molecule exhibits a wide and shallow minor groove owing to the considerable x-displacement (–2.5 Å on average) and the sugar puckers fall largely into the C3'-*endo* range. Also, the base pair stacking resembles that of A-DNA, while the intrastrand phosphate–phosphate distance (average 6.5 Å), the rise per residue (average 3.2 Å), and the inclination (average 2.6°) are reminiscent of the B-form. Analogously, the A-form duplex d(CTCTAGAG) (115) resembles the DNA structure in the *trp* repressor–operator complex (116). These analyses point out the difficulty of assigning a DNA, in complex with a protein, to a pure structural class and emphasize the apparent structural continuum of helical conformations. In the recently determined crystal structures of the TATA box-binding protein complexed to different TATA box elements (19,20), the DNA is found to be highly distorted, and resembles an A-DNA (short intrastrand phosphate–phosphate distances, low twist angles, high roll) with a very large inclination (average about 50°) (21). It is also intriguing that the founding member of the zinc finger family, TFIIIA, can recognize both the 5S RNA gene and the 5S RNA molecule, and that its cognate DNA element crystallizes in the A-form (65).

   The deformations detected in the duplexes of A-DNA crystal structures owing to lattice forces may also be achieved, with even greater severity, when ligands, like proteins, interact at particular sequences. The structure of the CCGG tetramer sequence, a restriction endonuclease recognition site, suggested a possible identification scheme based on such DNA deformability (11). The corresponding enzyme(s) could make out the overall A-form structure and, subsequent to binding, 'test' the deformability of the sequence, which could be followed by introduction of the cuts.

10. Comparison with solution studies

While the A-form is a commonly adopted conformation in crystals, the majority of the duplexes convert to the B-form in aqueous environments (2,62), as shown, for example, for a DNA octamer (113) or the TFIIIA binding site (65,117). The belief that B-DNA constitutes the biologically 'active' form of DNA is therefore strengthened by solution studies but not necessarily by single-crystal work. However, certain sequences, like poly (dG):poly (dC), clearly prefer the A-form even in solution (62), confirming X-ray crystallographic studies of GC-rich oligomers (60). It has also been suggested that the helical differences in solution are more subtle than in the crystal and that intermediate forms may exist (55). Possibly, there are also equilibria between different helical forms in solution (2,3). Furthermore, it has been reported that long DNA sequences can exhibit the A-form in solution in the presence of certain cations (63). In some cases it is found that cobalt hexammine is an essential ingredient promoting the formation of A-DNA crystals (12). Similarly, we may expect certain counterions *in vivo* to favour the A-form. It should also be kept in mind, that the cell nucleus is akin to a 'semicrystalline' environment, with highly compacted DNA, reminiscent of tightly packed A-DNA crystals.

Even though DNA may commonly adopt the B-form in solution, the A-form is the natural conformation of RNA and chimeric duplexes. While a large body of structural data is available for A-DNA, far less chimera and RNA structures have been determined to date, mostly because of initial problems with their synthesis and purification. It is therefore of interest to see whether the lessons learned from A-DNAs can be transferred to RNA and chimeric molecules.

11. Conclusions

Our understanding of the biophysics of nucleic acids has been enhanced by crysal structure analyses of oligonucleotides. The right-handed A-DNAs comprise a structural family separated from the B-like cluster by the bimodal distribution of the sugar pucker (C3'-*endo* in A-DNA, C2'-*endo* in B-DNA), as well as their large *x*-displacement and the inclination of the base pairs. In other words, they constitute a continuum of right-handed duplex conformations, i.e. the same calf thymus DNA fibre can be interconverted between the A- and B-forms, depending on the relative humidity. The existence of this structural continuum suggests that A-DNA represents a biologically relevant conformation, although transcriptionally silent. The A-form is characterized by a more hydrophobic shallow groove compared with B-DNA. Therefore, a DNA ligand, such as a protein, that demands such a groove architecture for complex formation could evoke a switch from B- to A-DNA. A-DNA also seems to be favoured over B-DNA in certain water-deficient crystalline environments. Genomic DNA must be intricately folded and compacted in order to fit into the tight space provided by eukaryotic nuclei. It is therefore not unreasonable to consider the aggregate state of this DNA as paracrystalline, in which the DNA may favour the A-form at particular sequences, possibly with packing similar to that observed in A-DNA crystals. Therefore, the observed sequence and packing effects in the crystal structures may have biological relevance.

Acknowledgements

Support from the National Institutes of Health (NIH grants GM-17378 and GM-49547), the Endowment from an Ohio Regents Eminent Scholar Chair, and an OSU Presidential Fellowship (to M.C.W) are gratefully acknowledged.

References

1. Franklin, R.E. and Gosling, R.G. (1953) *Acta Cryst*. **6**, 673.
2. Frederick, C.A., Quigley, G.J., Teng, M.K., Coll, M., van der Marel, G.A., van Boom, J.H., Rich, A. and Wang, A.H.J. (1989) *Eur. J. Biochem*. **181**, 295.
3. Kennard, O. and Hunter, W.N. (1989) *Biophys. J*. **22**, 327.
4. Wahl, M.C.and Sundaralingam, M. (1997) *Biopolymers* **44**, 45.
5. Fuller, W., Wilkins, M.H.F., Wilson, H.R. and Hamilton, L.D. (1965) *J. Mol. Biol*. **12**, 60.
6. Arnott, S., Dover, S.D. and Wonacott, A.J. (1969) *Acta Cryst*. **B25**, 2192.
7. Arnott, S. and Hukins, D.W.L. (1972) *Biochem. Biophys. Res. Commun*. **47**, 1504.
8. Sundaralingam, M. and Ban, C. (1993). In *Aspects of Crystallography in Molecular Biology*, New Aga International Limited, Publishers, New Delhi, India.
9. Shakked, Z., Rabinovich, D., Cruse, W.B., Egert, E., Kennard, O., Sala, G., Salisbury, S.A. and Viswamitra, M.A. (1981) *Proc. R. Soc. (London)*, **B213**, 479.
10. Wang, A.H.-J., Fujii, S., van Boom, J.H. and Rich, A. (1982) *Proc. Natl. Acad. Sci. USA* **79**, 3968.
11. Conner, B.N., Yoon, C., Dickerson, J.L. and Dickerson, R.E. (1984) *J. Mol. Biol*. **174**, 663.
12. Bingman, C.A., Zon, G. and Sundaralingam, M. (1992) *J. Mol. Biol*. **227**, 738.
13. Verdaguer, N., Aymami, J., Fernandez-Forner, D., Fita, I., Huynh-Dinh, T., Igolen, J. and Subirana, J.A. (1991) *J. Mol. Biol*. **221**, 623.
14. Mooers, B.H., Schroth, G.P., Baxter, W.W. and Ho, P.S. (1995) *J. Mol. Biol*. **249**, 772.
15. O''Brien, E.J. and MacEwan, A.W. (1970) *J. Mol. Biol*. **48**, 243.
16. Arnott, S., Hukins, D.W.L., Dover, S.D., Fuller, W. and Hodgson, A.R. (1973) *J. Mol. Biol*. **81**, 107.
17. Haran, T.E., Shakked, Z., Wang, A.H.-J. and Rich, A. (1087) *J. Biomol. Struct. Dynamics* **5**, 199.
18. Wahl, M.C. andSundaralingam, M. (1995) *Curr. Opin. Struct. Biol*. **5**, 282.
19. Kim, Y., Geiger, J.H., Hahn, S. and Sigler, P. B. (1993) *Nature* **365**, 512.
20. Kim, L.J., Nikolov, D.B. and Burley, S.K. (1993) *Nature* **365**, 520.
21. Guzikevich-Guerstein, G. and Shakked, Z. (1996) *Nature Struct. Biol*. **3**, 32.
22. Sundaralingam, M. (1973) *Jerus. Symp. Quant. Chem. Biochem*. **5**, 417.
23. Yathindra, N. and Sundaralingam, M. (1973) *Biopolymers* **12**, 297.
24. Wang, A.H.J., Fujii, S., van Boom, J.H., van der Marel, G.A., van Boeckel, S.A. and Rich, A. (1982) *Nature* **299**, 601.
25. Egli, M., Usman, N., Zhang, S. and Rich, A. (1992) *Proc. Natl. Acad. Sci. USA* **89**, 534.
26. Egli, M., Usman, N. and Rich, A. (1993) *Biochemistry* **32**, 3221.
27. Ban, C., Ramakrishnan, B. and Sundaralingam, M. (1994) *J. Mol. Biol*. **236**, 275.
28. Ban, C., Ramakrishnan, B. and Sundaralingam, M. (1994) *Nucl. Acids Res*. **22**, 5466.
29. Horton, N.C. and Finzel, B.C. (1996) *J. Mol. Biol*. **264**, 521.
30. Nunn, C.M. and Neidle, S. (1996) *J. Mol. Biol*. **256**, 340.
31. Ramakrishnan, B. and Sundaralingam, M. (1993) *Biochemistry* **32**, 11458.
32. Jain, S. and Sundaralingam, M. (1989) *J. Biol. Chem*. **264**, 12720.
33. Shakked, Z., Guerstein-Guzikevich, G., Eisenstein, M., Frolow, F. and Rabinovich, D. (1989) *Nature* **342**, 456.
34. Ramakrishnan, B. and Sundaralingam, M. (1993) *J. Biomol. Struct. Dynamics* **11**, 11.
35. Heinemann, U. (1991) *J. Biomol. Struct. Dynamics* **8**, 801.
36. Jain, S., Zon, G. and Sundaralingam, M. (1991) *Biochemistry* **30**, 3567.
37. Ramakrishnan, B. and Sundaralingam, M. (1993) *J. Mol. Biol*. **231**, 431.
38. Dock-Bregeon, A.C., Chevrier, B., Podjarny, A., Johnson, J., de Bear, J.S., Gough, G.R., Gilham, P.T. and Moras, D. (1989) *J. Mol. Biol*. **209**, 459.

39. Tippin, D.B. and Sundaralingam, M. (1996) *Acta Cryst.* **D52**, 997.
40. Michel, F. and Westhof, E. (1990) *J. Mol. Biol.* **216**, 585.
41. Jaeger, L., Michel, F. and Westhof, E. (1994) *J. Mol. Biol.* **236**, 1271.
42. Pley, H.W., Flaherty, K.M. and McKay, D.B. (1994) *Nature* **372**, 111.
43. Dickerson, R.E., Goodsell, D.S. and Neidle, S. (1994) *Proc. Natl. Acad. Sci. USA* **91**, 3579.
44. Spôner, J. and Kypr, J. (1991) *J. Mol. Biol.* **221**, 761.
45. Shakked, Z., Rabinovich, D., Kennard, O., Cruse, W.B., Salisbury, S.A. and Viswamitra, M.A. (1983) *J. Mol. Biol.* **166**, 183.
46. Calladine, C.R. (1982) *J. Mol. Biol.* **161**, 343.
47. Dickerson, R.E. (1983) *J. Mol. Biol.* **166**, 419.
48. Fratini, A.V., Kopka, M.L., Drew, H.R. and Dickerson, R.E. (1982) *J. Biol. Chem.* **257**, 14686.
49. Lavery, R. and Sklenar, H. (1988) *J. Biomol. Struct. Dynamics* **6**, 63.
50. Bhattacharyya, D. and Bansal, M. (1990) *J. Biomol. Struct. Dynamics* **8**, 539.
51. Babcock, M.S., Pednault, E.P.D. and Olson, W. (1993) *J. Biomol. Struct. Dynamics* **11**, 597.
52. Spôner, J. and Kypr, J. (1990) in *Theoretical Biochemistry and Molecular Biophysics, Vol. 1: DNA* (Beveridge, D.L. andLavery, L., eds), Vol. 1, pp. 271–284. Adenine Press, New York.
53. Thota, N., Li, X.H., Bingman, C. and Sundaralingam, M. (1993) *Acta Cryst.* **D49**, 282.
54. Olson, W.K. (1982) *Nucl. Acids Res.* **10**, 777.
55. Heinemann, U., Lauble, H., Frank, R. and Blöcker, H. (1987) *Nucl. Acids Res.* **15**, 9531.
56. Lauble, H., Frank, R., Blöcker, H. and Heinemann, U. (1988) *Nucl. Acids Res.* **16**, 7799.
57. Jain, S., Zon, G. and Sundaralingam, M. (1987) *J. Mol. Biol.* **197**, 141.
58. Heinemann, U., Alings, C. and Bansal, M. (1992) *EMBO J.* **11**, 1931.
59. Arnott, S. and Selsing, E. (1974) *J. Mol. Biol.* **88**, 551.
60. McCall, M., Brown, T. and Kennard, O. (1985) *J. Mol. Biol.* **183**, 385.
61. Langridge, R. (1969) *J. Cell. Physiol.* **74**(Suppl. 1), 1.
62. Benevides, J.M., Wang, A.H.J., Rich, A., Kyogoku, Y., van der Marel, G.A., van Boom, J.H. and Thomas, G.J.Jr. (1986) *Biochemistry* **25**, 41.
63. Robinson, H. and Wang, A.H.-J. (1996) *Nucl. Acids Res.* **24**, 676.
64. Pavletich, N.P. and Pabo, C.O. (1993) *Science* **261**, 1701.
65. McCall, M., Brown, T., Hunter, W.N. and Kennard, O. (1986) *Nature* **322**, 661.
66. Steitz, T.A. (1990) *Q. Rev. Biophys.* **23**, 205.
67. Rabinovich, D., Haran, T., Eisenstein, M. and Shakked, Z. (1988) *J. Mol. Biol.* **200**, 151.
68. Doucet, J., Benoit, J.-P., Cruse, W.B.T., Prange, T. and Kennard, O. (1989) *Nature* **337**, 190.
69. Selsing, E., Wells, R.D., Alden, C.J. and Arnott, S. (1978) *J. Biol. Chem.* **254**, 5417.
70. Wahl, M.C., Rao, S.T. and Sundaralingam, M. (1996) *Biophys. J.* **71**, 2857.
71. Bingman, C., Li, X., Zon, G. and Sundaralingam, M. (1992) *Biochemistry* **31**, 12803.
72. Bingman, C., Jain, S., Zon, G. and Sundaralingam, M. (1992) *Nucl. Acids Res.* **20**, 6637.
73. Ban, C., Ramakrishnan, B. and Sundaralingam, M. (1996) *Biophys. J.* **71**, 1215.
74. Ban, C. and Sundaralingam, M. (1996) *Biophys. J.* **71**, 1222.
75. Tippin, D.B., Ramakrishnan, B. and Sundaralingam, M. (1996) *J. Mol. Biol.* **270**, 247.
76. Wang, A.H.J., Quigley, G.J., Kolpak, F.J., Crawford, J.L., van Boom, J.H., van der Marel, G. and Rich, A. (1979) *Nature* **282**, 680.
77. Behe, M. and Felsenfeld, G. (1981) *Proc. Natl. Acad. Sci. USA* **78**, 1619.
78. Fujii, S., Wang, A.H.-J., van der Marel, G.A., van Boom, J.H. and Rich, A. (1982) *Nucl. Acids Res.* **10**, 7879.

79. Heinemann, U., Rudolph, L.N., Alings, C., Morr, M., Heikens, W., Frank, R. and Blocker, H. (1991) *Nucl. Acids Res.* **19**, 427.

80. Frederick, C.A., Saal, D., van der Marel, G.A., van Boom, J.H., Wang, A.H.J. and Rich, A. (1987) *Biopolymers* **26**, S145.

81. Ramakrishnan, B. and Sundaralingam, M. (1995) *Biophys. J.* **69**, 553.

82. Tippin, D.B. and Sundaralingam, M. (1997) *J. Mol. Biol.* **267**, 1171.

83. Hunter, W.N., Kneale, G., Brown, T., Rabinovich, D. and Kennard, O. (1986) *J. Mol. Biol.* **190**, 605.

84. Crick, F.H.C. (1966) *J. Mol. Biol.* **19**, 548.

85. Kneale, G., Brown, T., Kennard, O. and Rabinovich, D. (1985) *J. Mol. Biol.* **186**, 805.

86. Westhof, E. (1987) *Annu. Rev. Biophys.* **17**, 125.

87. Betzel, C., Lorenz, S., Furste, J.P., Bald, R., Zhang, M., Schneider, T.R., Wilson, K.S. and Erdmann, V.A. (1994) *FEBS Lett.* **351**, 159.

88. Cruse, W.B.T., *et al.* (1989) *Nucl. Acids Res.* **17**, 55.

89. Chen, X., Ramakrishnan, B., Rao, S.T. and Sundaralingam, M. (1994) *Nature Struct. Biol.* **1**, 169.

90. Jain, S., Zon, G. and Sundaralingam, M. (1989) *Biochemistry* **28**, 2360.

91. Wilcock, D.J., Adams, A., Cardin, C.J. and Wakelin, L.P.G. (1996) *Acta Cryst.* **D52**, 481.

92. Neidle, S., Berman, H. and Shieh, H.S. (1980) *Nature* **288**, 129.

93. Kopka, M.L., Fratini, A.V., Drew, H.R. and Dickerson, R.E. (1983) *J. Mol. Biol.* **163**, 129.

94. Westhof, E., Prange, T., Chevrier, B. and Moras, D. (1985) *Biochimie* **67**, 811.

95. Kennard, O., *et al.* (1986) *J. Biomol. Struct. Dynamics* **3**, 623.

96. Berman, H.M. (1986) *Ann. N. Y. Acad. Sci.* **482**, 166.

97. Westhof, E. (1988) *Int. J. Biol. Macromol.* **9**, 185.

98. Berman, H.M., Sowri, A., Ginell, S. and Beveridge, D. (1988) *J. Biomol. Struct. Dynamics* **5**, 1101.

99. Schneider, B., Cohen, D. and Berman, H.M. (1992) *Biopolymers* **32**, 725.

100. Schneider, B., Cohen, D.M., Schleifer, L., Srinivasan, A.R., Olson, W.K. and Berman, H.M. (1993) *Biophys. J.* **65**, 2291.

101. Eisenstein, M. and Shakked, Z. (1995) *J. Mol. Biol.* **248**, 662 (1995).

102. Wahl, M.C. and Sundaralingam, M. (1997) *TIBS* **22**, 97.

103. Tippin, D.B. and Sundaralingam, M. (1997) *Biochemistry* **26**, 536.

104. Eisenstein, M., Hope, H., Haran, T.E., Frolow, F., Shakked, Z. and Rabinovich, D. (1988) *Acta Cryst.* **B44**, 625.

105. Eisenstein, M., Frolow, F., Shakked, Z. and Rabinovich, D. (1990) *Nucl. Acids Res.* **18**, 3185.

106. Gao, Y.-G., Robinson, H., van Boom, J.H. and Wang, A.H.-J. (1995) *Biophys. J.* **69**, 559.

107. Saenger, W., Hunter, W.N. and Kennard, O. (1986) *Nature* **324**, 385.

108. Shakked, Z. Guzikevich-Guerstein, G., Frolow, F., Rabinovich, D., Joachimiak, A. and Sigler, P.B. (1994) *Nature* **368**, 469.

109. Brennan, R.G., Westhof, E. and Sundaralingam, M. (1986) *J. Biomol. Struct. Dynamics* **3**, 649.

110. Gao, Y.G., Sriram, M. and Wang, A.H. (1993) *Nucl. Acids Res.* **21**, 4093.

111. Tari, L.W. and Secco, A.S. (1995) *Nucl. Acids Res.* **23**, 2065.

112. Wemmer, D.E., Srivenugopal, K.S., Reid, B.R. and Morris, D.R. (1985) *J. Mol. Biol.* **185**, 457.

113. Clark, G.R., Brown, D.G., Sanderson, M.R., Chwalinski, T., Neidle, S., Veal, J.M., Jones, R.L., Wilson, W.D., Zon, G., Garman, E. and Stuart, D.I. (1990) *Nucl. Acids Res.* **18**, 5521.

114. Weston, S.A., Lahm, A. and Suck, D. (1992) *J. Mol. Biol.* **226**, 1237.
115. Hunter, W.N., Langlois D'Estaintot, B. and Kennard, O. (1989) *Biochemistry* **28**, 2444.
116. Otwinowski, Z., Schevitz, R.W., Zhang, R.G., Lawson, C.L. Joanchimiak, A., Marmorstein, R.Q., Luisi, B.F. and Sigler, P.B. (1988) *Nature* **355**, 321.
117. Aboul-ela, F., Varani, G., Walker, G.T. and Tinoco, Jr, I. (1988) *Nucl. Acids Res.* **16**, 3559.

6

Helix structure and molecular recognition by B-DNA

Richard E. Dickerson

Molecular Biology Institute, University of California at Los Angeles, Los Angeles, CA 90025-1570, USA

1. Introduction

One day in 1980, Francis Crick arrived in Pasadena from the Salk Institute to visit James Olds of the Biology Division at Caltech. While on campus, he came by my laboratory to see graduate student Horace Drew's new DNA dodecamer structure, CGC-GAATTCGCG, the first single-crystal X-ray structure analysis of B-DNA. By chance the only people in the laboratory that afternoon were Horace, my long-time scientific colleague Mary Kopka, and myself. The four of us huddled around contoured plexiglas sheets stacked atop a light box in a darkened room, in that pre-computer graphics age. Horace pointed out with pride both the similarities to the Watson–Crick–Arnott fibre structure for B-DNA, and the differences produced by local base sequence, which only a single-crystal analysis could show. Crick looked on with interest, and then made a casual remark that had far more impact on Horace than anything that I ever said to him, either before or since. When Horace concluded his presentation, Crick smiled and said to him, 'So *that's* what it looks like!' Metaphorically speaking, Horace inflated with pleasure and floated around the ceiling of the room like a balloon. Crick's remark was quoted endlessly to all comers for days.

At one level, Francis Crick was only being his usual courteous self, complimenting a graduate student who was thrilled to be in the same room with him, let alone receive kind words about his research. But there is another sense in which Crick's remark was prescient, and quite justified. Fibre diffraction, of necessity, yields the averaged helix structure, and does not show the effects of base sequence except when that sequence affects the overall averaged structure. Single-crystal structures can reveal the local, base-to-base variations in helix structure that are a consequence of the particular DNA sequence in question.

2. Early sequence–structure correlations

The first 10 years of single-crystal DNA structure analysis, roughly 1979–1988, were a euphoric search for the rules that governed sequence–structure relationships. For B-DNA this was the 'decade of the dodecamer', since virtually all the structures solved were variants of the Drew dodecamer, CGCGAATTCGCG (1–5). This was the case because the CG-rich ends kept the helix from unravelling, and hydrogen bonded overlap of the final two base pairs at each end with neighbouring helices provided a strong crystalline scaffolding while preserving the identity of the 10 base pair crystallographic and helical repeats. Early single-crystal A-DNA structures (see Chapter 5) tended to be octamers containing many C–C and G–G steps. Z-DNA structures (see Chapter 7) were limited to variants of the alternating pyrimidine–purine sequence: CGCG…, with a price being paid in helix stability for each deviation from pyrimidine–purine alternation, or even for substitution of A for G and T for C.

CGCxxxxxxGCG was the 'magic' sequence for B-DNA dodecamers. Providing that these outer C:G base pairs and their intermolecular overlap were preserved, the central six base pairs could be varied almost at will, to yield crystals that were isomorphous with the Drew structure and could be solved by simple molecular replacement. To date more than 75 dodecamer crystal structures isomorphous with Drew's have been solved (Table 6.1a), all in the same orthorhombic $P2_12_12_1$ space group, with essentially identical unit cell dimensions and crystal packing. In the vast majority of cases, these were solved by taking as a first approximation the phases of the Drew dodecamer or one of its cousins. Twenty-four of these structure analyses have been of DNA alone (Table 6.1a.1). But the rugged crystals of Drew-like sequences also proved to be excellent scaffolding for drug-binding studies; 39 different analyses have been carried out of complexes with 18 different drug molecules that bind within the minor groove (Table 6.1a.2), as well as 16 studies of base mismatches (Table 6.1a.3). In contrast to the richness and variety of these results, only *eight* dodecamers have been solved independently in space groups other than orthorhombic $P2_12_12_1$ (Table 6.1b).

During this first 'dodecamer decade', several attempts were made to draw up rules connecting local helix structure with local base sequence. The obvious parameters to consider were helical twist angle, the roll angle between successive base pairs, the rise from one base pair to another along the helix axis, the inclination of a base pair away from perpendicularity to the helix axis, the lateral displacement of the base pair away from that same axis, and the propeller twist between bases in a base pair (Fig. 6.1 and ref. 120; see also Chapter 2). Indeed, these parameters can be used to describe the distinguishing characteristics of the three DNA helix families: A, B, and Z. An idealized Arnott fibre-derived B-DNA (Chapter 1) has a mean 36° twist, 3.38 Å rise, zero roll, –6.0° inclination, +0.23 Å x-displacement, and –4.4° propeller twist.[1] In contrast, ideal A-DNA has a 33° twist, 2.56 Å rise, 6° roll, 21° inclination, –4.5 Å x-displacement, and –7.5° propeller twist. Similar values were observed in the early single-crystal analyses, except that inclination in A-DNA oligomers tended to be smaller, around 13°, and propeller twist for both A- and B-DNA turned out to be a highly variable and sequence-dependent quantity. The fact that base pairs in A-DNA are displaced off-axis by 4.5 Å means that the minor groove is shallower than the major, whereas in B-DNA, where base pairs sit squarely on the helix axis (0.23 Å mean displacement), major and minor grooves are of equal depth.

---

[1] By a misplaced attention to torsion angle sign consistency, the propeller shown in Fig. 6.1 was defined as positive by the 1988 Cambridge Convention (120) Unfortunately, propeller twist in DNA is almost always in the opposite direction, because this facilitates stacking of bases along each individual strand of helix. As a consequence of this 1988 sign decision, the field has been cursed with a steady stream of *negative* propeller values.

Table 6.1. X-Ray analyses of B-DNA helices and their complexes with minor groove binding drug molecules

Sequence	Space group	Z	Ubp	Date, Institution	Ref.	NDB No.	I.D. No.
(a) **Dodecamers, isomorphous with CGCGAATTCGCG, space group P2$_1$2$_1$2$_1$**							
1. Oligonucleotides alone							
CGCGAATTCGCG		4	12	1980 UCLA	1–5,6	BDL001	101
CGCGAATTCGCG		4	12	1982 UCLA	7	BDL002	102
CGCGAATTCGCG		4	12	1987 Strasbourg	8	BDL020	103
CGCGAATTCGCG		4	12	1985 Berkeley	9	BDL005	104
CGCGAATT^{br5}CGCG		4	12	1982 UCLA	10,11	BDLB03	105
CGCGAATT^{br5}CGCG		4	12	1982 UCLA	6,10,11	BDLB04	106
CGCGAm^6ATTCGCG		4	12	1988 MIT	12	BDLB13	107
CGCGAATTm^5CGCG		4	12	1997 Cambridge	13	BDLB73	108
CGCGAATTm^5CGCG		4	12	1997 Cambridge	13	BDLB74	109
CGm^5CGAATTm^5CGCG		4	12	1997 Cambridge	13	BDLB72	110
CGCGAAUUCGCG		4	12	1997 Cambridge	13	BDL075	111
CGCGAA$m^{6\alpha}$T$m^{6\alpha}$TCGCG		4	12	1997 Northwestern	14	BDLS79	112
CGCGAA$h^{6\alpha}$TOH$^{6\alpha}$TCGCG		4	12	1997 Northwestern	14	BDLS80	113
CGCGAASSCGCG		4	12	1996 Manchester	15	BDLS67	114
CGCGATATCGCG		4	12	1997 Weizmann	16	pending	114a
CGCAAAAAGCG		4	12	1987 Cambridge	6,17	BDL006	115
CGCAAAAATGCG		4	12	1989 Yale	6,18	BDL015	116
CGCAAATTTGCG		4	12	1992 Inst. Can. Res.	6,19	BDL038	117
CGCAIAITm^5CTGCG		4	12	1997 Weizmann	16	pending	117a
CGCATATATGCG		4	12	1988 UCLA	20	BDL007	118
CGCGTTAACGCG		4	12	1991 Ohio State	21,22	BDL059	119
CGTGAATTCACG		4	12	1991 UCLA	6,23	BDL029	120
CGTGAATTCACG		4	12	1991 Rutgers	24	BDL028	121
CGCGAAAACGCG/ CGCGTT/TTCGCG (nicked strand)		4	12	1990 MIT	25	BDL021,32	122

Table 6.1. *Continued*

Sequence	Space group	Z	Ubp	Date, Institution	Ref.	NDB No.	I.D. No.
2. DNA: Drug complexes							
Cisplatin							
CGCGAATTCGCG/Cis		4	12	1984 UCLA	26	DDL017	123
Netropsin: +Py–Py+							
CGCGAATTbr5CGCG/N		4	12	1985 UCLA	27,28	GDLB05	124
CGCGAATTbr5CGCG/N		4	12	1995 UCLA	29	GDLB31	125
CGCGAATTCGCG/N		4	12	1992 Illinois	30	GDL018	126
CGCe6GAATTCGCG/N		4	12	1992 Illinois	30	GDLB17	127
CGCAAATTTGCG/N		4	12	1993 MIT	31	GDL014	128
CGCGATATCGCG/N		4	12	1989 MIT	32	GDL001,4	129
CGCGTTAACGCG/N		4	12	1995 Ohio State	33	GDL030	130
Lexitropsin: +Im–Py+							
CGCGAATTCGCG/L		4	12	1995 UCLA	34	GDL037,8	131
Distamycin: °Py–Py–Py+							
CGCAAATTTGCG/D		4	12	1987 MIT	35	GDL003	132
CGCGAATTCGCG/D		4	12	1996 Cambridge	13	GDLB41	133
Hoechst 33258 (para –OH on phenyl ring A)							
CGCGAATTCGCG/H		4	12	1987 UCLA	36	GDL006	134
CGCGAATTCGCG/H		4	12	1988 MIT	37	GDL002	135
CGCGAATTCGCG/H		4	12	1991 UCLA	38	GDL010,11	136
CGCGAATTCGCG/H		4	12	1991 UCLA	38	GDL012	137
CGCGAATTCGCG/H		4	12	1991 UCLA	38	GDL013	138
CGCGAATTCGCG/H		4	12	1989 MIT	39	GDL007(?)	139
CGCAAATTTGCG/H		4	12	1994 Inst. Can. Res.	40	GDL028	140
CGCAAATTTGCG/H		4	12	1994 MIT	41	GDL026	141
CGCGAATTCGCG/H		4	12	1992 Illinois	42	GDL022	142
CGCe6GAATTCGCG/H		4	12	1992 Illinois	42	GDLB19	143

Table 6.1. *Continued*

Sequence	Space group	Z	Ubp	Date, Institution	Ref.	NDB No.	I.D. No.
Meta-OH(N) Hoecht 33258 (meta -OH on ring A)							
CGCGAATTCGCG/H "in"	4	12	1996 Inst. Can. Res.	43	GDL047	144
CGCGAATTCGCG/H "out"	4	12	1996 Inst. Can. Res.	43	GDL048	145
Hoechst 33342 (para -OEt on ring A)							
CGCGAATTCGCG/H	4	12	1992 Illinois	42	GDLB21	146
CGCGe^{6}GAATTCGCG/H	4	12	1992 Illinois	42	GDLB20	147
Bis-benzimidazole compound (imidazole for piperazine on Hoechst 33258)							
CGCGAATTCGCG/B	4	12	1995 Inst. Can. Res.	44	GDL033	148
Berenil							
CGCGAATTCGCG/B	4	12	1990 Inst. Can. Res.	45	GDL009	149
CGCGAATTCGCG/B	4	12	1992 Inst. Can. Res.	46	GDL016	150
CGCGAATTm^{5}CGCG/B	4	12	1997 Cambridge	13	GDLB42	151
DAPI							
CGCGAATTCGCG/D	4	12	1989 UCLA	47	GDL008	152
2,5-Bis(4-guanylphenyl)furan (berenil analogue)							
CGCGAATTCGCG/F	4	12	1996 Inst. Can. Res.	48	GDL036	153
2,5-Bis{[4-(N-isopropyl)amidino]phenyl}furan (berenil analogue)							
CGCGAATTCGCG/F	4	12	1996 Inst. Can. Res.	49	GDL044	154
2,5-Bis{[4-(N-cyclopropyl)amidino]phenyl}furan (berenil analogue)							
CGCGAATTCGCG/F	4	12	1997 Inst. Can. Res.	49	GDL045	155
Pentamidine							
CGCGAATTCGCG/P	4	12	1992 Inst. Can. Res.	50	GDL015	156
γ-Oxapentamidine							
CGCGAATTCGCG/P	4	12	1994 Inst. Can. Res.	51	GDL027	157
Propamidine							
CGCGAATTCGCG/P	4	12	1993 Inst. Can. Res.	52	GDL023	158
CGCGAATTCGCG/P	4	12	1995 Inst. Can. Res.	53	GDL032	159

Table 6.1. *Continued*

Sequence	Space group	Z	Ubp	Date, Institution	Ref.	NDB No.	I.D. No.
SN6999							
CGCe^6GAATTCGCG/S		4	12	1993 Illinois	54	GDLB24	160
Tribiz or tris-benzimidazole (extended Hoechst 33258 analogue)							
CGCAAATTTGCG/T		4	12	1996 Inst. Can. Res.	55	GDL039	161
3. Mismatch oligonucleotides (mismatches underlined)							
CGCGAATTGGCG		4	12	1993 Inst. Can. Res.	56	BDL046	162
CGCGAATTAGCG		4	12	1986 Cambridge	57,58	BDL012	163
CGCGAATTe^{6}AGCG		4	12	1994 Manchester	59	BDLB54	164
CGCGAATTo^{8}AGCG		4	12	1992 Manchester	60	BDLB33	165
CGCGAATTIGCG		4	12	1985 Cambridge	61	BDL009	166
CGCm^{6}GAATTTGCG		4	12	1990 Edinburgh	62	BDLB26	167
CGCAAATTGGCG		4	12	1989 Manchester	63,64	BDL014	168
CGCAAGCTGGCG		4	12	1990 Inst. Can. Res.	6,65	BDL022	169
CGCAAATTo^{8}GGCG		4	12	1994 Edinburgh	66	BDLB56	170
CGCAAATTCGCG		4	12	1986 Cambridge	67	BDL011	171
CGCAAATTIGCG		4	12	1992 Edinburgh	68	BDLB41	172
CGCIAATTAGCG		4	12	1987 Cambridge	69	BDLB10	173
CGCIAATTCGCG		4	12	1992 Thos. Jeff. U.	70	BDLB40	174
CGCm^{2}IAATTCGCG		4	12	1993 Illinois	71	–	175
CGAGAATTCm^{6}GCG		4	12	1994 Rutgers	72	BDLB53	176
CGTGAATTCm^{6}GCG		4	12	1995 Rutgers	73	BDLB58	177
(b) Dodecamers: other space groups							
|CCUCTGGTCTCC / GGAGACCAGAGG|	P1	1	12	1995 MIT	74	DDLB57	1
CGCTCTAGAGCG	P2$_1$	2	24	1996 Barcelona	75	BDL070	2
CGTAGATCTACG	C2	4	12	1993 Manchester	6,76	BDL042	3

Table 6.1. *Continued*

Sequence	Space group	Z	Ubp	Date, Institution	Ref.	NDB No.	I.D. No.
CGCGAAAAACG	P2₁2₁2	4	24	1993 Yale	6,77	BDL047	4
ACCGGCGCCACA	R3	9	12	1989 Strasbourg	78–80	BDL018(?)	5
ACCGGCGCCACA	R3	9	12	1989 Strasbourg	79,80	BDL034(?)	6
ACCGCCGGCGCC	R3	9	12	1989 Strasbourg	79,80	BDL035(?)	7
GCCGCCGGCGCC	R3	9	12	1989 Strasbourg	79,80	–	8
(c) Decamers							
CCAAGATTGG	C2	4	5	1987 UCLA	81,82	BDJ008	9
CCAAIATTGG	C2	4	5	1992 UCLA		–	10
CCAACGTTGG, Mg	C2	4	5	1991 UCLA	83,84	BDJ019	11
CCAACITTGG, Ca	C2	4	5	1992 UCLA	85	BDJB44	12
CCAGGCCTGG	C2	4	5	1989 Berlin	86	BDJ017	13
CCAGGCᵃʳᵃCTGG	C2	4	5	1991 MIT	87	BDJS30(?)	14
CCAo⁸GCGCTGG	C2	4	5	1995 MIT	88	BDJB57	15
CTCTCGAGAG	C2	4	10	1994 UCLA	89	BDJ060	16
CGCAATTGCG	C2	4	10	1997 Inst. Can. Res.	90	BDJ069	17
CAAAGAAAAG	C2	4	20	1997 UCLA	91	BDJ081	18
\CGACGATCGT\ \TGCTAGCAGC\	P2₁	2	10	1997 New York U.	92	UDJ060	19
CGATCGATCG, Mg	P2₁2₁2₁	4	10	1991 UCLA	93	BDJ025	20
CGATTAATCG, Mg	P2₁2₁2₁	4	10	1992 UCLA	94	BDJ031	21
CGATATATCG, Mg	P2₁2₁2₁	4	10	1992 UCLA	95	BDJ037	22
CGATATATCG, Ca	P2₁2₁2₁	4	10	1992 UCLA	95	BDJ036	23
CATGGCCATG, Ca	P2₁2₁2₁	4	10	1993 UCLA	96	BDJ051	24
CATGGCCATG, Ca (+ di-imidazole Lexitropsin)	P2₁2₁2₁	4	10	1997 UCLA	97	pending	24a
\GGCCAATTGG\ \GGTTAACCGG\	P2₁2₁2₁	4	10	1996 Cambridge	98	UDJ049	25

Table 6.1. Continued

Sequence	Space group	Z	Ubp	Date, Institution	Ref.	NDB No.	I.D. No.
CGCAATTGCG	$I2_12_12_1$	4	5	1995 Inst. Can. Res.	99	UDJ031	26
CCIIICCCGG	$P3_1$	3	10	1997 Weizmann	16	pending	26a
CGATCGm⁶ATCG	$P3_221$	6	10	1992 UCLA	100	BDJB48	28
CGATGCm⁶ATCG	$P3_221$	6	10	1993 UCLA	–	–	29
CCAACITTGG, Mg	$P3_221$	6	10	1992 UCLA	85	BDJB43	30
CCATTAATGG, Mg	$P3_221$	6	10	1994 UCLA	101	BDJ055	31
CCACTAGTGG	$P3_221$	6	10	1994 Weizmann	102	–	32
CCAACGTTGG/A (+ anthramycin)	$P3_221$	6	5	1993 UCLA	103	GDJB29	33
CCAGGCm⁵CTGG	$P6$	6	10	1992 Berlin	104,105	BDJB27	34
CCAGGCm⁵CTGG	$P6$	6	10	1993 Berlin	106	BDJB49	35
CCAGGCm⁵CTGG	$P6$	6	10	1993 Berlin	106	BDJB50	36
CCAAGCTTGG	$P6$	6	10	1993 UCLA	107	BDJ052	37
CCGGCGCCGG	$R3$	9	10	1992 Berlin	108	BDJ039	38
(d) Octamers							
CGCTAGCG	$P2_12_12_1$	4	16	1996 Barcelona	75	BDH071	51
GAAGCTTC/Act D (actinomycin D)	$C2$	4	8	1992 Kansas	109	DDH037(?)	52
Side-by-side distamycins							
ICICICIC/2Dst	$P4_122$	8	4	1994 Ohio State	110	GDHB25	53
IɛICICIC/2Dst	$P4_122$	8	4	1995 Ohio State	22	GDHB34	54
IɛICICIC/2Dst	$P4_122$	8	4	1995 Ohio State	22	GDHB35	55
ICITACIC	$P4_122$	8	4	1997 Ohio State	111	–	56
ICATATIC	$P4_122$	8	4	1997 Ohio State	111	–	57
ICATATIC	$C2$	4	4	1997 Ohio State	111	–	58

Table 6.1. *Continued*

Sequence	Space group	Z	Ubp	Date, Institution	Ref.	NDB No.	I.D. No.
(e) Other oligonucleotide lengths							
CGCGAAATTTACGCG	I222	8	14	1988 NCI/Weiz.	112	UDO009(?)	61
CGCAGAATTCGCG	C2	4	12	1988 Weizmann	113,114	UDM010	62
\GCGAATTCG\ \GCTTAAGCG\	$P2_12_12_1$	4	8	1996 Cambridge	115	UDI030	63
/GCGTACGCG/ /GCGCATGCG/	$P4_12_12_1$	8	8	1997 Cambridge	116	UDI047	64
\CGGTGG\ \CCACCG\	$P6_122$	12	6	1995 Manitoba	117	BDF062	65
CTCGAG	$P6_222$	12	3	1996 Ohio State	118	BDF068	66
GpsCGpsCGpsC	$P2_12_12_1$	4	6	1987 Cambridge	119	BDFP24	67

Z = number of asymmetric units per cell; Ubp = number of base pairs per asymmetric unit; NDB No. = Nucleic Acid Database serial number; br^5 = 5-bromo; m^5 = 5-methyl; e^6 = 6-ethyl; $e^{6\alpha}$ = 6'-α-methyl; $h^{6\alpha}$ = 6'-α-hydroxyl; o^8 = 8-oxo; ara = arabinosyl; ps = phosphorothioate; A, T, G, C = DNA; a, u, g, c = RNA; Py = pyrrole; Im = imidazole.

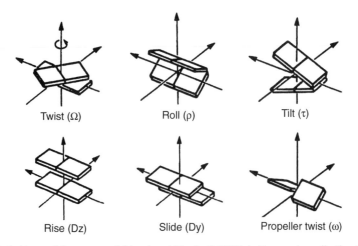

Fig. 6.1. Definitions of the most useful local variables in B-DNA helices: twist, roll, tilt, rise, and slide from one base pair to the next, and propeller within one base pair. (From ref. 120.)

In the simplest model, considering only the step between two base pairs, tables were constructed of 'the 10 twist angles of DNA', assuming that each of the 10 unique base pair steps in a DNA duplex would be associated with a unique value of twist (121). Calladine realized that the value of a helix parameter at a given step could be influenced by the steps preceding and following, and drew up a set of principles involving three adjacent base steps or four base pairs (122). Dickerson codified these principles into 'Calladine's Rules' and showed that the rules explained correctly the five A helices and lone B helix that were known at the time (123). The isolated base pair step was superseded by the tetrad as the fundamental unit of sequence and structure. All seemed for the best with this best of all possible helices.

However, this candid optimism crumbled during the following years, which can be termed the 'decamer decade', 1988 to the present (Table 6.1c). Advances in DNA synthesis methods made it easier to prepare many new sequences in crystallizable quantities. It was discovered that B-DNA decamers, with chains just one helical turn long, would stack atop one another in a manner that simulated repeating, endless helices running through the crystal (81). At first this appeared to open the way for rapid completion of the tetrad model of sequence–structure relationships. But as more sequences were examined, the same tetrad was observed to exhibit different helix parameters in different settings (84). Indeed, the same decamer sequence exhibited different local parameters in different crystalline environments (85,100). The tetrad model was evidently of little more validity than the old 'ten twist angle' model. What was going on? Did DNA have no fixed structure?

More structure analyses were required to suggest the answer. In the simple 'one sequence/one structure' picture that everyone had been assuming, a DNA duplex was a rigid object, its geometry fixed uniquely by its base sequence. The structural polymorphism observed in the newer crystal structures suggested an alternative and discouraging possibility: Perhaps the DNA duplex has no definite local structure, but is only a shapeless mass of duplex spaghetti. The individual structures that were observed by X-ray crystallography might then be no more than accidents produced by local crystal packing forces, without general applicability.[2]

But, as more crystal structures became available, a third and even more challenging pattern began to emerge. Some base pair steps such as C–A could exhibit twist angles of as little as 29° or as much as 54°, with many intermediate values. In contrast, A–A steps showed only a narrow range of variation around 36°, except when paired with a high twist C–A step in the sequence: CAA, in which case the A–A step could be forced down to around 29° (83). C–G and T–A steps could either be straight, or kinked sharply into the open major groove, whereas purine–purine steps generally displayed little or no bending. In short, different base sequences were found to display quite different degrees of variation with respect to twisting, bending, and other local helix deformations (84). It became apparent that the issue was not one of sequence-determined rigid local structure, but rather of a *sequence-based differential deformability* of the helix. This chapter deals with what we have learned about sequence-based differential deformability of the B-DNA helix, and how this deformability is utilized in the recognition of DNA sequence by drugs and proteins.

3. Molecular properties of B-DNA

The first B-DNA structure, that of the Drew dodecamer, CGCGAATTCGCG (1), illustrated several features of B-DNA that have become canonical from subsequent work.

1. The mean twist angle of base steps is centred around 36°, but with wide latitude at individual steps: from less than 20° to more than 55°.

2. Sugar pucker, although centred around the C2'-*endo* conformation expected from fibre diffraction models of B-DNA, is broadly distributed from C4'-*exo*, through O4'-*endo* and C1'-*exo*, to the expected C2'-*endo* and beyond. By contrast, sugar conformations in A-DNA oligomers are much more closely clustered around their expected centre, C3'-*endo* (see Figs 3 and 4 of ref. 124). This greater variability in B helix sugar pucker may reflect a greater malleability of the B helix, making it especially suitable for involvement in the molecular recognition process, and hence more suitable as a medium for storage and control of genetic information.

3. A:T base pairs exhibit greater variability of propeller twisting than do G:C pairs. This is undoubtedly a consequence of A:T pairs having only two hydrogen bonds rather than three, and hence offering less resistance to twisting of the base pair about its long axis.

4. Minor groove width is more variable in regions of successive A:T base pairs than in G:C regions. In the absence of groove-binding drugs, A:T regions of the minor

---

[2]This opinion has become almost a mantra in some circles, embodied in phrases such as, 'an artefact of crystal packing', usually preceded by 'only'. One major question to be addressed by this chapter is whether crystal packing is artefactual, or informative.

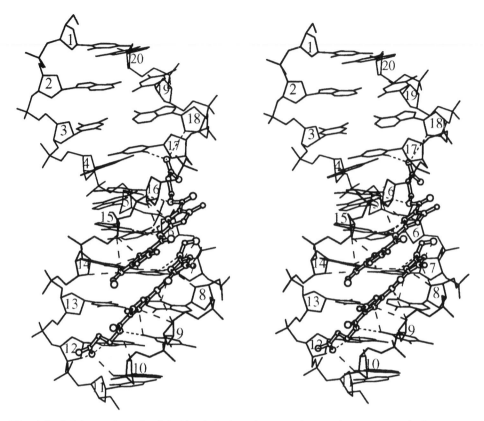

Fig. 6.2. A 2:1 complex of a di-imidazole lexitropsin, an analogue of netropsin, with the B-DNA decamer CATGGCCATG, illustrating widening of the minor groove by insertion of side-by-side drug molecules. (From ref. 97.)

groove are narrower than G:C regions. This follows immediately from the difference in propeller twist, since an increase in propeller magnitude brings C1′ atoms on opposite strands closer together and narrows the groove (see Fig. 9 of ref. 10). But regions of A:T base pairs are also capable of opening their minor groove enough to accommodate two planar drug molecules side by side (125, 126), as G:C regions can do without significant groove widening (Fig. 6.2).

5. Narrow A:T regions of the minor groove are filled with an ordered zigzag spine of hydration, in which a first hydration shell bridges adenine N3 and thymine O2 atoms diagonally across the groove, while a second hydration shell bridges these waters, giving each of them a local tetrahedral environment (see bottom of Fig. 6.3). Wider minor groove regions exhibit a somewhat less regular double ribbon of water molecules coordinated to base N and O atoms and to sugar O4′ atoms (see top of Fig. 6.3). These water molecules are displaced by a minor groove-binding drug, and the entropy of break-up of the spine contributes significantly to the free energy of binding (127).

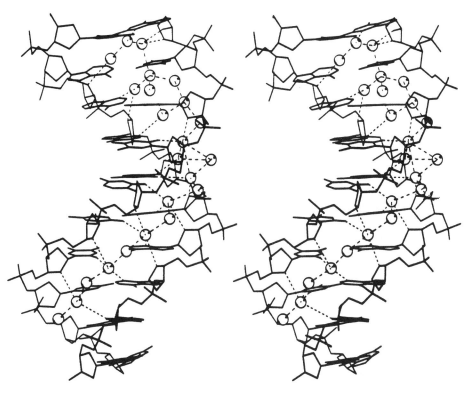

Fig. 6.3. Stereoview of the B-DNA decamer CAAAGAAAAG, showing a well-formed zigzag spine of hydration down the narrow AAAA region of the minor groove (bottom), and two less regular ribbons of hydration along the walls of the wider CAAA region of the groove (top, rear). (From ref. 91.)

6. Bending of the B-DNA duplex nearly always involves the easy deformation of roll, rather than the energetically unfavourable tilt (Fig. 6.1). As with stacked planks in a lumber yard, rolling the base pairs about their long axes is relatively easy, whereas pulling the two planks apart at one end requires more energy (128, 129). The easier roll deformation is that which compresses the more open major groove, and this is defined as positive roll. But in certain cases the observed bending involves negative roll, with compression of the minor groove instead.

B-DNA can be bent and twisted in ways that allow it to be wound around large proteins, to be supercoiled, and to be recognized by smaller proteins. But its susceptibility to bending and twisting is keyed to base sequence. Local physical properties of the helix are an expression of the underlying base sequence, in a manner that is useful for control. Base sequence does not define a fixed, static deformation, but rather a *differential deformability* of one region of helix vs. another. An analogy has been made to the human arm. At the elbow it can bend but not twist, at both the forearm and upper arm it can twist but not bend, and both bending and twisting are possible at the wrist. In combination, these possibilities allow a broad range of motion. But a determined effort to create a bend in the middle of the forearm would lead to disaster. Just as an ergonomic engineer needs to know the inherent structural capabilities of different regions of the human body, so we need to know the inherent deformabilities of different sequences of B-DNA.

4. Differences between individual base steps

Bases in DNA are of two types: single-ring pyrimidines (Y) and double-ring purines (R). Ten different base steps are possible in a helix composed of complementary hydrogen bonded chains. Three of these involve a Y–R step from a pyrimidine to a purine: T–A, C–G and C–A/T–G. (T–G is identical to C–A from the viewpoint of the other backbone chain.) Three purine–pyrimidine or R–Y steps exist: A–T, G–C and A–C/G–T, and four purine–purine (or pyrimidine–pyrimidine) steps: A–A/T–T, A–G/C–T, G–A/T–C, and G–G/C–C. Examples of each of these three classes from actual single-crystal B-DNA structures are shown in Figs 6.4–6.6.

To date, several generalizations have emerged from X-ray crystallography of DNA oligomers and their complexes with drugs and proteins. These of course will be subject to later verification and improvement; it takes more data to establish a trend or a probability than to discover an immutable law (provided that one exists). The following initial observations are presented as innate propensities or tendencies, not mandatory conformations. That is, the previously mentioned C–A step is more capable of exhibiting a large 55° twist angle than any other step, but this does not mean that every C–A will show such a large twist. However, one can be confident that such a large twist is unlikely at an A–T step.

(a) Y–R steps exhibit very little ring–ring overlap between adjacent base pairs (Fig. 6.4). Instead, outlying polar N or O atoms from one base pair are stacked against polarizable aromatic rings on the other. As a consequence, T–A, C–A, and C–G steps are weak, and are natural fracture points for the helix. They are especially susceptible to large twist and slide deformations, and to bending via positive roll. This is a useful feature in the bending of a B-DNA duplex by proteins such as the *Lac*I repressor (130), *Pur*R (131), TATA-binding protein (TBP) (132–135), γδ-resolvase (136), and others. In all these examples, the protein opens up the minor groove and forces the DNA to bend away from it, compressing the broad major groove. Lac and Pur accomplish this with the aid of an extremely large roll of more than +40° at a C–G step (Fig. 6.7). Human TATA-binding protein (TBP) opens the minor groove and induces a 100° bend in the sequence TATATATA by inserting phenylalanine rings into the two outermost T–A steps, giving them again a positive roll of more than 40° (Fig. 6.8a). Indeed, almost every local maximum of roll in γδ-resolvase is a Y–R step: T–A, C–A/T–G, or C–G (Fig. 6.8b).

(b) In contrast, R–Y and R–R steps exhibit extensive ring–ring overlap from one base pair to the next (Figs 6.5 and 6.6). Indeed, the base pairs in R–R steps seem almost to pivot around stacked purines as a hinge, with greater ring–ring separation at the pyrimidine end. Pyrimidine O2 atoms in the minor groove, and O4 or N4 atoms in the major groove, stack firmly over the six-membered ring of a neighbouring pyrimidine. These intimate stacking contacts all tends to keep base pairs parallel, and to give R–R steps smaller roll, slide, and twist deformations.

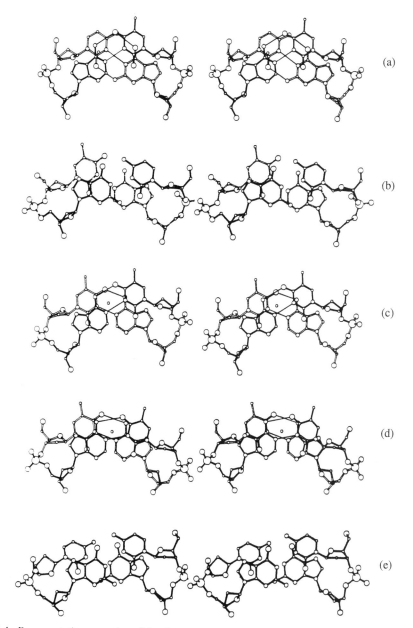

Fig. 6.4. Representative examples of the three pyrimidine–purine, or Y–R steps, showing little direct overlap between rings in adjacent base pairs. Individual roll (°), shift (Å) and twist (°) values are given as [*R*/*S*/*T*]. (a) C–A/T–G step from CCAACGTTGG, [−6.15/+2.59/50.8] (83). Large slide, with O2 of pyrimidines stacked against six-membered rings of neighbouring purines. (b) C–A/T–G step from CGCATATATGCG, [+3.65/+0.43/36.1] (20). Small slide, with pyrimidine O2 stacked against five-membered rings of purines. (c) T–A step from CGATATATCG, [−1.51/+1.05/43.7] (95). High twist but intermediate slide, with pyrimidine O2 stacked between purine rings. (d) T–A step from CGAT-TAATCG, [+9.11/+0.53/31.1] (94). Low twist. (e) C–G step from CGCATATATGCG, [−3.13/+0.34/37.8] (20).

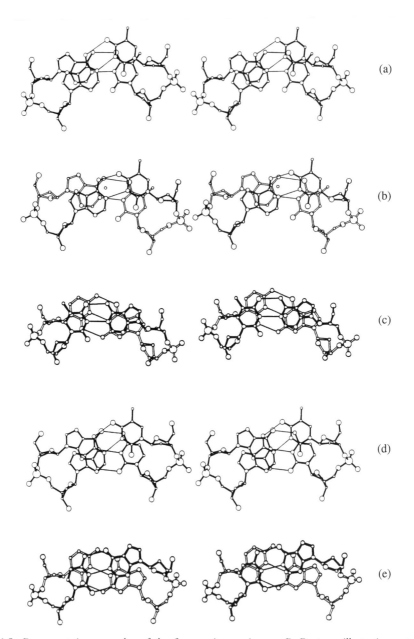

Fig. 6.5. Representative examples of the four purine–purine, or R–R steps, illustrating extensive ring–ring overlap, especially on the purine end. [*R*/*S*/*T*] as before. (a) A–A/T–T step from CGC-GAATTCGCG, [+0.31/−0.31/35.8] (3). (b) A–A/T–T step from CCAACITTGG, [−0.82/+0.08/34.7] (85). (c) A–G/C–T step from CCAGGCCTGG, [+5.55/+0.94/23.8] (86). (d) G–A/T–C step from CGCGAATTCGCG, [+2.67/−0.10/40.7] (3). (e) G–G/C–C step from CCAGGCCTGG, [+4.06/+0.74/36.9] (86).

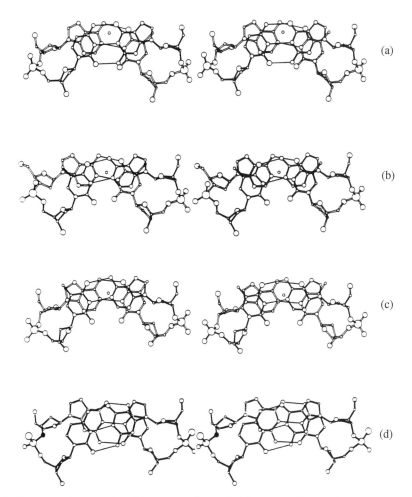

Fig 6.6. Representative examples of the three purine–pyrimidine, or R–Y steps, showing more ring–ring overlap than in Y–R steps. [*R/S/T*] as before. (a) A–C/G–T step from CCAAC<u>GT</u>TGG, [−1.99/−0.28/29.9] (83). (b) A–T step from CG<u>AT</u>TAATGC, [−1.84/−0.49/35.3] (94). (c) A–T step from CGAT<u>AT</u>ATCG, [+5.27/+0.04/25.0] (95). (d) G–C step from GC<u>GC</u>GC, [+1.31/+1.05/36.9] (119).

A–A steps are especially resistant to bending, probably because of the interlocking of base pairs with high propeller twist. To return to our lumber yard analogy, it is more difficult to push over a stack of nested sawhorses than a stack of planks. The unbent character of successive A–A steps is visible on both sides of the central C–G bend in the PurR complex (Fig. 6.7b), and will be encountered in several other protein:DNA complexes, as well as in all crystal structures of DNA alone (139,140).

(c) The tetrad concept, or the idea that behaviour of a central step is influenced by the steps flanking it to either side, still has a potential validity that has not been

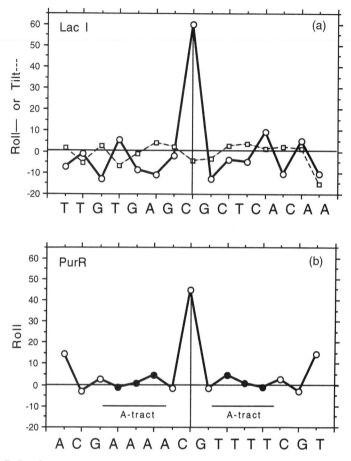

Fig. 6.7. Roll plots for two protein:DNA complexes illustrating bending at C–G steps: (a) Lac repressor (131), and (b) PurR (132). Tilt is also plotted in (a) to illustrate its negligible importance by comparison with roll. Short A-tracts in (b) are indicated by filled dots. Helix analysis using Richard Lavery's 'Curves' program (137) from the Nucleic Acid Database at Rutgers (138).

tested sufficiently because of the paucity of DNA structure data. For example, G–G–C–C steps are capable of large positive roll deformations at the G–C centre, even though G–C in general does not exhibit large roll. In particular, the sequence TGGCCA is observed to bend at the centre, with good parallel stacking of flanking TGG and CCA segments. The effect is so strong that it occurs both in the centre of the decamer CATGGCCATG (96), and across the gap between two stacked decamer helices in crystals of general sequence: CCAxxxxTGG/ CCAxxxxTGG, whether the central xxxx segment is ACGT (83), ACIT (85), AGAT (81), GGCC (86), TTAA (101), or CTAG (102). This illustrates the truism that *the structure of the DNA helix is determined primarily by the stacking of base pairs, with the role of the sugar–phosphate backbone being that of stabilizing the helix against disruption from outside.*

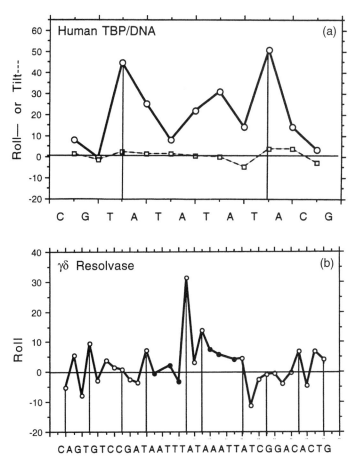

Fig. 6.8. Roll plots for two protein:DNA complexes illustrating bending at T–A steps: (a) human TATA-binding protein (135), and (b) γδ-resolvase (136). Tilt is also plotted in (a) to show its almost total insignificance when compared with roll. Short A-tracts in (b) are indicated by filled dots. Y–R steps in both (a) and (b) are marked by vertical lines.

5. DNA behaviour in crystals and in protein:DNA complexes

The sequence-induced behaviour of DNA in crystal structures has been analysed in detail by El Hassan and Calladine (141), who have studied 400 base pair steps from 24 A-DNA structures and 36 of the B-DNA structures of Table 6.1. They, like previous investigators (142,143), find that the most sequence-sensitive local helix parameters are roll (*R*), slide (*S*), and twist (*T*) from one base pair to the next, and propeller twist within a given base pair. Their conclusions from the systematic analysis of DNA conformations in the crystal will not be repeated here; but they generally confirm and extend the principles enunciated above. This chapter will build on their work, and carry it to another level.

An objection has been raised in the past as to the biological relevance of DNA crystal structures, especially when these appear to differ from results measured on

Table 6.2. Representative X-ray analyses of 63 protein: DNA complexes, indicating local roll, slide, and twist behaviour

Protein		DNA Binding Sequence	I.D. No.	Ref.
A. PROKARYOTIC H-T-H PROTEINS				
Lambda repressor (25°) concave		[A–**T**–**A**–C–**C**–**A**–C–**T**–**G**–G–**C**–**G**–G–**C**–**G**–G–**T**–**G**–A–**T**–**A**–T)	1	144
	R:	1 1 o 1 1 o -2 o -1 -1 1 2 2 o o 1 o		
	S:	o 1 -o -o 2 -2 -o **3** o -o 1 -2 -2 1 o -o o -o		
	T:	-1 1 -o -o -o -o -o 2 -1 o o -2 -o -o o -o 1 -o		
Lambda repressor (25°) concave		[A–**T**–**A**–C–**C**–**A**–C–**T**–**G**–G–**C**–**G**–**C**–**G**–**T**–**G**–A–G–A–T)	2	145
	R:	1 2 o 1 1 o 1 -1 o -1 -o 1 o **3** -o -o 1 o		
	S:	o 1 -o -o 1 -2 -o 2 o -o 1 -1 -1 o -o -o o -1		
	T:	-1 1 -o -1 -o -o 3 -1 o o -2 -o -o o -1 1 -o		
434 repressor (O_R1) (42°)[r_c=65Å] concave		[A–G–**T**–**A**–**C**–**A**–A–A–C–T–T–T–C–T–**T**–**G**–**T**–**A**–T)	4	146
	R:	-o -o -o o o 1 -1 o -1 -o 2 -1 1 1 1 -o-o o		
	S:	-o o -1 -1 o 1 o -1 -1 1 o -o o 1 o -o-o-o o		
	T:	-1 o -o -o o -1 -o 2 o o -1 o o-1 1 o		
434 repressor (O_R2) (35°) concave		[A–**T**–**A**–**C**–**A**–**A**–**T**–**G**–**T**–**A**–T–C–T–**T**–**G**–T–T–T)	5	147
	R:	-o 1 1 1 1 1 -o-o o 1 o o 1 -o o o		
	S:	-o o-1 -o o 1 1 -o-1 o 1 -o-1 -1 1 o		
	T:	-1 1 -o-o-1 o 2 -1 2 -o-o 2 -o-o-1 o-o		
434 repressor (O_R3) (42°) concave		[A–G–**T**–**A**–**C**–**A**–G–**T**–**T**–**T**–**T**–**T**–C–T–**T**–**G**–**T**–**A**–T)	6	148
	R:	1-o o o o o o o-o-o o o 1-o-o o		
	S:	-o-o o-o-o o-o-o-o o 1-o-o o-o o		
	T:	-o-o 1-o-o o o 1-o 1-o o-1-o o-o		
434 cro protein (O_R1) (27°) concave		[A–G–**T**–**A**–**C**–**A**–A–A–C–T–T–T–C–T–**T**–**G**–**T**–**A**–T)	7	149
	R:	o-o o 1-o-o o o o-1 o-o 2 o-o-o		
	S:	-o-o o-1 1 o o-1 o-o-o 1-o-o-o o		
	T:	-o-o o-2-o 1 o-o o o o o o-o-o-1 o o		

Table 6.2. *Continued*

Protein	DNA Binding Sequence	I.D. No.	Ref.
CAP protein 90° concave	[C-G-A-A-A-A-G-T-G-T-G-A-C-A-T-A-T|G-T-C-A-C-A-C-T-T-T-T-C-G) R: 1-o-o-o o o o-o-o **7**-o-o o 1-2 o 1-o-o **5**-o o o-o-o-1 1-1 S: 1 1-o-1-1-o-1 2-o o-2-o-o 2-o-2 2 2-o 1-2-1-o-o-o o 1 1 T: o 1 1-o o-o 1-o-2-o-o-1-1 2 1-o o-2-1-o o-o-o-o 1 o 1-1	8	150
CAP protein 90° concave	[C-G-A-A-A-A-G-T-G-T-G-A-C|A-T-A-T-G-T-C-A-C-A-C-T-T-T-T-C-G) R: o-o o-**4**-o-2 2-2 1 **9**-o-2-1 2-1-1 o o o **6**-o-1-o-1-o-**3** o 1 1 S: 1 1 o-1-o-1-1 2 o-o 2-1 o o o-1-o 1 2 1-o 2-2-o-1-o-o o 2 T: -1 o o o-o-o 2-1-2-o-o o-o-1-o o-2-2 1-**3**-1 1 1-2-o **3**-o-o-o-2	9	151
Lac repressor (60°) convex	[A-A-T-T-G-T-G-A-G-C-G-C-T-C-A-C-A-A-T-T) R: **-6-4** o o-1 1-2-2-1 **9**-2-o-1 2-1 1-1 **3** **3** S: **-5**-1 **4**-1-o 2 1-o-1 1-o-o 1-o-1-o 2-2-1 T: **-7-9** 2-2 2 o **3-5** 1-2-o-o 2-3-3 3 o 2-**8**	10	130
PurR repressor 45° convex	[A-C-G-A-A-A-A-C-G-T-T-T-T-C-G-T) R: 1-1 o-o-o-o **8**-o o-o-o-o o-1 1 S: -2-o o-o o-o-1 1-1-o o-o o-o-2 T: -2 2-1-o-o-o-1-o-1-o-o-o-1 2-2	11	131
Trp repressor (28°) concave	[G-T-A-C-T-A-G-T-T-A-A-C-T-A-G-T-A-C) R: -o-o-o 1 1 1-1-o o-1-1 1 1 1-o-1-o S: -o 1-1-o-1 o-1 1 2 1-1 o-1-o-1 1-o T: -1 1-o-o-o-o o o-o-o-o-o-o-o 1-o R: -o-o-o 1 2 o-o-o-o-o-o 2 1-o-o o S: -o 1-1-o-o o 1 1 2 1-o-o-1-1 1-o T: -1 1-o-o-o-o o o-o-o-o o-1-1 2-2	12	152
HIN recombinase straight	[G-T-T-T-T-G-A-T-A-A-G-A) R: **3** 2 2 o 1 2 1 o 1 o **3**-o S: -o o o o 1 1 o-1 1-1 1-2 T: -1-o o-o-o o 1-o 1-o-o-o	14	153

Table 6.2. *Continued*

Protein	DNA Binding Sequence	No.	Ref.
γδ resolvase 60° convex	[C-**A**-G-**T**-**G**-T-C-**C**-G-A-T-**A**-**A**-T-**A**-**A**|**T**-**A**-T-**C**-**G**-G-A-**C**-**A**-C-**T**-**G**]15	15	136

<pre>
γδ resolvase
60°   R: -1 1-1 1-o o o o-o-o 1-o   o-o 6 o 2 1 1   o o-2-o-o-o-o-o 1-o 1 o
convex   S: o-2-1 1 o-o-o 1-o-o-1 1   o o-o-1-o o o   o-o-o-o o-o-o o-1 o
         T: o-1-o-o-o-o-1 1 o-1-1 o   -o-o-o-o-o-1-o   -o o-o o o o-2 o 1-2 1-2 1
</pre>

B. EUKARYOTIC H-T-H PROTEINS

Protein	DNA Binding Sequence	No.	Ref.
Engrailed homeodomain straight	[**T**-T-**T**-**G**-C-**C**-**A**-**T**-**G**-**T**-**A**-**A**-**T**-**T**-**A**-C-C-**T**-**A**-**A**]	16	154

<pre>
R: o o o o-o-1-o o-o-o-o 0 1 o 1 o o o
S: o-o-o-1 o 2-2 2-o o o-o-1 o 1-o
T: o-1-o o-2 1-o 2-1 o o-o-o-o-o-1 1-2
</pre>

Protein	DNA Binding Sequence	No.	Ref.
MATα2 (Yeast) straight	[**C**-**A**-**T**-**G**-T-**A**-**A**-**T**-**T**-**C**-**A**-**T**-**T**-**T**-**A**-C-**A**-C-**G**-C]	17	155

<pre>
R: 1-o 1-o o o-1-o o o o-o-o-o 1-2 2 1
S: 1-o o-o o-o 1 o o 1-1 o 1 o-o-o-o o-o
T: 1 o-o-1 1-o o-o 1-1-o o o-o-o-1 1-1 1
</pre>

Protein	DNA Binding Sequence	No.	Ref.
MATa1/α2 (Yeast) 60° concave	[**C**-**A**-**A**-**T**-**G**-**T**-**A**-**A**-**T**-**T**-**T̲**-**A**-**T**-**T̲**-**A**-**C**-**A**-**T̲**-**C**-**A**]	18	156

<pre>
R: 1 o 2 o o 1-o o-2-1-1-o 1 o 2-o o 1
S: 1-o-o 1 o o o o-o-1-o 2-1 o-1-o 1
T: o-o o-2 1 o-1 1-o 2 1 o-o 1 o-o-o-o
</pre>

Protein	DNA Binding Sequence	No.	Ref.
Even-skipped straight	[**T**-**A̲**-**A**-**T̲**-**T̲**-**G**-**A̲**-**A**-**T̲**]	19	157

<pre>
R: o-o-o o-o-1 1-o 1
S: 2-o-1-1 1 1-o-o-o
T: 1-o-o 1 1-o o 1
</pre>

Protein	DNA Binding Sequence	No.	Ref.
Oct-1/POU straight	[G-**T**-**A**-**T**-**G**-**C**-**A**-**A**-**A**-**T̲**-**A**-**A**-A-G-G]	20	158

<pre>
R: o-o 1 1-2 2-1-o o 1 2-o
S: -1 1-1 1 o o o-o-1 1-o-o-o
T: -o 1-1 o-o-o o-o-o 1-o-o o
</pre>

Table 6.2. *Continued*

Protein	DNA Binding Sequence	No.	Ref.
Paired (prd) protein 20° concave	[C–**G**–T–**C**–**A**–**C**–**G**–G–T–**T**–**G**–A–C] R: -1-1 0-0-2 1 0 1-1-0-0-0 S: -0-1-0 1-1 2-2-1-0 2 2-1 T: 2-0-0 1-0 0-0-0-0 0 0-1	22	159
Paired (prd) protein 21° concave	[A–**T**–**A**–A–T–C–**T**–**G**–A–T–**T**–**A**–C] R: -0-0 0 0 0 0 2-0 1-0-0 0 S: 0 0-0-1-0-0 1-0-1-0 1-0 T: 0 0-0-1-0-1 0-0-0 0 0-0	23	160
	[G–**T**–**A**–A–T–C–**T**–**G**–A–T–**T**–**A**–C] R: -0-0-0 0 0 1 1 0 0-0-0-0 S: -0 1 0-1-0 0-0-0-1 0 1-0 T: -0 0 1-1-0-0-0-0-1 1 0-0	23	160
Pu.1 ETS-domain (40°?) concave	[<u>A</u>–A–A–<u>A</u>–G–G–G–G–A–A–G–**T**–**G**–G–G) R: -0-0-1 0 1 1 0 1 0-1 0 0 0 S: 0 0-0-0-1 0 1-0 0-1-1 1 0-0 T: 0 0 0 0-1-1-0-1 0 1-1 0-0 0 R: -0-0-1 0 1 0 1 1 0 0 0 0-0 S: 0 0-0-0-1 0 1-0 0-1-1 0-1 T: 0 0 0-0-1-0-1 0 1-1-0 1-1	24	161
RAP1 DNA domain 20° concave	[**C**–**G**–**C**–**A**–**C**–**A**–C–C–**A**–C–**C**–**A**–C–**C**–**A**–G) R: -1-1 0-0 0 1 1 0-1 0 0 0 0 0-0-0 S: 1 0 0-1-0-0-2 0 2-0 2-1-1-1-0 0 0 T: 0 1-0-0 1-0-2-1 1-0 1-1-1-0 1-0	25	162

Table 6.2. *Continued*

Protein	DNA Binding Sequence	No.	Ref.
C. ZINC-BINDING PROTEINS			
Zif268 (Cys₂His₂Zn) x 3 straight	[G–**C**–**G**–**T**–**G**–G–G–**C**–**G**–T]	26	163
	R: o–o o–o–o o–1–o–o		
	S: –o o–o o–1 o–o 1–o		
	T: –1 o–1 o–1 o–1		
YY1 zinc finger domain (Cys₂His₂Zn) x 4 straight	[A–G–G–G–T–C–T–C–C–**A**–T–T–**T**–**G**–A–A–G–**C**–**G**]	28	164
	R: o–o o–1 o o–1 o 1 o–o 1–o 2–o–o 1–1 o		
	S: o o–o–1 o–o o–o o–o–o o–1 o–o–o–2 2		
	T: –1 o–1–o–o o–1 o–o o–1 o–o 1–o–o–**3 4**		
Tramtrack (Cys₂His₂Zn) x 2 straight	[**T**–**A**–A–**T**–**A**–A–G–G–A–**T**–A–A–**C**–G–T–C–**C**–**G**]	29	165
	R: o–1–1 o o 1 1 o–o o o–1 1–o–o o 1		
	S: 1–o–o 1–o–o–1 o–o–o–o–1 1–o o–o–o		
	T: o o o 1–o–o–o–o–2 2 o–1 o–o o–o–o–o		
	R: o–o–1–o 1 o o–o–o o o–o 1–o o–o–o		
	S: o o–o–o 1–o–o–o o–o–o o 1 o–o–o–o		
	T: –o o–o 1–o o o–o–o–2 o o–1–o o o–o–o–1		
P53 tumor suppressor (Cys₃HisZn) straight	[T–T–C–C–**A**–T–**A**–C–T–**T**–**G**–C–C–C–**A**–**A**–T–**T**–**A**]	29a	166
	R: –1 1 1 2 o–1 1–1 1 o–1 o o–1 o o–o–o–o 2		
	S: –o–1 o–2 1 o–o–1–o o 1–1–o–o 1–o–1 o 1		
	T: –2–o–o–1–1 **4**–2–o o–o o–1 o–o 1–o–1 1–o		
Glucocorticoid receptor (Cys₄Zn) x 2 straight	[**C**–**A**–G–A–A–**C**–A–**T**–C–**G**–T–T–C–**T**–G	30	167
	R: –o o–o 1 1 o–1 1–o o–o–o–o–1 o–o		
	S: o–o–o o–o–o–1 o 1–o–1–o o o–o–o–o		
	T: o o–1 o–o o–2–o o–o–o–1 o o–o–o 1 1		
Glucocorticoid receptor (Cys₄Zn) x 2 straight	[T–C–**C**–**A**–G–A–A–**C**–**A**–T–**G**–T–T–C–**T**–**G**–G–A]	31	168
	R: –o 1–2 1 o–o–o 2 2 o–o–o 1–2 1–o		
	S: –1 o **3** o o–o–1–o–1–1–o–o o **3** o–1		
	T: –o–1 2–1 o o–o–o–1–o–o–o o 1–2 2–1–o		

Table 6.2. *Continued*

Protein		DNA Binding Sequence	No.	Ref.
Estrogen receptor (Cys$_4$Zn) × 2 straight		[C-**A**-G-G-T-**C**-**A**-**C**-**A**-G-T-**G**-A-C-C-**T**-**G**)	32	169
	R:	0 0 0 0-0-0-0 0 1-0 0-0 0 0-0-0		
	S:	0-0-1-1 1 2-2 0-0-2 2 1-1-1-0 1		
	T:	0-0 1-0 0-0-0 **3**-0-1 1 0-1 1 0-0		
	R:	0-0 0-0-0 0-0 1 1-0-0-0-0 1-0-0		
	S:	1-0-0-1 2-1 0-0-1 2 1-1-1-0 1		
	T:	0-0 0-0 1-0-1 2 0-1 0-0-0 1-1 0		
PPR1 (Cys$_6$Zn$_2$) straight		[T-**C**-**G**-G-**C**-**A**-A-T-**T**-**G**-C-**C**-**G**-A)	34	170
	R:	0-0-0 1 0-1 0 1 0 1-0-0 **3**		
	S:	-1 1-0-0 0-0-1-0 0-0 0 0-1		
	T:	0-0 0-2 1 2-1 1 0-0 0-1-1		

D. LEUCINE ZIPPER AND RELATED

Protein		DNA Binding Sequence	No.	Ref.
GCN4 (bZIP) straight		[G-G-A-G-A-**T**-**G**-A-C-G-T-C-**A**-T-C-T-C-C)	36	171
	R:	0 1 1 0 0 1 1-0 **3**-0 1 1 0 0 1 1 0		
	S:	1 0-0 0-1 1 0-1 1-1 0 1-0-0 0 1		
	T:	-0 0-0 1-1 1-1-1 0-1-1 1-1 1-0 0-0		
GCN4 (bZIP) straight		[G-G-A-G-A-**T**-G-A-C-G-T-C-**A**-T-C-T-C-C)	37	172
	R:	0 1 1 1-0 1 1-0 2-0 1 1-0 1 1 1 0		
	S:	-0 0-0-0-1 0-0-1 1-1-0 0-1-0-0 0-0		
	T:	-1 0-1 1-1 1-0-0-0-0 1-1-1 0-1		

Table 6.2. *Continued*

Protein	DNA Binding Sequence		No.	Ref.

Fos & Jun (bZIP) straight — No. 38, Ref. 173

```
        [A-T-G-G-A-G-T-C-A-T-A-G-G-A-G-A)
R:      -1 1 0 0 0 0 0-0 1 0 0-0-1-0 0-0-0 0-0
S:      -0 0 0 0 0-0 1-0-0-1-0 1-1 0 1 0 1-1 1
T:       1-0-0 0-1 1-0-0-0-0 1-1 0 1 0 0 0-1 3

R:       0-0-0-0 1 0 0-0-1 1 1 0 0 0 0-0-1
S:       0 0 0 0-1 1-0-0-1-0 0-1 1-0-0-1 0 0
T:       0-1 0 1-1 1-0 0-0 0-1 2-1 2-1-0 2
```

MyoD (bHLH) straight — No. 41, Ref. 174

```
        [T-C-A-A-C-A-G-C-T-G-T-T-G-A)
R:       1-1 0-0 1 0 1 0 1-0 0-0-0 0
S:      -1 2-0-1 1-1-1-1 1-1-1 2-0
T:      -1 1-0-0 1-0-1-0 1-0-0-0 1-0

R:       0-1 0-0 1 0 1 0-0-0-1 0
S:      -0 2-1-1 1-1-1-1 1-1-0 2-0
T:      -1 1-0-0 1-0-1-0 1-0-0 1-1
```

E. OTHER SPECIFIC PROTEIN/DNA COMPLEXES

EcoRI restriction enzyme straight — No. 44, Ref. 175

```
        [C-G-C-G-A-A-T-T-C-G-C-G)
R:       1 0 1-0 6-9 6-0 1 0 1
S:       0-0-0-0 0 0-0-0-0 0
T:      -1-0 0-0-2 0-2-0 0-0-1

R:       1 0 1-0 6-9 6-0 1 0 1
S:       0-0-0-0 0 0-0-0-0 0
T:      -1-0 0-0-2 0-2-0 0-0-1
```

Bov. papillomavirus-1 E2-DNA 68° concave — No. 45, Ref. 176

```
        [C-G-A-C-C-G-A-C-G-T-C-G-G-T-C-G)
R:       0-0-0 1 2-0-1-1-0 2 1-0-0-0 0
S:       1 1-1-1 1-1 0 2 0-1 1-1-1 1 1
T:      -1 1-0-0-0 1 0 1 0 1-0-0-0-0 1-1
```

Table 6.2. *Continued*

Protein	DNA Binding Sequence	No.	Ref.

Met J repressor straight (NDB) [2 × 25° (paper)] — No. 46, Ref. 177

[T-**A**-G-A-**C**-**G**-T-C-**T**-**A**-G-A-**C**-**G**-T-C-T]

```
R:   o 1-o  o o o o  1-o  o o-o  o
S:   1-1-o-o  o-o o-1  1-1-o-o  o-o  o-1
T:   1-1-o-o  o-1 o-1  1-1-o-o  o-1  o-1
```

Arc repressor 50° concave — No. 47, Ref. 178

[A-**T**-**A**-G-**T**-**A**-G-A-G-**T**-**G**-C-T-T-C-T-**T**-**A**-T-**C**-**A**-T]

```
R:   o 3 1 o-2 2 o-o-o-1-2 o-o 1 1-3 1 1 1 o
S:  -o-1 o-1 1 o-o-o-o 1 o-o-o-o o o-1 o-o-1
T:  -1-2 1-2 2-2-o-1-o o o-o o-2-o o-o-o-o-o
```

EcoRV restriction enzyme straight — No. 49, Ref. 179

[**C**-**G**-A-G-C-T-**C**-**G**]   (non-cognate sequence)

```
R:   2 o o-o-o-o-1
S:  -1-o o-o-o 1 o
T:   o-o-o-o-o-1

R:   1-1 o-o o o-o
S:  -o 1-o-o-o-1 2
T:  -o 1-1 o o-o-o
```

EcoRV restriction enzyme 50° convex — No. 48, Ref. 179

[G-G-G-A-**T**-**A**-T-C-C-C]   (cognate sequence)

```
R:  -o-o-o 2 9 1-1 1-o
S:   1 o-o-1-o-o-1 o 1
T:   o-o 1-3-2-3 o-o o

R:   o o-o 1 9 1-o o o
S:   1-o-o-o-o-o-o 1
T:   o-o 2-4-2-4 2-o o
```

Table 6.2. *Continued*

Protein	DNA Binding Sequence		No.	Ref.
EcoRV restriction enzyme (80°) convex		[A–A–G–A–T–**A**–T–C–T–T]	50	180
	R:	1 1–o 2 **9** 2 1 3 **4**		
	S:	1 0–o–1 0–o o 2 **4**		
	T:	–1–o o–1–**4**–2 1–1–o		
	R:	o 1–1 1 **9** 1 1 2 **3**		
	S:	1 0–o–1–o–o o 1 **3**		
	T:	–o–o 1–2–**3**–1 o–1–1		
IHF 160° concave	R: S: T:	[G–C–C–**A**–A–A–A–A–A–G–**C**–A–T–**T**–G–C–T–T–**A**–T–T–**T**–G–T–**T**–G–**C**–A–C–C] o 1 o–o–1–1–o–o–o–o–o 1 **9**–3–o–o–1–1–1–o **9** 1 1 o 2–1–1–**3** o–2 o–o –o–o 2–o–o o–o–1–o–o 1–o–1 **3**–o–o–1–o–1 2 1–o–1–o–o–o–2–o **3**–o 2–1–o –o–o o o–o–o–o 1–o o–**4** 1–1 o–o 1–o–o–o–1–1–o–o o–o 2–1 1–o o	51	181
TATA-binding protein (TBP) (Saccharomyces) 80° convex		[G–**T**–A–**T**–A–**T**–A–A–A–A–C–G . . . .]	52	132
	R:	–o **9** 3 1 **5** 2 1 **7**–o–o–o		
	S:	–2–1–2 2–o 1–o–o–1–**3**–1		
	T:	o–2–3–1–**5**–1–1–**3** o–o 1		
	R:	–o **9** 3 o **5** 2 1 **7**–o–o–o		
	S:	–2–1–1–2 o 1–o–1–1–**3**–1		
	T:	o–2–**3**–1–**4**–1–1–2 o–o 1		
TATA-binding protein (TBP) (Arabidopsis) 80° convex		[G–C–**T**–A–**T**–A–A–A–A–G–G–G–C–**A**]	53	133
	R:	–o–**4** **8** 2–1 2 1–o **6**–2–1–2–1		
	S:	–**3** 1–**3**–1 2 o–o–1–o–o–o 1		
	T:	–2 2–2–1–1–2–1–o 1–o–1–o o		
	R:	–2–2 **7** 1–1 2 1–o **6**–2–o–1–1		
	S:	–2 1–2–2 2 o o–1–o–o–2 2		
	T:	–1 1–1–1–2–1–o–o o o–**3** o		

Table 6.2. *Continued*

Protein	DNA Binding Sequence		No.	Ref.
TATA-binding protein (TBP) (Human) 80° convex		[C-T-G-C-T-A-T-A-A-A-G-G-C-T-G)	54	134
	R:	-1 o-1-1 8 2-1 2 2-o 6-1-2-o-1		
	S:	0 o-o-2-2 2 0 0 0-1 o-1-o 1		
	T:	0-o-o o-o-2-1-2-1-o-1 o-o-o 2		
TATA-binding protein (TBP) (Human) 80° convex		[C-G-T-A-T-A-T-A-T-A-C-G)	55	135
	R:	1-o 8 5 1 4 6 2 9 2 o		
	S:	2-o-1-1 2-2 3-1-2-1 2		
	T:	o 1-3-3-1-6-1-2-4-3-o		
TFIIA /TBP/DNA III° complex 80° convex		[T-G-T-A-T-G-T-A-T-A-T-A-A-A-A-C)	56	182
	R:	-2-2-o-o-o-2-o-2 7 1-1 1 o-1 3		
	S:	2 o-o-o-o o-o-o-2-2 3-o 1 o-4		
	T:	-1-o o-2-o o-1-2-o-o-o o		
TFIIB /TBP/DNA III° complex 80° convex		[G-G-C-T-A-T-A-A-A-A-G-G-G-C-T-G)	57	183
	R:	-o-2-1 6 1-o 2 1 2 4-o-1-2-o-4		
	S:	-1-1 1-2-1 3-o o-o-o-o-2-o-1 5		
	T:	o-1 o-1-1-o-2-1-1-o o-o-1 o o		
Kappa B P50 homodimer straight		[G-G-G-A-A-T-T-C-C-C)	61	184
	R:	o 1 o o o o 1 1 1		
	S:	2-1-o o-o o-o-o o		
	T:	-1-1-o o o o-1-1-1		

Table 6.2. *Continued*

Protein	DNA Binding Sequence											I.D. No.	Ref.
PVUII endonuclease straight	[G-A-C-**C-A**-G-C-**T-G**-G-T-C]											63	185
	R:	1	1	3	1-o	**4**-o	2	**3**	o	1			
	S:	-o	o-1	-o-1	o	-2-o	-1	o	2				
	T:	-2-o	-1-o	-2-o	-**3**-1	-o-o	o						

Notes:

Overall bend in DNA helix is given at left below the protein name, as well as whether the protein sits primarily on the concave or convex side of the bend produced in the DNA. Values in parentheses have been measured approximately from Figures, when no angle is cited explicitly in the text. Hence they are to be considered merely suggestive of the true bend angle. r_c = Radius of curvature of a smoothly bent DNA helix.

All **Y-R** steps are indicated in **bold face**. A-tracts, consisting of four or more successive AT base pairs without a disruptive T–A step, are underlined. 86 of the 157 A–A/T–T steps, or 55%, are found in A-tracts. For distributions of other step types, see Table 3.

Observed roll, slide, and twist, as calculated by the 'Curves' program, are indicated below each step of the sequence. As shown below. Roll is numbered in 5° intervals centered around zero, Slide is numbered in 0.5 Å intervals centered around zero, and Twist is numbered in 5° intervals centered around 35°.

Range of the given variable:

Roll:	−10°	–	−5°	–	0°	–	5°	–	10°	–	15°	–	20°	–	25°	–	30°
Slide:	−1.0 Å	–	−0.5 Å	–	0.0 Å	–	0.5 Å	–	1.0 Å	–	1.5 Å	–	2.0 Å	–	2.5 Å	–	3.0 Å
Twist:	25°	–	30°	–	35°	–	40°	–	45°	–	50°	–	55°	–	60°	–	65°
Symbol:	−1		−o		o		1		2		3		4		5

All steps greater than 9 are represented simply by a 9. Sequences followed by two sets of R/S/T values generally represent two independent molecules per asymmetric unit.

For a more extended analysis using the new FREEHELIX program, see references 191 and 192.

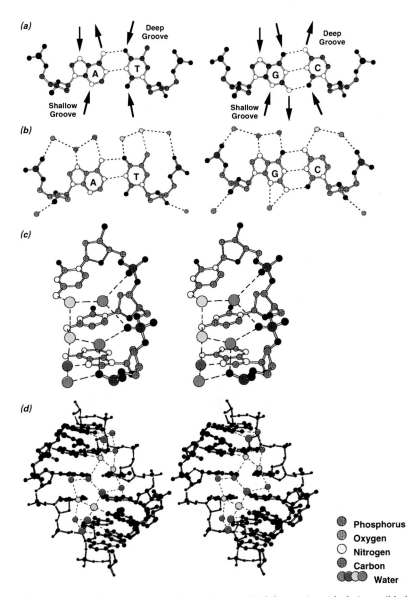

⊕	Phosphorus
⊕	Oxygen
○	Nitrogen
●	Carbon
◐◑○◯	Water

Plate I. Hydration of A-DNA. (a) G:C and A:T Watson–Crick base pairs with their possible hydration sites indicated as hydrogen bond donors or acceptors. (b) Observed hydration of the base pairs. Almost all hydrogen bonding sites on the bases are occupied except for the areas in the shallow groove that are blocked by crystal contacts. Waters in the shallow groove are shown in green. In the deep groove the waters contacting the phosphate anionic oxygens are blue, those hydrogen bonding to positions 6 or 7 in purines are red, and the water molecules hydrating position 4 of pyrimidines are shown in yellow. Note the fortuitous hydration of the 5-methyl group in thymine (thin broken line). (c) Water molecules bridging consecutive O1P phosphate oxygen atoms (blue) together with waters hydrating the O4/N4 position of pyrimidines (yellow) or the N7 position of purines (red) form pentagonal arrays. (d) Pentagonal water rings may originate in the deep groove of alternating sequences through the association of solvent molecules hydrogen bonding to the edges of the bases (red, waters hydrating N7 and O6 of G; yellow, waters hydrating N4 of C). The water networks are found in the structures of d(GGBr⁵UABr⁵UACC) (95), d(GGGTACCC) (105), and d(CCCGGCCGGG) (37).

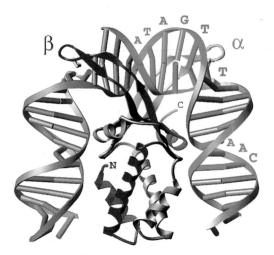

Plate II. Structure of integration host factor protein (IHF), complexed with B-DNA (181). The DNA is bent sharply at the top by insertion of loops into the minor groove, and has two mandatorily straight segments that pack along the protein to left and right. (From ref. 181.)

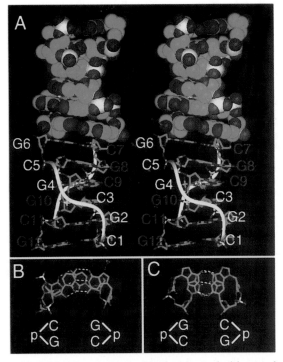

Plate III. Structure of d(CGCGCG) as Z-DNA. (A) The two stacked hexanucleotide duplexes in the crystal structure of d(CGCGCG) are shown as a stereodiagram. The upper duplex is shown as a CPK model using the van der Waals radii of each atom to define spheres for each atom. The lower duplex is shown as a stick model, with the backbone phosphates traced with a ribbon to show the zigzag nature of Z-DNA. The nucleotides are numbered from the 5′- to the 3′-terminus of each strand, 1–6 for one strand and 7–12 for the complementary strand. The d(CpG) dinucleotide in the *anti/p/syn* stacking arrangement (B) and the d(GpC) dinucleotide in the *syn/p/anti* stacking arrangement (C) are shown looking down the helix axis. Hydrogen bonds are shown as dashed lines connecting the bases of each base pair.

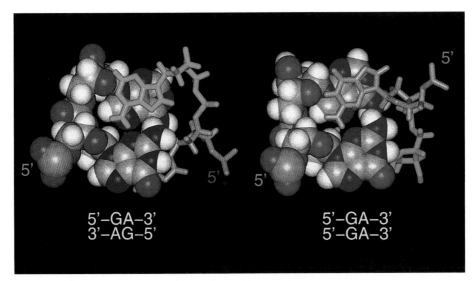

(a)

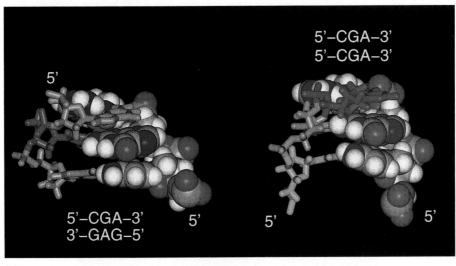

(b)

Plate IV. (a) Comparison between (right) the parallel (27) and (left) the antiparallel (40) 5'-(GA)$_2$-3' motifs. The guanosine residues of the stick-bond strand are coloured brown and the stick-bond strand adenosine residues are coloured blue. The 5'-termini towards the viewer are labelled in green, while the one 5'-terminus away from the viewer is coloured pink. (b) Comparison between the parallel-stranded 5'-(CGA)$_2$-3' motif (right) and the antiparallel 5'-(CGA):(GAG)-3' motif (left). The guanosine residues of the stick-bond strands are coloured brown, the adenosine residues are blue, and the cytidine residues are green.

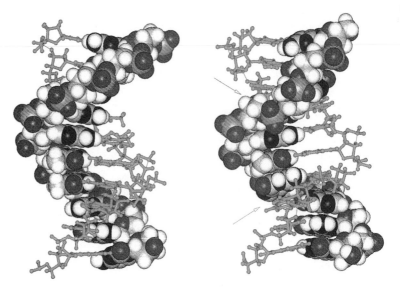

Plate V. Comparison between the unusual structure of (right) the d(GCGAATGAGC)$_2$ decamer duplex (40) containing two sheared (PYGAPU)$_2$ motifs and (left) a standard B-DNA duplex containing normal G:C and A:T Watson–Crick pairs. One strand is shown in a space-filling van der Waals form, while the other strand is shown in stick-bond form to illustrate better the interstrand stacking between the G and A bases (indicated by the red arrow in the bottom half). The unusual phosphate conformation of the GpA phosphodiester linkage is indicated by a blue arrow in the top half of the duplex.

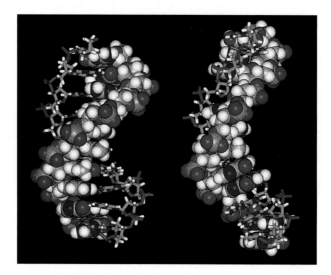

Plate VI. The structure of (right) the antiparallel self-paired centromeric pentamer repeat (GTG-GAATGGAAC)$_2$ (41,44) compared with (left) a normal B-DNA duplex (51). These two structures make the interesting point that DNA duplexes are sufficiently deformable or flexible to accommodate two such markedly different double helices that nevertheless have virtually identical thermodynamic stabilities. The grid of exposed hydrogen bond donors and acceptors (N2H, N1H, O6, and N7) of the four guanines in the (GGA)$_2$ motif is visible in the lower half of the right duplex. This motif, which can interact with itself, is repeated twice per turn on opposite sides of the duplex, and may be responsible for the condensation and mitotic/meiotic segregation of centromeres. The interstrand base stacking between the two unpaired guanosine residues from opposite strands can be seen at the interface of the space-filling and stick-bond strands.

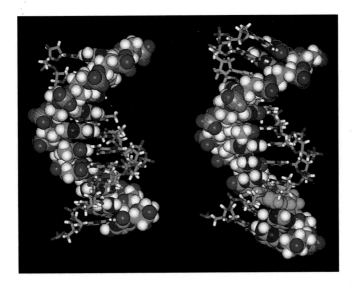

Plate VII. The structure (Chou *et al.*, in preparation) of the self-paired d(GCGAGTACGAAGC)$_2$ duplex (right) compared with normal B-DNA (left) (51). This sequence contains two GAA:GA motifs in which the 'extra A' from the GAA strand intercalates between the guanosine residues of two flanking sheared G:A pairs. The intercalated, unpaired adenosine residues are coloured blue in both the space-filling and stick-bond strands. The helix axis is bent ~20° at each of the GAA:GA sites because of the wedge-like insertion of the single adenosine.

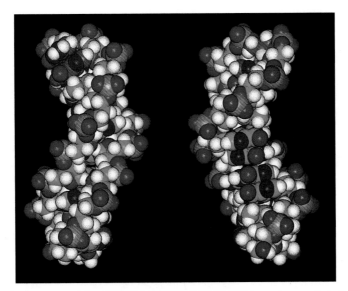

Plate VIII. The structure of the G(**TGGA**AT**GCA**AT**GGA**A)C heptadecamer hairpin containing a d(GCA) motif tight-turn loop and in intercalative (GGA)$_2$ motif in the hairpin stem (45). The carbon atoms of the tight-turn d(GCA) loop motif are coloured blue to differentiate them from the stem residues, the carbon atoms of which are coloured green. The loop motif, with the C base stacking on the G base of the sheared G:A pair, can be seen at the top of the left, minor groove view. The (GGA)$_2$ motif in the stem, with its grid of 16 hydrogen bonded donor/acceptor atoms, can be seen just below the centre of the right, major groove view. In the right view, the H4′ proton of the deoxycytidine sugar is coloured yellow to emphasize its stacking on the A base of the sheared G:A pair and its consequent large upfield shift (see text).

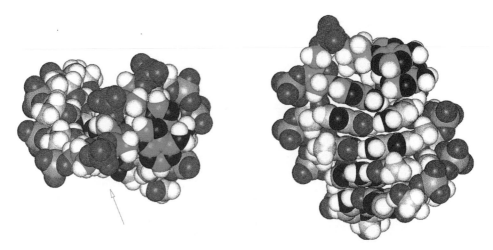

Plate IX. The structure of the single A-residue loop hairpin GTAC**AAA**GTAC (43) viewed from the major groove (right) and end-on from the top (left). The phosphorus atoms of the d(AAA) loop are coloured pink, while the stem phosphates are coloured orange. The kink in the phosphate backbone at the phosphodiester, connecting residue 6A to residue 7A, is indicated by a blue arrow; this kink is mostly a result of changes in the phosphate dihedral angles from $\zeta(g^-)$, $\alpha(g^-)$ to $\zeta(g^+)$, $\alpha(g^+)$ at the turn. Very good 5A–6A base stacking is observed and the stacking of the 6A-deoxyribose on the 7A base can be seen at the top of the right-hand figure.

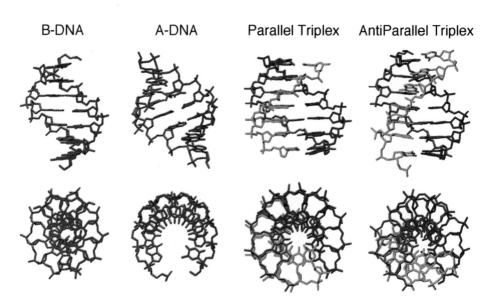

Plate X. Comparison of the parallel (YRY1) and antiparallel (RRY) triplexes to B- and A-DNA. The central purine strand, the duplex pyrimidine strand, and the third strand are coloured red, blue, and green, respectively. The views are: into the major groove illustrating the base pair inclinations and the groove widths; and down the helical axis illustrating the central 'hole'. The parallel triplex (YRY2) is after ref. 223, and the antiparallel triplex (RRY) is after ref. 244.

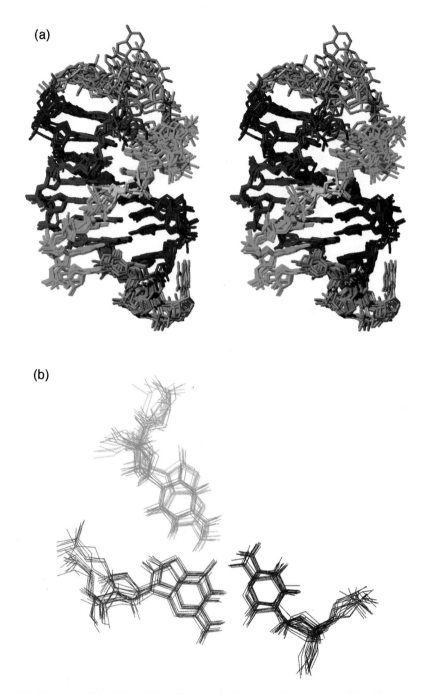

Plate XI. Structure of the N7G triplex. The strand colours are the same as in Plate X, except for the third strand guanine in the 7G:GC triplet, which is coloured yellow. The loops are coloured grey. (a) Stereoview looking into the major groove of the duplex for the family of the eight lowest energy structures. (b) Superposition of the 7G:GC triplet with a C$^+$:GC triplet from the same triplex [N7G]. After ref. 228.

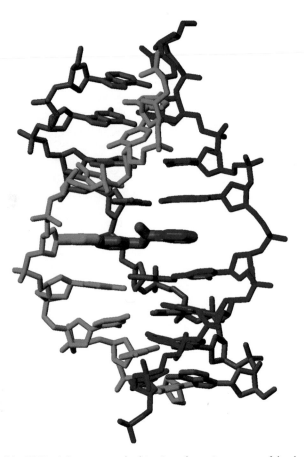

Plate XII. View of the DTA triplex structure looking into the major groove of the duplex. The D_3 base is drawn with thicker bonds. The strand colours are the same as in Plate X, except for the D_3 base. The A, B, and C rings of the D_3 base are green, red, and blue, respectively, to illustrate how the D_3 base mimics a triplet. After ref. 227.

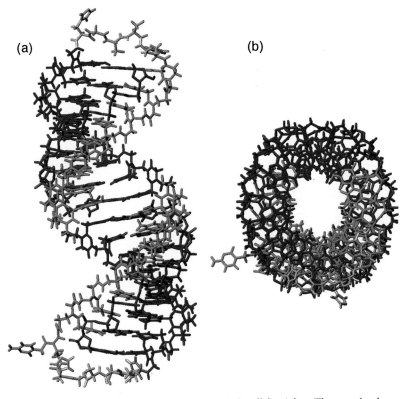

(a) (b)

Plate XIII. X-ray crystal structure of a PNA:DNA:PNA (parallel) triplex. The strand colours are the same as in Plate X. (a) View into the major groove of the duplex. (b) View down the helical axis. After ref. 193.

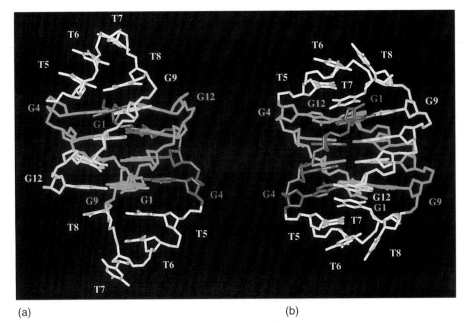

(a)   (b)

Plate XIV. A comparative colour display of: (a) the 2.5 Å X-ray crystal structure of the two-repeat *Oxytricha* telomere d(G₄T₄G₄) quadruplex containing lateral T₄ loops (45); and (b) the NMR-based solution structure of the two-repeat *Oxytricha* telomere d(G₄T₄G₄) quadruplex containing diagonal T₄ loops (47). The four G₄ segments around the quadruplex formed through dimerization of a pair of hairpins are coloured magenta, green, cyan, and yellow, while the T₄ loop segments are in white.

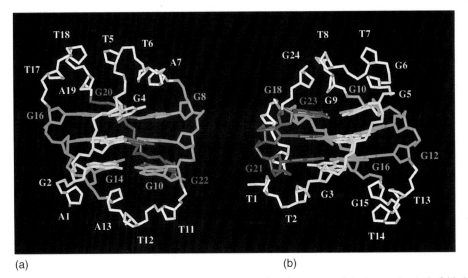

(a)   (b)

Plate XV. A comparative colour display of: (a) the NMR-based structure of the intramolecularly folded four-repeat human telomere d[AG₃(T₂AG₃)₃] quadruplex in Na⁺ solution containing a central T11–T12–A13 diagonal loop (59); and (b) the NMR-based solution structure of the intramolecularly folded four-repeat *Tetrahymena* telomere d(T₂G₄)₄ quadruplex containing a T19–T20 double chain reversal loop (77). The four G₃ segments around the quadruplex are coloured magenta, green, cyan, and yellow, while the loop segments are in white.

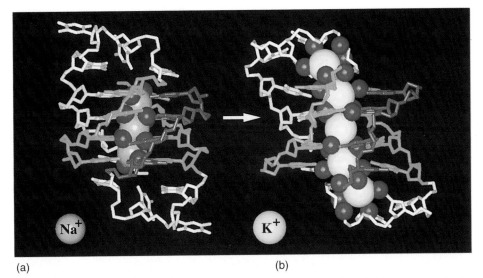

(a)   (b)

Plate XVI. A comparative colour display of the NMR-based structures of the d(G$_3$CT$_4$G$_3$C) quadruplex formed through head-to-tail dimerization of a pair of hairpins in: (a) Na$^+$ solution (116) and (b) K$^+$ solution (117). Each of the four strands involved in tetrad formation is shown in a separate colour. The T$_4$ loop residues are shown in white. Three Na$^+$ cations and their coordinated oxygens are shown in orange and red, respectively, in (a), while five K$^+$ cations and their coordinated oxygens are shown in yellow and red, respectively, in (b). The additional K$^+$ cation-binding sites in (b) are located within the symmetry-related T$_4$ hairpin loops.

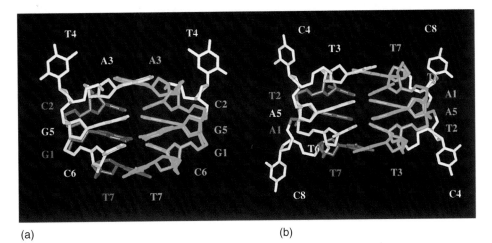

(a)   (b)

Plate XVII. A comparative colour display of the X-ray structures of: (a) the d(G$_3$CT$_4$G$_3$C) quadruplex containing G:C:G:C tetrads aligned through the minor groove edges of Watson–Crick G:C paris (118); and (b) the d(pAT$_2$CAT$_2$) quadruplex containing A:T:A:T tetrads aligned through the minor groove edges of Watson–Crick A:T pairs (119).

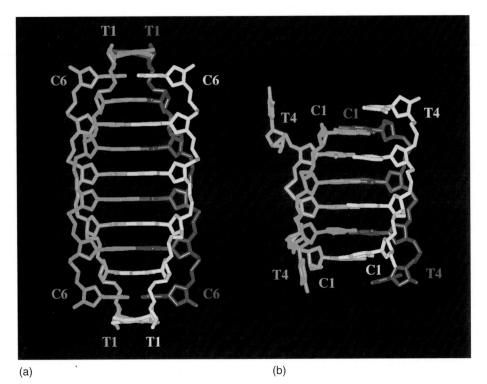

(a)   (b)

Plate XVIII. A comparative colour display of: (a) the NMR–based structure of the four-stranded d(TC$_5$) i-motif in acidic pH solution (15); and (b) the 1.4 Å crystal structure of the four-stranded d(C$_3$T) i-motif quadruplex (132). Each of the four strands involved in i-motif quadruplex formations is shown in a separate colour.

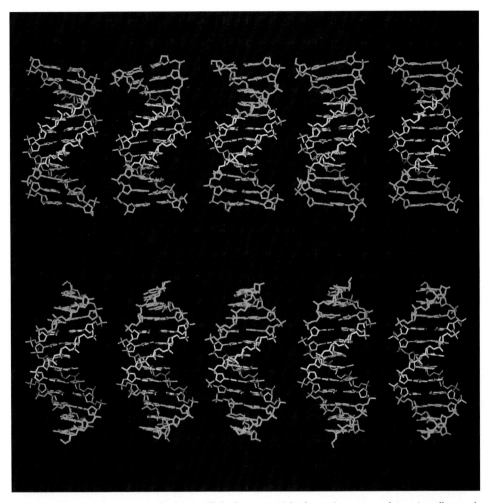

Plate XIX. A-tract-containing duplexes. All duplexes are 12 bp long, A-tracts are shown in yellow and the rest in cyan. From left to right: $(dA)_{12}$:$(dT)_{12}$ helix based on the fibre-derived structure of poly (dA):poly (dT) (47); GCCAAAAAAGCA from the crystal structure of the IHF protein:DNA complex (86); CAAGAAAAACTG from the crystal structure of the 434 repressor:DNA complex (81); crystal structure of CGCAAAAAAGCG (51); and fibre-based general sequence B-DNA (87). The view is perpendicular to the A-tract minor groove (top) and rotated by 90° about the helix axis (bottom).

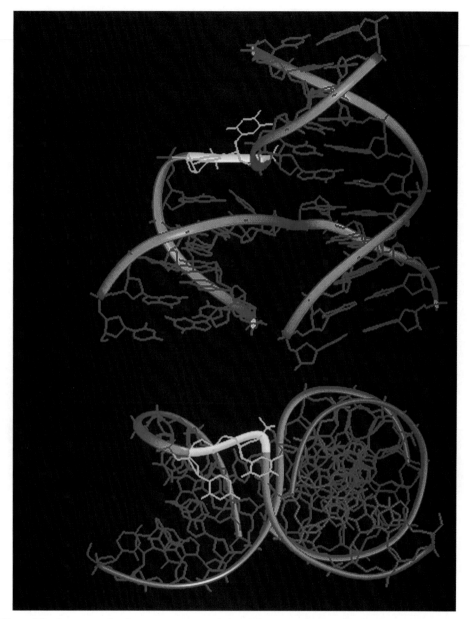

Plate XX. Structure of a three-way junction deduced from NMR data (70). This junction adopts a coaxially stacked structure in the isomer II conformation (refer to Fig. 16.8). Two views of one structure are shown, with a ribbon indicating the path of the deoxyribosephosphate backbone. The DNA forms a 3HS$_2$ junction with a m^{5}CpC bulge; the bulged cytosine nucleotides are indicated in cyan and the rest of the molecule in magenta. The upper image shows a view of the major groove side of the junction. The 'spacing' effect of the extra bases allowing the coaxial helical stacking to occur is very clear in this view. The lower image shows a view down the axis of the stacked helices. Note that the major distortion in helical geometry is largely localized to the extra bases. A loop closed one of the helices in the species studied experimentally, but has been removed from the structure presented for clarity. This figure was made from coordinates kindly made available by Dr D. Patel.

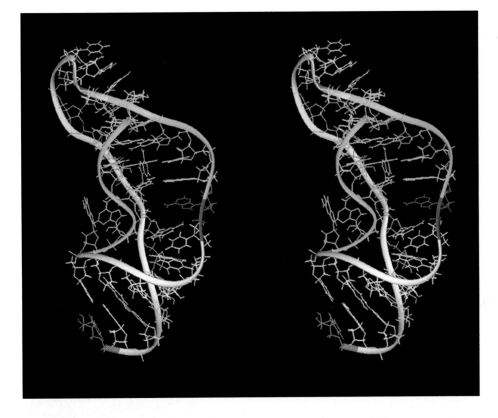

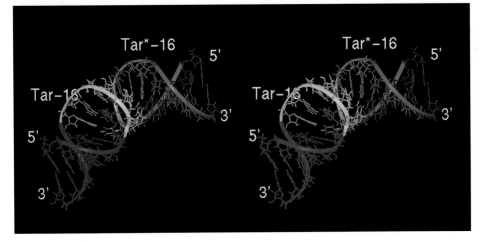

Plate XXI. Stereoview of: (top) a pseudoknot from mouse mammary tumour virus (MMTV) involved in ribosomal frameshifting (139), and (bottom) a kissing hairpin from the HIV TAR hairpin and its complement (145).

DNA in aqueous solution. The question then arises: Which DNA environment is more 'biological', the double helix isolated in dilute aqueous solution, or the helix packed against other helices in the low dielectric environment of a crystal? The 'natural' working environment of DNA, of course, is in close association with proteins of one kind or another. It seems useful, therefore, to examine in some detail the conformation of the B-DNA duplex when bound to proteins, and to compare its behaviour there with that in crystals of DNA alone. This is not intended as a systematic analysis of the mechanism of DNA recognition by proteins, but rather as an investigation of the conformations that B-DNA can undergo when subjected to the stresses of sequence-specific binding to proteins.

Table 6.2 lists 63 protein:DNA complexes for which X-ray crystal structures are available, and in which the protein recognizes and binds to a specific base sequence. These structures, involving 971 individual base steps, all are on deposit with the Nucleic Acid Database (138), and have been analysed using Richard Lavery's 'Curves' program (137; Chapter 2). Table 6.2 attempts to present this monumental archive of information in compact form, by representing the roll, slide, and twist by single characters below each step of a sequence. Roll angles are grouped into 5° intervals centred around zero roll, slide in 0.5 Å intervals to either side of zero slide, and twist in 5° intervals centred around 35°. A roll coefficient of $\pm R$ signifies a roll angle between $\pm[R5°]$ and $\pm[(R+1)5°]$. Similarly, slide coefficient $\pm S$ indicates a slide between $\pm[S0.5$ Å] and $\pm[(S+1)0.5$ Å]. Twist coefficient $\pm T$ designates a twist angle between $35°\pm[T5°]$ and $35°\pm[(T+1)5°]$. Hence, the zero for each coefficient is roughly the value expected in an ideal, undistorted B-DNA helix. To the left of the $R/S/T$ data are the overall DNA bend angles observed in the crystalline protein:DNA complex, and an indication of whether the protein sits on the outside (convex) surface of the bent DNA helix or the inside (concave). Where the original investigators do not quote a specific bend angle value, but the bend can be estimated from molecular drawings or normal vector plots, this approximate bend angle is enclosed in parentheses.

Four classes of protein:DNA complexes are listed in Table 6.2.

(A,B) Prokyotic and eukaryotic helix–turn–helix or HTH proteins, in which a packed bundle of two or three α helices presents one helix (usually the final one) towards the major groove so that its side chains can make hydrogen bonds with the floor of the groove.

(C) Zinc-binding proteins, in which a zinc finger or related complex involving Zn coordinated to cysteine and/or histidine side chains is used to present a recognition α helix towards the major groove of DNA.

(D) Basic leucine zipper (bZIP) and basic helix–loop–helix (bHLH) complexes that extend a pair of α helices in opposite directions along the major groove, holding the DNA in a scissor-like grip.

(E) Other DNA-binding proteins that do not fit the above three patterns, frequently involving an opening up of the minor groove by insertion of loops of polypeptide chain or intercalating phenylalanine side chains.

Most of the zinc and leucine zipper complexes insert their α helical 'reading heads' at parallel multiple sites along the DNA major groove, leaving the helix straight and unbent, as can be seen from Table 6.2. Bending is more common when the protein attacks the DNA at one particular site, or at two adjacent sites related by a real or pseudo-twofold symmetry axis. This is the pattern of HTH proteins (A,B) and of the groove-expanding proteins in category E.

Examination of the $R/S/T$ data in Table 6.2 shows a strong correlation between overall bending and large positive roll at key steps along the sequence, and a strong tendency for this large positive roll to occur at Y–R steps. The previously mentioned LacI, PurR, TBP, and $\gamma\delta$-resolvase are good examples (Figs 6.7 and 6.8). LacI has a large positive roll ($R = 9$) and positive slide ($S = 1$) at the central C–G step, and a lower than standard twist angle in the 20s ($T = -2$). PurR has very positive roll ($R = 8$) and positive slide ($S = 1$) at the central C–G step, with a less dramatic twist angle in the low 30s ($T = -0$). $\gamma\delta$-resolvase has one large roll step at T–A in the centre ($R = 6$), but many local Y–R roll maxima, as seen especially clearly in Fig. 6.8b. All of the TATA-binding proteins in Table 6.2e consistently exhibit large roll steps at the first

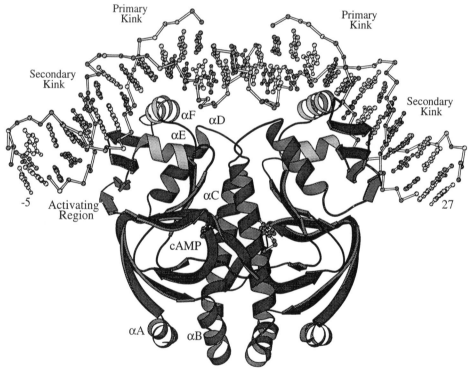

Half Complex A   Half Complex B

Fig. 6.9. Structure of catabolite activator protein (CAP), complexed with DNA (149,150). The DNA is kinked at two C–A/T–G steps in order to bend nearly 90° around the protein. Secondary kinks are problematical, being found in one structure analysis (151) but not the other (150). (From ref. 151.)

and the last step of the TATA box, with R between 6 and 9, and additional positive roll at intermediate steps (Fig. 6.8a). This pattern persists in complexes of TBP with TFIIA and TFIIB.

In contrast, the catabolite activator protein (CAP) (150,151) shown in Fig. 6.9 is an excellent illustration of roll bending in which the protein sits on the *concave* side of the bent DNA. This dimeric protein inserts helices into two successive major grooves of the B-DNA helix, inducing it to bend by nearly 90° around the protein. The bend occurs by compression of the major grooves, induced by large positive roll and small helical twist at C–A/T–G steps at the centre of each major groove half-site (Fig. 6.10). These bends appear in Table 6.2 as roll coefficients between 5 and 9, and are accompanied by unusually small local helical twist, with coefficient −2 (twist angles between 20 and 25°). Slide varies along the CAP DNA sequence in an apparently non-systematic manner between extrema of −2 and +2, or between −1.5 Å and +1.5 Å. The same information is conveyed by the three plots of Fig. 6.10, but in less compact form. The secondary kinks labelled in Fig. 6.9 are somewhat problematical. One of the two CAP:DNA analyses (151) has oscillating roll values at these two loci, but the other independent analysis (150) finds only two straight regions of low roll. This may be an example of working at the limit of resolution of the data, but the apparent inconsistency is puzzling.

CAP also illustrates the probabilistic nature of sequence–structure relationships. Although bending is produced by two C–A/T–G sites, four other such sites in the CAP sequence are straight, with roll angles less than 10° in magnitude. It is not that C–A/T–G steps are *required* to bend; only that they are *capable of bending* if outside circumstances dictate. Substitution of R–R steps for the two critical C–A steps would interfere with bending and probably would ruin the sequence as a CAP site. As an excellent illustration of the danger of unacceptable substitutions, Wobbe and Struhl (186) have examined all possible point mutations in the first six positions of the TATA box, TATAAA, for their ability to act as functional TBP-binding sites *in vivo* (see Table 1 of ref. 135). The most disastrous substitution was A for T in the third position, yielding the sequence TAAAAA. This led to less than 1% activity in both HeLa and yeast assay systems. Evidently the five base pair A-tract cannot be bent in the manner that is possible with TATAAA and is required for binding of TBP. Again, in the TBP system, note from Fig. 6.8a that, although the T–A step is capable of a roll bend of 45°, it does not inevitably adopt such a value. The T–A steps between the outer limits of the TATA box have rolls of less than 15°.

Integration host factor (IHF) (Plate II), illustrates roll kinking at other than Y–R steps (181). Bending is produced near the centre of the 35 base pair DNA duplex by insertion of loops of extended chain into the minor groove in two directions, rather like the insertion of α helices down two directions of the major groove in leucine zipper proteins. Spreading of the minor groove causes a bend towards the major groove, as with TBP, and the DNA helix is bent nearly 180° around the IHF protein. As seen in Fig. 6.11, bending occurs via positive roll and reduced twist at two ATTG/CAAT loci. Many C–A/T–G sites in IHF are not significantly bent, although all exhibit the strong positive slide that has been associated with a C–A step ever since the crystal structure analysis of the B-DNA decamer CCAACGTTGG (83).

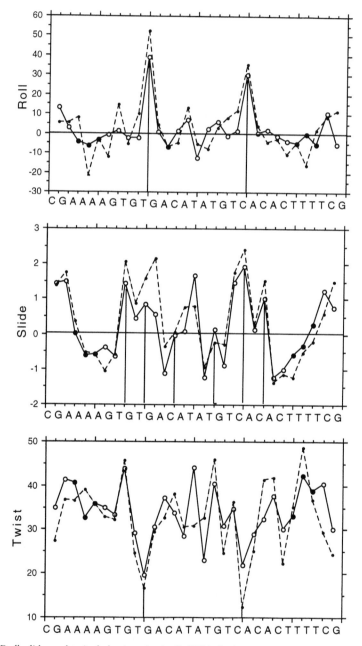

Fig. 6.10. Roll, slide, and twist behaviour in the B-DNA duplex as it wraps around CAP. Solid lines from ref. 150; dashed lines from ref. 151. Bending at the two C–A/T–G steps involves both large positive roll and small twist. Almost every significant positive slide occurs at a C–A/T–G step. Short A-tracts are indicated by filled dots.

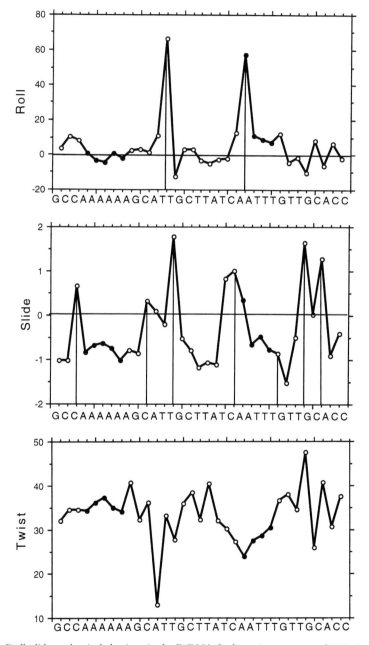

Fig. 6.11. Roll, slide, and twist behaviour in the B-DNA duplex as it wraps around IHF. Bending ocurs at two ATTG/CAAT regions. The six base pair A-tract in the initial arm of the DNA is straight and unbent. Short A-tracts are indicated by filled dots.

In the left arm as plotted in Plate II, IHF has a six-residue A-tract, defined as four or more successive AT base pairs without a disruptive T–A step. This A-tract in IHF is a region of essentially zero roll, uniform negative slide, and near-standard 36° twist. Both arms of the bent DNA duplex interact closely with the IHF protein, in a manner that requires straight helix segments displaying an unusually narrow minor groove (181). Many of the IHF-binding sites, as this one, already possess an A-tract in one of the two arms, others do not. But in the latter circumstance, binding is enhanced by insertion of an A-tract into the arm (187,188). Natural selection evidently has introduced A-tracts into loci where a straight, unbent DNA duplex is needed.

The *lambda*, *cro*, and *trp* repressors in Table 6.2a are all bent by 25–45° over ~18 base pairs, but produce this bending via many small deformations rather than one or two large roll steps, as in the proteins we have just been considering. Individual roll angles seldom fall outside the ±10° range. Similar comments apply to the eukaryotic HTH proteins in Table 6.2b. For reasons mentioned earlier, zinc-binding proteins and leucine zippers and their relatives, Table 6.2c and d, do not bend their DNA appreciably, and their $R/S/T$ coefficient entries in Table 6.2 are a sea of –1, –0, +0, and +1 values. This does not mean that these sequences lack C–A, T–A, and C–G steps, only that the bending capabilities of these steps are not being used.

Roll bending becomes important again in the proteins of Table 6.2e. EcoRI has a large central negative roll flanked by compensating positive rolls, apparently produced solely by the rolling of the central two base pairs about their long axes, caused by two α helices that point into the major groove (see Fig. 9 of ref. 189). Base pairs to either side of the central two are essentially unaffected by this interaction. EcoRV, like $\gamma\delta$-resolvase, produces a sharp bend at the central T–A step of its cognate sequences, again accompanied by diminished helical twist.

Short A-tracts of four to six base pairs are common in the sequences of Table 6.2, and are usually associated with small local roll angles, symbolized in Table 6.2 by o or –o. Individual roll angle plots generally show A-tracts to be regions of relative quiet within a more exaggerated roll milieu. This has been seen already for PurR (Fig. 6.7b), $\gamma\delta$-resolvase (Fig. 6.8b), CAP (Fig. 6.10a), and IHF (Fig. 6.11a). It also is true for 434 repressor, Engrailed, Mat $\alpha2$ and Even-skipped homeodomains, the Oct-1 POU domain, the Pu.1 ETS domain, the P53 tumour suppressor, and other instances where bending is not wanted.

Table 6.3 summarizes the behaviour of all 971 steps of the 63 helices in Table 6.2. The observed frequency of occurrence of the 10 possible dinucleotide steps is quite asymmetric, as evidenced by Table 6.4b. The A–A step is most common of all, occurring 16% of the time rather than its statistically expected 10%, and in 55% of these cases it occurs in A-tracts. Steps in which A is preceded or followed by C are also frequent. At the other end of the scale, steps with only G:C base pairs are strongly disfavoured, C–G and G–C most of all.

It is interesting to compare this distribution of steps with that encountered by El Hassan and Calladine (141) in their analysis of 400 steps from 60 crystal structures of A- and B-DNA (Table 6.4a). C–G was the most common step encountered in that study, and A–G was so rare as to have been excluded from subsequent analysis. Sharp differences were also seen between A- and B-DNA. However, these differences in

Table 6.3. Percentage of each step type in given ranges of roll, slide, and twist in the 63 protein:DNA complexes of Table 6.2

Step:	CA	TA	CG	AA	AG	GA	GG	AC	AT	GC	Total
No. of steps:	123	85	51	157	103	117	79	131	89	36	971
Roll (°)											
>15	4.0	**18.8**	6.0	7.0	5.8	0.9	2.5	1.5	7.9	2.8	V. pos.
5 to 15	**39.1**	**20.0**	**43.1**	18.5	31.1	25.6	29.1	7.6	27.0	13.9	Pos.
−5 to 5	40.7	45.8	35.2	**60.6**	**53.4**	**63.2**	**59.6**	**74.9**	**51.7**	38.9	Zero
−15 to −5	13.8	14.2	15.7	12.1	8.7	10.3	8.8	16.0	10.1	**44.4**	Neg.
< −15	2.4	1.2	–	1.8	1.0	–	–	–	3.3	–	V. neg.
Total	100	100	100	100	100	100	100	100	100	100	
Slide (Å)											
> 1.5	4.9	3.5	–	1.9	–	–	–	–	–	–	V. pos.
0.5 to 1.5	**48.0**	**42.3**	**52.9**	10.8	6.8	18.8	11.4	–	–	–	Pos.
−0.5 to 0.5	44.7	37.7	41.2	**75.9**	**68.9**	**70.0**	**62.0**	40.4	46.0	**66.7**	Zero
−1.5 to −0.5	2.4	15.3	5.9	10.8	24.3	11.2	26.6	**57.3**	**54.0**	30.5	Neg.
−1.5	–	1.2	–	0.6	–	–	–	2.3	–	2.8	V. neg.
Total	100	100	100	100	100	100	100	100	100	100	
Twist (°)											
> 50	2.4	–	2.0	0.6	1.0	1.7	–	–	–	–	V. high
40 to 50	34.1	31.7	15.7	10.2	8.7	23.1	6.4	4.6	1.1	8.3	High
30 to 40	**50.5**	**37.6**	**56.8**	**70.2**	**59.1**	**60.6**	**62.0**	**63.3**	39.4	**55.5**	Normal
20 to 30	11.4	25.9	25.5	17.2	29.2	14.6	31.6	30.6	**46.1**	30.6	Low
< 20	1.6	4.8	–	1.8	2.0	–	–	1.5	13.4	5.6	V. low
Total	100	100	100	100	100	100	100	100	100	100	

Table 6.4. Comparison of frequencies of the 10 base pair steps in this analysis of 63 protein:DNA complexes, and in the DNA analysis of El Hassan and Calladine

(a) By step type

Step type:	YR			RR = YY				RY			
Step:	CA	TA	CG	AA	AG	GA	GG	AC	AT	GC	Total
Protein: DNA	123	85	51	157	103	117	79	131	89	36	971
DNA crystals	26	21	83	64	9	25	56	28	32	56	400
A-DNA	2	14	16	0	4	4	40	20	4	11	115
B-DNA	24	7	67	64	5	21	16	8	28	45	285

(b) By frequency of occurrence in protein:DNA complexes

Step:	AA	AC	CA	GA	AG	AT	TA	GG	CG	GC	
Number	157	131	123	117	103	89	85	79	51	36	971
Percentage	16.2	13.5	12.7	12.0	10.6	9.2	8.8	8.1	5.2	3.7	100

DNA step frequencies are to be attributed mainly to the hazards of DNA syntheses and crystallization, not to any inherent distribution of steps in natural DNA. The preponderance of C–G and G–C steps came from multiple reuse of the 'magic' B-DNA sequence: CGCGxxxxCGCG. The high frequency of A–A steps in B-DNA crystal structures (22%) came from the interest of investigators in the A-tract bending question; in contrast, no A-DNA sequences were synthesized containing an A–A step! The preponderance of G–G steps in A-DNA came from the experimental finding that sequences containing these steps were easy to crystallize in the A-form.

In contrast to this behaviour with crystalline DNA, no investigator-induced selection is to be inferred in the 971 steps encountered in the 63 protein:DNA complexes of Table 6.2. Sequences were synthesized because they were recognized by the protein in question, not because they had worked well in other crystal structure analyses. Hence the distribution given in Table 6.4b can be taken as approximating the relative suitability of different steps for interaction with proteins. Evidently, steps from one GC base pair to another, regardless of base pair orientation, are not particularly compatible with interaction with a sequence-specific protein molecule.

Table 6.3 groups each kind of step into positive, neutral (effectively zero), and negative values of roll, slide, and twist. It shows that Y–R steps have a broad tendency towards positive roll and slide, whereas R–R steps even more strongly prefer near-zero roll and slide. Among Y–R steps, T–A has the greatest tendency towards major roll angles above 15°. Among R–Y steps, A–C and A–T favour zero roll but negative slide, whereas G–C tends towards the opposite behaviour: negative roll and zero slide. Twist angles centre around 36° for all except A–T, which favours lower twist.

6. Roll/slide/twist correlations in protein:DNA complexes

Local roll, slide, and tilt deformations are not mutually independent. Figure 6.12 compares slide vs. roll, twist vs. roll, and slide vs. twist plots for the 10 base steps. Among Y–R steps, C–A and T–A exhibit a negative correlation between slide and roll, and between twist and roll, with a positive correlation between slide and twist. This is easy to understand mechanically in terms of a finite length of sugar–phosphate backbone connecting one base pair with the next. As Fig. 6.13 demonstrates, a large positive roll separates base pair edges on their minor groove side, and works against either a large extension to the left, via slide, or to the right, via helical twist. Conversely, imposition of a large slide or twist would prevent the base step from simultaneously adopting a large positive roll. But slide and twist, as conventionally defined (120), act in opposite directions. A large slide to the left can be facilitated by a simultaneously large twist to the right; hence the correlations observed in Fig. 6.12a.

The relatively rare C–G steps do not exhibit a strong roll/slide/twist correlation, although, as Table 6.3 indicates, both positive roll and positive slide are favoured. Interestingly enough, G–A in the R–R category does show this roll/slide/twist correlation, which seems to be a general property of X–A steps. Among other R–R steps (Fig. 6.12b,c, and d), the very populous A–A cluster tightly around the origin of zero roll, zero slide, and standard 36° twist, consistent with the concept of the innate straightness of A-tracts (101,139,140). A–G and G–G show a similar clustering, with slight preferences for positive roll, and negative slide and twist, again visible in Table 6.3.

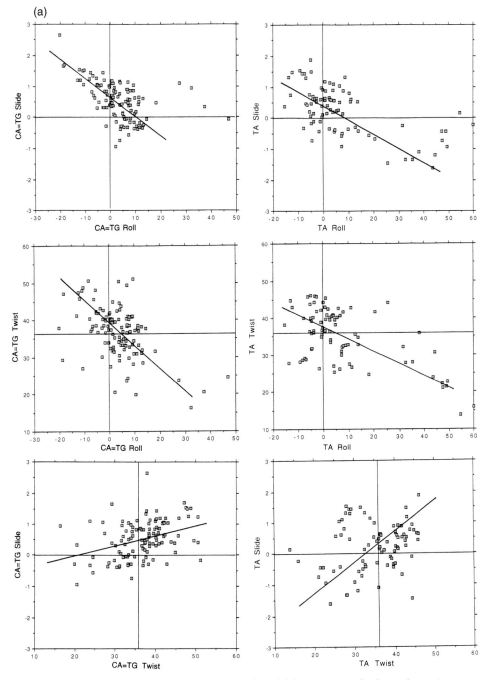

Fig. 6.12(a). Plots of slide versus roll, twist versus roll, and slide versus twist for the ten base pair steps as observed in the protein:DNA complexes of Table 6.2. In all, 968 steps are plotted. (a) C-A/T-G and T-A. (b) C-G and A-A/T-T. (c) A-G/C-T and G-A/T-C. (d) G-G/C-C and A-C/G-T. (e) A-T and G-C.

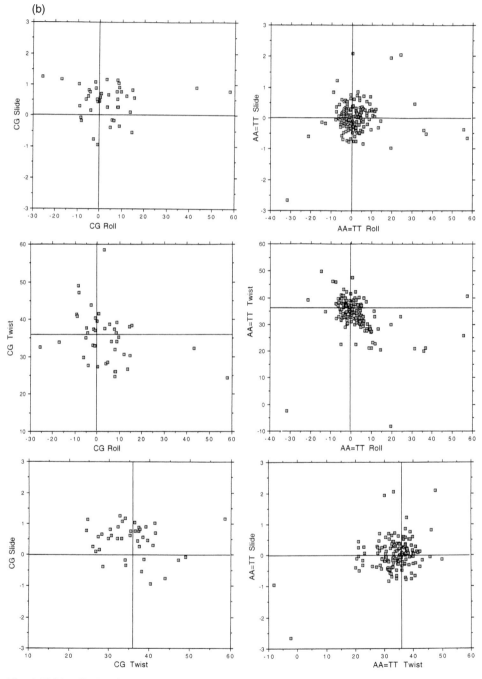

Fig. 6.12(b). *Continued*

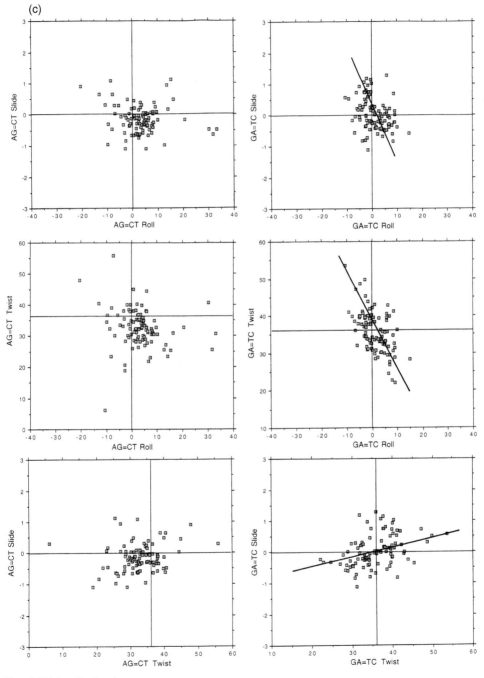

Fig. 6.12(c). *Continued*

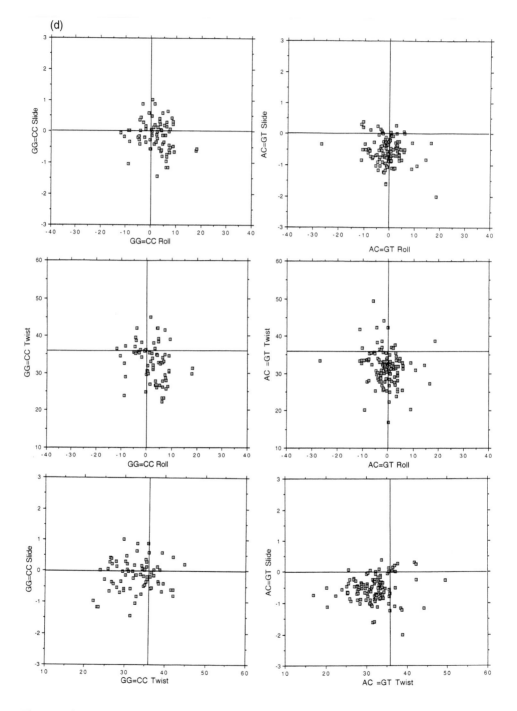

Fig. 6.12(d).   *Continued*

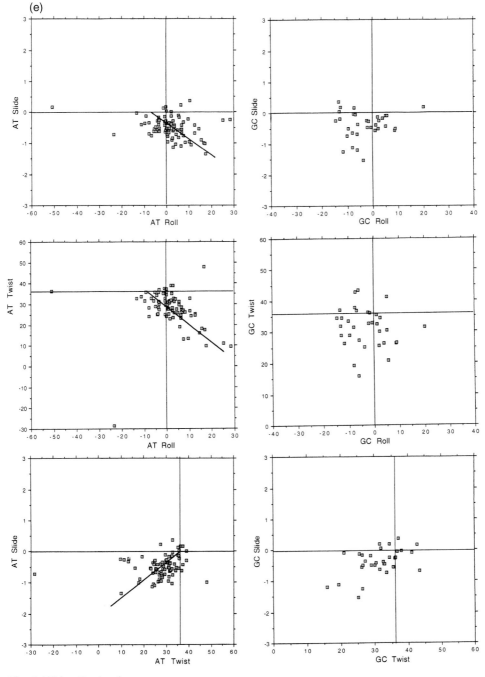

Fig. 6.12(e).   *Continued*

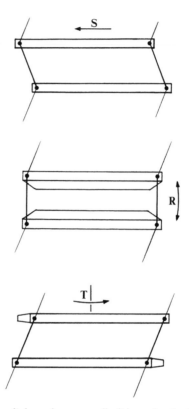

Fig. 6.13. Schematics explaining linkages between roll, slide, and twist. View into minor groove, helix axis vertical. Slabs are base pairs, dots are C1′ attachments to backbone, and heavy lines connecting dots are the sugar–phosphate backbone chain from one base pair to the next. Top: positive slide displaces the second base pair towards strand 1, thus stretching the backbone chain. Centre: positive roll opens the side of the base pairs to which the backbone chains are attached, forcing a diminution of slide. Hence, roll and slide are negatively correlated. Bottom: increased twist stretches the backbone chain in the opposite direction to positive slide, so twist and roll also are negatively correlated. But slide and twist are positively correlated because, with the present sign conventions, one variable relieves the backbone stress produced by the other.

The three R–Y steps show an interesting predilection for negative values of both slide and twist, leading to slide/twist plots populated mainly in the lower left quadrant. A–T exhibits a limited correlation of the type observed with C–A and T–A. As can be seen in Fig. 6.6, simultaneous negative slide and twist deformations tend, in the limit, to stack the six-membered rings of pyrimidine and purine of adjacent base pairs squarely over one another, and this apparently is a favourable situation.

These correlation plots for protein:DNA complexes are quite similar to those found by El Hassan and Calladine for crystals of DNA alone (Figs 6–8 of ref. 141). They illustrate the fact that DNA in crystals is behaving virtually identically to DNA complexed with proteins that have evolved to interact specifically with that DNA. Hence, DNA crystal structures offer a realistic and relevant snapshot of the possible conformational behaviour of the DNA duplex under biological conditions.

The behaviour seen in Fig. 6.12, and in Figs 6–8 of ref. 142, has not yet been explained theoretically. The most comprehensive attempt, by Hunter (190), calculates the stacking free energy of two base pairs as a function of roll and slide, considering two types of interactions: steric clash, because of van der Waals contacts, and electrostatic effects. The latter involve attractions and repulsions between: (a) partial ionic charges on individual atoms of different base pairs; (b) partial ionic charges on one base pair and the π electron cloud of the other; and (c) the π electron cloud of one base pair and the σ core of atoms on the other pair. Figure 20 of Hunter (190) presents contour plots of free energy vs. roll and slide for all 10 base step types. However, except in predicting essentially zero roll and slide for A–A steps, the potential minima on the Hunter plots *do not correspond* to the observed distribution of conformations for either isolated DNA or protein:DNA complexes. The correlation plots shown in Fig. 6.12 must remain as targets for any improved base stacking energy model.

7. Summary and conclusions

The 'take-home lesson' from this conformational analysis is that the B-DNA duplex in its working protein complexes behaves exactly like the DNA that has been observed in crystal structures. The tendencies of given sequences to roll, to slide, to twist, or not to undergo these local helix deformations, are just the same. In a sense this should not be surprising, as the local environments experienced by a DNA duplex in close proximity to proteins or to other DNA molecules are similar: low local dielectric constant, high ionic strength, low water activity, and partial dehydration. Max Perutz once made a then paradoxical, but now unsurprising, comment that haemoglobin molecules packed in erythrocytes had more in common with haemoglobin crystals than with haemoglobin in solution. For similar reasons, *DNA molecules packed against proteins in their normal biological environment have more in common with DNA crystals than with DNA in solution.* That perhaps is the most important single fact to be remembered from this data–crammed chapter.

Another significant conclusion about DNA sequence–structure relationships is that local conformations are statistical and probabilistic, not rigidly 1:1 cause and effect. Different base sequences have different conformational preferences, but with wide latitude in individual cases. Hence it must always be kept in mind that anecdotal isolated examples of sequence and structure can neither prove nor disprove a given generalization; only the examination of a large body of conformational examples can yield significant results. The flaw in the 'ten twist angles' approach and the Calladine/ Dickerson 'rules' was that they were based on too small a sample. The 400 base steps in the DNA crystals of El Hassan and Calladine (141) and the 971 base steps in the protein:DNA complexes of Table 6.2 are steps in the right direction.

With this warning in mind, the following conclusions appear to be valid from what is known to date.

1. B-DNA sugar pucker, although centred around the expected C2'-*endo* conformation, is more variable than in A-DNA.

2. Propeller twisting of base pairs is more variable, and in general larger in magnitude, with A:T base pairs than with G:C. As a consequence, the minor groove width is more variable with A:T than G:C, closing down around single drug molecules or opening up to accept two flat drug molecules side by side. The minor groove is hydrated in a regular manner, the pattern (spine vs. ribbons) depending on local groove width.

3. Bending of the DNA helix is almost always a consequence of roll, with tilt making a much smaller contribution. The bend can result from severe roll kinking at one or two loci (CAP, LacI, PurR, γδ-resolvase, IHF, TBP, etc.), or can be accomplished by smaller deformations spread out over a longer region (λ, 434, and Trp repressors, etc.). In most cases, roll bending occurs in a direction that compresses the major groove, whether the DNA is bent inwards by a protein that sits on the outside or convex surface of the bend, or whether the DNA wraps itself around a protein situated on the inside of the bend.

4. Y–R steps tend to stack polar N and O atoms over polarizable rings of the adjacent base pair. In contrast, ring-upon-ring stacking is more prevalent in R–R and R–Y steps. R–R steps, in particular, appear to rotate about their stacked purines, with their pyrimidine ends more widely separated so that the O2 atom of one base sits against the six-membered ring of the neighbour. In R–Y steps, purine N3 atoms stack against adjacent pyrimidine rings, but outlying O and N atoms contribute little.

5. Y–R steps are most prone to roll bending, particularly T–A and C–A/T–G. The latter also frequently exhibit a large positive slide. R–R steps are the least prone to roll and slide deformations, with the A–A step being the stiffest of all in both isolated DNA and in protein:DNA complexes.

6. Heterogeneous steps ending in A: C–A, T–A, and G–A, all display negative correlation between slide and roll, and between twist and roll, and positive correlation between slide and twist. This behaviour is explicable in terms of a finite backbone chain length from one base to the next. A–A has a faint tendency towards this behaviour, but only in its twist vs. roll plot. Other steps show no such correlation.

7. R–Y steps display a systematic preference for negative slide and twist values.

8. The frequency of occurrence of steps in the 63 protein:DNA complexes examined is quite asymmetric. Unlike the sequences of DNA alone, many of which were chosen for ease of synthesis or crystallization, these protein:DNA steps were largely self-selected by the protein:DNA complexes themselves. Hence they represent, not the absolute frequency of steps in DNA as a whole, but in that fraction of DNA that has evolved to interact closely and specifically with proteins. The A–A step is most common of all (16%), and in 55% of these cases it occurs in A-tracts, or runs of four or more successive As and Ts without an intrusive T–A step. In view of the known tendency of A-tracts towards straight and unbent character in both DNA crystal structures and in protein:DNA complexes, it is tempting to suggest that this preference for A–A is a consequence of natural selection for a stabilizing structural element.

9. By contrast, steps containing only G:C base pairs are rare, and seemingly are less compatible with complexes involving sequence-reading proteins.

The above are the 'rules' that any theory of base stacking will have to explain. They provide a vocabulary for the language of specific protein:DNA interactions: a language with many synonyms but also with a definite syntax. They demonstrate the near identity of behaviour of DNA when packed against other DNA molecules in the crystal, and when packed against protein molecules in its natural working environment. The 'odd man out' of DNA environments is the dilute aqueous solution, with low ionic strength and high dielectric constant. As has been remarked previously (91), the properties of DNA in dilute aqueous solution may be of interest to the physical chemist, but are of less interest to the biochemist.

In summary, B-DNA is a double helix that is uniquely capable of expressing its base sequence to the outside world, and hence of functioning as a regulatable or controllable carrier of genetic information.

References

1. Wing, R.M., Drew, H.R., Takano, T., Broka, C., Tanaka, S., Itakura, K. and Dickerson, R.E. (1980) *Nature* **287**, 755.
2. Dickerson, R.E. and Drew, H.R. (1981) *J. Mol. Biol.* **149**, 761.
3. Dickerson, R.E., Drew, H.R. and Conner, B.N. (1981) *Biomolecular Stereo-Dynamics*, (Sarma, R.H., ed.), Vol. 1, p. 1. Adenine Press, New York.
4. Drew, H.R. and Dickerson, R.E. (1981) *J. Mol. Biol.* **151**, 535.
5. H.R. Drew, R.M. Wing, T. Takano, C. Broka, S. Tanaka, K. Itakura and R.E. Dickerson (1981) Proc. Natl. Acad. Sci. USA 78, 2179.
6. Dickerson, R.E., Goodsell, D.S. and Neidle, S. (1994) *Proc. Natl. Acad. Sci. USA* **91**, 3579.
7. Drew, H.R., Samson, S. and Dickerson, R.E. (1982) *Proc. Natl. Acad. Sci. USA* **79**, 4040.
8. Westhof, E. (1987) *J. Biomol. Struct Dynamics* **5**, 581.
9. Holbrook, S.R., Dickerson, R.E. and Kim, S.-H. (1985) *Acta Cryst.* **B41**, 255.
10. Fratini, A.V., Kopka, M.L., Drew, H.R. and Dickerson. R.E. (1982) *J. Biol. Chem.* **257**, 14686.
11. Kopka, M.L., Fratini, A.V., Drew, H.R. and Dickerson. R.E. (1983) *J. Mol. Biol.* **163**, 129.
12. Frederick, C.A., Quigley, G.J., van der Marel, G.A., van Boom, J.H., Wang, A.H.-J. and Rich, A. (1988) *J. Biol. Chem.* **263**, 17872.
13. Partridge, B.L. and Salisbury, S.A. (1997) (unpublished).
14. Portmann, S., Altmann, K.-H., Reynes, N. and Egli, M. (1997) *J. Am. Chem. Soc.* **119**, 2396.
15. Boggon, T.J., Hancox, E.L., McAuley-Hecht, K.E., Connolly, B.A., Hunter, W.N., Brown, T., Walker, R.T. and Leonard, G.A. (1996) *Nucl. Acids Res.* **24**, 951.
16. Shatzky-Schwartz, M., Arbuckle, N.D., Eisenstein, M., Rabinovich, D., Bareket-Samish, A., Haran, T.E., Luisi, B.F. and Shakked, Z. (1997) *J. Mol. Biol.* **267**, 595.
17. Nelson, H.C.M., Finch, J.T., Luisi, B.F. and Klug, A. (1987) *Nature* **330**, 221.
18. DiGabriele, A.D., Sanderson, M.R. and Steitz, T.A. (1989) *Proc. Natl. Acad. Sci. USA* **86**, 1816.

19. Edwards, K.J., Brown, D.G., Spink, N., Skelly, J.V. and Neidle, S. (1992) *J. Mol. Biol.* **226**, 1161.
20. Yoon, C., Privé, G.G., Goodsell, D.S. and Dickerson, R.E. (1988) *Proc. Natl. Acad. Sci. USA* **85**, 6332.
21. Balendrian, J. and Sundaralingam, M. (1991) *J. Biomol. Struct. Dynamics* **9**, 511.
22. Chen, X., Ramakrishnan, B. and Sundaralingam, M. (1995) *Nature Struct. Biol.* **2**, 733.
23. Larsen, T.A., Kopka, M.L. and Dickerson, R.E. (1990) *Biochemistry* **30**, 4443.
24. Narayana, N., Ginell, S.L., Russu, I.M. and Berman, H.M. (1990) *Biochemistry* **30**, 4450.
25. Aymami, J., Coll, M., van der Marel, G.A., van Boom, J.H., Wang, A.H.-J. and Rich, A. (1990) *Proc. Natl. Acad. Sci. USA* **87**, 2526.
26. Wing, R.M., Pjura, P., Drew, H.R. and Dickerson, R.E. (1984) *EMBO J.* **3**, 1201.
27. Kopka, M.L., Yoon, C., Goodsell, D., Pjura, P. and Dickerson, R.E. (1985) *Proc. Natl. Acad. Sci. USA* **82**, 1376.
28. Kopka, M.L., Yoon, C.. Goodsell, D., Pjura, P. and Dickerson, R. (1985) *J. Mol. Biol.* **183**, 553.
29. Goodsell, D.S., Kopka, M.L. and Dickerson, R.E. (1995) *Biochemistry* **34**, 4983.
30. Sriram, M., van der Marel, G.A., Roelen, H.L.P.F., van Boom, J.H. and Wang, A.H.-J. (1992) *Biochemistry* **31**, 11823.
31. Tabernero, L., Verdaguer, N., Coll, M., Fita, I., van der Marel, G.A., van Boom, J.H., Rich, A. and Aymami, J. (1993) *Biochemistry* **32**, 8403.
32. Coll, M., Aymami, J., van der Marel, G.A., van Boom, J.H., Rich, A. and Wang, A.H.-J. (1989) *Biochemistry* **28**, 310.
33. Balendiran, K., Rao, S.T., Sekharudu, C.Y., Zon, G. and Sundaralingam, M. (1995) *Acta Cryst.* **D51**, 190.
34. Goodsell, D.S., Ng, H.L., Kopka, M.L., Lown, J.W. and Dickerson, R.E. (1995) *Biochemistry* **34**, 16654.
35. Coll, M., Frederick, C.A., Wang, A.H.-J. and Rich, A. (1987) *Proc. Natl. Acad. Sci. USA* **84**, 8385.
36. Pjura, P.E., Grzeskowiak, K. and Dickerson, R.E. (1987) *J. Mol. Biol.* **197**, 257.
37. Teng, M., Usman, N., Frederick, C.A. and Wang, A.H.-J. (1988) *Nucl. Acids Res.* **16**, 2671.
38. Quintana, J.R., Lipanov, A.A. and Dickerson, R.E. (1991) *Biochemistry* **30**, 10294.
39. Carrondo, M.A.A.F. de C.T., Coll, M., Aymami, J., Wang, A.H.-J., van der Marel, G.A., van Boom, J.H. and Rich, A. (1989) *Biochemistry* **28**, 7849.
40. Spink, N., Brown, D.G., Skelly, J.V. and Neidle, S. (1994) *Nucl. Acids Res.* **22**, 1607.
41. Vega, M.C., Garcia-Saez, I., Aymami, J., Eritja, R., Van Der Marel, G.A., Van Boom, J.H., Rich, A. and Coll, M. (1994) *Eur. J. Biochem.* **222**, 721.
42. Sriram, M., van der Marel, G.A., Roelen, H.L.P.F., van Boom, J.H. and Wang, A.H.-J. (1992) *EMBO J.* **11**, 225.
43. Clark, G.R., Squire, C.J., Gray, E.J., Leupin, W. and Neidle, S. (1996) *Nucl. Acids Res.* **24**, 4882.
44. Wood, A.A., Nunn, C.M., Czarny, A., Boykin, D.W. and Neidle, S. (1995) *Nucl. Acids Res.* **23**, 3678.
45. Brown, D.G., Sanderson, M.R., Skelly, J.V., Jenkins, T.C., Brown, T., Garman, E., Stuart, D.I. and Neidle. S. (1990) *EMBO J.* **9**, 1329.
46. Brown, D.G., Sanderson, M.R., Garman, E. and Neidle, S. (1992) *J. Mol. Biol.* **226**, 481.
47. Larsen, T.A., Goodsell, D.S., Cascio, D., Grzeskowiak, K. and Dickerson, R.E. (1989) *J. Biomol. Struct. Dynamics* **7**, 477.
48. Laughton, C.A., Tanious, F., Nunn, C.M., Boykin, D.W., Wilson, W.D. and Neidle, S. (1996) *Biochemistry* **35**, 5655.

49. Trent, J.O., Clark, G.R., Kumar, A., Wilson, W.D., Boykin, D.W., Hall, J.E., Tidwell, R.R., Blagburn, B.L. and Neidle, S. (1996) *J. Med. Chem.* **39**, 4554.
50. Edwards, K.J., Jenkins, T.C. and Neidle, S. (1992) *Biochemistry* **31**, 7104.
51. Nunn, C.M., Jenkins, T.C. and Neidle, S. (1994) *Eur. J. Biochem.* **226**, 953.
52. Nunn, C.M., Jenkins, T.C. and Neidle, S. (1993) *Biochemistry* **32**, 13838.
53. Nunn, C.M. and Neidle, S. (1995) *J. Med. Chem.* **38**, 2317.
54. Gao, Y.-G., Sriram, M., Denny, W.A. and Wang, A.H.-J. (1993) *Biochemistry* **32**, 9693.
55. Clark, G.R., Gray, E.J., Neidle, S., Li , Y.-H. nd Leupin, W. (1996) *Biochemistry* **35**, 13745.
56. Skelly, J.V., Edwards, K.J., Jenkins, T.C. and Neidle, S. (1993) *Proc. Natl. Acad. Sci. USA* **90**, 804.
57. Brown, T., Hunter, W.N., Kneale, G. and Kennard, O. (1986) *Proc. Natl. Acad. Sci. USA* **83**, 2402.
58. Hunter, W.N., Brown, T. and Kennard, O. (1986) *J. Biomol. Struct Dynamics* **4**, 173.
59. Leonard, G.A., McAuleyu-Hecht, K.E., Gibson, N.J., Brown, T., Watson, W.P. and Hunter, W.N. (1994) *Biochemistry* **33**, 4755.
60. Leonard, G.A., Guy, A., Brown, T., Teoule, R. and Hunter, W.N. (1992) *Biochemistry* **31**, 8415.
61. Hunter, W.N., Brown, T., Kneale, G., Anand, N.N., Rabinovich, D. and Kennard, O. (1987) *J. Biol. Chem.* **21**, 9962.
62. Leonard, G.A., Thomson, J., Watson, W.P. and Brown, T. (1990) *Proc. Natl. Acad. Sci. USA* **87**, 9573.
63. Brown, T., Leonard, G.A., Booth, E.D. and Chambers, J. (1989) *J. Mol. Biol.* **207**, 455.
64. Leonard, G.A., Booth, E.D. and Brown, T. (1990) *Nucl. Acids Res.* **18**, 5617.
65. Webster, G.D., Sanderson, M.R., Skelly, J.V., Neidle, S., Swann, P.F., Li, B.F. and Tickle, I.J. (1990). *Proc. Natl. Acad. Sci. USA* **87**, 6693.
66. McAuley-Hecht, K.E., Leonard, G.A., Gibson, N.J., Thomson, J.B., Watson, W.P., Hunter, W.N. and Brown, T. (1994) *Biochemistry* **33**, 10266.
67. Hunter, W.N., Brown, T., Anand, N.N. and Kennard, O. (1986) *Nature* **320**, 552.
68. Leonard, G.A., Booth, E.D., Hunter, W.N. and Brown, T. (1992) *Nucl. Acids Res.* **20**, 4753.
69. Corfield, P.W.R., Hunter, W.N., Brown, T., Robinson, P. and Kennard, O. (1987) *Nucl. Acids Res.* **15**, 7935.
70. Xuan, J.-C. and Weber, I.T. (1992) *Nucl. Acids Res.* **20**, 5457.
71. Yang, D., Gao, Y.-G., Robinson, H., van der Marel, G.A., van Boom, J.H. and Wang, A.H.-J. (1993) *Biochemistry* **32**, 8672.
72. Ginnell, S.L., Vojtechovsky, J., Gaffney, B., Jones, R. and Berman, H.M. (1994) *Biochemistry* **33**, 3487.
73. Vojtechovsky, J., Eaton, M.D., Gaffney, B., Jones, R. and Berman, H.M. (1995) *Biochemistry* **34**, 16632.
74. Takahara, P.M., Rosenzweig, A.C., Frederik, C.A. and Lippard, S.J. (1995) *Nature* **377**, 6749.
75. Urpi, L., Tereshko, V., Malinina, L., Huynh-Dinh, T. and Subirana, J.A. (1996) *Nature Struct. Biol.* **3**, 325.
76. Leonard, G.A. and Hunter, W.N. (1993) *J. Mol. Biol.* **234**, 198.
77. Di Gabriele, A.D. and Steitz, T.A. (1993) *J. Mol. Biol.* **231**, 1024.
78. Timsit, Y., Westhof, E., Fuchs, R.P.P. and Moras, D. (1989) *Nature* **341**, 459.
79. Timsit, Y., Vilbois, E. and Moras, D. (1991) *Nature* **354**, 167.
80. Timsit, Y. and Moras, D. (1991) *J. Mol. Biol.* **221**, 919.

81. Privé, G.G., Heinemann, U., Chandrasegaran, S., Kan, L.-S., Kopka, M.L. and Dickerson, R.E. (1987) *Science* **238**, 498.
82. Privé, G.G., Heinemann, U., Chandrasegaran, S., Kan, L.-S., Kopka, M.L. and Dickerson, R.E. (1988) *Structure and Expression, Vol. 2: DNA and its Drug Complexes*, (Sarma, R.H. and Sarma, M.H., eds), p. 27. Adenine Press, New York.
83. Privé, G.G., Yanagi, K. and Dickerson, R.E. (1991) *J. Mol. Biol.* **217**, 177.
84. Yanagi, K., Privé, G.D. and Dickerson, R.E. (1991) *J. Mol. Biol.* **217**, 201.
85. Lipanov, A., Kopka, M.L., Kaczor-Grzeskowiak, M., Quintana, J. and Dickerson, R.E. (1993) *Biochemistry* **32**, 1373.
86. Heinemann, U. and Alings, C. (1989) *J. Mol. Biol.* **210**, 369.
87. Gao, Y.-G., van der Marel, G.A., van Boom, J.H. and Wang, A.H.-J. (1991) *Biochemistry* **30**, 9922.
88. Lipscomb, L.A., Peek, M.E., Morningstar, M.L., Verghis, S.M., Miller, E.M., Rich, A., Essigmann, J.M. and Williams, L.D. (1995) *Proc. Natl. Acad. Sci. USA* **92**, 719.
89. Goodsell, D.S., Grzeskowiak, K. and Dickerson, R.E. (1995) *Biochemistry* **34**, 1022.
90. Wood, A.A., Nunn, C.M., and Neidle, S. (1997) *J. Mol. Biol.* **269**, 827.
91. Han, G.-W., Kopka, M.L., Cascio, D., Grzeskowiak, K. and Dickerson, R.E. (1997) *J. Mol. Biol.* **269**, 811.
92. Qiu, H., Dewan, J.C. and Seeman, N.C. (1997) *J. Mol. Biol.* **267**, 881.
93. Grzeskowiak, K., Yanagi, K., Privé, G.G. and Dickerson, R.E. (1991) *J. Biol. Chem* **266**, 8861.
94. Quintana, J.R., Grzeskowiak, K., Yanagi, K. and Dickerson, R.E. (1992) *J. Mol. Biol.* **225**, 379.
95. Yuan, H., Quintana, J.R. and Dickerson, R.E. (1992) *Biochemistry* **31**, 8009.
96. Goodsell, D.S., Kopka, M.L., Cascio, D. and Dickerson, R.E. (1993) *Proc. Natl. Acad. Sci. USA* **90**, 2930.
97. Kopka, M.L., Goodsell, D.S., Han, G.W., Chiu, T.K., Lown, J.W. and Dickerson, R.E. (1997) *Structure* **5**, 1033.
98. Vlieghe, D., Van Meervelt, L., Dautant, A., Gallois, B., Precigoux, G. and Kennard, O. (1996) *Science* **273**, 1702.
99. Spink, N., Nunn, C.M., Vojtechovsky, J., Berman, H.M. and Neidle, S. (1995) *Proc. Natl. Acad. Sci. USA* **92**, 10767.
100. Baikalov, I., Grzeskowiak, K., Yanagi, K., Quintana, J. and Dickerson, R.E. (1993) *J. Mol. Biol.* **231**, 768.
101. Goodsell, D.S., Kaczor-Grzeskowiak, M. and Dickerson, R.E. (1994) *J. Mol. Biol.* **239**, 79.
102. Shakked, Z., Guzlkevich-Guerstein, G., Frolow, F., Rabinovich, D., Joachimiak, A. and Sigler, P.B. (1994) *Nature* **368**, 469.
103. Kopka, M.L., Goodsell, D.S., Grzeskowiak, K., Baikalov, I., Cascio, D. and Dickerson, R.E. (1994) *Biochemistry* **33**, 13593.
104. Heinemann, U. and Alings, C. (1991) *EMBO J.* **10**, 35.
105. Heinemann, U. and Hahn, M. (1992) *J. Biol. Chem.* **267**, 7332.
106. Hahn, M. and Heinemann, U. (1993) *Acta Cryst.* **D49**, 468.
107. Grzeskowiak, K., Goodsell, D.S., Kaczor-Grzeskowiak, M., Cascio, D. and Dickerson, R.E. (1993) *Biochemistry* **32**, 8923.
108. Heinemann, U., Alings, C. and Bansal, M. (1992) *EMBO J.* **11**, 1931.
109. Kamitori, S. and Takusagawa, F. (1992) *J. Mol. Biol.* **225**, 445.
110. Chen, X., Ramakrishnan, B., Rao, S.T. and Sundaralingam, M. (1994) *Nature Struct. Biol.* **1**, 169.
111. Chen, X., Ramakrishnan, B. and Sundaralingam, M. (1997) *J. Mol. Biol.* **267**, 1157.

112. Miller, M., Harrison, R.W., Wlodawer, A., Appella, E. and Sussman, J.L. (1988) *Nature* **334**, 85.
113. Joshua-Tor, L., Rabinovich, D., Hope, H., Frolow, F., Appella, E. and Sussman, J.L. (1988) *Nature* **334**, 82.
114. Joshua-Tor, L., Frolow, F., Appella, E., Hope, H., Rabinovich, D. and Sussman, J.L. (1992) *J. Mol. Biol.* **225**, 397.
115. Van Meervelt, L., Vlieghe, D., Dautant, A., Gallois, B., Precigoux, G. and Kennard, O. (1995) *Nature* **374**, 742.
116. Mooers, B.H.M. and Ho, P.S. (1997) (to be published).
117. Tarim, L.W. and Secco, A.S. (1995) *Nucl. Acids Res.* **23**, 2065.
118. Wahl, M.C., Rao, S.T. and Sundaralingam, M. (1996) *Biophys. J.* **70**, 2857.
119. Cruse, W.B.T., Salisbury, S.A., Brown, T., Cosstick, R., Eckstein, F. and Kennard, O. (1986) *J. Mol. Biol.* **192**, 891.
120. Dickerson, R.E., *et al.* (1989) *EMBO J.* **8**, 1; *J. Biomol. Struct. Dynamics* **6**, 627; *Nucl. Acids Res.* **17**, 1797; *J. Mol. Biol.* **206**, 787.
121. Kabsch, W., Sander, C. and Trifonov, E.N. (1982) *Nucl. Acids Res.* **10**, 1097.
122. Calladine, C.R. (1982) *J. Mol. Biol.* **161**, 343.
123. Dickerson, R.E. (1983) *J. Mol. Biol.* **166**, 419.
124. Dickerson, R.E. (1990) in *Structure and Methods, Vol. 3: DNA and RNA*, (Sarma, R.H. and Sarma, M.H., eds), p. 1. Adenine Press, New York.
125. Pelton, J.G. and Wemmer, D.E. (1989) *Proc. Natl. Acad. Sci. USA* **86**, 5723.
126. Pelton. J.G. and Wemmer, D.E. (1990) *J. Am. Chem. Soc.* **112**, 1393.
127. Chalikian, T.V., Plum, G.E., Sarvazyan, A.P. and Breslauer, K.J. (1994) *Biochemistry* **33**, 8629.
128. Zhurkin, V.B., Lysov, Y.P. and Ivanov, V.I. (1979) *Nucl. Acids Res.* **6**, 1081.
129. Maroun, R.C. and Olson, W.K. (1988) *Biopolmers* **27**, 585.
130. Lewis, M., Chang, G., Horton, N.C., Kercher, M.A., Pace, H.C,. Schumacher, M.A., Brennan, R.G. and Lu, P. (1996) *Science* **271**, 1247.
131. Schumacher, M.A., Choi, K.Y., Zalkin, H. and Brennan, R.G. (1994) *Science* **266**, 763.
132. Kim, Y., Geiger, J.H., Hahn, S. and Sigler, P.B. (1993) *Nature* **365**, 512.
133. Kim, J.L., Nikolov, D.B. and Burley, S.K. (1993) *Nature* **365**, 520.
134. Nikolov, D.B., Chen, H., Halay, E.D., Hoffman, A., Roeder, R.G. and Burley, S.K. (1996) *Proc. Natl. Acad. Sci. USA* **93**, 4862.
135. Juo, Z.S., Chiu, T.K., Leiberman, P.M., Baikalov, I., Berk, A.J. and Dickerson, R.E. (1996) *J. Mol. Biol.* **261**, 239.
136. Yang, W. and Steitz, T.A. (1995) *Cell* **82**, 193.
137. Lavery, R. and Sklenar, H. (1988) *J. Biomol. Struct Dynamics* **6**, 63.
138. Berman, H.M., Olson, W.K., Beveridge, D.L., Westbrook, J., Gelbin, A., Demeny, T., Hsieh, S.-H., Srinivasan, A.R. and Schneider, B. (1992) *Biophys. J.* **63**, 751.
139. Dickerson, R.E., Goodsell, D.S. and Neidle, S.A. (1994) *Proc. Natl. Acad. Sci. USA* **91**, 3579.
140. Dickerson, R.E., Goodsell, D.S. and Kopka, M.L. (1996) *J. Mol. Biol.* **256**, 108.
141. El Hassan, M.A. and Calladine, C.R. (1997) *Phil. Trans. R. Soc. Lond. A* **355**, 43.
142. Calladine, C.R. and Drew, H.R. (1992, 1997) *Understanding DNA*. Academic Press, London and San Diego.
143. Calladine, C.R. and Drew, H.R. (1984) *J. Mol. Biol.* **178**, 773.
144. Beamer, L.J. and Pabo, C.O. (1992) *J. Mol. Biol.* **227**, 177.
145. Lim, W.A., Hodel, A., Sauer, R.T. and Richards, F.M. (1994) *Proc. Natl. Acad. Sci. USA* **91**, 423.
146. Aggarwal, A.K., Rodgers, D.W., Drottar, M., Ptashne, M. and Harrison, S.C. (1988) *Science* **242**, 899.

147. Shimon, L.J. and Harrison, S.C. (1993) *J. Mol. Biol.* **232**, 826.
148. Rodgers, D.W. and Harrison, S.C. (1993) *Structure* **1**, 227.
149. Mondragon, A. and Harrison, S.C. (1991) *J. Mol. Biol.* **219**, 321.
150. Schultz, S.C., Shields, G.C. and Steitz, T.A. (1991) *Science* **253**, 1001 .
151. Parkinson, G., Wilson, C., Gunasekera, A., Ebright, Y.W., Ebright, R.H. and Berman, H.M. (1996) *J. Mol. Biol.* **260**, 395.
152. Otwinowski, Z., Schevitz, R.W., Zhang, R.G., Lawson, C.L., Joachimiak, A., Marmorstein, R.Q., Luisi, B.F. and Sigler, P.B. (1988) *Nature* **335**, 321.
153. Feng, J.-A., Johnson, R.C. and Dickerson, R.E. (1994) *Science* **263**, 348.
154. Kissinger, C.R., Liu, B.S., Martin-Blanco, E., Kornberg, T.B. and Pabo, C.O. (1990) *Cell* **63**, 579.
155. Wolberger, C., Vershon, A.K., Liu, B., Johnson, A.D. and Pabo, C.O. (1991) *Cell* **67**, 517.
156. Li, T., Stark, M.R., Johnson, A.D. and Wolberger, C. (1995) *Science* **270**, 262.
157. Hirsch, J.A. and Aggarwal, A.K. (1995) *EMBO J.* **14**, 6280.
158. Klemm, J.D., Rould, M.A., Aurora, R., Herr, W. and Pabo, C.O. (1994) *Cell* **77**, 21.
159. Xu, W., Rould, M.A., Jun, S., Desplan, C. and Pabo, C.O. (1995) *Cell* **80**, 639.
160. Wilson, D.S., Guenther, B., Desplan, C. and Kuriyan, J. (1995) *Cell* **82**, 709 .
161. Kodandapani, R., Pio, F., Ni, C.Z., Piccialli, G., Klemsz, M., Mckercher, S., Maki, R.A. and Ely, K.R. (1996) *Nature* **380**, 456.
162. Koenig, P., Gioraldo, R., Chapman, L. and Rhodes, D. (1996) *Cell* **85**, 125.
163. Pavletich, N.P. and Pabo, C.O. (1991) *Science* **252**, 809.
164. Houbaviy, H.B., Ushjeva, A., Shenk, T. and Burley, S.K. (1996) *Proc. Natl. Acad. Sci. USA* **93**, 13577.
165. Fairall, L., Schwabe, J.W., Chapman, L., Finch, J.T. and Rhodes, D. (1993) *Nature* **366**, 483.
166. Cho, Y., Gorina, S., Jeffrey, P.D. and Pavletich, N.P. (1994) *Science* **265**, 346.
167. Luisi, B.F., Xu, W.X., Otwinowski, Z., Freedman, L.P., Yamamoto, K.R. and Sigler, P.B. (1991) *Nature* **352**, 497.
168. Gewirth, D.T. and Sigler, P.B. (1995) *Nature Struct. Biol.* **2**, 386.
169. Schwabe, J.W.R., Chapman, L., Finch, J.T. and Rhodes, D. (1993) *Cell* **75**, 567.
170. Marmorstein, R. and Harrison, S.C. (1994) *Genes Dev.* **8**, 2504.
171. Koenig, P. and Richmond, T.J. (1993) *J. Mol. Biol.* **233**, 139.
172. Keller, W., Konig, P. and Richmond, T.J. (1995) *J. Mol. Biol.* **254**, 657.
173. Glover, J.N.M. and Harrison, S.C. (1995) *Nature* **373**, 257.
174. Ma, P.C.M., Rould, M.A., Weintraub, H. and Pabo, C.O. (1994) *Cell* **77**, 451.
175. Kim, Y.C., Grable, J.C., Love, R., Greene, P.J. and Rosenberg, J.M. (1990) *Science* **249**, 1307.
176. Hegde, R.S., Grossman, S.R., Laimins, L.A. and Sigler, P.B. (1992) *Nature* **359**, 505.
177. Somers, W.S. and Phillips, S.E. (1992) *Nature* **359**, 387.
178. Raumann, B.E., Rould, M.A., Pabo, C.O. and Sauer, R.T. (1994) *Nature* **367**, 754.
179. Winkler, F.K., Banner, D.W., Oefner, C., Tsernoglou, D., Brown, R.S., Heathman, S.P., Bryan, R.K., Martin, P.D., Petratos, K. and Wilson, K.S. (1993) *EMBO J.* **12**, 1781.
180. Kostrewa, D. and Winkler, F.K. (1995) *Biochemistry* **34**, 683.
181. Rice, P.A., Yang, S.-W., Mizuuchi, K. and Nash, H.A. (1996) *Cell* **87**, 1295.
182. Tan, S., Hunziker, Y., Sargent, D.F. and Richmond, T.J. (1996) *Nature* **381**, 127.
183. Nikolov, D.B., Chen, H., Halay, E.D., Usheva, A.A., Hisatake, K., Lee, D.K., Roeder, R.G. and Burley, S.K. (1995) *Nature* **377**, 119.
184. Ghosh, G., van Duyne, G., Ghosh, S. and Sigler, P.B. (1995) *Nature* **373**, 303.

185. Cheng, X., Balendiran, K., Schildkraut, I. and Anderson, J.E. (1994) *EMBO J.* **13**, 3927.
186. Wobbe, C.R. and Struhl, K. (1990) *Mol. Cell. Biol.* **10**, 3859.
187. Hales, L.M., Gumport, R.I. and Gardner, J.F. (1994) *J. Bacteriol.* **176**, 2999.
188. Hales, L.M., Gumport, R.I. and Gardner, J.F. (1996) *Nucl. Acids Res.* **24**, 1780.
189. McClarin, J.A., Frederick, C.A., Wang, B.-C., Greene, P., Boyer, H.W., Grable, J. and Rosenberg, J.M. (1986) *Science* **234**, 1526.
190. Hunter, C.A. (1993) *J. Mol. Biol.* **230**, 1025.
191. Dickerson, R.E. (1998) *Nucl. Acids Res.* **26**, 1906.
192. Dickerson, R.E. and Chiu, T.K. (1998) *Biopolymers (Nucl. Acid Sci.)* **44**, 361.

7

The single-crystal structures of Z-DNA

Beth Basham, Brandt F. Eichman, and P. Shing Ho[1]

Department of Biochemistry and Biophysics, ALS 2011, Oregon State University, Corvallis, OR 97331, USA

1. Introduction

Z-DNA is a highly unique and unusual structure in biology. It is a left-handed double-helix, which, at the time of its discovery in 1979, was dramatically different from any of the known forms of DNA or RNA. Prior to the characterization of Z-DNA by crystallography, the fibre diffraction X-ray structures of naturally occurring and synthetic DNAs were all right-handed helices, with B-DNA, first described by Watson and Crick (1), and A-DNA, described by Franklin and Gosling (2) immediately afterwards, being the predominant models for the double-helix in solution and in the cell.

The discovery of Z-DNA itself was unusual in that it was the first detailed structure of any oligonucleotide to be determined by X-ray diffraction of single crystals. There had been prior spectroscopic evidence for a left-handed form of the synthetic sequence poly [d(GpC)] under high salt conditions (3), but the handedness of the structure could not be conclusively assigned until the crystal structure of d(CGCGCG) was determined by Wang *et al.* (4,5) in the laboratory of Professor Alexander Rich, and confirmed by the structure of d(CGCG) by Drew *et al.* (6) in Professor Richard Dickerson's laboratory. The single-crystal structures of A- and B-DNA were determined soon afterwards (7,8). Z-DNA was an unusual structure when first discovered in DNA crystals, since there had been no prior physical characterization, other than the circular dichroism results, or studies on the biology of this conformation.

The biology of Z-DNA is still widely debated. This problem of finding a function for a structure is difficult because it runs counter to the normal progression in biology, that is determining the structure that is responsible for a previously defined function. Since the crystal structures themselves cannot directly address this problem, we will not engage in this debate here.

Where the single-crystal structures become very important is in characterizing the physical properties of Z-DNA. Besides the crystal structures that will be described in detail here, this conformation has been extensively studied in solution by various spectroscopic and biochemical methods. Together, the results of these studies have yielded a highly detailed description of what is required to induce and stabilize Z-DNA. From the early studies, it was suggested that this unusual conformation can form only in alternating pyrimidine–purine (APP) sequences, specifically CG-rich APP sequences, and in the presence of extremely high salt concentrations. It is now known that not only are d(TA) base pairs accommodated by the structure, but non-alternating sequences can also adopt the Z conformation (Table 7.1). Furthermore,

Z-DNA can be induced to form under physiological conditions in the presence of cellular cations (e.g. polyamines) (14,30), by negative supercoiling in closed circular plasmids (31,32), or can be formed in the wake of a transcribing polymerase (33). Many of the sequence rules for the formation of Z-DNA were determined from crystallographic studies in concert with solution studies. The discussion in this chapter, therefore, will focus on these more tangible issues of how the structure of Z-DNA is affected by sequence and sequence modifications. Here, the structural effects include both the conformation of the DNA and the interactions of the DNA with the solvent.

2. The prototypical Z-DNA structure of d(CGCGCG)

To date, there are over 50 single-crystal structures of Z-DNA (Table 7.2), with lengths varying from 2 to 10 base pairs. This set includes structures with standard and modified bases, with standard Watson–Crick and mismatched base pairs, with standard and modified deoxyribose backbones, and with various cations. The data set is dominated by self-complementary APP hexanucleotide sequences that start with a dC or its methylated analogue dm^{5}C (Table 7.2) at the 5′-end of each strand. In all these structures, however, the overall conformation maintains the general features observed

Table 7.1. Conditions that affect Z-DNA crystallization and stability

Dinucleotide	$\Delta G°_T$ (kcal/mol-dn)[a]	Reference
d(CpG)	0.66	[9]
d(CpA)/d(TpG)	1.34	[10]
d(TpA)	≥2.4	[11,12]
d(CpC)/d(GpG)	2.4	[13]
d(TpC)/d(GpA)	2.5	[13]
Substituent[b]		
dBr^{5}C>dm^{5}C>dC>de^{5}C		[14–17][c]; [18,19]
dG>dI		[20,21][c]; [22]
dD>dA		[23]
dU>dT		[24,25]
Salts (cations and anions)		
Co(NH$_3$)$_6^{3+}$ >Ba^{2+} >Ca^{2+} >Mg^{2+} >Na$^+$ >Li$^+$ >NH$_4^+$		[14,26]
Spermine^{4+} >Spermidine^{3+}		[26]
Tetra-alkylammonium ion$^+$		[27]
Tetra-alkyl carboxylate$^-$		[27]
Solvents		
Alcohols: methanol, ethanol, propanol		[28]
Polyols: MPD, glycerol to stachyose		[22,28,29]

[a] B–Z transition free energies ($\Delta G°_T$) were determined in negatively supercoiled ccDNA.
[b] Modified nucleotide bases: dBr^{5}C = C5-bromodeoxycytosine; dm^{5}C = C5-methyldeoxycytosine; de^{5}C = C5-ethyldeoxycytosine; dI = deoxyinosine; dD = diaminodeoxypurine (2-aminodeoxyadenine); dU = deoxyuridine.
[c] These references refer to studies of synthetic polymers, as opposed to single crystals.

Table 7.2. Catalogue of Z-DNA crystal structures[a]

Sequence	NDB	PDB	Special features/ ions present	Space group	Resolution (Å)	R-Factor (%)	CS (molar)	Reference
d(CpG) family								
d(CpG)	zdb020	–		$P2_12_12_1$	0.8	13.6	–	[34]
d(CGCG)	zdd015	1zna	High salt, Cl⁻	$C222_1$	1.5	19.9	–	[6,35]
d(CGCG)[b]	zdd022	–	Spermine	$P6_5$	1.5	19.3	–	[36]
d(CGCG)[b]	zdd023	–		$P6_5$	1.5	21.0	–	[36]
d(CGCGCG)	zdf001	2dcg	Mg^{2+}, spermine	$P2_12_12_1$	0.9	14.0	2.19	[4]
d(CGCGCG)	zdf002	1dcg	Mg^{2+}	$P2_12_12_1$	1.0	17.5	2.00	[37]
d(CGCGCG)	zdf029	1d48	Spermine	$P2_12_12_1$	1.0	18.5	2.00	[38]
d(CGCGCG)	zdf035	131d	Spermine, –110°C	$P2_12_12_1$	1.0	18.0	2.00	[39]
d(CGCGCG)	zdf052	293d	Mg^{2+}, spermidine	$P2_12_12_1$	1.0	19.1	–	[40]
d(CGCGCG)	zdf007	–	$Ru(NH_3)_6^{3+}$	$P2_12_12_1$	1.2	20.0	–	[41]
d(CGCGCG)	zdf019	–	Mg^{2+}, $Co(NH_3)_6^{3+}$	$P2_12_12_1$	1.3	18.5	–	[42]
d(CGCGCG)	zdf044	–	Mg^{2+}, $Co^{(II)}Cl$	$P2_12_12_1$	1.5	19.6	–	[43]
d(CGCGCG)	zdf045	–	Spermine, $Co^{(II)}Cl$	$P2_12_12_1$	1.5	23.7	–	[43]
d(GCGCGCG)/ d(CCGCGCG)	zdg054	–	5'-overhang	$P2_12_12_1$	1.8	20.9	0.23	[44]
d(GCGCGCG)/ d(TCGCGCG)	zdg056	–	5'-overhang, Mg^{2+}, $Co(NH_3)_6^{3+}$	$P2_12_12_1$	1.9	19,1	0.23	[44]
d(CGCGCGCG)[b]	zdh017	–		$P6_5$	1.6	19.0	–	[45]
d(CGCGCGCGCG)[b]	zdj050	279d		$P6_522$	1.9	18.6	0.32	[46]
d(CCGCGG)	udf025	1d16		$C222_1$	1.9	18.5	0.14	[47]
Covalent modifications								
d(CGCGCG)	zdf028	1d39	$Cu^{(II)}$ Cl soaked	$P2_12_12_1$	1.2	19.8	2.00	[48]
d(Gm⁵CGCGCG)	zdgb55	–	5'-overhang, Mg^{2+}, $Co(NH_3)_6^{3+}$	$P2_12_12_1$	1.7	20.7	0.23	[44]
d(m⁵CGm⁵CGm⁵CG)	zdfb03	–		$P2_12_12_1$	1.3	15.6	0.57	[18]
d(Br⁵CGBr⁵CGBr⁵CG)	zdfb04	1dn4	291 K	$P2_12_12_1$	1.6	13.3	–	[19]

Table 7.2. *Continued*

Sequence	NDB	PDB	Special features/ions present	Space group	Resolution (Å)	R-Factor (%)	CS (molar)	Reference
d(Br⁵CGBr⁵CGBr⁵CG)	zdb05	1dn5	310 K	$P2_12_12_1$	1.4	12.5	–	[19]
d(CACGTG)	zdf008	–		$P2_12_12_1$	2.5	22.9	3.30	[49]
d(CGCACG)/	zdf038	–		$P2_1$	2.5	16.1	–	[50]
d(CGTGCG)								
d(CACGCG)/	zdf039	–		$P2_12_12_1$	1.6	19.9	0.60	[50]
d(CGCGTG)								
d(m⁵CGTAm⁵CG)	zdb06	–		$P2_12_12_1$	1.2	16.0	1.26	[51]
d(Br⁵CGATBr⁵CG)	zdb09	–		$P2_12_12_1$	1.5	19.3	–	[52]
d(m⁵CGUAm⁵CG)	zdb10	–	Cu(II)Cl soaked	$P2_12_12_1$	1.3	20.9	0.31	[53]
d(CDCGTG)	zdb11	–		$P2_12_12_1$	1.3	21.7	0.67	[23]
d(CGCm⁶GCG)	zdb21	1d24	O6-methylguanine	$P2_12_12_1$	1.9	19.0	0.71	[54]
d(m⁵CGUAm⁵CG)	zdb24	1d41		$P2_12_12_1$	1.3	20.8	0.31	[25]
d(CGCGm⁴CG)	zdb25	1da2	N4-methoxycytosine	$P2_12_12_1$	1.7	18.1	0.88	[55]
d(CGCUDCG)	zdb31	1d76		$P2_12_12_1$	1.3	13.8	0.60	[24]
d(CGCICG)	zdb34	–		$P2_12_12_1$	1.7	14.8	0.35	[56]
d(CGCGm⁵CG)	zdb36	133d	O4-methylcytosine	$P2_12_12_1$	1.8	18.9	2.29	[57]
d(m⁵CGGCm⁵CG)	–	–	Non-APP	$P2_12_12_1$	1.6	20.8	1.26	[58]
d(m⁵CGGGm⁵CG)/	zdb37	145d	Non-APP	$P2_12_12_1$	1.3	19.3	0.59	[59]
d(m⁵CGCCm⁵CG)								
d(CGCGm⁵CG)	–	–	Hemimethylated	$P2_12_12_1$	1.4	18.9	1.26	[60]
d(CGTDCG)	zdb41	210d		$P3_22_1$	1.4	17.4	1.35	[61]
d(CGTDC(Pt)G)	zdb42	211d	Platinated guanine	$P3_22_1$	1.6	17.0	1.60	[61]
d(CGCGOCG)	zdb43	223d	Oxydimethylene	$P2_12_12_1$	1.7	17.9	0.18	[62]
d(m⁵CGm⁵CGTG)	zdb48	–	BaCl₂ soaked		1.3	19.7	–	[43]
d(CGCGBr⁵CG)	zdb51	242d	Hemibrominated	$P2_12_12_1$	1.6	17.0	–	[63]
d(CGCATGCG)ᵇ	zdh016	–		$P6_5$	2.5	16.0	3.28	[45]
d(CGCICICG)	zdh030	1d53		$P6_5$	1.5	22.5	0.35	[64]
d(CGTACGTACG)	zdj018	1dn8	Co(NH₃)₆³⁺	$P6_5$	1.5	25.0	–	[65]

Table 7.2. *Continued*

Sequence	NDB	PDB	Special features/ions present	Space group	Resolution (Å)	R-Factor (%)	CS (molar)	Reference
d(CGCGCG)	zdf040		Racemic mixture	P1–	2.2	19.9	0.21	[66]
d(CGCGTG)	zdf046	–	Mg^{2+}, $Co^{(II)}Cl$	$P2_12_12_1$	1.5	21.5	–	[43]
d(CGCGTG)	zdf047	–	Mg^{2+}, $Cu^{(II)}Cl$	$P2_12_12_1$	1.5	18.0	–	[43]
Backbone modifications								
a(C)d(G)a(C)d(G)a(C)d(G)	zdfs33	–		$P6_522$	1.3	28.7	0.61	[67]
d(CG)r(CG)d(CG)	zhf026	–		$P2_12_12_1$	1.5	20.4	0.74	[68]
d(CG)a(C)d(GCG)	zdfs27	–		$P2_12_12_1$	1.5	16.7	0.61	[68]
Mismatches								
d(CGCGTG)	zdf013	–	G:T wobble base pair	$P2_12_12_1$	1.0	19.5	0.22	[69]
d(CGCGF⁵UG)	zdfb12	1dnf	F^5U:G wobble base pair	$P2_12_12_1$	1.5	17.2	2.42	[70]
d(Br⁵UGCGCG)	zdfb14	1da1	Br^5U:G base mismatch	$P2_12_12_1$	2.2	15.6	0.55	[71]

[a] NDB and PDB are the entry codes for the Nucleic Acid Database [72] and Protein Database [73], respectively. CS is the cationic strength of the solution used to crystallize the sequence, calculated as $CS = \Sigma Z_i^2 [M_i]$ (where Z is the charge and $[M]$ is the concentration of each cation species i in the crystallization setup as described in [22]).
[b] Disordered structures.

in the original Z-DNA structure of d(CGCGCG). A logical starting point in this chapter, therefore, is to discuss the conformation of d(CGCGCG) (4) as the prototypical Z-DNA structure to which all other structures will be compared. We will start by describing the gross morphology of the Z-DNA structure, which remains essentially identical in all structures of this conformation, followed by a detailed description of the helical parameters, where external and internal influences on the fine structure are observed.

2.1 The structure of Z-DNA

The two obvious features of Z-DNA that distinguish it from both A- and B-DNA are that it is a left-handed double helix and that its backbone has a characteristic zigzag pattern (Plate III). It is the zigzagged backbone that gives this form of DNA its name (4). This distinctive backbone pattern arises from an alternating conformation of the bases relative to the deoxyribose sugar, as defined by the rotation about the glycosidic bond (χ) (Fig. 7.1). In the two right-handed DNA conformations, all the bases along the chains adopt an *anti* conformation (defined as χ between 90 and 270°, but typically with values of $\chi \approx 210°$). The bases are thus extended out and away from the phosphoribose backbone. In Z-DNA, the nucleotides alternate between the standard *anti* conformation ($\langle \chi \rangle = 208°$), and the more compact *syn* conformation ($\langle \chi \rangle = 67°$) (Table 7.3) with the base essentially sitting on top of the deoxyribose ring (Plate III). The steric inhibition to pyrimidines adopting the *syn* conformation imposes the characteristic APP sequence motif commonly associated with Z-DNA. This alternating pattern of *anti/syn* nucleotide conformations, however, is strictly adhered to in all the crystal structures of Z-DNA, including those containing out-of-alternation sequences and non–Watson–Crick base pairs. Thus, it is the alternation in the backbone and not the sequence that defines Z-DNA.

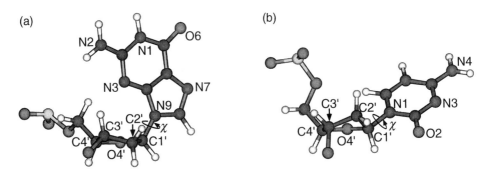

Fig. 7.1. Comparison of (A) the guanine nucleotide in the *syn* conformation and (B) the cytosine in the *anti* conformation of d(CGCGCG) as Z-DNA. The nitrogen and oxygens of each base, along with the atoms of the furanose ring of the 2′-deoxyribose sugar are labelled. The arrows show the carbons in the sugar rings that define the C3′-*endo* and C2′-*endo* sugar puckers for the *syn* guanine (A) and *anti* cytosine (B). The rotation about the glycosidic bonds that defines the *syn* and *anti* conformations of each nucleotide are labelled χ.

Table 7.3. Helical parameters of d(CGCGCG) crystallized in the presence of spermine[4+] and Mg[2+] [4]

Base	χ	Sugar pucker	Base	χ	Sugar pucker
C1	208.5	C2′-endo	G12	79.5	C2′-endo
G2	57.2	C3′-endo	C11	203.9	C2′-endo
C3	202.4	C2′-endo	G10	64.7	C3′-endo
G4	52.5	C3′-endo	C9	200.1	C2′-endo
C5	214.8	C2′-endo	G8	69.5	C3′-endo
G6	79.1	C2′-endo	C7	217.7	C2′-endo

$<\chi_{Cytosine}>$ 207.9±7.1 $<\chi_{Guanine}>$ 67.1±11.1

	Twist (Ω)	Rise (D_z)	Roll (ρ)	Tilt (τ)
d(CpG) step				
(C1:G12)/(G2:C11)	−8.5	3.8	−3.0	6.9
(C3:G10)/(G4:C9)	−9.1	3.8	3.6	1.1
(C5:G8)/(G6:C7)	−10.6	4.3	−2.1	0.7
Average	**−9.4±1.1**	**4.0±0.3**	**−0.5±3.6**	**2.9±3.5**
d(GpC) step				
(G2:C11)/(C3:G10)	−48.8	3.7	−0.8	−0.6
(G4:C9)/(C5:G8)	−51.4	3.6	0.4	0.2
Average	**−50.1±1.8**	**3.7±0.1**	**−0.2±0.8**	**−0.2±0.6**

					Displacement	
Base pair	Tip (θ)	Inclination (η)	Buckle (κ)	Propeller twist (ω)	d_x	d_y
C1:G12	3.0	6.9	0.3	0.8	3.3	2.5
G2:C11	2.1	7.5	4.8	2.1	3.1	1.9
C3:G10	−1.5	6.4	2.8	5.6	3.1	2.2
G4:C9	−1.1	6.6	5.9	3.4	3.1	2.4
C5:G8	1.0	7.3	0.1	0.6	3.5	2.0
G6:C7	0.9	7.7	4.4	3.2	3.4	1.9
Average	**0.7±1.8**	**7.1±0.5**	**3.1±2.4**	**2.6±1.9**	**−3.3±0.2**	**2.2±0.3**

All parameters were calculated with NASTE. All values are in degrees, except rise (D_z) and displacement (d_x, d_y), which are in Å.

In an antiparallel DNA double helix, there is a major groove and a minor groove. B-DNA has a deep major groove and shallow minor groove. Z-DNA has a minor groove that is a deep narrow crevice, which brings the phosphate groups of opposite strands closer together than in A- or B-DNA. In contrast, the major groove of Z-DNA is more a convex surface than a true groove and, consequently, exposes more atoms to solvent than would be expected for B-DNA (Plate III).

The conformations of the deoxyribose sugars along the phosphoribose backbone are strongly affected by the alternating *anti/syn* structure of Z-DNA. These sugar conformations are defined by how the furanose ring is distorted (or puckered) from planarity (Fig. 7.1). In B-DNA, the sugars adopt the C2'-*endo* conformation in which the C2' carbon sits above the plane (towards the base) formed by the C1', O1', and C4' atoms (see Chapter 1). In A-DNA, the sugar puckers are C3'-*endo*. The deoxyriboses of Z-DNA alternate between C3'-*endo* for nucleotides in a *syn* and C2'-*endo* for nucleotides in an *anti* conformation (Table 7.3). This alternating pattern is seen for all sequences, including non-APP sequences that place pyrimidines *syn*. In the crystal structure of d(CGCGCG), as well as other Z-DNA sequences, the 3'-terminal dG nucleotide has a C2'-*endo* sugar, even though the guanine is *syn* (Table 7.3). This end-effect and other exceptions to sugar pucker alternation most likely reflect distortions induced by the crystal lattice rather than any inherent sequence effect.

The phosphate backbone linking the sugars of each nucleotide shows two different conformations, Z_I and Z_{II}. The Z_I conformation is characterized by a pattern of alternating torsion angles along the backbone (α to ζ). The Z_{II}-form shows exceptions to this alternating pattern (most prominently at α, β, and γ), usually between the fourth and fifth base pairs of one strand of the hexamer duplex. The Z_{II} conformation rotates the phosphate out and away from the minor groove crevice at this nucleotide. This differentiates one strand from the other in the crystal for most of the structures in which the asymmetric unit is the DNA duplex; however, the bases are not dramatically affected by these deviations. The Z_{II} pattern has been attributed to crystal packing effects, and has been suggested to be stabilized by a specific pattern of waters at the interface between Z-DNA duplexes (24). The Z_I pattern is generally considered to be representative of the average structure of the Z-DNA backbone, while the existence of the Z_{II} pattern reflects the degree of flexibility in the backbone of an otherwise rigid structure.

This adherence to a characteristic zigzagged pattern in the backbone, with the nucleotides always in the alternating *anti/syn* conformation even for non-APP and non-Watson–Crick base pairs, suggests that Z-DNA is very rigid in its conformity to a structural and not to a sequence pattern. The repeating unit of Z-DNA is therefore a dinucleotide (dn) with very well-defined geometries. The zigzag pattern of the backbone results in the stacking of the base pairs in two different arrangements. A d(CpG) dinucleotide places the pyrimidines in the *anti* conformation 5' to *syn* purines along each chain (an *anti–p–syn* step), while the alternative d(GpC) dinucleotide has a *syn* purine stacked over an *anti* pyrimidine (a *syn–p–anti* step) (Plate IIIb, c). In the *anti–p–syn* dinucleotide, the bases of the pyrimidines from opposite strands are actually stacked, while the purines stack over the deoxyriboses of the adjacent pyrimidines along the same strands. It has been suggested that this latter stacking is stabilized by a favourable electrostatic interaction between the π electrons of the purine and the non-bonding electrons of the O4' oxygen of the sugar ring (74). The *syn–p–anti* step places the six-membered ring of the purine over the adjacent pyrimidine of the same strand. Thus, although an argument can be made that the *anti–p–syn* stacking arrangement is more stable, it is difficult to compare the base stacking interactions accurately because they are so different between the two stacking modes. There may be a difference imposed by the solvent interactions with Z-DNA, as will be discussed later. For now, we will treat the *anti–p–syn* dinucleotide as the repeating unit in Z-DNA. The sequence d(CGCGCG) can therefore be thought of as three stacked repeats of d(CpG) dinucleotides in the *anti–p–syn* conformation, with the interfaces being the *syn/anti* arrangements.

Thus, the overall shape of Z-DNA remains fairly consistent across all the Z-DNA crystal structures that have been determined. Factors such as sequence and solvent interactions affect the details of the structure, which are best described by the helical parameters.

2.2 The helix structure of d(CGCGCG)

We will compare the helical parameters of the various Z-DNA structures in order to elucidate the effect of any particular factor on the conformation. A set of standard definitions for helical parameters has previously been established by the Cambridge Workshop (75; see also Chapter 2); however, some of these parameters have a special meaning for Z-DNA. We have therefore developed an algorithm, NASTE (nucleic acid structure evaluation), to calculate various helical parameters specifically for Z-DNA. The algorithm first transposes all base pairs to a common frame of reference, defined by the helix axis and the normal from the helix axis to the base pair's long axis. The helical parameters of each base pair and each base step within the structures are calculated from this frame of reference. The effects of cations, sequence, base modifications, and crystal packing forces on these parameters will be discussed in this chapter.

The helical parameters for the *anti–p–syn* and *syn–p–anti* base steps of Z-DNA include the rise (D_z) and helical twist (Ω) between each base pair. The single-crystal structures of Z-DNA are long and narrow, with an average helical rise ($\langle D_z \rangle$) of 3.8 Å and a width of ~20 Å. By comparison, the $\langle D_z \rangle$ and width of B-DNA is 3.4 Å and ~24 Å, respectively. In Z-DNA , the average helical twist ($\langle \Omega \rangle$) is −30° per base pair; thus, each base pair of Z-DNA is underwound on average by −66° relative to B-DNA ($\langle \Omega \rangle$ = 36°). For the remainder of this discussion we will only be comparing Z-DNA structures, thus the terms 'overwound' and 'underwound' will refer to more and less left-handed twists (negative Ω), respectively. The repeat unit of Z-DNA, however, is the dinucleotide and thus it is more accurate to compare Ω for the distinct dinucleotide repeats in the structure. The d(CpG) step in d(CGCGCG) is characterized by $\langle \Omega \rangle$ = 9.4 ± 1.1°, while for the d(GpC) step $\langle \Omega \rangle$ = −50.1 ± 1.8° (Table 7.3), to give a total $\langle \Omega \rangle$ = −59.5° for the sum of the dinucleotide steps (or $\langle \Omega \rangle$ = 29.8° per base step).

Roll (ρ) and tilt (τ) describe the angles between adjacent base pairs along their long and short axes, respectively. Positive roll indicates that the bases open towards the major groove, and positive tilt indicates that the angle opens towards the leading strand, which is defined as the strand containing the first nucleotide (Plate III). NASTE's assignments of these parameters are consistent with the Cambridge conventions (75). The d(CpG) steps in Z-DNA have a greater average roll ($\langle \rho \rangle$ = −0.5 ± 3.6°) than the d(GpC) steps ($\langle \rho \rangle$ = −0.2 ± 0.8°) (Table 7.3). The average roll for all steps is −0.4 ± 2.6. Tilt differs slightly between d(CpG) and d(GpC) steps. The average tilt is $\langle \tau \rangle$ = 1.7 ± 2.9° for d(CGCGCG). By comparison, A-DNA tends to have a much larger roll ($\langle \rho \rangle$ = 6.3°) (76), and B-DNA tends to have only small degrees of roll ($\langle \rho \rangle$ = 0.6°) and tilt ($\langle \tau \rangle$ = 0.0°) (76).

The structures of DNA duplexes are additionally described by how each base pair is rotated along its long and short axes (the rotational helical parameters) and translated along these axes (the displacement) relative to the helix axis. The rotational helical parameters (tip and inclination) are calculated as the angle between the perpendicular to the base plane (the base normal) and the helix axis. Tip (θ) measures the rotation around the long axis of the base. In our comparisons, a positive value for tip indicates that the base pair is rotated towards the major groove. Inclination (η) measures the rotation around the short axis of the base pair, with a positive inclination reflecting a rotation towards the second strand. The average tip observed in Z-DNA (Table 7.3) is $\theta = 0.7 \pm 1.8°$ and is greater than that observed for B-DNA ($\theta = 0°$) (76), but much less than the average tip observed in A-DNA ($\theta = 11.0°$) (76). Likewise, Z-DNA has more inclination ($\eta = 7.1 \pm 0.5°$) (Table 7.3) than B-DNA ($\eta = 2.4°$) (76), but less than A-DNA ($\eta = 12.0°$) (76).

Like helical rise, displacement is a measure of translation, but in this case, translation of the position of the base pair relative to the helix axis. Displacement of the helix axis from the short axis at the centre of the base is the x-displacement (d_x), while that along the long axis is the y-displacement (d_y). Positive values of d_x reflect a translation towards the minor groove and positive d_y towards the leading strand. Base pairs in B-DNA are essentially centred on the helix axis and therefore show little or no displacement; however, in Z-DNA, the helix axis is displaced by ~4 Å into the minor groove. This can be separated into average values for $d_x = 3.3 \pm 0.2$ Å and $d_y = 2.2 \pm 0.3$ Å, respectively, for d(CGCGCG) (Table 7.3). By comparison, the helix axis of A-DNA is highly displaced towards the major groove (with $d_x = -4.5$ Å and $d_x < 0.3$ Å in the fibre structure).

Propeller twist (ω) and buckle (κ) describe the distortion from planarity of the two bases within each base pair, about the short and the long axes, respectively. In this analysis we report only the magnitude of these perturbations. The average propeller twist for Z-DNA (Table 7.3) ($\langle \omega \rangle = 2.6 \pm 1.9°$) is less than that of B-DNA ($\langle \omega \rangle = 11.0°$) (17) and A-DNA ($\langle \omega \rangle = 8.3°$) (76). The average buckle is similar between Z-DNA (Table 7.3) ($\langle \kappa \rangle = 3.1 \pm 2.4°$) and A-DNA ($\langle \kappa \rangle = 2.4°$) (76), but greater than in B-DNA ($\langle \kappa \rangle = 0.2°$) (76).

In summary, Z-DNA is a long, narrow, double helix in which the planes of the base pairs all lie essentially perpendicular to the helix axis, with the helix axis lying in the minor groove. The alternating helical twist angles reflect the distinct difference between the d(CpG) and d(GpC) dinucleotide steps.

2.3 The solvent structure of d(CGCGCG)

Both the deep narrow minor groove crevice and the convex major groove surface are important sites for Z-DNA interactions with solvent and with metal complexes. The most immediately obvious site of interaction at the major groove surface is the N7 nitrogen of the guanine bases. This is the most accessible nucleophilic group of the surface, and has been found to form covalent adducts with transition metals [e.g. copper(II) (48,53) and platinum(II) (61)]. Perhaps more important in terms of their effect on the stability of Z-DNA, however, are the hydrogen bonding interactions. The potential hydrogen bonding groups in Z-DNA are basically the same as those present in B-DNA, with the exception that the N3 nitrogens of the adenine and guanine bases are not normally accessible in the minor groove crevice of Z-DNA. The hydrogen bonding groups interact with water molecules, with ligands of solvated magnesium and sodium complexes [$Mg(H_2O)_6^{2+}$ and $Na(H_2O)_n^+$, for $n = 5$–7 (reviewed in refs 77 and 78)], with the hexammine complexes of cobalt and ruthenium [$Co(NH_3)_6^{3+}$ (42) and $Ru(NH_3)_6^{3+}$ (41)], and with the polyamines spermine (39) and spermidine (40). We will focus first on the water structure and then on cation interactions and their effects on Z-DNA structure.

The solvent organization at the major groove surface and minor groove crevice of Z-DNA has been extensively studied by Gessner *et al.* (79) for three crystal forms of d(CGCGCG) (the forms crystallized with only magnesium, only spermine, and mixed magnesium/spermine solutions). The features that are common to these three crystal structures probably represent the typical organization of solvent around the d(CG) base pairs of Z-DNA.

There are two conserved patterns of water interactions observed at the major groove surface (Fig. 7.2). These regular solvent motifs connect cytosines to cytosines and guanines to guanines across the strands. In the first motif, two waters bridge adjacent cytosines on opposite strands of the *anti–p–syn* steps [the d(CpG) dinucleotides]. This appears to be the more stable pattern of waters. The bridging water molecules are very well ordered, as indicated by their low temperature factors (average = 16.8 ± 5.1 Å^2), and fall into very well-defined geometries (with average water–cytosine hydrogen bond distances of 2.99 ± 0.15 Å, water–water distances of 2.94 ± 0.34 Å, and angles of 92.4 ± 6.5° for cytosine–water–water). The waters are not disrupted by either magnesiums or spermines in the crystal, even though hydrated magnesium complexes are located in close proximity to the connected cytosines.

The second motif at the major groove surface is formed by single waters that directly connect two guanine bases on the opposite strands of the *syn–p–anti* steps. These are less regular in structure than those that bridge the cytosines (with average hydrogen bond distances of 3.05 ± 0.36 Å between the waters and the O6 oxygen of the guanines, and guanine–water–guanine angles of 69.5 ± 5.7°). They are also readily displaced by hydrated magnesium complexes and spermine. Thus, this set of bridging waters at the *syn–p–anti* steps are less regular and apparently less stable than those at the *anti–p–syn* steps.

The minor groove crevice is lined by a continuous network of well-ordered water molecules. There are typically at least two water molecules lying in the plane of each d(CG) base pair (77). These form hydrogen bonds to the O2 keto oxygen of the cytosine and the N2 amino group of the guanine bases. The interconnected waters bound to the O2 oxygens of the cytosine bases form a continuous network, referred to as the spine of hydration. Similar spines of waters are observed in the minor grooves of B-DNA structures (80). The significance of regular networks of waters in the minor groove of DNA duplexes has previously been discussed for B-DNA (81) and is further analysed in Chapter 9. The basic conclusions from NMR studies on exchange between DNA-bound and bulk solvent were that these spines exist in solution in B-DNA (82,83) and thus can be treated as an integral part of the DNA structure (81). These same concepts are likely to apply to the hydration spine in the more rigid Z-DNA structures.

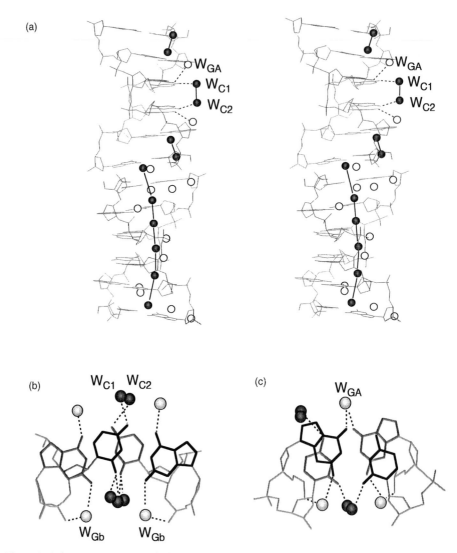

Fig. 7.2. Solvent interactions with d(CGCGCG) as Z-DNA. (A) Stereodiagram comparing the solvent structures at the major groove surface and the minor groove crevice of Z-DNA. The waters that interact at the major groove surface are shown on the upper duplex. Hydrogen bonds between each water are shown as solid lines, while hydrogen bonds from each water to the DNA surface are shown as dotted lines. Waters that bridge the stacked cytosines of the d(CpG) dinucleotide (through the N4 amino groups of the bases) are shown as dark spheres (labelled W_{C1} and W_{C2}), while waters that bridge the stacked guanines of the d(GpC) dinucleotide (through the O6 oxygen of the bases) are shown as open circles (labelled W_{GA}). Waters that interact with the minor groove crevice are shown in the lower duplex. Dark spheres represent the spine of hydration that links the cytosines (through interactions with the O2 oxygen of the bases), while those that link the guanine N2 amino groups to the phosphoribose backbone are shown as open circles. The solvent interactions with (B) the d(CpG) and (C) the d(GpC) dinucleotide steps are shown looking down the helix axis. In addition to the labels described above, W_{Gb} represents the waters that link the *syn* guanine bases to the phosphoribose backbone. Waters that are hydrogen bonded to cytosines are shown as dark spheres, while those to guanine are shown as light spheres.

The waters at the guanine bases are significant in that they bridge the N2 amino groups to the phosphate oxygens of the backbone (Fig. 7.2b). This interaction may be important for stabilizing the *syn* conformation of the guanine bases. Any perturbation to the solvent interactions in the major groove surface and minor groove crevice caused by various base substituent groups will affect the stability of Z-DNA.

2.4 Cation effects on the structure of d(CGCGCG)

Z-DNA has been crystallized in the presence of several different types of cations, including magnesium and the polyamines spermine and, most recently, spermidine. The effect that cations have on the structure of Z-DNA is significant because the cations help to stabilize the left-handed structure in solution by screening, and thus shielding, the negatively charged phosphates. The phosphate–phosphate distances are closer in Z-DNA than in either A- or B-DNA because of the narrow minor groove crevice. The effect on the stability of Z-DNA is dependent on both the concentration and the charge of the cation, with higher charged ions being more effective at stabilizing this conformation. The stabilization of Z-DNA in solution follows the trend spermine^{4+} > spermidine^{3+} > Mg^{2+} > Na^{+} (14). In particular, the stability of different sequences as Z-DNA is dependent on the the cation strength of a solution ($CS = \Sigma Z_i^2[M_i]$, where Z_i is the charge and (M_i) is the concentration of the cation type i). This relationship has been used as a quantitative method to predict the solutions for crystallizing different sequences as Z-DNA (22).

The polyamines have been extensively studied because they are known to aid in DNA condensation and to prevent thermal denaturation of the duplex (84). Levels of polyamines are highly dependent on the cell cycle, and are perturbed in cancer cells (reviewed in ref.85). For these reasons, polyamine binding has been of interest not only to biologists, but also to crystallographers because of the analogy between crystallization and condensation.

Four crystal structures of d(CGCGCG) have been analysed to determine the effect of polyamines and magnesium on the structure of Z-DNA. These structures include the magnesium only (MG) form (37), the spermine only (SP) form (38), the mixed magnesium and spermine (MGSP) form (4), and the mixed magnesium and spermidine (MGSD) form (40). Although all these crystals were grown in the presence of sodium ions, it is the interactions of the multivalent cations (specifically Mg^{2+} versus the two polyamines) that will be discussed here.

The reference d(CGCGCG) structure to which we have been referring is the original MGSP form (4). The structures of the DNA in both the MG, MGSP, and MGSD forms are nearly identical in all respects, except for the ligand interactions (Table 7.4). Thus, although the polyamines are more effective at stabilizing Z-DNA in solution, the crystal structures appear to be determined by the presence of magnesium. We will, therefore, treat the MG form as the reference for comparison, with the realization that the MGSP and MGSD forms are very similar to this.

The observed lattice of the SP crystal is different from that of the other d(CGCGCG) crystals, suggesting that this DNA structure is significantly different from the reference structure (Table 7.4). The DNA in the SP lattice is rotated by 70° around the helix axis, shifted by 3 Å along the helix axis, and rotated around the

Table 7.4. Effect of cations on the Z-DNA structure of d(CGCGCG)

	Crystal form			
	MG	MGSP	MGSD	SP
Twist (Ω)				
(C1:G12)/(G2:C11)	−9.2	−8.5	−9.0	−11.6
(G2:C11)/(C3?G10)	−48.9	−48.8	−48.8	−47.7
(C3:G10)/(G4:C9)	−9.4	−9.1	−8.7	−11.7
(G4:C9)/(C5:G8)	−50.8	−51.4	−51.6	−49.3
(C5:G8)/(G6:C7)	−12.2	−10.6	−11.6	−12.3
Average d(CpG)	−10.3 ± 1.7	−9.4 ± 1.1	−9.8 ± 1.6	−11.9 ± 0.4
Average d(GpC)	−49.9 ± 1.3	−50.1 ± 1.8	−50.2 ± 2.0	−48.5 ± 1.1
Rise (D_z)				
(C1:G12)/(G2:C11)	3.8	3.8	3.9	3.9
(G2:C11)/(C3:G10)	3.6	3.7	3.7	3.7
(C3:G10)/(G4:C9)	3.9	3.8	3.8	3.4
(G4:C9)/(C5:G8)	3.5	3.6	3.6	3.6
(C5:G8)/G6:C7)	4.1	4.3	4.1	3.4
Average	3.8 ± 0.2	3.8 ± 0.3	3.8 ± 0.2	3.6 ± 0.2
Roll (ρ)				
(C1:G12)/G2:C11)	−0.8	−3.0	−1.6	2.5
(G2:C11)/(C3:G10)	−1.5	−0.8	−0.9	−4.6
(C3:G10)/G4:C9)	−1.1	3.6	−2.1	1.2
(G4:C9)/C5:G8)	0.3	0.4	0.0	−1.9
(C5:G8)/(G6:C7)	3.6	−2.1	1.4	5.6
Average	0.1 ± 2.1	−0.4 ± 2.6	−0.6 ± 1.4	0.6 ± 3.9
Inclination (η)				
C1:G12	6.0	6.9	6.2	5.7
G2:C11	5.9	7.5	7.1	4.1
C3:G10	6.8	6.4	7.5	1.4
G4:C9	7.4	6.6	7.5	1.4
C5:G8	8.1	7.3	9.3	2.3
G6:C7	7.4	7.7	8.1	0.6
Average	6.9 ± 0.9	7.1 ± 0.5	7.6 ± 1.0	2.8 ± 1.9
x-displacement (d_x)				
C1:G12	−3.0	−3.3	−3.1	−4.8
G2:C11	−3.1	−3.1	−3.1	−4.8
C3:G10	−3.3	−3.1	−3.2	−3.9
G4:C9	−3.3	−3.1	−3.2	−3.4
C5:G8	−3.5	−3.5	−3.5	−3.8
G6:C7	−3.4	−3.4	−3.4	−4.2
Average	−3.3 ± 0.2	−3.3 ± 0.2	−3.3 ± 0.2	−4.2 ± 0.6

All values are in degrees, except rise (D_z) and displacement (d_x), which are in Å. MG refers to the crystal grown in the presence of Mg^{2+} only (37), MGSP is the crystal grown with MG^{2+} and spermine (4), MGSD is the crystal grown with Mg^{2+} and spermidine (40), and SP is the crystal grown with spermine only (38).

intramolecular pseudo-twofold axis compared with the DNA in the MG and MGSP crystals (38). The most obvious difference between the Z-DNA structure of the SP form is the shorter $\langle D_Z \rangle$ (3.6 Å) and larger $\langle d_X \rangle$ (−4.2 Å) compared with the MG structure (3.8 and −3.3 Å, respectively) (Table 7.4). As a result, the SP structure of Z-DNA is shorter and wider than the reference conformation. This is proabably a result of the binding of spermine to the major and minor grooves. The $\langle \Omega \rangle$ shows that the d(CpG) steps are overwound (by ~2°) while the d(GpC) steps are underwound (by ~1.0°) in the SP structure relative to the reference MG structure (Table 7.4). The compensating under- and over-winding of the dinucleotide steps renders the overall $\langle \Omega \rangle$ of the structure identical to that of the other Z-DNA structures. Finally, the SP form shows a slight increase in roll and a dramatic decrease in the inclination of the base pairs (Table 7.4).

In order to understand the effect of the cations on Z-DNA stability, we must first characterize the specific interactions that the cations and their ligands make with the DNA. Starting with the reference MG form, there are four unique hydrated magnesium clusters that were observed to bind the DNA duplex (Fig. 7.3). This

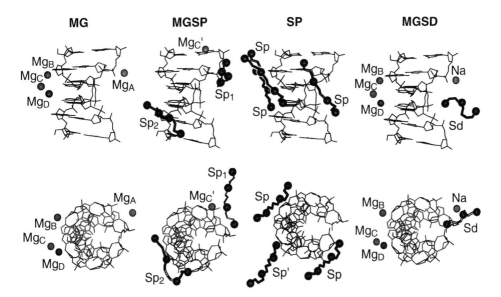

Fig. 7.3. Comparison of the cation interactions between the magnesium only (MG), mixed magnesium/spermine (MGSP), spermine only (SP), and mixed magnesium/spermidine (MGSD) forms of d(CGCGCG). Views perpendicular to (top) and down the helix axes (bottom) of each structure are shown. In the structures of the polyamines (spermine and spermidine), the nitrogen atoms are shown as spheres. In the MG form of the structure, each unique magnesium ion (waters not shown) is labelled as Mg_A, Mg_B, Mg_C, and Mg_D. The two unique spermine molecules of the MGSP form are labelled Sp_1 and Sp_2, while the single magnesium (which is symmetry related to Mg_C of the MG form) is labelled Mg_C'. Although there is only one unique spermine in the SP form, it makes three different interactions with each duplex. These three types of interactions are shown. Finally, the single unique spermidine (Sd), the three magnesiums (identical to Mg_B, Mg_C, and Mg_D of the MG form), and the cation identified as a sodium (labelled as Na, but similar in position to Mg_A of the MG form) are shown for the MGSD form of d(CGCGCG).

Table 7.5. Hydrogen bonding contacts of the four unique magnesium ions in the MG form of d(CGCGCG)

Cation	Residue	Atom
Mg_A	G6(s)	PO, O4'
	G8	PO, N2
	C9	PO
	G10	PO
Mg_B	G4(s)	O6
	C5(s)	PO
	G6	PO
	G8(s)	N7
	G10	PO
	C11	PO(w)
Mg_C	C1(s)	N4(w)
	C5	PO(w)
	G6(s)	N7, O6(w)
	C9(s)	N4, PO(w)
		Mg_D(w)
Mg_D	C1(s)	N4
	G6	O6
	C9(s)	N4, PO
	G10(s)	O6, N7(w)
	G12(s)	O6
		Mg_C(w)

Interactions of the ion with adjacent, symmetry-related residues is indicated by the designation (s), while (w) indicates that this contact is mediated through a coordinating water molecule.

does not entirely neutralize the net -10 charge of the phosphoribose backbone in d(CGCGCG), requiring either one additional magnesium or two sodium ions that cannot be observed in the crystal structure. The observed ions, however, represent the specific interactions. The hexahydrated Mg_A (Table 7.5) makes six hydrogen bonding contacts with the DNA. It interacts with a phosphate oxygen of G8, C9, G10, and of the G6 of a neighbouring duplex, as well as with the N2 of G8 and the O4' of the G6 of the neighbouring duplex. Mg_B (Table 7.5) is also hexahydrated and makes contacts with a phosphate oxygen of G6, G10, and C11, although the contact with the C11 oxygen is mediated by a water molecule. Three contacts are made with neighbouring duplexes. These are with the O6 of G4, the phosphate oxygen of C5, and the N7 of G8. Mg^{2+} complexes Mg_C and Mg_D (Table 7.5) are linked together and share two water ligands. One of these ligands binds to N4 of C1 (a neighbouring duplex) and additional water molecules mediate contacts with the N4 of C9 and the O6 of G6 on a neighbouring duplex. Additionally, an Mg_C ligand makes contact with the N7 of the same G6. Additional contacts of Mg_C ligands include water-mediated interactions with the phosphate oxygen of C5 and the N4 of C1 in a neighbouring duplex. Mg_D has additional interactions with the DNA, specifically with the O6 of G10 and a water-mediated contact with the C9 phosphate oxygen. It also makes contacts with two other duplexes, namely the O6 of G12 of one duplex and the O6 of G6 of another duplex. Thus, these Mg^{2+} complexes not only provide intramolecular stabilization of Z-DNA, but also stabilize the crystal through intermolecular interactions.

In the MGSD form, there is a single unique spermidine per duplex that displaces Mg_A, and the remaining three divalent cations are unperturbed (Fig. 7.3). The

Table 7.6. Hydrogen bonding contacts of polyamines and magnesium with the MGSP, MGSD, and SP forms of d(CGCGCG)[a]

Structure	Cation	Residue	Atom	MG contact equivalent
MGSP[b]	Mg_C	G6	N7	C(s)
	Sp_1	G4	N7	–
		C5(s)	PO	B,C
		G6(s)	PO, PO(w), O4'	A,B
		G8	O6	–
		G8(s)	PO(w)	A
		C9(s)	PO	A,C,D
		G10(s)	PO(w)	A,B
	Sp_2	C1(s)	5'-OH	–
		G2	N7	–
		G2(s)	N7, O6(w)	–
		C3	O6, N4(w)	–
		G10	O6, N7	D
		C11	N4	–
		C11(s)	PO	–
		G12	O6	D
MGSD[c]	Mg_B			B
	Mg_C			C
	Mg_D			D
	Sd	C3(s)	PO(w)	–
		G6	PO(w)	A
		G12	PO(w)	–
SP[d]	Sp	C3	PO	–
		G8	N7	B
		C9(s)	PO	A,C,D
		G10(s)	N7, O6	D
		C11(s)	PO	B
		G12(s)	PO	–

[a] Sp and Sd denote the polyamines spermine[4+] and spermidine[3+], respectively. PO denotes a phosphate oxygen. Interactions of the ion with adjacent, symmetry related residues is indicated by the designation (s), while (w) indicates that this contact is mediated through a coordinating water molecule. 'MG contact equivalent' refers to the analogous Mg^{2+} complex of the MG structure.
[b] MGSP refers to the crystal grown in the presence of Mg^{2+} and spermine.
[c] MGSD is the crystal structure of Z-DNA with Mg^{2+} and spermidine.
[d] SP is the structure that was crystallized only with spermine.

spermidine itself interacts with the phosphoribose backbone of two adjacent duplexes in the crystal lattice. These are all mediated by water bridges between the amino nitrogens of the ligand and the oxygens of the phosphates. Specifically, these interactions (Table 7.6) are with phosphate oxygens of C3, G6, and G12. The C3 interaction is with a neighbouring duplex. The interaction with G6 is equivalent to a contact made by Mg_A in the other structures. Interestingly, the amino groups of a truncated analogue [N-(2-amino-ethyl)-1,4-diaminobutane] of spermidine [spermindine is N-(2-amino-propyl)-1,4-diaminobutane] bind directly to the phosphates, and show direct interactions with the bases at the major groove surface (40).

Similarly, in the mixed MGSP form, one of the original Mg^{2+} clusters remains in place, but in this case the complexes Mg_A, Mg_B, and Mg_D are displaced by the polyamines (Fig. 7.3). There are two spermines per duplex in the asymmetric unit, each interacting with three DNA duplexes. Spermine$_1$ makes two contacts with the DNA (Table 7.6), one with the N7 of G4 and the other with the O6 of G8. The remainder of the interactions are with other duplexes. These interactions are with the phosphate oxygens of C5, G6, and C9 and the O4$'$ of G6. There are also water-mediated contacts to phoshate oxygens G6, G8, and G10. The amino groups of spermine$_2$ also make numerous contacts with the DNA and neighbouring duplexes (Table 7.6). Direct interactions include those with the N7 of G2, the O6 of G10 and G12, and the N4 of C3 and C11 (although the interaction with C3 is water mediated). Interactions with other duplexes include the 5$'$-OH of C1, the phosphate oxygen of C11, the N7 of G2, and a water-mediated contact with the O6 of G2. The interactions of spermine with the DNA are similar to those observed in the MG structure and, in fact, many of the spermine contacts are equivalent to those seen with Mg^{2+} in the MG form (Table 7.6).

In the SP crystal, there is one spermine per duplex. Each spermine interacts with three different DNA molecules, and three spermines interact with each DNA molecule (Fig. 7.3). This large number of interactions between the polyamine and the DNA is consistent with spermine's ability to condense DNA (38). The spermines interact with the DNA as follows (Fig. 7.3), one binds in the major groove of the DNA, the second interacts with the phosphates along the minor groove, and the third interacts with only the C9 and G10 of the DNA. Direct contacts are made between the phosphate oxygen of C3 and the N7 of G8. Interactions with neighbouring duplexes include hydrogen bonds with the phosphate oxygens of C9, C11, and G12 as well as interactions with the N7 and O6 of G10 (Table 7.6). These interactions are common with interactions observed between all four Mg^{2+} and the DNA in the MG structure (Table 7.6). However, the interactions between spermine and the DNA duplex in the SP form versus the mixed cation MGSP form are not identical.

In summary, the cations make similar contacts with the DNA across the different structures, and display both intra- and inter-duplex interactions, which often involve the coordination of bridging water molecules. When comparing the concentration of cations required to crystallize each of these forms of d(CGCGCG), it became evident that spermine had twice the effect expected relative to other cations. We therefore simply increased the effective *CS* for spermine on Z-DNA crystallization by a factor of 2 (this has already been incorporated into the *CS* values in Table 7.2). This is an empirical observation, but may be related to the base-specific interactions of this polyamine with Z-DNA.

The binding of spermine to supercoil-induced Z-DNA in closed circular DNA plasmids (86) appears to be consistent with that observed in the crystal. The association constant of spermine for Z-DNA [1.5×10^7 M^{-1} for d(CpG) and 1.2×10^8 M^{-1} for d(CpA/TpG) dinucleotides] is $\geq$ 100-fold greater than that for B-DNA (1.4×10^5 M^{-1}), consistent with the stabilizing effect that this polyamine has on the left-handed conformation. The size of the spermine-binding site for Z-DNA was determined to be 10.4 d(CG) base pairs. This is larger than that observed in the crystal structure of the SP form of d(CGCGCG) (1 spermine per duplex, or 6 bp/spermine). However, the crystal structure may overexaggerate the number of ligands actually bound to the DNA. The temperature factors for the spermines are about twice that observed for the DNA, even at $-100°$C. This suggests that the spermine may not be fully occupied and, therefore, the number of ligands bound per duplex is likely to be significantly less than 1. This would give an overall binding size that is more consistent with the results from the solution studies.

2.5 *Length effects on the structure of d(CpG) sequences as Z-DNA*

The structures of alternating d(CpG) dinucleotides as Z-DNA have been determined for five different lengths of duplexes, from a single dinucleotide in the structure of d(CpG) to five dinucleotides in d(GCGCGCGCGC) (Table 7.2). A comparison of lengths shorter than that of the hexamer d(CGCGCG) allows us to determine whether the conformation of the *anti–p–syn* stacking in d(CpG) is inherent to this dinucleotide in the absence of significant flanking base pairs. Comparisons of longer sequences address the questions of whether the structure of d(CGCGCG), or any hexanucleotide, can indeed be extrapolated to longer and even infinite lengths of Z-DNA, and whether the *anti–p–syn* dinucleotide of d(CpG) is the stable repeating unit of Z-DNA.

The overall conformations of the structures in this comparison are all very similar to one another and to d(CGCGCG), with just a few exceptions. One interesting feature that is common to all structures is the presence of the Z_{II} backbone conformation. The crystal lattice interactions that are associated with this perturbation in the reference d(CGCGCG) structure are not identical across this set of structures. It is unclear then which specific crystal lattice interactions are directly responsible for this backbone conformation.

Within this set of nine d(CpG)$_n$ Z-DNA sequences (Table 7.2), the heptamers d(GCGCGCG),d(CCGCGCG) and d(GCGCGCG),d(TCGCGCG) do not have blunt ends (44). These structures are essentially that of d(CGCGCG) as Z-DNA, with nucleotides dangling from the 5'-ends of each strand. These orphaned nucleotides pair between adjacent duplexes to form reverse Watson–Crick d(GC) and reverse wobble d(GT) base pairs that are sandwiched between two stacked d(CGCGCG) Z-DNA duplexes. These can therefore be treated as variations on the reference d(CGCGCG) structure in which the Z-DNA pattern is disrupted at the ends, serving as a true indicator of end-effects. In this case, the overall structure of the Z-DNA duplexed region is remarkably similar to that of the reference d(CGCGCG) in all respects (Table 7.7). The terminal base pairs of the duplex region (C1:G12 and G6:C7, where the nucleotides are numbered according to the duplex Z-DNA regions only, ignoring the 5′-overhangs) show a larger buckle than found in any of the d(CG) base pairs of d(CGCGCG); but, otherwise, all helical parameters are reproduced, including the average helical twist at each dinucleotide step and even the C2′-*endo* sugar pucker of the 3′-terminal guanine that breaks the alternating sugar conformation along each strand. The notable exceptions are the shorter rise and greater buckle in the corresponding terminal d(CG) base pairs in the heptamer structures. Thus, these appear to be true end-distortions associated with the lack of a Z-DNA-like d(GpC) step between duplexes in the crystal lattice.

Unlike d(CGCGCG), the short duplexes of d(CG) and d(CGCG) do not stack end-to-end to form pseudo-continuous helices. There are no internal d(CpG) steps in either structure, and only a single internal d(GpC) step in the tetramer structure. We would expect, therefore, that these structures are essentially 'all ends'. The dimer has a twist, rise, and tilt comparable with the first d(CpG) step of the heptamer, making it less left-handed and shorter than comparable steps in d(CGCGCG) (Table 7.7). Thus the structure of d(CpG) is that of unconstrained Z-DNA ends. There are additional distortions to the dimer, such as the significant roll and propeller twist between and within the base pairs. These, however, may be related to the unusual ammonium cation present in this crystal that is not present in other Z-DNA structures.

The d(CpG) steps in the tetramer structure of d(CGCG) are the most overwound of all the structures, with $\langle \Omega \rangle = -13°$. The single d(GpC) step, however, is significantly underwound, compensating for the overwound d(CpG) and resulting in an $\langle \Omega \rangle = -29.4°$ per base step that is almost identical to that of the d(CGCGCG) structures (Table 7.7). It should be noted that only the high salt, orthorhombic form of d(CGCG) (35) was available for this comparison. However, the structure of d(CCGCGG) has the 5′-terminal cytosine nucleotide flipped out to an extrahelical conformation and, thus, can be treated as a tetramer of four central Z-DNA base pairs, analogous to the treatment of the heptamers as six Z-DNA duplex base pairs with unusual ends. With respect to Ω, the tetramer within d(CCGCGG) is more similar to d(CGCGCG), particularly the MGSP form, than to that of d(CGCG). The similarity between the tetramer structures lies in the high negative roll of both the d(CpG) and d(GpC) base steps, a high negative tilt in the d(CpG) steps, large variations in the tip and inclination at each base pair, and large variations in propeller twist and buckle within each base pair (Table 7.7). These distortions are evidently associated with this short length of the duplex and, again, may reflect the structure of Z-DNA ends, as opposed to the internal dinucleotides that one would expect in longer sequences.

There are a number of octanucleotide Z-DNA structures that have been determined, including that of d(CGCGCGCG), but they are all in disordered lattices. The only reliable parameter that we can determine from this structure is the average helical rise per base pair (3.6 Å/bp), which was calculated from the length of the helical axis (the crystallographic *c*-axis) of 43.6 Å for six base pairs (45). This is shorter than the average for the alternating d(CG) tetramer and hexamer sequences.

The longest Z-DNA duplex crystal structure solved to date is that of the decamer d(GCGCGCGCGC). This sequence is unusual in that it starts with a guanine

Table 7.7. Helical base step and base pair parameters of d(CG)$_n$ sequences that crystallize as Z-DNA[a]

	d(CG)	d(CGCG)[b]	d(CCGCGG)	d(CGCGCG) MGSP	d(GCGCGCG)/ d(TCGCGCG)	d(GCGCGCG)/ d(CCGCGCG)	d(GCGGGCGCGC)[c]
d(CpG) steps							
Twist (Ω)							
(C1:G12)/(G2:C11)	−7.4	−12.3	−7.8	−8.5	−7.4	−6.5	−9.7
(C3:G10)/(G4:C9)		−13.6	−9.2	−9.1	−10.3	−10.5	
(C5:G8)/(G6:C7)				−10.6	−10.6	−9.8	
Average	−7.4	−13.0 ± 0.9	−8.5 ± 1.0	−9.4 ± 1.1	−9.4 ± 1.8	−8.9 ± 2.1	−9.7
Rise (D$_z$)							
(C1:G12)/(G2:C11)	3.2	3.8	3.8	3.8	3.4	3.4	3.9
(C3:G10)/(G4:C9)		3.7	3.8	3.8	4.1	3.9	
(C5:G8)/(G6:C7)				4.3	3.8	3.9	
Average	3.2	3.8 ± 0.1	3.8	4.0 ± 0.3	3.8 ± 0.4	3.7 ± 0.3	3.9
Roll (ρ)							
(C1:G12)/(G2:C11)	−13.7	−7.1	−9.0	−3.0	0.6	0.9	3.2
(C3:G10)/(G4:C9)		−0.7	−4.0	3.6	4.7	−3.7	
(C5:G8)/(G6:C7)				−2.1	2.2	2.0	
Average[d]	13.7	3.9 ± 4.5	6.5 ± 3.5	2.9 ± 0.8	2.5 ± 2.1	2.2 ± 1.4	3.2
Tilt (τ)							
(C1:G12)/(G2:C11)	−5.8	−10.1	−8.4	6.9	−5.8	−4.4	0.0
(C3:G10)/(G4:C9)		−3.4	0.7	1.1	0.2	0.6	
(C5:G8)/(G6:C7)				0.7	−2.7	−0.1	
Average[d]	5.8	6.8 ± 4.7	4.6 ± 5.4	2.9 ± 3.5	2.9 ± 2.8	1.7 ± 2.4	0.0
d(GpC) steps							
Twist (Ω)							
(G2:C11)/(C3:G10)		−45.8	−44.0	−48.8	−47.5	−47.5	−50.3
(G4:C9)/(C5:G8)				−51.4	−47.1	−48.0	
Average		−45.8	−44.0	−50.1 ± 1.8	−47.3 ± 0.3	−47.8 ± 0.4	−50.3

Table 7.7. *Continued*

	d(CG)	d(CG<u>CG</u>)[b]	d(CC<u>GC</u>GG)	d(CGCGCG) MGSP	d(GG<u>GCGCG</u>)/ d(T<u>CGGCGG</u>)	d(GG<u>CGCGCG</u>)/ d(CC<u>GCGCG</u>)	d(GCG<u>GCGCGCGC</u>)[c]
Rise (D_z)							
(G2:C11)/(C3:G10)		3.7	3.7	3.7	3.6	3.7	3.2
(G4:C9)/(C5:G8)				3.6	3.7	3.6	
Average		3.7	3.7	3.7±0.1	3.7±0.1	3.7±0.1	3.2
Roll (ρ)							
(G2:C11)/(C3:G10)		-4.2	-6.1	-0.8	-1.8	-1.8	-3.2
(G4:C9)/(C5:G8)				0.4	-3.7	0.1	
Average[d]		4.2	6.1	0.6±0.8	2.8±1.3	0.9±1.3	3.2
Tilt (τ)							
(G2:C11)/(C3:G10)		4.3	-1.5	-0.6	1.9	1.1	0.0
(G4:C9)/(C5:G8)				0.2	1.0	2.6	
Average[d]		4.3	1.5	0.4±0.6	1.5±0.6	1.9±1.1	0.0
Base pairs							
Tip (θ)							
C1:G12	13.7	7.1	9.0	3.0	-0.6	-0.9	-1.6
G2:C11	1.7	2.8	3.0	2.1	1.2	0.9	1.6
C3:G10		2.1	-1.0	-1.5	-3.6	-2.8	
G4:C9		-7.4	-6.2	-1.1	0.1	-2.9	
C5:G8				1.0	-2.1	-0.9	
G6:C7				0.9	4.6	5.5	
Average[d]	7.7±8.5	4.9±2.8	4.8±3.5	1.6±0.8	2.0±1.8	2.3±1.8	1.6
Inclination (η)							
C1:G12	5.8	10.1	8.4	6.9	5.8	4.4	-2.7
G2:C11	2.2	5.9	7.0	7.5	3.8	3.2	-2.7
C3:G10		9.2	6.2	6.4	4.0	3.8	
G4:C9		4.9	7.4	6.6	5.0	6.4	

Table 7.7. *Continued*

	d(CG)	d(CGCG)[b]	d(CCGCGG)	d(CGCGCG) MGSP	d(GCGCGCG)/ d(TCGCGCG)	d(GCGCGCG)/ d(CCGCGCG)	d(GCGCGCGCGC)[c]
C5:G8				7.3	7.7	6.4	
G6:C7				7.7	5.4	2.6	
Average[d]	4.0 ± 2.5	7.5 ± 2.5	7.3 ± 0.9	7.1 ± 0.5	5.3 ± 1.4	4.5 ± 1.6	2.7
Propeller twist (ω)							
C1:G12	10.6	1.9	3.2	0.8	-0.7	6.3	1.5
G2:C11	-15.7	2.0	-7.7	2.1	2.2	0.1	1.5
C3:G10		3.6	3.7	5.6	-0.9	-0.7	
G4:C9		4.2	0.9	3.4	1.2	-0.9	
C5:G8				0.6	-3.6	2.2	
G6:C7				3.2	5.4	0.1	
Average[d]	13.1 ± 3.6	2.9 ± 1.2	3.9 ± 2.8	2.6 ± 1.9	2.3 ± 1.8	1.7 ± 2.4	1.5
Buckle (κ)							
C1:G12	4.6	8.5	2.2	0.3	10.4	10.0	-10.5
G2:C11	1.2	-2.8	-6.6	-4.8	-5.9	-3.5	10.5
C3:G10		8.9	4.9	2.8	0.1	0.0	
G4:C9		-2.0	-1.5	-5.9	-0.7	-2.1	
C5:G8				0.1	8.2	3.8	
G6:C7				4.4	-8.3	-1.3	
Average[d]	2.9 ± 2.4	5.6 ± 3.7	3.8 ± 2.4	3.1 ± 2.4	5.6 ± 4.3	3.5 ± 3.5	10.5

[a] Underlined sequences denote the d(CpG) dinucleotides in the standard Z-DNA duplex. The numbering of residues refers only to those nucleotides in the duplex. All values are in degrees, except rise (D_z) and displacement (d_x), which are in Å. MGSP refers to the form of d(CGCGCG) crystallized in the presence of magnesium and spermine.

[b] d(CGCG) refers to the high salt orthorhombic form [6].

[c] Only one value for each type of base step in the d(GCGCGCGCGC) decamer is shown. These values are repeated throughout the decamer because of the dinucleotide asymmetric unit.

[d] Averages of the magnitudes of roll, tilt, tip, inclination, propeller twist, and buckle are shown, and were calculated by Ave = $(\Sigma |q_i|)/i$, where q_i is the value of that parameter at base step i.

nucleotide and thus there are more d(GpC) steps (5) than d(CpG) steps (4). Shorter alternating d(GpC) sequences that have been solved crystallographically [d(Gm⁵CGCGC) and d(Gm⁵CGm⁵CGCGC)] were in the A-form (87). The unmethylated versions of these hexamer and octamer sequences crystallize, but are highly disordered, with the octamer showing a strong Bragg reflection at 3.4 Å resolution, suggesting that it is probably in the B-form. Thus, it appears that Z-DNA is not the preferred form in alternating d(GC)$_n$ sequences until $n \geq 5$ dinucleotides. This is consistent with the solution studies of Quadrafoglio *et al.* (88),which showed that oligonucleotides of d(GC)$_n$ are left-handed only in longer sequences ($n > 7$ dinucleotides), while shorter sequences ($3 < n < 7$ dinucleotides) remain right-handed even under dehydrating conditions. Thus, it does appear that the d(CpG) step is the significant determinant for Z-DNA formation, and in oligonucleotides where the d(GpC) steps would be dominant, the left-handed conformation is not stable. In longer sequences, the number of destabilizing d(GpC) and stabilizing d(CpG) dinucleotides become equalized, allowing Z-DNA to form.

Unfortunately, the structure of d(GCGCGCGCGC) shows positional disorder and therefore end-effects could not be distinguished from the remainder of the structure. Still, the average values for the helical parameters can be compared to the averages for the shorter DNA lengths. For the most part, this decamer is very similar to the hexamers. The most interesting deviation is that the rise at the d(GpC) step is significantly shorter (by ~0.5 Å) than in the hexamer or tetramer structures, to give $\langle D_z \rangle = 3.2$ Å (Table 7.7). When compared with the $\langle D_z \rangle$ of the octamer structure, which is intermediate between the shorter (tetramer and hexamer sequences) and this longer sequence, the compressed rise appears to be length dependent and suggests that the shorter sequences have an elongated d(GpC) step. Alternatively, the shorter $\langle D_z \rangle$ of the octa- and deca-nucleotide structures may be related to the crystal lattice since both are in disordered hexagonal space groups. In support of this, the disordered d(CGCG) tetramers crystallize in hexagonal space groups and show shorter rises (3.61 to 3.67 Å) when determined from the lengths of the helix axes (36).

When comparing all these lengths of alternating d(CG) sequences as Z-DNA, the crystal structures of the hexanucleotides appear indeed to be a reasonable model for long, and perhaps even infinite, lengths of Z-DNA. The helical twist for Z-DNA is very consistent, at ~−30° per base step for all lengths. The rise at the Z-DNA stabilizing d(CpG) steps is ~3.8 Å, but may be slightly exaggerated in the shorter sequences at the d(GpC) step. The base pairs are all nearly perpendicular to the helix axis, with very little distortion to the base pair plane (as would be expected for this rigid structure). It is also clear that the base pairs at the ends of a Z-DNA stretch (as typified by the dimer and tetramer structures, and the ends of the heptamer structures) are more variable in structure.

3. Sequence and substituent efffects on the structure and stability of Z-DNA

The tendency of dinucleotides to form Z-DNA is as follows, d(m⁵CpG) > d(CpG) > d(CpA)/d(TpG) > d(TpA) (77). In order to understand the structural basis behind

d(C·G) type base pairs

R_A= H	=> Cytosine (C)
CH$_3$	=> 5-methyl Cytosine (m^{5}C)
Br	=> 5-bromo Cytosine (Br^{5}C)

| R_B = NH$_2$ | => Guanine (G) |
| H | => Inosine (I) |

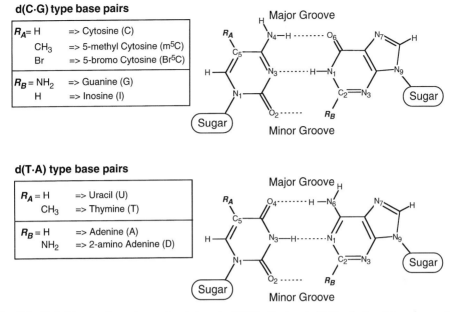

d(T·A) type base pairs

| R_A = H | => Uracil (U) |
| CH$_3$ | => Thymine (T) |

| R_B = H | => Adenine (A) |
| NH$_2$ | => 2-amino Adenine (D) |

Fig. 7.4. Definitions and structures of variations in d(CG)- (top) and d(TA)- (bottom) type base pairs. Substituents at the C5 carbon of the pyrimidine bases are labelled R_A, while those at the C2 carbon of the purine bases are R_B. In the d(CG)-type base pairs, substituents at the C5 carbon of the cytosine base form 5-methylcytosine and 5-bromocytosine. Removing the amino group at the C2 carbon of guanine forms the unusual inosine nucleotide. In the d(TA)-type base pairs, removing the methyl group at the C5 carbon of thymine forms the unusual deoxynucleotide uridine, while adding an amino group to the C2 carbon of adenine forms the unusual 2-aminoadenine (or diaminoadenine) nucleotide.

these trends, DNAs containing many different sequences and base modifications have been crystallized as Z-DNA. We will focus the discussion here on base modifications that both stabilize and destabilize the Z conformation in terms of the substituent groups that are added, deleted, or replaced in the standard bases of cytosine, thymine, guanine, and adenine (Fig. 7.4).

In this analysis of sequences that have been crystallized as Z-DNA, we compare hexanucleotide sequences that have been crystallized in the same crystal lattice to determine which structural features are sequence versus crystal packing effects. The impact that sequence has on the stability of Z-DNA will be addressed by considering two measures of its stability relative to B-DNA. These are the solvent free energies (SFEs) and the cationic strength (CS) of the crystallization solution. SFEs are estimated from the solvent-accessible surfaces (SAS) calculated for the DNA molecule and, therefore, reflect the energy associated with the DNA in an aqueous environment. The difference in SFE for a sequence in the Z-form versus the B form (ΔSFE_{Z-B}) is indicative of the sequence's stability as Z-DNA (89). The other measure of Z-DNA stability that is relevant to the sequences in a single crystal relies on the recognition that the amount of salt required to convert a sequence to Z-DNA from B-DNA depends on the sequence's inherent stability as Z-DNA relative to B-DNA.

Indeed, the quantity of salt (particularly the cations, as defined as the cation strength, or *CS*) required to crystallize sequences as Z-DNA was observed to be related to the relative stability of that sequence as Z-DNA (as estimated from ΔSFE_{Z-B}), and can be used to predict quantitatively the conditions for obtaining these crystals (22). This relationship can be attributed to the requirement that sequences undergo a transition from B- to Z-DNA along the crystallization pathway. Thus, in this analysis of sequence effects on DNA structure and stability, we will focus on the effects that various substituent groups have on the structure (both the DNA conformation and the solvent structure), ΔSFE_{Z-B}, and the crystallization conditions for the various sequences that have been crystallized as Z-DNA.

3.1 Effects of cytosine methylation on Z-DNA structure

Methylation of cytosine at the C5 carbon of the base (m^5C) (Fig. 7.4) has been studied extensively because of its effect on DNA transcription (90). The effect that methylation has on Z-DNA has been studied both in solution (14) and in various crystal structures. These studies have shown that methylation stabilizes the Z-DNA conformation relative to the B-form. Using circular dichrosim spectroscopy to monitor salt and alcohol titrations, Behe and Felsenfeld (14) showed that poly [d(m^5CpG)] requires less salt or alcohol to convert to Z-DNA than the unmethylated poly [d(CpG)]. This stabilization is associated with the effect that the methyl group has on the hydrophobicity of Z- and B-DNA, as reflected in the ability of cations of the Hofmeister series to induce Z-DNA in poly [d(CpG)] and poly [d(m^5CpG)] (26). In the Hofmeister series, the cations would be expected to follow the trend $Mg^{2+} > Li^+ > Na^+ > K^+ > NH_4^+$ in affecting the transition, if, indeed, the hydrophobic effect is significant (91).

The crystallization of methylated and unmethylated d(CGCGCG) reflects the stabilizing effect of cytosine methylation on Z-DNA. Methylated sequences require less salt to crystallize than unmethylated sequences. The sequence d(CGCGCG) was crystallized from a solution with a $CS = 2.0$ M, whereas the methylated sequence d($m^5Cm^5CGm^5CG$) required $CS = 0.57$ M cations (Table 7.2).

The crystal structure of d($m^5CGm^5CGm^5CG$) (18) showed that the methyl groups reside in protected and recessed pockets at the major groove surface formed by the base and sugar of the adjacent guanine nucleotide. Thus, by burying the methyl into a hydrophobic pocket, this group forms a hydrophobic patch that is less accessible to solvent in Z-DNA than in B-DNA. In addition, the methyl group is involved in favourable contacts with the base and sugar (18).

The methyl group, however, should not be viewed as simply a substituent added to the d(CGCGCG) structure. It also affects the structure of Z-DNA, as is evident from the analysis of sequences having different degrees of cytosine methylation. In this analysis, we compare the sequence d(CGCGCG), which contains the fully unmethylated d(CG) base pairs, to the sequences d($m^5CGm^5CGm^5CG$) and d($CGCGm^5CG$), in which the d(CG) base pairs are fully and hemimethylated. Each d(CpG) and d(GpC) dinucleotide step was analysed for helical twist, rise, roll, tilt, propeller twist, buckle, and *x*-displacement (Table 7.8).

Table 7.8. Comparisons of helical parameters for modified d(CpG) dinucleotides d(m⁵CpG), d(Br⁵CpG), and d(CpI)

	d(CGCGCG)	d(m⁵CGm⁵CGm⁵CG)	d(CGCGm⁵CG)	d(Br⁵CGBr⁵CGBr⁵CG)	d(CGCGBr⁵CG)	d(CGCICG)[a]
d(CpG) steps						
Twist (Ω)						
(C1:G12)/(G2:C11)	−9.2	−14.4	−10.8	−14.5	−8.9	−11.8
(C3:G10)/(G4:C9)	−9.4	−14.5	−12.1	−11.8	−11.6	−11.9
(C5:G8)/(G6:C7)	−12.2	−16.1	−14.6	−14.8	−13.9	−12.3
Average	−10.3 ± 1.7	−15.0 ± 1.0	−12.5 ± 1.9	−13.7 ± 1.7	−11.5 ± 2.5	−12.0 ± 0.2
Rise (D_z)						
(C1:G12)/(G2:C11)	3.8	3.9	3.9	3.9	3.8	3.6
(C3:G10)/(G4:C9)	3.9	3.7	3.8	3.5	4.0	2.1
(C5:G8)/(G6:C7)	4.1	3.9	3.9	3.9	3.8	3.4
Average	3.9 ± 0.2	3.8 ± 0.1	3.9 ± 0.1	3.8 ± 0.2	3.9 ± 0.1	3.0 ± 0.8
Roll (ρ)						
(C1:G12)/(G2:C11)	−0.8	0.1	1.5	0.0	−2.2	−4.6
(C3:G10)/(G4:C9)	−1.1	−2.6	−2.9	−2.1	2.1	−5.8
(C5:G8)/(G6:C7)	3.6	1.5	2.3	3.9	1.6	3.4
Average	0.6 ± 2.6	−0.3 ± 2.1	0.3 ± 2.8	0.6 ± 3.0	0.5 ± 2.4	−2.3 ± 5.0
Tilt (τ)						
(C1:G12)/(G2:C11)	−6.0	−7.1	−7.4	−9.2	−6.4	1.1
(C3:G10)/(G4:C9)	0.9	−0.1	0.7	0.6	−0.3	−1.7
(C5:G8)/(G6:C7)	−0.8	−1.0	1.5	0.3	−2.3	0.0
Average	−2.0 ± 3.6	−2.7 ± 3.8	−1.7 ± 4.9	−2.8 ± 5.6	−3.0 ± 3.1	−0.2 ± 1.4
d(GpC) steps						
Twist (Ω)						
(G2:C11)/(C3:G10)	−48.9	−43.6	−48.4	−45.4	−47.9	−49.2
(G4:C9)/(C5:G8)	−50.8	−44.5	−47.0	−46.0	−48.6	−48.2
Average	−49.9 ± 1.3	−44.1 ± 0.6	−47.7 ± 1.0	−45.7 ± 0.4	−48.3 ± 0.5	−48.7 ± 0.7

Table 7.8. *Continued*

	d(CGCGCG)	d(m⁵CGm⁵CGm⁵CG)	d(CGCGm⁵CG)	d(Br⁵CGBr⁵CGBr⁵CG)	d(CGCGBr⁵CG)	d(CGCICG)[a]
Rise (D_z)						
(G2:C11)/(C3:G10)	3.6	3.8	3.6	3.4	3.6	3.3
(G4:C9)/(C5:G8)	3.5	3.8	3.7	3.9	3.6	2.3
Average	3.6 ± 0.1	3.8 ± 0.0	3.7 ± 0.1	3.7 ± 0.4	3.6 ± 0.0	2.8 ± 0.7
Roll (ρ)						
(G2:C11)/(C3:G10)	−1.5	−4.6	−0.1	−4.3	−1.4	4.7
(G4:C9)/(C5:G8)	0.3	−2.4	1.1	0.0	0.3	4.4
Average	−0.6 ± 1.3	−3.5 ± 1.6	0.5 ± 0.8	−2.2 ± 3.0	−0.6 ± 1.2	4.6 ± 0.2
Tilt (τ)						
(G2:C11)/(C3:G10)	0.1	0.8	2.3	5.9	−0.5	−0.5
(G4:C9)/(C5:G8)	0.6	0.3	0.3	1.3	0.1	−0.6
Average	0.4 ± 0.4	0.6 ± 0.4	1.3 ± 1.4	3.6 ± 3.3	−0.2 ± 0.4	−0.6 ± 0.0
Base pairs						
Propeller twist (ω)						
C1:G12	1.1	2.0	0.6	6.6	2.2	1.5
G2:C11	3.2	3.4	0.9	6.5	0.4	1.7
C3:G10	0.9	4.8	1.2	4.4	0.5	5.0
G4:C9	1.5	1.2	0.4	0.7	5.7	1.0
C5:G8	0.5	0.3	2.6	5.9	0.2	1.9
G6:C7	2.7	2.1	2.1	0.7	0.4	2.0
Average	1.7 ± 1.1	2.3 ± 1.6	1.3 ± 0.9	4.1 ± 2.8	1.6 ± 2.2	2.2 ± 1.4

Table 7.8. *Continued*

	d(CGCGCG)	d(m⁵CGm⁵CGm⁵CG)	d(CGCGm⁵CG)	d(Br⁵CGBr⁵CGBr⁵CG)	d(CGCGBr⁵CG)	d(CGCICG)[a]
Buckle (κ)						
C1:G12	1.9	6.2	4.4	11.6	2.2	5.3
G2:C11	3.5	4.8	0.8	3.5	3.5	8.7
C3:G10	3.0	2.1	1.5	2.2	0.6	8.7
G4:C9	2.4	5.7	3.4	10.3	0.8	8.3
C5:G8	2.0	5.3	4.1	4.2	4.1	3.6
G6:C7	0.0	3.8	3.5	0.6	4.0	0.7
Average	2.1±1.2	4.7±1.5	3.0±1.5	5.4±4.5	2.5±1.6	5.9±3.3
x-displacement (d_x)						
C1:G12	−3.0	−3.6	−3.1	−3.7	−3.2	−3.8
G2:C11	−3.1	−3.6	−3.2	−3.6	−3.3	−3.3
C3:G10	−3.3	−3.4	−3.1	−3.4	−3.3	−2.1
G4:C9	−3.3	−3.5	−3.2	−3.4	−3.3	−2.3
C5:G8	−3.5	−3.8	−3.8	−3.7	−3.9	−3.4
G6:C7	−3.4	−3.6	−3.5	−3.7	−3.6	−3.4
Average	−3.3±0.2	−3.6±0.1	−3.3±0.3	−3.6±0.1	−3.4±0.3	−3.0±0.7

All values are in degrees, except rise (D_z) and displacement (d_x) which are in Å.
[a] Parameters for d(CGCICG) were taken directly from the tables in ref. 56.

The most significant effect of methylation on d(CpG) steps is on the helical twist (Table 7.8). In its fully unmethylated form, $\langle\Omega\rangle = -10.3 \pm 1.7,°$ whereas the fully methylated dinucleotide, d(m⁵CpG/m⁵CpG), is overwound by ~5° with an $\langle\Omega\rangle = -15.0 \pm 1.0°$. This has been attributed to unfavourable steric contacts between the methyl group and the C2′ carbon of the deoxyribose of the neighbouring guanine (18). In the d(GpC) steps, methylation affects both the twist and the roll, independent of the dinucleotide's location in the sequence. In the unmethylated d(GpC/GpC) dinucleotide, $\langle\Omega\rangle = -49.9 \pm 1.3°$, whereas the fully methylated d(Gpm⁵C/Gpm⁵C) is underwound by 5.8° with $\langle\Omega\rangle = -44.1 \pm 0.6°$. Thus, there is a compensating over- and under-winding of the d(CpG) and d(GpC) steps so that the overall structure of methylated Z-DNA remains relatively unperturbed [$\langle\Omega\rangle = -30.3°$ per base step in the unmethylated and $\langle\Omega\rangle = -29.8°$ per base step for the fully methylated d(CpG) sequences]. This suggests, once again, that the primary determinant of Z-DNA structure is the d(CpG) step, with the d(GpC) steps acting to compensate for perturbations to the structure.

At the base pair level, buckle and x-displacement are the only parameters that are significantly affected by methylation. In this case, d(CG) base pairs have an average buckle of $2.1 \pm 1.2°$ and an average x-displacement of 3.3 ± 0.2 Å. whereas d(m⁵CG) base pairs have an average buckle of $4.7 \pm 1.5°$ and an average x-displacement of 3.6 ± 0.1 Å (Table 7.8). This again is associated with steric interactions between the substituent and the neighbouring guanine nucleotide.

Studies on the hemimethylated dinucleotides d(m⁵CpG/CpG) and d(Gpm⁵C/GpC) show that each methyl group acts independently to affect the structure and stability of Z-DNA. The structures of the two hemimethylated d(m⁵CpG/CpG) and d(Gpm⁵C/GpC) steps in the sequence d(CGCGm⁵CG) are intermediate between those observed for the corresponding unmethylated and fully methylated dinucleotide steps (Table 7.8). This suggests that the hemimethylated form is a true conformational intermediate. This is evident when comparing, for example, the helical twist at the two d(m⁵CpG/CpG) to the average of the corresponding dinucleotides at each position. The helical twist between base pairs d(C1:G12) and d(G2:m⁵C11) is $-10.8°$, while the average for the corresponding base pairs in the unmethylated and methylated structures is $\langle\Omega\rangle = -11.8°$. Likewise, base pairs d(m⁵C5:G8) and d(G6:C7) at the opposite end of the duplex have $\Omega = -14.6°$ compared with $\langle\Omega\rangle = -14.2°$ for the unmethylated and methylated analogues. The position effects are probably a result of the crystal packing forces of the lattice. Similarly, $\langle\Omega\rangle$ for the d(Gpm⁵C/GpC) steps are intermediate between the analogous unmethylated and methylated structures.

We would expect hemimethylation of cytosines to have an intermediate effect on the stability of Z-DNA. Indeed, this is the case. The sequence d(CGCGm⁵CG) was crystallized from a solution with $CS = 1.26$ M, which is intermediate between that required for the unmethylated, d(CGCGCG) (2.0 M), and the fully methylated, d(m⁵CGm⁵CGm⁵CG), sequence (0.57 M). The CS we predict for crystallizing this hemimethylated sequence is 1.52 M, assuming equal contributions from each methyl group. A salt titration of 24 alternating base pairs of d(CG) showed that the midpoint for the transition from B- to Z-DNA in the unmethylated sequence occurred at ~1.25 M MgCl₂, while the fully methylated sequence was predominantly Z-DNA in

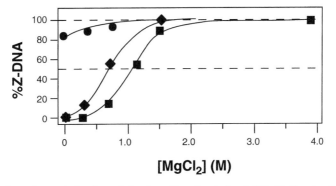

Fig. 7.5. Titration of unmethylated, methylated, and hemimethylated d(CpG) dinucleotides with MgCl$_2$ to induce the formation of Z-DNA (60). The unmethylated sequence d(CG)$_{12}$ (squares), fully methylated sequence d(m^{5}CG)$_{12}$ (circles), and hemimethylated sequence d(m^{5}CGCG)$_3$(CGCG)$_3$ (diamonds)were titrated with 0.0–4.0 M MgCl$_2$. The formation of Z-DNA was monitored by following the ratio of light absorbed at 260 nm versus 290 nm (A_{260}/A_{290} ratio). The conformations of the DNA at the beginning and end of the titration were confirmed to be that of B-DNA and Z-DNA, respectively, by circular dichroism spectroscopy.

the absence of added salt (Fig. 7.5). This salt-induced transition in the sequence d(m^{5}CGCGm^{5}CGCGm^{5}CGCGCGCGCGCGCGCG), which has six true hemi-methylated dinucleotide steps, had a midpoint for the transition at ~1 M MgCl$_2$ (60). Again, we would predict a midpoint at 0.9 M for equal contributions from each methyl group. Thus, the hemimethylated dinucleotides represent a true intermediate, both structurally and thermodynamically, between fully unmethylated and fully methylated dinucleotides.

The Z-DNA stabilizing effect from cytosine methylation is likely to be associated with the hydration of the DNA structure. The only notable effect of cytosine methylation on the solvent structure in the crystal, however, is that the water that is hydrogen bonded to the cytosine N4 nitrogen at the major groove surface is slightly displaced away from the methyl group (18). Otherwise, the arrangement of waters around the Z-DNA structure remains unperturbed. This is not entirely surprising because the methyl group actually sits recessed in a pocket of the major groove surface and thus is largely inaccessible to solvent.

The calculated solvent free energies show that methylation makes the Z-DNA surface more hydrophobic; however, the ΔSFE_{Z-B} indicates that methylation works to stabilize Z-DNA primarily by destabilizing B-DNA [by making its surface even more hydrophobic, (Table 7.9)]. Methylating the cytosines of Z-DNA increases the SFE_Z by 1.3 kcal/mol/bp and thus we would expect this to destabilize the left-handed conformation. In contrast, there is an even greater destabilization of B-DNA for these sequences (ΔSFE_{Z-B} = 0.29 kcal/mol/dn for the unmethylated sequence, but is −0.87 kcal/mol/dn for the methylated analogue). The SFE_Z and ΔSFE_{Z-B} for the hemimethylated dinucleotides in d(CGCGm^{5}CG) are again intermediate between the corresponding values for d(CGCGCG) and d(m^{5}CGm^{5}CGm^{5}CG) (Table 7.9).

Table 7.9. Solvent free energies of various dinucleotides as Z- and B-DNA

Dinucleotide anti–p–syn	SFE_Z (kcal/mol/dn)	SFE_B (kcal/mol/dn)	$\Delta SFE_{(Z-B)}$ (kcal/mol/dn)
d(CpG)	−12.97	−13.26	0.29
d(m⁵CpG)/(CpG)	−12.58	−12.08	−0.50
d(m⁵CpG)	−10.28	−9.41	−0.87
d(TpA)	−9.90	−8.53	1.35
d(UpA)	−11.50	−10.64	0.86
d(ApT)ᵃ	−8.17	−9.45	1.28
d(GpC)/d(Gpm⁵C)ᵃ	−12.08	−11.62	−0.46
d(CpC)/d(GpG)ᵃ	−12.26	−12.90	0.64

SFE_Z and SFE_B are the solvent free energies for the dinucleotide in the Z and B conformations, respectively. SFE_Z was calculated from the crystal structure containing that dinucleotide step and SFE_B was calculated from idealized B-DNA models. $\Delta SFE_{(Z-B)}$ is the free energy difference for the dinucleotide step in the Z form versus the B form.
ᵃ Dinucleotide out-of-alternation (e.g. the first base pair of the step is *anti*, followed by *syn*).

3.2 Effects of cytosine bromination on Z-DNA structure

The effective radius (~2 Å) and hydrophobicity of a bromine atom is very similar to that of a methyl group. We would therefore expect bromination of cytosines to have a similar effect in stabilizing Z-DNA, and on the structure of Z-DNA. The two Z-DNA sequences that have been crystallized that contain a brominated C5 of cytosine (Fig. 7.4) are the fully brominated sequence d(Br⁵CGBr⁵CGBr⁵CG) (19) and the hemibrominated sequence d(CGCGBr⁵CG) (63).

The effect of cytosine bromination on the stability of Z-DNA is equivalent to or greater than that of cytosine methylation. Poly (Br⁵CpG) is constituitively in the Z-form even in the absence of alcohols and high concentrations of added salts (15). Like d(m⁵CGm⁵CGm⁵CG), d(Br⁵CGBr⁵CGBr⁵CG) required very little salt to crystallize. The enhanced stability of brominated Z-DNA compared with even the methylated form may result from the smaller perturbation of the structure.

Comparison of base dinucleotide parameters reveals trends similar to those seen for methylation (Table 7.8). Specifically, the d(CpG/CpG) dinucleotide has an average twist of −10.3 ± 1.7°, while the d(Br⁵CpG/Br⁵CpG) dinucleotide is overwound by 3.4 to −13.7 ± 1.7°. Bromination, therefore, has an similar, but less dramatic, effect on Z-DNA structure than methylation. This smaller perturbation on the structure may be a result of differences in the interactions with adjacent nucleotides between the spherically shaped bromine atom compared with the tetrahedral methyl group. As with the methylation effect on twist, the overwinding of the *anti–p–syn* step in brominated steps is compensated at the d(GpC) step to give no net difference in the helical twist of the hexanucleotide structures. The $\langle \Omega \rangle = -30.3°$ per base step in d(CGCGCG) and the brominated structure has $\langle \Omega \rangle = -30.2°$ per base step. Unlike other Z-DNA structures, the helical twist of d(Br⁵CGBr⁵CGBr⁵CG) is not position dependent (Table 7.8), suggesting that the conformation of the fully brominated sequence is less affected by the crystal lattice. Additionally, only the Z_I backbone conformation is present in this fully brominated structure. Finally, bromination does not appear to have any effect on the base parameters of the tip, inclination, propeller twist, buckle, and *x*-displacement.

Unlike methylation, it is not clear if hemibromination represents an intermediate between fully brominated and fully unbrominated dinucleotides (Table 7.8). The helical twist ($\Omega = -8.9°$) for the hemibrominated dinucleotide at one end is similar to that of d(CpG/CpG) (−9.2°). However, $\Omega = -13.9°$ for this same hemibrominated dinucleotide at the opposite end, and is intermediate between $\Omega = -12.2°$ and −14.8° observed for the dinucleotides d(CpG/CpG) and d(Br⁵CpG/Br⁵CpG), respectively. Additionally, the $\langle \Omega \rangle = -48.3°$ for the hemibrominated d(GpBr⁵C/GpC) steps is identical to the average for d(GpC/GpC) and d(GpBr⁵C/GpBr⁵C). These observations are consistent with the hemibrominated sequence representing an intermediate conformation except at one terminal dinucleotide.

3.3 Effect of the N2 amino of guanine on the structure and stability of Z-DNA

Removing the amino group at the N2 position of guanine (to form inosine, dI) (Fig. 7.4) would be expected to destabilize Z-DNA. This would eliminate one hydrogen bond within the base pair but, perhaps more importantly, would affect the spine of hydration in the minor groove of the Z-DNA duplex. The minor groove water that bridges this N2 amino group to the phosphate oxygens of the backbone is thought to be important for stabilizing the *syn* conformation of the guanine bases (77). The two published structures of inosine-containing Z-DNA are for the octamer sequence d(CGCICICG) (64) and the hexamer d(CGCICG) (56). The coordinates of neither of these structures were available for analysis by our program, but some helical parameters could be gleaned from the data presented in the published papers.

The structure of d(CGCICICG) was disordered in the crystal and thus specific parameters for the d(CpI) and d(IpC) steps could not be distinguished from those of the d(CpG) and d(GpC) steps. The values reported are therefore averages for the respective dinucleotides. The average rise {3.6 Å for the d[CpG(I)] and 3.7 Å for d[G(I)pC]} and helical twist {16.5° for the d[CpG(I)] and 43.5° for d[G(I)pC]} of this structure are more similar to the tetramer d(CGCG) and the disordered octanucleotide d(CGCGCGCG) than to the parent hexanucleotide structures. This, again, may be related more to the hexagonal space group of these crystals than to any intrinsic structural property of Z-DNA.

The structure of d(CGCICG) shows a crystal lattice and conformation that is similar to the SP form of d(CGCGCG). The minor groove of the duplex is 0.6 Å narrower than the standard MGSP structure, but it was not clear whether this is primarily localized at the d(CI) base pairs, or averaged over the structure. The water structure of d(CGCICG) was said to be similar to that of the spermine form of d(CGCGCG), including the continuous spine connecting the O2 oxygens of the cytosines along the minor groove crevice. This suggests that the N2 amino group is not absolutely essential to ordering the waters in the crevice. Still, the bridge from the purine to the phosphate cannot be made. The SFEs calculated suggest that d(CI) base pairs are less stable as Z-DNA by 0.30 kcal/mol/bp compared with d(TA) base pairs. The sequence d(CICGCG) required $CS = 4.2$ to crystallize as Z-DNA (22), the highest salt concentration required for any APP sequence.

3.4 The structure and stability of d(TpA) dinucleotides in Z-DNA

The observation that d(TpA) dinucleotides can be incorporated into the structure of Z-DNA extends the range of sequences that can adopt the left-handed conformation. Although this is an APP dinucleotide, it does not promote the formation of Z-DNA, and must be flanked by methylated d(m^{5}CpG) dinucleotides to crystallize, as in the sequence d(m^{5}CGTAm^{5}CG). The structure of d(m^{5}CGm^{5}CGm^{5}CG) therefore serves as the reference when analysing this d(TpA)-containing structure. The destabilization of Z-DNA in the crystals by the d(TpA) dinucleotide is reflected in the CS for the crystallization of d(m^{5}CGTAm^{5}CG) (1.3 M) compared with that for d(m^{5}CGm^{5}CGm^{5}CG) (0.57 M) (Table 7.2).

The overall structure of d(m^{5}CGTAm^{5}CG) is indeed more similar to d(m^{5}CGm^{5}CGm^{5}CG) than to d(CGCGCG) in all respects (Tables 7.8 and 7.10). Differences in the structural details are attributed to replacing the central d(m^{5}CpG) dinucleotide with d(TpA). The helical twist is reduced by 1.7°, approaching that of d(CGCGCG). This is associated with a sliding of the d(TA) base pairs towards each other. This sliding is localized, however, to only the d(TpA) dinucleotide, since the d(GpT) and d(Apm^{5}C) steps show increases in $\langle \Omega \rangle = 0.5°$ each to compensate. The d(TpA) dinucleotide is also significantly compressed ($D_z = 3.3$ Å) compared with any d(CpG) dinucleotide.

The destabilization of Z-DNA by d(TpA) dinucleotides appears to be associated with the presence of the methyl group at the major groove surface of the thymine base and the absence of an N2 amino group from the minor groove crevice of the adenine base (Fig. 7.4), both of which perturb the solvation around the d(TA) base pairs (89). The cytosines of d(CpG) dinucleotides are bridged by a well-defined pattern of waters at the major groove surface (79). In contrast, the solvent structure at the d(TpA) dinucleotide major groove surface can best be described as a set of disordered waters and/or cation complexes, with no specific hydrogen bonding pattern to the thymine bases (51).

In comparison, the structure of d(m^{5}CGUAm^{5}CG) helps to pin-point the role of the thymine methyl group in the instability of d(TpA) dinucleotides as Z-DNA. In this deoxyuridine-containing structure of Z-DNA (Fig. 7.4), the twist angle between the central d(UA) base pairs approach the values of Ω for the d(m^{5}CpG) dinucleotides (Tables 7.8 and 7.10). This appears to result from the coupling of the two stacked uridine bases by a Mg(H$_2$O)$_6^{2+}$ complex. This complex is analogous to the waters that bridge the stacked cytosines at the major groove surface of the d(CpG) dinucleotides. The thymine methyls in the structure of d(m^{5}CGTAm^{5}CG) evidently disrupt these interactions. This was suggested by the lower ΔSFE_{Z-B} calculated for the d(UpA) compared with the d(TpA) dinucleotides (Table 7.9).

Interestingly, the solvent in the minor groove crevice is also perturbed by the C5 methyl of the thymines in d(m^{5}CGTAm^{5}CG). The two well-ordered waters typically observed at each d(CG) base pair in Z-DNA (Fig. 7.2) could not be located at either d(TA) base pair (51). Thus, the spine of hydration in the minor groove crevice is

Table 7.10. Helical parameters for d(A), d(T), d(U), and d(D)–containing sequences

	d(m⁵CGTAm⁵CG)	d(m⁵CGUAm⁵CG)	d(CGTDCG)	d(CDCGTG)	d(CDUDCG)
d(CpG) steps					
Twist (Ω)					
(C1:G12)/(G2:C11)	-16.1	-14.8	-13.5	-7.5	-13.5
(C3:G10)/(G4:C9)	-12.8	-13.8	-7.3	-12.4	-7.3
(C5:G8)/(G6:C7)	-14.9	-17.0	-13.5	-11.9	-13.5
Average	-14.6 ± 1.7	-15.2 ± 1.6	-11.4 ± 3.6	-10.6 ± 2.7	-11.4 ± 3.6
Rise (D_z)					
(C1:G12)/(G2:C11)	3.9	3.8	4.2	4.0	4.2
(C3:G10)/(G4:C9)	3.3	3.4	3.6	3.9	3.6
(C5:G8)/(G6:C7)	3.9	3.8	4.3	3.8	4.3
Average	3.7 ± 0.3	3.7 ± 0.2	4.0 ± 0.4	3.9 ± 0.1	4.0 ± 0.4
Roll (ρ)					
(C1:G12)/(G2:C11)	0.1	0.6	-9.4	0.5	-9.4
(C3:G10)/(G4:C9)	-1.4	-0.3	-6.6	2.6	-6.6
(C5:G8)/(G6:C7)	1.1	1.4	-5.8	0.3	-5.8
Average	-0.1 ± 1.3	0.6 ± 0.9	-7.3 ± 1.9	1.1 ± 1.3	-7.3 ± 1.9
Tilt (τ)					
(C1:G12)/(G2:C11)	7.3	6.2	-9.7	3.9	-9.7
(C3:G10)/(G4:C9)	-2.4	3.8	0.1	-0.1	0.1
(C5:G8)/(G6:C7)	-1.1	-1.7	1.7	1.7	1.7
Average	1.3 ± 5.3	2.8 ± 4.1	-2.6 ± 6.2	1.8 ± 2.0	-2.6 ± 6.2
d(GpC) steps					
Twist (Ω)					
(G2:C11)/(C3:G10)	-44.9	-44.7	-42.0	-49.7	-42.0
(G4:C9)/(C5:G8)	-44.2	-45.8	-42.0	-47.5	-42.0
Average	-44.6 ± 0.5	-45.3 ± 0.8	-42.0 ± 0.0	-48.6 ± 1.6	-42.0 ± 0.0

Table 7.10. *Continued*

	d(m⁵CGTAm⁵CG)	d(m⁵CGUAm⁵CG)	d(CGTDCG)	d(CDCGTG)	d(CDUDCG)
Rise (D_z)					
(G2:C11)/(C3:G10)	3.9	3.9	3.6	3.6	3.6
(G4:C9)/(C5:G8)	3.8	3.6	3.6	3.7	3.6
Average	3.9 ± 0.1	3.8 ± 0.2	3.6 ± 0.0	3.7 ± 0.1	3.6 ± 0.0
Roll (ρ)					
(G2:C11)/(C3:G10)	−5.3	4.0	−3.1	−0.6	−3.1
(G4:C9)/(C5:G8)	−5.2	6.5	−1.9	−3.1	−1.9
Average	-5.3 ± 0.1	5.3 ± 1.8	-2.5 ± 0.8	-1.9 ± 1.8	-2.5 ± 0.8
Tilt (τ)					
(G2:C11)/(C3:G10)	−0.6	−1.1	−1.8	1.8	−1.8
(G4:C9)/(C5:G8)	0.6	0.7	0.2	0.3	0.2
Average	0.0 ± 0.8	-0.2 ± 1.3	-0.8 ± 1.4	1.1 ± 1.1	-0.8 ± 1.4
Base pairs					
Propeller twist (ω)					
C1:G12	0.7	2.8	5.7	4.2	5.7
G2:C11	4.1	7.9	0.8	0.8	0.8
C3:G10	2.6	4.0	0.2	3.0	0.2
G4:C9	2.0	1.3	0.2	2.9	0.2
C5:G8	1.6	0.4	0.7	1.1	0.7
G6:C7	2.2	2.0	5.7	2.6	5.7
Average	2.2 ± 1.1	3.1 ± 2.7	2.2 ± 2.7	2.4 ± 1.3	2.2 ± 2.7

Table 7.10. *Continued*

	d(m⁵CGTAm⁵CG)	d(m⁵CGUAm⁵CG)	d(CGTDCG)	d(CDCGTG)	d(CDUDCG)
Buckle (κ)					
C1:G12	5.5	7.7	2.9	0.9	2.9
G2:C11	7.5	7.2	1.2	1.8	1.2
C3:G10	5.1	2.0	6.9	1.0	6.9
G4:C9	8.8	4.3	6.9	1.8	6.9
C5:G8	2.1	0.3	1.2	5.2	1.2
G6:C7	2.8	2.5	2.8	6.3	2.8
Average	5.3 ± 2.6	4.0 ± 3.0	3.7 ± 2.6	2.8 ± 2.3	3.7 ± 2.6
x-displacement (d_x)					
C1:G12	-3.8	-3.9	-3.7	-3.1	-3.7
G2:C11	-3.7	-3.6	-3.1	-3.1	-3.1
C3:G10	-3.4	-3.3	-2.8	-3.3	-2.8
G4:C9	-3.4	-3.2	-2.9	-3.3	-2.9
C5:G8	-3.8	-3.9	-3.3	-3.7	-3.3
G6:C7	-3.8	-3.9	-4.0	-3.4	-4.0
Average	-3.7 ± 0.2	-3.6 ± 0.3	-3.3 ± 0.5	-3.3 ± 0.2	-3.3 ± 0.5

All values are in degrees, except rise (D_z) and displacement (d_x), which are in Å. Parameters for the reference sequences d(CGCGCG) and d(m⁵CGm⁵CGm⁵CG) are shown in Table 7.8.

disrupted at each d(TA) base pair. This may contribute to the reduced stability of d(TpA) dinucleotides as Z-DNA. The water network in the minor groove of B-DNA is continuous even at the d(TA) base pairs (80). In this case, the waters are hydrogen bonded to the N3 nitrogen of the purine ring, which is largely inaccessible in Z-DNA. Thus, there are no waters that bridge the N2 amino group of the purine to the phosphate backbone to stabilize the *syn* conformation, as was observed with the d(CpG) dinucleotides in Z-DNA.

The ordered hydration in the minor groove, however, is restored to the d(CpG)-like spine at d(UA) base pairs of d(m⁵CGUAm⁵CG) (56). This apparently results from a widening of the minor groove caused by the coupled binding of the uridine bases by the magnesium–water complex at the major groove surface. There are two waters at each d(UA) base pair. One water is directly hydrogen bonded to the O2 of the uridine base, while the second connects this water to the phosphoribose backbone of the opposite strand. Thus, although no water directly connects the adenine base to the backbone, there may still be a degree of stabilization of the *syn* conformation conferred by the pyrimidine–water–water–phosphate bridge. This would suggest that d(UpA) dinucleotides are more stable as Z-DNA than d(TpA) dinucleotides. Indeed, the sequence d(m⁵CGUAm⁵CG) was crystallized in solutions having $CS = 0.31$ M, which is less than half of that required to crystallize d(m⁵CGTAm⁵CG).

The magnesium complex of d(m⁵CGUAm⁵CG) can be displaced from the major groove by binding copper ions to the purines (53). The result is that the minor groove crevice of the d(UpA) dinucleotide becomes narrower, although not as narrow as that of the d(TpA) dinucleotide. The effect of this on the spine of hydration is that the four waters at the d(UpA) dinucleotide are perturbed, but not displaced. One water remains hydrogen bonded to the O2 and in the plane of the uridine base. The second water for each base pair, however, is pushed out of plane and, therefore, cannot form the pyrimidine–water–water–phosphoribose bridge of the native d(UpA) structure. This displacement effectively isolates the cluster of four waters at the d(UpA) dinucleotides from those of the neighbouring d(CpG) dinucleotides. Thus, although the number of waters in the spine remains unchanged, its continuity along the minor groove and across the helix becomes disrupted by removing the magnesium complex at the major groove surface.

To see how perturbations to the major groove surface affect the stacking of the bases and the water structure of Z-DNA, we start with the d(UpA) dinucleotide of the copper-soaked structure, which has a minor groove crevice that is intermediate in width (Fig. 7.6). Introducing a magnesium complex at the major groove surface slides the base pairs to provide a wider crevice that can accommodate the four water molecules in the plane of the d(UA) base pairs. Methylating the uridine bases, on the other hand, prevents the binding of this magnesium complex and slides the base pairs in the opposite direction to narrow the crevice. The narrower crevice prevents the waters from forming a well-ordered network at the d(TA) base pairs. Thus, the major and minor grooves of Z-DNA cannot be treated as two isolated domains of the structure. Perturbations to one side are transmitted through the double helix to the other side of the duplex.

The other substituent that affects the stability of d(TpA) dinucleotides as Z-DNA is the N2 amino, or, more precisely, the lack of this group on the adenine bases. The

d(m⁵CGUAm⁵CG)   d(m⁵CGUAm⁵CG)*   d(m⁵CGTAm⁵CG)

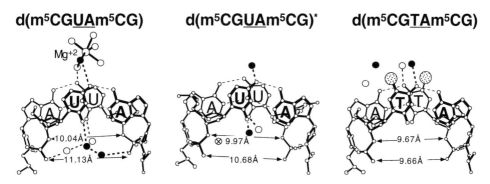

Fig. 7.6. Comparison of the solvent structures and widths of the minor groove crevice of d(U[T]pA) dinucleotides in Z-DNA. Shown are the structures of d(m⁵CGUAm⁵CG) (25), d(m⁵CGUAm⁵CG) soaked with copper [(53), d(m⁵CGUAm⁵CG)*], and d(m⁵CGTAm⁵CG) (51). The top base pair of each dinucleotide is shown with thick bonds and labelled in bold, while the lower base pairs are shown as thin bonds and labelled in standard type. Waters that interact with the top base pairs are shown as filled circles, while those interacting with the lower base pairs are open circles [the circle with a cross in the structure of d(m⁵CGUAm⁵CG)* sits between the two base pairs]. Widths of the minor groove crevice are measured between the O3' oxygens, and between the closest oxygens of the phosphate group of the dinucleotides. The methyl groups of the thymines in d(m⁵CGTAm⁵CG) are stippled.

unusual base 2-aminoadenine [or diamminopurine, d(D)] (Fig. 7.4) has been used to probe the effect of this group on Z-DNA structure and stability. Introducing this additional amino group to adenines apparently stabilizes Z-DNA. The sequence d(CGTDCG) was crystallized from solutions with $CS = 1.1$ M (61). These conditions are comparable to that of d(m⁵CGTAm⁵CG), even though the cytosines are not methylated. There are two potential means by which dD stabilizes Z-DNA. The first effect would be the introduction of an additional hydrogen bond to the base pair, making d(TD) more akin to d(CG) base pairs. Since Z-DNA is a more rigid helix than B-DNA (77,78), this would affect the difference in conformational entropy between the two DNA forms for the modified base pair. The second effect would be to place an additional hydrogen bonding function into the minor groove crevice to accommodate the waters of the hydration spine. This has been more extensively studied, and thus will be the focus of this discussion on d(TpD) dinucleotides.

The structure of the sequence d(CGTDCG) was solved in an unusual space group for Z-DNA, P3₂21 (61). Although in a completely different lattice arrangement from other Z-DNA hexanucleotides, its structure shows many of the same features as standard Z-DNA (Tables 7.8 and 7.10). It is, however, slightly underwound (the average helical twist is ~8° more positive) compared with d(CGCGCG), with most of this distortion associated with the d(GpC) steps [being approximately 7–9° less negative than comparable steps of d(CGCGCG)] and at one of the terminal d(CpG) steps (in this case 4.7° overwound in the left-handed direction). The minor groove is narrower as a result of appreciable negative roll at nearly all dinucleotide steps of the helix. These distortions may arise from the crystal lattice in that the terminal base pairs are not stacked end-to-end to form essentially continuous strands of Z-DNA as in the 'standard' hexamer crystals. The duplexes pack perpendicular to and against the major groove surface of the neighbouring duplex. This general lattice is similar to A-DNA packing modes, except that the ends of the duplexes pack against the minor groove in the crystals of A-DNA hexanucleotides (87). Thus, this structure may show more 'end-effects' than would normally be observed. There are, however, some sequence-dependent features.

Despite these distortions, the first hydration shell is again nearly identical to that of d(CGCGCG), if the waters at the interface between helices are ignored. The narrower minor groove crevice shifts the spine of hydration, but does not apparently 'squeeze' any water out as in the d(TpA) dinucleotides. Thus, it is clear that the N2 amino group of the purine does play a significant role in defining the regular pattern of this water network. Both the crystallization conditions and salt titrations followed by circular dichroism spectroscopy show that d(TpD) dinucleotides are more stable as Z-DNA than are d(TpA), but less so than d(CpG) (23). Under dehydrating conditions, however, the hexamer d(TDTDTD) forms A-DNA instead of Z-DNA, as measured by circular dichroism. The flanking d(CpG) dinucleotides in d(CGTDCG) are required to induce d(TpD) to form Z-DNA, although the cytosine bases do not need to be methylated.

All this taken together suggests that demethylating the thymine and adding an amino group to the adenine [as in a d(UpD) dinucleotide] would greatly enhance the stability of Z-DNA relative to the standard d(TpA) dinucleotide, to the point where it should behave more like a d(CpG) base pair. Indeed, the structure of d(CGUDCG) (24) most closely resembles that of the MG and MGSP forms of d(CGCGCG) in terms of the DNA conformation and the solvent interactions at the major groove surface and minor groove crevice. The CS for crystallization of this sequence as Z-DNA was identical to that of d(CGCGCG). It would be interesting to extrapolate from this to determine whether d(UDUDUD), as opposed to d(TDTDTD), would form Z-DNA in solution or in a crystal.

3.5 d(CpA)/d(TpG) dinucleotides in Z-DNA

One of the most prevalent simple, repeating sequences found in eukaryotic genomes is the alternating pattern of d(CpA)/d(TpG) dinucleotides (92–94). These APP sequences are thought to form Z-DNA. Studies on Z-DNA formed in negatively supercoiled plasmids indicate that the order of stability for APP dinucleotides is d(CpG) > d(CpA)/d(TpG) > d(TpA) (77,78). The thermodynamic propensity of a d(CpA)/d(TpG) dinucleotide to form Z-DNA is not, however, simply an average of the d(CpG) and d(TpA) dinucleotides. The first conversion of a d(CG) base pair in the standard d(CpG) dinucleotide to a d(TA) base pair is not as destabilizing as the second. Is this reflected in the crystal structure?

The single-crystal structure of the sequence d(CACGTG) has been solved to ~2.5 Å resolution (49), which is one of the lowest resolution structures of Z-DNA. The structure shows two features that may contribute to the lower propensity of d(CpA)/d(TpG) dinucleotides to form Z-DNA. One is that the lack of an N2 amino group on the adenine base reduces the stacking surface and thus results in poorer stacking interactions at the d(ApC) steps as opposed to the d(GpC) steps. This cannot be the major contributor, since only the d(ApC) step at the A8/C9 positions shows this poorer stacking. The d(ApC) step at A2/C3 compensates by placing the phosphate of C3 in the Z_{II} conformation. This displaces the A2 purine so that its six-membered ring lies directly on top of the cytosine base.

The other effect is observed in the solvent structure of the minor groove. Although the minor groove crevice of this sequence is identical in width to that of d(CGCGCG), there were no ordered solvent molecules located at or near the adenine bases in the groove (49). The suggestion here was that the N2 amino group that is missing from the adenine base contributes to the disruption of the spine of hydration. As with the d(TpA) dinucleotide, the bridge from the purine base to the phosphori-bose backbone, which appears to be important for stabilizing the purine in the *syn* conformation, is lost. In support of this proposition, the structure of d(CDCGTG) (23) shows the same organization of water molecules in the minor groove as does d(CGCGCG). In addition, the structure of d(CDCGTG) is identical to the MGSP form of d(CGCGCG) in all respects (Tables 7.8 and 7.10). This would contribute to the lower stability of d(CpA)/d(TpG) dinucleotides.

We had argued above with the structure of d(m⁵CGUAm⁵CG), however, that a wide minor groove, even in the absence of the N2 amino group on the purine, allows waters to organize into the well-ordered spine in Z-DNA. Why in this case, where the widths of the minor groove crevice of d(CpA)/d(TpG) are identical to those of d(CpG) and d(CpD)/d(TpG), were no ordered waters located near the adenines? It may be that the waters are less populated and thus could not be observed at the lower resolution of this structure. The structure of d(CACGCG)/d(CGCGTG) has been solved to 1.6 Å resolution (50), where one could expect to observe less populated solvent molecules. However, this asymmetric sequence shows orientational disorder about the dyad-axis of the duplex; therefore, it would be difficult to assign solvent structure definitively at the d(TA) base pairs. These base pairs effectively overlap in the electron density maps. The question therefore remains unanswered. If a higher resolu-tion structure of d(CpA)/d(TpG) dinucleotide does indeed show the same type of pyrimidine–water–water–phosphoribose bridge as was observed with the d(UpA) step, then we can start to understand why introducing the first d(TA) base pair into a dinu-cleotide is not as destabilizing to Z-DNA as the second.

3.6 Out-of-alternation structures

Z-DNA can tolerate dinucleotides that do not follow the APP rule for its formation (that is, they are out-of-alternation, and place pyrimidine bases in the disfavoured *syn* conformation). The crystal structures of d(Br⁵CGATBr⁵CG) and d(m⁵CGATm⁵CG) were the first to indicate that the APP rule could be violated (52), and the structure of the brominated sequence was the one reported. In the structure of d(Br⁵CGATBr⁵CG), both thymine bases of the central dinucleotide adopt the *syn* conformation while the complementary adenines are *anti*. Still, the backbone conformation is remarkably similar to that of d(CGCGCG) (Tables 7.8 and 7.11). The twist angle (Ω) for the *anti–p–syn* step of the d(ApT) dinucleotide is −9°, while all the *syn–p–anti* steps are −49°. All nucleotides in the *anti* conformation have C2′-*endo* sugar puckers, while a majority of those that are *syn* have C3′-*endo* puckers. Exceptions to this rule were at the guanines at the 3′-end of each strand. Thus, the alternating sugar conformations remain even when the pyrimidines are *syn*.

Table 7.11. The effects of out-of-alternation base steps on the helical structure of Z-DNA[a]

	d(m^5CGm^5CGm^5CG)	d(m^5CGG$\underline{C}m^5$CG)	d(m^5CGGGm^5CG)/ d(m^5CGC$\underline{C}m^5$CG)	d(m^5CGGGm^5CG)/ d(m^5CGC $\underline{m^5C}$CG)	d(Br^5CG$\underline{AT}Br^5$CG)[b]
d(CpG) steps					
Twist (Ω)					
(C1:G12)/(G2:C11)	−14.4	−13.6	−13.6	−13.2	−13.0
(C3:G10)/(G4:C9)	−14.5	−11.4	−12.4	−12.2	−9.0
(C5:G8)/(G6:C7)	−16.1	−14.8	−14.7	−14.7	−12.0
Average	−15.0 ± 1.0	−13.3 ± 1.7	−13.6 ± 1.2	−13.4 ± 1.2	−11.0 ± 2.0
Rise (D_z)					
(C1:G12)/(G2:C11)	3.9	4.0	3.9	4.0	
(C3:G10)/(G4:C9)	3.7	3.6	3.6	3.8	
(C5:G8)/(G6:C7)	3.9	3.6	3.8	4.0	
Average	3.8 ± 0.1	3.7 ± 0.2	3.8 ± 0.2	3.9 ± 0.1	
Roll (ρ)					
(C1:G12)/(G2:C11)	0.1	4.3	2.0	2.2	
(C3:G10)/(G4:C9)	−2.6	0.4	−0.9	0.2	
(C5:G8)/(G6:C7)	1.5	−0.3	−0.3	−2.0	
Average	−0.3 ± 2.1	1.5 ± 2.5	0.3 ± 1.5	0.2 ± 2.1	
Tilt (τ)					
(C1:G12)/(G2:C11)	−7.1	3.4	−1.9	5.1	
(C3:G10)/(G4:C9)	−0.1	5.1	2.0	1.8	
(C5:G8)/(G6:C7)	−1.0	−3.9	−1.7	2.5	
Average	−2.7 ± 3.8	1.5 ± 4.8	−0.5 ± 2.2	3.2 ± 1.7	
d(GpC) steps					
Twist (Ω)					
(G2:C11)/(C3:G10)	−43.6	−46.6	−46.8	−47.0	
(G4:C9)/(C5:G8)	−44.5	−46.8	−46.8	−47.3	
Average	−44.1 ± 0.6	−46.7 ± 0.1	−46.8	−47.2 ± 0.2	−49.0

Table 7.11. *Continued*

	d(m^5CGm^5CGm^5CG)	d(m^5CGGCm^5CG)	d(m^5CGGGm^5CG)/ d(m^5CGCCm^5CG)	d(m^5CGGGm^5CG)/ d(m^5CGC m^5CCG)	d(Br^{5}CGATBr^{5}CG)[b]
Rise (D_z)					
(G2:C11)/(C3:G10)	3.8	3.7	3.6	3.6	
(G4:C9)/(C5:G8)	3.8	3.7	3.8	3.8	
Average	3.8	3.7	3.7 ± 0.1	3.7 ± 0.1	
Roll (ρ)					
(G2:C11)/(C3:G10)	−4.6	5.6	0.2	1.3	
(G4:C9)/(C5:G8)	−2.4	1.3	−0.1	2.1	
Average	−3.5 ± 1.6	3.5 ± 3.0	0.0 ± 0.2	1.7 ± 0.6	
Tilt (τ)					
(G2:C11)/(C3:G10)	0.8	−1.4	−1.3	−0.1	
(G4:C9)/(C5:G8)	0.3	0.9	−0.1	−0.8	
Average	0.6 ± 0.4	−0.3 ± 1.6	−0.7 ± 0.8	−0.4 ± 0.5	
Base pairs					
Propeller twist (ω)					
C1:G12	2.0	5.0	0.9	2.2	
C2:G11	3.4	5.8	1.8	1.7	
C3:G10	4.8	2.4	3.0	4.5	
G4:C9	1.2	3.1	2.2	1.7	
C5:G8	0.3	1.1	2.3	1.0	
G6:C7	2.1	1.3	2.8	0.8	
Average	2.3 ± 1.6	3.1 ± 1.9	2.2 ± 0.7	2.0 ± 1.3	

Table 7.11. *Continued*

	d(m⁵CGm⁵CGm⁵CG)	d(m⁵CGGCm⁵CG)	d(m⁵CGGGGm⁵CG)/ d(m⁵CGCCm⁵CG)	d(m⁵CGGGGm⁵CG)/ d(m⁵CGC m⁵CCG)	d(Br⁵CGATBr⁵CG)[b]
Buckle (κ)					
C1:G12	6.2	4.7	2.6	1.5	
G2:C11	-4.8	-0.5	-0.8	-3.1	
C3:G10	2.1	14.1	14.8	13.9	
G4:C9	-5.7	-12.6	-5.4	-4.0	
C5:G8	5.3	0.6	2.3	1.9	
G6:C7	-3.8	-3.5	-3.2	-1.0	
Average[c]	4.7 ± 1.5	6.0 ± 5.9	4.8 ± 5.1	4.2 ± 4.9	
x-displacement (d_x)					
C1:G12	-3.6	-3.8	-3.7	-3.5	
G2:C11	-3.6	-3.7	-3.7	-3.4	
C3:G10	-3.4	-3.3	-3.3	-3.2	
G4:C9	-3.5	-3.4	-3.4	-3.4	
C5:G8	-3.8	-4.0	-3.9	-3.9	
G6:C7	-3.6	-4.0	-3.7	-3.6	
Average	-3.6 ± 0.1	-3.7 ± 0.3	-3.6 ± 0.2	-3.5 ± 0.2	

[a] Base step and base pair parameters are shown for crystallized Z-DNA structures containing out-of-alternation base pairs (underlined). All values are in degrees, except rise (D_z) and displacement (d_x), which are in Å.

[b] Values shown for d(Br⁵CGATBr⁵CG) are from ref. 52.

[c] Averages for base pair buckle were calculated from the magnitudes of the values listed ($<\kappa> = (\Sigma|\kappa_i|)/i$, where κ_i is the buckle at base pair i).

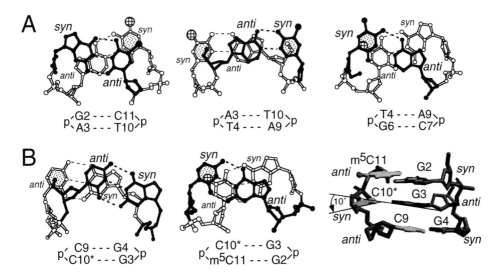

Fig. 7.7. Comparison of the out-of-alternation bases in the structures of (A) d(m⁵CGATm⁵CG) (52) and (B) d(m⁵CGGGm⁵CG)/d(m⁵CGm⁵CCm⁵CG) (59). Shown are the dinucleotide stacks of the out-of-alternation base pairs. The pyrimidine bases that are in the disfavored *syn* conformation are highlighted by the stippled rings. The top base pairs of the stacks are shown as solid atoms and bonds, while the bottom base pairs are in open atoms and bonds. (A) Views down the helix axis of the *syn–p–anti*, *anti–p–syn*, and *syn–p–anti* arrangements of the out-of-alternation base pairs in d(m⁵CGATm⁵CG) are shown. The structure shows that the *syn* thymine is unstacked and protrudes away from the major groove surface for the d(G2:C11)/d(A3:T10), d(A3:T10)/d(T4:A9), and d(T4:A9)/d(G6:C7) stacked base pairs. The guanines at the two out-of-alternation base pairs stacked on top of each other in the d(A3:T10)/d(T4:A9) stack. (B) The *anti–p–syn* and *syn–p–anti* stacking of d(CG) base pairs are shown down the helix axis and perpendicular to the axis of the d(m⁵CGGGm⁵CG)/d(m⁵CGCm⁵Cm⁵CG) structure. In the views down the axis, the single *syn* cytosine is shown to be unstacked also and protruding away from the major groove surface. The view along the helix shows the out-of-alternation d(G3:C10★) base pair. The C10★ base is buckled with respect to the G3 plane.

The significant effects of the out-of-alternation base pairs on the structure of Z-DNA are seen in the stacking of the bases (Fig. 7.7). The purine bases nearly completely overlap in the *anti–p–syn* stack, even more so than the pyrimidine bases of the standard APP sequences. The thymine bases, however, are completely unstacked in both the d(ApT) and d(GpA) steps and, therefore, protrude out from the major groove surface and into the solvent.

The organization of solvent in the minor groove is different from that of the APP d(m⁵CGTAm⁵CG) structure. In this latter case, no ordered waters were observed at the d(TA) base pairs. The d(AT) base pairs of the out-of-alternation structure do support ordered waters, but in a slightly different arrangement than in d(CGCGCG). In this case, the N3 nitrogen of adenine is accessible, as it is in B-DNA.

There are several questions that were left unanswered by this structure. Why are *syn* pyrimidines unstable in Z-DNA? The supercoil-induced B–Z transition free energy ($\Delta G°_T$) for the APP dinucleotide d(CpA)/d(TpG) is 1.3 kcal/mol (10), while that for the non-APP dinucleotide d(TpC)/d(GpA) is 2.5 kcal/mol (13) (Table 7.1). Thus, placing a single *syn* thymine requires 1.2 kcal/mol. The original explanation was that pyrimidines are sterically inhibited from adopting the *syn* conformation because of collisions between the base and the deoxyribose (95,96). The intramolecular distances from the thymine to the sugar in the d(Br⁵CGATBr⁵CG) structure, however, are only slightly shorter than those of guanines *syn* to their sugars. It is unclear as to whether the stacking of bases accounts for this destabilizing effect since, although the thymines are poorly stacked, the adenines show better stacking interactions. It is more likely that the protrusion of the out-of-alternation thymines into the solvent makes the difference. This will be discussed in greater detail later.

The other questions remaining are whether a single base pair that is out-of-alternation is more or less stable than two adjacent out-of-alternation base pairs in a non-APP dinucleotide? Finally, are out-of-alternation d(TA) base pairs more or less stable than d(GC)? These questions can potentially be addressed by studying the structures of non-APP d(GpC) sequences.

Only recently have structures of Z-DNA hexanucleotides been solved that place d(CG) base pairs out-of-alternation. The first was the non-self-complementary sequence d(m⁵CGGGm⁵CG)/d(m⁵CGCCm⁵CG) (59), which has a single *syn* cytosine base (underlined). Like the d(ApT)-containing structure, this cytosine protrudes into the major groove, but the base pair is significantly buckled (Fig. 7.7). This distortion to the base pair, which relieves the steric strain of placing the cytosine *syn*, appears to be induced by the methyl group of an adjacent cytosine. We had proposed that in the absence of methylation of the flanking d(CpG) dinucleotides, the pyrimidine base would slide away from the ribose to relieve the steric strain, much like the thymines do in the d(Br⁵mCGATBr⁵CG) structure (52). In the refined structure, the steric energy was calculated to be essentially identical between this out-of-alternation structure and the standard structure of d(m⁵CGm⁵CGm⁵CG). The structure of d(m⁵CGGGm⁵CG)/d(m⁵CGCm⁵CCG), however, shows the out-of-alternation d(CG) base pair with essentially the same high buckle, even in the absence of the methyl group of the adjacent cytosine. Similarly, both base pairs that are out-of-alternation in the structure of d(m⁵CGGCm⁵CG) show this same buckling (Table 7.11). Thus this distortion to the base plane is inherent to out-of-alternation base pairs, regardless of the flanking base pairs. It may simply be that the *syn* pyrimidine base is not sandwiched by the base and deoxyriboses of the two flanking base pairs, as is the standard guanine base.

The *syn* cytosine affects the solvent structure at both the major groove surface and minor groove crevice. In the minor groove of the out-of-alternation d(CG) base pair, a water is hydrogen bonded to the N2 amino group of the guanine base and no waters are observed bound to the now inaccessible O2 oxygen of the cytosine, as in the non-APP d(ApT) dinucleotides. In addition, there is no pattern of ordered waters around this d(CG) base pair. This may, however, be associated with the orientational disorder of this non-self-complementary sequence.

The difference in stability between a standard d(CG) base pair and an out-of-alternation d(GC) base pair was estimated from supercoiled ccDNA studies to be 1.7 kcal/mol/bp [$\Delta G°_T$ for an APP d(CpG) dinucleotide (dn) is 0.7 kcal/mol/dn (9), while that for a d(CpC)/d(GpG) dinucleotide is 2.4 kcal/mol/dn (13)]. We believe that these solvent rearrangements play a role in this destabilization of Z-DNA.

Perhaps the two most dramatic examples of the out-of-alternation structures are the sequences d(CCCGGG) and d(m⁵CGGCm⁵CG). Both resemble sequences that one might expect to form A-DNA instead of Z-DNA. Indeed, the reverse of the latter sequence, as in the hexamers d(GCCGGC) and its methylated analogue d(Gm⁵CCGGC), have been crystallized as A-DNA (87) . The structure of d(CCCGGG) has not been published in detail and thus we cannot discuss it in this review (47). We have recently completed the structure of d(m⁵CGGCm⁵CG) and find it to be nearly identical to the structure of d(CGCGCG) at the level of the DNA (Table 7.11). The one major exception is in the high buckle of the base pairs that are out-of-alternation (as discussed above). The other important structural perturbation is found in the solvent structure. At the major groove surface, the waters that bridge each cytosine are not as apparent, even at the flanking *anti–p–syn* d(CpG) dinucleotides. At the central out-of-alternation *anti–p–syn* d(GpC) dinucleotide, the two stacked guanines, however, show analogous solvent structures to those of the standard stacked cytosines in d(CGCGCG). For the flanking *anti–p–syn* d(CpG) dinucleotides in the minor groove, the waters that link the guanine N2 amino groups to the phosphoribose backbone were still observed and thus help to stabilize these in the *syn* conformation. The spine of hydration that links the cytosines in the minor groove, however, is no longer present. At the two central out-of-alternation d(GC) base pairs, the *syn* cytosines are not at all accessible to solvent in the minor groove crevice. The two stacked guanines, however, are bridged by two waters that are analogous to the waters that normally form the spine that bridges the central cytosine bases in d(CGCGCG). This may help to increase the stability of the two base pairs that are out-of-alternation if they occur adjacent to each other as opposed to being separated in a sequence. Thus, although the DNA structure is not dramatically affected in this very unlikely Z-DNA sequence, the water interactions are.

4. Summary: sequence effects on the structure and stability of Z-DNA

The nucleotide sequence affects not only the structure, but also the stability of Z-DNA. We have concentrated on how the major and minor grooves are affected, as well as the related solvent rearrangments at these surfaces because these are the classical explanations given for whether a DNA duplex conformation is stable or not. The characterization of Z-DNA sequences that contain d(m⁵CG), d(CG), d(CI), d(UA), d(TA), and d(TD) base pairs in various combinations suggests that there are several distinct factors important for Z-DNA stability. Amidation of the purine base at the C2 helps to stabilize Z-DNA. Removing the N2 amino group from guanine destabilizes Z-DNA in d(CG) sequences, while adding this group to adenine helps to stabilize the structure in d(TA)-containing sequences. Methylation at the C5 position of pyrimidine bases has both a stabilizing and destabilizing effect on Z-DNA. Z-DNA is stabilized by methylation of cytosines, as in d(m⁵CG), and also when thymines are demethylated to form deoxyuridine. This apparent contradictory effect of methylation depends on its position relative to the amino and keto groups of the base pairs in the major groove.

We should stress, however, that comparisons of Z-DNA structures alone cannot provide an accurate account of the factors that stabilize a sequence in this form. These same parameters must be compared with the reference B-DNA structures of these sequences. Even then, however, it is not entirely clear how all these various factors contribute to the ability or inability of certain sequences to adopt the left-handed form of the duplex. For example, if one simply compares the spine of hydration in the narrow minor groove across the various Z-DNA structures, we see that this spine is disrupted by narrow minor grooves, the lack of an amino group contributed by the purine, and base pairs that violate the alternating pyrimidinepurine sequence motif for the *anti–p–syn* dinucleotide stacking. It is also clear that solvent interactions at the major groove (e.g. cation complexes that bridge stacked adjacent bases) will also affect the structure of the minor groove and its hydration spine. Whether this facilitates or hinders the formation of Z-DNA depends on ones point of view. One can argue that solvent interactions are stabilizing since waters can form a direct bridge from the N2 amino group (if present) of the purine base to the DNA backbone, which would help to hold the base in the *syn* conformation. Furthermore, all this discussion says nothing about the effect of this amino group on the spine of hydration in the minor groove of B-DNA. However, a well-structured water network can be argued to be destabilizing to either B- or Z-DNA from the perspective of the reduced entropy of the solvent structure (24). Thus, although the large data set of single-crystal structures for different sequences and substituent groups as Z-DNA provides a wealth of structural information, the details may not tell us much about the stability of this unusual conformation if we are confined to these qualitative comparisons.

One approach that does utilize the crystal structures to study and predict the effects of sequence of substituent groups on Z-DNA stability is to calculate solvent free energies (SFEs) from the structures, and to compare these to SFEs for the same sequences as B-DNA (ΔSFE_{Z-B}). In this case, the reference B-DNA state is treated explicitly. Unfortunately, not all the various substituent modifications are well represented in B-DNA crystal structures; however, the SFEs calculated from B-DNA models constructed using idealized parameters appear to represent accurately the free energy for hydrating this form, even when compared with the conformations of sequences in single crystals (89,97).

For the standard APP sequences, we can derive a thermodynamic cycle (26) to elucidate how each base substituent affects the hydration and stability of Z-DNA. For example, deamination of the guanine in d(CG) base pairs to form d(CI) has an energetic cost of +1.6 kcal/mol, whereas amination of d(TA) to form d(TD) favours Z-DNA by −1.4 kcal/mol (Fig. 7.8). This underscores the importance of the amino group in the minor groove and is consistent with its role in coordinating water molecules to form the spine of water molecules that traverse the minor goove (51). It also explains the apparent contradictory effect of methylation on the stability of Z-DNA, with methylation of cytosines favouring the left-handed form and the thymine methyl disfavouring this form. This is not intuitive, but when the SASs of each surface type are compared for B- and Z-DNA, they become more apparent (Table 7.12). Methylation of cytosine does have the effect of increasing the overall exposed hydrophobic surface for the d(CpG) dinucleotides; however, this increase is significantly greater for B-DNA than for Z-DNA and thus increases the relative stabil-

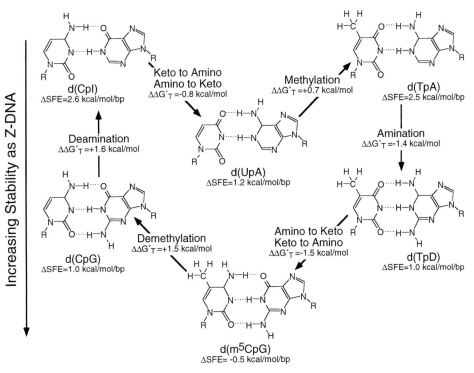

Fig. 7.8. Effect of substituent groups on the differences in solvent free energies (ΔSFE) and the stability (ΔΔG°$_T$) of dinucleotides in Z-DNA versus B-DNA. A thermodynamic cycle is shown for the addition, removal, or replacement of various substituent groups, starting with the most stable dinucleotide as Z-DNA d(m⁵CpG) to the least stable [d(CpI) and d(TpA)] and back to d(m⁵CpG). The ΔSFE are shown for each dinucleotide, while the effects of the change in the substituent on the stability of Z-DNA (ΔΔG°$_T$) are shown for each modification step.

ity of the left-handed form. It is now becoming evident that cytosine methylation destabilizes B-DNA, allowing the formation of A-DNA in crystals (87), and increasing the frequency for cytosine deamination in solution (98).

We can also make some predictions concerning the base pairs that are out-of-alternation. From the SFE calculations (Table 7.9), we can see that the d(ApT) as an out-of-alternation *anti–p–syn* dinucleotide is predicted to be less stable as Z-DNA compared with the analogous d(GpC) out-of-alternation dinucleotide. This appears to be associated primarily not with the out-of-alternation steps themselves [in this case, the d(ApT) step is actually more stable], but with how each out-of-alternation base pair affects the flanking base pairs. Finally, a single d(CG) that is out-of-alternation is predicted to be only slightly less destabilized as Z-DNA compared with the d(GpC) dinucleotide. Thus, we would expect that placing the cytosines of two adjacent base pairs in a *syn* conformation is more favourable than having them separated.

Upon putting all of this together in the context of the crystallography of Z-DNA, it became evident that the SFE calculations are useful as an analytical tool for predicting the target salt concentrations for obtaining crystals of this conformation (22). A

Table 7.12. Solvent accessible surface areas (Å^2) of dinucleotides steps as B- and Z-DNA

Conformation (B/Z)	Dinucleotide	Base atoms					Ribose atoms			Total
		C	CH$_3$(C5)	O	N	N2	C'	O'	P	
B	d(TpA)	43.6	44.8	32.6	55.6	–	182.8	51.4	132.8	543.6
Z	d(TpA)	46.0	46.2	27.0	50.8	–	188.2	42.0	133.6	533.8
B	d(TpD)	28.8	44.8	29.8	46.8	26.0	183.8	43.8	132.8	536.6
Z	d(TpD)	33.3	46.3	27.4	49.7	21.5	170.9	41.8	133.2	524.1
B	d(CpG)	49.4	–	31.0	59.1	23.6	185.5	47.2	132.6	528.4
Z	d(CpG)	56.3	–	44.2	48.7	19.6	184.0	47.4	132.1	532.3
B	d(CpG)	64.8	–	38.0	65.2	–	197.7	52.3	132.6	550.6
Z	d(CpG)	71.4	–	47.2	48.8	–	199.4	46.8	133.0	546.6
B	d(UpA)	63.2	–	39.2	57.6	–	190.0	51.4	133.6	535.0
Z	d(UpA)	68.6	–	37.8	57.0	–	194.6	42.0	133.8	533.8
B	d(m^{5}CG)	26.9	48.3	31.1	46.6	30.3	195.7	42.3	127.1	548.2
Z	d(m^{5}CG)	31.4	50.5	37.7	43.8	20.2	180.8	40.6	141.9	547.0
B	d(ApT)[a]	33.0	55.1	35.1	58.4	–	185.4	48.0	127.7	542.7
Z	d(ApT)[a]	55.4	80.3	24.6	37.2	–	171.8	48.9	131.0	549.2
B	d(GpC)[a]	34.3	–	28.3	57.9	28.1	187.8	40.3	128.8	505.6
Z	d(GpC)[a]	75.3	–	21.6	70.5	20.8	177.2	47.1	132.5	544.9
B	d(GpG)/ d(CpC)[a]	32.8	–	28.5	60.0	27.7	191.3	41.5	128.3	510.1
Z	d(GpG)/ d(CpC)[a]	67.2	–	25.5	62.0	20.6	183.6	40.9	135.9	535.7

[a] Out-of-alternation dinucleotide step.

comparison of the *CS* for crystallization of the current Z-DNA sequences (Table 7.2) shows a strong corelation to the ΔSFE_{Z-B} for these sequences (Fig. 7.9). This relationship apparently arises for the stabilization of the Z- versus the B-form as both the salt and alcohol concentrations in the crystallization set-ups are increased. The pathway for crystallization, therefore, directs the DNA to the left-handed form in solution, while avoiding various amorphous precipitant forms along the way.

The shortcoming of the SFE approach in studying stability is that we do not utilize any of the detailed information on solvent interactions gleaned from the high resolution single-crystal structures. The general hydration parameters from the SFE calculations should somehow be related to these specific patterns of water structure. This is perhaps where Z-DNA may play its most significant role in physical biochemistry. The accumulated structural and thermodynamic data for all these various sequences can provide a benchmark for the development of molecular force fields. It serves much the same function as the hydrogen atom to physical chemists. The properties of

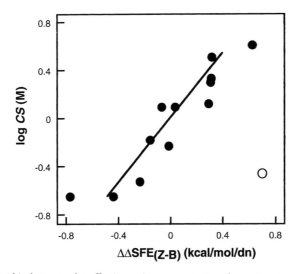

Fig. 7.9. Relationship between the effective cation concentration (log cation strength or logCS) of the crystallization solutions and the difference in solvent free energy between Z-DNA and B-DNA ($\Delta\Delta SFE_{Z-B}$) for sequences crystallized as Z-DNA. The logCS that could be determined (Table 7.2) for all sequences (that contain base pairs of the type defined in Fig. 7.4) are plotted relative to the $\Delta\Delta SFE_{Z-B}$ calculated (Table 7.9) for these sequences. The open circle represents the sequence d(CGCICG), which was crystallized by the hanging drop method of vapour diffusion. The line represents the best linear fit of the data for sequences with SFE between -0.4 and $+0.4$ kcal/mol/dn (slope = 1.36, y-intercept = 0.05, R = 0.93). The plot asymptotes at both high and low values for logCS. At the high end, the salt concentrations reach the point of saturation in the crystallization solutions, while the low end represents the minimum amount of cations required to crystallize the DNAs (approximately equal to the concentration of mononucleotide equivalents in the DNA).

Z-DNA are now very well understood; next, we need to develop the theories to explain the properties. Once developed, these same principles should be generally applicable to the study and prediction of all classes of biological macromolecules.

Acknowledgements

This work has been supported by grants from the National Science Foundation (MCB972824), the National Institutes of Health (R05GM54538A), and the Environmental Health Sciences Center at Oregon State University (NIEHS ES00210). We would like to thank Mason Kwong and Christine Nguyen for their help with this project.

References

1. Watson, J.D. and Crick, F.H.C. (1953) *Nature* **171**, 737.
2. Franklin, R.E. and Gosling, R.G. (1953) *Nature* **172**, 156.
3. Pohl, R.M. and Jovin, T.M. (1972) *J. Mol. Biol.* **647**, 375.
4. Wang, A.H.-J., Quigley, G.J., Kolpak, F.J., Crawford, J.L., van Boom, J.H., van der Marel, G. and Rich, A. (1979) *Nature* **282**, 680.

5. Wang, A.H.-J., Quigley, G.J., Kolpak, F.J., van der Marel, G., van Boom, J.H. and Rich, A. (1979) *Science* **211**, 171.
6. Drew, H.R., Takano, T., Tanaka, S., Itakura, K. and Dickerson, R.E. (1980) *Nature (London)* **286**, 755.
7. Shakked, D., Rabinovich, D., Cruse, W.B.T., Egert, E., Kennard, O., Sals, G., Salisbury, S.A. and Viswamitra, M.A. (1981) *Proc. R. Soc. (London) B* **213**, 479.
8. Wing, R.M., Drew, H.R., Takano, T., Broka, C., Tanaka, S., Itakura, K., and Dickerson, R.E. (1980) *Nature* **287**, 755.
9. Peck, L.J. and Wang, J.C. (1983) *Proc. Natl. Acad. Sci. USA* **80**, 6206.
10. Vologodskii, A.V. and Frank-Kamenetskii, M.D. (1984) *J. Biomol. Struct. Dynamics* **1**, 1325.
11. Ellison, M.J., Feigon, J., Kelleher, R.J., III, Wang, A.H.-J., Habener, J.F. and Rich, A. (1986) *Biochemistry* **25**, 3648.
12. McLean, M.J., Lee, J.W. and Wells, R.D. (1988) *J. Biol. Chem.* **263**, 7378.
13. Ellison, M.J., Kelleher, R.J., III, Wang, A.H.-J., Habener, J.F. and Rich, A. (1985) *Proc. Natl. Acad. Sci. USA* **82**, 8320.
14. Behe, M. and Felsenfeld, G. (1981) *Proc. Natl. Acad. Sci. USA.* **78**, 1619.
15. Moller, A., Nordheim, A., Kozlowski, S.A., Patel, D. and Rich, A. (1984) *Biochemistry* **23**, 54.
16. Jovin, T.M., McIntosh, L.P., Arndt-Jovin, D., Zarling, D.A., Robert-Nicoud, M., van de Sande, J.H. and Jorgenson, K.F. (1983) *J. Biomol. Struct. Dynamics* **1**, 21.
17. Sagi, J., Szemzo, A., Otvos, L., Vorlikckova, M. and Kypr, J. (1991) *Int. J. Biol. Macromol.* **13**, 329.
18. Fujii, S., Wang, A.H.-J., van der Marel, G., van Boom, J.H. and Rich, A. (1982) *Nucl. Acids Res.* **10**, 7879.
19. Chevrier, B., Dock, A.C., Hartmann, B., Leng, M., Moras, D., Thuong, M.T. and Westhof, E. (1986) *J. Mol. Biol.* **188**, 707.
20. Vorlickova, M. and Sagi, J. (1991) *Nucl. Acids Res.* **21**, 2343.
21. Wang, L. and Keiderling, T.A. (1993) *Nucl. Acids Res.* **21**, 4127.
22. Ho, P.S., Kagawa, T.F., Tseng, K., Schroth, G.P. and Zhou, G. (1991) *Science* **254**, 1003.
23. Coll, M., Wang, A.H.-J., van der Marel, G.A., van Boom, J.H. and Rich, A. (1986) *J. Biomol. Struct. Dynamics* **4**, 157.
24. Schneider, B., Ginell, S.L., Jones, R., Gaffney, B. and Berman, H.M. (1992) *Biochemistry* **31**, 9622.
25. Zhou, G. and Ho, P.S. (1990) *Biochemistry* **29**, 7229.
26. Kagawa, T.F., Howell, M.L., Tseng, K. and Ho, P.S. (1993) *Nucl. Acids Res.* **21**, 5978.
27. McDonnell, N.B. and Preisler, R.S. (1989) *Biochem. Biophys. Res. Commun.* **164**, 426.
28. Preisler, R.S., Chen, H.H., Colombo, M.F., Choe, Y., Short, B.J.J. and Rau, D.C. (1995) *Biochemistry* **34**, 14400.
29. Tereshko, V. and Milinina, L. (1990) *J. Biomol. Struct. Dynamics* **7**, 827.
30. Feuerstein, B.G., Williams, L.D., Basu, H.S. and Marton, L.J. (1991) *J. Cell. Biochem.* **46**, 37.
31. Thomas, T.J., Gunnia, U.B. and Thomas, T. (1991) *J. Biol. Chem.* **266**, 6137.
32. Thomas, T.J. and Thomas, T. (1994) *Biochem. J.* **298**, 485.
33. Rahmouni, A.R. and Wells, R.D. (1989) *Science* **246**, 358.
34. Ramakrishnan, B. and Viswamitra, M.A. (1988) *J. Biomol. Struct. Dynamics* **6**, 511.
35. Drew, H.R. and Dickerson , R.E. (1981) *J. Mol. Biol.* **152**, 723.
36. Crawford, J.L., Kolpak, F.J., Wang, A.H.–J., Quigley, G.J., van Boom, J.H., van der Marel, G.A. and Rich, A. (1980) *Proc. Natl. Acad. Sci. USA* **77**, 4016.

37. Gessner, R.V., Frederick, C.A., Quigley, G.J., Rich, A. and Wang, A.H.-J. (1989) *J. Biol. Chem.* **264**, 7921.

38. Egli, M., Williams, L.D., Gao, Q. and Rich, A. (1991) *Biochemistry* **30**, 11388.

39. Bancroft, D., Williams, L.D., Rich, A. and Egli, M. (1994) *Biochemistry* **33**, 1073.

40. Ohishi, H., Nakanishi, I., Inubushi, K., van der Marel, G.A., van Boom, J.H., Rich, A., Wang, A.H.-J., Hakoshima, T. and Tomita, K. (1996) *FEBS Lett.* **391**, 153.

41. Ho, P.S., Frederick, C.A., Saal, D., Wang, A.H.-J. and Rich, A. (1987) *J. Biomol. Struct. Dynamics* **4**, 521.

42. Gessner, R.V., Quigley, G.J., Wang, A.H.-J., van der Marel, G.A., van Boom, J.H. and Rich, A. (1985) *Biochemistry* **24**, 237.

43. Gao, Y.G., Sriram, M. and Wang, A.H.-J. (1993) *Nucl. Acids Res.* **21**, 4093.

44. Mooers, B.H.M., Eichman, B.F. and Ho, P.S. (1997) *J. Mol. Biol.* **269**, 796.

45. Fujii, S., Wang, A.H.-J., Quigley, G.J., Westerink, H., van der Marel, G., van Boom, J.H. and Rich, A. (1985) *Biopolymers* **24**, 243.

46. Ban, C., Ramakrishnan, B. and Sundaralingam, M. (1996) *Biophys. J.* **7**, 1215.

47. Malinina, L., Urpi, L., Salas, X., Huynh-Dinh, T. and Subirana, J.A. (1994) *J. Mol. Biol.* **243**, 484.

48. Kagawa, T.F., Geierstanger, B.H., Wang, A.H.-J. and Ho, P.S. (1991) *J. Biol. Chem.* **266**, 20175.

49. Coll, M., Fita, I., Lloveras, J., Subirana, J.A., Bardella, F., Huynh-Dinh, T. and Igolen, J. (1988) *Nucl. Acids Res.* **16**, 8695.

50. Sadsivan, C. and Gautham, N. (1995) *J. Mol. Biol.* **248**, 918.

51. Wang, A.H.-J., Hakoshima, T., van der Marel, G., van Boom, J.H. and Rich, A. (1984) *Cell* **37**, 321.

52. Wang, A.H.-J., Gessner, R.V., van der Marel, G.A., van Boom, J.H. and Rich, A. (1985) *Proc. Natl. Acad. Sci. USA* **82**, 3611.

53. Geierstanger, B.H., Kagawa, T.F., Chen, S.-L., Quigley, G.J. and Ho, P.S. (1991) *J. Biol. Chem.* **266**, 20185.

54. Ginell, S.L., Kuzmich, S., Jones, R.A. and Berman, H.M. (1990) *Biochemistry* **29**, 10461.

55. van Meervelt, L., Moore, M.H., Lin, P.K.T., Brown, D.M. and Kennard, O. (1990) *J. Mol. Biol.* **216**, 773.

56. Kumar, V.D. and Weber, I.T. (1993) *Nucl. Acids Res.* **21**, 2201.

57. Cervi, A.R., Guy, A., Leonard, G.A., Téoule, R. and Hunter, W.N. (1993) *Nucl. Acids Res.* **21**, 5623.

58. Eichman, B.F., Basham, B., Schroth, G.P. and Ho, P.S. (submitted).

59. Schroth, G.P., Kagawa, T.F. and Ho, P.S. (1993) *Biochemistry* **32**, 13381.

60. Bononi, J. (1994) *MS Thesis*, Oregon State University, Corvallis.

61. Parkinson, G.N., Arvanitis, G.M., Lessinger, L., Ginell, S.L., Jones, R., Gaffney, B. and Berman, H.M. (1995) *Biochemistry* **34**, 15487.

62. Moore, M.H., van Meervelt, L., Salisbury, S.A., Kong Thoo Lin, P. and Brown, D.M. (1995) *J. Mol. Biol.* **251**, 665.

63. Peterson, M.R., Harrop, S.J., McSweeney, S.M., Leonard, G.A., Thompson, A.W., Hunter, W.N. and Helliwell, J.R. (1996) *J. Synch. Rad.* **3**, 24.

64. Kumar, V.D., Harrison, R.W., Andrews, L.C. and Weber, I.T. (1992) *Biochemistry* **31**, 1541.

65. Brennan, R.G., Westhof, E. and Sundaralingam, M. (1986) *J. Biomol. Struct. Dynamics* **3**, 649.

66. Doi, M., Inoue, M., Tomoo, K., Ishida, T., Ueda, Y., Akagi, M. and Urata, H. (1993) *J. Am. Chem. Soc.* **115**, 10432.

67. Zhang, H., van der Marel, G., van Boom, J. and Wang, A.H.-J. (1992) *Biopolymers* **32**, 1559.

68. Teng, M., Liaw, Y.-C., van der Marel, G.A., van Boom, J.H. and Wang, A.H.-J. (1989) *Biochemistry* **28**, 4923.

69. Ho, P.S., Frederick, C.A., Quigley, G.J., van der Marel, G.A., van Boom, J.H., Wang, A.H.-J. and Rich, A. (1985) *EMBO J.* **4**, 3617.

70. Coll, M., Saal, D., Frederick, C.A., Aymami, J., Rich, A. and Wang, A.H.-J. (1989) *Nucl. Acids Res.* **17**, 911.

71. Brown, T., Kneale, G., Hunter, W.N. and Kennard, O. (1986) *Nucl. Acids Res.* **14**, 1801.

72. Berman, H.M., Olson, W.K., Beveridge, D.L., Westbrook, J., Gelbin, A., Demeny, T., Hsieh, S.-H., Srinivasan, A.R. and Schneider, B. (1992) *Biophys. J.* **63**, 751.

73. Bernstein, F.C., Koetzle, T.F., Williams, G.J.B., Meyer, E.F., Jr., Brice, M.D., Rodger, J.R., Kennard, O., Shimanouchi, T. and Tasumi, M. (1977) *J. Mol. Biol.* **112**, 535.

74. Egli, M. and Gessner, R.V. (1995) *Proc. Natl. Acad. Sci. USA* **92**, 180.

75. Diekmann, S. (1989) *EMBO J.* **8**, 1.

76. Dickerson, R.E. (1992) *Meth. Enzymol.* **211**, 67.

77. Rich, A., Nordheim, A. and Wang, A.H.-J. (1984) *Annu. Rev. Biochem.* **53**, 791.

78. Jovin, T.M., Soumpasis, D.M. and McIntosh, L.P. (1987) *Annu. Rev. Phys. Chem.* **38**, 521.

79. Gessner, R.V., Quigley, G.J. and Egli, M. (1994) *J. Mol. Biol.* **236**, 1154.

80. Drew, H.R. and Dickerson, R.E. (1981) *J. Mol. Biol.* **151**, 535.

81. Berman, H.M. (1994) *Curr. Opin. Struct. Biol.* **4**, 345.

82. Kubinec, M.G. and Wemmer, D.E. (1992) *J. Am. Chem. Soc.* **114**, 8739.

83. Liepinsh, E., Otting, G. and Wuthrich, K. (1992) *Nucl. Acids Res.* **20**, 6549.

84. Morgan, J.E., Blankenship, J.W. and Matthews, H.R. (1986) *Arch. Biochem. Biophys.* **246**, 225.

85. Tabor, C.W. and Tabor, H. (1984) *Annu. Rev. Biochem.* **53**, 749.

86. Howell, M.L., Schroth, G.P. and Ho, P.S. (1996) *Biochemistry* **35**, 15373.

87. Mooers, B.H.M., Schroth, G.P., Baxter, W.W. and Ho, P.S. (1995) *J. Mol. Biol.* **249**, 772.

88. Quadrifoglio, F., Manzini, G. and Yathindra, N. (1984) *J. Mol. Biol.* **175**, 419.

89. Kagawa, T.F., Stoddard, D., Zhou, G. and Ho, P.S. (1989) *Biochemistry* **28**, 6642.

90. Futscher, B.W., Rice, J.C., Ho, P.S. and Dalton, W.S. (submitted).

91. Melander, W. and Horvath, C. (1977) *Arch. Biochem. Biophys.* **183**, 200.

92. Hamada, H. and Kakunaga, T. (1982) *J. Cell. Biochem.* **3**, 333.

93. Hamada, H., Petrino, M.G. and Kakunaga, T. (1982) *Proc. Natl. Acad. Sci. USA* **79**, 6465.

94. Schroth, G.P., Chou, P.J. and Ho, P.S. (1992) *J. Biol. Chem.* **267**, 11846.

95. Davies, D.B. (1978) *Progress in NMR Spectroscopy*, Vol. 12, p. 135. Pergamonn, Oxford.

96. Haschmeyer, A.E.V. and Rich, A. (1967) *J. Mol. Biol.* **27**, 369.

97. Basham, B., Schroth, G.P. and Ho, P.S. (1995) *Proc. Natl. Acad. Sci. USA* **92**, 6464.

98. Zhang, X. and Mathews, C.K. (1994) *J. Biol. Chem.* **269**, 7066.

8

Standard DNA duplexes and RNA:DNA hybrids in solution

Uli Schmitz, Forrest J. H. Blocker, and Thomas L. James

Department of Pharmaceutical Chemistry, University of California, San Francisco, San Francisco, CA 94143-446, USA

1. Introduction

Following the current paradigm that biological function is encoded in three-dimensional molecular structure, the last 10 years have seen great efforts in the determination of well-defined solution structures of DNA using high resolution nuclear magnetic resonance (NMR) methods and modelling tools. Since the 1970s, NMR has emerged as the method of choice for the study of biomolecules under more physiological conditions than can be created in the crystalline state. Initially, it was successfully applied to the study of nucleic acids to yield only 'low-resolution' structural insights, such as secondary structure information or qualitative differences between similar samples. With large improvements in magnetic field strengths and development of sensitive two-dimensional (2D) NMR experiments, as well as the development of large-scale synthetic methods for nucleic acids, high resolution structures of DNA could finally be tackled in the mid to late 1980s (see refs 1–3 for review). Consequently, most of the research in that period was directed towards establishing adequate methodologies for: (i) achieving complete assignments of solvent-exchangeable and non-exchangeable protons in nucleic acids (4); (ii) converting spectral observables, i.e. intensities from quantitative 1D and 2D NOE spectroscopy, or coupling constants from correlated spectroscopy into structural information (5–8); and (iii) building adequate models using the NMR-derived structural information (ref. 5 and references therein).

A spate of high resolution DNA duplex structures appeared in the early 1990s, and it became conceivable that sequence-specific structural rules could be established readily if enough high quality duplex structures could be solved. However, prompted by DNA crystal structure results, it became evident that a sequence-specific code for structure is more complex than anticipated. On the other hand, it has become clear that the particular structure of a dinucleotide step depends on neighbouring sequences (9). This also led to the idea of a specific flexibility or malleability of certain DNA sequences. This, in turn, causes difficulties in the interpretation of NMR-derived structural data, making the conformational analysis of a particular sequence in terms of rigid helical parameters problematic.

Not surprisingly, some of the early excitement stimulated cautionary studies analysing the accuracy and limitations of the NMR-based approach (10–14). Our discussion of structures will be preceded by a critical summary of methodologies and inherent limitations, as well as their impact on the structures derived.

Despite the intrinsic limitations of NMR-derived structures, which are still not fully understood, some promising results emerged from a statistical analysis of a handful of the most accurately defined DNA duplex structures (15,16). This analysis revealed that the average helical twist of a particular set of structures is virtually identical with that measured by independent biophysical solution measurements (17,18) lending some long-awaited credibility to the NMR approach.

There have been a large variety of applications of NMR to DNA. Over 1000 studies are listed in the Medline® database (19) for the years 1990–1996 in which NMR was applied to some kind of sample containing DNA. A comparatively small number of publications report high resolution structures of unmodified, standard DNA duplexes (4%), but over 50% of all studies focus on the interaction of DNA with other molecules, e.g. proteins and peptides (9%), drugs (covalently attached or as a non-covalent complex) (40%), and cations (5%). This chapter will emphasize the structures of standard duplexes.

NMR is also an excellent tool for elucidating the dynamic properties of nucleic acids (20), since many of the NMR observables either encode conformational flexibilites, owing to their nature as time-average parameters, e.g. coupling constants and NOE intensities, or reflect global and local dynamic properties directly, e.g. line widths and other relaxation parameters.

2. Data and methods for high resolution structure determination

Before discussing detailed differences between high resolution NMR structures of DNA or RNA:DNA hybrids, a brief review of the methods involved in structure determination is helpful to understand some of the intrinsic problems.

2.1 NMR data and restraints for the determination of high resolution structures

Structural information used for DNA structure determination is typically in the form of distance restraints extracted from multidimensional homonuclear NOE spectra; $\{^1H,^1H\}$NOE cross-peak volumes are converted into distance restraints in a more or less quantitative fashion (21). Distance data are often augmented by scalar coupling constant-derived $^1H^1H$ or $^1H^{31}P$ torsion angles, available from various correlated spectroscopy experiments (e.g. COSY, HETCOR). Detailed reviews are available describing acquisition of suitable data sets, proton assignments, and the generation of adequate restraint lists for the structure refinement procedure (4,5,22,23). For the present purpose, it is important to realize that the NMR-derived parameters largely represent local information, where coupling constants usually extend over three bonds and NOEs reflect distances of >7 Å, with the higher values only in the more auspicious case of methyl groups.

It is clear that there is not enough information available from NMR data alone to define precisely the structure of a DNA duplex without making assumptions about base pairing geometries and without inclusion of an empirical chemical force field. The precision of a structure, i.e. the reproducibility of the structures generated from the experimental data will improve with an increasing number of distance and torsion angle restraints, as well as with the precision of these restraints. Consequently, decreased precision in the restraints, equivalent to wider distance error bounds, will require a larger number of restraints to define a structure with the same precision. For the increasingly encountered situation of isotope-labelled proteins and RNA, heteronuclear NOE spectra require a very conservative quantitative interpretation of distance restraints, e.g. using only an upper bound of 6 Å (24). However, the huge number of NOEs in multidimensional spectra of proteins and RNA compensates well for this loss in precision. Also, in these cases the focus is not usually on small sequence-specific structural details.

For DNA duplexes, however, where isotope-labelled samples and multidimensional heteronuclear NMR is just on the horizon, one can currently assign 10–15 NOEs per residue, 20 in the most favourable cases. However, the distribution is not ideal; there are relatively few restraints involving the proton-poor base moieties. These restraints are not sufficient to derive a precise structure with high accuracy unless the tightness of the distance restraints is considerably improved compared with the conservative approach in the case of isotope-labelled samples. One possibility, commonly used in protein structural work, is to group the NOE intensities semi-quantitatively (e.g. weak, medium, and large, corresponding to distances of 4–7, 3–5, and 1.8–3.5 Å, respectively) (25). Since, in DNA duplexes, the restraints are not equally distributed over the molecule and there are typically few 'long-range distances', i.e. non-sequential, cross-strand distances, the semi-quantitative interpretation is even more prone to cause artefacts resulting from multispin effects, commonly termed spin diffusion (26,27). An approximate way of coping with spin diffusion is to extrapolate NOE intensities towards very small mixing times using NOE build-up curves (28). This procedure requires a number of NOE datasets to be analysed and still cannot properly account for spin diffusion effects (26). A systematic improvement in the restraints' accuracy requires the consideration of all structure-dependent relaxation pathways of the entire proton system, which can be done with complete relaxation matrix methods (29). With this approach, the theoretical NOE spectra, including spin diffusion effects, can be calculated for a given structure with the assumption of a motional model. In reverse, this strategy can be used to compute accurate distances from NOE intensities for unknown structures using a hybrid matrix approach, where all experimentally unobserved NOEs are taken from a model structure that is similar to the target structure (30–32). Exact details of methods differ, but the elements of the relaxation matrix are varied until a consistent fit to the experimentally observed NOEs is obtained. While programs like IRMA (32) and MORASS (31) integrate this process directly into the conformational search for the final structure via restrained molecular dynamics (rMD), the program MARDIGRAS (30) simply varies cross-relaxation rates until the best solution is found as indicated by a minimum residual index. Interproton distances are readily obtained from the converged relaxation matrix. This latter route, where defining the target and meeting it are separate steps, offers the methodological advantage that no structural assumptions are implicit via the chemical force field of the search engine or the constraint imposed by the requirement that all distances must be satisfiable by a single DNA model. Therefore, a set of MARDIGRAS-derived distances might even exhibit some mutually inconsistent restraints, a situation that can result from conformational flexibility leading to dynamically averaged, and therefore possibly inconsistent, distances. Consequently, examination of these distances may aid the recognition of conformational flexibility.

Complete relaxation matrix methods have also been implemented in different molecular dynamics programs to allow refinement directly against NOE intensities, e.g. AMBER (33), XPLOR (34), GROMOS (35), CHARMM (36), and some commercial packages such as DISCOVER (37). Back-calculation methods lead to spin diffusion-adjusted structures, when preliminary, distance geometry-derived models (38,39) or canonical starting models (40) are refined by adjusting local model geometries to match the appearance of the actual NOESY spectra. In a related approach, the program NUCFIT varies local structural parameters to match NOE build-up curves by taking into account spin diffusion and anisotropic molecular motion (12,41).

Another important difference between the various complete relaxation matrix implementations is the way they deal with the assessment of errors and propagation of those errors into the structure derived. Integrated NOE cross-peak volumes have limited accuracy, owing to peak overlap, spectral noise levels, distortion from solvent or diagonal peaks, exchange processes, incomplete proton relaxation, or baseline problems.

To obtain reliable NMR structures, it is important to use distance restraints with bounds as tight as possible, without ignoring the inherent experimental and methodological errors (26). The subsequent refinement process should then yield a final structure along with a conformational envelope reflecting the intrinsic limitations in the original data and the method. Therefore, it is important to use a refinement procedure that utilizes upper and lower restraint bounds for obtaining DNA structures, such that the degree of accuracy is not lost in the process. The MARDIGRAS procedure uses an absolute noise error, e.g. conservatively using the size of the smallest NOE, and an additional relative error, e.g. 10–20% in distance bounds determination. Furthermore, different motional models with varying correlation times are used (e.g. the isotropic correlation time may be varied from 2 to 4 ns for a short oligonucleotide) in combination with different starting geometries. Typically, the difference between upper and lower MARDIGRAS-derived distance bounds encountered in the DNA structural work discussed below is around 0.3 Å. However, distance determinations for a covalently modified DNA exhibiting a number of unusual, long fixed distances revealed that NOE intensity error propagation, especially for the weak NOEs, needed to be accounted for in order to reproduce all the fixed distances (42). An alternative procedure was devised, entailing MARDIGRAS calculations repeated 30–100 times using NOE intensity sets randomly perturbed with user-selected noise and relative error limits. With this modified MARDIGRAS procedure (42), the average distance restraint width increased to 0.6–1.0 Å, reflecting the significantly wider bounds for weak NOEs (e.g. 3–5 Å instead of 1–2 Å). The compensation for the broader bounds is that distances up to 7 Å could be reliably determined (42).

Even using coupling constant-derived torsion angles as structural restraints is not as straightforward as it seemed in the early 1990s (see refs 3 and 7 for review). With respect to $^3J_{HH}$ coupling constants, which are commonly employed to restrain the deoxyribose moieties, it appears that one cannot rely blindly on the modified Karplus equation (7,43), which arose from the empirical correlation of hundreds of measured coupling constants and torsion angles of small cyclic molecules. For macromolecules, dipolar effects can influence the scalar coupling constants (44,45); the effects become significant at larger correlation times. On a practical level, coupling constants are extracted from COSY-type experiments, often involving extensive peak shape analysis via simulation (46–50). Typical errors range from ±0.3 Hz for the best-defined coupling constants, i.e. $J_{H1'H2'}$ and $J_{H1'H2''}$, to ±1 Hz or more for $J_{H3'H4'}$. Model calculations have shown that dipolar contributions are small for correlation times below 5 ns (45,51), with effects smaller than experimental errors. Furthermore, the sum of the vicinal coupling constants for a particular proton, e.g. $\Sigma H1'$, is not affected by dipolar relaxation (50,52) because of compensatory effects for individual coupling constants, $J_{H1'H2''}$ and $J_{H1'H2'}$. Altona and coworkers (6) showed that $\Sigma H1'$ is a very useful parameter for assessing whether or not a particular sugar moiety adopts a rigid conformation; even the relative occupation of S- and N-type conformations can be inferred from $\Sigma H1'$. Overall, it seems that deoxyribose coupling constants can be used safely for DNA structural work when the length does not exceed 12–14 base pairs and the temperature is above 15–20°C. Lane and coworkers (50) demonstrated that a reliable deoxyribose conformational analysis can be accomplished even for a 16-mer duplex if the temperature is chosen appropriately (e.g. 50°C, yielding a correlation time of 5 ns).

^{31}P–^{1}H coupling constants can be used to define parts of the DNA backbone (3,8). However, limited by experimental accuracy and the relative insensitivity of this coupling constant on the dihedral angle, backbone torsion angles restraints are typically implemented in a rather qualitative fashion with restraint widths from 60 to 180°. If the NOE restraints define the relative geometry between sequential nucleotides sufficiently, extant chemical force fields, e.g. AMBER4.1 (53), (see also Chapter 4) have proven very robust in defining the phosphodiester moitiey. Thus, most DNA duplex structures have been determined without explicit backbone restraints, except for those defining the sugar geometry.

2.2 Structure refinement procedure

To obtain high resolution DNA duplex structures, NMR-derived restraints (i.e. distances or NOE intensities, and torsion angles) are used to drive a conformational search tool. This conformational search should enable all conformations consistent with the experimental data to be defined. This leads to a conformational envelope defining the structure—a source of consternation for scientists accustomed to the single structure typically reported in X-ray studies. Most commonly employed is the rMD approach [e.g. AMBER (33) , GROMOS (35), XPLOR (34)], where the match of a theoretical structure with the restraint target is translated into a penalty term in the empirical force field. Hence, the conformational search ideally leads to a structure that agrees with both the chemical force field and the restraints. Similar strategies have been implemented for restrained Monte Carlo (rMC) [e.g. DNAminiCARLO, (54,55)] and restrained molecular mechanics [e.g. JUMNA (56)], which become particularly powerful when internal coordinates are substituted for Cartesian coordinates, typically used for rMD. While distance geometry has also been applied to DNA structure determination, it typically produces structures that need subsequent refinement against a chemical force field to obtain energetically feasible structures.

For all methods, successful refinement requires convergence, which means that essentially the same target structure can be reached from different starting models using a reasonable protocol (for a detailed discussion see ref. 5). The similarity between converged structures obtained from different starting structures (the atomic rms deviation should be below ≈1 Å for a short DNA duplex), expresses the precision of the whole refinement process. It is not to be confused with the accuracy of the structure (57). Since NMR-derived restraints do not completely define a DNA duplex structure, the conformational search results will always depend somewhat on the refinement protocol, especially the relative weighting of chemical force field and NMR restraints. However, it should be noted that the structure of one duplex refined with rMC agreed within 0.5 Å of atomic rms deviation with that refined with rMD (55).

The accuracy of a refined structure can be partially inferred from comparison of the theoretical spectral data (i.e. NOE intensities and coupling constants) with the experimental data, which is similar to the procedure in X-ray crystallography that produces *R*-factors. Similar figures of merit can be used for NMR-derived structures as well (26,29,58,59). However, such figures of merit for NMR-derived structures also depend on the number of observables, such that it is difficult, for example, to compare NMR *R*-factors of different structures. None the less, NMR *R*-factors have become valuable parameters to indicate the progress of a refinement process.

An improved representation of final conformational parameters involves the depiction of their relative error bars. Such error bars can be inferred from analysing larger clusters of structures obtained from different refinement runs or even from longer rMD trajectories. The latter provides a simple way to discern whether small differences in helical parameters or backbone parameters for certain residues are, indeed, significant.

2.3 Limitations in NMR-derived high resolution structures

The issues discussed thus far are closely linked to the question of how well defined are the structural parameters in DNA solution structures published to date. Although this question cannot be considered solved, some clarification is emerging.

It is known that high precision distance restraints can reproduce fine structural features fairly well (60). When a DNA crystal structure was used to simulate 2D NOE cross-peak volumes, which were subsequently used as input data for a relaxation matrix/rMD protocol, the resulting structures exhibited almost identical structural parameters, including most helical parameters. However, this study underestimated errors, especially for small NOE intensities, and assumed no peak overlap.

We note that an earlier study (10) showed that the accuracy of a back-calculation-based refinement procedure to reproduce helical parameters was more limited than in the case above (60). It has been noted (61), however, that some DNA structures calculated with back-calculation methods exhibit extreme helical features, e.g. high helical twist. A statistical analysis of a number of high resolution duplex structures (61) revealed that one of the structures determined by back-calculation methods (62) was highly overwound, exhibiting large structural deviations from what could be considered the 'average B-DNA solution structure' (61). For the rest of the structures in the analysis, the average twist angle for all relaxation matrix/rMD-refined structures (see below) is virtually identical with independent solution data (17,18).

A reassuring result from another study was that two independent structure determinations, one with rMD and the other with rMC, for a Pribnow box duplex carried out with the same relaxation matrix-derived distances (12 restraints per residue on average), converged to essentially the same structure (see below), despite use of different refinement tools and force fields (55,63).

With respect to the number of relaxation matrix-derived distance restraints necessary to generate structures reproducibly, calculations with a particularly well-defined structure (20 distance restraints per residue on average) (64) demonstrated that practically the same final structure could be obtained when up to 20% of the distance restraints had been randomly omitted from the protocol. A 'free R-factor' analysis also suggested that those structures exhibited the same degree of accuracy. Therefore, an average number of 12–15 restraints per residue should be considered adequate, if they are reasonably accurate and well distributed over the molecule. The latter is important, especially for the sparse cross-strand restraints. An earlier study of a self-complementary DNA duplex (13) revealed significant overall structural changes when the central, cross-strand restraint between two adenine H2 [A(H2)] protons was removed. This result exemplifies the way NMR-derived restraints enforce local structure to which the rest of the molecule must adjust. Therefore, structural features associated with such constraints must be interpreted with care. These latter results also explain why certain sequences are better defined by NMR data than others: AT-rich sequences provide NOEs involving the A(H2) protons and the thymine methyl group, not available in GC pairs; sequences with runs of the same nucleotide or strict alternation, e.g. $(AC)_4(GT)_4$ (47,65), exhibit too much peak overlap and therefore provide a smaller number of restraints.

The most elusive question is how well the individual conformational parameters are defined through NMR restraints. Several studies have tried to answer this question by characterizing the distribution in the structural parameters when a series of refinement protocols produced a number of structures for a particular restraint data set (10,11,13,60). Other approaches included evaluation of changes in simulated NOE intensities when structural parameters were systematically scanned (12,14) or analysis of average helical parameters and their standard deviations obtained from rMD trajectories with dramatically different weights for the NMR restraints (66).

Together, these studies indicate that intraresidue parameters, i.e. sugar pucker and glycosidic torsion angle, are well defined by NMR restraints, especially when deoxyribose coupling constants are available. On the other hand, the dependence of the base pair parameters on the NMR restraints is not clear, because the derived base pair geometry is a structural compromise between the NMR-determined structure of each strand and ideal Watson–Crick base pairing. NMR structures of DNA are typically modelled with explicit Watson–Crick base pairs (5) connecting the strands, which, other than this, are only connected through few A(H2) cross-strand restraints. Since the balance between such holonomic restraints and the NMR restraints is defined arbitrarily, it is often difficult to interpret the actual values for parameters that describe deviations from the ideal flat Watson–Crick base pair. In general, larger weights for the NMR restraints drive the base pair parameters towards a larger deviation from 0, the idealized B-DNA value. For example, values for opening of −5 to −15°, or values for shear and stretch of >0.5 Å are significant distortions of the hydrogen bonding scheme and can be structural artefacts from overfitting (5,66). (See ref. 67 and Chapter 2 for definitions of helical parameters.)

Among the helical parameters, x-displacement and inclination [as calculated by the program 'Curves' (68)] describe the position of a base pair with respect to the global helix axis and most readily distinguish between A- and B-DNA. Interestingly, x-displacement and inclination show a strong correlation and a pronounced dependence upon the weight of the NMR restraints (66); when NMR restraint weights are increased, x-displacement decreases but absolute values for inclination increase. The result is a leaner, less hollow, double helix, which deviates significantly from canonical B-DNA. Unusual values for these two parameters should be interpreted with great caution, as they also are sensitive to overfitting.

For large NMR restraint weights, some reduction in the range derived for step parameters slide and shift, and, to a smaller extent, twist, are noticeable. However, values for the step parameters exhibited a more pronounced NMR restraint weight dependence, which seems to be stronger for the central part of the Pribnow box octamer. Systematic changes in all steps were apparent for tilt and roll, whereas the most dramatic changes for individual steps could be seen for the parameters slide, twist, and roll. Thus, even relaxation matrix-based rMD refinement will render structures with limited precision, which can be translated into a conformational envelope spanning a considerable range of helical parameter values. Nevertheless, sequence-specific patterns for the absolute values of the helical parameters seem to be independent of refinement protocol, even when individual errors in the helical parameters are quite large (66).

Qualitatively similar effects were seen in the refinement of the nonamer, d(GCAAAAACG):d(CGTTTTTC) (69),where three different rMD protocols (two different charges for the phosphate groups under *in vacuo* conditions, −0.3 and −1.2, versus utilization of explicit solvent with neutralized phosphates) yielded reasonable structures with low energies and NOE R-factors but significant deviations for some of the helical parameters. Despite the fact that some of these discrepancies might be a result of extreme conditions with respect to treating phosphate charges in rMD simulations, most of the sequence-dependent patterns of the helical parameters are similar for the different structures. Some gross helical features are clearly affected by use of explicit water molecules (69). Besides reduced values for twist and propeller twist, the minor groove was more compressed when no explicit solvent was used. This is in marked contrast to an earlier study, where rMD trajectories of another AT-rich duplex (70) obtained with and without explicit solvent yielded a narrower minor groove with explicit solvent. Helical parameter values in this latter study, especially with explicit solvent, were similar to other structural studies in which 'a spine of hydration' in the narrowed minor groove of AT-rich sequences was postulated (71,72). Furthermore, fluctuations of the minor groove width were reduced dramatically for rMD trajectories with explicit solvent (66,70). Therefore, the minor groove width is not well defined by NMR data and absolute values are more reliable when explicit solvent is used.

All of the above considerations regarding the accuracy of structural parameters apply only in the case where a DNA molecule exists in only one conformation. In reality, all molecules are more or less flexible, and this flexibility presents the largest systematic problem for defining high resolution solution structures. Since NOE-derived distance restraints are calculated from time-averaged data, which maybe severely biased towards shorter distances, even relatively small fractions of minor conformations can distort NOE intensities. In more pronounced situations, conformational flexibility will be manifest in mutually exclusive NOE-derived distances (16,70) or coupling constants (7,43,73). Since we barely have enough NMR data to define a single structure unambiguously, generating a well-defined description of a flexible molecule is a formidable task. Nevertheless, despite the underdefined nature of the problem, some partial solutions have been offered. These will be discussed in the context of DNA:RNA hybrid structures (see below).

3. DNA duplex structures

The number of published NMR structures of DNA is still quite modest compared with proteins. As this chapter's focus is on double helical DNA, Table 8.1 lists those high resolution structures where the DNA contains solely standard nucleotides with Watson–Crick base pairs and no modifications. For purposes of comparison, only 39 studies that reported a description of the structure determination process along with the discussion and depiction of the full structure are included in the table. A number of excellent structural studies that focus on certain structural details, e.g. sugar pucker determination, could not be included.

Along with DNA sequence, Table 8.1 also lists type and number of experimental restraints used and gives a synopsis of refinement methods employed. Methods for the generation of distance restraints are listed, i.e. isolated spin pair approximation (ISPA) and NOE build-up curves versus a hybrid relaxation matrix methods which is either integrated into the rMD refinement (int hyb, rMD) or is independent (hyb dd); the conformational search tool is also listed, including the name of the program, i.e. distance geometry (DG), restrained MD (rMD), restrained Monte Carlo (rMC), restrained energy minimization with internal helical parameters (rEMhp), and back-calculation methods (bkcalc). If methods other than bkcalc used NOE volumes as restraints, they are listed explicitly (vol ref).

In general, it seems that the vast majority of structures from the 1990s were generated such that spin diffusion has been accounted for, either when the distance restraints were generated, or in a final refinement step against NOE volumes. The amount of experimental information used in the refinement differs enormously from structure to structure, ranging from just a few to 20 distance restraints per nucleotide. About half of the newer structures were derived with additional torsion angle restraints, most of which constrain only the deoxyribose moiety. To enable further exploration of the structures in Table 8.1, we also provide accession numbers of the coordinates deposited at the Protein Data Bank (PDB) (URL: http://pdb.bnl.gov) (100). However, disappointingly, only nine of the listed structures are actually available at PDB. [Note that DNA crystal structures are available at the Nucleic Acid Databank as well, URL: http://ndbserver.rutgers.edu (101), see Chapter 3.] Before taking a more detailed look at the structures in Table 8.1, we note that several extensive reviews are available covering the earlier years of this field (1,2,4,61).

Table 8.1. Standard duplex DNA structures solved using NMR data

Entry	Sequence	Exp.[a] restraints NOE/tor/nuc	Restraint generation and refinement methods[b] (names of programs)	Access number PDB	Reference
1	5'-AAGTGTGACAT 3'-TTCACACTGTA	150/N/7	ISPA dd; rEM	na	74
2	5'-CGTACG 3'-GCATGC	190/N/16	ISPA dd; rEM	na	75
3	5'-CGTACG 3'-GCATGC	192/N/16	ISPA dd; rMD (CHARMM)	na	76
4	5'-CTGGATCCAG 3'-GACCTAGGTC	160/N/8	ISPA dd; rMD (CHARMM)	na	77
5	5'-GCATGC 3'-CGTACG	158/N/13	ISPA dd; rMD (CHARMM)	na	78
6	5'-GTTTTAAAAC 3'-CAAAATTTTG	200/N/10	vol ref	na	79
7	5'-GAAAATTTTC 3'-CTTTTAAAAG	200/N/10	vol ref	na	80
8	5'-GCGTATGTTGCG 3'-CGCATACAACGC	212/N/9	ISPA dd, DG; bkcalc	na	81
9	5'-CGCGAATTCGCG 3'-GCGCTTAAGCGC	155/N/7	ISPA dd, DG; bkcalc (DSPACE; BKCALC)	171D	38
10	5'-GCCTGATCAGGC 3'-CGGACTAGTCCG	162/N/7	ISPA dd, DG; bkcalc (DSPACE; BKCALC)	na	82
11	5'-TCTATCACCG 3'-AGATAGTGGC	330/Y/17	ISPA dd, rMD; vol ref (GROMOS)	1D20	83
12	5'-GTACGTAC 3'-CATGCATG	244/Y/15	ISPA dd, rMD; vol ref (GROMOS)	1D19	84

Table 8.1. *Continued*

Entry	Sequence	Exp.[a] restraints NOE/tor/nuc	Restraint generation and refinement methods[b] (names of programs)	Access number PDB	Reference
13	5'-CATGCATG 3'-GTACGTAC	208/Y/13	ISPA dd, rMD; vol ref (GROMOS)	1D18	84
14	5'-CGCCTAATCG 3'-GCGGATTAGC	293/N/15	ISPA dd, DG; bkcalc (DSPACE; BKCALC)	na	10
15	5'-CGTCACGCGC 3'-GCAGTGCGCG	290/N/15	ISPA dd, DG; bkcalc (DSPACE; BKCALC)	na	10
16	5'-CGCTTAAGCG 3'-GCGAATTCGC	138/N/7	int hyb dd; rMD (MORASS; AMBER)	na	85
17	5'-GGAAATTTCC 3'-CCTTTAAAGG	182/Y/9	ISPA dd; rMD (AMBER)	na	86
18	5'-ACACACAC 3'-TGTGTGTG	220/Y/14	hyb dd; rMD (MARDIGRAS; AMBER)	na	65
19	5'-ATATATAUAT 3'-TATATATATA	166/N/88	hyb dd; rMD (MARDIGRAS; AMBER)	na	87
20	5'-GTATATAC 3'-CATATATG	197/N/12	hybrid: MARDIGRAS rMD	1D42	13
21	5'-CCTTAAGG 3'-GGAATTCC	89/Y/6	ISPA dd; rMD (AMBER) w/ Raman torsions	na	88
22	5'-CGATCG 3'-GCTAGC	703v/N/na	vol ref; rMM SPEDREF, XPLOR	na	40
23	5'-GTACTGCAGTAC 3'-CATGACGTCATG	73/Y/3	ISPA dd; rHP (JUMNA)	na	56
24	5'-CATGACGTCATG 3'-GTACTGCAGTAC	95/Y/4	ISPA dd; rHP (JUMNA)	na	56

Table 8.1. *Continued*

Entry	Sequence	Exp.[a] restraints NOE/tor/nuc	Restraint generation and refinement methods[b] (names of programs)	Access number PDB	Reference
25	5'-GCGTATACGC 3'-CGCATATGCG	na/Y/na	ISPA dd, DG; bkcalc (DSPACE; BKCALC)	1D68	89
26	5'-GCCGTTAACGGC 3'-CGGCAATTGCCG	80/Y/3	ISPA dd, DG; bkcalc (DSPACE; BKCALC)	132D	62
27	5'-GTATAATG 3'-CATATTAC	184/Y/12	hyb dd; rMD (MARDIGRAS; AMBER) MDtar (AMBER) rMC (DNAminiCarlo) prob assess (PDQPRO)	1D70 na na na	63 70 90 61
28	5'-GGATCC 3'-CCTAGG	na/N/na	hyb dd; rHP	na	91
29	5'-AGCTTGCCTTGAG 3'-TCGAACGGAACTC	267/Y/13	hyb dd; rMD (MARDIGRAS; AMBER)	142D	92
30	5'-GAATTTAAATTC 3'-CTTAAATTTAAG	250/N/10	ISPA dd, rMD;bkcalc (DISCOVER, AMBER; BKCALC)	na	93
31	5'-GGTATACC 3'-CCATATGG	149/N/9	int hyb dd; rMD (IRMA; GROMOS)	na	94
32	5'-AATGGAATGGAATGG 3'-GGTAAGGTAAGGTAA	200/Y/9	hyb dd; rMD (AMBER)	na	95
33	5'-CATTTGCATC 3'-GTAAACGTAG	398/Y/20	hyb dd; rMD (MARDIGRAS; AMBER)	na	64
34	5'-ACCGTTAACGGT 3'-TGGCAATTGCCA	190/N/10	ISPA dd, rMD; vol ref (XPLOR)	na	96
35	5'-CGGACAAGAAG 3'-GCCTGTTCTTC	217/N/10	hyb dd; rMD (MARDIGRAS; AMBER)	na	97

Table 8.1. *Continued*

Entry	Sequence	Exp.[a] restraints NOE/tor/nuc	Restraint generation and refinement methods[b] (names of programs)	Access number PDB	Reference
36	5'-GCAAAAACG 3'-CGTTTTTC	189/Y/11	hyb dd; rMD (MARDIGRAS; AMBER)	na	69
37	5'-TTTCTCCTTTCT 3'-AAAGAGGAAAGA	149/N/6	int hyb dd; rMD (IRMA; GROMOS)	na	98
38	5'-CGCAAAAATGCG 3'-GCGTTTTTACGC	na/Y/na	vol ref; rMD (CHARMM)	na	36
39	5'-CGAGGTTTAAACCTCG 3'-GCTCCAAATTTGGAGC	460/Y/14	ISPA dd, rMD;bkcalc (DISCOVER, AMBER; BKMAT)	na	99

[a] Number of distance restraints; use of torsion angle restraints for sugars or backbone (yes = Y, no = N); number of distance restraints per residue.
[b] Abbreviations for methods used to generate distance restraints and type of refinement procedure: DG, distance geometry; rMD, restrained molecular dynamics; rEM, restrained energy minimization; dd, distance derivation; ISPA, isolated spin–pair approximation; rMM, restrained molecular mechanics; int, integrated; vol ref, NOE volume refinement; MDtar, MD with time-averaged restraints; bkcalc, backcalculation; hyb dd, hybrid matrix approach for distance derivation; rHP, restrained energy minimization with internal helical parameters.

3.1 The average DNA structure in solution as seen by NMR

The least surprising, but not at all trivial, result from Table 8.1, is that all DNA structures are reported to be in the B family. To date, no structural studies of unmodified DNA using near physiological solution conditions have found any other duplex topology. Note the many studies of the transition from B- to the left-handed Z-form of DNA, which might be involved in some regulation of DNA transcription *in vivo* (102). In general, alternating purine–pyrimidine (RY) sequences, in particular GC-rich sequences, especially with specific methylations (103), could adopt a Z-form geometry. Although some solution studies suggested that even an alternating CG hexamer could undergo the B to Z transition (104), high resolution NMR studies (105) could not corroborate this for d(CGCGCGTATACGCGCG)$_2$. In general it seems, that fairly drastic, non-physiological conditions, e.g. ethanol, high concentrations of divalent cations, or perchlorate (106–108) are necessary to drive DNA towards a left-handed form.

Before considering detailed differences between the structures in Table 8.1, it is instructive to picture the overall similarities of the average B-form in solution deduced by NMR methods. Such a picture emerged from a quantitative analysis (16) of the helical parameters of nine of the structures listed in Table 8.1 (entries 11, 12, 13, 20, 25, 26, 27, 29, 33). Note that a meaningful analysis of a pool of structures requires re-examining them with the same tool for derivation of helical parameters. The analysis was also limited by the availability of coordinates for published structures. Nevertheless, some interesting features could be extracted, even for specific traits of certain nucleotide steps. Table 8.2 lists results for a number of helical parameters in juxtaposition to average solid-state data derived from crystal structures.

One of the most important results of this statistical analysis (16) was that the average local helical twist (35.3°) is in good agreement with the independent values (17,18) for the solution state (34.0–34.3°). The match is nearly perfect when a highly over-wound structure derived with a particular back-calculation method (entry 26) is dropped from the average. The large twist values of that structure might be a result of refinement artefacts, since the back-calculation method did not use a full empirical force field. Several distinctions can be made between B-DNA in solution and in the solid state. First, the parameter 'roll' is positive in solution and slightly negative in the crystalline state. Positive roll, also encountered in A-form structures, is correlated with minor groove widening. Secondly, the average value for slide is slightly negative for the NMR structures, while it is positive for B-form crystal structures. Both of these features show that the NMR average tends slightly towards the A-form.

Table 8.2 also indicates increased bending anisotropy of the double helix which should be increased in the absence of crystal packing forces. Both the mean value and standard deviation for roll of B-DNA are bigger in solution than for the X-ray average. The tilt averages to zero regardless of state, but the standard deviation is clearly smaller for the NMR average. Rolling towards the minor groove is easier to accomplish in solution than tilting, which results from unfavourable compressing of one strand and stretching of the other. A similar rationale can explain the reduced values for twist and the associated standard deviation. For the crystalline state, differen-

Table 8.2. Structural parameters and standard deviations for average DNA from solution and solid state data and individual nucleotide steps

Source[a] (number of values in dataset)	Twist[b] (°)	Tilt (°)	Roll (°)	Shift (Å)	Slide (Å)	Rise (Å)	Propeller twist (°)
B-DNA[c]	36.1	0.0	−0.2	0.00	0.21	3.35	−13.8
	4.2	3.6	5.6	0.55	0.75	0.24	6.6
A-DNA[d]	30.8	0.0	7.9	0.00	−1.57	3.32	−9.8
	4.8	3.3	5.6	0.52	0.38	0.31	5.6
NMR[e]	35.3	−0.1	4.6	0.00	−0.34	3.15	−10.8
(42)	4.2	2.5	7.0	0.28	0.46	0.21	7.8
RR:YY[e]	34.8	0.0	3.5	0.00	−0.41	3.16	
(12)	2.8	2.7	3.9	0.29	0.39	0.16	
RY:RY[e]	37.3	−0.2	1.4	0.01	−0.29	3.20	
(15)	3.6	2.0	6.1	0.28	0.30	0.23	
YR:YR[e]	34.2	0.0	9.8	−0.01	−0.24	3.10	
(15)	5.4	2.0	7.7	0.24	0.59	0.20	

[a] Data are cited from ref. 61; first line of each entry gives the mean value, the second line gives the standard deviation in each individual set of data.
[b] Local helical parameters reported here conform to the Cambridge convention (67) and were calculated with an algorithm by Zhurkin and coworkers (109).
[c] Data of the survey of high resolution X-ray structures in the B-form (110).
[d] Data of the survey of high resolution X-ray structures in the A-form (111).
[e] Data of the survey of nine high resolution NMR structures (see text for description of dataset).

tial helical twist is an efficient means to reduce steric clashes, whereas increased roll can more efficiently reduce clashes in solution.

Another feature of the solution B-form is the pronounced non-planarity of individual base pairs. Mean values for the parameters propeller twist, buckle, and stagger deviate from solid-state values (data not shown). An interesting feature of the NMR average is the slightly compressed helix rise, for which no other systematic explanation can be offered beyond the idea that non-flat base pairs in conjunction with appropriate step parameters might stack in a more compressed way. Taken together, the rise and slide values suggest that DNA in solution is slightly shorter and fatter than anticipated from canonical B-DNA.

3.2 Sequence-dependent structural variation

Sequence-dependent structural features are difficult to determine since for both solid and solution states the structure determination methods are not devoid of artefacts, e.g. the role of crystal packing forces in crystal structures and the empirical force fields of the refinement tools in both methods should not be underestimated. But beyond these limitations, it is not clear what unit of a double helix uniquely defines the structural equivalent of the DNA sequence code. Although there are pronounced differences between the 10 unique complementary dinucleotide segments and the three groups of purine–pyrimidine sequences, a host of crystal structure studies have suggested that at least the adjacent steps exert a distinct structural influence, which would yield 136 structurally unique tetramers (9). Analysis of this problem with experimental structures is not possible in the foreseeable future, but we can focus on unique features apparent in dinucleotide steps.

Earlier NMR studies (65,87) tried to elucidate such properties by studying strictly alternating sequences, assuming that the dinucleotide step features do not get swamped by the effects of the flanking base pairs. For example, in the case of d(ATATATATATAT)$_2$ (87) alternating patterns for some helical parameters were found. These and subsequent studies also made clear that ultimate and penultimate base pairs are affected by their terminal position and hardly bear sequence-dependent features.

More definite insights came from the statistical analysis of non-terminal base pairs cited above (61), where the 10 unique dinucleotide steps were analysed utilizing 2–7 data points for each step. Despite this paucity of statistical data, a few themes emerged, especially for the three A–T steps that were the best defined. Instead of listing helical parameters for the latter, Fig. 8.1 depicts the average structures of the dyads computed from the average helical parameters (61). The parameters roll and slide constitute the biggest differences between the three sequences. For both parameters, maximum values are seen for TpA, and a minimum for ApT, with ApA assuming an intermediate value. The large slide value for TpA results in poor stacking, which can be seen easily in Fig. 8.1. This trend can be rationalized by the so-called Calladine

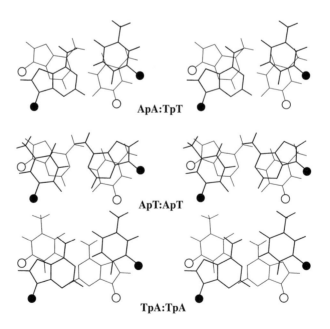

ApA:TpT

ApT:ApT

TpA:TpA

Fig. 8.1. Stereoviews of average conformations for the three unique A–T base steps according to the local, average helical parameters, as described in ref. 61. (The top base pair is bold; circles mark glycosidic bonds, with filled circles belonging to the top and open circles to the bottom base pair.)

rules (112), according to which purines on opposite strands relieve steric clashes by moving away from each other. The TpA step exhibits the only positive slide value among the 10 analysed dyads, while ApT shows the only positive roll value. These trends are also confirmed by independent modelling studies (113,114) and hold qualitatively for a number of B-DNA crystal structures.

TpA has the largest value for helical twist, followed by ApA and ApT. Although the absolute differences are fairly small (≈ 1–2°) and significant sequence heterogeneity is found for this parameter, the tendency agrees well with conformational calculations and various crystal data. Some sequences (62,89) showing huge differences between TpA and ApT steps (>10°) have been reported. Considering all results for alternating AT sequences, it becomes clear that the parameters interact to minimize base stacking for TpA and maximize stacking for ApT steps, which is in accord with the so-called 'alternating B-form' model (115,116). Most of these observations also apply to broader classes of sequences of purines and pyrimidines, with larger roll and slide values for YpR compared with RpY steps (see Table 8.2). However, the trend for twist is reversed from alternating AT with larger values for RpY than for YpR and RpR steps. The YpR step is more compressed on average than the other two and also exhibits the largest value for the parameter cup. (For a definition of this non–standard helical parameter, see Chapters 2 and 6, and ref. 117). These parameters lead to a unique situation for the TpG:CpA dyad, where especially the large positive roll, associated with a compressed major groove, causes a fairly localized bend. This effect is obvious for d(CATTTGCATC):d(CTAAACGTAG) depicted in Fig. 8.2 and to a lesser extent also for a trisdecamer, entry 29 in Table 8.1.

Fedoroff *et al.* (118) took a different approach to elucidate sequence-dependent effects in the groups of sequences of purines and pyrimidines, by extracting various sequential interproton distances via back-calculation methods. Although the dependence between specific distances and the overall helical parameters is fairly complex,

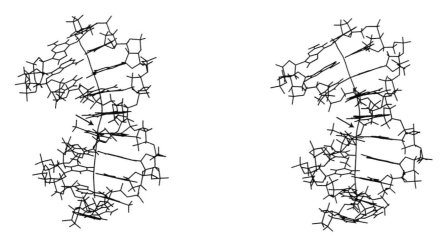

Fig. 8.2. Stereoview of d(CATTTGCATC):d(GTAAACGTAG) (Table 8.1, entry 33) with a superposition of the global helix axis calculated with the program 'Curves' (68). (The 5′-end of the first strand is at the bottom in the back.) The arrow points at the centre of the TpG step, which is the center of a bend.

the sequential H1′–H8/6 and H2′–H8/6 distances seem to follow a pattern. H1′–H8/6 distances are shorter (>3.5 Å) for RpY and RpR, and longer (>3.5 Å) for YpR and YpY steps, while shorter H2′–H8/6 distances (>2.7 Å) ensue for YpY and RpY steps and larger distances (>2.7 Å) for YpR and RpR steps. Distances involving A(H2) protons have been recognized as a good structure indicator (13,119). The cross-strand A(H2)(n)–H1′(m+1) distances, in particular, appear to depend on sequence: for CpA, TpG, and TpA, it is larger than 4.5 Å; in GA steps, it varies between 3.8 and 4.5 Å; and in AA steps it is between 3.7 and 4.2 Å. However, for other sequences both cross–strand and sequential A(H2)(n)–H1′(n+1), the distances appear to be affected by factors other than the dyad identity and the details of these complicating factors are not yet understood entirely (119). Nevertheless, decreased A(H2)(n)–H1′(m+1) distances are usually indicative of a narrowed minor groove, found in most AT-rich NMR structures.

In general, it is clear that many more NMR structures will be needed to confirm and extend the results described in this section.

3.3 Bending in NMR structures of DNA

It has been recognized for two decades now that DNA can be bent. Much research has been dedicated to describing this phenomenon structurally, to finding physical explanations for the underlying mechanics, and to understanding the biological consequences. A detailed introduction of all of the issues is beyond the scope of this section, since Chapter 14 deals with this topic. Also, a review summing up 20 years of research into DNA bending can be found in ref. 120. However, since a large number of NMR studies were geared towards understanding DNA bending, we will summarize some interesting NMR-derived results.

In general, it seems clear that DNA bending, which can span several residues or be more localized, is a special case of sequence-dependent DNA features, especially those that involve the mechanics of the intrinsic flexibility of the double helix. Several models have been proposed for sequence-dependent bending, and no general agreement has been achieved. The so-called 'static wedge models' (121,122) assume that the origin of curvature lies solely in the localized properties of the particular dinucleotide steps, where a propensity for certain roll, tilt, and twist parameters is thought to govern bending.

An example that would be in good agreement with this notion has already been discussed in the section above, where a particular combination of helical parameters, tilt, twist, and, foremost, roll for the TpG:CpA dyad, give rise to a distinct bend in several structures. However, a host of other experimental data, including gel mobility studies, X-ray data, nucleosome positioning data, and energy calculations gave the motivation for several 'context-dependent' models (120), which postulate cooperative interaction and distant neighbour conformations in the duplex. These latter models have evolved around the intrinsic curvature of longer runs of adenines. The 'A-tract' model (123) postulates a negative roll for (A)$_n$: (T)$_n$ segments, which causes bending into the minor groove. The effect of the roll motion is assumed to be more pronounced for ApA steps in the (A)$_n$: (T)$_n$ environment compared with other surrounding sequences. In contrast, the alternative 'non-A-tract' model assumes that ApA steps have zero roll, as seen in known B-DNA crystal structures, but postulates that other steps, on average, exhibit a positive roll value (122,124). The crucial difference between the latter models seems to lie in the precise roll values for the ApA step in $(A)_n:(T)_n$ segments and mixed sequences. It has also been pointed out that when addressing these issues attention must be paid to the general conditions, e.g. concentration of ions and other materials. Definitive answers are hardly available from macroscopic techniques such as gel mobility studies, and much hope has been invested in NMR solution structures to give a more precise and more accurate picture.

However, we must bear in mind the intrinsic limitations of NMR-derived structures (see above). The combination of the underdetermined nature of the experimental NMR data and the lack of experimental long distance restraints makes it very difficult to derive a smooth curvature for a DNA NMR structure. Several studies have demonstrated that sets of NMR data can simultaneously be fitted by a curved or straight structure (14). However, the result could be dependent on sequence and on number and distribution of experimental restraints. It is important to bear in mind that some of the global helical features depend also on the non-NMR restraining information. It has been pointed out in several studies (see above) that the treatment of electrostatic interactions in particular, i.e. dealing with phosphate charges or solvent models, can have a significant influence on the structural results (5,69,125); this is pronounced when the number of restraints is low as one would find in badly overlapped spectra from A-tract sequences. Thus, it appears that curvature and some of its corollaries, like a certain major or minor groove width, can be an artefact of the force field used in the refinement process, and not necessarily implied by the NOE data. For example, Chuprina *et al.* (93) studied d(GAATTTAAATTC)$_2$ (Table 8.1, entry 30) and suggested, on the basis of energetics, that A-tract NOE-derived distance restraints alone were not sufficient to distinguish between structures with a narrowed minor groove arising from a large propeller twist with a small inclination or from a small propeller twist with a large negative inclination.

A consistent picture of A-tract bending cannot be drawn from the 16 $(A)_n$-rich structures listed in Table 8.1. Several refined NMR structures appear to be bent (Table 8.1: entries 7, 17, 30, and 38), but straight structures can also be found in Table 8.1 (entries 6, 33, 34, 37, and 39). Most of the A-tract sequences structures are reported to be in the B family but with a narrower minor groove (36, 62, 64, 69, 86, 96, 119). A larger propeller twist value than standard B-DNA seems to be another feature of A-tracts, regardless of the presence of bending (46, 63, 64, 79, 86, 96, 99).

A bent structure was observed for d(CGCAAAAATGCG):d(GCGTTTTTACGC) (entry 38) (36). In this case, a detailed comparison of two crystal structures obtained in the same lattice (126), a well-defined NMR structure, and extensive free molecular dynamics calculations with explicit solvent was presented. On the basis of detailed correlation analysis of parameters the roll and tilt, both crystal structures exhibited fairly straight A-tract geometries, with bends at the junctions with the flanking sequences towards minor or major groove, respectively. The NMR structure, however, exhibited a slight, but nevertheless concerted, bend towards the major groove for the entire A-tract on top of a more pronounced bend at one of the junctions. The ambiguity of the bending direction at the crystal structures' junctions was reflected by the MD simulations, which showed an overall bend towards the major groove, besides considerable oscillations between major and minor groove directions. Interestingly, the NMR structure was clearly more similar to one of the two forms observed in the crystal lattice. This form also exhibited a characteristic narrowing of the minor groove similar to the NMR structure. The values for propeller twist were very similar for the A-tracts of the NMR and both the crystal structures. The free MD simulations revealed extensive buckle and propeller twist dynamics, slightly larger for the A-tract than for flanking sequences. Furthermore, enhanced backbone dynamics involving torsion angles α and γ are seen within the A-tract and the junctions, which on a qualitative level rationalizes small but distinct differences in sugar pucker at the end of the A-tract indicated by interproton coupling constants. The sugar repuckering, as seen in the MD trajectories, coincides with local bending. A detailed sugar pucker analysis (46) for $[d(A)_5(T)_5]_2$ also found unique structural features at the end of the A-tract, suggesting some distortion of the regular B-form geometries. A sharp drop for the pseudo-rotation angle of the sugar moieties was observed when going from the last two As ($150° > P > 180°$) to the first two Ts ($100° > P > 130°$) at the junction. It was also noted that intraresidue H1′–H4′ NOEs were much stronger for the T junction compared with all other As, which is consistent with an even lower pseudo-rotation angle ($60° > P > 120°$). A number of NOEs for the A at the junction were reported as being different from the other As. None of these sharp differences were observed for d(GCAAAAACG):d(CGTTTTTGC) (Table 8.1, entry 36), where all sugars were predominately C2′-*endo*, typical of B-form DNA (69).

Discontinuities in helical parameters and apparent kinks have also been detected in a number of other structures with ApT junctions: d(CGCGAATTCGCG)$_2$ (Table 8.1, entry 9) appears to be kinked at the ApT step (38) and d(CCTAAATTTGCC): d(GGCAAATTTAGG) appears to be distorted at the ApT and TpA steps (127).

Gel electrophoresis (128) and crystallization studies (129) suggested a difference between the curvature of A_n:T_n tracts and T_n:A_n tracts (130). Several T_n:A_n NMR studies are available involving a Trp promoter sequence (131) and endonuclease cleavage sites (62,132). For the latter, high negative propeller twist, large rise, negative buckle, and large opening values were found for the TpA step. In the sequence d(GTTTTAAAAC)$_2$ (Table 8.1, entry 6), there is no finite distinguishable discontinuity at the TpA junction (79), whereas d(GAAAATTTTC)$_2$ (Table 8.1, entry 7) was reported to be bent by approximately 10° with a discontinuity at the ApT junction (80). d(CCTTAAGG)$_2$ (Table 8.1, entry 21) was found, using NMR and Raman spectroscopy, to bend towards the major groove at the TpA step. It was suggested that this TpA junction bending is the result of a hydrophobic interaction between the methyl groups of the thymines (88). Another duplex, d(GAATTTAAATTC)$_2$ (Table 8.1, entry 30) (93), was found to be bent locally into the major groove at the TpA step.

The role of water in DNA bending has been studied recently by NMR. Large NOEs were observed between A(H2) of d(GTGGAATTCCAC)$_2$ and hydration water (133), consistent with the presence of a 'spine of hydration' in the minor groove. In contrast, no such NOEs were detected in the d(TTAA)$_2$ segment of d(GTGGT-TAACCAC)$_2$, indicating no tightly bound water molecules. These results can be correlated with the larger width of the minor groove in d(TTAA)$_2$ segments relative to d(AATT)$_2$ segments. The spine of hydration in the minor groove of d(CGC-

GAATTCGCG)$_2$ was found to be particularly stable, emphasizing the potential structural significance of bound water (133–135) (see also Chapter 9).

3.4 Conformational flexibility in DNA duplex structures

NOEs, coupling constants, line widths, and relaxation parameters are dependent upon conformational fluctuations of the molecule in solution, which may entail time-scales ranging from subnanosecond processes for small amplitude vibrations to large-scale millisecond conformational transitions. A great deal of research has been dedicated to unravelling the above issues, as many reviews have noted (20,136,137). Here, we will focus on some practical structural manifestations of conformational flexibility related to NMR-derived structures in general.

In general, the fastest, subnanosecond motions are of relatively low amplitude such that cross-relaxation rates and associated NOE intensities are not very different from those for a rigid body (137), where overall molecular tumbling governs the relaxation processes. Some NMR relaxation studies have focused on local differences in DNA flexibility indicated by measuring ^{1}H–^{1}H, ^{13}C–^{1}H, or ^{12}C–^{2}H relaxation parameters (137–139). The results of those and other studies are not completely consistent. For example, a proton relaxation study reported the same correlation time for the deoxyribose proton H1′, and base protons H6 and H8, based on proton spin–lattice relaxation time values (T_1) (140). Some other studies (137,139) also found no significant differences between the proton relaxation behaviour of bases and deoxyriboses, implying little or no internal motion on a nanosecond time-scale. However, a more recent natural abundance ^{13}C relaxation study (138) with DNA hexamers and octamers found small, but significant, differences between base and sugar moieties. For non-terminal residues, the 'order parameters' were around 0.8 for protonated base carbons and 0.6 for sugar carbons. [These 'order parameters' describe the relative mobility with values from 0 to 1, going from absolute disorder to a rigid body (141).] The lowest order parameters were observed for terminal residues, with values as low as 0.2–0.3 for the HO–^{13}C5′/3′ positions. In general terms, the anisotropy of molecular motion in short DNA oligonucleotides does not seem have a significant effect on the determination of accurate distance restraints and the ensuing high resolution structures (137). Nevertheless, for NMR structures that involved complete relaxation matrix methods, it has been shown that the match of calculated NOEs with experimental data can be improved by appropriate treatment of the aforementioned fast internal dynamics and anisotropic rotations (142,143) without invoking significant structural changes.Therefore, larger DNA fragments might require a different treatment of the relaxation as the molecular tumbling becomes more anisotropic owing to the increasingly rod-like shape.

Slower molecular motions lead to averaging of coupling constants and NOEs. For example, averaging is observed for vicinal coupling constants when the rate of torsional fluctuations exceeds roughly 10^2 s^{-1}. The time-scale of fast exchange, leading to averaging on the chemical shift scale, which also involves averaging of NOEs, depends on the actual chemical shift differences of the conformers involved, which, on the other hand, is expected to be no more than one ppm in DNA (137). It is clear that, with interchanging conformations, the averaging might lead to a 'virtual' NMR structure of limited value. Evidence of such averaging has been reported by several groups (6,12,48) for the sugar moieties in nucleic acid structures.

Conformational exchange processes in an intermediate realm might be manifest in unusual line widths of specific resonances. Such a situation was found for many sequences with TpA steps, which seems to be a unique example where flexibility might be a sequence-dependent feature.

3.4.1 Specific flexibility of the TpA step

Beyond the structural differences between ApT and TpA steps discussed in Section 3.1, the TpA step exhibits a unique, enhanced line broadening for the adenine base protons, especially for the usually very sharp A(H2) resonance (62,63,131,132). The line width of A(H2) is dependent upon the magnetic field strength and the temperature: the line width increases with temperature to a maximum, after which it shows the usual narrowing upon further heating. This behaviour was elucidated by $T_1\rho$ measurements (63), which revealed that the adenine base of the TpA step is involved in a relatively slow conformational exchange process on the submillisecond time-scale: 10^{-4} s (63), 5×10^{-5} s (144), and 10^{-6} to 10^{-2} s (132). The A(H2) chemical shift also shows a clear temperature dependence, typical for fast exchange. Kennedy *et al.* (132) concluded that the effects are most likely owing to enhanced mobility of the adenine base plane with an amplitude range of 20–50° degrees. Since large chemical shift differences between the rapidly exchanging conformations are required for rationalizing the experimental data, it seems logical that fluctuations in the ring current contributions to the chemical shifts are the main reason for the observed effects. In this vein, Kennedy *et al.* made a convincing case for the TpA junction in [d(CGAGGTT-TAAACCTCG)]$_2$, where A(H2) at the TpA junction is indeed positioned closely enough beneath the aromatic plane of the following A.

Additional evidence for unique flexibility involving adenine bases came from a high resolution NMR structure of d(GTATAATG):d(CATATTAC) (Table 8.1, entry 27). Unusually, short inter-residue distance restraints were observed between the two H8 protons and between the two H2 protons of the central, stacked adenines. This was rationalized by movement of the two bases relative to one another such that a short H8–H8 distance exists in one conformation and a short H2–H2 distance in the other conformation. As measured NOEs strongly reflect the shorter distance with motional averaging (see above), the measured H8–H8 and H2–H2 distances could not be satisfied by a single structure (63); this was part of the motivation for a more flexible structure refinement (see below). Note that the unique flexibility is very much a property of the TpA step and does not seem to require specific flanking sequences (145). This idea is also supported by the fact that N6-methylation of the junction adenine removes the line broadening effects completely (99).

3.4.2 Accounting for sugar flexibility in the refinement process

A known limitation of methods using NMR-derived distances and torsion angles is that they produce static models of DNA even though the structure may be dynamic. Dynamic averaging of distances and torsion angles yields single values of the restraints that are used to derive the static structure. However, the dynamic averaging is non-linear, so the precise values of the restraints reflect the structure and population of each of the interchanging conformations. A particular situation exists for deoxyribose rings where both NOE restraints and vicinal coupling constants can be used to describe the ring conformation independently. The interproton distances that are most sensitive to sugar conformation are the H8/6–H2′ and H8/6–H3′ distances, which, in conjunction with the intrasugar H1′–H4′ and H2″–H4′ distances, usually lead to a reasonably well-defined sugar pucker. An even more precise description can be obtained from intradeoxyribose vicinal coupling constants, which relate to sugar pucker via the modified Karplus equation, parameterized according to Altona and coworkers (6,7,146).

It was noted early on that some experimentally determined sets of coupling constants were not compatible with one rigid sugar conformation in DNA duplexes (6,43,46). This led to the simplest, non-rigid model, a quickly interconverting mixture of the two energetically most favourable and crystallographically most frequently observed conformations, C2′-*endo* (S) and C3′-*endo* (N). No two-state mixture can be defined unambiguously by four or five experimental coupling constants, but with a couple of reasonable assumptions the so-called S/N mixture approximates a dynamic sugar ring. Significant effort has been devoted to extracting accurate coupling constants using various simulation procedures. Different approaches have been undertaken, where the fitting procedure was either manual (46–48) or iterative (147,148). Such methods typically yield relatively precise values for $J_{H1'H2''}$, and $J_{H1'H2'}$, but even the more elusive $J_{H3'H2''}$ and $J_{H3'H2'}$ can be extracted with error bounds of up to ± 1 Hz.

Most studies of pucker entailing three or more coupling constants per deoxyribose have found that the coupling constants cannot be fitted simultaneously with one rigid conformation. Instead, most researchers report results (see ref. 5 and references therein) in the form of a two-state model where the populations of the two conformers and the pseudo-rotation angle of the major conformer, so far always within the S range, are varied to fit the data. Most DNA duplex studies to date indicate a small percentage of an N-type conformer (0–30% for non-terminal residues). Occasional repuckering has been observed in virtually all of the reliable free MD simulations that approach the nanosecond time-scale (149, 50) (see also Chapter 4). From such simulations and other theoretical work (151), it is clear that the S range is energetically very shallow and another local energy minimum is found near the O4′-*endo* conformation. This shows how oversimplified the classical two-state interpretation is. On the other hand, a small amount of sugar repuckering does not seem to cause big problems in deriving an average NMR structure.

Several studies have assessed the structural implications of sugar repuckering in a DNA duplex containing the Pribnow box (Table 8.1, entry 27) by adopting flexible refinement schemes. First, the rigid, average structure of the Pribnow box octamer was determined by conventional rMD and by rMC methods with virtually the same result (see Fig. 8.3a). It was noted that the measured H8/6–H3′ distances for most nucleotides were too short for the S-type pucker, and coupling constant analysis suggested the presence of a small percentage of the minor N-form conformer (63). A shorter H8/6–H3′ distance (*r*) in a minor N-conformer will strongly skew a measured distance to a shorter value owing to the non-linear (r^{-6} weighted) averaging of the interconverting conformers. The sugar conformations of the octamer were thus

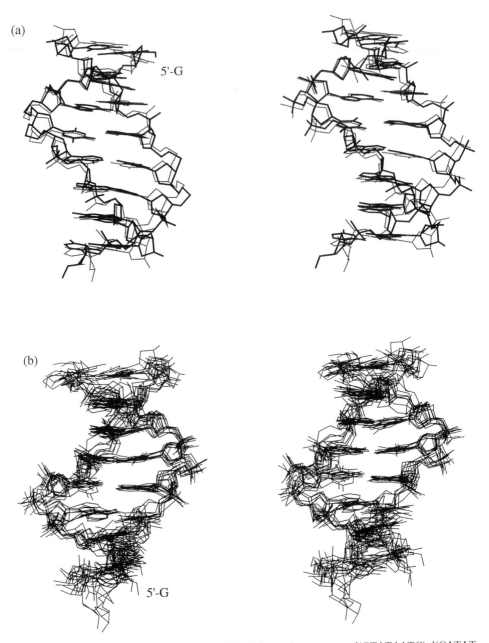

Fig. 8.3. High resolution NMR structures of the Pribnow box octamer d(GTATAATG):d(CATAT-TAC) (Table 8.1, entry 27). (The 5′-end of the first strand is labelled. Hydrogen atoms are omitted.) (a) Stereoview of best fit heavy atom superposition of two structures determined using rMC (bold) and rMD (thin) refinement methods (see text). (The atomic rms deviation for the heavy atoms of the inner six nucleotides is 0.5 Å.) The view is into the major groove. (b) Stereoview of nine structures representing a MDtar ensemble of the Pribnow box octamer according to ref. 70. Nine snapshots are shown covering the last 90 ps of a 120 ps simulation. Note that the backbone is more flexible than the bases. Terminal residues are more disordered than inner nucleotides. The view is into the minor groove.

restrained to the range determined from the coupling constants for the major conformers.

The rigid NMR structure was then subjected to MD with time-averaged restraints
(152), where the restraints are enforced as an average over the course of the trajectory
rather than at each step of the MD calculations (70,153). Depending on the averaging
time window, this method allows the exploration of extensive local dynamics, as several
low energy conformations can be sampled driven by the restraints. In the case of the
Pribnow box octamer, MDtar simulations utilized only distance restraints, including the
discrepant H8/6–H3′ distances. The ensuing structural ensemble could easily satisfy
both distance restraints, especially H8/6–H2′ and H8/6–H3′ distances, as well as the
coupling constants. This seems to be a direct effect of the sugar repuckering, which is
feasible when using MDtar, but not conventional rMD. The conformational envelope
produced with MDtar is wider than with rMD or even free MD simulations, as the
restraints encode structural averaging over a time-scale much longer than an MD trajectory. We emphasize that the average conformational parameters for the MDtar
ensemble are very similar to those obtained by conventional rMD refinement. This
means that the overall structure of the DNA octamer does not change significantly
when the sugars undergo occasional repuckering to the N-form. A detailed discussion
of the conformational parameters is beyond the scope of this chapter and can be found
elsewhere (70). However, the overall structural effect of the MDtar refinement can be
gleaned from Fig. 8.3b, which shows a representative ensemble covering 100 ps. Note
that the backbone is more disordered than the bases; also, terminal residues experience
a much larger conformational range than non-terminal residues. On the other hand, it
must be noted that the discrepancy between adenine sequential H8–H8 and H2–H2
restraints (see above) could not be remedied completely by the MDtar approach, suggesting that the motion required to satisfy all the restraints is not achievable in the time
window for averaging presented by the MDtar simulations. Another example of this
type of refinement is discussed in Section 4.1.

A different approach to dynamic refinement, termed PARSE (probability assessment via relaxation rates of a structural ensemble), was applied to the same Pribnow
box octamer (16). A large pool of conformers was created through a series of
restrained Monte Carlo refinements, where restraints were variously excluded or
included. The modifications in the restraint sets were made to allow individual sugars
to assume non-S-form puckers. In all, 60 different permutations of the distance
restraint file were used. The pool of conformers (>500) was created such that at least
one member had N-type pucker for each nucleotide; the pool also contained the final
rMC-refined NMR structure and other similar structures. Then, probabilities for all
conformers were computed using PARSE to yield the best match with all experimental cross-relaxation rates, which were derived from NOE intensities. The resulting
PARSE ensemble contained 13 conformers with non-zero probabilities. All 13 conformers contain at least one deoxyribose that is flipped at least partially to the N
domain. Here also, a better match with the experimental data could be achieved by
allowing flexibility in the sugar region. The average structural parameters were very
similar to those from the conventional rMD (63) or rMC (55) structures.

In summary, it seems that allowing flexibility in the refinement leads to better
agreement with experimental data. Although the sugar dynamics ultimately lead to

Table 8.3. RNA:DNA hybrid high resolution structural studies since 1990

Entry	Sequence	Exp.[a] restraints NOE/tor/nuc	Restraint generation refinement methods[a] (names of programs)	Access number PDB	Reference
1	5'-d(GTCACATG) 3'-r(CAGUGUAC)		ISPA dd; DG, rMD,backcalc. (DISCOVER, BIRDER)	124d	162
2	5'-d(GTGAACTT) 3'-r(CACUUGAA)	na	int.vol.ref.– internal par. rMD;NUCFIT, DISCOVER	na	163
3	5'-d(GCTATAA$_{ps}$TGG) 3'-r(CGAUAUU ACC) ps= (S) phosphorothioate	322/Y/16	hyb matr; rMD, MDtar, (MARDIGRAS,AMBER4.1)	219d	73,164
4	5'-d(GCTATAAAprTGG) 3'-r(CGAUAUU ACC) pr= (R) phosphorothioate	320/Y/16	hyb matr; rMD, MDtar, (MARDIGRAS,AMBER4.1)	219d	73,164
5	5'-d(CGCGTTTTGCGC) 3'-r(GCGCAAAACGCG)	Pucker analysis	Qualitative	na	165
6	5'-d(GGGTATA CGC) 3'-d(CCCATAT)r(GCG)	na	ISPA dd; DG, rMD,backcalc. (DISCOVER, BIRDER)	169d	166
7	5'-r(GCCA)d(CTGC) 3'-d(CGGT GACG)	205/Y/13	ISPA dd; DG, rMD,backcalc. (DISCOVER, BIRDER)	1gtc	167,168
8	5'-d(GGGT TTACT) 3'-r(CCCA)d(AATGA)	ISPA dd; DG, rMD,backcalc.	1oka (DISCOVER, BIRDER)	52	
9	5'-d(GGAGA)r(UGAC) 3'-d(GTCAT CTCC)	260v[b]	ISPA dd;rMD,vol ref (XPLOR)	1dm	169
10	5'-d(CG)r(CG)d(CG) 3'-d(GC)r(GC)d(GC)	592v[b]	int.vol.ref., rEM (SPEDREF)	na	170
11	5'-d(CG)r(C)d(TA G CG) 3'-d(GC G AT)r(C)d(GC)	789v[b]	int.vol.ref., rEM (SPEDREF)	na	170

Table 8.3. *Continued*

Entry	Sequence	Exp.[a] restraints NOE/tor/nuc	Restraint generation refinement methods[a] (names of programs)	Access number PDB	Reference
12	5'-r(CGCG)d(TATA)r)CGCG) 3'-r(CGCG)d(ATAT)r(GCGC)	na	ISPA dd; DG, rMD,backcalc. (DISCOVER, BIRDER)	104d	171
13	5'-d(CG)r(AGAU)d(GAC) 3'-d(CC TCTA CTG)	256v[b]	ISPA dd;rMD,vol ref (XPLOR)	1dhh	169
14	5'-d(CGTTATAATGCG) 3'-r(GCAAUAUUACGC)	Chem. shift comparison	Qualitative	na	172
15	5'-d(CGCG)r(AAUU)d(CGCG) 3'-d(GCGC)r(UUAA)d(GCGC)	Chem. shift comparison	Qualitative	na	173
16	5'-d(CGCG)r(AUAU)d(CGCG) 3'-d(GCGC)r(UAUA)d(GCGC)	Chem. shift comparison	Qualitative	na	173
17	5'-d(CGTT)r(AUAA)d(TGCG) 3'-d(GCAA)r(UAUU)d(AAGC)	Chem. shift comparison	Qualitative	na	173

[a] For further explanation see legend to Table 18.1.
[b] Volume refinement may involve above and below diagonal peaks plus some diagonal peaks; a restraints–per–nucleotide value is not readily available.

higher disorder for the backbone, the overall structural features remain the same. On the other hand, neither of the above approaches can lead to a unique solution of the problem, which makes dynamic refinement only desirable for situations where averaging artefacts are obvious (see below).

4. RNA:DNA hybrid structures

RNA:DNA hybrids are formed during essential biological processes such as transcription of DNA into RNA and the reverse transcription of viral RNA code into DNA sequences. Another strong motivation for structural studies of hybrids is to understand 'antisense' pharmaceuticals which are generally modified DNA oligonucleotides targeted to mRNA or viral RNA. Such modified RNA:DNA hybrids are thought to be hydrolysed by largely sequence-independent, hybrid-specific RNAases, e.g. RNAase H (154,155). Understanding the interaction between hybrids and associated enzymes on a structural level should aid 'antisense' drug design.

While sequence-dependent structural features are of interest for DNA duplexes, for RNA:DNA hybrids even the gross helical structure was not established until a few years ago. The agreement between different methods was poor in earlier work (156–159), but it had already been suggested that hybrid structure in solution is different from that in the solid state. In crystal structures the introduction of ribonucleotides into short DNA oligomers drives the structure from the typical B-form to the A-form. Just one 5'-ribonucleotide was enough to drive an octamer hybrid into an A-form crystal structure (160,161). This is not the case for RNA:DNA hybrids in solution.

High resolution NMR structures are available for different constructs, where either a part of one or both strands contains ribonucleotides, or a whole strand is RNA. A compilation of NMR structural studies for hybrids is presented in Table 8.3. Note that the selection criteria were less restrictive than for the DNA duplexes in Table 8.1 since we wanted to gather high resolution structural information on all the different types of hybrids mentioned above.

4.1 RNA:DNA hybrids with one complete RNA strand

Hybrids with one complete RNA strand have been studied most extensively with, however, conflicting results. From fibre diffraction data, it became clear early on (158) that the structure of poly r(A):poly d(T) depends on the relative humidity. An A-type diffraction pattern changed with increasing humidity to one representing the RNA strand with C3'-*endo* sugars, but the DNA strand with B-like C3'-*exo* sugars. Similar results were found for poly d(A): poly r(U) and poly d(I):poly r(C) (159). More evidence for different backbone conformations for the two strands, termed 'heteronomous', came from solid state ^{31}P NMR (174), circular dichroism (175), and Raman spectroscopy (176). However, some solution studies came to a different conclusion, with both strands assuming loose B-form geometries (156,157). A more detailed high resolution NMR study (172), where several non-exchangeable protons had been assigned, resolved most of the old discrepancies, revealing that the deoxyriboses in d(CGTATAATGCG):r(CGCAUUAUAACG) assume sugar puckers in the general C2′-*endo* region, whereas the RNA strand definitely adopts the C3′-*endo* conformation. A detailed comparison between the hybrid and the all–DNA analogue of chemical shifts and NOE connectivity patterns in DNA versus RNA strands demonstrated clearly that the two strands assume different geometries. At the time, a high resolution structure was not determined for the dodecamer since the sequence was too long for complete and unambiguous assignments of all the required protons, especially H3′ and H4′ protons. A few years later, high resolution structures were presented by several groups (73,162–164) establishing the heteronomous character of the solution structure. A detailed structural analysis of the d(GTCACTATG):r(CAUGU-GAC) hybrid (see Fig. 8.4a) was presented by Salazar *et al.* (162), including a discussion of the interaction of this hybrid with RNAase H (177). From NOESY and COSY data, the authors could clearly establish the A-type character of the RNA strand through small $J_{H1'H2'}$ coupling constants, typical for C3′-endo pucker, and strong sequential H6/8–H2′ NOEs. For the DNA strand, the NMR data clearly showed that it is neither A- nor B-form, but rather something intermediate. Strong H1′–H4′ NOEs, very similar $J_{H1'H2'}$ and $J_{H1'H2''}$ coupling constants ($\approx$ 6–7 Hz), and strong H3′–H4′ and medium H2′–H3′ COSY peaks indicated a deoxyribose conformation in the O4′-*endo* range (168). Before turning to the broader helical structural features, it is interesting to compare the results for the deoxyribose conformation with other studies (73,163,165), as the sugar moieties can be well defined by NMR data even without deriving a complete model structure.

In general it seems that most spectra are consistent in all studies. However, a different interpretation for the sugar conformation has been proposed. A more consistent interpretation of all the data suggests a flexible deoxyribose pucker model involving S- and N-type conformations, largely substantiated through additional information that could not be reconciled with the O4′-*endo* conformation (164).

As mentioned, the sugar ring cannot be defined with high precision by NOEs alone since the largest sugar pucker change-induced distance fluctuation is for H1′–H4′, being 3.3 Å for C2′-*endo* and C3′-*endo*, and 2.5 Å for O4′-*endo*; i.e. barely larger than the typical accuracy of the experimentally determined distance ($\pm$0.2 to $\pm$0.4 Å). We note that a quickly interconverting S/N pucker mixture with only 10 or 20% of O4′-*endo* conformers would give rise to a significantly stronger H1′–H4′ NOE, with a sixth-root-weighted average distance of 3.0 or 2.9 Å, respectively. Pucker conformations close to O4′-*endo* have been found to be relatively stable with theoretical studies (151) and dynamics structure refinement of DNA (70).

Coupling constants typically augment the structural description. Here also, the accuracy of the picture increases not only with the accuracy of the coupling constants but also with how many of them have actually been determined. To distinguish clearly between a flexible pucker model and the rigid O4′-*endo* conformation, more coupling constants are necessary than just values for $J_{H1'H2'}$ and $J_{H1'H2''}$ and semi-quantitative assessment of $J_{H2''H3'}$. González *et al.* (73) extracted $J_{H1'H2'}$, $J_{H1'H2''}$, $J_{H2'H3'}$, and $J_{H2''H3'}$ with error bounds from $\pm$0.3 Hz to$\pm$1 Hz via simulation of COSY cross-peaks. Typical experimental values for $J_{H1'H2'}$, $J_{H1'H2''}$, and $J_{H2''H3'}$ (6.1–8.5, 6.0–6.5, >3–4.0 Hz) are indeed compatible with a single sugar geometry around O4′-*endo*. However,

(a)

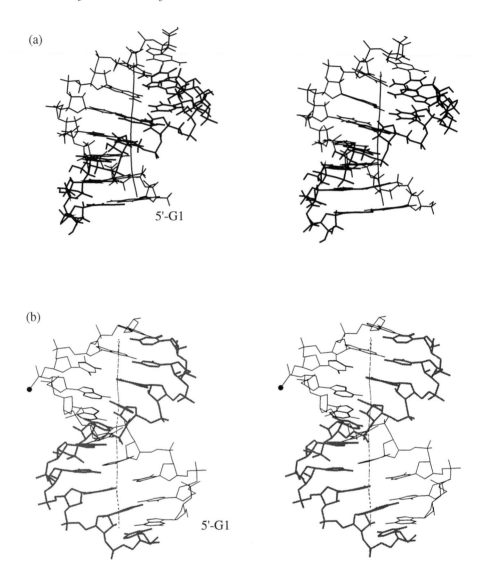

5'-G1

(b)

5'-G1

values for $J_{H2'H3'}$ (5.5–6.6 Hz) are clearly not in agreement with the above geometry as this coupling constant assumes a maximum value of $\approx$ 9.5 Hz for O4'-*endo*. (Minimum values of 5–6 Hz arise for C2'-*endo* and C3'-*endo* conformations.) It seems clear that such a substantial deviation cannot simply be caused by dipolar contributions to coupling constants (45) because the correlation times were >4 ns (see Section 2.1 above). The most difficult coupling to determine is $J_{H3'H4'}$ because of the lack of fine structure in the COSY peak. Gao and Jeffs overcame this obstacle by acquiring {H2',H2''}-decoupled COSY spectra (165). The reported $J_{H3'H4'}$ values for non-terminal residues (6.1–7.5 Hz) are compatible with both a single O4'-*endo* geometry

(c)

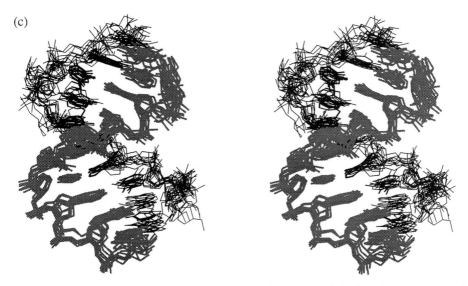

Fig. 8.4. Stereoviews of high resolution NMR structures of RNA:DNA hybrids with one entire strand being RNA . (The 5′-end of the DNA strand is marked. The RNA strand is shown in bold.) (a) Structure of d(GTCACATG):r(CAGUGUAC) (Table 8.3, entry 1) with superposition of the relatively straight global helix axis, calculated with the program 'Curves' (68). (b) Structure of d(GCTATAA$_{pR}$TGG):r(CGAUAU-UACC) (Table 8.3, entry 4) with superposition of the relatively straight global helix axis, calculated with the program 'Curves' (68). 'pR' indicates a chirally pure *R*-phosphorothioate modification, which is indicated by a small closed circle in the structure. (c) Heavy atom best fit superposition of 10 snapshots from a MDtar ensemble for d(GCTATAA$_{pR}$TGG):r(CGAUAUUACC) (Table 8.3, entry 4) showing the relative flexibilities of the two strands. Note the less flexible RNA strand in the foreground. The overall geometry is not changed compared with depiction of the conventional rMD average structure in (b). (The 10 snapshots cover the last 100 ps of 120 ps in 10 ps steps.)

($J_{H3'H4'} \approx 7$ Hz) and a S/N-mixture ($J_{H3'H4'} \approx 1$ Hz for C2′-*endo*; $J_{H3'H4'} \approx 8$ Hz for C3′-*endo*). Furthermore, the COSY data of Salazar *et al.* (168) showed very strong H3′H4′ COSY peaks for the RNA and significantly weaker ones for the DNA, which also argues against a rigid O4′-*endo* conformation in light of the small coupling constant difference between C3′-*endo* and O4′-*endo*.

Another indication of flexibility can be the incompatibility of NOE-derived distances (16). The intranucleotide H6/8–H3′ and H6/8–H2′ distances assume their shortest values for C3′-*endo* and C2′-*endo*, respectively, assuming that the glycosidic torsion angle is adjusted to the change in sugar pucker. In the case of an S/N mixture, the average values of these two distances become short and potentially unsatisfiable by a single conformer (16). For a decamer hybrid containing a single, chirally pure phosphorothioate modification (73), the above distances for the RNA strand assume typical C3′-*endo* values for the majority of residues. For the DNA strand, however, H6/8–H2′ distances are closer to C2′-*endo*, while H6/8–H3′ distances are closer to C3′-*endo*. The indiscrepancies were not dramatic, however, owing to the relatively high error bounds on some of those distances.

In light of the above, it is not surprising that rMD refinement of the hybrid yielded different results when NOE-derived distances alone were used versus refining with additional coupling constant restraints. In the first case, a structure was obtained exhibiting good agreement with the experimental data and sugar puckers for the DNA strand around O4′-*endo* (see Fig. 8.4b). Refinement with both types of restraints indicates that both cannot be satisfied equally well in one model; interestingly, DNA sugar puckers were still mostly in the O4′-*endo* range. These findings are consistent with a flexible molecule where the average structure cannot satisfy all data. Furthermore, it must be noted that the O4′-*endo* conformation is the most probable compromise between satisfying all restraints and keeping a reasonable conformational energy for a single structure as the restrained portion of the sugar ring is indeed flat. The restraints for the physically unachievable average of an equally populated S/N mixture encode a conformation with all sugar carbon atoms in one plane.

Besides conventional rMD refinement of two phosphorothioate hybrids (164), differing only in the chirality of the single phosphorothioate (Table 8.3, entries 3 and 4), González *et al.* also employed the more flexible refinement strategy using time-averaged distance and coupling constant restraints (MDtar). Such a MDtar ensemble (see Fig. 8.4c) was shown to satisfy both the coupling constants and the distance restraints equally well. Structural parameters were calclulated for the conventional rMD structures, as well as for long trajectories using conventional (rMD) and time-averaged restraints (MDtar). Both refinement methods have been applied to the two related, (R) and (S) chiral forms of the hybrid. Besides the reassuring result that the completely independently determined (R)- and (S)-form rMD structures are virtually identical, with slight differences only for the thioate step, the most striking result was that the average values for the helical parameters were very similar for rMD and MDtar ensembles (164). Despite the larger standard deviations for MDtar parameters, indicating a wider conformational envelope, all the sequence-specific patterns were reproduced very well compared with the rMD data.

With regard to sugar conformations, rMD and MDtar ensembles exhibited the same tight distribution around C3′-*endo* for the RNA strand. For the DNA strand, however, the results were different. The tight distribution for the rMD ensemble, often centred in the lower S-range, became a complex distribution pattern for the MDtar ensemble, with a widely populated S-range (O4′-*endo* to C2′-*endo*) and a significant population in the C3′-*endo* region (20–47%). González *et al.* concluded that the overall helical appearance of the hybrid does not change significantly when going from the rMD ensemble, which only satisfies the NOE distances, to the MDtar ensemble, which also satisfies coupling constants (see Fig. 8.4b and c). This implies that forcing the deoxyribose moieties into compromise average conformations does not distort the overall structure.

For both of the high resolution hybrid structures (164,177) helical parameters were reported. Unfortunately, the sequences are quite different and different definitions for the helical parameters were used. Nevertheless, in both structures twist and rise are low, more similar to the A-form. For the thioate hybrid, the *x*-displacement for most steps is around −3 Å, roughly between the values of the A- and B-forms. Fedoroff *et al.* report some of their helical parameters independently for each strand (177), which leads to the interesting observation that the DNA strand is more susceptible to rotations about the long base pair axis (large fluctuations for roll and tip), while the RNA strand seems to be more prone to undergo rotations about the short axis (large fluctuations for tilt and inclination) The most important double helical feature is probably the minor groove width, because it is thought to be the primary locus for specific interactions with proteins such as RNAase H. Both studies (164,177) agree, reporting a minor groove width of 7.5–9 Å for the hybrid, compared with 11 Å for the A-form and 6 Å for the B-form.

Manual docking of RNAase H and the d(GTCACATG):r(CAGUGUAC) hybrid structures led to the interesting idea that the overall helix geometry of the hybrid, and especially the intermediate groove width, is the basis for the discrimination of RNAase H against double helical RNA or DNA (177). For the authors, the propensity of the DNA to adopt O4'-*endo* sugar pucker constitutes one of the key elements for the interaction. However, it is easy to see that a hybrid model with essentially the same helical geometry but a more flexible DNA strand (164) should certainly fit into the binding area of RNAase H equally well.

4.2 Okazaki-like fragments

These fragments are hybrids where only part of the one strand is RNA. Okazaki fragments form during DNA transcription and reverse transcription of viral genomes. RNAases ultimately remove the RNA part by cleaving exactly at the junction or in the case of *E. coli* RNAase HI, just before the last step. This implies that structural discontinuities at the junction are available to guide the RNAases. For synthetic oligomers modelling Okazaki fragments, solution and solid state results differ. Crystal structures of [r(GGC)d(TATAGCC)$_2$] (178) and r(GCG)d(TATACCC):d(GGGTATACGC) (179) revealed A-form geometries without large disruptions at the junctions, whereas high resolution NMR results (52,167), similar to those described for hybrids above, suggested the sugar conformations to be in a an 'heteronomous' structure. Indeed, for the r(GCCA)d(CTGC):d(GGTGACG) hybrid (162), which represents a substrate for the RNAase H activity of HIV-1 reverse transcriptase, the four ribonucleotides adopt C3'-*endo* conformations while the deoxyribonucleotides cover a wide range of the pseudo-rotation wheel (54°> P >144°). The most unusual sugar pucker is reported for the deoxyribose at the junction (54°> P >90°), which was also observed for the 3'-ends. On the other hand, DNA puckers in the hybrid part are not very different from other DNA sugars. Nevertheless, the last two base pairs of the hybrid part appeared to be the most 'heteronomous'. Federoff *et al.* recently published two well-defined high resolution structures of Okazaki fragments from HIV-1 (167) and Moloney Murine leukaemia virus (52) (see Fig. 8.5a). In general, both structures exhibit the previously described heteronomous features for the hybrid section and more regular B-form properties for the DNA duplex part. Interestingly, the discontinuities produced a clear bend associated with the junction (≈ 16° in ref. 52, ≈ 18° in ref. 167). Distinct changes are seen for some helical parameters at the junction, while other parameters describing the general helical appearance change more gradually. For example, a large negative *x*-displacement and a small inclination for the hybrid part change concertedly to a more pronounced positive inclination and a reduction in the *x*-displacement. Although, the values for *x*-displacement give the hybrid segment some A-like appearance, inclination and other helical parameters depict a double helix

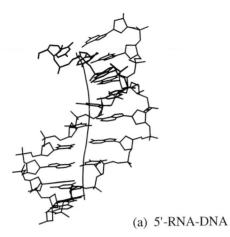
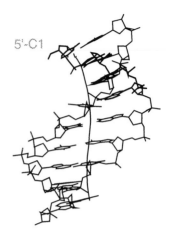

(a) 5'-RNA-DNA

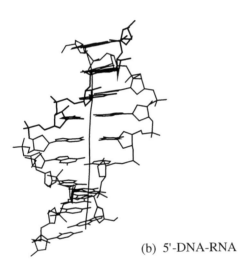
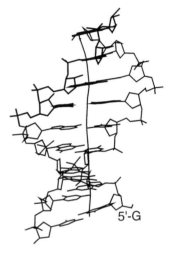

(b) 5'-DNA-RNA

(c) 5'-DNA-RNA-DNA

Fig. 8.5. Stereoviews of high resolution NMR structures of RNA:DNA hybrids where only a part of one strand is RNA. The global helix axis was calculated with the program 'Curves' (68). (The 5'-end of the DNA strand is marked. The RNA part is shown in bold.) (a) RNA at the 5'-end: Okazaki-like fragment r(CCCA)d(AATGA):d(GGGTTACT) (Table 8.3, entry 8). Note the clear bend associated with the RNA:DNA junction. (b) RNA at the 3'-end: d(GGAGA)r(UGAC):d(GTCATCTCC) (Table 8.3, entry 9). The distortion in the global helix axis is minimal. (c) RNA in the middle of a DNA strand: d(CG)r(AGAU)d(GAC):3'-d(CCTCTACTG) (Table 8.3, entry 13). Note the curvature in the global helix axis at both junctions.

with a narrow minor groove, which decreases gradually from the hybrid segment towards the DNA part.

Intrastrand distances between phosphate groups assume values between the A- and B-form only for the RNA segment, whereas the entire DNA part exhibits fairly B-like values. Fedoroff and coworkers offer a side-by-side comparison of some of the helical parameters for the two hybrids (52,167). Besides obvious sequence-dependent structural effects, similar discontinuities around the DNA:RNA junction can be seen for parameters roll, tilt, rise, and buckle. Especially for the latter, a large negative value for the junction base pair(s) seems to be a unique feature, as a similar behaviour is not only found for two other hybrids, [r(GCG)dTATACCC):d(GGGTATACGC)] and [r(CGCG)d(TATACGCG)$_2$] (166), but also for the A-form crystal structure (179). For rise, a distinct increase towards the junction is apparent for both of the recent NMR structures (52,167), although the decrease in the DNA segment towards the A-form values is somewhat surprising. Most backbone torsion angles are close to typical values for either the A- or B-form. The ribonucleotides exhibit values different from the deoxyribonucleotides only for δ, ε and ζ, which describe the 3'-proximity of the sugar moieties. Furthermore, χ angles for ribonucleotides reflect the expected A-form values, whereas, for the deoxyribonucleotides, intermediate values are found, reflecting the adjustment to pucker values in the lower S-range. The structural results for the self-complementary hybrid, [r(CGCG)d(TATACGCG)$_2$] paint a very similar picture, where an all-DNA TATA box is flanked by GC-rich hybrid segments on both sides (171). This structure was analysed as three independent segments, which on the one hand brings out the grossly different helical features, but on the other hand creates the impression that discontinuities, especially for parameters rise, twist, buckle, x-displacement, and inclination, occur strictly at the junctions. Nevertheless, a distinct bend ($\approx$ 23°) results, similar to the other hybrid structures. DNA sequences with alternating A:T pairs are known for their compressed minor groove (13,180). Since the hybrid segments seem to induce small groove widths for the DNA moieties as seen in the structures above, it is not surprising that for the three-segment hybrid, a seriously compressed minor groove is found for the TATA segment. The two strands get close enough in the middle of this part (closest cross-strand interphosphate distance > 5 Å) that strong interstrand H2–H1' NOEs were observed.

A different structural situation exists when the hybrid segment follows the DNA in the 3'-direction (169). A hybrid segment at the 3'-end does not have nearly the same structural impact as the 5'-counterpart (see Fig. 8.5b). Since the riboses at the junction and the 3'-end adopt some intermediate pucker values ($J_{H1'H2'} \approx 6$ Hz) and the minor groove width is largely the same for the entire hybrid, the heteronomous character is hardly tangible. Nevertheless, helical parameters rise, roll, tilt, and slide show unique trends for the 3′-hybrid segment.

4.3 RNA inserted into DNA sequences

Although, no biological role has been assigned to hybrids where a short RNA segment is inserted into DNA sequences, such constructs are interesting to complete a systematic structural picture. Qualitative studies on hybrids with double helical RNA inserts (172) found the DNA segments in B-form geometries, while the RNA section is essentially A-form, based on the sugar pucker criterion and a chemical shift comparison between the hybrid and the all-DNA analogue. The RNA base pair at the junction was described as heteronomous since the 5′-ribose does not assume typical A-form pucker, in contrast to the 3′-ribose. Also, the DNA base pair at the junction is heteronomous in that the 3′-deoxyribose exhibits B-form pucker and the 5′-sugar assumes some intermediate state. A high resolution structure of this type of hybrid confirmed some of the above observations. However, it is questionable if the results for [d(GC)r(GC)d(GC)]$_2$ (170) can really be compared with the data for the above dodecamers, since every base pair must be considered terminal or at the junction. Nevertheless, the structure of the GC hybrid is reported to be between the A- and B-form, with large negative x-displacement but small tilt values. All riboses adopt C3′-endo pucker whereas only the 5′-deoxyribose at the junction is C2′-endo, and all others assume an intermediate S/N value. $J_{H1'H2'}$ and $J_{H1'H2''}$ were interpreted as S/N mixtures with 40–75% S population. The same authors also report the structure of a self-complementary octamer where only one sugar is changed into a ribose (170). Whereas the deoxyriboses are all in the S regime, the ribonucleotide adopts A-form pucker, which does not seem to perturb the overall B-form geometry.

Nihizaki *et al.* reported a detailed structure of a hybrid nonamer (169) where four ribonucleotides are placed in the middle of one strand (see Fig. 8.5c). In accord with the earlier qualitative interpretation (173) it seems that overlapped H2′H2″ proton resonances for the 5′-deoxyribose and non-C3′-endo puckers for the first ribose are unique features for the DNA to RNA transition. Overall, the structure of the one-strand insertion hybrid (169) is reported to be closer to the A-form than to a B-form geometry, although the helical parameters presented show many fluctuations without giving clear A- or B-form tendencies. The minor groove width, however, is clearly larger for the hybrid section, which is very different from all other hybrid structures discussed here. In this RNA insert hybrid structure, bends can be seen for both junctions.

5. Outlook for the future

The discussion of DNA and DNA:RNA hybrid structures above has shown the potential and the limitations of the NMR-based approach for high resolution structures. Even the most accurate DNA duplex structures are somewhat dependent on the choice of refinement method, and this suggests caution in using specific values of the derived structural parameters, especially helical parameters, and comparing them with the results of other studies. Nevertheless, when features can be compared, the structural results of different studies are generally found to be in accord. In our opinion, most of the recent DNA and DNA:RNA hybrid structures represent good structural models that capture a great deal of the sequence-dependent structural traits. With current NMR computational methods, reliable insights about structural features should be expected for nucleic acid systems that deviate from a standard duplex. The influence on structural features exerted by a distinct, localized modification should be revealed readily through NMR methods. Such systems might include modifications in backbone and nucleotides, mismatches, unusual base pairs, and bulged nucleotides; some of these are reviewed in Chapters 10–13. For standard DNA duplex structures, we can expect improvements via isotopic labelling techniques similar to developments in protein and RNA structure determination. ^{15}N- and ^{13}C -labelled precursors are becoming available now for synthesizing DNA chemically (181) and enzymatic preparation of samples (182). Furthermore, selective deuteration at several positions of the sugar moieties (183,184) is creating unique possibilities for the observation of only a portion of a larger system without losing the full structural context.

With respect to the accuracy of any NMR-derived structure, the biggest limitation comes from the potential flexibility of biological macromolecules, which might lead to average structures with artefacts. The availability of isotope-labelled DNA samples should also provide a handle for addressing the flexibility problem by measuring relaxation properties of the heteronuclei similar to earlier studies with very concentrated natural abundance samples (138,185). Although most cases of conformational flexibility present an underdefined system in terms of the information available from NMR, future refinement methods need to address these issues more systematically. Besides the tools mentioned in this chapter, other methods for generating multiple conformers in accord with NMR data or finding the combination of conformers best satisfying NMR data have already been reported.

References

1. van den Ven, F.J.M. and Hilbers, C.W. (1988) *Eur. J. Biochem.* **178**, 1.
2. Patel, D.J., Shapiro, L. and Hare, D. (1987) *Q. Rev. Biophys.* **20**, 35.
3. Wijmenga, S.S., Mooren, M.M.W. and Hilbers, C.W. (1994) in *NMR Macromolecules*, (ed. Roberts, G.C.K.), p. 217. Oxford University Press.
4. Feigon, J., Sklenar, V., Wang, E., Gilbert, D.E., Macaya, R.F. and Schultze, P. (1992) *Meth. Enzymol.* **211**, 235.
5. Schmitz, U. and James, T.L. (1995) *Meth. Enzymol.* **261**, 1.
6. Rinkel, L.J. and Altona, C. (1987) *J. Biomol. Struct. Dynamics* **4**, 621.
7. van Wijk, J., Huckriede, B.D., Ippel, J.H. and Altona, C. (1992) *Meth. Enzymol.* **211**, 286.
8. Kim, S.-G., Lin, L.-J. and Reid, B.R. (1992) *Biochemistry* **31**, 3564.
9. Dickerson, R. (1992) *Meth. Enzymol.* **211**, 67.
10. Metzler, W.J., Wang, C., Kitchen, D.B., Levy, R.M. and Pardi, A. (1990) *J. Mol. Biol.* **214**, 711.
11. Pardi, A., Hare, D.R. and Wang, C. (1988) *Proc. Natl. Acad. Sci. USA* **85**, 8785.
12. Lane, A.N. (1990) *Biochim. Biophys. Acta* **1049**, 189.
13. Schmitz, U., Pearlman, D.A. and James, T.L. (1991) *J. Mol. Biol.* **221**, 271.
14. Ulyanov, N.B., Gorin, A.A., Zhurkin , V.B., Chen, B., Sarma, M.H. and Sarma, R.H. (1992) *Biochemistry* **31**, 3918.
15. Ulyanov, N.B. and James, T.L. (1994) *Appl. Magn. Reson.* **7**, 21.

16. Ulyanov, N.B., Schmitz, U., Kumar, A. and James, T.L. (1995) *Biophys. J.* **68**, 13.
17. Peck, L.J. and Wang, J.C. (1981) *Nature* **292**, 375.
18. Rhodes, D. and Klug, A. (1981) *Nature* **292**, 378.
19. Melvyl (1984) *Registered Trademark of the Regents of the University of California.* Internet address: melvyl.ucop.edu.
20. Lane, A.N. (1994) *Meth. Enzymol.* **261**, 413
21. Macura, S. and Ernst, R.R. (1980) *J. Mol. Phys.* **41**, 95.
22. Gorenstein, D.A. (1992) *Meth. Enzymol.* **211**, 254.
23. James, T. L. (1991) *Curr. Opin. Struct. Biol.* **1**, 1042.
24. Allain, F.H.T., Gubser, C.C., Howe, P.W.A., Nagai, K., Neuhaus, D. and Varani, G. (1996) *Nature* **380**, 646.
25. Wüthrich, K. (1986) *NMR of Proteins and Nucleic Acids.* Wiley, New York.
26. Thomas, P.D., Basus, V.J. and James, T.L. (1991) *Proc. Natl. Acad. Sci. USA* **88**, 1237.
27. Borgias, B.A. and James, T.L. (1989) *Meth. Enzymol.* **176**, 169.
28. Kumar, A., Ernst, R.R. and Wüthrich, K. (1981) *J. Am. Chem. Soc.* **103**, 3654.
29. Keepers, J.W. and James, T.L. (1984) *J. Magn. Reson.* **57**, 404.
30. Borgias, B.A. and James, T.L. (1990) *J. Magn. Reson.* **87**, 475.
31. Post, C.B., Meadows, R.P. and Gorenstein, D.G. (1990) *J. Am. Chem. Soc.* **112**, 6796.
32. Boelens, R., Koning, T.M.G., van der Marel, G.A., van Boom, J.H. and Kaptein, R. (1989) *J. Magn. Reson.* **82**, 290.
33. Pearlman, D. A., Case, D. A., Caldwell, J. C., Seibel, G. L., Singh, U. C., Weiner, P. and Kollman, P. A. (1990) *AMBER, version 4.0.* University of San Francisco, San Francisco.
34. Brünger, A. T. (1992) *X-PLOR, Version 3.1: A System for X-ray Crystallography and NMR.* Yale Universiy Press, New Haven.
35. de Vlieg, J., Boelens, R., Scheek, R. M., Kaptein, R. and van Gunsteren, W. F. (1986) *Isr. J. Chem.* **27**, 181.
36. Young, M.A., Srinivasan, J., Goljer, I., Kumar, S., Beveridge, D.L. and Bolton, P.H. (1995) *Meth. Enzymol.* **261**, 121.
37. Molecular Simulations Inc. (1995) *Discover, InsightII.* San Diego, CA.
38. Nerdal, W., Hare, D.R. and Reid, B.R. (1989) *Biochemistry* **28**, 10008.
39. Nibedita, R., Kumar, R.A., Majumdar, A. and Hosur, R.V. (1992) *J. Biomol. NMR* **2**, 477.
40. Robinson, H. and Wang, A.H.J. (1992) *Biochemistry* **31**, 3524.
41. Lane, A.N. (1990) *Biochim. Biophys. Acta* **1049**, 205.
42. Liu, H., Spielmann, H.P., Ulyanov, N.B., Wemmer, D.E. and James, T.L. (1995) *J. Biomol. NMR* **6**, 390.
43. Altona, C. (1982) *Rec. Trav. Chim. Pays-Bas* **101**, 413.
44. Harbison, G.S. (1993) *J. Am. Chem. Soc.* **115**, 3026.
45. Zhu, L., Reid, B.R., Kennedy, M. and Drobny, G.P. (1994) *J. Mag. Res. Ser. A* **111**, 195.
46. Celda, B., Widmer, H., Leupin, W., Chazin, W.J., Denny, W.A. and Wüthrich, K. (1989) *Biochemistry* **28**, 1462.
47. Gochin, M., Zon, G. and James, T.L. (1990) *Biochemistry* **29**, 11161.
48. Schmitz, U., Zon, G. and James, T.L. (1990) *Biochemistry* **29**, 2357.
49. Macaya, R., Wang, E., Schultze, P., Sklenar, V. and Feigon, J. (1992) *J. Mol. Biol.* **225**, 755.
50. Conte, M.R., Bauer, C.J. and Lane, A.N. (1996) *J. Biomol. NMR* **7**, 190.
51. Schmidt, P. and Griesinger, C. (1994) unpublished data.
52. Salazar, M., Fedoroff, O.Y. and Reid, B.R. (1996) *Biochemistry* **35**, 8126.
53. Cornell, W., Cieplak, P., Bayly, C. I., Gould, I.R. and Kollman, P.A. (1996) *J. Am. Chem. Soc.* **118**, 2309.

54. Zhurkin, V.B., Ulyanov, N.B., Gorin, A.A. and Jernigan, R.L. (1991) *Proc. Natl. Acad. Sci. USA* **88**, 7046.
55. Ulyanov, N., Schmitz, U. and James, T. (1993) *J. Biomol. NMR* **3**, 547.
56. Mauffret, O., Hartmann, B., Convert, O., Lavery, R. and Fermandjian, S. (1992) *J. Mol. Biol.* **227**, 852.
57. James, T.L. (1994) *Meth. Enzymol.* **239**, 416.
58. González, C., Rullmann, J.A.C., Bonvin, M.J.J., Boelens, R. and Kaptein, R. (1991) *J. Magn. Reson.* **91**, 659.
59. Withka, J.M., Srinivasan, J. and Bolton, P.H. (1992) *J. Magn. Reson.* **98**, 611.
60. Kaluarachchi, K., Meadows, R.P. and Gorenstein, D.G. (1992) *Biochemistry* **30**, 8785.
61. Ulyanov, N.B. and James, T.L. (1995) *Meth. Enzymol.* **261**, 90.
62. Kim, S.-G. and Reid, B.R. (1992) *Biochemistry* **31**, 12103.
63. Schmitz, U., Sethson, I., Egan, W.M. and James, T.L. (1992) *J. Mol. Biol.* **227**, 510.
64. Weisz, K., Shafer, R.H., Egan, W. and James, T.L. (1994) *Biochemistry* **33**, 354.
65. Gochin, M. and James, T.L. (1990) *Biochemistry* **29**, 11172.
66. Schmitz, U. and James, T.L. (1993) in *Structural Biology: The State of the Art*, (Sarma, R.H. and Sarma, M.H., eds), Vol. 2, p. 251. Adenine Press, Schenectady.
67. Diekmann, S. (1989) *EMBO J.* **8**, 1.
68. Lavery, R. and Sklenar, H. (1990) *CURVES 3.0, Helical Analysis of Irregular Nucleic Acids.* Laboratory for Theoretical Biochemistry CNRS, Paris, France 1990.
69. Leijon, M., Zdunek, J., Fritzsche, H., Sklenar, H. and Graslund, A. (1995) *Eur.J. Biochem.* **234**, 832.
70. Schmitz, U., Ulyanov, N.B., Kumar, A. and James, T.L. (1993) *J. Mol. Biol.* **234**, 373.
71. Kopka, M.L., Fratini, A.V., Drew, H.R. and Dickerson, R.E. (1983) *J. Mol. Biol.* **163**, 129.
72. Chuprina, V.P. (1987) *Nucl. Acids Res.* **15**, 293.
73. González, C., Stec, W., Kobylanska, A., Hogrefe, R.I., Reynolds, M. and James, T.L. (1994) *Biochemistry* **33**, 11062.
74. Clore, G.M. and Gronenborn, A.M. (1985) *EMBO J.* **4**, 829.
75. Clore, G.M., Gronenborn, A., Moss, D. and Tickle, I. (1985) *J. Mol. Biol.* **185**, 219.
76. Nilsson, L., Clore, G.M., Gronenborn, A.M., Brünger, A.T. and Karplus, M. (1986) *J. Mol. Biol.* **188**, 455.
77. Nilges, M., Clore, G.M., Gronenborn, A.M., Brünger, A.T., Karplus, M. and Nilsson, L. (1987) *Biochemistry* **26**, 3734.
78. Nilges, M., Clore, G.M. and Gronenborn, A.M. (1987) *Biochemistry* **26**, 3718.
79. Gupta, G., Sarma, M.H. and Sarma, R.H. (1988) *Biochemistry* **27**, 7909.
80. Sarma, M.H., Gupta, G. and Sarma, R.H. (1988) *Biochemistry* **27**, 3423.
81. Nerdal, W., Hare, D.R. and Reid, B.R. (1988) *J. Mol. Biol.* **201**, 717.
82. Banks, K.M., Hare, D.R. and Reid, B.R. (1989) *Biochemistry* **28**, 6996.
83. Baleja, J.D., Pon, R.T. and Sykes, B.D. (1990) *Biochemistry* **29**, 4828.
84. Baleja, J.D., Germann, M.W., van de Sande, J.H. and Sykes, B.D. (1990) *J. Mol. Biol.* **215**, 411.
85. Powers, R., Jones, C.R. and Gorenstein, D.G. (1990) *J. Biomol. Struct. Dynamics* **8**, 253.
86. Katahira, M., Sugeta, H. and Kyogoku, Y. (1990) *Biochemistry* **29**, 7214.
87. Kerwood, D.J., Zon, G. and James, T.L. (1991) *Eur. J. Biochem.* **197**, 583.
88. Ito, N., Nakamura, H., Sumikawa, H. and Nagashima, N. (1991) *J. Mol. Struct.* **242**, 119.
89. Cheng, J.-W., Chou, S.-H., Salazar, M. and Reid, B.R. (1992) *J. Mol. Biol.* **228**, 118.
90. Ulyanov, N.B., Sarma, M.H., Zhurkin, V.B. and Sarma, R.H. (1993) *Biochemistry* **32**, 6875.
91. Ulyanov, N.B., Gorin, A.A., Zhurkin, V.B., Chen, B.C., Sarma, M.H. and Sarma, R.H. (1992) *Biochemistry* **31**, 3918.

92. Mujeeb, A., Kerwin, S.M., Kenyon, G.L. and James, T.L. (1993) *Biochemistry* **32**, 13419.
93. Chuprina, V.P., Sletten, E. and Fedoroff, O. (1993) *J. Biomol. Struct. Dynamics* **10**, 693.
94. Shapiro, L., Nilges, M. and Eriksson, M. (1993) *Acta Chem. Scand.* **47**, 43.
95. Catasti, P., Gupta, G., Garcia, A.E., Ratliff, R., Hong, L., Yau, P., Moyzis, R.K. and Bradbury, E.M. (1994) *Biochemistry* **33**, 3819.
96. Radha, P.K., Madan, A., Nibedita, R. and Hosur, R.V. (1995) *Biochemistry* **34**, 5913.
97. Feng, B. and Stone, M.P. (1995) *Chem. Res. Toxicol.* **8**, 821.
98. Sodano, P., Hartmann, B., Rose, T., Wain-Hobson, S. and Delepierre, M. (1995) *Biochemistry* **34**, 6900.
99. Lingbeck, J., Kubinec, M.G., Miller, J., Reid, B.R., Drobny, G.P. and Kennedy, M.A. (1996) Biochemistry **35**, 719.
100. Bernstein, F.C., Koetzle, T.F., Williams, G.J., Meyer, E.E., Brice, M. D., Rodgers, J. R., Kennard, O., Shimanouchi, T. and Tasumi, M. (1977) *J. Mol. Biol.* **112**, 535.
101. Berman, H. M., Olson, W. K., Beveridge, D. L., Westbrook, J., Gelbin, A., Demeny, T., Hsieh, S.-H., Srinivasan, A. R. and Schneider, B. (1992) *Biophys. J.* **63**, 751.
102. Herbert, A., Lowenhaupt, K., Spitzner, J., Berger, I. and Rich, A. (1995) in *Biological Structure and Dynamics*, (Sarma, R.H. and Sarma, M.H., eds), Vol. 2, p. 189. Adenine Press, Schenectady.
103. Orbons, L.P. and Altona, C. (1986) *Eur. J. Biochem.* **160**, 141.
104. Klysik, J., Stirdivant, S.M., Larson, J., Hart, P.A. and Wells, R.D. (1981) *Nature* **290**, 672.
105. Patel, D.J., Kozlowski, S.A., Hare, D.R., Reid, B., Ikuta, S., Lander, N. and Itakura, K. (1985) *Biochemistry* **24**, 926.
106. Ikuta, S. and Wang, Y.S. (1989) *Nucl. Acids Res* **17**, 4131.
107. Vorlickova, M. (1995) *Biophys. J.* **69**, 2033.
108. Riazance-Lawrence, J.H. and Johnson, W.C.J. (1992) *Biopolymers* **32**, 271.
109. Ulyanov, N.B., Gorin, A.A. and Zhurkin, V.B. unpublished results.
110. Gorin, A.A., Zhurkin, V.B. and Olson, W.K. (1995) *J. Mol. Biol.* **247**, 34.
111. Gorin, A.A., Zhurkin, V.B. and Olson, W.K. (unpublished results, cited from Ulyanov and James, *Meth. Enzymol.* **261**, 90).
112. Calladine, C.R. (1982) *J. Mol. Biol.* **161**, 343.
113. Ulyanov, N.B. and Zhurkin, V.B. (1984) *J. Biomol. Struct. Dynamics* **2**, 361.
114. Poncin, M., Piazzola, D. and Lavery, R. (1992) *Biopolymers* **32**, 1077.
115. Yoon, C., Privé, G.G., Goodsell, D.S. and Dickerson, R.E. (1988) *Proc. Natl. Acad. Sci. USA* **85**, 6332.
116. Yuan, H., Quintana, J. and Dickerson, R. (1992) *Biochemistry* **31**, 8009.
117. Yanagi, K., Privé, G.G. and Dickerson, R.E. (1991) *J. Mol. Biol.* **217**, 201.
118. Fedoroff, O.Y., Reid, B.R. and Chuprina, V.P. (1994) *J. Mol. Biol.* **235**, 325.
119. Chuprina, V.P., Lipanov, A.A., Fedoroff, O.Y., Kim, S.-G., Kintanar, A. and Reid, B.R. (1991) *Proc. Natl. Acad. Sci. USA* **88**, 9087.
120. Olson, W.K. and Zhurkin, V.B. (1996) *Biological Structure and Dynamics*, (Sarma, R. H. and Sarma, M. H., eds), Vol. 2, p. 341. Adenine Press, Schenectady.
121. Bolshoy, A., McNamara, P., Harrington, R.E. and Trifonov, E.N. (1991) *Proc. Natl. Acad. Sci. USA* **88**, 2312.
122. Calladine, C.R., Drew, H.R. and McCall, M.J. (1988) *J. Mol. Biol.* **201**, 127.
123. Crothers, D.M., Haran, T.E. and Nadeau, J.G. (1990) *J. Biol. Chem.* **265**, 7093.
124. Goodsell, D.S., Kaczor-Grzeskowiak, M. and Dickerson, R.E. (1994) *J. Mol. Biol.* **239**, 79.
125. Allain, F.H.T. and Varani, G. (1995) *J. Mol. Biol.* **250**, 333.

126. DiGabriele, A.D., Sanderson, M.R. and Steitz, T.A. (1989) *Proc. Natl. Acad. Sci. USA* **86**, 1816.

127. Fawthrop, S.A., Yang, J.C. and Fisher, J. (1993) *Nucl. Acids Res.* **21**, 4860.

128. Hagerman, P.J. (1986) *Nature* **321**, 449.

129. Koo, H.-S. and Crothers, D.M. (1988) *Proc. Natl. Acad. Sci. USA* **85**, 1763.

130. Sanghani, S.R., Zakrzewska, K., Harvey, S.C. and Lavery, R. (1996) *Nucl. Acids Res.* **24**, 1632.

131. Lefévre, J.F., Lane, A.N. and Jardetzky, O. (1987) *Biochemistry* **26**, 5076.

132. Kennedy, M.A., Nuutero, S.T., Davis, J.T., Drobny, G.P. and Reid, B.R. (1993) *Biochemistry* **32**, 8022.

133. Liepinsh, E., Leupin, W. and Otting, G. (1994) *Nucl. Acids Res.* **22**, 2249.

134. Kubinec, M.G. and Wemmer, D.E. (1992) *J. Am. Chem. Soc.* **114**, 8739.

135. Liepinsh, E., Otting, G. and Wüthrich, K. (1992) *Nucl. Acids Res.* **20**, 6549.

136. Kearns, D.R. (1984) *Crit. Rev. Biochem.* **15**, 237.

137. Lane, A. (1993) *Progr. NMR Spectrosc.* **25**, 481.

138. Borer, P.N., LaPlante, S.R., Kumar, A., Zanatta, N., Martin, A., Hakkinen, A. and Levy, G.C. (1994) *Biochemistry* **33**, 2441.

139. Alam, T.M., Orban, J. and Drobny, G.P. (1991) *Biochemistry* **30**, 9229.

140. Reid, B.R., Banks, K., Flynn, P. and Nerdal, W. (1989) *Biochemistry* **28**, 10001.

141. Lipari, G. and Szabo, A. (1982) *J. Am. Chem. Soc.* **104**, 4546.

142. Withka, J.M., Swaminathan, S., Srinivasan, J., Beveridge, D.L. and Bolton, P.H. (1992) *Science* **255**, 597.

143. Koning, T.M.G., Boelens, R., van der Marel, G.A., van Boom, J.H. and Kaptein, R. (1991) *Biochemistry* **30**, 3787.

144. Lane, A., Bauer, C.J. and Frenkiel , T.A. (1993) *Eur. Biophys. J.* **21**, 425.

145. McAteer, K., Ellis, P.D. and Kennedy, M.A. (1995) *Nucl. Acids Res.* **23**, 3962.

146. Altona, C. and Sundaralingam, M. (1972) *J. Am. Chem. Soc.* **94**, 8205.

147. Macaya, R.F., Schultze, P. and Feigon, J. (1992) *J. Am. Chem. Soc.* **114**, 781.

148. Emsley, L., Dwyer, T.J., Spielmann, H.P. and Wemmer, D.E. (1993) *J. Am. Chem. Soc.* **115**, 7765.

149. Beveridge, D., Swaminathan, S., Ravishanker, G., Withka, J., Srinivasan, J., Prevost, C., Louise-May, S., Langley, D., DiCapua, F. and Bolton, P.H. (1993) in *Water and Biological Macromolecules*, (Westhof, E., ed.), p. 143.CRC Press, Boca Raton.

150. Cheatham III, T.E. and Kollman, P.A. (1996) *J. Mol. Biol.* **259**, 434.

151. Gorin, A.A., Ulyanov, N.B. and Zhurkin, V.B. (1990) *Mol. Biol.* **24**, 1036.

152. Torda, A.E., Scheek, R.M. and van Gunsteren, W.F. (1991) in *Computational Aspects of the Study of Biological Macromolecules by Nuclear Magnetic Resonance Spectroscopy*, (Hoch, J.C., ed.), p.219. Plenum Press, New York.

153. Schmitz, U., Kumar, A. and James, T.L. (1992) *J. Am. Chem. Soc.* **114**, 10564.

154. Nakamura, H., Oda, Y., Iwai, S., Inoue, H., Ohtsuka, E., Kanaya, S., Kimura, S., Katsuda, C., Katayanagi, K., Morikawa, K., Miyashiro, H. and Ikehara, M. (1991) *Proc. Natl. Acad. Sci. USA* **88**, 11535.

155. Oda, Y., Iwai, S., Ohtsuka, E., Ishikawa, M., Ikehara, M. and Nakamura, H. (1993) *Nucl. Acids Res.* **21**, 4690.

156. Reid, D.G., Salisbury, S.A., Brown, T., Williams, D.H., Vasseur, J.J., Rayner, B. and Imbach, J.L. (1983) *Eur. J. Biochem.* **135**, 307.

157. Gupta, G., Sarma, M.H. and Sarma, R.H. (1985) *J. Mol. Biol.* **186**, 463.

158. Zimmerman, S.B. and Pheiffer, B.H. (1981) *Proc. Natl. Acad. Sci. USA* **78**, 78.

159. Arnott, S., Chandrasekaran, R., Millane, R.P. and Park, H.S. (1986) *J. Mol. Biol.* **188**, 631.

160. Egli, M., Usman, N. and Rich, A. (1993) *Biochemistry* **32**, 3221.
161. Ban, C., Ramakrishnan, B. and Sundaralingam, M. (1994) *J. Mol. Biol.* **236**, 275.
162. Salazar, M., Fedoroff, O.Y., Miller, J.M., Ribeiro, N.S. and Reid, B.R. (1993) *Biochemistry* **32**, 4207.
163. Lane, A.N., Ebel, S. and Brown, T. (1993) *Eur. J. Biochem.* **215**, 297.
164. González, C., Stec, W., Reynolds, M. and James, T.L. (1995) *Biochemistry* **34**, 4969.
165. Gao, X. and Jeffs, P.W. (1994) *J. Biomol. NMR* **4**, 367.
166. Salazar, M., Fedoroff, O., Zhu, L. and Reid, B.R. (1994) *J. Mol. Biol.* **241**, 440.
167. Fedoroff, O., Salazar, M. and Reid, B.R. (1996) *Biochemistry* **35**, 11070.
168. Salazar, M., Champoux, J.J. and Reid, B.R. (1993) *Biochemistry* **32**, 739.
169. Nishizaki, T., Iwai, S., Ohkubo, T., Kojima, C., Nakamura, H., Kyogoku, Y. and Ohtsuka, E. (1996) *Biochemistry* **35**, 4016.
170. Jaishree, T.N., van der Marel, G.A., van Boom, J.H. and Wang, A.H. (1993) *Biochemistry* **32**, 4903.
171. Zhu, L., Salazar, M. and Reid, B.R. (1995) *Biochemistry* **34**, 2372.
172. Chou, S.-H., Flynn, P. and Reid, B.R. (1989) *Biochemistry* **28**, 2435.
173. Chou, S.-H., Flynn, P., Wang, A. and Reid, B. (1991) *Biochemistry* **30**, 5248.
174. Shindo, H. and Matsumoto, U. (1984) *J. Biol. Chem.* **259**, 8682.
175. Steely, H.T., Gray, D.M. and Ratcliff, R.L. (1986) *Nucl. Acids Res.* **24**, 10071.
176. Benevides, F.C., Koetzle, T.F., Williams, G.J.B., Meyer, E.F., Brice, M.D., Rodgers, J.R., Kennard, O., Shimanouchi, T. and Tasume, M. (1988) *Biochemistry* **27**, 3868.
177. Fedoroff, O.Y., Salazar, M. and Reid, B.R. (1993) *J. Mol. Biol.* **233**, 509.
178. Wang, A.H.J., Fujii, S., Van Boom, J.H., Van der Marel, G.A., Van Boeckel, C.A.A. and Rich , A. (1982) *Nature* **299**, 601.
179. Egli, M., Usman, N., Zhang, S. and Rich, A. (1992) *Proc. Natl. Acad. Sci. USA* **89**, 534.
180. Arnott, S., Chandrasekaran, R., Puigjaner, L.C., Walker, J.K., Hall, I.H., Birdsall, D.L. and Ratcliff, R.L. (1983) *Nucl. Acids Res.* **11**, 1457.
181. Tate, S., Ono, A. and Kainosho, M. (1994) *J. Am. Chem. Soc.* **116**, 5977.
182. Zimmer, D.P. and Crothers, D.M. (1995) *Proc. Natl. Acad. Sci. USA* **92**, 3091.
183. Yamakage, S.I., Maltseva, T.V., Nilson, F.P., Foldesi, A. and Chattopadhyaya, J. (1993) *Nucl. Acids Res.* **21**, 5005.
184. Agback, P., Maltseva, T.V., Yamakage, S.I., Nilson, F.P., Foldesi, A. and Chattopadhyaya, J. (1994) *Nucl. Acids Res.* **22**, 1404.
185. LaPlante, S.R., Zanatta, N., Hakkinen, A., Wang, A.H. and Borer, P.N. (1994) *Biochemistry* **33**, 2430.

9
Nucleic acid hydration

Helen M. Berman[1] and Bohdan Schneider[2]

[1]*Department of Chemistry, Rutgers University, Piscataway, NJ 08854–8087, USA*
[2]*J. Heyrovsky Institute of Physical Chemistry, Academy of Sciences of the Czech Republic, 18223 Prague, Czech Republic*

1. Introduction

It is perhaps only a small exaggeration to say that the timing of the birth of modern molecular biology was dependent on selecting the DNA sample with the correct water content. Franklin and Gosling (1) first observed that as the humidity of the sample increased, the characteristics of the fibre diffraction pattern changed. The low humidity A-form was apparently more crystalline and was, therefore, the initial focus of their attention. However, the high humidity B-form was more interpretable because it yielded the characteristic helical diffraction pattern. Once attention was given to this form, the double helical structure of DNA was discovered (2).

Fibre diffraction (3,4) studies established that the B-form of DNA is the long, slender, righthanded helix that has now become an icon of biology. A-DNA is shorter and squatter, with the bases inclined to the helix axis. These studies also confirmed that the presence of ions and solvent plays a very strong role in determining which conformation a given DNA will adopt. The challenge is to find out why this is so.

Early solution studies introduced the concept of hydration shells and differentiated this water from the bulk solvent (5,6). Numerous experimental and theoretical studies have given further insight into the effects of sequence and environment on DNA structure. The ability to crystallize short, defined sequences of nucleic acids has made it possible to visualize the bound water using the methods of X-ray crystallography. Newly developed NMR techniques have allowed us to obtain a dynamic picture of the waters and their interactions with nucleic acids.

The importance of water in macromolecular recognition is well appreciated, if not fully understood. The careful balance between the enthalpic contribution of hydrogen bonding and the entropic consequences of disrupting those bonds drives the interactions between nucleic acids and other molecules including drugs and proteins (7).

This chapter will summarize the results of some recent studies of the behaviour of nucleic acids in solution, and then present the current state of our knowledge about the structure of water around nucleic acids as derived from X-ray, NMR, and theoretical analyses. Reviews of some of the earlier work can be found in several sources (8–15).

2. Macroscopic studies

Many different studies on the behaviour of DNA and RNA in solution and in fibres have provided data about how water influences the behaviour of these molecules. The results of thermodynamic studies of nucleic acid duplexes (16) and their complexes with drugs (7,17) have been interpreted in terms of the influence of hydration on the structure and interactions of nucleic acids. Changes in entropy in particular have been correlated with transition or binding events that induce ordered solvent to be released to the bulk medium. Magnetic densimetric techniques have been used to measure ligand binding-induced volume changes, which have been found to correlate with entropy changes (18). In these studies, volume increases correspond to entropy increases which, in turn, are thought to relate to the release of bound water; volume contractions are proposed to be associated with net hydration.

By judicious choice of samples, it is possible to relate the macroscopic behaviour to specific microscopic properties, such as sequence and conformation. Calorimetric and densimetric measurements (19) suggest that B-form homoduplexes are more hydrated than their A-form counterparts. The formation of bulged DNA is accompanied by volume contractions, indicating that there is more coulombic hydration (20). These experiments suggest that favourable changes in the thermodynamics of hydration compensate for the otherwise destabilizing effect of the bulge. The observation that the volume contracts more when distamycin binds to alternating AT polymers than when it binds to homopolymers was interpreted as an indication of higher hydration of homopolymeric duplexes. In addition, it was suggested that drug complexation is accompanied by an increase in hydration, which may result from the strengthening of the hydrogen bonded water network by the hydrophobic groups of the ligand (21). Other thermodynamic experiments have led to the conclusions that parallel DNA is less hydrated than antiparallel DNA (22), and that AT homopolymers have different hydration properties than their GC counterparts (23).

Both the partial molar volume and the partial molar adiabatic compressibility are extremely sensitive to solute hydration (24). These have been measured for DNA alone and complexed with netropsin using densimetric and newly developed ultrasonic techniques (25,26). It was found that the coefficient of adiabatic compressibility of the first hydration shell is significantly different from that of bulk water. It was also shown that duplexes with 55–60% AT compositions exhibit the weakest hydration; increases or decreases in AT content from this range lead to enhanced hydration. Although all of the B-DNA sequences studied showed the same total quantity of water, the fact that poly (dA):poly (dT) homopolymers are thought to be more hydrated than alternating poly (dAdT):poly (dAdT) copolymers may be a consequence of the fact that they have stronger DNA–water interactions. However, this possibility is not confirmed by the results of the volumetric measurements; both types of AT duplexes exhibit similar values for both the partial molar volume and the partial molar compressibility. Thus, further studies are required to understand better the hydration properties of these DNA polynucleotides.

Osmotic stress measurements have given another view about the role of water in intermolecular interactions (27,28). Osmotic stress is applied to an array of DNA molecules and their intermolecular separations are measured by diffraction techniques. The results of the experiments are interpreted to mean that cations between helices reorganize the water in such a way as to balance repulsive forces with long-range attractive hydration forces. The concept of positive hydration forces may lead to a simpler physical model than the one implied by the more traditional concepts about hydrophobic interactions.

Osmotic stress has also been used to study the interactions of *Eco*RI and DNA (29). It has been known for some time that under some conditions, the enzyme has reduced sequence specificity; this has been called 'star' activity. When the DNA cleavage reaction was measured in the presence of several osmolytes, the 'star' activity was demonstrated to be directly related to osmotic pressure. At high pressures, the water activity is lowered and, with it, the specificity. The interpretation of these results is that the bound water at the DNA–protein interface is key to the molecular specificity.

3. Structural analyses of nucleic acid hydration

3.1 Early studies and methods of analysis

The early crystal structure determinations of dinucleoside phosphates (30–32) demonstrated the presence of ordered water in crystals and led to the concept of water involvement in the recognition process (33). The structure of a dinucleoside phosphate complexed to the small molecule drug proflavine showed an elegant pentagonal network of water molecules reminiscent of clathrate structures around hydrophobic small molecules (34). This structure provided a test bed for many subsequent theoretical studies (35–37). The results of these early structural analyses gave some important insights into the hydration of nucleic acids. However, it was the observation of the spine of hydration in the first B-DNA structure (38) that made it necessary to consider seriously the concept of water as an integral part of nucleic acids.

The first structural studies of the hydration of nucleic acids were done using X-ray crystallographic methods, which allow us to observe a time-averaged view of the atoms in a crystal. Therefore, if a water molecule is exchanging between one site on a molecule and the bulk solvent, it will be observed on the electron density map; the rate of exchange does not affect its observation. On the other hand, if a water molecule occupies multiple sites it will be difficult to observe. In recent years, NMR methods have been developed that not only allow the determination of macromolecular structures, but also provide a view of hydration structure around these molecules. Unlike diffraction methods, which give a time-averaged view, NMR methods provide a dynamic view and thus depend on the rate of exchange of the water molecule with the bulk solvent. If the rate of exchange is slow, the water molecule will be observed; if the water molecule is rapidly exchanging even between a single site and the surrounding water, it will not be detectable. Both methods are thus complementary and offer us insight into the characteristics of the hydration structure of nucleic acids. In the following discussion, we give the results of both methods for the various types of nucleic acid structures.

3.2 B-DNA

There are a few distinctive hydration motifs observed in B-DNA helices. The spine of hydration first seen in the crystal structure of a dodecamer containing the *Eco*R1 restriction site sequence d[GAATTC]$_2$ (BDL001) (39) has been observed in many

(a)

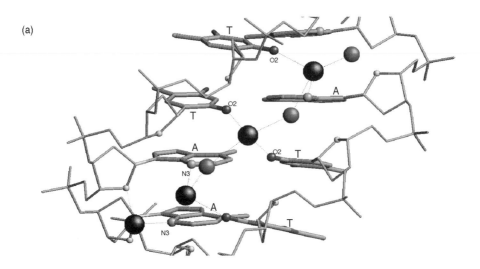

(b)

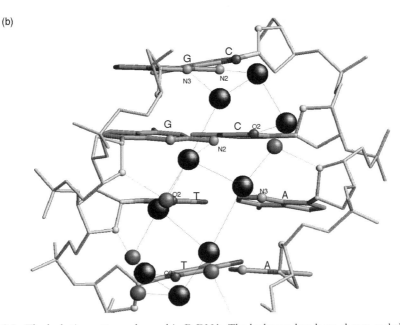

Fig. 9.1. The hydration patterns observed in B-DNA. The hydrogen bonds are shown as dashed lines. (a) The spine of hydration in the minor groove of d[CGCGAATTCGCG]$_2$ (BDL001) (39). The first shell waters are hydrogen bonded to the purine N3 and pyrimidine O2 atoms and are shown as large spheres. The second shell waters bridge the first shell waters and are shown as small spheres. (b) The double row of waters in the minor groove of d[CCAACGTTGG]$_2$ (BDJ019) (40). The waters making links between the base-attached waters are shown as smaller spheres. (c) Major groove hydration in d[CGATCGATCG]$_2$ (BDJ025) (41). Two waters attached to guanine make hydrogen bonds to a water attached to thymine.

(c)

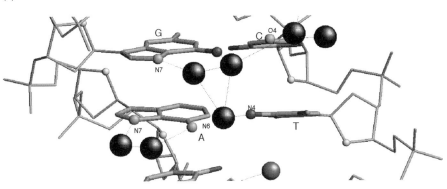

(d)

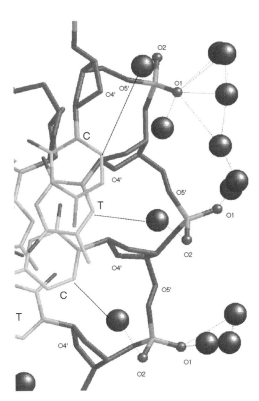

When the helical twist is lower than that shown, another hydrogen bond forms between the waters attached to guanine N7 and adenine N6, thus forming a more extensive network. (d) Phosphate hydration in d[CTCTCGAGAG]₂ (BDJ060) (47). Hydration spheres of the top and middle phosphates are linked by a hydrogen bond, while waters hydrating O1P and O2P of the same phosphate are far apart. Dotted lines show ~3.6 Å long contacts between waters hydrogen bonded to O2P and the C6 atoms of adjacent pyrimidine bases.

other structures of B-DNA. The spine is found in the minor groove of AATT regions. First shell waters hydrogen bonded to the purine N3 or pyrimidine O2 atoms are bridged by second shell waters (Fig. 9.1a). Additional contacts are made with the O4′ atoms of the sugars. In crystal structures of oligonucleotides with wider minor grooves, such as d[CCAACGTTGG]$_2$ (BDJ019) (40), double rows of waters can be accommodated into the minor groove (Fig. 9.1b).

There are also distinctive hydration patterns seen in the major groove. In d[CGATCGATCG]$_2$ (BDJ025) (41) (Fig. 9.1c), for example, a spine of first shell waters interconnects the hydrophilic atoms in the CGAT sequence. In this particular example, the hydration patterns around the two GA steps show differences that are directly correlated with the differences in their helical twist.

Systematic analyses of the patterns of hydration around the bases, sugars, and phosphates (42–44) contained in DNA crystal structures have led to a much clearer under-

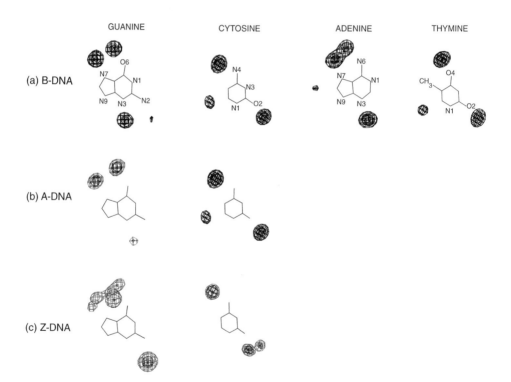

Fig. 9.2. Pseudo–electron densities of water around bases in B-, A-, and Z-type DNA conformations. The major groove is on the upper left side of each base; the minor groove on the lower right side. (a) The purines have two principal hydration sites in the major groove and one in the minor groove; pyrimidines have one such site in each groove. Low but significant densities near pyrimidine atom C6 reflect the existence of waters trapped between phosphate atom O2P and pyrimidine C6 (see Fig. 9.1d). (b) A-DNA base hydration is similar to that in B-DNA. Hydration in the major groove of A-DNA is more extensive than in the minor groove. The difference is quite striking for guanine. (c) Hydration sites of bases in the Z conformation are different from both right-handed conformations. In the minor groove, guanine N2 rather than N3 is hydrated, and cytosine has two hydration sites. In the major groove, hydration of guanine is quite distinct with four non-planar hydration sites, two hydrating O6, and two N7.

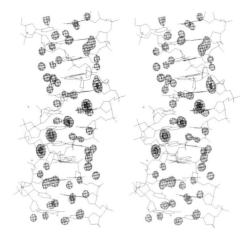

Fig. 9.3. A stereogram of the pseudo-electron density for the d[CGCGAATTCGCG]$_2$ (BDL001) (39) dodecamer structure. The five strongest densities in the front reproduce the spine of hydration.

standing of the basis of the networks in DNA double helices. When all bases of a single type, with their associated water molecules, are superimposed to create hydrated building blocks, there are clusters of water associated with each polar atom (Fig. 9.2a). If these building blocks are modelled into known B-DNA structures, the hydration patterns in the grooves are reproduced (43). Furthermore, building blocks created from decamer structures alone could be used to model the hydration patterns in a dodecamer sequence (45) (Fig. 9.3). This means that the hydration patterns around the bases are local. In that same study, it was demonstrated that, in principle, the spine of hydration can be formed by both AATT and GGCC sequences. If GGCC could form the same conformation as AATT, resulting in the narrow minor groove dimensions, then it too could nucleate a spine. Thus, the hydration pattern of the central AATT in the dodecamer structure is as much a function of the local base conformation as it is of the hydrogen bonding potential of the base.

In contrast to the hydration geometry around the bases, which shows very strong clustering, the hydration around phosphates is more variable and conformation dependent. Analyses of known crystal structures and theoretical analyses suggest that each charged oxygen can be surrounded by up to three water molecules which can be arranged in a cone of hydration (46). An example of the types of water patterns that can be formed around the phosphate backbone is shown in the structure of d[CTCTCGAGAG]$_2$ (BDJ060) (47) (Fig. 9.1d).

Of considerable interest is whether the hydration seen in crystal structures can be observed in solution (48). NMR methods using a combination of nuclear Overhauser (NOESY) and rotating frame nuclear Overhauser spectroscopies (ROESY) (49) have been successfully employed to study hydrated DNA (50–52). These studies confirm the existence of the spine of hydration in the minor groove of DNA containing AATT segments. A very recent study showed that while in some sequences minor groove hydration of TTAA segments is kinetically destabilized (51), there are sequences where this is not the case (52). It has been assumed that the width of the minor groove is directly related to the stability of the spine and that the TTAA segment would have a wide groove. However, the authors point out that NMR methods do not give accurate information about groove width and there are not enough X-ray structures to be able to predict the groove width of a particular sequence because, among other things, we do not, as yet, know the effects of flanking sequences. More studies are needed to 'confirm this putative connection between hydration lifetimes, minor groove width, and nucleotide sequence' (52).

Theoretical studies of DNA hydration have been reviewed elsewhere (9,53). One very recent molecular dynamics simulation gives the very stimulating result that in over half the trajectory, a sodium ion, rather than a water molecule, is found in the A–T step of the minor groove (54). This type of geometry was observed in the high resolution crystal structure of ApU (32) but it has never been seen, as yet, in DNA oligomer crystals. The lower resolution of these structures makes it difficult to distinguish sodium ions from water molecules, especially if the sites are not fully occupied. The results from the theoretical analysis strongly suggest that at least some of the water molecules found in X-ray structures are actually ions and/or the hydration sites are partially occupied by ions. Further experimental analyses at much higher resolution are needed to resolve this issue.

3.3 A-DNA

One of the earliest oligonucleotide crystal structures to be reported, d[GGBr⁵UABr⁵ UACC]₂ (ADHB11) (55) contains fused pentagonal rings of water in the major groove

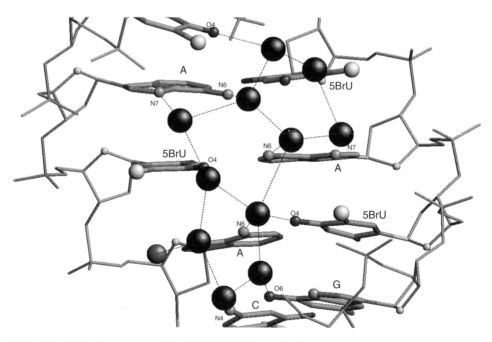

Fig. 9.4. Ordered water network in d[GGBr⁵UABr⁵UACC]₂ (ADHB11) (55). Water molecules associated with bridges are shown as large spheres; other waters as smaller spheres.

(Fig. 9.4). Since this first observation, numerous A-DNA structures have been determined and many exhibit complex and interesting hydration patterns.

In some cases, as in d[GGGTACCC]$_2$ (ADH030, ADH031), temperature strongly affects the hydration pattern (56). On the other hand, in d[GGGCGCCC]$_2$ (ADH057), this is not the case. Here, the structure of the duplex was determined at three different temperatures and the hydration patterns were very similar (57). Only the hydration of the phosphate backbone is less conserved.

In an analysis of five A-DNA crystal structures that contain CG in the central part of their sequence, certain common features were observed (57). The most striking is a chain of water molecules in the minor groove that interconnect the central CG to the backbone atoms of symmetry-related molecules (Fig. 9.5a). In these same structures, there are water-mediated groove–groove and groove–backbone interactions (Fig. 9.5b). In two cases, the second type of water bridge is also involved in pentagonal networks in the crystal.

Systematic analysis of the base hydration in A-DNA duplexes shows the same type of tight clustering of waters around the base heteroatoms exposed to solvent (43) (Fig. 9.2b). The individual hydration sites for bases in the A and B conformations are very similar, with the major difference being in the relative occupancies of the waters in the

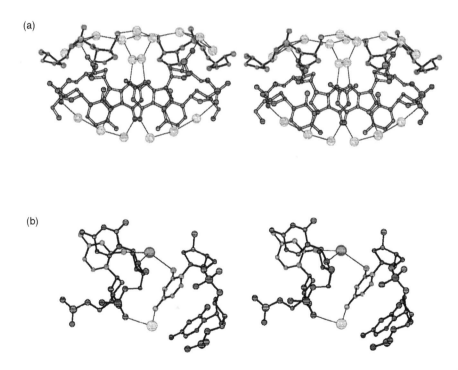

(a)

(b)

Fig. 9.5. (a) Water network at the CG step in an A-DNA structure d[GGGGCCCC]$_2$. (b) Intermolecular water-mediated groove–groove and groove–backbone interactions in A-DNA structures. (From ref. 57 with permission.)

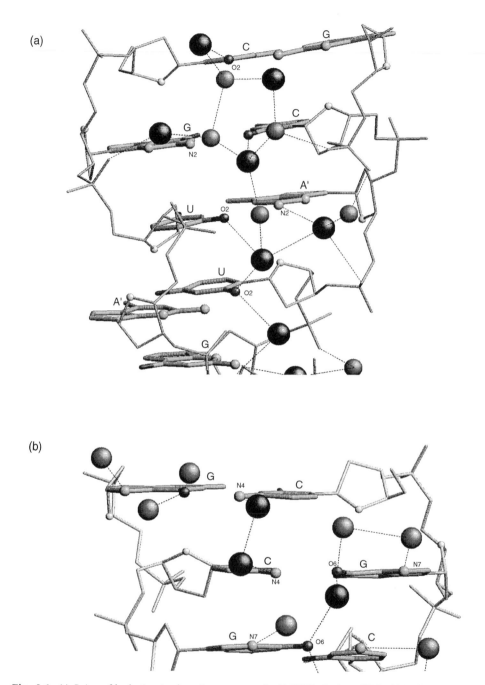

Fig. 9.6. (a) Spine of hydration in the minor groove of a Z-DNA duplex, d[CGU′ACG]$_2$ (ZDFB31) (60). Water molecules hydrogen bonded to the bases are drawn as large spheres; other waters as smaller spheres. Note that some waters hydrogen bond to phosphate oxygens. (b) Cross-strand O6–w–O6 and N4–w–w–N4 water bridges in the major groove of a Z-DNA helix d[CGCGCG]$_2$ (59). Water molecules associated with bridges are shown as large spheres; other waters as smaller spheres.

major and minor grooves. In B-DNA, the waters are localized equally well in both grooves, whereas in A-DNA more waters are found in the major groove than in the minor groove. Some studies of the phosphate hydration show that, as in B-DNA, the hydration sites are less well conserved (57). In many, but certainly not all, A-DNA structures, there are water bridges between adjacent phosphates in a strand. This feature led to the concept of the 'economy of hydration' (58) which is suggested as a driving force in the B to A transition when the humidity is lowered.

3.4 Z-DNA

Z-DNA duplexes show very distinctive hydration patterns. A spine of hydration is formed in the very deep minor groove. A network of water molecules is formed by water molecules connected to O2 atoms of cytosines from opposite strands which are further hydrogen bonded to second shell water molecules (Fig. 9.6a). In a detailed analysis of the crystal structure of d[CGCGCG]$_2$ (59), it was shown that in the convex major groove there are bridges between the two guanine O6 atoms at GpC steps from opposite strands and between the two N4 cytosines in CpG steps (Fig. 9.6b).

In addition to the intrahelix networks found in Z duplexes, there are bridges that connect the helices in the crystal. An analysis of these bridges shows that their presence may be related to the Z_I/Z_{II} conformation found in step 4–5 of many Z-DNA structures (Fig. 9.7) (60).

As in the other DNA helix types, the hydration sites around the bases are tightly clustered. The structure of the hydration shell around the bases in Z-DNA is very different and more complex than that in B- or A-DNA structures. In the minor groove,

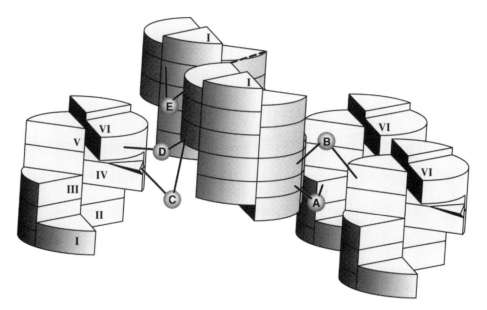

Fig. 9.7. Interhelical water bridges between Z helices. The water bridges (labelled A–E) occur between the phosphates in the Z_{II} conformation at step 4–5 and adjacent helices. (From ref. 60 with permission.)

the primary hydration site of guanine is N2, rather than N3 as in the A- and B- conformations. In the major groove, cytosine has one localized hydration site in the base plane (in a position similar to B-DNA). Hydration of purines is concentrated into three major and one lesser sites, all of which lie outside the base plane (43).

3.5 RNA

In DNA it is clear that the water molecules bonded to the bases are an integral part of the structure. In RNA the picture is far more complex, with the hydration patterns of the sugar–phosphate backbone playing a dominant structural role.

In a comparative study of four tRNA crystal structures, it has been shown that the sugar groups are much more hydrated than those in DNA (61). Although the structures are at relatively low resolution, more than 40% of the water sites are the same in all four structures. The helical stems have repetitive hydration patterns, many of which involve the O2′ hydroxyl groups. The unusual base pairs found in abundance in tRNA exhibit water bridges between the base and the backbone atoms. Of most interest is the fact that water sites are conserved in the loop areas and at the sites of the tertiary interactions. The authors of the study conclude that water molecules may indeed be related to the stabilization of these interactions.

More recent high resolution studies of RNA duplexes also demonstrate the diverse roles that water plays in these structures. In one study of two RNA octamers (62,63), the 2′-OH groups are hydrogen bonded to water molecules and form a repetitive hydration pattern in the minor groove (Fig. 9.8). The major groove also has a network of hydrogen bonds that involves the water molecules, the phosphate oxygens, and the hydrophilic base atoms. The authors suggest that the hydration of the 2′-OH group may contribute to the greater rigidity of A-RNA duplexes compared with A-DNA. In many ways the water patterns seen in this structure are analogous to

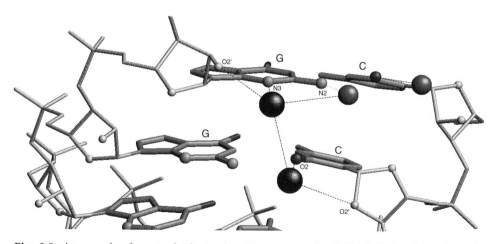

Fig. 9.8. An example of a water bridge in the minor groove of an RNA helix involving the O2′ hydroxyl of the ribose sugar (62).

those observed in the highly hydrated structure of collagen in which the hydroxyl group on the hydroxyproline (64) appears to play a synergistic role with the water molecules in stabilizing the conformation.

Water is also involved in the base mismatches that are seen in RNA structures that contain internal loops. In the G:U pairs, waters bridge the N2 of the guanine and the O2′ hydroxyl in the minor groove (65). In G:T pairs in DNA, the water bridges the N2 and the O2 atoms (66,67). The U:C pairs are even more unusual because there is only one hydrogen bond between the base atoms in the pair. The second base–base link is mediated by a water bridge.

3.6 Drug–nucleic acid complexes

The structure of d(CpG)–proflavine provided the first example of an ordered water network in a DNA drug complex (34) (Fig. 9.9). The water molecules hydrogen bonded with the base and drug heteroatoms in each complex associate with water molecules in symmetry-related complexes to form the pentagonal arrays characteristic of this crystal (68). The intriguing quality of these networks, as well as the high reso-

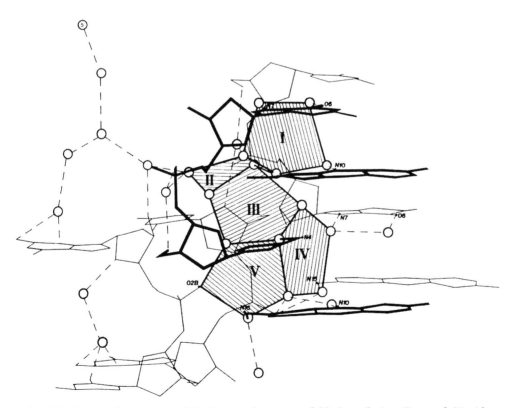

Fig. 9.9. Pentagonal water network in the crystal structure of dCpG–proflavine. (From ref. 34 with permission.)

lution of the structure analysis, made this particular structure a benchmark for several theoretical analyses that, to varying degrees, were able to reproduce the experimental results (35–37).

While the water molecules observed in the d(CpG)–proflavine structure are perhaps more important in the crystalline interactions, there are now examples in which water

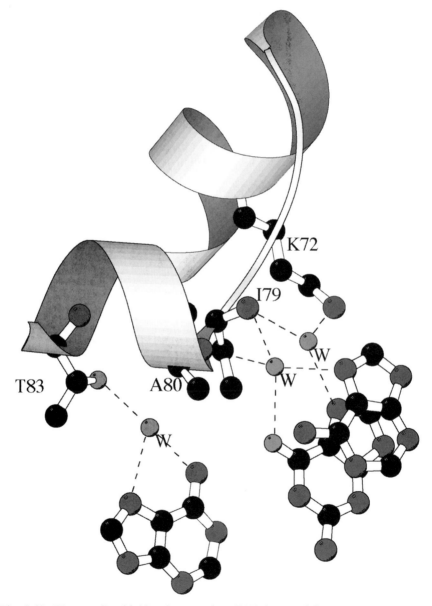

Fig. 9.10. Water-mediated bridges between three DNA bases and four amino side chains in the *trp* repressor–operator complex (73).

is situated at the interface between the drug and the DNA. This is seen in intercalated complexes with daunomycin analogues (69,70) as well as in complexes between DNA and groove binders (71,72).

What is not clear at this point is whether the presence of these water molecules at the interaction site is fortuitous or whether they play a role in recognition and specificity. Further solution studies, as well as theoretical calculations, will be needed to determine this.

3.7 Protein–DNA complexes

The importance of water in mediating protein–DNA interactions was first demonstrated in the crystal structure of a complex between the *trp* repressor and its target DNA (73). There is only one direct base contact but there are four water-mediated contacts involving three base pairs (base pairs 5, 6, 7), four amino acid residues, and three water molecules in each of the symmetrical half-sites (Fig. 9.10). Recent mutagenesis studies of this system (74) show that if the G6 is changed to A, affinity is diminished. However, this is reversed if A5 is simultaneously changed to G. This is explained by consideration of hydrogen bonding patterns involving the water (Fig. 9.11).

In a comparative study of the crystal structures of an uncomplexed decamer containing the six base pair recognition site and the DNA found in the *trp* repressor–DNA complex (75), it was found that there are 10 conserved water molecules in the major groove. These conserved water molecules include the three that are involved in the protein interactions, and the authors conclude that these waters are an integral part of the DNA.

Since the first observation of water-mediated protein–DNA interactions, others have been observed (76,77) in crystals of DNA–protein complexes. NMR investigations have also indicated that water-mediated interactions between DNA and proteins exist in solution (78). A combination of NMR and molecular dynamics simulation of an *Antennapedia* homeodomain–DNA complex provides further insight into the role of

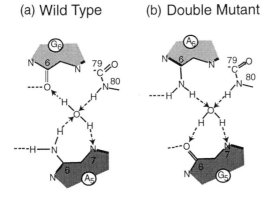

Fig. 9.11. (a) Water mediating the interactions between the amino acid amide group and G6 and A5 in the wild type *trp* repressor–operator complex. (b) The same bridge in the G5:A6 double mutant. (From ref. 74 with permission.)

water at the protein–DNA interface (79). In this case, the situation is more complex than in the *trp* system in that water-mediated contacts coexist with direct contacts, leading to several different contact geometries. The authors suggest that the specificity is a result of the rapid interconversion of the ensemble of structures. This has the interesting consequence of reducing the entropic cost of complex formation.

4. Summary

There is an accumulating body of evidence suggesting that water plays a key role in modulating the conformations, interactions, and recognition properties of nucleic acids. Physical and biochemical methods continue to be developed that provide strong circumstantial evidence that the hydration characteristics of nucleic acids must always be taken into account in trying to understand their macroscopic behaviour.

Structures of nucleic acids derived by X-ray methods have displayed a variety of networks of waters associated with the molecules and their complexes. NMR methods have confirmed that at least some of these water molecules are kinetically stable. Systematic analysis of hydrated structures has allowed us to determine the likely positions of the waters associated with the bases, as well as the locus of those associated with phosphate groups. The fact that it is possible to predict the positions of waters in the grooves of nucleic acids with known conformations is yet another indication that water should be considered an integral part of nucleic acids.

The next challenge is to use this knowledge about hydration, which has been derived from structural studies, to produce physical models for the solution and biochemical behaviour of nucleic acids. Although, it is not possible to do this now, it is not overly optimistic to think that this goal is achievable in the foreseeable future.

Acknowledgements

We wish to acknowledge the wonderful and stimulating discussions we have had over the years with David Beveridge, Ken Breslauer, and Stephen Neidle who continue to influence our thinking about hydration. We thank Christine Zardecki for her help with this manuscript and Eric Plum, T.V. Chalikian, and Rachel Kramer for reviewing the manuscript. We are also grateful for the continued funding of this work by the NIH.

References

1. Franklin, R.E. and Gosling, R.G. (1953) *Nature* **171**, 740.
2. Watson, J.D. and Crick, F.H.C. (1953) *Nature* **171**, 737.
3. Arnott, S. (1970) *Progr. Biophys. Mol. Biol.* **21**, 267.
4. Arnott, S., Campbell Smith, P.J. and Chandrasekaran, R. (1976) in G.D. Fasman (ed.), *CRC Handbook of Biochemistry and Molecular Biology: Nucleic Acids*, (Fasman, G.D., ed.), pp. 411–422. CRC Press, Cleveland.
5. Saenger, W. (1983) *Principles of Nucleic Acid Structure*, Springer Advanced Texts in Chemistry, (Cantor, C.R., ed.). Springer-Varlag, Berlin.
6. Texter, J. (1978) *Progr. Biophys. Mol. Biol.* **33**, 83.

7. Breslauer, K.J., Remeta, D.P., Chou, W.-Y., Ferrante, R., Curry, J., Zaunczkowski, D., Snyder, J. and Marky, L.A. (1987) *Proc. Natl. Acad. Sci. USA* **84**, 8922.

8. Westhof, E. (1987) *Int. J. Biol. Macromol.* **9**, 186.

9. Westhof, E. and Beveridge, D.L. (1989) *Water Sci. Rev.* 24.

10. Westhof, E. (1993) *Water and Biological Macromolecules. Topics in Molecular and Structural Biology*, (Westhof, E., ed.). CRC Press, Boca Raton.

11. Westhof, E. (1988) *Annu. Rev. Biophys. Biophys. Chem.* **17**, 125.

12. Berman, H.M. (1986) in *Computer Simulation of Chemical and Biomolecular Systems*, (Beveridge, D.L. and Jorgensen, W.L., eds), pp. 166–178. New York Academy of Science, New York.

13. Berman, H.M. (1991) *Curr. Opin. Struct. Biol.* **1**, 423.

14. Berman, H.M. (1994) *Curr. Opin. Struct. Biol.* **4**, 345.

15. Jeffrey, G.A. and Saenger, W. (eds) (1991) *Hydrogen Bonding in Biological Structures*. Springer-Verlag, New York.

16. Breslauer, K.J. (1991) *Curr. Biol.* **1**, 416.

17. Marky, L.A. and Breslauer, K.J. (1987) *Proc. Natl. Acad. Sci. USA* **84**, 4359.

18. Rentzeperis, D., Marky, L.A. and Kupke, D.W. (1992) *J. Phys. Chem.* **96**, 9612.

19. Rentzeperis, D., Kupke, D.W. and Marky, L.A. (1993) *Biopolymers* **33**, 117.

20. Zieba, K., Chu, T.M., Kupke, D.W. and Marky, L.A. (1991) *Biochemistry* **30**, 8018.

21. Rentzeperis, D., Kupke, D.W. and Marky, L.A. (1992) *Biopolymers* **32**, 1065.

22. Rentzeperis, D. and Marky, L.A. (1993) *J. Am. Chem. Soc.* **115**, 1645.

23. Remeta, D.P., Mudd, C.P., Berger, R.L. and Breslauer, K.J. (1993) *Biochemistry* **32**, 5064.

24. Chalikian, T.V., Sarvazyan, A.P. and Breslauer, K.J. (1994) *Biophys. Chem.* **51**, 89.

25. Chalikian, T.V., Sarvazyan, A.P., Plum, G.E. and Breslauer, K.J. (1994) *Biochemistry* **33**, 2394.

26. Chalikian, T.V., Plum, E.G., Sarvazyan, A.P., and Breslauer, K.J. (1994) *Biochemistry* **33**, 8629.

27. Rau, D.C. and Parsegian, V.A. (1992) *Biophys. J.* **61**, 246.

28. Rau, D.C. and Parsegian, V.A. (1992) *Biophys. J.* **61**, 260.

29. Robinson, C.R. and Sligar, S.G. (1993) *J. Mol. Biol.* **234**, 302.

30. Rosenberg, J.M., Seeman, N.C., Kim, J.J.P., Suddath, F.L., Nicholas, H.B. and Rich, A. (1973) *Nature* **243**, 150.

31. Rosenberg, J.M., Seeman, N.C., Day, R.O. and Rich, A. (1976) *J. Mol. Biol.* **104**, 145.

32. Seeman, N.C., Rosenberg, J.M., Suddath, F.L., Kim, J.J.P. and Rich, A. (1976) *J. Mol. Biol.* **104**, 109.

33. Seeman, N.C., Rosenberg, J.M. and Rich, A. (1976) *Proc. Natl. Acad. Sci. USA* **73**, 804.

34. Neidle, S., Berman, H. and Shieh, H.S. (1980) *Nature* **288**, 129.

35. Swaminathan, S., Beveridge, D.L. and Berman, H.M. (1990) *J. Phys. Chem.* **92**, 4660.

36. Kim, K.S., Corongiu, G. and Clementi, E. (1983) *J. Biomol. Struct. Dynamics* **1**, 263.

37. Hummer, G., Garcia, A.E. and Soumpasis, D.M. (1995) *Biophys. J.* **68**, 1639.

38. Drew, H.R. and Dickerson, R.E. (1981) *J. Mol. Biol.* **151**, 535.

39. Drew, H.R., Wing, R.M., Takano, T., Broka, C., Tanaka, S., Itakura, K. and Dickerson, R.E. (1981) *Proc. Natl. Acad. Sci. USA* **78**, 2179.

40. Privé, G.G., Yanagi, K. and Dickerson, R.E. (1991) *J. Mol. Biol.* **217**, 177.

41. Grzeskowiak, K., Yanagi, K., Privé, G.G. and Dickerson, R.E. (1991) *J. Biol. Chem.* **266**, 8861.

42. Schneider, B., Cohen, D. and Berman, H.M. (1992) *Biopolymers* **32**, 725.

43. Schneider, B., Cohen, D.M., Schleifer, L., Srinivasan, A.R., Olson, W.K. and Berman, H.M. (1993) *Biophys. J.* **65**, 2291.

44. Umrania, Y., Nikjoo, H. and Goodfellow, J.M. (1995) *Int. J. Radiat. Biol.* **67**, 145.

45. Schneider, B. and Berman, H.M. (1995) *Biophys. J.* **69**, 2661.
46. Westhof, E. (1993) in *Water and Biological Macromolecules*, (Westhof, E., ed.), pp. 226–243. CRC Press, Boca Raton.
47. Goodsell, D.S., Grzeskowiak, K. and Dickerson, R.E. (1995) *Biochemistry* **34**, 1022.
48. Kochoyan, M. and Leroy, J.L. (1995) *Curr. Opin. Struct. Biol.* **5**, 329.
49. Otting, G., Liepinsh, E. and Wüthrich, K. (1991) *Science* **254**, 974.
50. Kubinec, M.G. and Wemmer, D.E. (1992) *J. Am. Chem. Soc.* **114**, 8739.
51. Liepinsh, E., Leupin, W. and Otting, G. (1994) *Nucl. Acids Res.* **22**, 2249.
52. Jacobson, A., Leupin, W., Liepinsh, E. and Otting, F. (1996) *Nucl. Acids Res.* **24**, 2911.
53. Jayaram, B. and Beveridge, D.L. (1996) *Annu. Rev. Biophys. Biomol. Struct.* **25**, 367.
54. Young, M., A., Jayaram, B. and Beveridge, D.L. (1997) *J. Am. Chem. Soc.* **119**, 59.
55. Kennard, O., Cruse, W.B.T., Nachman, J., Prange, T., Shakked, Z. and Rabinovich, D. (1986) *J. Biomol. Struct. Dynamics* **3**, 623.
56. Eisenstein, M., Frolow, F., Shakked, Z. and Rabinovich, D. (1990) *Nucl. Acids Res.* **18**, 3185.
57. Eisenstein, M. and Shakked, Z. (1995) *J. Mol. Biol.* **248**, 662.
58. Saenger, W., Hunter, W.N. and Kennard, O. (1986) *Nature* **324**, 385.
59. Gessner, R.V., Quigley, G.J. and Egli, M. (1994) *J. Mol. Biol.* **236**, 1154.
60. Schneider, B., Ginell, S.L., Jones, R., Gaffney, B. and Berman, H.M. (1992) *Biochemistry* **31**, 9622.
61. Westhof, E., Dumas, P. and Moras, D. (1988) *Biochimie* **70**, 145.
62. Egli, M., Portmann, S. and Usman, N. (1996) *Biochemistry* **35**, 8489.
63. Portmann, S., Usman, N. and Egli, M. (1995) *Biochemistry* **34**, 7569.
64. Bella, J., Brodsky, B. and Berman, H.M. (1995) *Structure* **3**, 893.
65. Holbrook, S.R., Cheong, C., Tinoco, Jr, I. and Kim, S.-H. (1991) *Nature* **353**, 579.
66. Hunter, W.N., Brown, T., Kneale, G., Anand, N.N., Rabinovich, D. and Kennard, O. (1987) *J. Biol. Chem.* **262**, 9962.
67. Kneale, G., Brown, T., Kennard, O. and Rabinovich, D. (1985) *J. Mol. Biol.*, **186**, 805.
68. Schneider, B., Ginell, S.L. and Berman, H.M. (1992) *Biophys. J.* **63**, 1572.
69. Moore, M.H., Hunter, W.N., d'Estaintot, B.L. and Kennard, O. (1989) *J. Mol. Biol.* **206**, 693.
70. Wang, A.H.-J., Ughetto, G., Quigley, G.J. and Rich, A. (1987) *Biochemistry* **26**, 1152.
71. Brown, D.G., Sanderson, M.R., Skelly, J.V., Jenkins, T.C., Brown, T., Garman, E., Stuart, D.I. and Neidle, S. (1990) *EMBO J.* **9**, 1329.
72. Sriram, M., van der Marel, G.A., Roelen, H.L.P.F., van Boom, J.H. and Wang, A.H.-J. (1992) *Biochemistry* **31**, 11823.
73. Otwinowski, Z., Schevitz, R.W., Zhang, R.-G., Lawson, C.L., Joachimiak, A., Marmorstein, R.Q., Luisi, B.F. and Sigler, P.B. (1988) *Nature* **335**, 321.
74. Joachimiak, A., Haran, T. and Sigler, P. (1994) *EMBO J.* **13**, 367.
75. Shakked, Z., Guzikevich-Guerstein, G., Frolow, F., Rabinovich, D., Joachimiak, A. and Sigler, P.B. (1994) *Nature* **368**, 469.
76. Hirsch, J.A. and Aggarwal, A.K. (1995) *EMBO J.* **14**, 6280.
77. Wilson, D.S., Guenther, B., Desplan, C. and Kuriyan, J. (1995) *Cell* **82**, 709.
78. Qian, Y.Q., Otting, G. and Wuthrich, K. (1993) *J. Am. Chem. Soc.* **115**, 1189.
79. Billeter, M., Guntert, P., Luginbuhl, P. and Wuthrich, K. (1996) *Cell* **85**, 1057.

10

Single-crystal X-ray diffraction studies on the non-Watson–Crick base associations of mismatches, modified bases, and non-duplex oligonucleotide structures

William N. Hunter[1],★ and Tom Brown[2]

[1]*Department of Biochemistry, University of Dundee, Dundee, DD15EH, UK*
[2]*Department of Chemistry, University of Southampton, Southampton, SO17 1BJ, UK*

1. Introduction

The replication of DNA must occur with a high degree of precision in order for genetic information to be faithfully transmitted from one generation to the next. Watson and Crick recognized that a complementary base pairing scheme in duplex DNA could contribute to such a mechanism (1). In this way, purines interact with pyrimidines so that guanine (G) pairs with cytosine (C) and adenine (A) pairs with thymine (T) to form what are termed Watson–Crick base pairs (Fig. 10.1).

The very specific manner in which the Watson–Crick base pairs are formed contributes stability to an oligonucleotide structure and a particular arrangement of functional groups for interaction with enzymes and proteins by, for example, specific hydrogen bonding patterns (2). However, given that the human genome is estimated to contain around 10^9 base pairs it is hardly surprising that mistakes can and do occur during the replication process. Given the redundancy in the genetic code, not every alteration of the DNA sequence will lead to a change in the gene product but a single error in a triplet may be carried through and eventually lead to a serious mutation. Errors can be introduced via non-Watson–Crick base pairs, termed mismatches or mispairs. Alternatively, damage to DNA can produce bases with altered chemical properties capable of scrambling the genetic code (3). Some mutations may confer an evolutionary advantage, but in general the propagation of such mistakes must not be allowed and a complicated protein recognition and repair system plays a key role in maintaining the fidelity of replication (4).

Structural investigations of the proteins involved in this recognition of mistakes in DNA, and subsequent repair, represent one of the most exciting subjects in structural biology (5 and references therein). Studies on these enzymes follow on from research in a number of laboratories directed towards the biophysical characterization of mismatches and modified bases in DNA and RNA and their biological implications. Crystallographic studies have provided stru ral detail to complement thermo-

---

★Corresponding author.

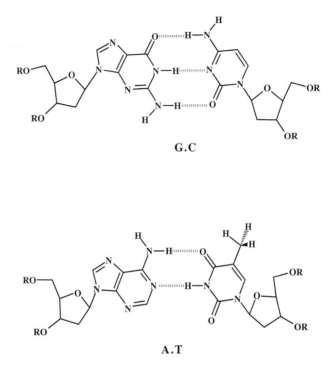

G.C

A.T

Fig. 10.1. The Watson–Crick base pairs G:C (top) and A:T (bottom). In all figures the hydrogen bonds are represented by dashed lines.

dynamic studies on the stability of the mismatches or base pairs involving chemically modified components (6). In addition to a description of mismatch pairings in DNA, a number of studies on RNA fragments, triplexes, quadruplexes, and a novel loop assembly have highlighted the important role of non-Watson–Crick base associations (see Chapter 17). This can involve an extension from two bases, interacting with each other using a specific pattern of hydrogen bonds, to three- and four-base assemblies. Our aim in this chapter is to highlight the crystallographic results on base associations (NMR studies are the subject of Chapter 11). Although we concentrate on mismatches in duplex DNA some mention is made of other examples involving RNA, triplexes, and quadruplexes. However, the reader is directed elsewhere in this volume for more detailed coverage of RNA (Chapter 17) and higher order DNA structures (Chapters 12 and 13).

2. Mismatches

There is a competition between the Watson–Crick A:T or G:C pairs and eight non-Watson–Crick alternatives that are called mismatches or mispairs. These are the purine–pyrimidine G:T and A:C pairings, the purine–purine G:G, A:A, and G:A pairings, and, finally, the pyrimidine–pyrimidine C:C, T:T, and C:T mismatches. The incorporation of non-Watson–Crick base pairs in duplex DNA is one of the most common errors that occurs during the replication process. Mutagenic pathways are

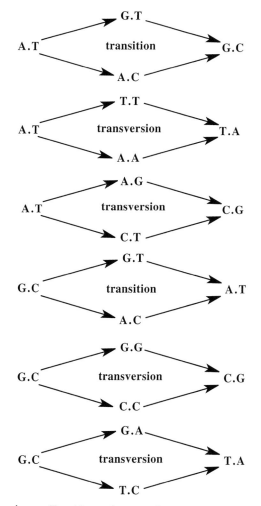

Fig. 10.2. Mutagenic pathways. Transition and transversion mutations starting from A:T or G:C base pairs.

divided into transition and transversion paths. The former invokes purine–pyrimidine mismatches, the latter purine–purine or pyrimidine–pyrimidine mispairs. Figure 10.2 presents the mutagenic pathways starting from both A:T and G:C pairs.

The theory of mispair formation, initially proposed by Watson and Crick (7), extended by Topal and Fresco (8), and reviewed by Strazewski and Tamm (9) relies on the involvement of rare tautomer forms of the bases. The mismatches involving these tautomers could be sterically equivalent to Watson–Crick base pairs and unlikely to distort or perturb the duplex into which they are formed.

The crystallographic study of mispairs cannot give any information on the occurrence of rare tautomers during the replication process. However, these studies do define the structure of the oligonucleotide hosting the mispair, thus serving to characterize any localized perturbations of structure, the hydrogen bonding patterns linking the bases, the influence of neighbouring bases, and clues about how recognition and subsequent repair of mismatches may occur. One of the main conclusions from mismatch studies is that there is no need to invoke the presence of rare tautomers in mismatch formation and stability.

The crystallographic study of mismatches has in general used complementary sequences known to form well-ordered systems into which the mispairs have been engineered. The most common framework has been the Drew–Dickerson dodecamer duplex (10). This sequence, which crystallizes readily in the B-form, is d(CGC-GAATTCGCG). Other templates have been A-form DNA octamers and Z-form hexamers (11). In each case a duplex containing two mispairs has been formed. There are two main benefits in this approach. It maximizes the likelihood of getting well-ordered single crystals for the analysis and it means that there is a native Watson–Crick structure that can be used for comparative purposes.

2.1 Purine–pyrimidine base pairs

The first mismatch pair to be characterized was the G:T in an A-form octamer (12,13). Subsequently, this was studied in different sequence environments and in different DNA forms (14–16). This type of purine–pyrimidine pairing adopts what is termed the wobble configuration, which was first proposed by Crick to explain G:U pairing at the third codon position during codon–anticodon interactions (17). The purine is shifted towards the DNA minor groove and the pyrimidine towards the major groove. The bases maintain the major tautomeric forms and create two interbase hydrogen bonds (Fig. 10.3a). Well-ordered solvent molecules bridge functional groups on the bases in both major and minor grooves and confer additional stability to the pairing. G:Br^{5}U and G:F^{5}U pairs (where uracil contains a bromine or fluorine at the 5 position) have also been characterized in Z-form hexamers (18,19) and wobble G:U pairs, plus attendant solvent molecules observed in a fragment of 5S rRNA (20).

Inosine (I) is a guanine analogue that lacks the 2-amino group. This base is commonly found in tRNA where it is able to pair with A, C, and U in codon–anticodon interactions. It is an important base since the ability to pair with three other bases contributes to the degeneracy of the genetic code. Inosine occurs rarely in DNA, as a result of deamination of deoxyguanosine, where it is potentially mutagenic. A specific glycosylase is available to remove it from DNA. The I:T pair (21) assumes a similar structure to the G:T pair, although the loss of N2 on the minor groove side of the duplex removes the possibility of a stabilizing water bridge between the bases in that groove.

A:C pairing also displays a similar structure to the G:T, but there are two arrangements that could be invoked to explain the formation of two hydrogen bonds linking the bases (22,23; Fig. 10.3b, c). A solvent molecule can link the bases on the major groove side to aid stability, but not on the minor groove side. The adenine is either protonated or in a rare tautomeric form. Energetic considerations suggested the former and biophysical characterization of A:C mispairs using NMR and UV melting methods over a wide pH range subsequently supported this proposal (24). It is perhaps more appropriate to denote this base pair as A$^+$:C.

(a)

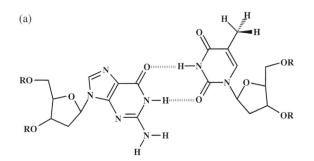

G.T

(b)

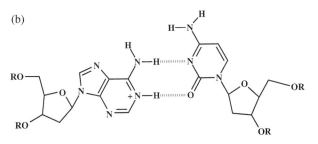

A⁺.C

(c)

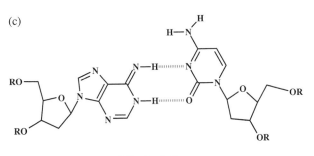

A(imino).C

Fig. 10.3. (a) The G:T 'wobble' pair; (b) the A⁺:C pair; and (c) the A:C pair with the purine in the imino form.

2.2 Purine–purine base pairs

Both A:G and G:G pairs have been characterized in duplex B-DNA. The A:A pairing will be discussed in the context of non-duplex DNA later. The G:A pairing has attracted particular interest since biochemical studies have identified such mismatches as being repaired with much less efficiency than other mispairs (25). A structural explanation has been sought.

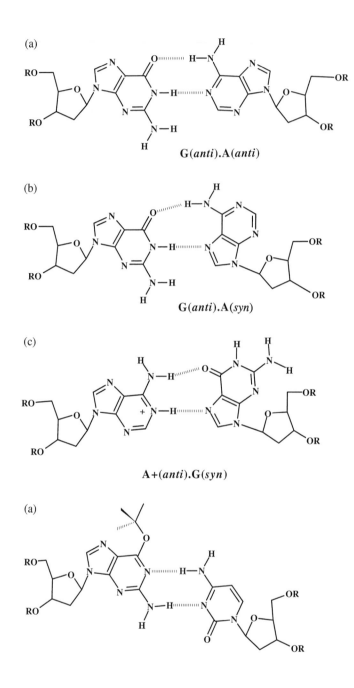

Fig. 10.4. Four examples of a G:A pair highlight the variability of this mismatch. (a) G(*anti*):A(*anti*); (b) G(*anti*):A(*syn*); (c) A⁺(*anti*):G(*syn*); (d) G(*anti*):A(*anti*) amino.

Crystallographic and NMR studies have identified four G:A configurations in DNA (26–29, Fig. 10.4). The form of the mispair that is observed has been shown to depend on a number of factors such as pH, salt concentration, and, in particular, the sequence environment in which the mismatch is located. The dependence of the G:A conformation on the adjacent sequence can be rationalized in part by dipole–dipole interactions with adjacent bases (28). Hydrogen bonding using a functional group provided by an adjacent base can also be important and this is clear in the example of the G(*anti*):A(*anti*) pairing. The presence of an intrabase pair hydrogen bond between the amino N2 of guanine and the O2 of an adjacent thymine on the opposing strand has been noted (26). Presumably, without an O2 in this position that is ready to participate in hydrogen bonding some other G:A conformation could be preferred. The G(*anti*):A(*anti*) mismatch also produces a bulge in the duplex structure as the backbone is forced apart to accommodate the purine–purine pair in which each base adopts the *anti* conformation. When one of the bases is in the *syn* conformation this bulging effect is not observed. The key point about studies on the G:A mispair is that the variablility of conformations that can be observed would present quite a challenge to an enzyme recognition and repair system and this may be an important factor in the poor recognition and repair of the G:A mismatch.

In the RNA duplex r(CGCGAAUUAGCG) there are two A(*anti*):G(*anti*) base pairs and evidence to suggest the same degree of variability as that observed in DNA (30). A careful investigation of the hydrogen bonding possibilities suggests that the A(*anti*):G(*anti*) pairing uses a conventional hydrogen bond formed between N6 and O6 and what is termed a reverse, three-centre hydrogen bond in which the lone pair on N1 is shared with the N–H groups of the guanine N1 and N2. In this way the destabilizing effects of having unsatisfied hydrogen bonding functional groups can be avoided.

The structural variation observed for the G:A mismatch also applies to I:A pairs (31–33). This variability may help explain the mutagenicity of inosine. UV melting studies indicate that inosine-containing mismatches are surprisingly stable (33). Most other mismatches have a tendency to destabilize the DNA duplex and produce local melting effects that can open up the duplex. Repair enzymes can use this physical property of the mismatch duplex to recognize incorrect base pairing. Local destabilization could also assist the flipping out of mismatched bases for excision. The phenomenon of base flipping as part of the protein recognition and repair process has been noted on the basis of crystallographic studies (5).

There has only been a single structure for the homopurine G:G mismatch. It shows a G(*anti*):G(*syn*) arrangement (34). The details are slightly different for the two mispairs in the DNA duplex and two hydrogen bonding schemes have been put forward (Fig. 10.5). G:G transversion mismatches are readily repaired and in this case the authors note that the sugar–phosphate backbone is distorted in comparison to the native duplex.

2.3 Pyrimidine–pyrimidine base pairs

These mismatches have proven difficult to characterize when they are incorporated in duplex DNA, but there are some examples of C:U and U:U associations in duplex

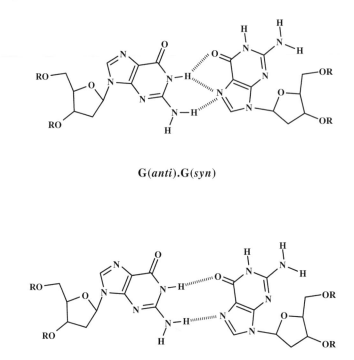

G(*anti*).G(*syn*)

G(*anti*).G(*syn*)

Fig. 10.5. Two slightly different G:G mismatches have been observed in a B-form dodecamer duplex. Although they are both G(*anti*):G(*syn*) the details of the hydrogen bonding vary.

RNA. The C:U mispair has been observed in r(GGACUUCGGUCC) (35). In this case there is a single hydrogen bond between the bases involving C(N4) and U(O4) and a bridging solvent linking the two N3 groups (Fig. 10.6).

The U:U pair is polymorphic. What are called *cis* U:U wobble pairs have been observed in two RNA dodecamer structures (36,37). These are also discussed in Chapter 17. The U:U pairs are held together with two hydrogen bonds (Fig. 10.7a), and although an ordered solvent is not observed in both crystal structures, this pair has

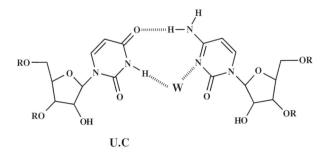

U.C

Fig. 10.6. The U:C mispair observed in RNA. W represents a water molecule that bridges the pyrimidines.

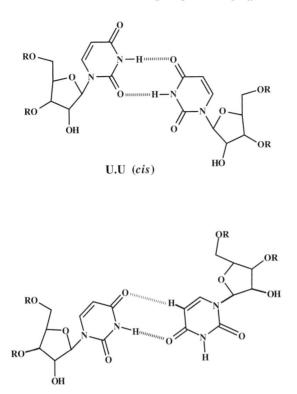

U.U (*cis*)

U.U (*trans*)

Fig. 10.7. Two forms (*cis* and *trans*) of the U:U pairing as observed in RNA structures.

what appears to be an attractive site to bring in a water molecule in both the major and minor groove sides. This would be similar to the G:T mismatch discussed above. The nonameric sequence r(GCUUCGGC)d(Br⁵U) has a similar U:U pair at the end of one of the helices, which is disordered (38). The hexanucleotide sequence r(UUCGCG) crystallizes with a tetranucleotide duplex involving C:G pairs and two U:U pairs formed by the overhanging bases (39). There is a conventional hydrogen bond between N3 and O4 but also a C–H···O hydrogen bond between C5 and O4 (Fig. 10.7b). The importance and occurrence of C–H···O hydrogen bonds in nucleic acid structure has been discussed recently (40). This type of interaction has been invoked in this particular type of U:U pair and occurs in a number of base–base interactions involving modified bases, and also in triplex formation.

3. Pairings with modified bases

In addition to the pressures of carrying out replication involving a large number of bases, the genetic code is constantly pressured by chemical and physical forces in the environment or generated in cells during the normal course of metabolism. Carcinogenic chemicals, ultraviolet light, ionizing radiation, and reactive oxygen species are all capable of inducing modifications to DNA (3,4). Of particular interest are alterations to the purines.

Guanine can be methylated by alkylnitrosoureas to form O6-methylguanine (O6MeG), which is potentially very damaging since it alters the hydrogen bonding potential of the base, thereby promoting G to A transition mutations. The O6MeG:T mispair could then be selected during replication in preference to a O6MeG:C pair.

(a)

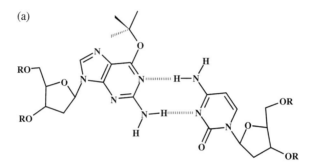

O(6)-MeG.C

(b)

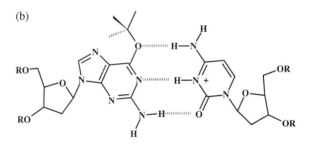

O(6)-MeG.C⁺

(c)

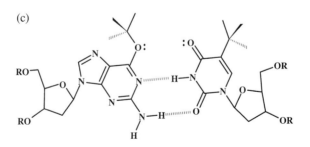

O(6)-MeG.T

The structure of a O6MeG:C pair has been determined at physiological pH (41–43) and is shown to adopt a wobble conformation (Fig. 10.8a). A highly specific enzyme, O6-methylguanine methyltransferase, which is able to repair this particular alteration by excising the methyl group, has evolved to control this aspect of damage to DNA.

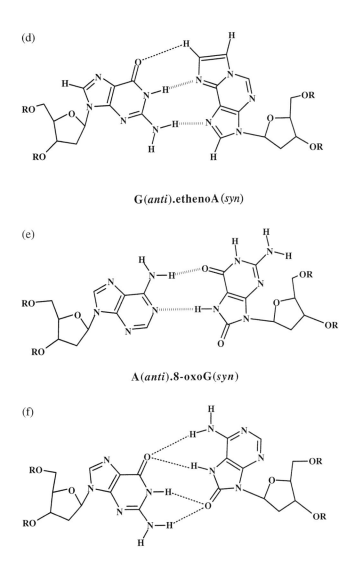

Fig. 10.8. (a) The O6MeG:C pair, which resembles the G:T mismatch. (b) The O6MeG:C⁺ pairing, which resembles a Watson–Crick base pair. (c) O6MeG:T mismatch, which also resembles a Watson–Crick pair. (d) The G (*anti*):edA pair where edA is ethenoA. (e) The A(*anti*):O8G(*syn*) and (f) G(*anti*):O8A(*syn*) pairings, where O8G and O8A represent 8-oxoG and 8-oxoA, respectively.

Chemical damage is not induced solely by alkylating agents, but by many other carcinogens as well. Adenine can react with vinyl chloride to produce 1,N6-ethenoadenosine (edA). The structure of the G:edA pairing has been determined (44) and the association is depicted in Fig. 10.8d. There are two obvious hydrogen bonds and a C–H⋯O hydrogen bond has been invoked between the H8 and O6 of G to alleviate the destabilization of an unsatisfied hydrogen bond acceptor in the pair. Unlike other non-Watson–Crick pairings, there is significant alteration in the conformation of the sugar–phosphate backbone when edA is incorporated into the duplex. Such perturbation could represent a signal for the recognition and repair of this modified base by 3-methyladenine–DNA glycosylase.

Both purines can undergo oxidation at the 8 position to produce 8-oxoadenine (O8A) and 8-oxoguanine (O8G). The bases are predominantly in the keto form. Whilst modification at the 8 position does not affect the hydrogen bonding patterns on functional groups used in G:C and A:T pairs, the presence of the O8 and N7(H) does promote other hydrogen bonding possibilities and a *syn* conformation about the glycosidic bond. This is noted in the structures of O8G:A and O8A:G pairings (45,46; Fig. 10.8e). The presence of the highly mutagenic O8G lesion in genomic DNA can produce a G to T transversion mutation via an intermediate O8G:A base pair. The thermodynamic stability of this pair, in addition to the psuedo-symmetry about the glycosidic bonds, perhaps explains why it is not readily recognized by proof-reading enzymes. O8A is not particularly mutagenic and the O8A:G pairing, whilst again showing a *syn/anti*, pair is asymmetric about the glycosidic bonds, a structural feature that may make it easier to recognize and repair. This pairing is held together by four bifurcated hydrogen bonds resulting from two reverse, three-centred hydrogen bonding systems. Such an arrangement helps to stabilize the duplex, since it allows all functional groups in the mismatched pair to fulfil their hydrogen bonding capacity.

The structural studies on duplexes containing mismatches or modified bases have clearly indicated that DNA has sufficient flexibility to incorporate these with ease. The sugar–phosphate backbone makes small adjustments as required and any distortions are highly localized. Biophysical characterization including UV melting studies indicate that when non-Watson–Crick associations are involved there is more often than not a reduction in T_m. This can be ascribed to localized destabilization of the duplex structure. The recognition and repair of mistakes in the DNA duplex is thus likely to occur at a very localized level. It will involve a combination of structural and thermodynamic effects such as distortions to the furanose–phosphate backbone, the disposition of functional groups able to participate in hydrogen bonding interactions with specific enzyme residues, and localized melting effects.

4. Non-Watson–Crick associations stabilize higher order structures

There is a requirement for non-Watson–Crick base interactions in some aspects of nucleic acid structure, in particular where large assemblies are involved. Such interactions are important in the stabilization of large RNA structures, for example, tRNA (reviewed in Chapter 19) and more recently shown in ribozyme structures (Chapter 17; 47 and references therein, 48). RNA structures are detailed in Chapter 17 and we shall confine ourselves to some comments on DNA triplexes, quadruplexes, and two loops.

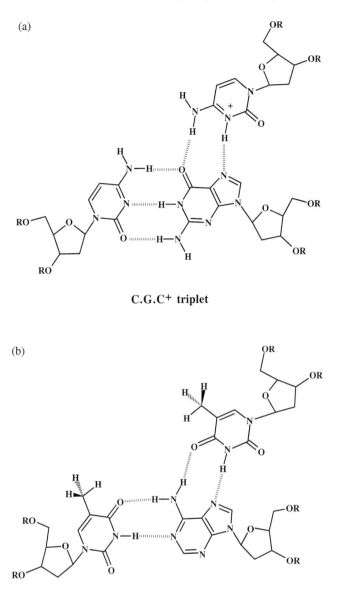

(a)

C.G.C⁺ triplet

(b)

T.A.T triplet

Fig. 10.9. Two examples of base triplets that involve one Watson–Crick base pair interacting with a third base. (a) CGC⁺ and (b) TAT triads.

4.1 Triple helices

A triplex is a duplex on to which a third strand is attached, for example by binding in the major groove. The three-stranded structure has been implicated in genetic recombination, and the design of molecular fragments able to form and stabilize a designated triplex is an area of interest with prospects for antigene therapy.

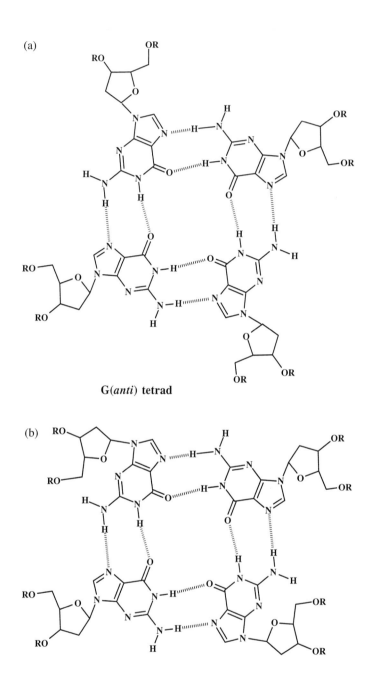

(a)

G(*anti*) tetrad

(b)

G(*anti*). G(*syn*). G(*anti*). G(*syn*) tetrad

Fig. 10.10. Two examples of G tetrads: (a) the G(*anti*) and (b) the G(*anti*):G(*syn*) tetrad.

Crystallographic studies of a nonamer (49) then a decamer (50) with a sequence designed to form an overhanging base have produced models for both parallel and antiparallel triplexes. Two types of C:G:C triplet are formed by crystal lattice contacts, which on the basis of model building can be extended to provide two distinct types of triplex (50).

A fully formed triplex structure has been characterized by the crystallographic analysis of a peptide nucleic acid–DNA complex (51). This molecule utilizes both T:A:T and C:G:C triplets to create a unique triplex called the P-form helix. The use of a nuclease-resistant backbone, as in this case, in combination with a design strategy targeting triplex formation, opens up new possibilities in the area of antisense therapeutic agents. In this example Watson–Crick pairs are supplemented by a Hoogsteen base pair involving the purine interacting with a pyrimidine in the major groove. Two types of triplet associations are depicted in Fig. 10.9.

4.2 Quadruplexes

The terminal segments of eukaryotic chromosomes are called telomeres. These sections of the chromosome have been implicated in replication processes and in stability (52–55). They have an unusual sequence which involves repeating tracts of guanines. The guanines are able to self-associate as tetrads or quartets (Fig. 10.10) and, under the influence of specific cations, this type of G-rich DNA is able to form a range of parallel and antiparallel quadruplexes. The structures of d(GGGGTTTTGGGG) (56) and d(TGGGGT) (57) have been determined. In the first case, each strand forms an intramolecular hairpin stabilized by G:G pairs. Two hairpins associate in an antiparallel manner to create a stack of four guanine tetrads. The glycosyl bonds alternate between *syn* and *anti*. In the case of d(TGGGGT), the strands in the tetraplex are all parallel to each other and the glycosyl bonds are all in an *anti* conformation. Each quadruplex binds a cation, the antiparallel stucture binds potassium, and the parallel quadruplex binds sodium, either at the centre of or between the G quartets.

A series of crystal structures has been determined that are stabilized by intercalating hemiprotonated C:C$^+$ pairs. This pairing is shown in Fig. 10.11 and involves three hydrogen bonds linking the cytosines. The sequences that provide these structures

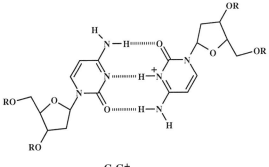

C.C$^+$

Fig. 10.11. The C:C$^+$ pairing.

include d(CCCC) (58), d(CCCT) (59), d(CCCAAT) (60), and d(TAACCC). (61). This last example also involves Hoogsteen A(*syn*):T pairs. In each case a tetraplex is formed that can be thought of as a combination of two parallel duplexes, intercalated with opposite polarity.

4.3 A unique loop structure

The structure of d(GCATGCT) has been determined to high resolution (62). The asymmetric unit is a single strand that folds back upon itself to create a loop structure not previously seen in structures of DNA. The stem of the loop is formed from the two GpC steps. However, the hydrogen bond donor and acceptor groups used in Watson–Crick G:C base pairs are positioned end-on rather than in the normal head-to-head fashion noted in hairpin loop structures (63,64). Dimerization using a crystallographic twofold axis leads to the formation of an extensive network of hydrogen bonds formed by Watson–Crick pairing and, in addition, by the G:C base pairs interacting with each other on what can be termed the minor groove side of the pair. A:A and T:T base pairs are clearly important for the stability of this unusual DNA structure. The A:A pair is formed using a symmetric N6–N7 amino hydrogen bonded conformation, similar to that observed in yeast tRNA[phe] (2). This purine–purine pairing assists dimerization of the loop through the hydrogen bonds and it also assists the association of two loop dimers by the base stacking of adjacent A:A pairs.

The T:T pair contributes mainly to stabilizing the crystal lattice. It is formed by a symmetric hydrogen bonding of the extruding thymine N3 and O2 atoms with an equivalent thymine of a symmetry-related loop. The crystal structure of a Z-form stem hairpin loop has also created a T:T pair owing to lattice interactions (64), the sequence is d(CGCGCGTTTTCGCGCG). The loop contains four thymines. The T:T pair formed between one loop with a symmetry-related loop is such that the rare enol tautomeric form must be present for one of the bases.

It remains unclear whether the quadruplex and loop structures that have been characterized are of direct biological relevance. What has been shown to be important is the use of non-Watson–Crick base associations, base pairs, triplets, and quartets, to help create such structures. It is tempting to suggest that the tight packaging of nucleic acids in, for example, viral genomes and chromosomes could well utilize similar structural motifs.

Acknowledgements

Financial support was provided by the Biotechnology and Biochemistry Science Research Council (BBSRC), the Engineering and Physical Sciences Research Council (EPSRC), and, in particular, the Wellcome Trust.

References

1. Watson, J.D. and Crick, F.H.C. (1953) *Nature* **171**, 737.
2. Saenger, W. (1984) *Principles of Nucleic Acid Structure*. Springer-Verlag, New York.
3. Loft, S. and Poulsen, H.E. (1996) *J. Mol. Med.* **74**, 297.
4. Modrich, P. (1987) *Annu. Rev. Biochem.* **56**, 435.

5. Vassylyev, D.G. and Morikawa, K. (1997) *Curr. Opin. Struct. Biol.* **7**, 103.

6. Brown, T., Hunter, W.N. and Leonard, G.A. (1993) *Chem. Brit.* **6**, 484.

7. Watson, J.D and Crick, F.H.C. (1953) *Nature* **171**, 964.

8. Topal, M.D. and Fresco, J.R. (1976) *Nature* **263**, 290.

9. Strazewski, P. and Tamm, C. (1990) *Angew. Chem. Intl. Ed. Engl.* **29**, 36.

10. Wing, R.M., Drew H.R., Takano, T., Broka, C., Takana, S., Itakura, K. and Dickerson, R.E. (1980) *Nature* **287**, 755.

11. Kennard, O. and Hunter, W.N. (1991) *Angew. Chem. Intl. Ed. Engl.* **30**, 1254.

12. Brown, T., Kennard, O., Kneale, G. and Rabinovich, D. (1985) *Nature* **315**, 604.

13. Hunter, W.N., Kneale, G., Brown, T., Rabinovich, D. and Kennard, O. (1986) *J. Mol. Biol.* **190**, 605.

14. Kneale, G., Brown, T., Kennard, O. and Rabinovich, D. (1985) *J. Mol. Biol.* **186**, 805.

15. Hunter, W.N., Brown, T., Kneale, G., Anand, N.N., Rabinovich, D and Kennard, O. (1987) *J. Biol. Chem.* **262**, 9962.

16. Ho, P.S., Frederick, C.A., Quigley, G., van der Marel, G.A. van Boom, J.H., Wang, A.H-J. and Rich, A. (1985) *EMBO J.* **4**, 3617.

17. Crick, F.H.C. (1966) *J. Mol. Biol.* **19**, 548.

18. Brown, T., Kneale, G., Hunter, W.N. and Kennard, O. (1986) *Nucl. Acids Res.* **14**, 1801.

19. Coll, M., Saal, D., Frederick, C.A., Aymami, J., Rich, A., Wang, A.-H. J. (1989) *Nucl. Acids Res.* **17**, 911.

20. Betzel, C., Lorenz, S., Furste, J.P., Bald, R., Zhang, M., Schneider, T., Wilson, K.S. and Erdmann, V.A. (1994) *FEBS Lett.* **351**, 159.

21. Cruse, W.B.T., Aymami, J., Kennard, O., Brown, T., Jack, A.G.C. and Leonard, G.A. (1989) *Nucl. Acids Res.* **17**, 55.

22. Hunter, W.N., Brown, T., Anand, N.N. and Kennard, O. (1986) *Nature* **320**, 552.

23. Hunter, W.N., Brown, T. and Kennard, O. (1987) *Nucl. Acids Res.* **15**, 6589.

24. Brown, T., Leonard, G.A., Booth, E.D. and Kneale, G. (1990) *J. Mol. Biol.* **221**, 437.

25. Fersht, A.R., Knill-Jones, J.W. and Tsui, W.C. (1982) *J. Mol. Biol.* **156**, 37.

26. Privé, G.G., Heinemann, U., Kan, L.S., Chandrasegaran, S., and Dickerson, R.E. (1987) *Science* **238**, 498.

27. Brown, T., Hunter, W.N., Kneale, G.G. and Kennard, O. (1986) *Proc. Natl. Acad. Sci. USA* **83**, 2402.

28. Brown, T., Leonard, G.A., Booth, E.D. and Chambers, J. (1989) *J. Mol. Biol.* **207**, 455.

29. Hunter, W.N., Brown, T. and Kennard, O. (1986) *J. Biolmol. Struct. Dynamics* **4**, 173.

30. Leonard, G.A., McAuley-Hecht, K., Abel, S., Lough, D.M., Brown, T. and Hunter, W.N. (1994) *Structure* **2**, 483.

31. Corfield, P.W.R., Hunter, W.N., Brown, T., Robinson, P and Kennard, O (1987) *Nucl. Acids Res.* **15**, 7935.

32. Webster, G.D., Sanderson, M.R., Skelly, J.V., Neidle, S., Swann, P.F., Li, B.F. and Tickle, I. (1990) *Proc. Natl. Acad. Sci. USA* **87**, 6693.

33. Leonard, G.A., Booth, E., Hunter, W.N. and Brown, T. (1992) *Nucl. Acids Res.* **20**, 4753.

34. Skelly, J.V., Edwards, K.J., Jenkins, T.C. and Neidle, S. (1993) *Proc. Natl. Acad. Sci. USA* **90**, 804.

35. Holbrook, S.R., Cheong, C., Tinoco, I. and Kim, S. H. (1991) *Nature* **353**, 579.

36. Baeyens, K.J., De Bondt, H.L. and Holbrook, S.R. (1995) *Nature Struct. Biol.* **2**, 56.

37. Lietzke, S.E., Barne, C.L., Bergland, J.A. and Kundrot, C.E. (1996) *Structure* **4**, 917.

38. Cruse, W.B.T., Saludjian, P., Biala, E., Strazewski, P., Prange, T. and Kennard, O. (1994) *Proc. Natl. Acad. Sci. USA* **91**, 4160.

39. Wahl, M.C. Rao, S.T. and Sundaralingam, M. (1996) *Nature Struct. Biol.* **3**, 24.

40. Leonard, G.A., McAuley-Hecht, K., Brown, T. and W.N. Hunter., W.N. (1995) *Acta Cryst.* **D51**, 136.
41. Leonard, G.A., Thomson, J.B., Watson, W.P. and Brown, T. (1990) *Proc. Natl. Acad. Sci. USA* **87**, 9573.
42. Ginell, S.L., Vojtechovsky, J., Gaffney, B., Jones, R. and Berman, H.M. (1994) *Biochemistry* **33**, 3487.
43. Vojtechovsky, J., Eaton, M.D., Gaffney, B., Jones, R. and Berman, H.M. (1994) *Biochemistry* **34**, 16632.
44. Leonard, G.A., McAuley-Hecht, K.E., Gibson, N.J., Brown, T., Watson, W.P. and Hunter, W.N. (1994) *Biochemistry* **33**, 4755.
45. Leonard, G.A., Guy, A., Brown, T., Teoule, R. and Hunter, W.N. (1992) *Biochemistry* **31**, 8415.
46. McAuley-Hecht, K.E., Leonard, G.A., Gibson, N.J., Thomson, J.B., Watson, W.P., Hunter, W.N. and Brown, T. (1994) *Biochemistry* **33**, 10266.
47. Scott, W.G. and Klug, A. (1996) *TIBS* **21**, 220.
48. Cate, J.H., Gooding, A.R., Podell, E., Zhou, K., Golden, B.L., Kundrot, C.E., Cech, T.R. and Doudna, J.A. (1996) *Science* **273**, 1678.
49. van Meervelt, L., Dautant, A., Gallois, B., Precigoux, G. and Kennard, O. (1995) *Nature* **374**, 742.
50. Vlieghe, D., van Meervelt, L., Dautant, A., Gallois, B., Precigoux, G. and Kennard, O. (1996) SCIENCE **273**, 1702.
51. Betts, L., Josey, J.A., Veal, J.M. and Jordan, S.R. (1995) *Science* **270**, 1838.
52. Sen, D. and Gilbert, W. (1988) *Nature* **334**, 364.
53. Sunquist, W. I. and Klug, A. (1989) *Nature* **342**, 825.
54. Williamson, J. R., Raghuraman, M. K. and Cech, T. R. (1989) *Cell* **59**, 871.
55. Smith, F. W. and Feigon, J. (1992) *Nature* **356**, 164.
56. Kang, C., Zhang, X., Ratcliff, R., Moyzis, R. and Rich, A. (1992) *Nature* **356**, 126.
57. Laughlin, G., Murchie, A.I.H., Norman, D.G., Moore, M.H., Moody, P.C.E., Lilley, D.M.J. and Luisi, B. (1994) *Science* **265**, 520.
58. Chen, L., Cai, L., Zhang, X. and Rich, A. (1994) *Biochemistry* **33**, 13540.
59. Kang, C., Berger, I., Lockshin, C., Ratcliff, R., Moyzis, R. and Rich, A. (1994) *Proc. Natl. Acad. Sci. USA* **91**, 11636.
60. Berger, I., Kang, C., Fredian, A., Ratcliff, R., Moyzis, R. and Rich, A. (1995) *Nature Struct. Biol.* **2**, 416.
61. Kang, C., Berger, I., Lockshin, C., Ratcliff, R., Moyzis, R. and Rich, A. (1995) *Proc. Natl. Acad. Sci. USA* **92**, 3874.
62. Leonard, G. A., Zhang, S., Peterson, M. R., Harrop, S. J., Helliwell, J. R., Cruse, W.B.T., Langlois d'Estaintot, B., Kennard, O., Brown T. and Hunter, W. N. (1995) *Structure* **3**, 335.
63. Chattopadhyaya, R., Ikuta, S., Grzeskowiak, K. and Dickerson, R.E. (1988) *Nature* **334**, 175.
64. Chattopadhyaya, R., Grzeskowiak, K. and Dickerson, R. E. (1990) *J. Mol. Biol.* **211**, 189.

11

DNA mismatches in solution

Shan-Ho Chou[1] and Brian R. Reid[2]

[1]*Institute of Biochemistry, National Chung-Hsing University, Taichung, 40227, Taiwan*
[2]*Department of Chemistry and Biochemistry, University of Washington, Seattle WA 98195, USA*

1. Introduction

The DNA double helix, with its complementary G:C and A:T Watson–Crick base pairing, is a remarkably efficient device for the storage and expression of information and the stable transmission of this information through successive generations. Although normal, or Watson–Crick, base pairing is mediated through hydrogen bonding between A and T residues and between G and C residues, the double helix is also stabilized by a variety of other 'stacking' interactions which obviously differ from one sequence to the next. In the process of copying each of the two strands of DNA to produce two identical double helices, i.e. daughter cells with the same genetic composition, incorrect or mismatch pairings (G or C with T or A, or with themselves) inevitably occur. Such errors are detected and corrected first by proof-reading at the replicative DNA polymerase level and, secondly, by DNA mismatch repair systems that operate *in vivo* to excise and correct, post-replicatively, those nucleotide misincorporations that have escaped proof-reading (1). This double line of defence serves to reduce the overall level of error propagation between generations to one in about 10^{11} base pairs.

In addition to mismatch pairing of standard bases produced by enzymatic errors, abnormal pairing involving non-standard bases that have been modified by chemical agents, or by ionizing radiation, are also excised and corrected by the post-replicative repair enzyme system. Failure to repair such aberrant mismatches leads to the introduction of mutations in the progeny cell DNA molecules, with potentially fatal consequences that include cancer and genetic diseases. Neglecting for the moment protonation at acidic pH and base pair orientation, there are eight possible mismatch pairings, each of which is equally likely to pass on a mutation to a daughter duplex, yet these different mismatches are repaired with quite different efficiencies. The efficiency of correction/repair depends on whether the mispairing is of the Pu:Pu, Py:Py or Pu:Py type (2,3), as well as on the sequence of the flanking base pairs (4), implying the recognition of discrete structural features of the duplex surrounding the error. It would therefore appear obvious that an understanding of the repair mechanism and the recognition of mispaired bases by post-replicative repair enzymes (5) at the molecular level will require reasonably detailed studies of the structures of the corresponding mismatched base pairs in a variety of sequence contexts.

Although standard Watson–Crick pairing tends to optimize hydrogen bonding, the possibility of other energetically equivalent, non-standard hydrogen bonding schemes between two bases has long been recognized (6) and theoretical calculations have estimated that several such mismatch pairings should be energetically favourable as isolated base pairs (7), thus suggesting that not all abnormal or mismatch pairings should be assumed, *a priori*, to be destabilizing. Such 'isolated base pair' calculations obviously ignore important nearest neighbour stacking effects, such as dipole–dipole and van der Waals interactions, and it is to be expected that any given mismatch will be uniquely sensitive to the surrounding sequence context. It is therefore tempting to speculate whether the more stable 'mismatch' sequences should be considered 'abnormal' and whether non-standard base pairing might actually exist *in vivo* and carry out important biological functions. Particularly intriguing in this respect are the long stretches of tandemly repeated simple oligonucleotide sequences, known as 'satellite DNAs', found in eukaryotic chromosomes (8). The telomeres of chromosomes are another example; telomeric tandem repeats occur at the covalent ends of chromosomes and form 'abnormal' tetrad structures involving G:G pairing (for reviews see ref. 9 and Chapter 13).

Several studies on the solution structure of different DNA sequences containing a variety of isolated single mismatch base pairs have been carried out using NMR methods (for a recent review, see ref. 10). However, several of these attempts failed to obtain detailed NMR structures of the mismatch site because of the fact that the particular mismatch frequently caused destabilization of the DNA duplex, and often led to the formation of equilibrium mixtures of multiple interconverting structures. The latter problem is particularly troublesome in NMR structure determination and, with improvements in force field parameters, may be better investigated by molecular dynamics methods to probe rapid transitions between metastable states. In this chapter we will restrict ourselves to a discussion of well-defined, non-interconverting, stable base pairs involving non-complementary (in the Watson–Crick sense) bases. Particular emphasis will be placed on purine–purine mispairing and, where possible, we will also attempt to discuss the possible biological implications of these unusual structural motifs.

2. Mismatch pairing in antiparallel GA, GGA, and GGGA repeats

Tandem polypurine repeat sequences of the type $d(G_{1-3}A)_n$ are highly represented and widely distributed throughout mammalian genome satellite DNA sequences (11). Such sequences have been implicated in gene regulation as well as genetic recombination (12). Binding proteins specific for the complementary single-stranded $d(TC)_n$ and for $d(GA)_n$ DNA sequences have also been identified recently (13,14). DNA sequences with the purines on one strand and the pyrimidines on the other are structurally polymorphic and there is increasing evidence that they can form unusual structures that differ markedly from normal B-form DNA. Alternating $d(GA)_n$ sequences are perhaps the best studied example and there are reports that such sequences, in the absence of the complementary strand, form antiparallel duplexes with themselves (15), as well as parallel-stranded, self-paired duplexes (16) and tetraplexes (17). Furthermore, $d(TC)_n:d(GA)_n$ repeat sequences appear to serve as pause or arrest signals in DNA replication and amplification (18,19), perhaps as a result of forming non-canonical structures.

Although the formation of parallel-stranded (16) or anti-parallel (15) double helical structures for self-paired d(GA)$_n$ sequences has been inferred from native gel electrophoresis studies, the precise base pairing geometry of these proposed structures has been difficult to determine unambiguously. To date, no detailed NMR structural studies have been reported on this tandem repeat—probably because its small dinucleotide repeat nature produces highly overlapped proton spectra which may be further broadened by interconversion between multiple conformations. Two different types of pairing geometry have been indirectly deduced for d(GA)$_n$ sequences under different solution conditions using accessibility to chemical modification by DEPC (diethylpyrocarbonate) and DMS (dimethyl sulfate) as structural probes (15,16). The former reagent is used to distinguish between single-stranded and base paired regions of polynucleotides since the predominant reaction of DEPC is to carbethoxylate the N7 atoms of unpaired purine residues, with adenines being much more susceptible than guanines. Conversely, DMS methylates the N7 position of paired and unpaired guanines but can be used to probe the type of base pairing, since the guanine N7 is unreactive towards DMS when it participates in hydrogen bonding—as it does, for example, in a Hoogsteen base pair (20,21).

Using 52-residue DNAs in which the first (5′) 11 residues and the last (3′) 11 residues were autocomplementary and were separated by an intervening 30-residue stretch of 15 GA repeats, Huertas *et al.* (15) were able to show that such sequences formed fold-back hairpin structures, with hyperreactivity to DEPC (single-strandedness) confined to a hexanucleotide loop at the centre of the (GA)$_{15}$ run. They therefore concluded that the first (GA)$_6$ dodecanucleotide must be base paired to the last (GA)$_6$ dodecanucleotide in an antiparallel fashion. All guanines were found to be DMS-susceptible, indicating no Hoogsteen pairing to G(N7) atoms. Because the adenines in the descending arm of the stem, i.e. the second (3′) (GA)$_6$ run, were somewhat less reactive towards DEPC than those in the first (5′) ascending arm (GA)$_6$ run, a pairing scheme consisting of G$_{anti}$: A$_{syn}$ pairs alternating with G$_{anti}$:A$_{anti}$ pairs was proposed for the (GA)$_n$:(GA)$_n$ repeats in the double-stranded stem of this hairpin (15). Based on these conclusions, a molecular model of an antiparallel, right-handed duplex containing G$_{anti}$:A$_{anti}$ base pairs interleaved with G$_{anti}$:A$_{syn}$ base pairs was constructed by computer modelling; all the deoxyribose sugars could be successfully incorporated into this model in the normal C2′-*endo* conformation.

Using similar DEPC and DMS probes of N7 accessibility, as well as excimer fluorescence of 5′ pyrene-labelled shorter oligonucleotides of the type (GA)$_{7.5}$ and (GA)$_{12.5}$, Rippe *et al.* (16) came to completely different conclusions about GA:GA pairing in bimolecular homoduplexes of GA repeats. The excimer fluorescence studies indicated that the 5′-pyrene labels were at the *same end* of the presumed bimolecular duplexes, thus indicating parallel-stranded structures; the formation and stability of these structures did not require acidic conditions and thus did not appear to involve protonated bases. Based on this information, the authors succeeded in constructing a right-handed, parallel-stranded duplex with a register in which G$_{syn}$:G$_{syn}$ base pairs alternated with symmetrical A$_{anti}$:A$_{anti}$ base pairs (16). However, the sugar puckers could only be incorporated into this model in the less usual (for DNA) C3′-*endo* con-

(a)

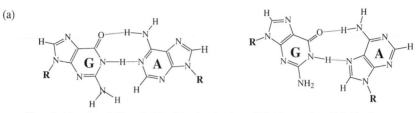

G(*anti*)•**A**(*anti*) and **G**(*anti*)•**A**(*syn*) in the Antiparallel Alternating d(GA)$_n$ Sequence

(b)

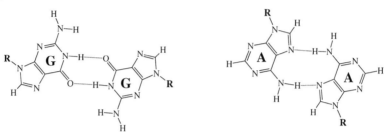

G(*syn*)•**G**(*syn*) and **A**(*anti*)•**A**(*anti*) in the Parallel Alternating d(GA)$_n$ Sequence

(c)

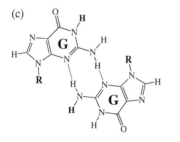

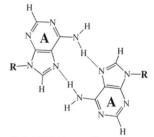

G(*anti*)•**G**(*anti*) and **A**(*anti*)•**A**(*anti*) in the Parallel d(CGA) motif

(d)

e

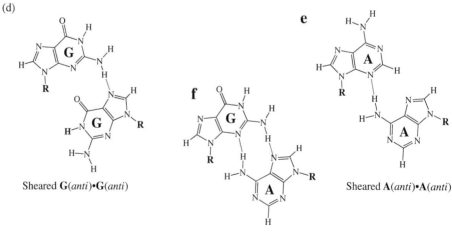

Sheared **G**(*anti*)•**G**(*anti*)

f

Sheared **A**(*anti*)•**A**(*anti*)

Sheared **G**(*anti*)•**A**(*anti*)

Fig. 11.1. The various types of purine–purine pairings discussed in this chapter.

formation. The quite different purine–purine pairing schemes in these two models for self-paired $(GA)_n$:$(GA)_n$ homoduplexes are shown in Fig. 11.1; the validity of these models remains to be tested at atomic resolution by more powerful techniques such as solution NMR or X-ray crystallography, with the former being preferable by virtue of the avoidance of possible lattice packing artefacts.

The structural properties of $d(GGA)_n$ and $d(GGGA)_n$ repeat sequences are even less well understood. The formation of four-stranded tetraplex structures by $d(GGA)_n$ repeat sequences has been proposed on the basis of cation-stabilization thermal melting studies (17). Recently, the demonstration of intramolecular hairpin formation by the *Drosophila* centromeric dodeca-satellite DNA sequence (22) and by $d(GGA)_n$ repeats, as well as $d(GGGA)_n$ repeat sequences, has shed some light on their pairing (23).

In the evolutionarily conserved dodeca-satellite 5′-$d(GTACGGGACCGA)_n$ repeats of *Drosophila* centromeres, the G-rich strand alone has been shown to form a fold-back structure, based on non-denaturing gel electrophoresis, electron microscopy, accessibility to chemical modification, and thermal denaturation studies (22). The central GGGA tract of the 12-mer repeat, and particularly the formation of G:A pairs, was found to be critical for the stability of the intramolecular hairpin forms. However, the alignment and precise geometry of the purine–purine pairing is not known and three different registers for a $d(GGGA)$ tract interacting with a second $d(GGGA)$ tract were proposed, namely a -2 register with a $(GA)_2$ motif, a $(GGA)_2$ motif register, and even a $(GGGA)_2$ motif alignment (22).

By the same token, in $d(GGA)_n$ and $d(GGGA)_n$ direct tandem repeat sequences, the types of purine–purine pairings and their precise geometry are also unclear. However, an interesting conclusion has been drawn by Huertas and Azorin (23) on the basis of chemical modification studies; namely, that pairing between $d(GGA)_n$ sequences is stabilized by G:A pairing of some kind, while pairing between $d(GGGA)_n$ repeats involves only G:G and A:A pairs, and not G:A pairs.

3. Mismatches between parallel-stranded CGA triplets and their repeats

The above discussions suggest that two different conformations for self-paired $d(GA)_n$ repeat sequences may exist under different conditions. Unfortunately, detailed three-dimensional structures, determined either by X-ray crystallographic or NMR solution methods, are not yet available for these repeats. An interesting related observation is that cytidine residues, which can pair with themselves under mildly acidic conditions to form stable C^+:C pairs, have been found to help in aligning GA:GA pairing to form the parallel-stranded $(CGA)_2$ motif, first reported by Wang and co-workers (24–26) and later confirmed by Patel's group (27).

The first experimental demonstration of parallel alignment between strands in DNA duplexes was the X-ray structure of crystals of $d(CG)_2$ grown under acidic conditions; the nucleotide residues in these crystals were found to pair via C^+:C and G:G homobase pair mismatches (28,29). More recently, Gueron and co-workers (30) have shown that in mildly acidic solution even simple dC_n-containing sequences also form parallel-stranded duplexes containing C^+:C mismatches; these parallel-stranded duplexes dimerize intercalatively in an antiparallel orientation to form the four-stranded structure known as the 'i-motif'. Sequences containing one or more CGA triplets were found to adopt distinct structures at pH values below 5.0, which are in a slow exchange equilibrium with the neutral pH form (24). NMR studies of the self-complementary oligodeoxyribonucleotides CGATCG, TCGATCGA, and CGATCGATCG revealed that the neutral forms of these oligomers adopt an antiparallel canonical B-form DNA structure, while their acidic forms are right-handed, *parallel-stranded* duplexes containing symmetrical C^+:C, G:G, A:A, and T:T homobase pairs instead of Watson–Crick pairs (24). The parallel-stranded $(CGA)_2$ motif is crucial to the formation of such structures, which were proposed to be stabilized by strong interstrand GA stacking, as well as by hemiprotonated C^+:C pairing.

In fact, the simple tetranucleotide d(TCGA) had been studied by NMR much earlier (31); based on the observation of several shifted proton resonances, it was suggested that this sequence forms a non-B-form DNA duplex at low temperature. The temperature-dependent transitions of this non-B-form structure could be duplicated reversibly by titration to acidic pH values, and a protonated antiparallel $(TCGA)_2$ duplex model containing a $G:C^+$ Hoogsteen base pair was proposed (32). However, the NMR data, which can readily distinguish between the *syn* and *anti* orientations of the glycosidic bond, do not support the *syn* conformation proposed for the guanosines in this model.

The structure of the $d(TCGA)_2$ duplex was solved more recently by Patel's group using more extensive NMR data (27), and was found to form hemiprotonated C^+:C pairs, as well as G:G and A:A homopurine pairs, as expected in a parallel-stranded $d(CGA)_2$ motif. The one-dimensional imino proton spectrum exhibited a resonance at approximately 15 ppm, which is a characteristic of C^+:C pairing and reflects the mildly acidic pH conditions. In addition, imino proton resonances were observed at 10.2 and 11.3 ppm. The resonance at 10.2 ppm is characteristic of a non–hydrogen bonded, but slowly exchanging, guanosine imino proton, as occurs, for example, in G:A pairs that are in the sheared, or side-by-side, geometry (33–35). Similarly, the imino proton at 11.3 ppm is characteristic of a thymidine imino proton that is not hydrogen bonded but exchanges slowly with water (36), either as a result of restricted solvent accessibility or the reduced hydroxide/buffer exchange catalysis at lower pH values.

The terminal thymidine residues in the $d(TCGA)_2$ sequence thus may not be paired via hydrogen bonding, since the imino proton is the only potential hydrogen bond donor in deoxythymidine; and, indeed, in the NMR structure reported for this parallel-stranded duplex, the thymine base is oriented towards the sugar moiety of the thymidine on the opposite strand (27). It is not clear whether the lack of T:T pairing is an inherent property of the parallel-stranded $(TCGA)_2$ duplex or merely a reflection of the thymines being terminal residues, especially since mismatched T:T pairing has been proposed when the TCGA tetranucleotide is embedded in the centre of a longer sequence (24).

Another important structural feature of sequences forming such parallel-stranded duplexes is strong *interstrand* G/A stacking, which has some similarities to the interstrand G/G and A/A stacking in antiparallel-stranded tandem sheared G:A pairs in d(PyGAPu)$_2$ motifs (see below). A comparison between parallel-stranded 5'-(GA)$_2$ stacking and antiparallel-stranded 5'-(GA)$_2$ stacking is shown in Plate IVa. One GpA strand is shown as a space-filling, van der Waals representation, while the second GpA strand is represented in stick-bond form; the space-filling strands have the same orientation, with the guanosine residues on top and the adenosine residues below. The glycosidic bonds of the bases in both duplexes all have the same *anti* conformation, while the phosphate backbone conformation is quite different for the parallel and antiparallel cases. The torsion angles of the sugar phosphate backbone connecting the guanosine and adenosine residues exhibit a $\varepsilon(g^-)\zeta(t)$ configuration in the antiparallel duplex, while in the parallel-stranded duplex they are both *trans* in a $\varepsilon(t)\zeta(t)$ configuration (27).

Important differences between the parallel and antiparallel structures occur in the strands shown in stick-bond form. While the guanosine (coloured brown) is located on the bottom in the antiparallel duplex, it is on top in the parallel-stranded duplex. Furthermore, in the parallel-stranded duplex the H8 protons of the purines point into the narrow or 'minor' groove, whereas in the antiparallel duplex they are located in the wide, major groove. It is clear from this Plate that excellent interstrand stacking of both the G/G and A/A type occurs in the antiparallel motif, while only the G/A type of interstrand stacking is observed in the parallel motif. Plate IVb compares the parallel 5'-CGA-3':5'-CGA-3' motif with an antiparallel 5'-CGA-3':3'-GAG-5' duplex containing a (GA)$_2$ motif, viewed from the side instead of end-on. The reference 5'-d(CGA)-3' strands are again shown in van der Waals representation in the same orientation for ease of comparison. The parallel and antiparallel nature of the two motifs are apparent from this figure. In the parallel motif, the excellent intrastrand C/G and *interstrand* G/A stacking can easily be seen, while in the antiparallel-stranded (GA)$_2$ motif, the interstrand stacks are of the G/G and A/A type, even though the intrastrand C/G stacking is similar.

It is also worthwhile to compare the G:G and A:A pairing geometry in the two parallel-stranded duplexes, namely the proposed d(GA)$_n$ tandem repeat (16) and the parallel-stranded d(TCGA) duplex (27). As can be seen from Figure 11.1b and c, while the A:A pairings between the two parallel duplexes are similar (they superimpose when flipped over horizontally), the G:G pairings are quite different. In the parallel-stranded d(GA)$_n$ repeat, the G:G bases pair symmetrically *via* their N1H and O^6 atoms and both guanosine residues adopt the *syn* glycosidic conformation, while in the d(CGA) motif, the G:G bases pair through their N^{2}H and N3 atoms and are in the *anti* conformation.

Another point worth noting is that it is now well established (see below) that CGA sequences in CGAG contexts (or TGA sequences in TGAA contexts) form antiparallel duplexes containing two tandem sheared G:A pairs [the (PyGAPu)$_2$ motif] flanked by Watson–Crick pairs (33–35). It is therefore interesting that, in complete contrast to CGAG, a single change to CGAT with a 3'-pyrimidine should result in a parallel duplex with C$^+$:C and homopurine base pairs under acidic conditions (24). This argues for an important structural role for the purine following the tandem G:A pairs in the antiparallel (GA)$_2$ motif. It is not known at this point whether CGAG or TGAA sequences can form parallel duplexes under acidic conditions, but experiments are in progress to investigate this point.

4. Tandem sheared G:A mismatches separated by Watson–Crick base pairs

The sequences described above in Sections 2 and 3 all form duplexes containing continuous runs of adjacent mismatched base pairs, i.e. no intervening normal Watson–Crick base pairs are involved. We will now discuss small, stable mismatched motifs containing tandem sheared G:A base pairs that are quite stable when flanked by, and embedded in, normal Watson–Crick base-paired duplexes. Perhaps the most remarkable and unusual feature of these duplexes is that the 'destabilization effect' of each mismatched G:A pair does not accumulate progressively. Instead, they contribute significantly to the stabilization of adjacent G:A mismatches and flanking normal base pairs to form very stable duplexes (33,37), but only in certain sequence contexts (35).

4.1 Tandem sheared G:A mismatches in the [Py(GA)Pu]₂ motif: sequence dependence

Non-standard base pairing, including sheared or side-by-side G:A pairing, has long been recognized as a theoretical alternative to standard Watson–Crick pairing in nucleic acids (6,7,38). However, interest in the actual existence and remarkable stability of G:A pairing came from the finding of Wilson and colleagues (39) that certain purine–rich oligodeoxynucleotide sequences could pair with themselves to form duplexes of similar stabilities to those formed in the presence of the complementary pyrimidine-rich strand. Based on sequence alignment, the self-paired homoduplex was proposed to consist of two adjacent G:A pairs separated from another two G:A pairs by two intervening Watson–Crick pairs, and flanked by two Watson–Crick pairs at each end. NMR studies, combined with the effects of replacing guanosine residues with inosines, led to a model in which guanine paired with adenine *via* G(N²H) to A(N7) and A(N⁶H) to G(N3) hydrogen bonds (33). Such sheared tandem 5′-GA:GA-3′ pairs could be incorporated into an antiparallel duplex model with little distortion from a standard B-form DNA backbone configuration (33).

In the five years following this pioneering study, several structural and thermodynamic investigations of tandem G:A mismatched pairs have been reported (34,35, 40–49). Using characteristic chemical shift signatures for the sheared geometry, it was shown from 1D NMR studies that the formation of tandem sheared G:A pairs was sensitive to the orientation of the flanking Watson–Crick pairs, requiring a PyGAPu context on each of the antiparallel strands (35). Some quite unusual (for DNA) cross-strand NOEs (33,34,40,49) produced a set of distance restraints that led to the determination of fairly high precision structures for DNA duplexes containing one or more [Py(GA)Pu]₂ motifs; for example, 15 refined structures exhibiting pairwise rmsd values of 0.96 ± 0.34 Å (40).

To illustrate the gross structural differences between duplexes containing tandem sheared G:A pairs and normal duplexes containing Watson–Crick pairs, these two structures are presented in simplified ring-and-arrow form in Fig. 11.2. As can be seen from the side view into the major groove, the DNA containing the tandem sheared G:A pairs has two kinks in the backbone; part of the cause of this kinking is the result of changes in the backbone torsion angles from the B₁ conformation of

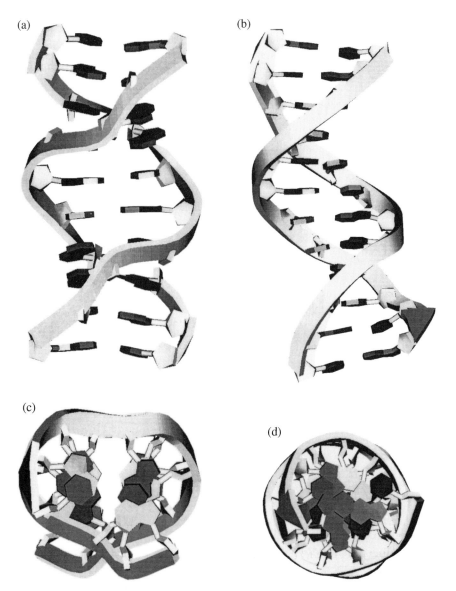

Fig. 11.2. The unusual structure of the d(GCGAATGAGC)$_2$ decamer duplex (a and c) containing two sheared (PYGAPU)$_2$ motifs (40) compared with the B-DNA (51) crystal structure (b and d) containing normal G:C and A:T Watson–Crick pairs. The phosphate backbones are represented by an arrow ribbon to illustrate the two kinks caused by the B$_I$→B$_{II}$ phosphodiester transition between the two sheared G:A base pairs. (a) Also shows the other unusual structural characteristic of DNAs containing sheared (PyGAPu)$_2$ motifs, namely the bases do not follow the backbone spiral but stack vertically down each side of the cylinder owing to the cross-over between interstrand and intrastrand stacking, with none of the base twisting that occurs in B-DNA (b). This phenomenon is more easily seen in the end-on views in (c) and (d), where the clear-cut separation of the stacking down the right and left sides can be seen in (c), while the bases are radially distributed in the B-DNA helix shown in (d). This view also demonstrates that the cross-section of the DNA duplexes containing (PyGAPu)$_2$ motifs is not circular, as in B-DNA, but has an elongated oval shape.

B-form DNA to a B_{II} configuration at the GpA phosphodiester link (40,50). Furthermore, the adenosine now swings away from intrastrand stacking on the preceding guanine and participates in cross-strand stacking with the adenine of the opposite strand. The change in phosphodiester conformation results in a downfield shift of the GpA ^{31}P resonance (35,50). Interestingly, this backbone rearrangement to produce cross-strand Pu/Pu stacking is restricted to juxtaposed GpA dinucleotides and does not occur with juxtaposed ApG, ApA, or GpG sequences (S.-M. Chou and B.R. Reid, unpublished observations).

The tandem sheared G:A DNA also has a wider minor groove than B-DNA (Fig. 11.2) (51), which has a very smooth backbone trace. Figures 11.2c and d show end-on views of these duplexes and the base stacking patterns are obviously dramatically different for the two duplexes. While the B-DNA duplex (Fig. 11.2d) adopts the usual intrastrand base stacking, with *c.* 30–40° of twist per step, the tandem sheared G:A-containing DNA exhibits a combination of *intrastrand* and *interstrand* stacking, resulting in two clear-cut sides to the base stacking. The high stability of the tandem sheared G:A pair-containing DNA is obviously a result of the extensive cross-strand purine/purine base stacking and the increased intrastrand stacking of the GpA:GpA dinucleotides with their Watson–Crick nearest neighbours. An increase in diameter is also associated with the tandem sheared G:A-containing DNA duplex (Fig. 11.2c). Plate V shows a comparison between the B-DNA duplex d(GCGAATTCGC)$_2$ (51) and the d(GCGAATGAGC)$_2$ duplex containing tandem G:A pairs (40). To better compare the two, the terminal base pairs of the B-form crystal structure have been removed so that both sequences now have 10 base pairs. From the figure, it can be seen that the B-DNA structure (left) has regular base stacking with a smooth backbone trace, while the d(GCGAATGAGC)$_2$ duplex containing the (PyGAPu)$_2$ motif, although still a right-handed double helix, has a quite different appearance; residue 8A has swung from intrastrand stacking to an interstranded stack with residue 4A of the opposite strand and, in a similar fashion, 7G now stacks with residue 3G of the opposite strand. This excellent cross-stranded G/G and A/G stacking is indicated by a red arrow in the bottom half of Plate V. The accompanying change in the phosphate backbone from B_I to a B_{II} conformation causes a kink in the otherwise smooth backbone trace, as indicated by the blue arrow in the top half of Plate V.

Major effects on the stability and structure of G:A mismatch pairs as a result of changing the immediately adjacent Watson–Crick pairs have been revealed by thermodynamic (33–37) and ^{31}P NMR studies (35). While thermodynamic studies reveal that DNA sequences containing the [PyGAPu]$_2$ motif have stabilities comparable to those of fully Watson–Crick base-paired duplexes (32,37), NMR studies indicate that changes to a different, non-sheared, G:A pairing geometry (or even to non-paired GA bulges) occur when antiparallel GpA dinucleotides are juxtaposed in PuGAPy contexts (35). Thus, the head-to-head $G_{anti}:A_{anti}$ geometry, with a hydrogen bonded G imino proton, that occurs in the (AGAT)$_2$ context switches to the more stable sheared tandem G:A pairing in either (CGAG)$_2$ or (TGAA)$_2$ contexts, while no duplex is formed at all in a (GGAC)$_2$ context. This dramatic change in G:A pairing from head-to-head to side-by-side geometry is clearly revealed in the NMR spectrum by a characteristic shifting of the guanosine imino proton resonance from 12.4 ppm (hydrogen bonded) to 10.1 ppm (not hydrogen bonded) (35). The ^{31}P resonances connecting the 5′-GpA -3′

residues are also found to shift from −2 ppm to −3 ppm in the B_{II} conformation (35). It is also interesting to note that tandem adjacent GA:GA pairs in the non self-complementary DNA sequences 5′GGA*CGAC*ATC:GAT*GGAG*TCC-3′ were also found to adopt sheared pairing geometry (48). The flanking neighbour stacking interactions PyGA:GAPu were thus proposed to contain the minimal essential elements for the formation and stabilization of the sheared $(GA)_2$ motif (48). However, the full structure of the CGAC:GGAG duplex was not determined and further studies are still needed to clarify and explain fully the context requirements of this motif.

Interest in the relevance and biological function of the $(GA)_2$ motif stems from the finding that the sequence requirements for the replication origin of the single-stranded DNA virus ϕX174 suggest that it contains two adjacent sheared G:A pairs in a $(GA)_2$ motif (52). A unique sequence within a hairpin region in the ϕX174 genome was found to be the binding site for the protein n′, which is a pre-priming DNA replication enzyme of *E. coli* (52). It has been suggested that recognition of this hairpin sequence is the signal that leads to the initiation of ϕX174 replication. Interestingly, the secondary structure proposed for the hairpin stem recognition locus of protein n′ contains juxtaposed CGAA:TGAG sequences of the archetypal PyGAPu type and would be expected to form an antiparallel $(GA)_2$ motif with adjacent sheared G:A geometry. Preliminary NMR data of a DNA duplex containing juxtaposed CGAA:TGAG sequences in the same flanking context as the ϕX174 replication origin does in fact show the formation of tandem G:A pairs in the sheared pairing geometry (S.-H. Chou, L. Zhu, and B.A. Reid, unpublished results). If, as seems likely, protein n′ recognizes this special segment of DNA, it will bind it in a quite different way to that proposed in the literature (53,54), because of the previously unsuspected major groove backbone kink and the presence of unusual hydrogen bond donors and acceptors in the major groove (Plate V). The N7, O6, and N1H atoms of the guanosine residues are now exposed and available for ligand binding in the major groove, while the N1 and N6H atoms of the adenine residues have moved to the minor groove where they are available for ligand binding, if indeed this atypical minor groove is wide enough for peptide interaction.

The extent to which the sheared antiparallel $(GA)_2$ motif participates in other biological systems is not yet established at present, but we believe it will have quite widespread application, because of its great stability and its complete compatibility with flanking Watson–Crick base pairs (in the appropriate orientation) in forming a continuous duplex. In fact, the recently solved X-ray crystal structures of an RNA:DNA hybrid ribozyme (55) and a pure RNA hammerhead ribozyme (56) both contain juxtaposed CGAA:UGAG sequences in a highly conserved loop region (57). Furthermore, the backbone linking the GpA residues of the adjacent G:A pairs in these hammerhead ribozymes (55,56) adopts the B_{II} $[\varepsilon(g^-)\zeta(t)]$ conformation observed earlier in NMR studies of the isolated $(GpA)_2$ motif (34,50).

4.2 *Sheared G:A mismatches in the pPy(GGA)Pu]₂ motif*

Over 95% of the human genome consists of so-called 'junk' DNA, which does not code for functional RNA or protein molecules (58). Much of this DNA consists of highly repeated, relatively simple core sequences, the possible functions of which are only beginning to emerge. In humans, the satellite DNA repeat (TTAGGG)$_n$ has been mapped to the telomere regions at the ends of chromosomes (59, 60). The purine-rich strand alone has been found to adopt an unusual G tetrad tetraplex structure involving G:G:G pairing (61). Satellite II and satellite III DNAs, which comprise approximately 5% of the human genome, were found to contain (ATTCC)$_n$, or alternatively (GGAAT)$_n$, repeats (8) that have recently been shown to be localized at the centromeres of human chromosomes (62). This repeat, which can also be considered to be a (TGGAA)$_n$ repeat by simple phase-shifting on the purine strand, is highly conserved among all eukaryotic species and is a high affinity ligand for specific nuclear proteins—the affinity is comparable to other highly selective protein–DNA interactions, such as the *lac* repressor–operator DNA interaction (62). These observations have led to the suggestion that the (TGGAA)$_n$ repeat may be a component of the functional human centromere. An extremely interesting aspect of this repeat is the fact that the purine-rich strand alone forms homoduplexes that have the same thermal stability as the Watson–Crick duplex formed in the presence of the complementary pyrimidine-rich strand. Several groups have subsequently investigated the structure of the unusual duplex formed by this self-paired repeat. Jaishree and Wang (64) used the phase-shifted variant C(AATGG) sequence as a model of the (AATGG)$_n$ tandem repeat. Unfortunately, the additional C residue (which does not occur in the natural repeat) at the 5′-terminus reset the pairing register and forced the C(AATGG) duplex into a configuration with two *non-adjacent* head-to-head G$_{anti}$:A$_{anti}$ pairs separated by two Watson–Crick A:T pairs. In a separate NMR structural study of this repeating pentamer, Catasti *et al* (64) carried out NMR studies on the self-paired duplexes formed by (AATGG)n sequences (where n = 2 or 3) and derived a solution structure in which the repeating motif contained a G:G base pair sandwiched between two sheared G:A pairs. However the G$_{syn}$:G$_{anti}$ mismatch pair that they proposed is not compatible with their own NMR data in that there were no strong intranucleotide G(H8) to G(H1′) NOEs that would be diagnostic of their proposed Gsyn conformation; a fast flip-flop interconversion between G$_{anti}$:G$_{syn}$ and G$_{syn}$:G$_{anti}$ pairing had to be proposed as an *ad hoc* rationalization for this discrepancy. Furthermore, their assignment of the critical guanosine H3′ and H4′ protons (64), which are actually upfield shifted by *c.* 2 ppm owing to an unusual stacking arrangement in the (GGA)$_2$ motif (41,44), also appear to be incorrect. The structure of this repeat was finally solved by Chou *et al.* (41) using a tandem repeat of the pentamer sequence with a TGGAA phase, i.e. (GTGGAATGGAAC)$_2$. The fact that the 'G:G pair' was, in fact, not paired at all but was intercalated, was established by guanosine to inosine substitutions. This led to the detection of many unusual and informative NOEs from the inosine H2 proton; for example, in the 3G–4I–5A:8G–9I–10A segment of the duplex, the detection of 4I(H2) ↔ 9I(H8) and 9I(H2) ↔ 4I(H8) NOEs, together with the absence of 'nearest neighbour' 4I(H8) ↔ 5A(H8) and 9I(H8) ↔ 10A(H8) NOEs, is incompatible with I:I pairing and establishes the intercalative stacking arrangement of the two I residues on each other. This conclusion was also complemented by many additional and unexpected types of NOEs in this region, as summarized in Fig. 11.3. The unusual H4′ chemical shifts and the C3′-*endo* sugar conformations of the unpaired guanosine residues were confirmed by DQF-COSY and ^{31}P-^{1}H correlation experiments (44).

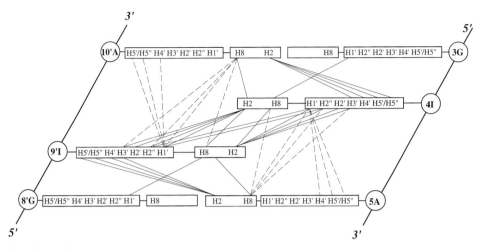

Fig. 11.3. The NOE connectivity pattern for the antiparallel d(GIA)₂ motif. Guanosine residues were replaced by inosines at the central unpaired purine to exploit the extra I(H2)-related, through space connectivities, which were found to be critical in solving the structure of this unusual motif (41). The detectable NOE connectivities are indicated by solid lines, while those expected in normal DNA, but not detectable experimentally, are indicated by dashed lines. These data are only consistent with 4I/9I intercalating and stacking on each other between the bracketing sheared G:A pairs.

Plate VI compares a standard B-form (GCGAATTCGC)₂ crystal structure (left) with the structure of the NMR-derived (TGGAATGGAA)₂ duplex containing the pericentromeric TGGAA repeat (right). In this figure, one strand is shown in space-filling display and the other strand in stick-bond form. This emphasizes the excellent cross-strand stacking between the unpaired intercalated guanosine residue and the guanosine residue of the sheared G:A pair, as shown by the parallel interface between the space-filling strand and the stick-bond strand in the bottom half of the duplex. Owing to its obviously different gross morphology compared with standard B-DNA (the major groove is much wider and the minor groove is much narrower), the mode of interaction of this novel duplex with proteins, i.e. isolated HeLa cell nuclear extracts (62), can also be expected to be quite different to that of normal DNA. The d(GGA)₂ motif contains a grid of 16 hydrogen bond donors and acceptors, i.e. the N2H-N1H-O6-N7 atoms of the four co-stacked guanine residues, that are exposed to the exterior in the major groove (41,44).

Whether or not the self-paired (TGGAA)₂ repeat is actually formed *in vivo* is not yet established but the exposed four-guanine, 'sticky patch' is repeated twice per turn, facing opposite sides of the duplex, and could perhaps be responsible for the highly condensed nature of DNA at the centromere, and may even participate in the capture of chromosomes by means of the centromere during mitosis.

The participation of the analogous r(GGA)₂ motif in RNA function is less clear, but it should be noted that in the folding of tRNA, the G57 base of the rT loop intercalates between the G19:C56 and G18:ξ55 tertiary base pairs to form a continuous G19-G57-G18-m¹A58 stack (65). This four-purine stack is one of the most important stabilizing interactions in tRNA folding.

4.3 Sheared G:A mismatches in the [Py(GAA)Pu]:[Py(GA)Pu] motif

After the discovery of sheared tandem G:A pairing in the antiparallel $(GA)_2$ and double-guanine intercalative $(GGA)_2$ motifs, it became of interest whether single G intercalation between sheared G:A pairs could also occur, and whether adenosine residues could replace guanosine residues in these intercalations. The biological relevance of this question stems from the fact that a potential antiparallel GAA:GA motif could occur in a highly conserved region at the 3'-termini of single-stranded rodent parvovirus genomes (66,67). A Y-shaped double hairpin fold-back structure was proposed for this conserved sequence that juxtaposes a GA dinucleotide opposite an antiparallel GAA triplet in all four parvovirus sequences, suggesting some essential function for this element, which is located in the region of the genome where initiation of DNA replication occurs. An unpaired bubble structure was originally proposed for this mismatch region, but it is interesting to note that it is 'constrained' in that it is resistant to mung bean endonuclease, which is a single-stranded DNA-cleaving enzyme (67).

We have carried out NMR studies of this potentially important motif that indicate that it does not form an unpaired bubble, but instead forms a G:A-bracketed single-A stack intercalated motif in solution (S.-H. Chou, L. Zhu and B.R. Reid, unpublished results), which explains the resistance of this motif to mung bean endonuclease cleavage. The structure of the 5'-(CGAGTACGAAG)$_2$ 11-mer duplex, containing two GAA:GA motifs separated by four Watson–Crick pairs, has been determined (S.-H. Chou, L. Zhu and B.R. Reid, in preparation), and is shown alongside the B-DNA crystal structure of 5'-(GCGAATTCGC)$_2$ in Plate VII. The unpaired adenosine that is intercalated between antiparallel sheared G:A pairs is shown in blue (on both strands) and can be seen to stack very well with the guanine residues of both of flanking G:A pairs. Interestingly, since the adenine following the sheared G:A guanine now stacks on it, there is now no need for this GpA phosphodiester to switch into a B_{II} configuration to permit cross-strand G/G and A/A stacking. The backbone has now reverted to the B-DNA type with no kink and, unlike the $(GA)_2$ motif, no unusually shifted phosphorus resonances are observed in the ^1H-^{31}P correlation spectrum of the GAA:GA motif.

5. Sheared G:A mismatches closing single-residue hairpin loops

5.1 The (GCA) motif and (GNA) motifs

There has been considerable interest in the structure of small hairpin loops in connection with the discovery of the expansion of tandem triplet repeats in the target genes of several genetic diseases that show anticipation (68,69). The formation of hairpin fold-back structures by either the pyrimidine-rich strand or the purine-rich strand (or both) of these repeated triplets has been suggested to be part of a proposed replicative slippage mechanism for the expansion of the triplet repeats (70). Several gel electrophoretic studies on the formation of fold-back hairpins by such repeats have been reported recently (71–75), including proposals for the formation of loops putatively closed by

A:A or G:G pairs, but the type and register of the mismatch pairings and the actual structure of the base pairs in the stem, as well as the structure of the loop in such hairpins, remains unclear at this point. Earlier structural and thermodynamic studies led to the original conclusion that oligonucleotide hairpins containing less than three nucleotides in the loop were sterically impossible (76) and the optimal hairpin loop size in DNA hairpins was considered to be 4–5 residues (77). However, these conclusions were found to require revision when later studies established that the stability of DNA hairpins increased as the size of the loop was reduced, with trinucleotide loops (especially TTT or AAA) being the most stable (78). The nature of the closing pair at the top of the base paired stem has a major influence on loop stability, and in 1994 Hirao *et al.* (79) reported that the DNA heptanucleotide d(GCGAAGC) forms an extraordinarily stable fold-back structure that is resistant to nucleases and heat. Based on NMR studies, the authors reported a compact hairpin model with a three-base pair stem closed by a sheared G:A pair and a 'mobile' loop consisting of a single adenosine (79).

In completely separate studies on variants of the d(GGA)$_2$ motif in d(TGGAA)$_n$ repeat sequences, the present authors, together with Leiming Zhu, investigated the solution structure of d(TGCAA) sequences—expecting them to form intercalative (GCA)$_2$ motifs analogous to the (GGA)$_2$ motifs described above, since thermal denaturation studies of (GCAAT)$_6$ had shown that it has almost the same melting temperature as the (GGAAT)$_6$ sequence (62). However, to our surprise, they did not form intercalative motifs, and the decamer CAAT**GCA**ATG instead formed an unusual stable hairpin with a four-base paired stem and a single-cytidine loop closed by a sheared G:A pair (80). Studies of the remaining two NAATGNAATG variants, namely AAAT**GAA**ATG and TAAT**GTA**ATG, revealed that neither formed a single stable structure. Instead, they both established an equilibrium mixture of hairpins containing a single-residue 'tight-turn' loop closed by a sheared G:A pair [the d(GNA) motif] and bimolecular duplexes containing intercalative d(GNA)$_2$ motifs (42); these two quite different conformations were found to be in slow exchange on the NMR time-scale for both decamer sequences. Thus, GNA triplets exhibit remarkably different folding and interaction properties that depend on the identity of the N residue. When N = G, d(NAATGNAATG) sequences have a strong propensity to form duplexes containing an intercalative d(GGA)$_2$ motif (*c.* 80% of the population). However, when N = C, such decamers form exclusively hairpins containing a single-C tight-turn loop, i.e. the d(GCA) loop motif. Finally, when N = T or A, the decamers both exhibit slow exchange hairpin-duplex equilibria, with a stronger tendency to form single-residue, tight-loop hairpins (*c.* 80%) than bimolecular intercalative duplexes (*c.* 20%) under NMR conditions (42).

The fact that the GCA triplet exclusively forms tight-turn hairpins with single C loops may well be of biological relevance in modulating the folding of pericentromeric DNA since TGCAA is the most common variant in (TGGAA)$_n$ runs (81). We have recently shown that, while (G)TGGAATGGAATGGAA(C) sequences form antiparallel duplexes containing three intercalative (GGA)$_2$ motifs, a single change to (G)TGGAATGCAATGGAA(C) results in the exclusive formation of hairpins containing a (GGA)$_2$ motif in the stem and a (GCA) motif tight-turn loop (45).

This extraordinary hairpin-promoting capability of (GCA) triplets in the middle of (TGGAA)$_n$ runs would be expected to form multi-arm fold-back structures which may be related to the condensation of human centromeres. Plate VIII shows the hairpin structure of such a d(TGGAATGCAATGGAA) sequence in two different views (45). In the major groove view, the grid of 16 hydrogen bond donors and acceptors of the four well-stacked guanosines can be clearly seen just below the centre of the right view, while in the minor groove view, on the left, the excellent base stacking in the GCA tight-loop (in which the carbons are blue) is evident. Furthermore, the deoxyribose of residue 8C and the base of residue 9A (in which the carbons are blue) are also 'stacked'—as shown at the top of the right view. The deoxyribose H4′ proton of the residue 8C is coloured yellow in this Figure to reveal its direct stacking over the 9A base, which explains its unusually upfield chemical shift of c. 1.8 ppm (45,80). The stacking interaction of the deoxyribose of the loop cytidine residue with the adenine base of the closing sheared G:A pair now explains how this motif can form such small hairpin loops containing only one nucleotide. In a loop closed by a normal Watson–Crick pair, the C5′ atoms of the ascending strand and the C3′ atom of the descending strand are too far apart to be bridged by a single nucleotide, and loops bridging the ends of such stems require a minimum of two nucleotides (77). However, the sheared geometry of the closing G:A pair swings the ends of the two stem strands closer together and this, combined with the interaction of the loop residue sugar ring with the closing G:A pair, is sufficient to permit bridging by a single nucleotide.

5.2 *(AAA) and (GAG) motifs*

Given the requirement for sheared G:A pairs in closing single-nucleotide loops, an interesting question became whether this function could be carried out by other Pu:Pu combinations in sheared geometry. We have now extended this closing pair motif to A:A and G:G pairs. The DNA sequences d(GTAC**AAA**GTAC) and d(GTAC**GAG**GTAC) also form hairpins with analogous tight-turn loops, containing a single adenosine residue, that are closed by sheared A:A and G:G pairs, respectively; the solution structure of the d(GTAC**AAA**GTAC) 11-mer hairpin has been rigorously determined by NMR distance geometry methods (43). Because of the small molecular size of this undecamer, its well-resolved NMR spectra, and abundant distance constraints from the A(H2) protons and stereospecifically assigned H5′/H5″ protons, the rmsd between 30 distance geometry structures was only a. 1.15 Å before energy minimization. The backbone ε, β and γ torsion angles were also constrained from ^{31}P-^{1}H correlation experiments combined with the in-plane 'W' rule (82,83). The ζ and α dihedral angles were excluded from the *trans* domain, based on the observation of no unusually upfield-shifted ^{31}P resonances (84). These backbone torsion angle restraints were found to be quite useful for converging the distance geometry structures for this single-A loop hairpin. The structure of the d(GTAC**AAA**GTAC) hairpin is shown in two different views in Plate IX. In the left view, the kink in the backbone of the loop region is indicated by a blue arrow and is brought about mainly by a change in torsion angles from $\zeta(g^-)$, $\alpha(g^-)$ to $\zeta(g^+)$, $\alpha(g^+)$ at the turn, and by a change in the γ torsion angle of residue 7A from *gauche$^+$* to *trans*. The excellent 5A/6A base stacking and 6A deoxyribose/7A stacking in the major groove is clear in the right view, while the clear distinction between the major and minor grooves is evident in the left view.

Owing to the unusual stacking interaction between the 3'-adenine base and the deoxyribose rings of the central nucleotide in the d(GNA) and d(AAA) tight-turn motifs, as well as in the d(GNA)$_2$ intercalative motifs, the H4' proton of this sugar experiences extraordinary upfield shifts of *c.* 2 ppm into the H2'/H2'' spectral region (41–43). This is due to the 'stacking' of the unpaired N residue sugar directly over the strong ring current of the following adenine base, producing a maximal upfield ring current shift on this H4' proton (see the right views in Plates VIII and IX). This special stacking of deoxyribose sugar rings with their O4' and H4' atoms directly on top of an adjacent base is not without precedent, and has been observed at CpG steps in Z-DNA (89), in the crystal structures of self-intercalated dimers of cyclic d(ApAp) (85), and in the crystal structures of DNA complexes with drugs (86–88). The stabilizing forces in such interactions has been discussed recently and is believed to arise from O4'-CH hydrogen bonding and an $n \leftrightarrow \pi^*$ interaction (89).

5.3 Sheared pairs and single-residue loops in biology

It should be noted that the single-residue loop hairpins discussed above are required for promoter recognition by the N4 virion RNA polymerase (90,91) and are extensively used by the single-stranded DNA parvoviruses in forming the special 'rabbit's ears' fold-back structure at their 3'-terminus that is necessary for their replication (66,67). In the coliphage N4, the phage-coded RNA polymerase is unable to transcribe normal double-stranded DNA. However, it can transcribe, accurately and efficiently, single-stranded, promoter-containing templates. It was later proposed that a hairpin is formed at the promoter region which, after forming an 'activated promoter' with *E. coli* single-stranded-binding proteins, can be recognized by virion RNA polymerase. Analysis of a large series of mutant promoters revealed that a particular template secondary structure, specifically a hairpin with a 5–7 bp stem and a three-base loop, was required for recognition of the promoter by the RNA polymerase. These hairpin structures contain a core sequence consisting of either 5'-GC**GAA**GC or 5'-GC**GGA**GC. The sequence 5'-GC**GAA**GC has already been found to be extraordinarily stable (79), existing predominantly as a compact hairpin with a single-A residue loop that is closed by a sheared G:A pair (42,79). The 5'-GC**GGA**GC sequence should also be able to form a similar hairpin structure with a single-G loop, or at least a hairpin–duplex equilibrium, if it is embedded in a relatively long sequence. By itself, this heptamer is more likely to form a duplex structure with an intercalated d(GGA)$_2$ motif (41,42,44).

The single-residue hairpin loop is even more prevalent in the single-stranded DNA parvoviruses (66,67). The double-hairpin secondary structure proposed on the basis of sequence data contains a GAA single-A loop motif at the end of one hairpin and a proposed CAG loop at the end of the other hairpin of the Y-shaped 'rabbit's ear' at the 3'-terminus—as well as an AAA loop capping the proposed hairpin at the 5'-end of the genome, as shown at the top of Fig. 11.4. Careful examination of the parvovirus sequence reveals that the proposed hairpin containing the CAG loop can actually be rearranged into a hairpin with a tight-turn d(GCA) motif that contains the same number of G:C pairs in the stem, as shown at the bottom of Fig. 11.4. A consequence of this rearrangement is that there will now be two unpaired nucleotides

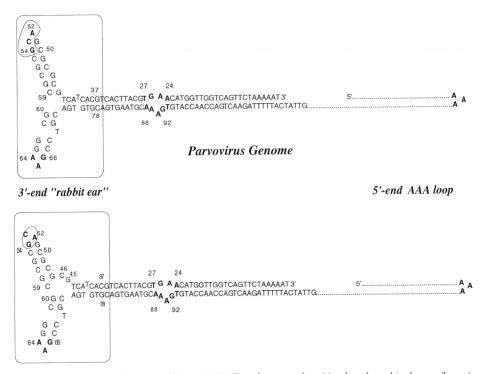

Fig. 11.4. The originally proposed (upper) (66,67) and suggested revision based on this chapter (lower) secondary structure of the parvovirus genome. Note the reversed numbering system (starting from the 3′-end) used by the authors. The 3′-end of the genome contains a Y-shaped 'rabbit ear' structure, in which a triplet d(GAA) loop motif is present (shown in bold). The d(CAG) loop proposed in the upper diagram can be rearranged into an alternative hairpin containing a d(GCA) loop and two unpaired residues at the three-way junction (lower figure) that we believe would be a more stable form, because of the strong hairpin-promoting ability of the d(GCA) motif and the reported stabilizing effect of two unpaired bases at the three-way junction (93,94). A conserved GAA:GA motif is present at positions 91, 90, 89 and positions 26, 25. The structure of this motif has recently been solved and contains an unpaired A residue intercalated between two sheared G:A pairs. A d(AAA) loop motif is also present near the 5′-end of the genome. The parvovirus genome thus contains several variants of sheared PU:PU pairing motifs, which may serve important biological recognition functions in this single-stranded DNA virus.

(G45–C46 or G70–C71—note the inverted numbering from the 3′-end) in the three-way junction region. We believe that the latter folding will be the more stable form because of the strong hairpin-promoting ability of the GCA triplet motif and furthermore, the two unpaired residues would stabilize the DNA three-way junction (92,93). If this rearranged model is correct, then the fold-back secondary structure of single-stranded DNA parvoviruses has made great use of sheared mismatch G:A and A:A pairing by forming a single-C loop d(GCA) motif, a single-A loop d(GAA) motif, a single-A loop d(AAA) motif, and an intercalated GAA:GA motif in the stem of the Y-shaped 3′-fold-back (see Fig. 11.4). Because these motifs are so well conserved in all parvovirus genomes, these special structures and their motifs are probably important for the autonomous DNA replication of these viruses.

6. Sheared GA mismatches closing two–residue hairpin loops

Since the sugar–sugar geometry in sheared Pu:Pu pairs allows them to be easily bridged by a single nucleotide loop, then one would surmise that two-residue loops closed by sheared G:A pairs should be at least as a stable, and, indeed, such loop structures in DNA do occur. Hirao *et al.* (94) first reported a rather stable structure for the short single-stranded DNA sequence GC*GAAA*GC that moves unexpectedly faster than other DNA octamers during electrophoresis in denaturing polyacrylamide gels containing 7M urea. At that time a hairpin structure consisting of a two-G:C base paired stem with a GAAA loop was proposed. A related, similar result was obtained later when Sandusky *et al.* (95) reported that the DNA 15-mer d(TGC **GGCA**GCAACAGC) reverse transcript of T-cell leukaemia virus 2 was found not to be suitable as a DNA sequencing primer. This pentadecamer was found to have the electrophoretic mobility of a nonamer on denaturing acrylamide gels in 8M urea. An NMR-derived structural model indicated that this sequence (with a melting temperature *c.* 75°C) forms a hairpin structure with three Watson–Crick pairs in a 4 bp stem, a two-residue loop closed by a sheared G:A pair, and five dangling bases (95). Studies on a separate series of DNA oligomers indicated that 5′-C**GXYA**G sequences have melting temperatures 18–20°C higher than DNA oligomers with 5′-C**AXYG**G sequences (95). The identities of the XY loop residues were found to play a negligible role in determining the stability of these DNA hairpins. The orientation dependence of the flanking Watson–Crick context was implicated by the observation that 5′-**G**GXYA**C** hairpins were found to have *c.* 20°C lower melting temperatures than 5′-**C**GXYA**G** sequences.

The two-residue loop GC*GAAA*GC sequence has also been found at the replication origin of the single-stranded DNA phage G4 (96). However, the hairpin structure proposed by the authors contains a five-residue AAAGC loop closed by a G:C pair, instead of the two-residue AA loop closed by a G:A pair that is formed by the isolated GCGAAAGC octamer sequence. Hirao *et al.* (97) constructed a series of hairpins containing different stem lengths, and found that at least eight base pairs in the stem were required to maintain the AAAGC loop hairpin structure proposed by Sims *et al.* (96). When the number of base pairs in the stem was less than eight, the secondary structure rearranged to form the hairpin with a two-residue AA loop closed by a sheared G:A pair (97). In fact, they found that the 'GAAA loop' hairpin with only four Watson–Crick pairs in the stem was even more stable than the AAAGC loop hairpin containing eight Watson–Crick pairs in the stem. This suggests that the sheared G:A-closed, two-residue loop is a very stable motif that readily forms stable hairpin structures when flanked by self-complementary sequences. In the case of RNA, similar stable, two-residue r(GNRA) loop hairpins closed by sheared G:A pairs have also been found both in solution (98) and in the crystalline state (99).

7. Conclusion

From the above discussion, it is clear that d(Py**GNA**Pu) motifs containing sheared G:A pairs have established a special niche of their own in the Watson–Crick based B-DNA world. When there is no N residue between adjacent G:A pairs, such sequences form the rather stable d(Py*GA*Pu)$_2$ motif. A similar stable sheared d(Py*GA*Pu): d(Py*AA*Pu) antiparallel duplex motif has also been found (100). When N = C, T, or A, such sequences form d(Py*GCA*Pu), d(Py*GTA*Pu), or d(Py*GAA*Pu) tight-turn single-residue loop motifs, although bimolecular d(Py*GTA*Pu)$_2$ and d(Py*GAA*Pu)$_2$ intercalative duplex forms are also detected when N = T or A. The d(Py*GCA*Pu) case is special in that no intercalative d(Py*GCA*Pu)$_2$ duplex is observed. For the N = A case, stable d(Py*AAA*Pu) and d(Py*GAG*Pu) tight-loop motifs can also be formed. The two-residue, G:A-closed d(Py*GAAA*Pu) hairpin loop motif has also been found to be stable. In contrast, when N = G, the stable, symmetrical bimolecular d(Py*GGA*Pu)$_2$ motif is formed, with two unpaired guanosine residues intercalated between adjacent sheared G:A pairs. When N = A on one strand only, a stable non-symmetrical d(Py*GAA*Pu):d(Py*GA*Pu) motif with only a single unpaired adenosine residue inter-calated between adjacent sheared G:A pairs can be formed. It is important to note that, except for closing single- or double-residue hairpin loops, no single isolated sheared G:A pairs have been found in duplexes or hairpin stems. Only adjacent pairs of sheared G:A pairs in (Py*GA*Pu)$_2$ motifs are found to be stable and compatible with flanking B-DNA geometry. Although G:A mismatches with other types of geometry can coexist with Watson–Crick B-DNA, they all introduce some sort of destabiliza-tion to the parent B-form duplexes. The cross-strand stacking, either in the form of base-base or base-deoxyribose stacking, is a characteristic of these tandem pairings and is thus of critical importance to the stabilization of these mismatched DNA sequences. It will be of interest in the future to ascertain to what extent, if any, these Pu:Pu mismatches participate in the structure and expansion of GCA repeats in spin-obulbar muscular atrophy (101) myotonic dystrophy (102) and Huntington's disease (103), in the structure and expansion of GCG repeats in Fragile X mental retardation (104), and in the structure and expansion of GAA repeats in Friedreich's ataxia (105).

Acknowledgements

The National Chung-Hsing University, Taiwan is thanked for financial support (to S.-H. C.) and B.R.R. gratefully acknowledges the support of NIH grants GM32681 and GM52883.

References

1. Loeb, L.A. and Kunkel, T.A. (1982) *Ann. Rev. Biochem.* 52, 429.
2. Kramer, B., Kramer, W. and Fritz, H.-J. (1984) *Cell* **38**, 879.
3. Dohet, C., Wanger, R. and Radman, M. (1985) *Proc. Natl. Acad. Sci USA* **82**, 503.
4. Fazakerley, G.V., Quignard, E., Woisard, A., Guschlbauer, W., van der Marel, G.A., van Boom, J.H., Jones, M. and Radman, M. (1986) *EMBO J.* **5**, 3697.
5. Su, S.-S., Lahue, R.S., Au, K.G. and Modrich, P. (1988) *J. Biol. Chem.* **263**, 6829.
6. Donohue, J. (1956) *Proc. Natl. Acad. Sci. USA* **42**, 60.
7. Poltev, V.I. and Shulyupina, N.V. (1986) *J. Biomol. Struct. Dynamics* **3**, 739.
8. Prosser, J., Frommer, M., Paul, C. and Vincent, P.C. (1986) *J. Mol. Biol.* **187**, 145.
9. Zakian, V.A. (1995) *Science* **270**, 1601.
10. Fazakerley, G.V. and Boulard, Y. (1995) *Meth. Enzymol.* **261**, 145.

11. Manor, H., Rao, B.S. and Martin, R.G. (1988) *J. Mol. Evol.* **27**, 96.

12. Palecek, E. (1991) *CRC Crit. Rev. Biochem. Mol. Biol.* **26**, 151.

13. Yee, H.A., Wong, A.K., van de Sande, J.H. and Ratter, J.B. (1991) *Nucl. Acids. Res.* **19**, 949.

14. Aharoni, A., Baran, N. and Manor, H. (1993) *Nucl. Acids. Res.* **21**, 5221.

15. Huertas, D., Bellsolell, L., Casasnovas, J.M., Coll, M. and Azorin, F. (1993) *EMBO J.* **12**, 4029.

16. Rippe, K., Fritsch, V., Westhof, E. and Jovin, T.M. (1992) *EMBO* **11**, 3777.

17. Lee, J.S. (1990) *Nucl. Acids Res.* **18**, 6057.

18. Baran, N., Lapidot, A. and Manor, H. (1991) *Proc. Natl. Acad. Sci. USA* **88**, 507.

19. Lapidot, A., Baran, N. and Manor, H. (1989) *Nucl. Acids Res.* **17**, 883.

20. Sen, D. and Gilbert, W. (1988) *Nature* **334**, 364.

21. Sundquist, W.I. and Klug, A. (1989) *Nature* **342**, 825.

22. Ferrer, N., Azorin, F., Villasante, A., Gutierrez, C. and Abad, J.P. (1995) *J. Mol. Biol.* **245**, 8.

23. Huertas, D. and Azorin, F. (1996) *Biochemistry* **35**, 13125.

24. Robinson, H., van der Marel, G.A., van Boom, J.H. and Wang, A.H.-J. (1992) *Biochemistry* **31**, 10510.

25. Robinson, H. and Wang, A.H.-J. (1993) *Proc. Natl. Acad. Sci. USA* **90**, 5224.

26. Robinson, H., van Boom, J.H. and Wang, A.H.-J. (1994) *J. Am. Chem. Soc.* **116**, 1565.

27. Wang, Y. and Patel, D.J. (1994) *J. Mol. Biol.* **242**, 508.

28. Cruse, W.B.T., Egert, E., Kennard, O., Sala, G.B., Salisbury, S.A. and Viswamitra, M.A. (1983) *Biochemistry* **22**, 1833.

29. Coll, M., Solans, X., Font-Altaba, M. and Subirana, J.A. (1987) *J. Biomol. Struct. Dynamics* **4**, 797.

30. Gehring, K., Leroy, J.-L. and Gueron, M. (1993) *Nature* **363**, 561.

31. Reid, D.G., Salisbury, S.A., Brown, T. and Williams, D.H. (1985) *Biochemistry* **24**, 4325.

32. Topping, R.T., Stone, M.P., Brush, C.K. and Harris, T.M. (1988) *Biochemistry* **27**, 7216.

33. Li, Y., Zon, G. and Wilson, W.D. (1991) *Proc. Natl. Acad. Sci, USA* **88**, 26.

34. Chou, S.-H., Cheng, J.-W. and Reid, B.R. (1992) *J. Mol. Biol.* **228**, 138.

35. Cheng, J.-W., Chou, S.-H. and Reid, B.R. (1992) *J. Mol. Biol.* **228**, 1037.

36. Hare, D.R. and Reid, B.R. (1986) *Biochemistry* **25**, 5341.

37. Ebel, S., Lane, A.N. and Brown, T. (1992) *Biochemistry* **31**, 12083.

38. Saenger, W. (1984) *Principles of Nucleic Acid Structure.* Springer, New York.

39. Wilson, W.D., Dotrong, M.H., Zuo, E.T. and Zon, G. (1988) *Nucl. Acids Res.* **16**, 5137.

40. Chou, S.-H., Cheng, J.-W., Fedoroff, O. and Reid, B.R. (1994) *J. Mol. Biol.* **241**, 467.

41. Chou, S.-H., Zhu, L. and Reid, B.R. (1994) *J. Mol. Biol.* **244**, 259.

42. Chou, S.-H., Zhu, L. and Reid, B.R. (1996) *J. Mol. Biol.* **259**, 445.

43. Chou, S.-H., Zhu, L., Gao, Z., Cheng, J.-W. and Reid, B.R. (1996) *J. Mol. Biol.* **264**, 981.

44. Zhu, L., Chou, S.-H. and Reid, B.R. (1995) *J. Mol. Biol.* **254**, 623.

45. Zhu, L., Chou, S.-H. and Reid, B.R. (1996) *Proc. Natl. Acad. Sci. USA* **93**, 12159.

46. Lane, A., Martin, S.R., Ebel, S. and Brown, T. (1992) *Biochemistry* **31**, 12087.

47. Lane, A., Ebel, S. and Brown, T. (1994) *Eur. J. Biochem.* **220**, 717.

48. Katahira, M., Sato, H., Mishima, K., Uesugi, S. and Fujii, S. (1993) *Nucl. Acids Res.* **21**, 5418.

49. Green, K.L., Jones, R.L., Li, Y., Robinson, H., Wang, A.H., Zon, G. and Wilson, W.D. (1994) *Biochemistry* **33**, 1053.

50. Chou, S.-H., Cheng, J.-W., Fedoroff, O.Y., Chuprina, V.P. and Reid, B.R. (1992) *J. Amr. Chem. Soc.* **114**, 3114.

51. Wing, R., Drew, H., Takano, T., Broka, C., Tanaka, S., Itakura, K. and Dickerson, R.E. (1980) *Nature* **287**, 755.
52. Shlomai J. and Kornberg, A. (1980) *Proc. Natl. Acad. Sci. USA* **77**, 799.
53. Freemont, P.S., Lane, A.N. and Sanderson, M.R. (1991) *Biochem. J.* **278**, 1.
54. Steitz, T.A. (1990) *Q. Rev. Biophys.* **23**, 205.
55. Pley, H.W., Flaherty, K.M. and McKay, D.B. (1994) *Nature* **372**, 68.
56. Scott, W.G., Finch, J.T. and Klug, A. (1995) *Cell* **81**, 991.
57. Uhlenbeck, O.C. (1987) *Nature* **328**, 596.
58. Nowak, R. (1994) *Science* **263**, 608.
59. Blackburn, E.H. (1991), *Nature* **350**, 569.
60. Moyzis, R.K., Torney, D.C., Meyne, J., Buckingham, J.M., Wu, J.R., Burks, C., Sirotkin, K.M. and Goad, W.B. (1989) *Genomics* **4**, 273.
61. Williamson, J.R., Raghuraman, M.K. and Cech, T.R. (1989) *Cell* **59**, 871.
62. Grady, D.L., Ratliff, R.L., Robinson, D.L., McCanlies, E.C., Meyne, J. and Moyzis, R.K. (1992) *Proc. Natl. Acad. Sci. USA* **89**, 1695.
63. Jaishree, T.N. and Wang, A.H.-J. (1994) *FEBS Lett.* **347**, 99.
64. Catasti, P., Gupta, G., Garcia, A.E., Ratliff, R., Hong, L., Yau, P., Moyzis, R.K. and Bradbury, E.M. (1994) *Biochemistry* **33**, 3819.
65. Rich, A., Quigley, G.J. and Wang, A.H.-J. (1979) in *Stereodynamics of Molecular Systems*, (Sarma, R.H. ed.), pp. 315–330. Pergamon Press, New York.
66. Astell, C.R., Chow, M.B. and Ward, D.C. (1985) *J. Virol.* **54**, 171.
67. Astell, C.R., Smith, M., Chow, M.B. and Ward, D.C. (1979) *Cell* **17**, 691.
68. Bates, G. and Lehrach, H. (1994) *BioEssays* **16**, 277.
69. Sutherland, G.R. and Richards, R.I. (1995) *Proc. Natl. Acad. Sci. USA* **92**, 3636.
70. Richards, R.I. and Sutherlands, G.R. (1994) *Nature Genet.* **6**, 114.
71. Gacy, A.M., Goellner, G., Juranic, N., Macura, S. and McMurray, C.T. (1995) *Cell* **81**, 533.
72. Yu, A., Dill, J., Wirth, S.S., Huang, G., Lee, V.H., Haworth, I.S. and Mitas, M. (1995) *Nucl. Acids Res.* **23**, 2706.
73. Yu, A., Dill, J. and Mitas, M. (1995) *Nucl. Acids Res.* **23**, 4055.
74. Chen, F.-M. (1991) *Biochemistry* **30**, 4472.
75. Mitas, M., Yu, A., Dill, J. and Haworth, I.S. (1995) *Biochemistry* **34**, 12803.
76. Tinoco, I.J., Uhlenbeck, O.C. and Levine, M.D. (1971) *Nature* **230**, 362.
77. Haasnoot, C.A.G., Hilbers, C.W., van der Marel, G.A., van Boom, J.H., Singh, U.C., Pattabiraman, N. and Kollman, P.A. (1986) *J. Biomol. Struct. Dynamics* **3**, 843.
78. Germann, M.W., Kalisch, B.W., Lundberg, P., Vogel, H.J. and van de Sande, J.H. (1990) *Nucl. Acids Res.* **18**, 1489.
79. Hirao, I., Kawai, G., Yoshizawa, S., Nishimura, Y., Ishido, Y., Watanabe, K. and Miura, K. (1994) *Nucl. Acids Res.* **22**, 576.
80. Zhu, L., Chou, S.-H., Xu, J. and Reid, B.R. (1995) *Nature Struct. Biol.* **2**, 1012.
81. Vissel, B., Nagy, A. and Choo, K.H.A. (1992) *Cytogenet. Cell Genet.* **61**, 81.
82. Sarma, R.H., Mynott, R.J., Wood, D.J. and Hruska, F.E. (1973) *J. Am. Chem. Soc.* **95**, 6457.
83. Altona, C. (1982) *Rec. Trav. Chim. Pays-Bas* **101**, 413.
84. Gorenstein, D.G., Schroeder, S.A., Fu, J.M., Metz, J.T., Roongta, V. and Jones, C.R. (1988) *Biochemistry* **27**, 7223.
85. Frederick, C.A., Coll, M., van der Marel, G.A., van Boom, J.H. and Wang, A.H.-J. (1988) *Biochemistry* **27**, 8350.
86. Teng, M.-K., Usman, N., Frederick, C.A. and Wang, A.H.-J. (1988) *Nucl. Acids Res.* **16**, 2671.

87. Kopka, M.L., Yoon, C., Goodsell, D., Pjura, P. and Dickerson, R.E. (1985) *Proc. Natl. Acad. Sci. USA* **82**, 1376.
88. Coll, M., Frederick, C.A., Wang, A.H.-J. and Rich, A. (1987) *Proc. Natl. Acad. Sci. USA* **84**, 8385.
89. Egli, M. and Gessner, R.V. (1995) *Proc. Natl. Acad. Sci. USA* **92**, 180.
90. Glucksmann-Kuis, M.A., Malone, C., Markiewicz, P. and Rothman-Denes, L.B. (1992) *Cell* **70**, 491.
91. Glucksmann-Kuis, M.A., Dai, X., Markiewicz, P. and Rothman-Denes, L.B. (1996) *Cell* **84**, 147.
92. Leontis, N.B., Kwok, W. and Newman, J.S. (1991) *Nucl. Acids Res.* **19**, 759.
93. Leontis, N.B., Hills, M.T., Piotto, M., Ouporov, I.V., Malhotra, A. and Gorenstein, D.G. (1994) *Biophys. J.* **68**, 251.
94. Hirao, I., Nishimura, Y., Naraoka, T., Watanabe, K., Arata, Y. and Miura, K. (1989) *Nucl. Acids Res.* **17**, 2223.
95. Sandusky, P., Wooten, E.W., Kurochkin, A.V., Kavanaugh, T., Mandecki, W. and Zuiderweg, E.R. (1995) *Nucl. Acids Res.* **23**, 4717.
96. Sims, J., Capon, D. and Dressler, D. (1979) *J. Biol. Chem.* **254**, 12615.
97. Hirao, I., Ishida, M., Watanabe, K. and Miura, K. (1990) *Biochim. Biophy. Acta*, **1087**, 199.
98. Heus, H.A. and Pardi, A. (1991) *Science* **253**, 191.
99. Pley, H.W., Flaherty, K.M. and McKay, D.B. (1994) *Nature* **372**, 111.
100. Maskos, K., Gunn, B.M., LeBlanc, D.A. and Morden, K.M. (1993) *Biochemistry* **32**, 3583.
101. La Spada, A.R., Wilson, E.M., Lubahn, D.B., Harding, A.E. and Flschbeck, K.H. (1991) *Nature* **352**, 77.
102. Brook, J.D., McCurrach, M.E., Harley, H.G., Buckler, A.J., Church, D., Aburatani, H., Hunter, K., Stanton, V.P., Thirion, J.P., Hudson, T. *et al.* (1992) *Cell*, **68**, 799.
103. The HD Collaborative Research Group (1993) *Cell* **72**, 971.
104. Verkerk, A.J.M.H., Pieretti, M., Sutcliffe, J.S., Fu, Y.H., Kuhl, D.P., Pizzuti, A., Reiner, O., Richards, S., Victoria, M.F., Zhang, F.P. *et al.* (1991) *Cell* **65**, 905.
105. Campuzano V., Montermini, L., Molto, M.D., Pianese, L., Cossee, M., Cavalcanti, F., Monros, E., Rodius, F., Duclos, F., Monticelli, A. *et al.* (1996) *Science* **271**, 1423.

12

Structures of nucleic acid triplexes

Edmond Wang and Juli Feigon

Department of Chemistry and Biochemistry, University of California, Los Angeles, CA 90095, USA

1. Introduction

1.1 Historical background

During the past 10 years, interest in triple-stranded nucleic acids (triplexes) has been considerable. Hundreds of research papers have been published on various aspects of triple-stranded nucleic acids: these include studies ranging from stability and sequence requirements to potential biological roles and pharmaceutical uses. Although the idea of triple-stranded nucleic acids (triplexes) seems contemporary, it actually dates back to 1953 when Linus Pauling proposed a three-stranded model for standard DNA that placed the phosphate backbones in the centre of a helix and the bases facing outwards (1,2). This model was considered unlikely for electrostatic reasons, and the correct double-stranded structure was proposed by Watson and Crick (3). A few years later, Felsenfeld and coworkers discovered that nucleic acids actually could form a three-stranded structure (4,5). Using RNA homopolymers, they demonstrated that poly r(A) and poly r(U), besides forming a duplex (presumably with Watson–Crick-type base pairs), could also form a complex with a stoichiometry of 2:1 [poly r(U)]:[poly r(A)]. They proposed that the second poly r(U) strand was binding in one of the grooves of the duplex and base pairing to either the adenine strand, the uracil strand, or both strands, but was most likely Hoogsteen base pairing (6) to the adenine strand.

Additional triple-stranded RNAs were soon discovered. Poly r(C) was shown to form a 1:2 complex with poly r(G) (7) or with guanosine oligonucleotides (8,9). Lipsett demonstrated that a duplex of Watson–Crick base pairs was formed initially, and that the second guanine strand bound to this duplex. Lipsett proposed that the second guanine strand bound to the duplex in the major groove via Hoogsteen pairing to the first guanine strand, forming G:GC triplets.[1] Interestingly, if the pH was lowered to pH 5 or 6, poly r(C) was shown to form a 2:1 complex with guanosine oligonucleotides preferentially (9,10), rather than a 1:2 complex. Lipsett proposed that the second poly r(C) strand was protonated at the N3 (imino) positions, enabling it to form Hoogsteen base pairs with the poly r(G) strand, to form C^+:GC triplets (9).

The formation of triple-stranded DNA structures was subsequently demonstrated for poly d(I):poly d(I):poly d(C) (11,12) and poly d(T):poly d(T):poly d(A) (13).

---

[1] In the triplet convention used here, the base preceding the ':' is the third strand base. The base following the ':' is the Watson–Crick strand base that is involved in the (reverse) Hoogsteen pairing with the third strand base. The last base is the other Watson–Crick base.

Mixtures of RNA and DNA polymers were also shown to form triplexes, such as poly d(T):poly r(A):poly d(T) (14) and poly d(I):poly d(I):poly r(C) (12). Many additional triplexes were formed from polymers of modified bases of both DNA and RNA (reviewed in refs 15 and 16).

These studies, although limited to homonucleotide polymers, demonstrated that triplex formation was sequence dependent [e.g. poly d(T) would form a triplex with poly d(A):poly d(T), but not with poly d(G):poly d(C)]. It was not initially recognized that many of these triplexes could be isomorphous structures. In 1968, Morgan and Wells showed that a stable triplex could be constructed from the mixed sequence polynucleotide poly r(UC):poly d(GA):poly d(TC) (17). This important result led to the realization that the sequence requirements for triplex formation could be generalized to homopurine–homopyrimidine sequences (at least in the case of a homopyrimidine third strand). Furthermore, these results indicated that at least two of the triplets (U:AT and C+:GC) were likely to be isosteric. Morgan and Wells also found they could inhibit RNA polymerase by using an exogenous RNA strand to target a duplex DNA sequence, suggesting a possible biological role for triple-stranded nucleic acids.

1.2 Overview of triplex motifs and triplet base-pairing schemes

A triplex is formed by the binding of a third nucleic acid strand in the major groove of a duplex nucleic acid. The duplex must generally be composed of a homopurine–homopyrimidine sequence (for reviews, see refs 18–24). There are two types of triplexes that can be distinguished by the orientation and composition of their third strand. In this review, we define the two types of triplexes as: (i) parallel motif triplexes, also known as the pyrimidine, or YRY, motif triplexes; and (ii) the antiparallel motif triplexes, also known as the purine, or RRY, motif triplexes.

The parallel motif is generally characterized by a homopyrimidine third strand that binds parallel to the homopurine strand of the duplex (central strand of the triplex). This motif has two canonical triplets: a T:AT triplet, which is formed when a thymine in the third strand Hoogsteen base pairs with an adenine in the duplex (Fig. 12.1a), and a C$^+$:GC triplet, which is formed when a protonated cytosine in the third strand Hoogsteen base pairs with a guanine in the duplex (Fig. 12.1b). The third strand cytosine is protonated at the N3(imino) position; thus parallel triplex formation has a pH dependence and is favoured by low pH (9,25,26). Triplexes of the parallel motif will be referred to as PTs (parallel triplexes).

The antiparallel motif is characterized by a homopurine third strand that binds antiparallel to the homopurine strand of the duplex (central strand of the triplex). This motif has three canonical triplets: a G:GC triplet, which is formed when a guanine in the third strand reverse Hoogsteen base pairs with an guanine in the duplex (Fig. 12.2a); an A:AT triplet, which is formed when an adenine in the third strand reverse Hoogsteen base pairs with an adenine in the duplex (Fig. 12.2b); and a T:AT triplet (which is different from the T:AT triplet in the parallel motif), which is formed when a thymine in the third strand reverse Hoogsteen base pairs with an adenine in the duplex (Fig. 12.2c). Unlike the canonical triplets in the parallel motif, the three canonical triplets in the antiparallel motif are not isosteric (Fig. 12.3), leading to possible backbone distortions when the triplets are intermixed. The antiparallel triplexes,

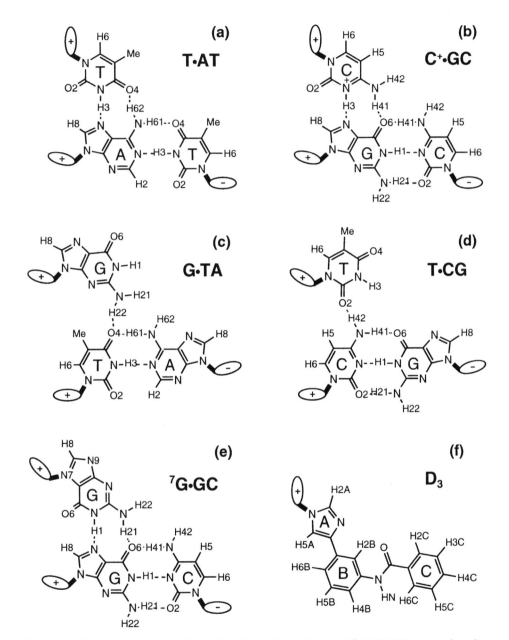

Fig. 12.1. Triplet base-pairing schemes for the parallel triplex motif: (a) T:AT canonical triplet, (b) C⁺:GC canonical triplet, (c) G:TA mismatch triplet, (d) T:CG mismatch triplet, (e) ⁷G:GC triplet, and (f) D₃ base. The mismatch triplets are the ones for which high resolution triplex structures containing them have been determined.

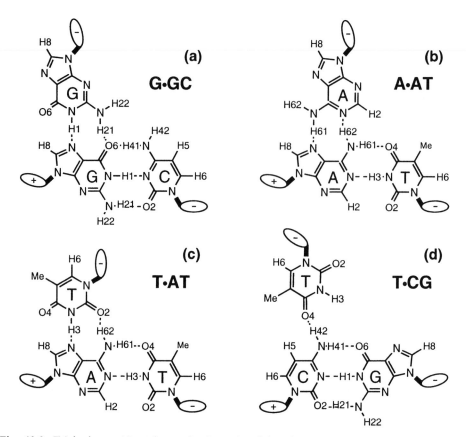

Fig. 12.2. Triplet base-pairing schemes for the antiparallel triplex motif: (a) G:GC canonical triplet, (b) A:AT canonical triplet, (c) T:AT canonical triplet, and (d) T:CG mismatch triplet.

unlike the parallel triplexes, are not pH dependent. Triplexes of the antiparallel motif will be referred to as APTs (antiparallel triplexes).

The sequence definition of the parallel and antiparallel motifs is somewhat complicated by the fact that a third strand composed of a mixture of guanines and thymines can switch polarity from antiparallel to parallel depending on the ratio of guanines to thymines and on the number of GpT and TpG steps, (27–29). Also, antiparallel sequences can sometimes be forced parallel (30), and parallel sequences can sometimes be forced antiparallel (21,31).

1.3 Biological significance of triplex formation

Triplexes readily form under physiological conditions, but it remains unclear what biological roles triplexes play *in vivo*, if any. In this section, we give a brief overview of possible biological roles and evidence for the formation of triplexes *in vivo*. More extensive discussions can be found in other reviews (24,32).

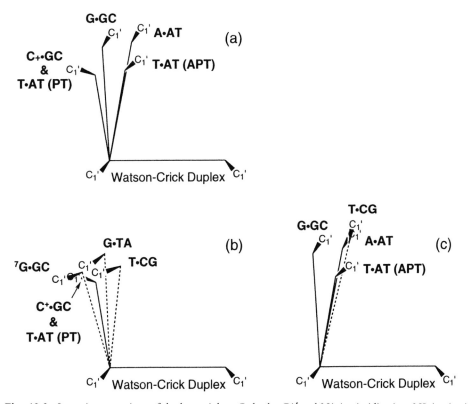

Fig. 12.3. Isosteric comparison of the base triplets. Only the C1′ and N1 (pyrimidines) or N9 (purines) atoms are shown as the tails and heads of small arrows, indicating the orientation of the glycosidic bond. The Watson–Crick base pairs for each triplet have been superimposed to illustrate the relative position of the third strand base. The mismatch triplets are connected by dashed lines. (a) Superposition of the canonical triplets for both the parallel and antiparallel triplex motifs. (b) Superposition of the canonical and mismatch triplets for the parallel triplexes. (c) Superposition of the canonical and mismatch triplets for the antiparallel triplexes.

Transcriptional regulation is a possible and obvious role for triplexes. In fact, from the first discovery of triple-stranded nucleic acids, it was suggested that a biologically important three-stranded complex could be constructed from single-stranded RNA and duplex DNA (4). Also, early on, triplexes were shown to be stable under physiological conditions and to inhibit various enzymes such as RNA polymerase (17), DNAase I (33), and RNAase (33). An early proposal by Miller and Sobell ingeniously hypothesized that certain repressors may be ribonucleoproteins where the sequence specificity is conferred by a complementary mRNA capable of forming a triple-stranded complex with DNA (34).

One important point to consider about the biological relevance of triplexes is that triplex formation requires a run of purines in one strand (and pyrimidines in the other). This requirement would appear to restrict triplexes to a minor role *in vivo*. However, homopurine–homopyrimidine tracts turn out to be statistically three to four times over-represented in eukaryotic (35) and eukaryotic viral genomes (36) (for review see ref. 18). Homopurine–homopyrimidine tracts are *not*, however, over-represented in prokaryotic (35) or bacteriophage genomes (36), implying that triplexes may have a biological role in eukaryotes, but not in prokaryotes. Many homop-urine–homopyrimidine tracts are found upstream of genes (for examples, see refs 37–40) or within genes (41,42), consistent with the hypothesis that triplexes play a role in transcriptional regulation. These homopurine–homopyrimidine tracts are often hypersensitive to single-stranded nucleases (43–47), indicating that they may adopt a non-B-DNA conformation.

In 1986, Frank-Kamenetskii and coworkers made a discovery that had important implications for the *in vivo* existence of triplexes, and consequently sparked much of the renewed interest in triplexes. They showed that an intramolecular triplex could be formed at homopurine–homopyrimidine mirror repeat sequences in negatively super-coiled plasmids (25,48,49). Their proposed triple-stranded structure also explained the S1 nuclease hypersensitivity of these sequences (25,48,49). Many of the homopurine–homopyrimidine sequences discovered so far are in fact mirror repeats (50), strongly suggesting that triplexes do form *in vivo*. These triple-stranded structures, dubbed H-DNA,[2] are created when one half of the mirror repeat dissociates into sep-arate homopurine and homopyrimidine single strands, followed by the homopyrimi-dine strand folding back on to the remaining duplex half of the mirror repeat and binding in the major groove to from a parallel triplex (Fig. 12.4). The remainder of the homopurine strand is single-stranded and accounts for the S1 nuclease sensitivity. It is also possible for the homopurine strand to fold back, forming an antiparallel triplex (★H-DNA) (51–53). In fact, there are numerous possible H-DNA-related structures (for reviews see refs 24 and 54). Divalent cations appear to be required for

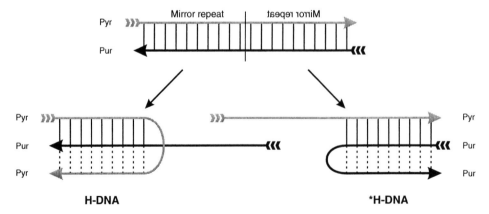

Fig. 12.4. Intramolecular folding scheme for H-DNA and ★H-DNA illustrating the parallel and anti-parallel orientation of the third strands. The homopyrimidine strand is in grey and the homopurine strand is in black. The thin solid lines represent Watson–Crick base pairing and the thin dashed lines represent (reverse) Hoogsteen base pairing.

[2]The 'H' indicates a proton because the original sequences required low pH, or, the 'H' stands for 'Hoogsteen' or 'hinged'.

the favourable formation of ★H-DNA over H-DNA (51–53). *In vivo* footprinting results support the hypothesis that H- and ★H-DNA exist in living cells (55,56).

The requirement for protonation at N3 of cytosines in the third strand of PTs suggests that only small amounts of triplex may exist at neutral pH. However, several groups have demonstrated that parallel triplexes can be formed at physiological pH (57–60), and can be further stabilized at neutral pH by replacing the third strand cytosines with the naturally occurring 5-methylcytosine (^{m5}C) (61–64). H-DNA can also be stabilized at physiological pH by increasing the negative superhelical density (25,49,58), and has been shown to form at physiological pH and superhelical density (60). Although the antiparallel triplexes are not pH dependent, their base triplets are not isosteric (Fig. 12.3), which make them less stable.

The identification of triplex-binding proteins provides some evidence that triplexes are actually used *in vivo*. Two triplex-binding proteins with apparent molecular weights of 55 kDa have been purified from HeLa cells (65,66). Both proteins preferentially bind triple-stranded DNA over duplex DNA, but they have differing binding affinities and sequence specificities, indicating that they are different proteins (65,66).

Immunostaining of mouse and human chromosomes with monoclonal antibodies that specifically recognize triplexes revealed a strong correlation with chromosome banding patterns, which suggests that triplex formation is cell cycle dependent and may play a role in chromosome condensation and organization (67–69). Direct binding of the two antibodies to nuclei was also shown to inhibit cell growth, specifically at the end of S phase and during G2 (70), while control antibodies had no effect. This further suggests a role for triplexes in chromosome condensation.

Several investigations have addressed the question of whether triplexes are involved in transcriptional regulation. For example, a homopurine–homopyrimidine sequence was constructed within a β-galactosidase gene without altering the amino acid sequence by taking advantage of codon degeneracy (71). In *E. coli*, the total enzyme activity was reduced roughly 80% relative to the wild type sequence. Truncated transcripts were also isolated that were of the length predicted between the start site and the putative intramolecular triplex site. Another set of experiments indicate that triplexes may regulate transcription via a *trans*-acting factor. In mouse cells, poly d(G) sequences upstream of a gene were found to act as enhancers (72). However, the enhancement was strongly dependent on the length of the poly d(G) tract; d(G)$_{27-30}$ enhances transcription, whereas d(G)$_{35}$ does not. *In vitro*, when poly d(G) tracts are inserted into a mildly supercoiled plasmid, tracts 32 bp or longer form H-DNA, while tracts 30 bp or shorter do not. Furthermore, if another plasmid with a poly d(G) tract is cotransformed with the first, then the second plasmid can reduce expression of the first if the poly d(G) tract is 30 bp, but not 35 bp (72). This suggests that *in vivo* the longer poly d(G) tracts are forming intramolecular triplexes and blocking a *trans*-acting transcription factor.

Triplexes may also play a role in homologous recombination. For example, homologous recombination was induced between two direct repeats by active transcription *in vivo* when a poly d(G):poly d(C) sequence was inserted between them (73). This effect was proposed to be caused by formation of ★H-DNA, which may then bring two remote sequences together to stimulate homologous recombination (73). Similarly, using an *in vivo* plasmid–plasmid recombination assay, it has been demonstrated that plasmids containing potential H-DNA (or *H-DNA)-forming sequences undergo increased recombination, while those plasmids containing nearly identical sequences that are unlikely to form intramolecular triplexes, have no effect (74). The single strand produced by H-DNA formation may be acting as an invading single strand in homologous recombination (74).

Recently, a palindromic homopurine–homopyrimidine sequence required for the lytic replication of the Epstein–Barr viral genome has been studied and shown to be capable of forming *H-DNA (75). Mutations in the sequence inhibit both replication and *H-DNA formation. Surprisingly, complementary mutations that restore the palindrome also restore replication and *H-DNA formation. This result suggests that it is not the sequence, but the palindrome (and its resulting structure) that is important for replication. This is the strongest evidence yet of a biological role for triple-stranded nucleic acids.

1.4 Triplexes as therapeutics

The second area of research that has caused the resurgence of interest in nucleic acid triplexes is in the use of triplexes as potential therapeutics. This work has been the motivation for many of the studies on sequence specificity and alternate triplets in the structures reviewed here. Here we present a brief overview of efforts to target particular sequences of duplex DNA through triplex formation, with an emphasis on modifications used to help extend and improve sequence specificity (for more complete reviews, see refs 76–78).

The general idea behind most potential pharmaceutical applications of triplexes is the targeting of sequences within or upstream of a particular gene via triplex formation, in order to block transcription and thus repress protein production at the DNA level. This strategy is sometimes called the antigene strategy and is analogous to the antisense strategy, except that the antigene strategy operates at the transcriptional level instead of the translational level. Many researchers have proven the feasibility of the antigene strategy *in vitro* by inhibiting transcription of specific genes (for examples, see refs 79–81). In addition to blocking RNA polymerase, a number of other DNA-binding proteins can also be inhibited, such as DNA polymerase (82–84), various endonucleases (85–88), methylase (89), NF-κB (88), and other transcription factors (89–91). The antigene strategy has also been shown to work *in vivo* (92–96). An antigene RNA oligonucleotide has been constitutively expressed from a vector and shown to reach a high steady state concentration *in vivo* (97).

Another pharmaceutical application is to create artificial nucleases by coupling triplex-forming oligonucleotides to DNA cleaving reagents, such as Cu^{II}–phenanthroline (98,99), Fe^{II}–EDTA (57,100), or an azidoproflavine derivative (101). A photoactive nuclease has also been produced using ellipticine (102,103). These artificial nucleases are much more specific than naturally occurring nucleases because their recognition sequences are potentially much longer, although the cleavage position is less precise. They have been used to cleave a single site in the bacteriophage λ genome (104) and a yeast chromosome (105). Because of their ability to generate large DNA fragments, these artificial nucleases are potentially useful in chromosome mapping.

In vitro triplex applications include uses as an artificial ligase (106), as a sequence-specific mutagen (107–109), as an agent for the purification and isolation of specific double-stranded DNA sequences (110–114), and as an agent to purify PCR products (115). Triplex-forming oligonucleotides have also been used as sequence probes, as in Southern blotting, except that they hybridize to double-stranded DNA (116). By covalently linking proteins to single-stranded oligonucleotides, triplex formation has been used to target proteins to specific DNA sequences (117).

A major shortcoming of triplexes as therapeutics is that their formation requires a homopurine–homopyrimidine sequence. The usefulness of the antigene strategy would be greatly expanded if one could form triplets with all four base pairs. In an effort to target base pair inversions within a homopurine–homopyrimidine sequence, several groups have investigated the stability and selectivity of aternate triplets in both the parallel motif (118–125) and the antiparallel motif (126,127). Even *in vitro* selection techniques have been employed to identify mismatch triplets (117).

For the parallel motif, every investigation found that the canonical T:AT and C^+:GC triplets are the most stable. Non-canonical bases in the third strand have essentially the same effect on triplex stability that mismatches do on DNA triplexes. Thus, an AT base pair is most effectively recognized by a T, although an A can also form a reasonably stable A:AT triplet (117,118,122,123). Similarly, a GC base pair is most effectively recognized by a protonated C, although A^+:GC, T:GC, and G:GC triplets can all form depending on the conditions, but are considerably less stable (117,118,120–123). Several studies have investigated the stability of triplets formed by third strand recognition of TA and CG base pairs (where the pyrimidine base is in the 'homopurine' strand). It was found that to recognize a TA base pair, a G in the third strand forms the most stable triplet (119,122–125) under most conditions, but in some cases a C:TA triplet is more stable (120). A CG base pair is recognized by both a T and a C, although the resulting triplets are not very stable, and results vary as to which triplet is more stable (120,124,125). Thus, for the parallel motif, all four base pairs can be targeted, but at a cost in triplex stability when TA and, especially, CG base pairs are involved. In addition, some specificity is lost since a T can recognize both an AT and a CG base pair, and a C can recognize both a GC and a CG base pair. The effect of these alternate triplets on triplex stability depends on the sequence context, i.e. which triplets are neighbouring, and the pH (128–131).

For the antiparallel motif, the canonical G:GC, A:AT, and T:AT triplets were found to be by far the most stable (126,127). However, an A can also bind to a GC base pair, and T can bind to a CG base pair. For a TA base pair, there are no stable triplets (a T:TA triplet is the least destabilizing) (126,127). Thus, the lack of a stable TA-containing triplet means that the antiparallel motif is more restricted than the parallel motif in the sequences that can be targeted. In addition, a large amount of specificity is lost since a T can recognize both an AT or a CG base pair and an A can recognize both an AT and a GC base pair. This means that although T:AT and A:AT are the most stable triplets, both the T and the A can form reasonably stable aternate triplets.

Another tactic to avoid the sequence restrictions is to simply bypass a homopurine–homopyrimidine inversion site by inserting an abasic residue in the third strand. Abasic substitutions generally yield stable triplexes in both the parallel (124,132) and the antiparallel motif (133), but decrease specificity (124). An imidazole has also been used at an inversion site with some success (134). Several nucleotide derivatives and synthetic bases have also been studied. One purine derivative, 7-deaza-2′-deoxyxanthosine (dzaX), was used as a T analogue in APTs (135). Unlike a T:AT triplet, a dzaX:AT triplet is isosteric with a G:GC triplet. The dzaX:AT-containing triplex was found to be 100-fold more stable than the equivalent T:AT-containing triplex. Another purine derivative, deoxynebularine, was found to recognize both CG and AT base pairs in APTs (136). Two synthetic bases, 3-(2-deoxy-β-D-ribofuranosyl)-2-methyl-8-(N′-n-butylureido)naphthyl[1,2]imidazole and 1-(2-deoxy-β-D-ribofura-nosyl)-4-(3-benzamidophenyl)imidazole (D₃), have been designed to recognize CG base pairs in PTs by forming specific hydrogen bonds with both the guanine and cyto-sine (137,138). The latter base (D₃) was shown to intercalate and mimic a complete triplet instead of hydrogen bonding to a Watson-Crick base pair (139,140).

For target sequences that consist of a homopurine tract followed by a homopyrimi-dine tract, one could envision binding two oligonucleotides: one to each homopurine tract on opposite strands, linked together at the junction where they meet. This would effectively allow triplexes to be targeted to a wider range of sequences. This alternative strand-targeting strategy has been tested by several groups, using either a 5′–5′ linkage (141) or a 3′–3′ linkage (142), and has been found to be effective in forming stable triplexes. Another method for alternate strand triplex formation comes from the real-ization that third strands of PTs and APTs bind to their homopurine strands in oppo-site orientations. Therefore, alternate strands can be targeted by alternating the triplex motif, without changing the third strand polarity, and without the use of unnatural 5′–5′ or 3′–3′ linkages. Such triplexes do indeed form stable complexes (143–146), greatly enhancing the sequence space that can be targeted by triplexes. An interesting variation of the alternate strand triplexes takes advantage of the fact that APTs can switch polarity depending on the TpG and GpT content (27–29). Thus, certain sequences can be targeted on alternate strands using solely the antiparallel motif (27). A significant limitation of all of the cross-over triplexes is that a longer total sequence is required to form a stable triplex.

Although the parallel motif can more successfully and specifically target a larger sequence space than the antiparallel motif, the parallel motif has the disadvantage of being considerably less stable at physiological pH owing to the need to protonate the cytosines. Much research has been devoted to reducing the pH dependence of PTs. The most common method of increasing triplex stability at neutral pH is the substitu-tion of ^{m5}C for cytosine (28,61,62,89,147,148). An alternative is to use uncharged C$^+$ analogues, such as pseudo-isocytidine (149,150), 1-(2-deoxy-β-D-ribofuranosyl)-3-methyl-5-amino-1H-pyrazolo[4,3-d]pyrimidin-7-one (P1) (151–153), N7-glycosy-lated guanine (154), 8-oxoadenine (155,156), or 4-amino-5-methyl-2,6-pyrimidone (157). A G:GC triplet in a parallel motif has also been used instead of a C$^+$ (28).

Triplexes can also be stabilized by the use of intercalating agents (158). However, mismatches are also stabilized (159), so some specificity may be lost. Most of these intercalating agents also bind preferentially to triplexes over duplexes. Some intercala-tors, such as echinomycin and actinomycin D, have no effect on triplex stability and may even destabilize triplexes (160). When conjugated to the end of a triplex-forming oligonucleotide, the intercalating agent anchors the oligonucleotide, which greatly improves triplex stability in both parallel (147,159–161) and antiparallel (161) motifs.

The minor groove binding drugs netropsin and berenil destabilize triplexes (162–164), which suggests that the minor groove environment may be significantly different in triplexes and duplexes.

Another method of stabilizing triplexes is to attach a cross-linking reagent to the third strand and induce a covalent linkage to the duplex. Several alkylating agents have been shown to cross-link to one of the duplex strands (165,166). For example, psoralen, an intercalator, can dramatically stabilize triplexes by cross-linking to both strands of the duplex (167–169). Psoralen-cross-linked triplexes have even been demonstrated *in vivo*, and shown to inhibit transcription more effectively than the free oligonucleotide. However, the psoralen–triplex-mediated inhibition may be abolished in only a few hours by cellular DNA repair systems (170).

Finally, in order to improve various pharmaceutical characteristics, such as resistance to degradation, increased stability, or increased cellular uptake, researchers have investigated triplex formation using modified backbones (171). The most common target for modification is the phosphates, which have been replaced with phosphorothioates (172–175), methyl phosphonates (176,177), or guanidinium groups (178). Oligonucleotides containing phosphorothioate or methyl phosphonate linkages form triplexes, but are less stable than their DNA counterparts (179–182). Mixed guanidinium–DNA triplexes are extremely stable (with melting temperatures as high as 100°C) because of the favourable interactions between the positively charged guanidinium moieties and the negatively charged phosphates (178). Other modifications involve the ribose sugars, such as replacing the 2′-hydroxyl in RNA with a 2′-O-methyl (149,183,184), or replacing the ribose with a riboacetal group (185), or bicyclic ring structure (186). These modifications result in triplexes that are more stable than their DNA counterparts (185–188). The entire backbone has been replaced by a peptide-like structure, a so-called peptide nucleic acid or PNA (189–191) (for review, see ref. 192). When targeted to duplex DNA, PNAs bind to their complementary DNA strand via strand invasion, followed by a second PNA strand binding in the major groove, to form a PNA:DNA:PNA triplex (189,193). These complexes are extremely stable.

2. Structures of parallel triplexes

2.1 Background

Of the two triplex motifs, the PTs are the best characterized. Early researchers had speculated on the base-pairing schemes of the canonical T:AT and C$^+$:GC triplets (9,10,34,194). In 1973, the first direct structural information about PTs was provided by X-ray fibre diffraction experiments on poly rU:poly rA:poly rU and its DNA equivalent (195–197). The X-ray diffraction data illuminated the correct base-pairing scheme of the U:AU and T:AT triplets (Fig. 12.1a). The T:AT (U:AU) triplet consists of a standard Watson–Crick AT base pair with a second thymine binding in the major groove via Hoogsteen hydrogen bonding to the adenine in a parallel orientation. The third strand has an *anti* glycosidic conformation.

The fibre diffraction data indicated that the triple helices had a helical rise similar to B-DNA but a low twist and a deep major groove similar to A-DNA (Table 12.1; ref.

Table 12.1. Structural data on parallel triplexes

Triplex sequence[a]	Method	DNA:RNA composition	χ angle	Sugar pucker[b]	Rise[c] (Å)	Twist[c] (°)	x-disp[c] (Å)	Inclin[c] (°)	Reference (see notes)
B-DNA[d]	–	–	anti	S	3.4	36	-0.7	-6.0	–
A-DNA[d]	–	–	anti	N	2.6	33	-5.4	19.1	–
YRY1	NMR	D:DD	anti	S[e]	3.1 [3.3]	31 [31]	-4.0	2.4	1
YRY2	NMR	D:DD	anti	S[f]	3.3	31	-2.8	4.9	2
GTA	NMR	D:DD	anti	S[g]	3.5 [3.4]	31 [31]	-1.8 [-1.9]	3.8	3
TCG	NMR	D:DD	anti	S[h]	3.5 [3.4]	32 [32]	-2.2 [-2.1]	2.8	4
N7G	NMR	D:DD	anti	S	3.1 [3.1]	29 [29]	-3.1 [-2.9]	5.3	5
DTA	NMR	D:DD	anti	S[i]	3.2 [3.2]	31 [31]	-1.2 [-1.4]	-1.9	6
PAT	NMR	D:DD[j]	anti	S[k]	3.0	30	-3.0	13.9	7
(T:AT)₁₂	X-ray	D:DD	anti	S	3.3 [3.3]	28 [28]	[2.6][l]	[-5.0]	8
Mixed	X-ray	D:DD	anti	S	3.2 [3.2]	30 [30]	[2.5][l]	[5.0]	8
Poly(T:AT)[d]	X-ray	D:DD	anti	N[m]	3.3 [3.3] [3.2]	30 [30] [33]	-3.6	5.4 [8.5] [10.0]	9
Poly(C+:IC)	X-ray	D:DD	anti	N[m]	–	–	–	–	9
Mixed	FTIR	D:DD[n]	–	S & N[o]	–	–	–	–	10
(AG)₃	FTIR/Raman	D:DD[p]	anti	S & N[q]	–	–	–	–	11
Poly(C+:GC)	FTIR	D:DD	–	S & N[r]	–	–	–	–	12
Poly(C+:GC)	FTIR	D:DD	–	S & N[r]	–	–	–	–	13
Poly(T:AT)	FTIR	D:DD	–	S	–	–	–	–	14
Poly(T:AT)	FTIR	D:DD	–	S	–	–	–	–	15
(GA)ₙ	Gel	D:DD	–	–	–	[32]	–	–	16
UAT	NMR	R:DD	anti	S	3.2	29	-2.0	-3.2	17
Mixed	FTIR	R:DD[y]	–	S & N[t]	–	–	–	–	10
Mixed	FTIR	R:DD	–	S & N[u]	–	–	–	–	18
Poly(C+:GC)	FTIR	R:DD	–	N	–	–	–	–	13
Poly(U:AT)	FTIR	R:DD	–	S & N	–	–	–	–	15
(GA)ₙ	Gel	R:DD	–	–	–	[32]	–	–	16
Poly(T:AT)	FTIR	D:RD	–	S & N	–	–	–	–	15
Poly(U:AU)	X-ray	R:DR	anti	N[m]	[3.0]	[33]	–	[12.0]	9
Poly(U:AT)	FTIR	R:RD	–	s & N	–	–	–	–	15
Poly(C+:GC)	FTIR	D:RR	–	N	–	–	–	–	13
Poly(T:AU)	FTIR	D:RR	–	N	–	–	–	–	15
Mixed	NM	R:RR	anti	N	–[r]	–	–	–	19
Mixed	NMR/FTIR	R:RR	anti	N	–[r]	–	–	–	20,21

Table 12.1. Continued

Triplex sequence[a]	Method	DNA:RNA composition	χ angle	Sugar pucker[b]	Rise[c] (Å)	Twist[c] (°)	x-disp[c] (Å)	Inclin[c] (°)	Reference (see notes)
Poly(U:AU)[w]	X-ray	R:RR	anti	N[m]	[3.0]	[30]	—	[12.0]	9
Poly(U:AU)[x]	X-ray	R:RR	anti	N[m]	[3.0]	[33]	—	[12.0]	9
Mixed	FTIR	R:RR	—	Z	—	—	—	—	18
Poly(C+:GC)	FTIR	R:RR	—	Z	—	—	—	—	13
Poly(U:AU)	FTIR	R:RR	—	Z	—	—	—	—	15

[a] The table is grouped by DNA, DNA:RNA hybrid, and RNA triplexes. The high resolution NMR structures are listed in bold by the names used in the text; Otherwise, the composition of the triplexes are given as triplets [e.g. (T:AT)n] or as the sequence in the purine strand [e.g. (AG)n] or as 'mixed' for more complex sequences.

[b] S-type sugar pucker is a C2'-endo conformation. N-type sugar pucker is a C3'-endo conformation.

[c] Helical parameters are calculated using 'Curves' v5.1 with a linear helical axis for the duplex alone. Values in [ ] are the parameters and may have been calculated using a different method and/or helical axis.

[d] The standard A-DNA, B-DNA, and triplex parameters were calculated from structures created in Insight 95 (Biosym). The Biosym parameters are from X-ray fibre diffraction data (refs. 22–24).

[e] The pyrimidine strands have some N-type character.

[f] The pyrimidine strands have some N-type character, especially the cytosines.

[g] The guanine in the G:TA mismatch triplet is N-type.

[h] The thymine in the T:CG mismatch triplet is N-type.

[i] The sugar puckers are generally S-type, except for one thymine which is adjacent to the interca-lation site. The thymine methyl groups in the third strand have been replaced by propyne groups.

[k] The cytosines in the third strand have some N-type character.

[l] The authors report a positive displacement from the helical axis, which probably represents a neg-ative x-displacement in the standard helical parameter convention.

[m] The sugar conformations were assumed based on the helical structure.

[n] The cytosines in the third strand are either unmodified or have been replaced with 5-MeC.

[o] The sugars in the T:AT triplets are S-type, while the sugars in the C+:GC triplets have a ratio of 1:2 S-:N-type.

[p] The cytosines in neither, either, or both pyrimidine strands have been replaced with 5-MeC.

[q] The purines are all S-type, but the overall triplex has a ratio of 2:1 S-:N-type.

[r] The guanines are all S-type and the cytosines in both pyrimidine strands are N-type.

[s] The riboses in the third strand are 2'O-methylated.

[t] The sugar pucker in the duplex pyrimidine strand is S-type, while the purine strand and the RNA third strand are N-type.

[u] The sugar puckers are mostly N-type with some S-type. Binding of the RNA third strand changes the DNA duplex sugar puckers from all S-type to mostly N-type.

[v] NOE cross-peak patterns in the NMR data are typical of A-form helices (ref. 25).

[w] Data were collected at 92% relative humidity.

[x] Data were collected at 75% relative humidity.

1. Bornet, O. and Lancelot, G. (1995) J. Biomol. Struct. Dyn. 12, 803–14.
2. Tarköy, M., Phipps, A.K., Schultze, P. and Feigon, J. (1998) Biochemistry 37, 5810–19.
3. Radhakrishnan, I. and Patel, D.J. (1994) Structure 2, 17–32.
4. Radhakrishnan, I. and Patel, D.J. (1994) J. Mol. Biol. 241, 600–19.
5. Koshlap, K.M., Schultze, P., Brunar, H., Dervan, P.B. and Feigon, J. (1997) Biochemistry 36, 2659–68.
6. Wang, E., Koshlap, K.M., Gillespie, P., Dervan, P.B. and Feigon, J. (1996) J. Mol. Biol. 257, 1052–69.
7. Phipps, A.K., Tarköy, M., Schultze, P. and Feigon, J. (1998) Biochemistry 37, 5820–30.
8. Liu, K., Sasisekharan, V., Miles, H.T. and Ragunathan, G. (1996) Biopolymers 39, 573–89.
9. Arnott, S., Bond, P.J., Selsing, E. and Smith P.C.J. (1976) Nucleic Acids Res. 3, 2459–70.
10. Dagneaux, C., Liquier, J. and Taillandier, E. (1995) Biochemistry 34, 16618–23.
11. Fang, Y., Bai, C., Wei, Y., Lin, S.B. and Kan, L. (1995) J. Biomol. Struct. Dyn. 13, 471–82.
12. Ouali, M., Letellier, R., Adnet, F., Liquier, J., Sun, J.-S., Lavery, R. and Taillandier, E. (1993) Biochemistry 32, 2098–103.
13. Akhebat, A., Dagneaux, C., Liquier, J. and Taillandier, E. (1992) J. Biomol. Struct. Dyn. 10, 577–88.
14. Howard, F.B., Miles, H.T., Liu, K., Frazier, J., Raghunathan, G. and Sasisekharan, V. (1992) Biochemistry 31, 10671–7.
15. Liquier, J., Coffinier, P., Firon, M. and Taillandier, E. (1991) J. Biomol. Struct. Dyn. 9, 437–45.
16. Shin, C. and Koo, H.S. (1996) Biochemistry 35, 968–72.
17. Gotfredsen, C.H., Schultze, P. and Feigon, J. (1998) J. Am. Chem. Soc. 120, 4281–9.
18. Liquier, J., Taillandier, E., Klinck, R., Guittet, E., Gouyette, C. and Huynh-Dinh, T. (1995) Nucleic Acids Res. 23, 1722–8.
19. Holland, J.A. and Hoffman, D.W. (1996) Nucleic Acids Res. 24, 2841–8.
20. Klinck, R., Guittet, E., Liquier, J., Taillandier, E., Gouyette, C. and Huynh-Dinh, T. (1994) FEBS Lett. 355, 297–300.
21. Klinck, R., Liquier, J., Taillandier, E., Gouyette, C., Huynh-Dinh, T and Guittet, E. (1995) Eur. J. Bioch. 233, 544–53.
22. Arnott, S., Hukins, D.W. and Dover, S.D. (1972) Biochem. Biophys. Res. Comm. 48, 1392–9.
23. Arnott, S. and Hukins, D.W. (1972) Biochem. Biophys. Res. Comm. 47, 1504–9.
24. Arnott, S. and Selsing, E. (1974) J. Mol. Biol. 88, 509–21.
25. Heus, H.A. and Pardi, A. (1991) J. Am. Chem. Soc. 113, 4360–1.

9). Consequently, it was concluded that the structure was A-DNA-like, and, therefore, the sugars must adopt a C3′-*endo* confromation (195–197).

Attempts to crystallize oligomer (instead of polymer) DNA have been unsuccessful to date, and have also yielded fibre-type diffraction results (198,199). The data indicate similar helical parameters, but with a significantly lesser x-displacement of the duplex from the helical axis (Table 12.1; ref. 8).

Arnott *et al.* also collected data on a poly dC:poly dI:poly dC triplex (197). The base-pairing scheme for a C:IC triplet presumably involves a protonated third strand C and would be predicted to be similar to a C^+:GC triplet. It was already known that the T:AT and C^+:GC triplets were likely to be isosteric (17). Some early NMR experiments provided the first direct evidence for the existence of the protonated cytosine in the C^+:GC triplet (200,201). However, more recent NMR studies have provided the first definitive proof for not only the protonated cytosine, but also the details of the C^+:GC base-pairing scheme (26) (Fig. 12.1b). Rajagopal and Feigon (202) were able to observe the protonated cytosine imino proton directly and to define the base-pairing scheme from magnetization transfer pathways. This and subsequent NMR studies (26,59,202,203) also confirmed the base-pairing scheme of the T:AT triplet.

2.2 DNA parallel triplexes

2.2.1 Helix morphology

NMR studies have provided the few high resolution structures of PTs to date (see Section 2.5). Although many of these structures contain a mismatch triplet or modified bases, they still have several features in common.

(a) The sugar puckers are generally S-type (with the exception of some sugars of bases involved in alternate/mismatch triplets). However, the cytosines have some N-type character.

(b) The base pair axial rises are in the typical range of B-DNA.

(c) The triplexes are generally slightly underwound as indicated by the low helical twists, even when compared with A-DNA.

(d) The x-displacement from the helical axis is intermediate between A- and B-DNA.

(e) The inclination of the base pairs is small, similar to B-DNA.

(f) The third strand nucleotides are in the *anti* conformation.

An important distinction between the more recent NMR results and the early X-ray results is that the conformations of the sugars are generally C2′-*endo* (S-type) (121,204–206) as opposed to the early assumption that they were C3′-*endo* (N-type) (195–197). This is supported by IR data as well (Table 12.1). Of the helical features, only the helical twist is similar to A-DNA. The x-displacement is roughly −2 Å, which is greater than that of B-DNA but is not as dramatic as the −5.4 Å x-displacement found in A-DNA. Visually, the triplexes resemble B-DNA more than A-DNA (Plate X). The structural characteristics of triplexes can probably best be interpreted within the context of a B-form DNA duplex with a nucleic acid ligand binding in the major groove. For example, the x-displacement from the helical axis is quite significant and can be considered a consequence of having to accommodate the third strand in the major groove. If one assumes that the sugar pucker of DNA prefers the S-type conformation and that the Hoogsteen base pairing of a DNA third strand tends to maintain the B-DNA-like rise and base pair inclination, then the increased x-displacement can only be accommodated by an unwinding of the helix. These are precisely the results observed experimentally.

A comment on helical parameters[3] is in order here (see Chapter 2 for a detailed discussion of this topic). Depending on the program and analysis method used, the helical parameters can vary significantly. Table 12.2 illustrates how the helical parameters can vary depending on how the global axis is calculated. For this review, the program 'Curves', version 5.1, (208,209) was used to evaluate the helical parameters for all triplexes for which coordinates were available. The Watson–Crick duplex of each triplex is used as the reference point, with the third strand essentially being considered a ligand; this makes direct comparison to A- and B-DNA most meaningful. Therefore, the helical parameters reported in represent only the duplex portions of the triplexes. A linear helical axis was used for the analysis in Tables 12.1 and 12.3 to allow a direct comparison between triplexes.

2.3 RNA parallel triplexes

Since RNA duplexes are A-form, it seems likely that RNA triplexes will also adopt an A-form conformation. Although an RNA triplex structure has yet to be solved, there is much data to support this prediction. The original triplex structure of poly r(U):poly r(A):poly r(U) by Arnott and coworkers has a lower rise and a greater base pair inclination than their DNA triplex, i.e. more A-form-like (195–197) (Table 12.1; ref. 9). Chemical cleavage with Fe^{II}–EDTA produces slightly different cleavage patterns for A- or B-form helices (210). The cleavage pattern for an all-RNA triplex is consistent with an A-form structure (210). Results from IR studies of PTs composed entirely of RNA (Table 12.1) show that only N-type sugar puckers are present (211,212). NMR studies on RNA triplexes have also found only N-type sugar puckers (213,214). More importantly, certain NMR cross-peak patterns are diagnostic of A-form helices (215,216) and these patterns are observed in the NMR spectra of these RNA triplexes (213,214).

2.4 DNA:RNA hybrid parallel triplexes

Relatively little is known about the structure of triplexes formed from combinations of RNA and DNA. X-ray fibre diffraction data on poly r(U):poly d(A):poly r(U) indicates that its structure has more in common with the all-RNA polymer triplexes than with the all-DNA polymer triplexes (197) (Table 12.1). IR data of mixed

---

[3]The helical parameters conform to the conventions defined at the 1988 EMBO workshop on DNA curvature and bending (207).

Table 12.2. Effect of axis calculation on helical parameters

Structure[a]	Rise[b] (Å)		Twist[b] (°)		x-Disp[b] (Å)		Incl[b] (°)	
B-DNA[c]	3.4	(3.4)	36	(36)	−0.7	(−0.7)	−5.9	(−5.9)
A-DNA[c]	2.6	(2.6)	33	(33)	−5.4	(−5.4)	19.1	(19.1)
Triplex[c]	3.3	(3.3)	30	(30)	−3.6	(−3.6)	5.4	(5.4)
Parallel D:DD								
YRY1	3.1	(3.3)	31	(31)	−4.0	(−3.2)	2.4	(−3.6)
YRY2	3.3	(3.4)	31	(31)	−2.8	(−2.5)	4.9	(0.9)
GTA	3.5	(3.4)	31	(31)	−1.8	(−2.0)	3.8	(4.1)
TCG	3.5	(3.4)	32	(32)	−2.2	(−2.1)	2.8	(−0.2)
N7G	3.1	(3.6)	29	(28)	−3.1	(−1.3)	5.3	(−12.5)
DTA	3.2[d]	(3.2)[d]	31[e]	(31)[e]	−1.2	(−1.4)	−1.9	(−1.6)
PAT	3.0	(3.3)	30	(29)	−3.0	(−1.6)	13.9	(0.2)
Parallel R:DD								
UAT	3.2	(3.2)	29	(29)	−2.0	(−1.9)	−3.2	(−6.1)
Antiparallel D:DD								
RRY	3.6	(3.6)	30	(30)	−2.1	(−1.9)	−1.3	(−4.4)

[a] High resolution NMR structures discussed in the text.
[b] Helical parameters are calculated using 'Curves' v5.1 with a linear helical axis applied to the duplex alone. Values in parenthesis were calculated using a best-fit curved axis (refs 1 and 2 below).
[c] The standard A-DNA, B-DNA, and triplex parameters were calculated from structures created in Insight 95 (Biosym). The Biosym parameters are from X-ray fiber diffraction data (refs 3–5 below).
[d] Calculated excluding the base step at the D3 intercalation site (ref. 6 below).
[e] Calculated excluding the 5′ base step (with respect to the purine strand) and the base step at the D3 intercalation site (ref. 6 below).

1. Lavery, R. and Sklenar, H. (1988) *J. Biomol. Struct. Dyn.* **6**, 63–91.
2. Lavery, R. and Sklenar, H. (1989) *J. Biomol. Struct. Dyn.* **6**, 655–67.
3. Arnott, S. and Hukins, D.W. (1972) *Biochem. Biophys. Res. Comm.* **47**, 1504–9.
4. Arnott, S., Hukins, D.W. and Dover, S.D. (1972) *Biochem. Biophys. Res. Comm.* **48**, 1392–9.
5. Arnott, S. and Selsing, E. (1974) *J. Mol. Biol.* **88**, 509–21.
6. Wang, E., Koshlap, K.M., Gillespie, P., Dervan, P.B. and Feigon, J. (1996) *J. Mol. Biol.* **257**, 1052–69.

DNA:RNA triplexes show mixtures of S- and N-type sugar puckers (211,212,217) (Table 12.1), indicating that these triplex structures may be a mixture of B- and A-forms, depending on RNA content. Chemical cleavage of mixed DNA:RNA triplexes with Fe[II]–EDTA produces two families of cleavage patterns corresponding to B- and A-form helices (210). Both the IR and chemical cleavage studies find a general trend towards increasing A-form characteristics with increasing RNA content. The latter study also finds that the identity of the purine strand, DNA or RNA, correlates with the helical conformation, B- or A-form, respectively. This finding is also supported by a recent NMR structure of an R:DD intramolecular triplex (UAT triplex in Table 12.1), which is most similar to B-form with S-type sugar puckers for the DNA strands (218). Interestingly, several studies have found that the D:RD and D:RR strand combinations do not form stable triplexes (184,219–221). This result, combined with the idea that the identity of the purine strand determines the helical con-

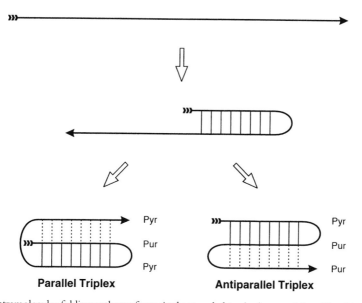

Fig. 12.5. Intramolecular folding pathway from single-stranded to duplex to triplex. The thin solid lines represent Watson–Crick base pairing and the thin dashed lines represent (reverse) Hoogsteen base pairing. The arrows are drawn 5′ to 3′ from tail to head.

formation, suggests that the A-form triplex typical of all-RNA triplexes cannot accommodate a DNA third strand.

2.5 High resolution structures

All of the high resolution DNA triplex structures solved to date are composed of a single strand that folds to form an intramolecular triplex, as first demonstrated by Sklenár and Feigon (59) (Fig. 12.5). This provides a convenient model system for studying triplex structures, since it guarantees the correct stoichiometry and eliminates most potential problems caused by the formation of alternative structures. The first NMR-based model structure of an intramolecular triplex was published in 1992 (206). This model structure used distance restraints derived from NMR data, but was refined from a starting structure based on the Arnott fibre diffraction structures (197). Subsequently published high resolution structures of triplexes have also been calculated from starting structures (A- and/or B-DNA), or have been calculated from distance geometry generated starting structures.

There are currently five published high resolution parallel motif DNA triplex structures, all of which were solved using NMR (Table 12.1). Three additional triplex structures have recently been solved in our laboratory. All the sequences are given in Fig. 12.6. Two of the structures are composed entirely of canonical T:AT and C^+:GC triplets (222,223) and one contains P:AT and C^+:GC triplets (224), where the P is a thymine with a propyne group at the 5 position instead of a methyl group. These structures are hereafter referred to as YRY1, YRY2, and PAT, respectively. Other triplexes incorporate a modified base or a single mismatch triplet, such as a G:TA

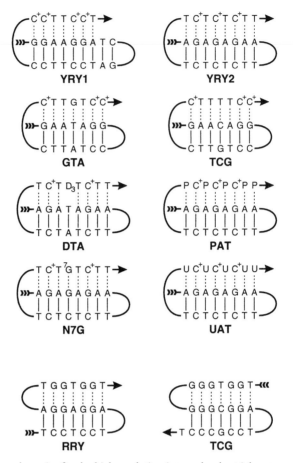

Fig. 12.6. Sequence schematics for the high resolution intramolecular triplex structures. The thin solid lines represent Watson–Crick base pairing and the thin dashed lines represent (reverse) Hoogsteen base pairing. The arrows are drawn 5′ to 3′ from tail to head. The sequences for the eight parallel triplexes are on the top and the two antiparallel triplexes are on the bottom.

triplet (GTA) (225), a T:CG triplet (TCG) (226), a 1-(2-deoxy-D-ribofuranosyl)-4-(3-benzamido)phenyl-imidazole (D$_3$) targeted to a TA base pair (DTA) (227), and an N7-glycosylated guanine targeted to a GC base pair (N7G) (228). The four structures currently in the Brookhaven PDB databank (229) (http://www.pdb.bnl.gov) are GTA (accession #149d), TCG (#177d), DTA (#1wan), and N7G (#1gn7).

While none have been published to date, we have recently determined a high resolution DNA:RNA hybrid triplex structure by NMR. This triplex (UAT triplex in Table 12.1) is composed of an RNA third strand bound to a DNA duplex (218).

Before discussing the structures, we note that NMR refinement has certain strengths and weaknesses. NMR is very good at determining sugar pucker and distinguishing between a *syn* or *anti* glycosidic conformation. Also, in theory, NMR can accurately define the backbone angles β, γ, δ, and ε. However, in practice, β and γ require assignment of the H5′ and H5″ protons and ε is often ambiguous owing to the periodicity of the coupling constants. The α and ζ backbone angles cannot be determined reliably by NMR, except qualitatively through the phosphorus chemical shift (230). Therefore, backbone angles determined by NMR may be influenced more by the energy potentials used in the structure calculations than by real data and must be taken with the proverbial grain of salt. Some helical parameters are reasonably well defined by NMR, such as rise and twist. Other parameters are probably poorly defined by NMR, such as propeller twist.

The *method* used for NMR structure determination is also an important consideration. Starting from model structures tends to lead to smaller root mean squared deviation (rmsd) than distance geometry. Precision (manifested as rmsd) is often mistaken for accuracy. The rmsd can be manipulated by refining 1000 structures and showing only the three lowest energy structures. There are other factors, such as the inclusion of explicit water molecules or ions during the calculation. Since methods for structure determination of nucleic acids have evolved during the time period in which these structures were solved, the reader is advised to consider the methodology used when evaluating the finer details of the structures. (See also Chapter 8).

2.5.1 Canonical triplexes, YRY1 and YRY2

The two 'canonical' triplexes, YRY1 and YRY2 (Plate X and Fig. 12.6), are the basis of comparison for the other mismatch/modified triplexes. The triplexes form regular helices whose general parameters have already been described (Table 12.1). It should be noted that these helical parameters are measured for the duplex portions of the triplexes. Interestingly, the sugar puckers of YRY1 and YRY2 are not completely S-type. Only the purines of YRY1 are completely S-type, while many of the pyrimidines have a partial N-type character (222). In studies of YRY2 (223) and in previous NMR studies of a related triplex (205,206), the cytosines in both pyrimidine strands have significant N-type character. This observation is supported by an IR study in which the T:AT triplets are S-type, but the C$^+$:GC triplets have a 1:2 ratio of S- to N-type sugar pucker (217). Another IR study has found that the purines are completely S-type and that there is an overall 2:1 ratio of S- to N-type sugar puckers (231).

2.5.2 The GTA triplex

Two of the high resolution triplex structures, GTA and TCG, contain a single mismatch triplet. The G:TA triplet was identified by chemical probing and UV spectroscopy as the most stable mismatch triplet (119,122,123). The detailed base-pairing scheme was first ascertained by NMR (204,232) and is illustrated in Fig. 12.1c. The guanine is base paired to the thymine via a single hydrogen bond between the G [H2(2)] amino proton and the T(O4) oxygen. When the G:TA triplet is compared to either the T:AT and C$^+$:GC triplets (Figs 12.1a, b, c and 12.3b), it can be seen that it is not isosteric with the canonical triplets. The positions of the third strand sugars are all roughly the same in the three triplets, but the orientation of the guanine glycosidic bond in the G:TA triplet is quite different from the orientation of the corresponding glycosidic bonds in the canonical triplets.

In the GTA triplex structure (225) (Fig. 12.6), this difference induces several localized structural perturbations. Measurement of the twist in the third strand (as opposed to the duplex) reveals that the 5′-base step is dramatically overwound while the

3′-base step is dramatically underwound. This is a direct effect of the orientation of the guanine glycosidic bond relative to those of the neighbouring third strand bases. This altered twist affects the base stacking in the third strand. The guanine is stacked completely over the 5′-base and has almost no overlap with the 3′-base. The twist of the 5′-base step in the duplex[4] is also overwound but the 3′-base step is unaffected. The G:TA guanine sugar adopts an N-type pucker, apparently to reduce the backbone distortion caused by the unusual orientation of the guanine glycosidic bond (204,232). The G:TA guanine is tilted out of plane with the TA base pair towards the 3′-triplet, possibly forming a weak hydrogen bond between the guanine amino H2(1) proton and the O4 oxygen of the duplex thymine of the 3′-triplet (225), although there is no direct NMR evidence for such a hydrogen bond.

2.5.3 The TCG triplex

The other mismatch-containing triplex is the TCG triplex (Fig. 12.6), which contains a single T:CG triplet centred within canonical T:AT and C⁺:GC triplets (226). A thymine in the third strand was found to form the most stable triplet with a CG base pair (122,125). The detailed base-pairing scheme of the T:CG triplet was shown to involve a single hydrogen bond between the third strand thymine O2 oxygen and the Watson–Crick cytosine H2(2) amino proton (226) (Fig. 12.1d). Surprisingly, both the position and orientation of the thymine glycosidic bond are very similar to the guanine glycosidic bond of the G:TA triplet (Fig. 12.3b). As a consequence, the structural perturbations observed in the TCG triplex are very similar to those observed in the GTA triplex. Both triplexes have the same third strand twist perturbations: a large overwinding in the base step that is 5′ to the third strand mismatch base, and a large underwinding in the 3′-base step (225,226). All the sugar puckers are generally S-type with the exception of the thymine of the T:CG triplet, which has an N-type sugar pucker, exactly like the guanine in the G:TA triplet. In fact, the two triplexes have nearly identical backbone positions (not shown) and have remarkably similar helical parameters (Table 12.1).

2.5.4 The N7G triplex

The N7G triplex (228) (Fig. 12.6) contains a guanosine that is glycosylated at the N7 position (⁷G) instead of at the N9 position (Fig. 12.1e). This base was designed to be an uncharged analogue of a protonated cytosine, thereby allowing a C⁺:GC triplet to be replaced with a ⁷G:GC triplet, and increase the stability of PTs at physiological pH (154). The designed base-pairing scheme of the ⁷G:GC triplet (154) (Fig. 12.1e) was confirmed by the NMR structure (228). The helical parameters for the N7G triplex are similar to the other PTs, except it has a slightly smaller twist and a slightly larger x-displacement. These two changes may be related since a larger x-displacement will produce a smaller twist (all else being equal). Interestingly, the twist and rise between base pairs are inversely correlated, and display an alternating high and low pattern (228). The ApG and ApA base pair steps have a low twist and a large rise, and the

---

[4]References to the 5′-end of the triplex (or duplex) as a whole are with respect to the central purine strand, e.g. a reference to the triplet at the 5′-end of a triplex refers to the triplet containing the base at the 5′-end of the purine strand.

GpA base pair steps have a high twist and a small rise. This is the same sequence effect observed in duplex DNA (233,234). There are no unusual helical parameters associated with the ^{7}G:GC triplet. Overall the N7G triplex is a very regular structure (Plate XIa).

When the ^{7}G:GC triplet is superimposed with a C$^+$:GC triplet from the N7G triplex, the C1$'$ carbons of the ^{7}G and C$^+$ occupy similar positions (Plate XIb and Fig. 12.3b). However, as in the GTA and TCG triplexes, the third strand glycosidic bond of the mismatch triplet is oriented differently from the canonical third strand glycosidic bonds, although the difference is smaller for the ^{7}G:GC triplet. This perturbation may be partially responsible for the lower stability at pH 5.2 of the N7G triplex relative to a triplex containing a C$^+$:GC triplet (228). However, at pH 7, the distortion caused by the ^{7}G:GC triplet apparently is less destabilizing than the inability to protonate a C$^+$:GC triplet (154).

2.5.5 The DTA triplex

The DTA triplex (227) (Fig. 12.6) has a novel synthetic base, D$_3$, designed to form specific hydrogen bonds with a CG base pair (138). However, chemical footprinting studies found that the D$_3$ base recognizes both TA and CG base pairs, forming D:TA and D:CG triplets (138). NMR experiments revealed that the D$_3$ base was not base pairing via hydrogen bonds, but was intercalating instead (139) (Plate XII).

As expected, a large rise is observed at the intercalation site to accommodate the D$_3$ base. Concomitantly, a large unwinding of the helix is also found at the intercalation site (227). If these distortions caused by the D$_3$ base intercalation are disregarded, then the DTA triplex has helical parameters that are similar to the other PTs (Table 12.1). Another similarity is that the DTA triplex generally has S-type sugar puckers, except for some cytosines in the third strand, as has been observed for other PTs (205,217,222). The third strand thymine that is 5$'$ to the D$_3$ base also adopts an N-type sugar pucker. This conformation positions the D$_3$ base directly over the 3$'$ triplet at the intercalation site, allowing the D$_3$ base to mimic a triplet (Plate XII). The ability of the D$_3$ base to mimic a triplet was unexpected, and suggests that greater triplex stability can be achieved by designing a synthetic base that more closely mimics a triplet.

2.5.6 The propyne triplex

The final DNA triplex, PAT (Fig. 12.6), is one in which the thymine methyl groups in the third strand have been replaced with propyne groups (224). These propyl bases have been shown to enhance both duplex and triplex stability (235). The helical parameters of the PAT triplex conform, in general, to other DNA PTs (Table 12.1). However, the 5$'$-end of the triplex has a significant inclination (6.2°) and makes this end of the triplex resemble A-DNA visually. An opposite inclination (−5.3°) is observed in the 3$'$-end of the triplex, making this end of the triplex resemble B-DNA visually. The unusual variation in inclination might to be an effect of the propyne groups in combination with the triplex sequence. The sequence of the third strand is 5$'$-PCPCPCPP-3$'$. The increase of hydrophobicity resulting from the propyne groups may dehydrate one of the grooves and/or alter the stacking interactions, possibly inducing an A-form inclination, while the two sequential propyne groups at the

3'-end of the helix apparently can produce some steric clash, which induces a reverse inclination.

2.5.7 *The DNA:RNA triplex*

The structure of a triplex (UAT) (Fig. 12.6) composed of a DNA duplex and an RNA third strand has been solved by NMR (218). Surprisingly, the replacement of a DNA third strand with an RNA third strand appears to have very little effect on the triplex structure. The helical parameters are very much like the DNA PTs (Table 12.1). In fact, the inclination may be more B-form then other PTs studied. The sugar pucker conformations for the RNA strand were difficult to evaluate and could only be qualitatively determined (218). Normally, A-form sugar puckers in RNA are indicated by small H1' to H2' coupling constants leading to weak or non-existent cross-peaks in a correlation experiment. In the UAT triplex, the RNA cross-peaks were fairly intense, indicating significant S-type sugar pucker, but the sugar conformation could not be unambiguously determined from the single coupling constant. This apparently contradicts IR studies on R:DD triplexes (212,217), where the sugars have significant N-type sugar puckers (Table 12.1). However, distinctly weaker cross-peaks were observed for the RNA cytosines in the UAT triplex, indicating that they had a more N-type sugar pucker. This may be the preferred conformation for the cytosines in a triplex, since NMR studies on DNA triplexes have also found to be of greater N-type character (204–206), and an IR study on an R:DD triplex composed solely of C^+:GC triplets also finds only N-type sugar puckers (211) (Table 12.1).

3. Structures of antiparallel triplexes

3.1 *Background*

Much less is known about the structure of APTs than PTs. Early UV studies of poly r(C) combined with poly r(G) or oligo r(G) detected a 1:2 complex (7–9). In hindsight, these complexes were undoubtedly forming APTs. APTs were not recognized as a separate triplex motif until recently. In fact, poly r(A):poly r(A):poly r(U) was not discovered until 1987 (236), and the corresponding DNA polymers were not shown to form a triplex until 1995 (237). The A:AU triplex has been the most difficult to characterize and so far has only been found when the poly r(A) strand is ~30–150 bases long (236). The discovery of the three canonical triplets (G:GC, A:AT, and T:AT) sparked the realization that APTs constituted a separate triplex motif (238). The antiparallel strand orientation was also only recently determined (238–241).

For triplexes in which the third strand is antiparallel to the duplex purine strand, there are at least two possible base-pairing schemes for the canonical triplets (238). NMR studies have defined the actual base-pairing scheme as being reverse Hoogsteen with an *anti* glycosidic bond conformation (242,243) (Fig. 12.2). An important difference between the PT canonical triplets and the APT canonical triplets is that the PT canonical triplets are isosteric, while the APT canonical triplets are not (Fig. 12.3a).

3.2 DNA antiparallel triplexes

3.2.1 Helix morphology

There are currently two high resolution NMR structures of DNA APTs (244,245) and an X-ray structure of two stacked G:GC triplets (246). Although the number of structures is limited, some generalities can still be ventured. Overall, the DNA APTs greatly resemble the DNA PTs both in their helical parameters (Tables 12.1 and 12.3) and in their appearance (Plate X). The base pair rise and x–displacement are similar to that of the PTs. The helical twists of one NMR structure (244) and the X-ray structure (246) are comparable to the PTs, but the NMR structure with the T:CG mismatch (245) has a much larger twist (Table 12.3). Results from gel migration studies, which apparently can measure the twist with high precision, support the smaller twist (247) (Table 12.3).

The inclination for the DNA APTs is small and slightly more negative than the DNA PTs and more similar to B–DNA. However, there are only two values of inclination, and only one that we could verify (Table 12.3). The sugar puckers in these structures are predominantly S–type. This is confirmed by IR studies on DNA APTs (30,248) (Table 12.3). One IR study on an APT composed solely of G·GC triplets suggests that the guanine duplex strand may be N–type (248).

3.3 RNA antiparallel triplexes

Very little is known about the structure of RNA or hybrid RNA:DNA APTs. The only structural information comes from an IR/Raman study of poly r(G):poly d(G):poly d(C), which finds that both guanine strands adopt N–type sugar puckers while the cytosine strand appears to be S–type (248).

3.4 High resolution structures

The two high resolution NMR DNA APT structures are composed of (1) canonical G:GC and T:AT triplets (hereafter referred to as RRY) (244) or (2) G:GC, T:AT triplets, plus a single T:CG triplet (TCG) (245). The crystal structure of a triplex (GGC) was extrapolated from two G:GC triplets formed at the ends of a DNA duplex with two overhanging guanines that bind to two GC base pairs from a symmetry-related duplex (246). The coordinates for RRY (accession # 134d-136d) and GGC (#272d) are in the PDB database. A high resolution structure of an A:AT triplet has not yet been solved, although the details of the base-pairing scheme have been determined by NMR (243) (Fig. 12.2b).

3.4.1 Canonical triplex, RRY

The RRY triplex (244) (Plate X and Fig. 12.6) has helical parameters (Table 12.3) very similar to the DNA PTs. However, a number of sequence effects are found in the RRY triplex originating from the fact that the G:GC and T:AT triplets are not isosteric (Fig. 12.3). If twist and rise are measured for the third strand, then the TpG base steps are underwound and have a large rise. Conversely, the GpT base steps are overwound and have a small rise (244). The GpG base steps have an average twist and rise, typical of other DNA triplexes. No TpT base step occurs in the RRY triplex, but

Table 12.3. Structural data on antiparallel triplexes

Triplex sequence[a]	Method	DNA:RNA composition	X angle	Sugar pucker[b]	Rise[c] (Å)	Twist[c] (°)	x-disp[c] (Å)	Inclin[c] (°)	Reference (see notes)
B-DNA[d]	–	–	anti	S	3.4	36	-0.7	-6.0	–
A-DNA[d]	–	–	anti	N	2.6	33	-5.4	19.1	–
RRY	NMR	D:DD	anti	S[e]	3.6 [3.6]	30 [30]	-2.1 [-1.9]	-1.3	1
Mixed[f]	NMR	D:DD	anti	S					2
TCG	NMR	D:DD	anti	S	[3.3][g]	[38]	[-2.9]	–	3
GGC	X-ray	D:DD	anti	S[h]	[3.3]	[30]	[-1.5]	[-4.8]	4
(A:AT)₁₀	FTIR	D:DD	–	S	–	–	–	–	5
Poly(G:GC)	FTIR/Raman	D:DD	anti[i]	S & N[j]	–	–	–	–	6
(GA)ₙ	Gel	D:DD	anti	–	–	[32]	–	–	7
Poly(G:GC)	FTIR/Raman	R:DD	anti[i]	S & N[k]	–	–	–	–	6

[a] The table is grouped by DNA and DNA:RNA hybrid triplexes. The high resolution NMR structures are listed in bold by the names used in the text; Otherwise, the composition of the triplexes are given as triplets [e.g. $(T:AT)_n$] or as the sequence in the purine strand [e.g. $(AG)_n$] or as 'mixed' for more complex sequences.

[b] S-type sugar pucker is a C2'-endo conformation. N-type sugar pucker is a C3'-endo conformation.

[c] Helical parameters are calculated using 'Curves' v5.1 with a linear helical axis for the duplex alone. Values in [ ] are the parameters given in the references and may have been calculated using a different method and/or helical axis.

[d] The standard A-DNA and B-DNA parameters were calculated from structures created in Insight 95 (Biosym). The Biosym parameters are from X-ray fibre diffraction data (refs 8 and 9 below).

[e] The sugar puckers in the final structures are generally S-type, although no data on the sugar conformations were obtained and no direct restraints were used.

[f] Contains a T:CG mismatch triplet.

[g] The helical parameters for the TCG triplex were not independently calculated because the triplex coordinates were not available.

[h] Five of the six sugars are S-type. One cytosine is N-type.

[i] Only the glycosidic angle of the guanines could be determined.

[j] Both the poly (dC) duplex strand and the poly (dG) third strand are S-type, while the poly (dG) duplex strand is N-type.

[k] The poly (dC) strand is S-type, while both the poly (dG) and poly (rG) strands are N-type.

1. Radhakrishnan, I. and Patel, D.J. (1993) *Structure* **1**, 135–52.
2. Dittrich, K., Gu, J., Tinder, R., Hogan, M.E. and Gao, X. (1994) *Biochemistry* **33**, 4111–20.
3. Ji, J., Hogan, M.E. and Gao, X. (1996) *Structure* **4**, 425–35.
4. Vlieghe, D., Van Meervelt, L., Dautant, A., Gallois, B., Précigoux, G. and Kennard, O. (1996) *Science* **273**, 1702–5.
5. Dagneaux, C., Gousset, H., Shchyolkina, A.K., Ouali, M., Letellier, R., Liquier, J., Florentiev, V.L. and Taillandier, E. (1996) *Nucleic Acids Res.* **24**, 4506–12.
6. Ouali, M., Letellier, R., Sun, J.-S., Akhebat, A., Adnet, F., Liquier, J. and Taillandier, E. (1993) *J. Amer. Chem. Soc.* **115**, 4264–70.
7. Shin, C. and Koo, H.S. (1996) *Biochemistry* **35**, 968–72.
8. Arnott, S. and Hukins, D.W. (1972) *Biochem. Biophys. Res. Comm.* **47**, 1504–9.
9. Arnott, S., Hukins, D.W. and Dover, S.D. (1972) *Biochem. Biophys. Res. Comm.* **48**, 1392–9.

since such a base step would consist of like triplets, the twist and rise are expected to be typical of other DNA triplexes. Similar sequence effects might be expected for mixtures of G:GC and A:AT triplets.

The *x*-displacement of the duplex base pairs is larger for the GC base pairs than for the AT base pairs, which correlates with the size of the third strand base. Evidently, the increased size of a G:GC triplet with respect to a T:AT triplet is partially accommodated by displacing the duplex strand.

The sugar conformations of the RRY triplex are unclear from the NMR data. The NOE cross-peak intensities for the H6,H8–H3′ cross-peaks are quite strong, indicating partial N-type sugar pucker. However, the coupling patterns for the H1′–H2′,2″ cross-peaks are neither standard S-type nor N-type (244).

3.4.2 The TCG triplex

Studies on alternate triplets within an APT motif have revealed that a T:CG triplet is the most stable mismatch triplet, although it is significantly less stable than the canonical triplets (126,249). The TCG triplex (Fig. 12.6) is comprised of canonical G:GC and T:AT triplets and a single T:CG triplet (245). The triplex has helical parameters similar to other APTs and PTs (Table 12.3), except for local distortions about the T:CG triplet. The NMR structure reveals that the thymine of the T:CG triplet interacts with the CG base pair via a single hydrogen bond from the C[H4(2)] amino proton to the T(O4) oxygen (Fig. 12.2d), rather than the T(O2) oxygen, as previously predicted (127).

Incorporation of the T:CG triplet has some effect on the third strand conformation. The width of the groove that is formed by the third strand and the (predominantly) purine strand is much wider near the T:CG triplet (245). The helical twist of the third strand is generally the same as the duplex except at the two base steps that involve the thymine of the T:CG triplet. The base step that is 5′ to the thymine, a GpT step, is extremely underwound (5.8°), and the base step that is 3′ to the thymine, a TpG step, is extremely overwound (67.4°) (245). This sequence effect on the twist is the reverse of what is observed in the RRY triplex, where GpT steps are overwound and TpG steps are underwound (244). We can speculate that had the thymine O2 been involved in the hydrogen bond instead of the O4, then the sequence effect on the twist would match the trend found in the canonical triplex. However, by using the O4 oxygen, the thymine sugar is placed further from the helical axis, which more closely matches the position of a guanine sugar in a G:GC triplet (Figs 12.2d and 12.3c).

Comparison of the T:CG triplet in this APT to the T:CG triplet in the PT shows that they differ in their thymine sugar puckers. The thymine sugar in the APT remains in the same conformation as the other sugars, S-type (245), while the thymine sugar in the PT is N-type (226). The two triplets also differ in the atoms involved in the hydrogen bond. The APT thymine utilizes the O4 oxygen, while the PT thymine utilizes the O2 oxygen. Remarkably, if the two triplets are superimposed, the O4 atom of the APT thymine perfectly superimposes on the O2 atom of the PT thymine. In fact, the thymine base in the APT is perfectly related to the PT thymine by a twofold rotation about a pseudo-symmetry axis that runs through the N3 and C6 atoms (245). This symmetry operation places the C1′ atom of the APT thymine in the same position as the PT methyl group, which places the APT sugar further from the helical axis and closer to the analogous position of the guanine sugar in a G:GC triplet. Thus, the

APT and PT T:CG triplets adopt the same base-pairing scheme, except that their thymines utilize pseudo-symmetry-related carbonyl oxygens for the hydrogen bonding, and each thymine sugar is positioned most favourably for the particular triplex motif.

3.4.3 The GGC triplex

The final triplex is extrapolated from a 2.0 Å resolution crystal structure of two tandem G:GC triplets on the end of a duplex (246). The helical parameters generally conform to the helical parameters of the RRY and TCG triplexes (Table 12.3). In the two base triplets from which the triplex structure was calculated, all of the sugars are S-type, except for one of the Watson–Crick paired cytosines. However, the structure of the triplex may be influenced by the duplex/triplex junction and a triplex/triplex junction, where a pair of antiparallel G:GC triplets interact with parallel G:GC triplets. The GGC structure provides very precise detail of the G:GC triplet.

4. PNA triplex structures

A crystal structure of a triplex composed of a homopurine DNA strand and a hairpin homopyrimidine peptide (or polyamide) nucleic acid (PNA) strand (193) has been solved (Plate XIII). PNAs are nucleic acids in which the phosphodiester backbone has been replaced with a peptide backbone (189). When targeted to duplex DNA, they form more stable triplexes than their DNA counterparts (189). In addition, strand invasion by the PNA displaces the homopyrimidine strand of the duplex, forming a 2:1 PNA:DNA triplex (189,250). The crystal structure of the PNA:DNA triplex is composed of both T:AT and C⁺:GC canonical triplets (193). Based on the triplet composition, this triplex is of the parallel motif, where the N-terminus is analogous to the 5′-end of a DNA strand.

The most striking characteristic of the PNA:DNA triplex is the cavity down the centre of the helix caused by the large *x*-displacement (−6.8 Å) (Plate XIII). This structure differs significantly from both A- and B-DNA. The sugar puckers of the DNA strand are all N-type as in A-DNA, the rise and inclination are both similar to B-DNA, the *x*-displacement is larger than either A- or B-DNA, the twist is much smaller than either A- or B-DNA, and the glycosidic bonds are all in the *anti* conformation, as in both A- and B-DNA. Since the helix is neither A- nor B-form, the helix has been called P-form (193).

Interestingly, the Hoogsteen PNA strand and the DNA strand are extremely close together and share extensive van der Waals contacts (193). There is a series of hydrogen bonds between the amides of the PNA backbone and the O1P phosphate oxygens of the DNA backbone. These hydrogen bonds probably account for the increased stability of the PNA triplex.

5. Conclusion

In summary, triplex structure can be viewed in the context of a duplex structure that has been perturbed to accommodate the binding of a third strand 'ligand' in the major groove. For DNA triplexes, the duplex structure is a B-DNA structure, and the perturbations that are observed are an increased (negative) x-displacement and an unwinding of the helix. For RNA triplexes, there is less direct structural information. However, IR and fibre diffraction data indicate that the structures are A-form with some perturbations. Both the parallel and antiparallel triplexes adopt similar structures. However, the parallel triplexes have a more regular backbone in the third strand because their canonical triplets are isosteric.

Acknowledgments

The authors thank Charlotte Gotfredsen, A. Kathryn Phipps, Markus Tarköy, and Peter Schultze for unpublished work discussed here. This work was supported by NIH grant GM 37254 (to J.F).

References

1. Pauling, L. and Corey, R.B. (1953) *Nature* **171**, 346.
2. Pauling, L. and Corey, R.B. (1953) *Proc. Natl. Acad. Sci. USA* **39**, 84.
3. Watson, J.D. and Crick, F.H. (1953) *Nature* **171**, 737.
4. Felsenfeld, G., Davies, D.R. and Rich, A. (1957) *J. Am. Chem. Soc.* **79**, 2023.
5. Felsenfeld, G. and Rich, A. (1957) *Biochim. Biophys. Acta* **26**, 457.
6. Hoogsteen, K. (1959) *Acta Cryst.* **12**, 822.
7. Fresco, J.R. (1963) in *Some Investigations on the Secondary and Tertiary Structure of Ribonucleic Acids*, (Fresco, J.R., ed.), pp. 121. Academic Press, Inc.,New York.
8. Lipsett, M.N. (1963) *Biochem. Biophys.* Res. *Commun.* **11**, 224.
9. Lipsett, M.N. (1964) *J. Biol. Chem.* **239**, 1256.
10. Howard, F.B., Frazier, J., Lipsett, M.N. and Miles, H.T. (1964) *Biochim. Biophys. Res. Commun.* **17**, 93.
11. Inman, R.B. (1964) *J. Mol. Biol.* **10**, 137.
12. Chamberlin, M.J. and Patterson, D.L. (1965) *J. Mol. Biol.* **12**, 410.
13. Riley, M., Maling, B. and Chamberlin, M.J. (1966) *J. Mol. Biol.* **20**, 359.
14. Rich, A. (1960) *Proc. Natl. Acad. Sci. USA* **46**, 1044.
15. Felsenfeld, G. and Miles, H.T. (1967) *Annu. Rev. Biochem.* **36**, 407.
16. Michelson, A.M., Massoulié, J. and Guschlbauer, W. (1967) *Progr. Nucl. Acid Res. Mol. Biol.* **6**, 83.
17. Morgan, A.R. and Wells, R.D. (1968) *J. Mol. Biol.* **37**, 63.
18. Wells, R.D., Collier, D.A., Hanvey, J.C., Shimizu, M. and Wohlrab, F. (1988) *FASEB J.* **2**, 2939.
19. Cheng, Y.K. and Pettitt, B.M. (1992) *Progr. Biophys. Mol. Biol.* **58**, 225.
20. Sun, J.S. and Hélène, C. (1993) *Curr. Opin. Struct.* **3**, 345.
21. Lu, G. and Ferl, R.J. (1993) *Int. J. Biochem.* **25**, 1529.
22. Radhakrishnan, I. and Patel, D.J. (1994) *Biochemistry* **33**, 11405.
23. Plum, G.E., Pilch, D.S., Singleton, S.F. and Breslauer, K.J. (1995) *Annu. Rev. Biophys. Biomol. Struct.* **24**, 319.
24. Frank-Kamenetskii, M.D. and Mirkin, S.M. (1995) *Annu. Rev. Biochem.* **64**, 65.
25. Lyamichev, V.I., Mirkin, S.M. and Frank-Kamenetskii, M.D. (1987) *J. Biomol. Struct. Dynamics* **5**, 275.
26. Rajagopal, P. and Feigon, J. (1989) *Nature* **339**, 637.
27. Sun, J.S., De Bizemont, T., Duval-Valentin, G., Montenay-Garestier, T. and Hélène, (1991) *C. C. R. Acad. Sci. III* **313**, 585.

Oxford Handbook of Nucleic Acid Structure

28. Giovannangeli, C., Rougée, M., Garestier, T., Thuong, N.T. and Hélène, C. (1992) *Proc. Natl. Acad. Sci.USA* **89**, 8631.

29. de Bizemont, T., Duval-Valentin, G., Sun, J.S., Bisagni, E., Garestier, T. and Hélène, C. (1996) *Nucl. Acids Res.* **24**, 1136.

30. Dagneaux, C., Gousset, H., Shchyolkina, A.K., Ouali, M., Letellier, R., Liquier, J., Florentiev, V.L. and Taillandier, E. (1996) *Nucl. Acids Res.* **24**, 4506.

31. Dagneaux, C., Liquier, J. and Taillandier, E. (1995) *Biochemistry* **34**, 14815.

32. Soyfer, V.N. and Potaman, V.N. (1996) *Triple-Helical Nucleic Acids*. Springer-Verlag, New York.

33. Murray, N.L. and Morgan, A.R. (1973) *Can.J. Biochem.* **51**, 436.

34. Miller, J.H. and Sobell, H.M. (1966) *Proc. Natl. Acad. Sci. USA* **55**, 1201.

35. Behe, M.J. (1995) *Nucl. Acids Res.* **23**, 689.

36. Beasty, A.M. and Behe, M.J. (1988) *Nucl. Acids Res.* **16**, 1517.

37. Gillies, S.D., Folsom, V. and Tonegawa, S. (1984) *Nature* **310**, 594.

38. Fowler, R.F. and Skinner, D.M. (1986) *J. Biol. Chem.* **261**, 8994.

39. de Martynoff, G., Pohl, V., Mercken, L., van Ommen, G.J. and Vassart, G. (1987) *Eur.J. Biochem.* **164**, 591.

40. Gee, J.E., Yen, R.L., Hung, M.C. and Hogan, M.E. (1994) *Gene* **149**, 109.

41. Belland, R.J. (1991) *Mol. Microbiol.* **5**, 2351.

42. Vasquez, K.M., Wensel, T.G., Hogan, M.E. and Wilson, J.H. (1995) *Biochemistry* **34**, 7243.

43. Elgin, S.C. (1981) *Cell* **27**, 413.

44. Mace, H.A., Pelham, H.R. and Travers, A.A. (1983) *Nature* **304**, 555.

45. Nickol, J.M. and Felsenfeld, G. (1983) *Cell* **35**, 467.

46. Cantor, C.R. and Efstratiadis, A. (1984) *Nucl. Acids Res.* **12**, 8059.

47. Evans, T. and Efstratiadis, A. (1986) *J. Biol. Chem.* **261**, 14771.

48. Lyamichev, V.I., Mirkin, S.M. and Frank-Kamenetskii, M.D. (1986) *J. Biomol. Struct. Dynamics* **3**, 667.

49. Mirkin, S.M., Lyamichev, V.I., Drushlyak, K.N., Dobrynin, V.N., Filippov, S.A. and Frank-Kamenetskii, M.D. (1987) *Nature* **330**, 495.

50. Schroth, G.P. and Ho, P.S. (1995) *Nucl. Acids Res.* **23**, 1977.

51. Kohwi, Y. and Kohwi-Shigematsu, T. (1988) *Proc. Natl. Acad. Sci. USA* **85**, 3781.

52. Kohwi, Y. (1989) *Nucl. Acids Res.* **17**, 4493.

53. Bernués, J., Beltrán, R., Casasnovas, J.M. and Azorín, F. (1989) *EMBO J.* **8**, 2087.

54. Mirkin, S.M. and Frank-Kamenetskii, M.D. (1994) *Annu. Rev. Biophys. Biomol. Struct.* **23**, 541.

55. Karlovsky, P., Pecinka, P., Vojtiskova, M., Makaturova, E. and Palecek, E. (1990) *FEBS Lett.* **274**, 39.

56. Kohwi, Y., Malkhosyan, S.R. and Kohwi-Shigematsu, T. (1992) *J. Mol. Biol.* **223**, 817.

57. Moser, H.E. and Dervan, P.B. (1987) *Science* **238**, 645.

58. Htun, H. and Dahlberg, J.E. (1988) *Science* **241**, 1791.

59. Sklenár, V. and Feigon, J. (1990) *Nature* **345**, 836.

60. Collier, D.A. and Wells, R.D. (1990) *J. Biol. Chem.* **265**, 10652.

61. Lee, J.S., Woodsworth, M.L., Latimer, L.J. P. and Morgan, A.R. (1984) *Nucl. Acids Res.* **12**, 6603.

62. Povsic, T.J. and Dervan, P.B. (1989) *J. Am. Chem. Soc.* **111**, 3059.

63. Xodo, L.E., Manzini, G., Quadrifoglio, F., van der Marel, G.A. and van Boom, J.H. (1991) *Nucl. Acids Res.* **19**, 5625.

64. Hanvey, J.C., Williams, E.M. and Besterman, J.M. (1991) *Antisense Res. Dev.* **1**, 307.

65. Kiyama, R. and Camerini-Otero, R.D. (1991) *Proc. Natl. Acad. Sci. USA* **88**, 10450.

66. Guieysse, A.L., Praseuth, D. and Hélène, C. (1997) *J. Mol. Biol.* **267**, 289.

67. Lee, J.S., Burkholder, G.D., Latimer, L.J. P., Haug, B.L. and Braun, R.P. (1987) *Nucl. Acids Res.* **15**, 1047.

68. Burkholder, G.D., Latimer, L.J. P. and Lee, J.S. (1988) *Chromosoma* **97**, 185.

69. Agazie, Y.M., Lee, J.S. and Burkholder, G.D. (1994) *J. Biol. Chem.* **269**, 7019.

70. Agazie, Y.M., Burkholder, G.D. and Lee, J.S. (1996) *Biochem. J.* **316**, 461.

71. Sarkar, P.S. and Brahmachari, S.K. (1992) *Nucl. Acids Res.* **20**, 5713.

72. Kohwi, Y. and Kohwi-Shigematsu, T. (1991) *Genes Dev.* **5**, 2547.

73. Kohwi, Y. and Panchenko, Y. (1993) *Genes Dev.* **7**, 1766.

74. Rooney, S.M. and Moore, P.D. (1995) *Proc. Natl. Acad. Sci. USA* **92**, 2141.

75. Portes-Sentis, S., Sergeant, A. and Gruffat, H. (1997) *Nucl. Acids Res.* **25**, 1347.

76. Chubb, J.M. and Hogan, M.E. (1992) *Trends Biotechnol.* **10**, 132.

77. Gee, J.E. and Miller, D.M. (1992) *Am.J. Med. Sci.* **304**, 366.

78. Hélène, C. (1991) *Anticancer Drug Des.* **6**, 569.

79. Cooney, M., Czernuszewicz, G., Postel, E.H., Flint, S.J. and Hogan, M.E. (1988) *Science* **241**, 456.

80. Young, S.L., Krawczyk, S.H., Matteucci, M.D. and Toole, J.J. (1991) *Proc. Natl. Acad. Sci. USA* **88**, 10023.

81. Duval-Valentin, G., Thuong, N.T. and Hélène, C. (1992) *Proc. Natl. Acad. Sci. USA* **89**, 504.

82. Hacia, J.G., Dervan, P.B. and Wold, B.J. (1994) *Biochemistry* **33**, 6192.

83. Samadashwily, G.M. and Mirkin, S.M. (1994) *Gene* **149**, 127.

84. Krasilnikov, A.S., Panyutin, I.G., Samadashwily, G.M., Cox, R., Lazurkin, Y.S. and Mirkin, S.M. (1997) *Nucl. Acid Res.* **25**, 1339.

85. François, J.C., Saison-Behmoaras, T., Thuong, N.T. and Hélène, C. (1989) *Biochemistry* **28**, 9617.

86. Maher, L.J. d., Dervan, P.B. and Wold, B.J. (1990) *Biochemistry* **29**, 8820.

87. Hanvey, J.C., Shimizu, M. and Wells, R.D. (1990) *Nucl. Acids Res.* **18**, 157.

88. Grigoriev, M., Praseuth, D., Robin, P., Hemar, A., Saison-Behmoaras, T., Dautry-Varsat, A., Thuong, N.T., Hélène, C. and Harel-Bellan, A. (1992) *J. Biol. Chem.* **267**, 3389.

89. Maher, L.J. D., Wold, B. and Dervan, P.B. (1989) *Science* **245**, 725.

90. Gee, J.E., Blume, S., Snyder, R.C., Ray, R. and Miller, D.M. (1992) *J. Biol. Chem.* **267**, 11163.

91. Reddoch, J.F. and Miller, D.M. (1995) *Biochemistry* **34**, 7659.

92. Orson, F.M., Thomas, D.W., McShan, W.M., Kessler, D.J. and Hogan, M.E. (1991) *Nucl. Acids Res.* **19**, 3435.

93. Postel, E.H., Flint, S.J., Kessler, D.J. and Hogan, M.E. (1991) *Proc. Natl. Acad. Sci. USA* **88**, 8227.

94. Lu, G. and Ferl, R. (1992) *J. Plant Mol. Biol.* **19**, 715.

95. Helm, C.W., Shrestha, K., Thomas, S., Shingleton, H.M. and Miller, D.M. (1993) *Gynecol. Oncol.* **49**, 339.

96. Ing, N.H., Beekman, J.M., Kessler, D.J., Murphy, M., Jayaraman, K., Zendegui, J.G., Hogan, M.E., O'Malley, B.W. and Tsai, M.J. (1993) *Nucl. Acids Res.* **21**, 2789.

97. Noonberg, S.B., Scott, G.K., Garovoy, M.R., Benz, C.C. and Hunt, C.A. (1994) *Nucl. Acids Res.* **22**, 2830.

98. François, J.C., Saison-Behmoaras, T., Chassignol, M., Thuong, N.T. and Hélène, C. (1989) *J. Biol. Chem.* **264**, 5891.

99. François, J.C., Saison-Behmoaras, T., Barbier, C., Chassignol, M., Thuong, N.T. and Hélène, C. (1989) *Proc. Natl. Acad. Sci. USA* **86**, 9702.

100. Boidot-Forget, M., Chassignol, M., Takasugi, M., Thuong, N.T. and Hélène, C. (1988) *Gene* **72**, 361.
101. Le Doan, T., Perrouault, L., Praseuth, D., Habhoub, N., Decout, J.L., Thuong, N.T., Lhomme, J. and Hélène, C. (1987) *Nucl. Acids Res.* **15**, 7749.
102. Perrouault, L., Asseline, U., Rivalle, C., Thuong, N.T., Bisagni, E., Giovannangeli, C., Le Doan, T. and Hélène, C. (1990) *Nature* **344**, 358.
103. Le Doan, T., Perrouault, L., Asseline, U., Thuong, N.T., Rivalle, C., Bisagni, E. and Hélène, C. (1991) *Antisense Res. Dev.* **1**, 43.
104. Strobel, S.A., Moser, H.E. and Dervan, P.B. (1988) *J. Am. Chem. Soc.* **110**, 7927.
105. Strobel, S.A. and Dervan, P.B. (1990) *Science* **249**, 73.
106. Luebke, K.J. and Dervan, P.B. (1992) *Nucl. Acids Res.* **20**, 3005.
107. Havre, P.A. and Glazer, P.M. (1993) *J. Virol.* **67**, 7324.
108. Havre, P.A., Gunther, E.J., Gasparro, F.P. and Glazer, P.M. (1993) *Proc. Natl. Acad. Sci. USA* **90**, 7879.
109. Wang, G., Levy, D.D., Seidman, M.M. and Glazer, P.M. (1995) *Mol. Cell Biol.* **15**, 1759.
110. Roberts, R.W. and Crothers, D.M. (1991) *Proc. Natl. Acad. Sci. USA* **88**, 9397.
111. Ito, T., Smith, C.L. and Cantor, C.R. (1992) *Nucl. Acids Res.* **20**, 3524.
112. Ito, T., Smith, C.L. and Cantor, C.R. (1992) *Proc. Natl. Acad. Sci. USA* **89**, 495.
113. Ito, T., Smith, C.L. and Cantor, C.R. (1992) *Genet. Anal. Tech. Appl.* **9**, 96.
114. Sonti, S., V., Griffor, M.C., Sano, T., Narayanswami, S., Bose, A., Cantor, C.R. and Kausch, A.P. (1995) *Nucl. Acids Res.* **23**, 3995.
115. Vary, C.P. (1992) *Clin. Chem.* **38**, 687.
116. Olivas, W.M. and Maher, L.J. R. (1994) *Biotechniques* **16**, 128.
117. Pei, D.H., Ulrich, H.D. and Schultz, P.G. (1991) *Science* **253**, 1408.
118. Letai, A.G., Palladino, M.A., Fromm, E., Rizzo, V. and Fresco, J.R. (1988) *Biochemistry* **27**, 9108.
119. Griffin, L.C. and Dervan, P.B. (1989) *Science* **245**, 967.
120. Belotserkovskii, B.P., Veselkov, A.G., Filippov, S.A., Dobrynin, V.N., Mirkin, S.M. and Frank-Kamenetskii, M.D. (1990) *Nucl. Acids Res.* **18**, 6621.
121. Macaya, R.F., Gilbert, D.E., Malek, S., Sinsheimer, J. and Feigon, J. (1991) *Science* **254**, 270.
122. Sun, J.S., Mergny, J.L., Lavery, R., Montenay-Garestier, T. and Hélène, C. (1991) *J. Biomol. Struct. Dynamics* **9**, 411.
123. Mergny, J.L., Sun, J.S., Rougée, M., Montenay-Garestier, T., Barcelo, F., Chomilier, J. and Hélène, C. (1991) *Biochemistry* **30**, 9791.
124. Horne, D.A. and Dervan, P.B. (1991) *Nucl. Acids Res.* **19**, 4963.
125. Yoon, K., Hobbs, C.A., Koch, J., Sardaro, M., Kutny, R. and Weis, A.L. (1992) *Proc. Natl. Acad. Sci. USA* **89**, 3840.
126. Beal, P.A. and Dervan, P.B. (1992) *Nucl. Acids Res.* **20**, 2773.
127. Greenberg, W.A. and Dervan, P.B. (1995) *J. Am. Chem. Soc.* **117**, 5016.
128. Kiessling, L.L., Griffin, L.C. and Dervan, P.B. (1992) *Biochemistry* **31**, 2829.
129. Colocci, N., Distefano, M.D. and Dervan, P.B. (1993) *J. Am. Chem. Soc.* **115**, 4468.
130. Volker, J. and Klump, H.H. (1994) *Biochemistry* **33**, 13502.
131. Colocci, N. and Dervan, P.B. (1995) *J. Am. Chem. Soc.* **117**, 4781.
132. Ebbinghaus, S.W., Gee, J.E., Rodu, B., Mayfield, C.A., Sanders, G. and Miller, D.M. (1993) *J. Clin. Invest.* **92**, 2433.
133. Mayfield, C. and Miller, D. (1994) *Nucl. Acids Res.* **22**, 1909.
134. Gee, J.E., Revankar, G.R., Rao, T.S. and Hogan, M.E. (1995) *Biochemistry* **34**, 2042.

135. Milligan, J.F., Krawczyk, S.H., Wadwani, S. and Matteucci, M.D. (1993) *Nucl. Acids Res.* **21**, 327.

136. Stilz, H.U. and Dervan, P.B. (1993) *Biochemistry* **32**, 2177.

137. Zimmerman, S.C. and Schmitt, P. (1995) *J. Am. Chem. Soc.* **117**, 10769.

138. Griffin, L.C., Kiessling, L.L., Beal, P.A., Gillespie, P. and Dervan, P.B. (1992) *J. Am. Chem. Soc.* **114**, 7976.

139. Koshlap, K.M., Gillespie, P., Dervan, P.B. and Feigon, J. (1993) *J. Am. Chem. Soc.* **115**, 7908.

140. Wang, E., Koshlap, K.M., Gillespie, P., Dervan, P.B. and Feigon, J. (1996) *J. Mol. Biol.* **257**, 1052.

141. Ono, A., Chen, C.N. and Kan, L.S. (1991) *Biochemistry* **30**, 9914.

142. Horne, D.A. and Dervan, P.B. (1990) *J. Am. Chem. Soc.* **112**, 2435.

143. Jayasena, S.D. and Johnston, B.H. (1992) *Biochemistry* **31**, 320.

144. Jayasena, S.D. and Johnston, B.H. (1992) *Nucl. Acids Res.* **20**, 5279.

145. Beal, P.A. and Dervan, P.B. (1992) *J. Am. Chem. Soc.* **114**, 4976.

146. Washbrook, E. and Fox, K.R. (1994) *Biochem. J.* **301**, 569.

147. Sun, J.S., François, J.C., Montenay-Garestier, T., Saison-Behmoaras, T., Roig, V., Thuong, N.T. and Hélène, C. (1989) *Proc. Natl. Acad. Sci. USA* **86**, 9198.

148. Collier, D.A., Thuong, N.T. and Hélène, C. (1991) *J. Am. Chem. Soc.* **113**, 1457.

149. Ono, A., Tso, P.O. P. and Kan, L.S. (1991) *J. Am. Chem. Soc.* **113**, 4032.

150. Ono, A., Tso, P.O. P. and Kan, L.S. (1992) *J. Org. Chem.* **57**, 3225.

151. Koh, J.S. and Dervan, P.B. (1992) *J. Am. Chem. Soc.* **114**, 1470.

152. Radhakrishnan, I., Patel, D.J., Priestly, E.S., Nash, H.M. and Dervan, P.B. (1993) *Biochemistry* **32**, 11228.

153. Priestley, E.S. and Dervan, P.B. (1995) *J. Am. Chem. Soc.* **117**, 4761.

154. Hunziker, J., Priestley, E.S., Brunar, H. and Dervan, P.B. (1995) *J. Am. Chem. Soc.* **117**, 2661.

155. Krawczyk, S.H., Milligan, J.F., Wadwani, S., Moulds, C., Froehler, B.C. and Matteucci, M.D. (1992) *Proc. Natl. Acad. Sci. USA* **89**, 3761.

156. Jetter, M.C. and Hobbs, F.W. (1993) *Biochemistry* **32**, 3249.

157. Xiang, G.B., Soussou, W. and McLaughlin, L.W. (1994) *J. Am. Chem. Soc.* **116**, 11155.

158. Thuong, N.T. and Hélène, C. (1993) *Angew. Chem. Int. Ed. Eng.* **32**, 666.

159. Stonehouse, T.J. and Fox, K.R. (1994) *Biochim. Biophys. Acta* **1218**, 322.

160. Collier, D.A., Mergny, J.L., Thuong, N.T. and Hélène, C. (1991) *Nucl. Acids Res.* **19**, 42(19.

161. Fox, K.R. (1994) *Nucl. Acids Res.* **22**, 2016.

162. Durand, M., Thuong, N.T. and Maurizot, J.C. (1992) *J. Biol. Chem.* **267**, 24394.

163. Park, Y.W. and Breslauer, K.J. (1992) *Proc. Natl. Acad. Sci. USA* **89**, 6653.

164. Durand, M., Thuong, N.T. and Maurizot, J.C. (1994) *J. Biomol. Struct. Dynamics* **11**, 1191.

165. Fedorova, O.S., Knorre, D.G., Podust, L.M. and Zarytova, V.F. (1988) *FEBS Lett.* **228**, 273.

166. Povsic, T.J. and Dervan, P.B. (1990) *J. Am. Chem. Soc.* **112**, 9428.

167. Takasugi, M., Guendouz, A., Chassignol, M., Decout, J.L., Lhomme, J., Thuong, N.T. and Hélène, C. (1991) *Proc. Natl. Acad. Sci. USA* **88**, 5602.

168. Giovannangeli, C., Thuong, N.T. and Hélène, C. (1992) *Nucl. Acids Res.* **20**, 4275.

169. Grigoriev, M., Praseuth, D., Guieysse, A.L., Robin, P., Thuong, N.T., Hélène, C. and Harel-Bellan, A. (1993) *Proc. Natl. Acad. Sci. USA* **90**, 3501.

170. Degols, G., Clarenc, J.P., Lebleu, B. and Leonetti, J.P. (1994) *J. Biol. Chem.* **269**, 16933.

171. Nielsen, P.E. (1995) *Annu. Rev. Biophys. Biomol. Struct.* **24**, 167.

172. Latimer, L.J., Hampel, K. and Lee, J.S. (1989) *Nucl. Acids Res.* **17**, 1549.

173. Kim, S.G., Tsukahara, S., Yokoyama, S. and Takaku, H. (1992) *FEBS Lett.* **314**, 29.

174. Tsukahara, S., Kim, S.G. and Takaku, H. (1993) *Biochem. Biophys. Res. Commun.* **196**, 990.

175. Hacia, J.G., Wold, B.J. and Dervan, P.B. (1994) *Biochemistry* **33**, 5367.

176. Callahan, D.E., Trapane, T.L., Miller, P.S., Ts'o, P.O. and Kan, L.S. (1991) *Biochemistry* **30**, 1650.

177. Reynolds, M.A., Arnold, L.J., Jr., Almazan, M.T., Beck, T.A., Hogrefe, R.I., Metzler, M.D., Stoughton, S.R., Tseng, B.Y., Trapane, T.L., Ts'o, P.O. and Woolf, T.M. (1994) *Proc. Natl. Acad. Sci. USA* **91**, 12433.

178. Browne, K.A., Dempcy, R.O. and Bruice, T.C. (1995) *Proc. Natl. Acad. Sci. USA* **92**, 7051.

179. Kibler-Herzog, L., Kell, B., Zon, G., Shinozuka, K., Mizan, S. and Wilson, W.D. (1990) *Nucl. Acids Res.* **18**, 3545.

180. Kibler-Herzog, L., Zon, G., Whittier, G., Mizan, S. and Wilson, W.D. (1993) *Anticancer Drug Des.* **8**, 65.

181. Alunni-Fabbroni, M., Manfioletti, G., Manzini, G. and Xodo, L.E. (1994) *Eur. J. Biochem.* **226**, 831.

182. Xodo, L., Alunni-Fabbroni, M., Manzini, G. and Quadrifoglio, F. (1994) *Nucl. Acids Res.* **22**, 3322.

183. Shimizu, M., Koizumi, T., Inoue, H. and Ohtsuka, E. (1994) *Bioorg. Med.* **4**, 1029.

184. Wang, S. and Kool, E.T. (1995) *Nucl. Acids Res.* **23**, 1157.

185. Jones, R.J., Swaminathan, S., Milligan, J.F., Wadwani, S., Froehler, B.C. and Matteucci, M.D. (1993) *J. Am. Chem. Soc.* **115**, 9816.

186. Tarköy, M., Bolli, M. and Leumann, C. (1994) *Helv. Chim. Acta* **77**, 716.

187. Escudé, C., Sun, J.S., Rougée, M., Garestier, T. and Hélène, (1992) *C. C. R. Acad. Sci. III* **315**, 521.

188. Shimizu, M., Konishi, A., Shimada, Y., Inoue, H. and Ohtsuka, E. (1992) *FEBS Lett.* **302**, 155.

189. Nielsen, P.E., Egholm, M., Berg, R.H. and Buchardt, O. (1991) *Science* **254**, 1497.

190. Egholm, M., Buchardt, O., Christensen, L., Behrens, C., Freier, S.M., Driver, D.A., Berg, R.H., Kim, S.K., Norden, B. and Nielsen, P.E. (1993) *Nature* **365**, 566.

191. Kim, S.K., Nielsen, P.E., Egholm, M., Buchardt, O., Berg, R.H. and Norden, B. (1993) *J. Am. Chem. Soc.* **115**, 6477.

192. Nielsen, P.E., Egholm, M. and Buchardt, O. (1994) *Bioconjug. Chem.* **5**, 3.

193. Betts, L., Josey, J.A., Veal, J.M. and Jordan, S.R. (1995) *Science* **270**, 1838.

194. Miles, H.T. (1964) *Proc. Natl. Acad. Sci. USA* **51**, 1104.

195. Arnott, S. and Bond, P.J. (1973) *Nature New Biol.* **244**, 99.

196. Arnott, S. and Selsing, E. (1974) *J. Mol. Biol.* **88**, 509.

197. Arnott, S., Bond, P.J., Selsing, E. and Smith, P.J. C. (1976) *Nucl. Acids Res.* **3**, 2459.

198. Liu, K., Miles, H.T., Parris, K.D. and Sasisekharan, V. (1994) *Nature Struct. Biol.* **1**, 11.

199. Liu, K., Sasisekharan, V., Miles, H.T. and Raghunathan, G. (1996) *Biopolymers* **39**, 573.

200. Kallenbach, N.R., Daniel, Jr, W.E., and Kaminker, M.A. (1976) *Biochemistry* **15**, 1218.

201. Geerdes, H.A. M. and Hilbers, C.W. (1977) *Nucl. Acids Res.* **4**, 207.

202. Rajagopal, P. and Feigon, J. (1989) *Biochemistry* **28**, 7859.

203. de los Santos, C., Rosen, M. and Patel, D. (1989) *Biochemistry* **28**, 7282.

204. Radhakrishnan, I., Patel, D.J., Veal, J.M. and Gao, X.L. (1992) *J. Am. Chem. Soc.* **114**, 6913.

205. Macaya, R.F., Schultze, P. and Feigon, J. (1992) *J. Am. Chem. Soc.* **114**, 781.

206. Macaya, R., Wang, E., Schultze, P., Sklenár, V. and Feigon, J. (1992) *J. Mol. Biol.* **225**, 755.

207. Anonymous (1989) *EMBO J.* **8**, 1.

208. Lavery, R. and Sklenar, H. (1988) *J. Biomol. Struct. Dynamics* **6**, 63.

209. Lavery, R. and Sklenar, H. (1989) *J. Biomol. Struct. Dynamics* **6**, 655.

210. Han, H. and Dervan, P.B. (1994) *Nucl. Acids Res.* **22**, 2837.

211. Akhebat, A., Dagneaux, C., Liquier, J. and Taillandier, E. (1992) *J. Biomol. Struct. Dynamics* **10**, 577.

212. Liquier, J., Taillandier, E., Klinck, R., Guittet, E., Gouyette, C. and Huynh-Dinh, T. (1995) *Nucl. Acids Res.* **23**, 1722.

213. Klinck, R., Liquier, J., Taillandier, E., Gouyette, C., Huynhdinh, T. and Guittet, E. (1995) *Eur.J. Biochem.* **233**, 544.

214. Holland, J.A. and Hoffman, D.W. (1996) *Nucl. Acids Res.* **24**, 2841.

215. Heus, H.A. and Pardi, A. (1991) *J. Am. Chem. Soc.* **113**, 4360.

216. Wüthrich, K. (1986) *NMR of Proteins and Nucleic Acids*. John Wiley & Sons, New York.

217. Dagneaux, C., Liquier, J. and Taillandier, E. (1995) *Biochemistry* **34**, 16618.

218. Gotfredsen, C.H., Schultze, P. and Feigon, J. (1998) *J. Am. Chem. Soc.* **120**, 4281.

219. Roberts, R.W. and Crothers, D.M. (1992) *Science* **258**, 1463.

220. Escudé, C., François, J.C., Sun, J.S., Ott, G., Sprinzl, M., Garestier, T. and Hélène, C. (1993) *Nucl. Acids Res.* **21**, 5547.

221. Han, H. and Dervan, P.B. (1993) *Proc. Natl. Acad. Sci. USA* **90**, 3806.

222. Bornet, O. and Lancelot, G. (1995) *J. Biomol. Struct. Dynamics* **12**, 803.

223. Tarköy, M., Phipps, A.K., Schultze, P. and Feigon, J. (1998) *Biochemistry* **37**, 5810.

224. Phipps, A.K., Tarköy, M., Schultze, P. and Feigon, J. (1998) *Biochemistry* **37**, 5820.

225. Radhakrishnan, I. and Patel, D.J. (1994) *Structure* **2**, 17.

226. Radhakrishnan, I. and Patel, D.J. (1994) *J. Mol. Biol.* **241**, 600.

227. Wang, E., Koshlap, K.M., Gillespie, P., Dervan, P.B. and Feigon, J. (1996) *J. Mol. Biol.* **257**, 1052.

228. Koshlap, K.M., Schultze, P., Brunar, H., Dervan, P.B. and Feigon, J. (1997) *Biochemistry* **36**, 2659.

229. Bernstein, F.C., Koetzle, T.F., Williams, G.J., Meyer, E.E., Jr., Brice, M.D., Rodgers, J.R., Kennard, O., Shimanouchi, T. and Tasumi, M. (1977) *J. Mol. Biol.* **112**, 535.

230. Roongta, V.A., Jones, C.R. and Gorenstein, D.G. (1990) *Biochemistry* **29**, 5245.

231. Fang, Y., Bai, C., Wei, Y., Lin, S.B. and Kan, L. (1995) *J. Biomol. Struct. Dynamics* **13**, 471.

232. Wang, E., Malek, S. and Feigon, J. (1992) *Biochemistry* **31**, 4838.

233. Yanagi, K., Prive, G.G. and Dickerson, R.E. (1991) *J. Mol. Biol.* **217**, 201.

234. Quintana, J.R., Grzeskowiak, K., Yanagi, K. and Dickerson, R.E. (1992) *J. Mol. Biol.* **225**, 379.

235. Froehler, B.C., Wadwani, S., Terhorst, T.J. and Gerrard, S.R. (1992) *Tetrahedron Lett.* **33**, 5307.

236. Broitman, S.L., Im, D.D. and Fresco, J.R. (1987) *Proc. Natl. Acad. Sci. USA* **84**, 5120.

237. Howard, F.B., Miles, H.T. and Ross, P.D. (1995) *Biochemistry* **34**, 7135.

238. Beal, P.A. and Dervan, P.B. (1991) *Science* **251**, 1360.

239. Durland, R.H., Kessler, D.J., Gunnell, S., Duvic, M., Pettitt, B.M. and Hogan, M.E. (1991) *Biochemistry* **30**, 9246.

240. Chen, F.M. (1991) *Biochemistry* **30**, 4472.

241. Pilch, D.S., Levenson, C. and Shafer, R.H. (1991) *Biochemistry* **30**, 6081.

242. Radhakrishnan, I., de los Santos, C. and Patel, D.J. (1991) *J. Mol. Biol.* **221**, 1403.

243. Radhakrishnan, I., de los Santos, C. and Patel, D.J. (1993) *J. Mol. Biol.* **234**, 188.

244. Radhakrishnan, I. and Patel, D.J. (1993) *Structure* **1**, 135.
245. Ji, J., Hogan, M.E. and Gao, X. (1996) *Structure* **4**, 425.
246. Vlieghe, D., Van Meervelt, L., Dautant, A., Gallois, B., Precigoux, G. and Kennard, O. (1996) *Science* **273**, 1702.
247. Shin, C. and Koo, H.S. (1996) *Biochemistry* **35**, 968.
248. Ouali, M., Letellier, R., Sun, J.S., Akhebat, A., Adnet, F., Liquier, J. and Taillandier, E. (1993) *J. Am. Chem. Soc.* **115**, 4264.
249. Durland, R.H., Rao, T.S., Revankar, G.R., Tinsley, J.H., Myrick, M.A., Seth, D.M., Rayford, J., Singh, P. and Jayaraman, K. (1994) *Nucl. Acids Res.* **22**, 3233.
250. Nielsen, P.E., Egholm, M. and Buchardt, O. (1994) *J. Mol. Recogn.* **7**, 165.

NOTE added in proof: This review covers the published literature and work from the Feigon laboratory through May, 1997.
References to unpublished work from that time have been updated.

13

Structures of guanine-rich and cytosine-rich quadruplexes formed *in vitro* by telomeric, centromeric, and triplet repeat disease DNA sequences

Dinshaw J. Patel, Serge Bouaziz, Abdelali Kettani, and Yong Wang
Cellular Biochemistry and Biophysics Program, Memorial Sloan-Kettering Cancer Center, New York, NY 10021, USA

1. Introduction

DNA sequences can adopt higher order architectures beyond duplex alignments, and research in this area is increasingly addressing the structural and energetics issues related to DNA triplexes (reviewed in refs 1–3 and Chapter 12), quadruplexes (4,5), and junctions (6,7, and Chapter 15). The range of strand directionalities and pairing alignments within these multistranded structures provides novel DNA architectures associated with molecular recognition and function.

The structure of DNA quadruplexes formed by guanine-rich DNA segments is of great interest currently, since it affects processes ranging from the architecture of telomeric and centromeric sites, to the potential pairing alignments during genetic recombination events. The initial efforts in this area have focused on monovalent cation-coordinated G quadruplexes formed by the stacking of planar G:G:G:G tetrads (8–12; reviewed in 13). This chapter focuses on recent structural insights into G quadruplex architecture that have emerged from crystallographic and solution NMR studies. The observed structural polymorphism is related to the relative strand directionality and to the distribution of *syn/anti* guanines along individual strands and around G tetrads in G quadruplexes (for earlier structural reviews see refs 4 and 14). This chapter also summarizes recent structures of quadruplexes containing G:C:G:C tetrads adopted by triplet repeat disease and related sequences. It also discusses recent structural efforts that have defined the role of monovalent cations sandwiched between G tetrads in stabilizing the G quadruplex fold and, in addition, identified the molecular basis associated with monovalent cation-dependent folding of loop domains of quadruplexes.

Cytosine-rich sequences have been shown to form quadruplexes at acidic pH, designated i-motifs, through antiparallel alignment of a pair of mutually intercalated parallel-stranded $C:CH^+$ mismatch-containing duplexes (15). This chapter outlines the range of i-motif structures adopted by telomeric and centromeric sequences and the role of flanking sequences in directing the overall folding topology of the i-motif quadruplex.

2. Telomeric sequence G quadruplexes

Telomeres are nucleic acid:protein complexes found at the ends of linear chromosomes. They are involved in chromosomal 3′-end replication without truncation, in chromosomal organization and in protection of chromosomal terminii against degradation, and in the anchoring of chromosomes to the nuclear envelope (reviewed in ref. 16). They contain tandem repeats of guanines and cytosines on partner strands together with guanine-rich segment overhangs at the 3′-ends, in species as divergent as ciliates, yeast, and humans. The critical functional role of telomere sequence follows directly from the observation that mutated telomeric sequences induce telomere length instability and subsequent death of the organism (17). The folding topologies of such G rich tandem repeats are of considerable interest since they have the potential to form G quadruplexes *in vitro*.

The fundamental unit of the G quadruplex is the G tetrad (18–20) which involves a cyclized, hydrogen bonded, square planar alignment of four guanines, as shown in Fig. 13.1. Adjacent guanines around the G tetrad are paired through their Watson–Crick and Hoogsteen edges, resulting in four electronegative carbonyl groups being directed towards the interior of the tetrad. G quadruplex formation has an absolute requirement for monovalent K^+ and Na^+ cations (21–24), with the monovalent cation-binding sites presumably positioned in the interior of the quadruplex between stacked G tetrads (11). G quadruplex architecture is to some extent dependent on the nature of the monovalent cation (25–27), with K^+ cations generating the most stable G quadruplexes (reviewed in 28). Early efforts at determining the folding topologies of G quadruplexes based on chemical modification, base analogue substitution, and cross-linking experiments (9–12,29) have been supplemented by X-ray and NMR approaches that provide atomic resolution views of the folding architecture in the crystalline and solution states, respectively. These G quadruplex structures are pre-

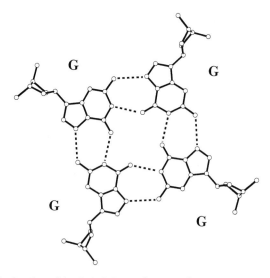

Fig. 13.1. A schematic drawing of the G:G:G:G tetrad pairing alignment.

sented below and categorized according to the relative strand directionalities and *syn/anti* distribution around individual G tetrads.

2.1 G quadruplexes containing anti:anti:anti:anti G tetrad alignments and parallel strand directionalities

Some of the earliest efforts at structure determination of G quadruplexes stabilized by G tetrads focused on sequences containing single dG_n repeats. Such dG_n sequences with non-guanine flanking bases have provided the necessary structural information on the architecture of parallel-stranded G quadruplexes in solution and crystalline states.

2.1.1 Solution structure of d(NG₄N) quadruplexes

NMR-based studies of single guanine-rich repeat $d(T_2AG_3T)$, $d(T_2G_4T)$ (30), $d(TG_4T)$ (31), and $d(TG_3T)$ (32) sequences lacking 3'-terminal guanines, provided the initial evidence for formation of parallel-stranded G quadruplexes containing only *anti* -glycosidic torsion angles in K^+-containing solution. These studies also established that the guanine imino protons of the internal G tetrads exchanged very slowly with solvent water (30). By contrast, sequences ending with 3'-terminal guanines tend to aggregate by forming higher order multistranded structures, as probed by gel mobility and methylation protection experiments (33,34) and NMR spectral parameters (30).

The solution structures of all parallel-stranded G quaduplexes have been solved through combined NMR and molecular dynamics studies of the sequences $d(T_2G_4T)$ (35), $d(T_4G_4)$ (36), and $d(TG_4T)$ (37). The structures are well defined within the guanine-rich segments, but underdefined at the thymine segments. A view looking into one of the four equivalent grooves of the solution structure of the G_4 segment of the $d(T_2G_4T)$ quadruplex is shown in Fig. 13.2a (35). The structure is right-handed with all residues adopting *anti* glycosidic torsion angles and S-type (C2'-*endo*) sugar pucker conformations. The four G tetrads, which approach coplanarity, are stacked on each other, with the overlap of the central tetrads shown in Fig. 13.2b.

2.1.2 Energetics of the d(TG₃T) quadruplex in solution

The energetics for the order–disorder transition of $d(TG_nT)$ quadruplexes in monovalent cation solution have been measured using optical (38) and calorimetric (32) experiments. The calorimetric studies on the $d(TG_3T)$ quadruplex in K^+ solution yield values of $\Delta G° = -9.6$ kJ/mol of tetrad, $\Delta H° = -87.8$ kJ/mol of tetrad, and $\Delta S° = 259$ J/K mol of tetrad at 25°C (32). These data establish that the stability of G quadruplexes reflects a favourable enthalpic contribution to formation.

2.1.3 Crystal structure of the d(TG₄T) quadruplex

The crystal structure of the $d(TG_4T)$ sequence in the presence of Na^+ cation was solved initially at 1.2 Å resolution (39) and refined further to 0.95 Å (40). There are four parallel-stranded G quadruplexes in the asymmetric unit of this crystallographic structure, with pairs of G quadruplexes stacked end-to-end in a head-to-head (5' to 5') orientation through their terminal G tetrads. The crystal structure of the G_4

(a)

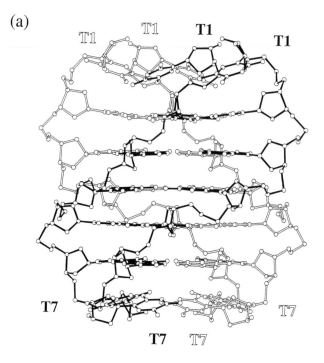

(b)

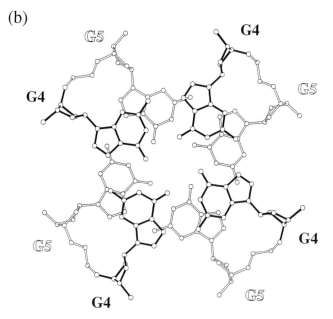

Fig. 13.2. (a) A view of the NMR-based solution structure of the four-stranded d(T₂G₄T) quadruplex (35). Two of the strands, directed towards the viewer, are shown with filled bonds and the other two, directed away from the viewer, are shown with open bonds. (b) Stacking between adjacent internal G:G:G:G tetrads in the solution structure of the d(T₂G₄T) quadruplex (35).

(a)

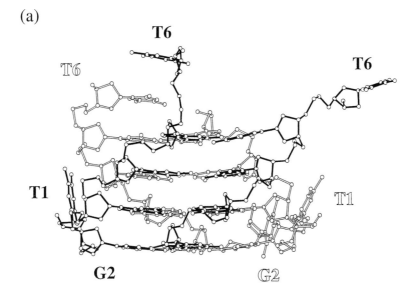

(b)

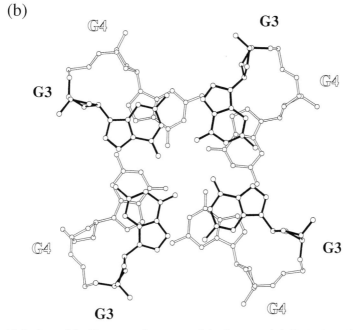

Fig. 13.3. (a) A view of the X-ray crystal structure of the four-stranded d(TG$_4$T) quadruplex (39,40). Two of the strands, directed towards the viewer, are shown with filled bonds and the other two, directed away from the viewer, are shown with open bonds. (b) Stacking between adjacent internal G:G:G:G tetrads in the crystal structure of the d(TG$_4$T) quadruplex (39,40).

segment of the d(TG$_4$T) G quadruplex is shown in Fig. 13.3a together with the overlap geometry between stacked central G tetrads, in Fig. 13.3b. The terminal thymines are less well defined and not involved in the stacking with the G tetrads of the G quadruplex. The Na$^+$ cations are well defined in this 0.95 Å high resolution crystal structure, and their positioning ranges from coordination sites associated with inwardly directed guanine O6 atoms located between G tetrad planes, to sites located within G tetrad planes (40). Bound water molecules can also be identified at this high resolution and are clustered around the backbone phosphates in the helical grooves. The basic architecture of the parallel-stranded G quadruplex segments are the same in the crystal (Fig. 13.3a) (40) and in solution (Fig. 13.2a) (35), as are the base pair overlaps between adjacent G tetrads in the crystal (Fig. 13.3b) and in solution (Figure 13.2b).

2.1.4 Solution structure of the r(UG$_4$U) quadruplex

Guanine-rich sequences are also detected in RNA, suggesting the potential for RNA G quadruplex formation. Indeed, guanine-rich sequences have been identified in *E. coli* 5S RNA, where they are known to aggregate into a tetrameric form in the presence of K$^+$ cation (41). An NMR and molecular dynamics-based characterization of the r(UG$_4$U) sequence in K$^+$ solution established formation of a right-handed G quadruplex (Fig. 13.4a) containing all *anti*-glycosidic torsion angles and stabilized by four stacked G tetrads (42). The majority of the sugar puckers adopted N-type (C3'-*endo*) or partially N-type sugar pucker conformations. This structural study also identified formation of a U tetrad (shown schematically in Fig. 13.4b) which stacks on the adjacent G-tetrad (42).

2.1.5 Self-assembly of guanine-rich telomeric sequences into larger superstructures

The *Tetrahymena* telomere d(G$_4$T$_2$G$_4$) sequence has been shown by gel electrophoresis to assemble spontaneously into larger superstructures in monovalent cation solution (43). These superstructures, called G wires, have been imaged by scanning probe microscopy (44) and exhibit characteristics of long, linear polymers of G tetrad-stabilized, parallel-stranded DNA (43).

2.2 G quadruplexes containing syn:anti:syn:anti G tetrad alignments and antiparallel directionalities of adjacent strands

The structure of the d(G$_4$T$_4$G$_4$) sequence, which contains two tandem guanine-rich segments within the sequence context of the *Oxytricha* telomeric d(T$_4$G$_4$)$_n$ repeat has been solved in both crystalline (45) and solution (46,47) states. The folding architecture of the G quadruplex formed through dimerization of a pair of d(G$_4$T$_4$G$_4$) segments is distinct between the X-ray (45) and NMR (46,47) structures, as defined by the relative alignment of adjacent strands, the *syn/anti* distribution of guanine glycosidic bonds around individual G tetrads, and the loop connectivities (lateral versus diagonal). The results of the X-ray structure of the d(G$_4$T$_4$G$_4$) G quadruplex are reported in this section.

(a)

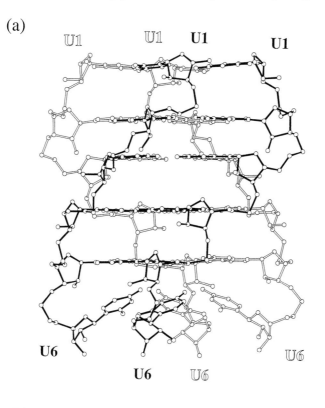

(b)

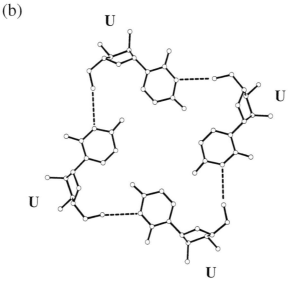

Fig. 13.4. (a) A view of the NMR-based solution structure of the four-stranded r(UG$_4$U) quadruplex (42). Two of the strands, directed towards the viewer, are shown with filled bonds and the other two, directed away from the viewer, are shown with open bonds. (b) Alignment around the U:U:U:U tetrad involving the U1 residue in the solution structure of the r(UG$_4$U) quadruplex (42).

2.2.1 Crystal structure of the Oxytricha telomere d(G₄T₄G₄) quadruplex

The X-ray structure of crystals of d(G₄T₄G₄) grown from K⁺ solution and solved at 2.5 Å resolution establishes formation of a pair of hairpins oriented in a head-to-tail alignment, with G₄ segments connected by lateral loops, as shown schematically in Fig. 13.5a (45). Adjacent strands are aligned antiparallel to each other with alternating *syn–anti–syn–anti* alignments of guanines along individual G₄ segments and *syn:anti:syn:anti* alignments of guanines around individual G tetrads. A view of the structure of this G quadruplex is shown in Fig. 13.6. A twofold axis of symmetry relates the two halves of the G quadruplex, resulting in two symmetric wide grooves and two symmetric narrow grooves. A K⁺ cation-binding site was associated with electron density between the two central G tetrads of the G quadruplex (45). The stacking patterns between adjacent G tetrads at G(*syn*)–G(*anti*) and G(*anti*)–G(*syn*) steps are shown in Fig. 13.7a,b, respectively.

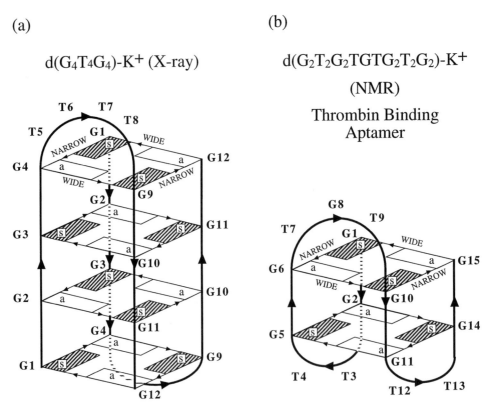

Fig. 13.5. (a) Schematic of the X-ray crystal structure-based folding topology adopted by the quadruplex formed through head-to-tail dimerization of the two-repeat *Oxytricha* telomere d(G₄T₄G₄) sequence in K⁺ solution (45). The T₄ loops are of the lateral type. The *syn* guanines are shown as hatched rectangles while *anti* guanines are shown as open rectangles. (b) Schematic of the NMR solution structure-based folding topology adopted by the intramolecular quadruplex formed by the d(G₂T₂G₂TGTG₂T₂G₂) sequence in K⁺ solution (49,50). All three loops are of the lateral type.

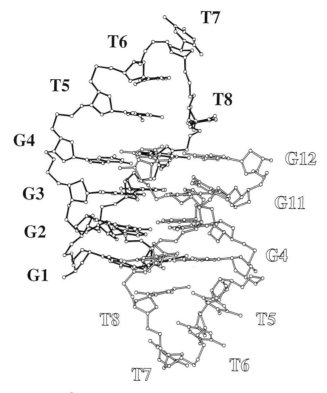

Fig. 13.6. A view of the 2.5 Å X-ray crystal structure of the two-repeat *Oxytricha* telomere d(G$_4$T$_4$G$_4$) quadruplex formed through head-to-tail dimerization of a pair of hairpins in K$^+$ solution (45). One d(G$_4$T$_4$G$_4$) hairpin is shown with filled bonds while the other is shown with open bonds. The T$_4$ loops are of the lateral loop type.

2.2.2 *Solution structure of the thrombin-binding* d(G$_2$T$_2$G$_2$TGTG$_2$T$_2$G$_2$) *DNA aptamer quadruplex*

A DNA aptamer with the consensus d(G$_2$T$_2$G$_2$TGTG$_2$T$_2$G$_2$) sequence was identified through *in vitro* selection based on its ability to bind α-thrombin (48). This thrombin-binding aptamer contains four G$_2$ steps with the potential to form an intramolecular G quadruplex in monovalent cation solution. Indeed, two groups independently established that the NMR parameters of the d(G$_2$T$_2$G$_2$TGTG$_2$T$_2$G$_2$) sequence in K$^+$ solution (49,50) were consistent with formation of a G quadruplex with antiparallel alignment of adjacent strands, alternating *syn–anti* alignments along individual G$_2$ steps and *syn:anti:syn:anti* alignments of guanines around individual G tetrads, as shown schematically in Fig. 13.5b. The T$_2$, TGT, and T$_2$ loops were all of the lateral type with a T:T wobble mismatch formed between the second thymines in the two T$_2$ loops (51). The thrombin-binding G quadruplex in K$^+$ solution is sufficiently stable, despite containing only two stacked G tetrads, to permit the single inosine for guanine substitutions necessary for distinguishing between alternative folding topologies (49). The NMR data have been quantitatively analysed to provide the solution structure of

(a)

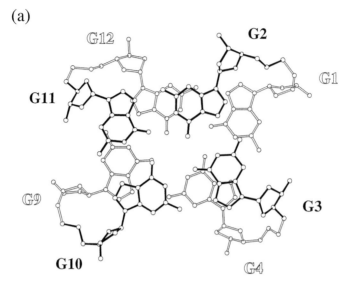

(b)

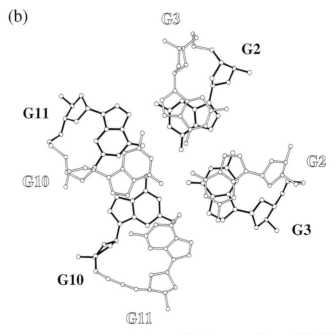

Fig. 13.7. Views down the helix axis showing stacking between adjacent G:G:G:G tetrads at (a) G(*syn*)–G(*anti*) and (b) G(*anti*)–G(*syn*) steps in the crystal structure of the *Oxytricha* telomere d(G$_4$T$_4$G$_4$) quadruplex (45). Individual G tetrads are drawn with either filled or open bonds.

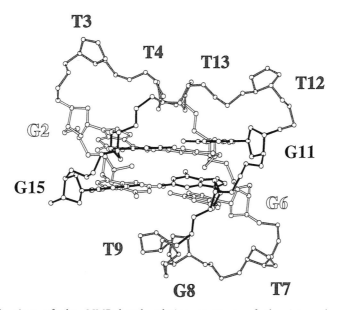

Fig. 13.8. A view of the NMR-based solution structure of the intramolecularly folded d($G_2T_2G_2TGTG_2T_2G_2$) quadruplex in Na^+ solution (51). Two of the guanine-containing G_2 steps, directed towards the viewer, are shown with filled bonds and the other two guanine-containing G_2 steps, directed away from the viewer, are shown with open bonds. The three loop segments (T3–T4, T7–G8–T9, and T12–T13) are shown with hatched bonds and the bases in these segments have been deleted in the interests of clarity. All three loops are of the lateral type.

the d($G_2T_2G_2TGTG_2T_2G_2$) G quadruplex (51,52). The solution structure of this G quadruplex structure is shown in Fig. 13.8 (51). The overlaps between adjacent G tetrads is shown in Fig. 13.9a and between the T:T mismatch and the G tetrad in Fig. 13.9b (51).

A combination of NMR and electron spin resonance (ESR) methods have been used to identify paramagnetic manganese divalent cation-binding sites on the thrombin-binding d($G_2T_2G_2TGTG_2T_2G_2$) G quadruplex (53). These divalent Mn cation-binding sites are located one per minor groove of the G quadruplex.

2.2.3 Crystal structure of the thrombin-binding d($G_2T_2G_2TGTG_2T_2G_2$) DNA aptamer quadruplex bound to thrombin

The crystal structure of d($G_2T_2G_2TGTG_2T_2G_2$) with Na^+ as counterion and bound to thrombin has been solved at 2.9 Å resolution (54). The bound DNA in the crystalline complex forms a G quadruplex (54) with an architecture where the strand runs in an opposite direction to that shown schematically in Fig. 13.5b. The intermolecular interface in the complex involves the heparin-binding site and fibrinogen-binding exosite on two different thrombins, and the loop segments in the G quadruplex (54). Interestingly, even though both X-ray (54) and NMR (51,52) methods have identified the same structure for the G quadruplex core containing two stacked G tetrads, they disagree with respect to the orientation of the connecting loops, as pointed out

(a)

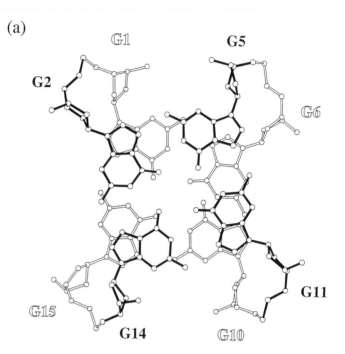

(b)

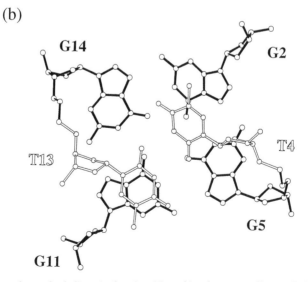

Fig. 13.9. Views down the helix axis showing (a) stacking between adjacent G:G:G:G tetrads at the G(syn):G($anti$) step and (b) stacking between the G:G:G:G tetrad and a T:T mismatch in the solution structure of the intramolecularly folded d($G_2T_2G_2TGTG_2T_2G_2$) quadruplex in Na$^+$ solution (51).

recently (55). This may reflect ambiguities in the X-ray structure of the complex in identifying the less well-defined electron densities in the loop-connecting segments.

2.2.4 Solution structures of insulin-linked polymorphic d(G₄TGTG₄) and d(G₄TGTG₄ACAG₄TGTG₄) quadruplexes

The human insulin gene contains a guanine-rich region that contains tandem repeats of the d(ACAG₄TGTG₄) sequence (56). The solution structures of both d(G₄TGTG₄) and d(G₄TGTG₄ACAG₄TGTG₄) sequences have been characterized by NMR in Na⁺-containing aqueous solution (57). The authors concluded that d(G₄TGTG₄) forms a G quadruplex through head-to-tail dimerization of hairpins containing TGT lateral loops, while d(G₄TGTG₄ACAG₄TGTG₄) forms an intramolecular G quadruplex containing TGT, ACA, and TGT lateral loops (57). There is reason to reserve judgement on these conclusions since the authors did not undertake inosine for guanine substitutions to identify individual guanine residues involved in quadruplex formation definitively (see refs 49,58,59), an approach that has proved invaluable in distinguishing between G(*syn*):G(*anti*):G(*syn*):G(*anti*) and G(*syn*):G(*syn*):G(*anti*):G(*anti*) tetrad alignments. The concern outlined above related to the proposed solution structures of the G quadruplexes formed by the insulin-linked polymorphic region sequences (57) could be resolved following completion of inosine for guanine substitution experiments along the lines reported earlier for the structure of the *Oxytricha* telomeric d(G₄T₄G₄T₄G₄T₄G₄) sequence in Na⁺-containing solution, where a proposed model of the folding topology (60) had to be corrected following inosine substitution experiments (61,62).

2.2.5 Dimeric RNA G quadruplex models

It has been suggested that guanine-rich regions may be involved in dimerization of retroviral RNAs through G quadruplex formation (63). Presumably, the proposed quadruplex involves formation of an intramolecular hairpin within G:A-rich segments, which can then dimerize through intermolecular association. This quadruplex model has been challenged subsequently, since an alternative dimerization site has been identified in HIV-1 (64,65) which does not involve G quadruplex formation.

2.2.6 Intramolecular RNA G quadruplex models

An intramolecular G quadruplex fold has also been postulated for a guanine-rich segment adjacent to an endonucleolytic cleavage site in insulin-like growth factor II mRNA (66). Chemical and enzymatic probing experiments have been interpreted in terms of the formation of a unimolecular G quadruplex conformation in Na⁺ and K⁺, but not in Li⁺, cation-containing solution.

2.3 G quadruplexes containing syn:syn:anti:anti G tetrad alignments and both parallel and antiparallel directionalities of adjacent strands

The relative alignment of strands around dimeric G quadruplexes is defined by the type of connecting loop linking the Gₙ segments. A key discovery was the identification of diagonal loops: initially in the *Oxytricha* telomere d(G$_4$T$_4$G$_4$) dimeric G quadruplex (46), and subsequently in the human telomere unimolecular G quadruplex d[AG$_3$(T$_2$AG$_3$)$_3$] (59) and the *Oxytricha* telomere unimolecular G quadruplex d[G$_4$(T$_4$G$_4$)$_3$] (46,61,62).

2.3.1 *Diagonal loops in G quadruplexes*

An NMR study of the *Oxytricha* telomere d(G$_4$T$_4$G$_4$) sequence in Na$^+$-containing solution identified formation of a G quadruplex with a folding topology (46) distinctly different from the corresponding topology for the same sequence observed in the crystalline state (45). This folding topology was verified from additional NMR measurements, including the inosine for guanine substitutions necessary for unambiguous spectral assignments (67).

The folding topology of the d(G$_4$T$_4$G$_4$) quadruplex in Na$^+$ solution involves head-to-tail alignment of a pair of d(G$_4$T$_4$G$_4$) segments containing diagonal connecting T$_4$ loops, as shown schematically in Fig. 13.10a (46). The formation of diagonal connecting loops affects both the directionality of adjacent strands around the G quadruplex and the *syn/anti* distribution of guanines around individual G tetrads. Specifically, individual strands have both a parallel and an antiparallel neighbour around the G quadruplex, *syn–anti–syn–anti* orientations are observed for guanines along individual G$_4$ segments and *syn:syn:anti:anti* alignments are observed for guanines around individual G tetrads (46). The hydrogen bond directionalities alternate between clockwise and anticlockwise orientations between adjacent stacked G tetrads in the quadruplex. The diagonal loop G quadruplex contains a twofold symmetry axis with one wide, one narrow, and two medium grooves. This diagonal loop-containing d(G$_4$T$_4$G$_4$) G quadruplex architecture is quite stable since the imino protons from the internal G tetrads exhibit very slow exchange rates on transfer from H$_2$O to D$_2$O solution (46).

2.3.2 *Solution structure of the human telomere d[AG$_3$(T$_2$AG$_3$)$_3$] quadruplex*

The sequence of the human telomere repeat d(T$_2$AG$_3$)$_n$ contains one less guanine than the corresponding d(T$_4$G$_4$)$_n$ *Oxytricha* and d(T$_2$G$_4$)$_n$ *Tetrahymena* telomeric repeats. The odd number of guanines in the human telomere repeat raises interesting questions about its folding topology and these have been addressed in a solution structure determination of the four AG$_3$ repeat human telomere d[AG$_3$(T$_2$AG$_3$)$_3$] quadruplex in Na$^+$-containing solution (59). These structural efforts have been complemented by chemical footprinting and base substitution studies on d(T$_n$AG$_3$)$_4$ sequences, where $n = 2$ and 4, which fold into intramolecularly folded G quadruplexes (68).

This structural characterization, which reported the first high resolution solution structure of a diagonal loop-containing G quadruplex, was undertaken on the d[AG$_3$(T$_2$AG$_3$)$_3$] sequence, since the d(T$_2$AG$_3$)$_4$ sequence gave poor quality NMR spectra, presumably owing to conformational heterogeneity. The resonance assignments in the d[AG$_3$(T$_2$AG$_3$)$_3$] 22-mer sequence were assigned after an in-depth analysis of NOE connectivities and on the basis of dU for T and partially successful inosine for guanine substitutions (59). The solution structure was solved by a combined NMR and molecular dynamics study including intensity-based refinement.

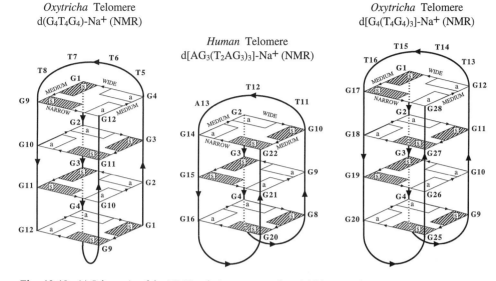

Fig. 13.10. (a) Schematic of the NMR solution structure-based folding topology adopted by the quadruplex formed through head-to-tail dimerization of the two-repeat *Oxytricha* telomere d(G₄T₄G₄) sequence in Na⁺ solution (46). The T₄ loop is of the diagonal type. The *syn* guanines are shown as hatched rectangles, while *anti* guanines are shown as open rectangles. (b) Schematic of the NMR solution structure-based folding topology adopted by the intramolecular quadruplex formed by the four-repeat human telomere d[AG₃(T₂AG₃)₃] quadruplex in Na⁺ solution (59). The central T₂A loop is of the diagonal type. (c) Schematic of the NMR solution structure-based folding topology adopted by the intramolecular quadruplex formed by the four-repeat *Oxytricha* telomere d[G₄(T₄G₄)₃] quadruplex in Na⁺ solution (61,62). The central T₄ loop is of the diagonal type.

The folding topology of the d[AG₃(T₂AG₃)₃] quadruplex in Na⁺ solution is shown schematically in Fig. 13.10b and the solution structure is shown in Fig. 13.11 (59). The solution structure contains three stacked G tetrads involving all 12 guanine residues in the sequence. The first and third TTA loops are of the lateral type, while the critical central TTA loop is of the diagonal type. These loop connectivities define the strand orientations such that individual strands have both a parallel and an antiparallel neighbour, as seen schematically in Fig. 13.10b. There is one wide, two medium, and one narrow groove in this quadruplex (59).

The guanine glycosidic torsion angles alternate between *anti* and *syn* (starting with an *anti* alignment at G2) along the entire length of the d[AG₃(T₂AG₃)₃] sequence, and the alternation remains in registry despite the intervening TTA loop segments. The guanines adopt *syn:syn:anti:anti* glycosidic torsion angles around individual G tetrads with the hydrogen bonding directionalities alternating between clockwise and anticlockwise orientations between adjacent stacked G tetrads, as seen schematically in Fig. 13.10b (59).

The overlap geometries between adjacent G tetrads at G(*syn*)–G(*anti*) and G(*anti*)–G(*syn*) steps in the solution structure are shown in Fig. 13.12a,b, respectively

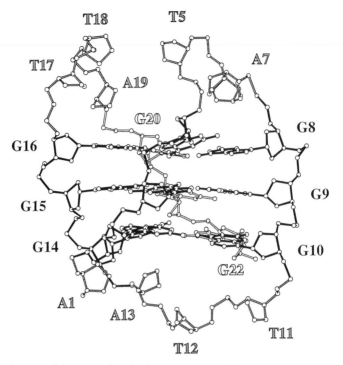

Fig. 13.11. A view of the NMR-based solution structure of the intramolecularly folded four-repeat human telomere d[AG₃(T₂AG₃)₃] quadruplex in Na⁺ solution (59). Three of the guanine-containing G₃ steps are shown with filled bonds while the remaining guanine-containing G₃ steps are shown with open bonds. The three loop segments (T5–T6–A7, T11–T12–A13, and T17–T18–A19) are shown with hatched bonds and the bases in these segments have been deleted in the interests of clarity. The central T11–T12–A13 loop is of the diagonal type.

(59). Base overlap between stacked G tetrads primarily involves the guanine five-membered rings at G(*syn*)–G(*anti*) steps (Fig. 13.12a) and the guanine six-membered rings at G(*anti*)–G(*syn*) steps (Fig. 13.12b).

Three of the four adenines in the sequence are stacked on adjacent G tetrads, while the fourth is tilted relative to the G tetrad plane. These stacking alignments involving loop adenine residues must contribute to the stabilization of the tertiary fold of the d[AG₃(T₂AG₃)₃] quadruplex. None of the thymines or adenines are involved in base pairing in the structure of the G quadruplex. The very slow exchange observed for imino protons of the internal G tetrads in the d(G₄T₄G₄) (46) and d[G₄(T₄G₄)₃] (46,61,62) G quadruplexes are not observed for the internal G tetrad in the d[AG₃(T₂AG₃)₃] G quadruplex (59), reflecting its marginal stability.

2.3.3 Energetics of the human telomere d(T₂AG₃)₄ quadruplex in solution

The thermodynamic parameters for human telomere G quadruplex formation have been determined from the concentration dependence of optical melting curves for d(T₂AG₃)₄ in Na⁺ and K⁺ solution (69). The estimated values are $\Delta G° = -3.3$ (−7.1) kJ/mol of tetrad and $\Delta H° = -54.3$ (−66.9) kJ/mol of tetrad in Na⁺ (K⁺) solution. These model-dependent thermodynamic parameters for the d(T₂AG₃)₄ G quadruplex

(a)

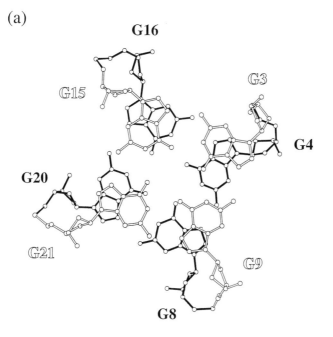

(b)

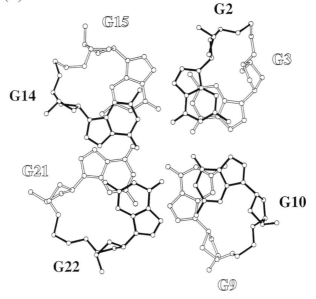

Fig. 13.12. Views down the helix axis showing stacking between adjacent G:G:G:G tetrads at (a) G(*syn*)–G(*anti*) and (b) G(*anti*)–G(*syn*) steps in the NMR-based solution structure of the intramolecularly folded four-repeat human telomere d[AG$_3$(T$_2$AG$_3$)$_3$] quadruplex (59). Individual G:G:G:G tetrads are drawn with either filled or open bonds.

(69) are a factor of two lower than their model-independent calorimetric counterparts for the d(TG$_3$T) (32) and d(G$_2$T$_5$G$_2$) (70) G quadruplexes. The origin of this discrepancy is not clear. It should be kept in mind that sequences such as d(T$_2$AG$_3$)$_4$ can, potentially, adopt a distribution of intramolecular folding topologies, and a meaningful evaluation of the energetics must be accompanied by a rigorous characterization of the conformational state(s) under consideration.

2.3.4 Solution structure of the Oxytricha telomere d(G$_4$T$_4$G$_4$) quadruplex

The details of the solution structure of the d(G$_4$T$_4$G$_4$) quadruplex in Na$^+$ solution have emerged from a combined NMR and molecular dynamics analysis of the spectral data (47). A view of the high resolution d(G$_4$T$_4$G$_4$) G quadruplex solution structure containing diagonal loops is shown in Fig. 13.13 (47). The guanine sugar puckers are of the S-type in the d(G$_4$T$_4$G$_4$) G quadruplex. The symmetry-related T$_4$ loop conformations are well defined, with the first and third thymines stacked over the terminal G tetrad planes, the second thymine stacked over the first thymine, and the last thymine looped out and somewhat disordered. The G tetrad overlaps at G(*syn*)–G(*anti*) and

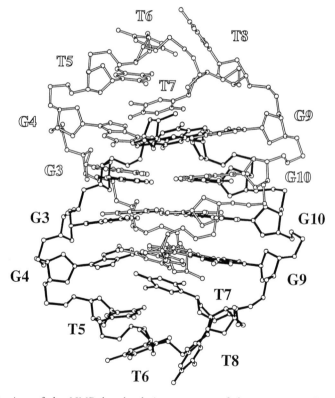

Fig. 13.13. A view of the NMR-based solution structure of the two-repeat *Oxytricha* telomere d(G$_4$T$_4$G$_4$) quadruplex formed through head-to-tail dimerization of a pair of hairpins in Na$^+$ solution (47). One d(G$_4$T$_4$G$_4$) hairpin is shown with filled bonds while the other is shown with open bonds. The T$_4$ loops are of the diagonal type.

G(*anti*)–G(*syn*) steps in the refined solution structure of the *Oxytricha* d(G$_4$T$_4$G$_4$) G quadruplex (47) are similar to those reported for these steps in the refined solution structure of the human d[AG$_3$(T$_2$AG$_3$)$_3$] G quadruplex (59). The structure of the lateral loop-containing d(G$_4$T$_4$G$_4$) G quadruplex in the crystalline state (45) and the diagonal loop-containing d(G$_4$T$_4$G$_4$) G quadruplex in solution (46) are directly compared Plate XIV.

More recent studies have established that the diagonal loop-containing fold of the d(G$_4$T$_4$G$_4$) G quadruplex is observed both in Na$^+$ and K$^+$-containing solution (71). Specific proton markers associated with the diagonal loop-linked d(G$_4$T$_4$G$_4$) G quadruplex underwent small shifts as average resonances on proceeding from Na$^+$ to K$^+$-containing solution. The monovalent cation selectivity of the diagonal loop-containing d(G$_4$T$_4$G$_4$) G quadruplex was assigned to the greater energetic cost of Na$^+$ dehydration relative to K$^+$ dehydration (71).

2.3.5 *Solution structure of the Oxytricha telomere d[G$_4$(T$_4$G$_4$)$_3$] quadruplex*

The *Oxytricha* telomere d[G$_4$(T$_4$G$_4$)$_3$] has the potential to fold into an intramolecular G quadruplex stabilized by four stacked G tetrads and three connecting T$_4$ loops. The d[G$_4$(T$_4$G$_4$)$_3$] sequence in Na$^+$ solution gives a surprisingly well-resolved imino proton spectrum corresponding to one predominant conformation (46). The first attempt at determining the solution structure of the d[G$_4$(T$_4$G$_4$)$_3$] quadruplex claimed to differentiate a folding topology favouring a lateral central loop over the alternative possibility of a diagonal central loop (60). This conclusion appeared to be questionable given the tentative nature of key guanine proton assignments and the paucity of details related to the computational protocols.

These uncertainties were resolved independently by two groups who solved the solution structure of the d[G$_4$(T$_4$G$_4$)$_3$] quadruplex in Na$^+$ solution based on an in-depth NMR and molecular dynamics computational approach (61,62). One of the groups incorporated six individual inosine for guanine substitutions (61), while the other used one inosine for guanine substitution and extensive comparison with related data on the d(G$_4$T$_4$G$_4$) quadruplex (62). These studies identified key assignment errors in the earlier NMR study (60) and ruled out the proposed central lateral loop in the intramolecularly folded G quadruplex.

The folding topology of the intramolecularly folded d[G$_4$(T$_4$G$_4$)$_3$] quadruplex in Na$^+$ solution is shown schematically in Fig. 13.10c (61,62) and its solution structure is shown in Fig. 13.14 (61). The structure is stabilized by four stacked G tetrads with a central diagonal T$_4$ loop and two lateral T$_4$ loops. The strand directionalities, guanine *syn/anti* alignments along individual strands and around G tetrads, and groove dimensions are the same in the G quadruplexes formed through dimerization of d(G$_4$T$_4$G$_4$) hairpins (Fig. 13.10a) (46) and through intramolecular folding of the d[G$_4$(T$_4$G$_4$)$_3$] sequence (Fig. 13.10c) (61,62).

A comparison of the folding schematics of the four guanine repeat *Oxytricha* telomere d[G$_4$(T$_4$G$_4$)$_3$] (Fig. 13.10c) (61,62) and human telomere d[AG$_3$(T$_2$AG$_3$)$_3$] (Fig. 13.10b) (59) quadruplexes establishes common elements in the folding topologies. Indeed, the three lower G tetrads in the d[G$_4$(T$_4$G$_4$)$_3$] quadruplex (Fig. 13.10c) exhibit the same structural features as the three tetrads in the d[AG$_3$(T$_2$AG$_3$)$_3$] quadruplex (Fig. 13.10b). These studies emphasize the importance of this folding topology

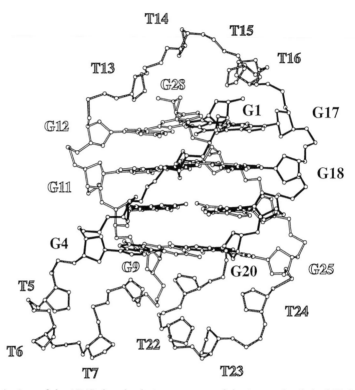

Fig. 13.14. A view of the NMR-based solution structure of the intramolecularly folded four-repeat *Oxytricha* telomere d[G$_4$(T$_4$G$_4$)$_3$] quadruplex in Na$^+$ solution (61). Two of the guanine-containing G$_4$ steps, directed towards the viewer, are shown with filled bonds while the other two guanine-containing G$_4$ steps, directed away from the viewer, are shown with open bonds. The three loop segments (T5–T6–T7–T8, T13–T14–T15–T16, and T21–T22–T23–T24) are shown with hatched bonds and the bases in these segments have been deleted in the interests of clarity. The central T13–T14–T15–T16 loop is of the diagonal type.

(Fig. 13.10b,c) for the solution structures of intramolecularly folded G quadruplexes, which is defined by a central diagonal loop (46).

2.3.6 Solution structure of the d(G$_2$T$_4$CG$_2$) quadruplex

The NMR parameters characteristic of G quadruplex formation were initially identified from a heteronuclear NMR study of the d(G$_2$T$_4$CG$_2$) sequence in Na$^+$ solution (58). The structure of this sequence, which contains a pair of G$_2$ repeats has been solved recently, with G quadruplex formation through dimerization of a pair of antiparallel d(G$_2$T$_4$CG$_2$) hairpins (72). The T$_4$C loops are of the diagonal type, which in turn defines the strand directionalities and the guanine *syn/anti* alignments along individual G$_2$ segments and around G tetrads. Thus, the d(G$_2$T$_4$CG$_2$) quadruplex containing two G tetrads (58,72) and the d(G$_4$T$_4$G$_4$) quadruplex containing four G tetrads (Fig. 13.10a) (46,47) adopt the same folding topology.

2.3.7 Energetics of the d(G₂T₅G₂) quadruplex in solution

The corresponding energetics for the order–disorder transition of the $d(G_2T_5G_2)$ quadruplex in Na⁺ solution have been measured calorimetrically and yield values of $\Delta G° = -15.9$ kJ/mol of tetrad and $\Delta H° = -117.0$ kJ/mol of tetrad at 25°C (70). These calorimetric parameters once again stress the importance of enthalpic contributions to the stability of G quadruplexes formed through alignment of a pair of diagonal loop-containing segments.

2.3.8 Solution structure of the d(G₃T₄G₃) quadruplex

The $d(G_3T_4G_3)$ sequence shows well-resolved NMR spectra in monovalent cation solution (73), with the spectral properties indicative of formation of an asymmetric G quadruplex through dimerization of a pair of $d(G_3T_4G_3)$ segments. Detailed NMR studies by two groups (74,75), including a molecular dynamics-based refinement (76), have identified the folding topology of the $d(G_3T_4G_3)$ quadruplex, which is shown schematically in Fig. 13.15. The solution structure of this G quadruplex contains several unusual features which are discussed below.

This G quadruplex, which contains three stacked G tetrads, forms through head-to-tail dimerization of a pair of $d(G_3T_4G_3)$ segments, with the directionality of the four strands defined by the diagonal alignment of the T_4 loops (74–76). The dimer is asymmetric as reflected in the 5′-*syn–syn–anti*–(loop)–*syn–anti–anti* alignments along one strand and 5′-*syn–anti–anti*–(loop)–*syn–syn–anti* alignments along the other, as shown

d(G₃T₄G₃)-Na⁺ (NMR)

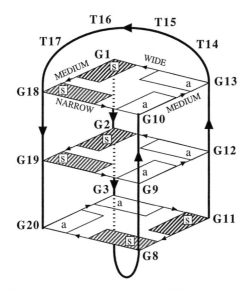

Fig. 13.15. Schematic of the NMR solution structure-based folding topology adopted by the quadruplex formed through head-to-tail dimerization of the two-repeat $d(G_3T_4G_3)$ sequence in Na⁺ solution (74–76). The T_4 loop is of the diagonal type.

schematically in Fig. 13.15. Each strand has both a parallel and an antiparallel neighbour and this is accompanied by *syn:syn:anti:anti* alignments around individual G tetrads (74–76).

It is interesting that one of two possible arrangements of diagonal loop folds for segment dimerization is favoured for formation of both the d($G_3T_4G_3$) quadruplex (74–76) and the d($G_4T_4G_4$) quadruplex (46,47). This preference has been attributed to the predominance of a specific intermediate in the folding pathway to G quadruplex formation (46).

2.3.9 Energetics of the d($G_3T_4G_3$) quadruplex in solution

The thermodynamic parameters for bimolecular G quadruplex formation have been determined from the concentration dependence of optical melting curves for d($G_3T_4G_3$) in Na^+ and K^+ solutions (73). The estimated values are $\Delta G° = -10.9\ (-16.7)$ kJ/mol of tetrad, $\Delta H° = -96\ (-133)$ kJ/mol of tetrad and $\Delta S° = -288\ (-393)$ J/K mol of tetrad in Na^+ (K^+) solution. These model-dependent thermodynamic parameters for the d($G_3T_4G_3$) G quadruplex (73) compare favourably with their model-independent calorimetric counterparts for the d(TG_3T) (32) and d($G_2T_5G_2$) (70) G quadruplexes.

2.4 A G quadruplex containing a double chain reversal loop, syn:syn:syn:anti and anti:anti:anti:syn G tetrads, and unequal strand directionalities

The G quadruplex structures presented above contained either lateral or diagonal central loops which defined the strand directionalities and the G(*syn*)/G(*anti*) distribution along given strands and around individual G tetrads. These quadruplexes contained even numbers of G(*syn*)/G(*anti*) residues around a given G tetrad and equal numbers of strands pointing in opposite directions. An exception to these rules has emerged following structure determination of the *Tetrahymena* telomere d(T_2G_4)$_n$ G quadruplex.

2.4.1 Solution structure of the Tetrahymena telomere d(T_2G_4)$_4$ quadruplex

The *Tetrahymena* telomere d(T_2G_4)$_n$ sequence differs from its *Oxytricha* telomere d(T_4G_4)$_n$ counterpart in having two fewer thymines per repeat that can potentially influence the loop topology involved in chain reversal. It is also conceivable that some of the guanines could participate in chain reversal, making it unclear as to the number of G tetrads stabilizing *Tetrahymena* telomere d(T_2G_4)$_n$ quadruplexes.

Initial efforts to address this issue focused on the four repeat *Tetrahymena* d(T_2G_4)$_4$ sequence in Na^+-containing solution, which was studied by non-denaturing gel electrophoresis, chemical footprinting, UV cross-linking, and NMR experiments (29). The data were interpreted in terms of an intramolecularly folded G quadruplex stabilized by three G tetrads, three T_2G lateral loops, and between 4 and 6 *syn*-guanines in the folded structure (29).

The same sequence has been investigated further based on additional NMR characterization, combined with intensity-restrained molecular dynamics computations (77). The study focused on the predominant conformation exhibiting narrow NMR resonances in the presence of a broad spectral envelope indicative of aggregated species.

Tetrahymena Telomere
d(T₂G₄)₄-Na⁺ (NMR)

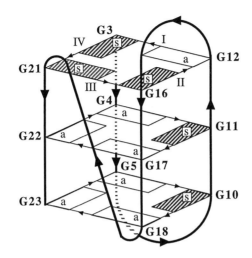

Fig. 13.16. Schematic of the NMR solution structure-based folding topology adopted by the intramolecular four-repeat *Tetrahymena* telomere d(T₂G₄)₄ quadruplex in Na⁺ solution (77). The T19–T20 loop segment forms a double chain reversal. The *syn* guanines are shown as hatched rectangles while *anti* guanines are shown as open rectangles.

The folding topology of the *Tetrahymena* telomere d(T₂G₄)₄ quadruplex is shown schematically in Fig. 13.16 and its solution structure shown in Fig. 13.17. The structure is unprecedented in terms of the *syn/anti* distribution along individual guanine stretches and around individual G tetrads, the directions of the four strands around the G quadruplex, and the presence of a loop involved in a double chain reversal (77).

The solution structure contains three G tetrads connected by three loop segments. The first, four-base GT₂G lateral loop is followed by a second, three-base T₂G lateral loop, and then by a third, two-base T₂ loop involved in double chain reversal, as shown schematically in Fig. 13.16. Since the double chain reversal T₂ loop connects two strands that are aligned in parallel, the overall G quadruplex contains three of the four strands aligned in one direction and the remaining strand aligned in the opposite direction (77). This results in four unique grooves around the G quadruplex, one of which is spanned by the T₂ loop. The two lateral loops are stabilized through formation of a wobble G:T base pair which stacks over the adjacent G tetrad in the structure of the G quadruplex.

Furthermore, the guanines adopt either *syn–anti–anti* or *syn–syn–anti* patterns along individual strands and *syn:syn:syn:anti* and *anti:anti:anti:syn* patterns around individual G tetrads within the G quadruplex, as shown schematically in Fig. 13.16 (77). There are two unique G–G steps in the solution structure of the *Tetrahymena* telomere d(T₂G₄)₄ G quadruplex with distinct stacking patterns. The G(*syn*)–G(*anti*) steps have an overlap pattern (Fig. 13.18a) that is similar to what has been observed for related steps in other

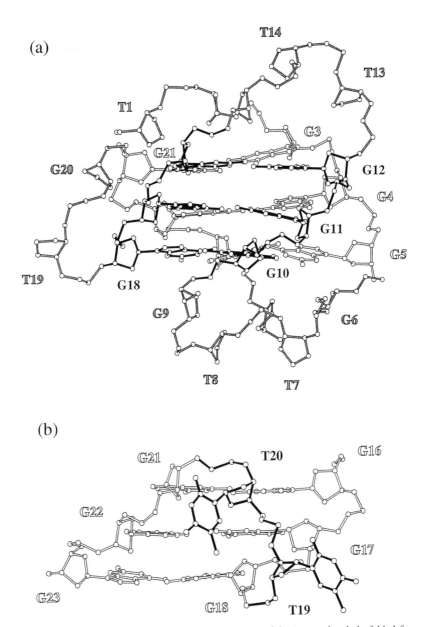

Fig. 13.17. (a) A view of the NMR-based solution structure of the intramolecularly folded four-repeat *Tetrahymena* telomere d(T$_2$G$_4$)$_4$ quadruplex in Na$^+$ solution (77). Two of the guanine-containing G$_3$ steps, directed towards the viewer, are shown with filled bonds while the other two guanine-containing G$_3$ steps, directed away from the viewer, are shown with open bonds. The three loop segments (G6–T7–T8–G9, T13–T14–G15 and T19–T20) are shown with hatched bonds and the bases in these segments have been deleted in the interests of clarity. The T19–T20 loop is of the double chain reversal type. (b) A close-up of the double chain reversal loop involving T19–T20 which connects G16–G17–G18 and G21–G22–G23 segments that are aligned in parallel in the solution structure of the intramolecularly folded four-repeat *Tetrahymena* telomere d(T$_2$G$_4$)$_4$ quadruplex (77).

(a)

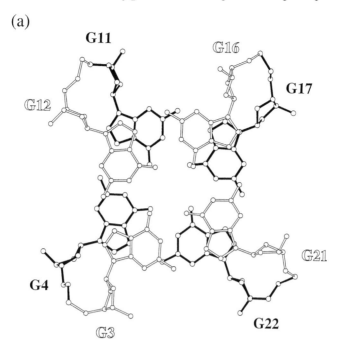

(b)

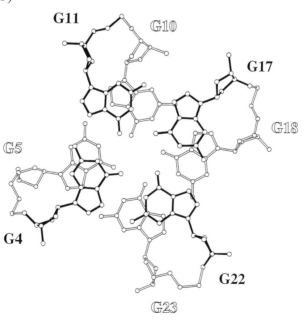

Fig. 13.18. Views down the helix axis showing stacking between adjacent G:G:G:G tetrads at (a) G(*syn*)–G(*anti*) and (b) G(*syn*)–G(*syn*) steps in the NMR-based solution structure of the intramolecularly folded four-repeat *Tetrahymena* telomere d(T$_2$G$_4$)$_4$ quadruplex (77). Individual G:G:G:G tetrads are drawn with either filled or open bonds.

G quadruplexes, while the G(*anti*)–G(*anti*) or G(*syn*)–G(*syn*) steps exhibit a stacking pattern (Fig. 13.18b) similar to that previously observed in an all parallel-stranded G quadruplexes (35). The two adjacent *anti:anti:anti:syn* G tetrads have the same clockwise hydrogen bond directionalities, in contrast to the anticlockwise hydrogen bond directionality of the *syn:syn:syn:anti* G tetrad, as shown schematically in Fig. 13.16 (77).

The solution structure of the d(T$_2$G$_4$)$_4$ in Na$^+$ solution (77) is in good agreement with the footprinting and cross-linking experiments reported previously (12,29). The *Tetrahymena* telomere d(T$_2$G$_4$)$_n$ sequence differs from its human telomere d(T$_2$AG$_3$)$_n$ counterpart in that a single G in the former sequence is replaced by an A in the latter sequence. This small difference results in distinctly different folding topologies for the *Tetrahymena* (Fig. 13.16) and human (Fig. 13.10b) G quadruplexes, with differences in strand directionalities, guanine *syn/anti* distributions along strands and around G tetrads, and in the number of bases and orientations of the connecting loop segments. The solution structures of the human (59) and *Tetrahymena* (77) telomere G quadruplexes are compared directly in Plate XV.

2.5 Telomeric sequence G quadruplexes containing G tetrads and base triads

The terminal G tetrads of a G quadruplex can potentially serve as templates for the stepwise annealing of novel stacked, multistranded pairing alignments. Such alignments could be unusual base mismatches, base triples and tetrads, and, as is shown in

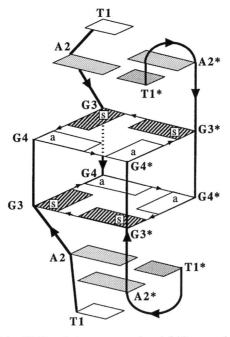

Fig. 13.19. Schematic of the NMR solution structure-based folding topology adopted by the four-stranded single repeat *Bombyx mori* telomere analogue d(TAG$_2$) quadruplex in Na$^+$ solution (78). This folding topology contains stacked A:(A:T) triads and G:G:G:G tetrads.

an example below, base triads. Such a concept provides an approach for the construction of novel multistranded structures emanating from a G tetrad foundation.

2.5.1 Solution structure of the Bombyx mori telomere d(T₂AG₂) quadruplex

The *Bombyx mori* telomere $d(T_2AG_2)_n$ sequence differs from the human telomere $d(T_2AG_3)_n$ sequence in having one less guanine in the repeat. The single repeat $d(T_2AG_2)$ sequence and its truncated $d(TAG_2)$ version give exceptionally well-resolved NMR spectra in Na^+-containing solution, exhibiting imino proton resonances between 11 and 12 ppm characteristic of G tetrad formation (78).

Single guanine-rich repeat segments are known to form parallel-stranded G quadruplexes containing *anti*-glycosidic torsion angles at the guanine residues. By contrast, both $d(TAG_2)$ and $d(T_2AG_2)$ contain a *syn*-guanine at the 5′-G residue, ruling out formation of a parallel-stranded G quadruplex. The $d(TAG_2)$ sequence [also the $d(T_2AG_2)$ sequence] forms a twofold, symmetric, four-stranded G quadruplex which is shown schematically in Fig. 13.19 (78). This G quadruplex contains two stacked *syn:syn:anti:anti* G tetrads, with individual strands having both a parallel and antiparallel neighbour around the quadruplex (Fig. 13.19). The solution structure of the G quadruplex is shown in Fig. 13.20.

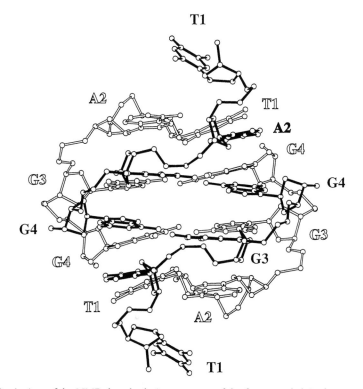

Fig. 13.20. A view of the NMR-based solution structure of the four-stranded single repeat *Bombyx mori* telomere analogue $d(TAG_2)$ quadruplex in Na^+ solution (78). Two of the strands, directed towards the viewer, are shown with filled bonds while the other two strands, directed away from the viewer, are shown with open bonds.

(a)

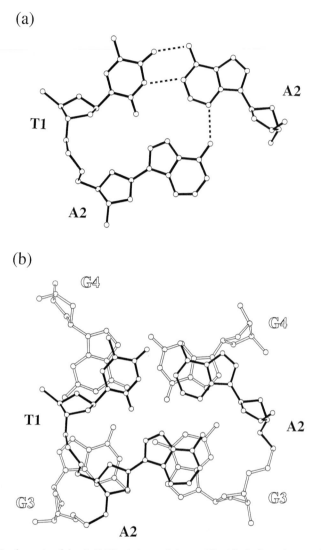

(b)

Fig. 13.21. (a) A schematic of the A:(A:T) triad containing an T1–A2 platform that was identified in the solution structure of the *Bombyx mori* telomere analogue d(TAG₂) quadruplex (78). (b) A view down the helix axis showing the overlap between the A:(A:T) triad and the G:G:G:G tetrad in the solution structure of the *Bombyx mori* telomere analogue d(TAG₂) quadruplex (78).

The two G tetrads are capped by novel (T:A):A triads, shown schematically in Fig. 13.21a, where an A residue hydrogen bonds to the minor groove edge of a Watson–Crick T:A base pair. The (T:A):A triad (Fig. 13.21a) contains a T–A base platform, where two sequential bases are aligned in the same plane (78). The concept of base triads had been postulated earlier on the basis of modelling studies (79), while base platforms were initially observed experimentally at three A–A steps in the crystal structure of the P4–P6 domain of the *Tetrahymena* self-splicing group I ribozyme (80). The overlap geometry between the (T:A):A triad and the G tetrad is shown in Fig. 13.21b.

2.6 G quadruplex recognition

The unique folding topologies associated with individual families of G quadruplex architectures make them attractive targets for ligands ranging from small organic molecules to proteins. There is a limited literature on small molecule recognition and a more extensive literature on protein recognition of G quadruplexes and these results are presented below from a structural perspective.

2.6.1 Small molecules complexed to G quadruplexes

There has been considerable interest in identifying small molecules that target G quadruplexes and are capable of forming site-specific stable complexes. Both ethidium bromide (81) and carbocyanine dyes (82) bind to G quadruplexes, but these efforts have not provided specific complexes necessary for structural characterization. More recently, DNA aptamers containing guanine-rich repeats capable of G quadruplex formation have been identified based on their ability to target anionic porphyrin ligands (83,84). The structure of this family of complexes will be of considerable interest given that the dimensions of the porphyrin ligand are comparable to that of the G tetrad.

2.6.2 Therapeutic potential of G quadruplexes

Three examples point to the potential of G quadruplex-based therapeutics, as reflected by the ability of this architecture to target functional proteins. Thus, a combinatorially selected, parallel-stranded G quadruplex was shown to be a potent inhibitor of HIV envelope-mediated cell fusion (85). The molecular basis of this recognition remains undefined at present.

The crystal structure of the thrombin-binding intramolecularly folded $d(G_2T_2G_2TGTG_2T_2G_2)$ DNA aptamer complexed to thrombin has been solved to 2.9 Å resolution (54). Molecular recognition involves ionic and hydrophobic interactions between loop segments of the G quadruplex fold and distinct regions (putative heparin-binding site and fibrinogen exosite) on two different thrombin molecules.

A DNA oligomer containing tandem guanine repeats and capable of intramolecularly folded G quadruplex formation in K^+ solution has been shown to be amongst the most active inhibitors of HIV integrase (86–88). The molecular characterization of this G quadruplex in the absence and presence of bound HIV integrase will be of great interest since the K^+ cation-folded loop domain of the G quadruplex has been shown to be involved in targeting the binding site on the HIV integrase (89,90).

2.6.3 Proteins that target G quadruplexes

Recent studies have identified a number of proteins that either facilitate DNA G quadruplex formation (91–93) or bind to parallel-stranded DNA G quadruplexes (94–96), including a nuclease that cleaves DNA 5′ to the G quadruplex fold (97,98). In addition, a cytoplasmic exoribonuclease has recently been shown to target RNA G quadruplexes preferentially (99). Currently, nothing is known about the molecular basis of G quadruplex–protein recognition in these systems. Several of these complexes represent attractive and challenging structural characterization projects.

2.7 Biological relevance of G tetrad-containing G quadruplexes

Sequences other than telomeres contain guanine repeats. These include immunoglob-ulin switch regions (10), insulin-linked polymorphic regions associated with diabetes mellitus (56), retinoblastoma susceptibility genes (100), and the control region of c-*myc* (101). These sequences form G quadruplexes *in vitro* but it remains to be estab-lished whether such quadruplexes play a biological role *in vivo*. There is some indirect evidence suggesting a potential biological role for G quadruplexes. Thus, both the β subunit of the *Oxytricha* telomere-binding protein (91,92,102) and the yeast Rap1 protein (93) exhibit molecular chaperone function in their ability to accelerate G quadruplex formation. Similarly, mutations in the yeast *KEM1* gene, which encodes a nuclease specific for G quadruplex DNA, have been shown to affect meiosis and mitosis (97). More research is needed to address definitively the issues related to potential biological roles for G quadruplexes.

3 G:C:G:C tetrad-containing quadruplexes

3.1 Triplet repeat disease sequence quadruplexes containing G:C:G:C tetrads formed through alignment of major groove edges of Watson–Crick G:C pairs

The discovery of the expansion of d(CGG)$_n$:d(CCG)$_n$ repeats associated with the fagile X syndrome (103–106) has stimulated spectroscopic and footprinting efforts to delineate the potential folding topologies adopted by such sequences. Indeed, it has been shown that the d(CGG)$_n$ repeat (n = 7) forms a stable quadruplex structure which is suggested to be of the all-parallel-stranded type, and that this process is facili-tated by methylation of the cytosine residues (107).

3.1.1 Solution structure of the d(GCG$_2$T$_3$GCG$_2$) quadruplex containing CG$_2$ fragile X syndrome triplet repeats

The d(GCG$_2$T$_3$GCG$_2$) sequence contains both guanines and cytosines with the potential of forming tetrads containing a mixture of G and C residues. The d(GCG$_2$T$_3$GCG$_2$) sequence in Na$^+$ solution exhibits exceptionally well-resolved, narrow resonances corresponding to formation of a single conformation (108). The NMR resonances were assigned definitively with the aid of inosine for guanine and uracil for thymine substitutions and the structure was solved by molecular dynamics calculations including intensity-based refinements. The resulting quadruplex forms through head-to-tail dimerization of a pair of d(GCG$_2$T$_3$GCG$_2$) hairpins, as shown schematically in Fig. 13.22a. The structure of this quadruplex in shown in Fig. 13.23 (108). The twofold symmetry in this quadruplex required the use of a sum-averaging protocol in the XPLOR molecular dynamics program (109) to overcome uncertainties associated with intramolecular versus intermolecular NOE contributions between pairs of protons (110,111) The connecting T$_3$ loops are of the lateral type, with adja-cent strands aligned in an antiparallel orientation around the quadruplex (108). The outer tetrads are of the G(*syn*):G(*anti*):G(*syn*):G(*anti*) type (see Fig. 13.1), while the inner tetrads are of the G(*anti*):C(*anti*):G(*anti*):C(*anti*) type, as shown schematically in

(a)

d(G-C-G-G-T₃-G-C-G-G)

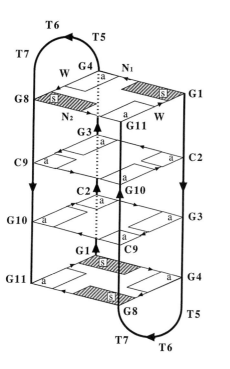

(b)

d(G-G-G-C-T₄-G-G-G-C)

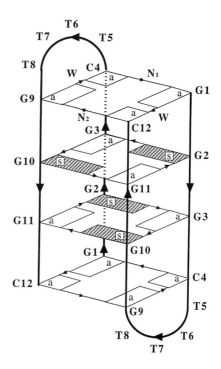

Contains C-G-G Triplet Repeats of
Fragile X Syndrome Disease Gene

Contains G-G-G-C Repeats of
Adeno-associated Virus Gene

Fig. 13.22. (a) Schematic of the NMR solution structure-based folding topology adopted by the quadruplex formed through head-to-tail dimerization of the d(GCG₂T₃GCG₂) sequence in Na⁺ solution (108). This topology contains outer G:G:G:G tetrads and inner G:C:G:C tetrads. The T₃ loop is of the lateral type. The *syn* guanines are shown as hatched rectangles while *anti* guanines are shown as open rectangles. (b) Schematic of the NMR solution structure-based folding topology adopted by the quadruplex formed through head-to-tail dimerization of the d(G₃CT₄G₃C) sequence in Na⁺ solution (116). This topology contains outer G:C:G:C and inner G:G:G:G tetrads. The T₄ loop is of the lateral type.

Fig. 13.24a. The stacking between the outer G:G:G:G and inner G:C:G:C terads is shown in Fig. 13.25a. This result represented the first experimental demonstration of a G:C:G:C tetrad involving pairing along the major groove edges of Watson–Crick G:C base pairs (108) (for earlier models, see refs 112–114). Both cytosine exocyclic amino protons are hydrogen bonded in this major groove-aligned G:C:G:C tetrad (Fig. 13.24a), which is consistent with both cytosine amino protons resonating at *c.* 9 ppm in the NMR spectrum. Furthermore, the observed NOEs between the

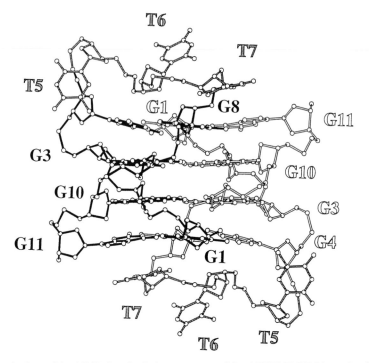

Fig. 13.23. A view of the NMR-based solution structure of the d(GCG$_2$T$_3$GCG$_2$) quadruplex formed through head-to-tail dimerization of a pair of hairpins in Na$^+$ solution (108). The tetrad segments of one d(GCG$_2$T$_3$GCG$_2$) hairpin is shown with filled bonds while the other is shown with open bonds. The T$_3$ loops are of the lateral type and are shown by hatched bonds.

guanine H8 and cytosine H5 protons across the Watson–Crick G:C base pairs of the tetrad provide key restraints defining the alignment in the central G:C:G:C tetrads in the solution structure of the d(GCG$_2$T$_3$GCG$_2$) quadruplex (108).

3.1.2 Solution structure of the d(G$_3$CT$_4$G$_3$C) quadruplex formed by G$_3$C repeats observed in adeno-associated viral DNA

The adeno-associated virus, a human parvovirus, is unique amongst eukaryotic DNA viruses in its ability to integrate site specifically into a defined region of chromosome 19 (reviewed in ref. 115). The G$_3$C sequence has been identified both in adeno-associated virus (as islands) and in chromosome 19 (as tandem repeats) and could play a role in the mechanism of site-specific integration. The NMR spectrum of the d(G$_3$CT$_4$G$_3$C) sequence, which contains two G$_3$C segments separated by a T$_n$ linker (*n* = 3 or 4), exhibits a set of resonances corresponding to a predominant conformation in Na$^+$ solution (116). The NMR resonances in the d(G$_3$CT$_4$G$_3$C) sequence were assigned unambiguously with the aid of site specifically incorporated ^{15}N-labelled guanines [inosine for guanine substitutions did not work in this case owing to destabilization of the d(G$_3$CT$_4$G$_3$C) structure on inosine substitution] and the structure was solved using the sum-averaging routine during both distance and intensity refined molecular dynamics calculations (116). The folding topology of the d(G$_3$CT$_4$G$_3$C)

(a)

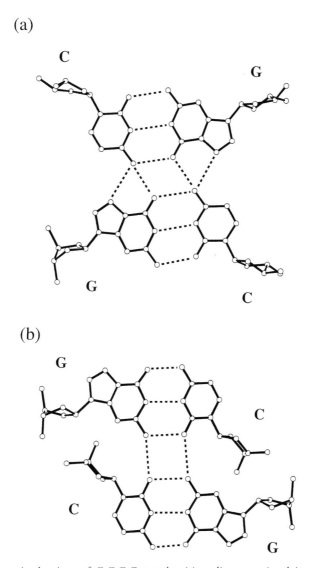

(b)

Fig. 13.24. Schematic drawings of G:C:G:C tetrad pairing alignments involving dimerization of Watson–Crick G:C base pairs along (a) their major groove edges (108) and (b) their minor groove edges (118).

quadruplex structure in Na$^+$ solution is shown schematically in Fig. 13.22b. This quadruplex forms through head-to-tail dimerization of a pair of d(G$_3$CT$_4$G$_3$C) hairpins involving connecting T$_4$ lateral loops and individual strands running antiparallel to each other around the quadruplex. This quadruplex also contains a pair of separated G:C:G:C tetrads formed through major groove alignment of a pair of Watson–Crick G:C base pairs, as shown previously in Fig. 13.24a. The structure of this quadruplex is shown in Fig. 13.26 (116). The base overlaps between the outer

(a)

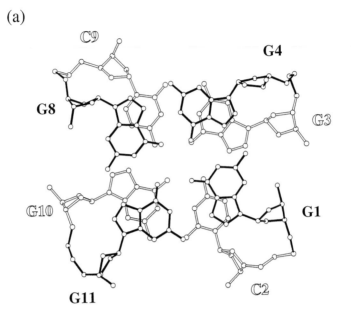

(b)

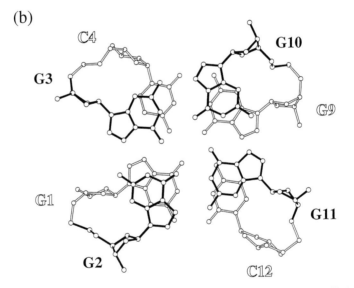

Fig. 13.25. Views down the helix axis showing stacking between adjacent G:G:G:G (filled bonds) and G:C:G:C (open bonds) tetrads in: (a) the solution structure of the d(GCG$_2$T$_3$GCG$_2$) quadruplex in Na$^+$ solution (108); and (b) the solution structure of the d(G$_3$CT$_4$G$_3$C) quadruplex in Na$^+$ solution (116).

G(*anti*):C(*anti*):G(*anti*):C(*anti*) and inner G(*syn*):G(*anti*):G(*syn*):G(*anti*) tetrads is shown in Fig. 13.25b.

The above studies on the folding topologies of quadruplexes formed through dimerization of the d(GCG$_2$T$_3$GCG$_2$) (Fig. 13.22a) (108) and d(G$_3$CT$_4$G$_3$C)

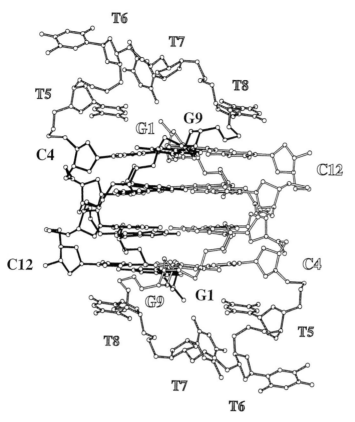

Fig. 13.26. A view of the NMR-based solution structure of the d(G$_3$CT$_4$G$_3$C) quadruplex formed through head-to-tail dimerization of a pair of hairpins in Na$^+$ solution (116). The tetrad segment of one d(G$_3$CT$_4$G$_3$C) hairpin is shown with filled bonds while the other is shown with open bonds. The T$_4$ loops are of the lateral type and are shown by hatched bonds.

(Fig. 13.22b) (116) sequences establish the prevalence of G:C:G:C tetrad formation (Fig. 13.24a) and that such tetrads can be either adjacent (Fig. 13.22a) or separated (Fig. 13.22b) from each other in the quadruplex, depending on sequence.

3.1.3 A Na$^+$ to K$^+$ cation-dependent conformational switch in the loop-spanning segment of a
       G$_3$C repeat-containing quadruplex

The role of Na$^+$ versus K$^+$ in stabilizing DNA quadruplexes has been one of considerable interest. The most favourable situation for a structural analysis of monovalent cation-dependent conformations would be one where distinct NMR spectra were observable for a quadruplex in Na$^+$ solution on the one hand and in K$^+$ solution on the other, and, in addition, interconversion between these distinct quadruplex conformations were slow on the NMR time-scale. The NMR spectrum of the d(G$_3$CT$_4$G$_3$C) sequence in K$^+$ solution exhibits a set of resonances corresponding to a predominant conformation (117) that is distinct from its predominant conformational counterpart in Na$^+$ solution (116). Furthermore, the distinct conformations of the d($G_3CT_4G_3C$) sequences in Na$^+$ and K$^+$ solutions are in slow exchange in solutions containing a mixture of these monovalent cations.

The solution structure of the d($G_3CT_4G_3C$) quadruplex in K$^+$ solution has been solved (Plate XVIb) (117) and, together with the corresponding quadruplex structure in Na$^+$ solution (Plate XVIa) (116), defines the molecular basis of the Na$^+$ to K$^+$ cation-dependent conformational switch. Both Na$^+$ and K$^+$ cation-dependent conformations of the d($G_3CT_4G_3C$) quadruplexes exhibit certain common structural features, which include head-to-tail dimerization of symmetry-related hairpins, antiparallel alignment of adjacent strands, and stacked adjacent G(*syn*):G(*anti*):G(*syn*):G(*anti*) tetrads in the central core of the quadruplexes. The two quadruplex conformations differ in the conformations of the T$_4$ loops (Fig. 13.27a,b for Na$^+$ and K$^+$ conformations, respectively), the relative alignment of opposing Watson–Crick G:C base pairs across

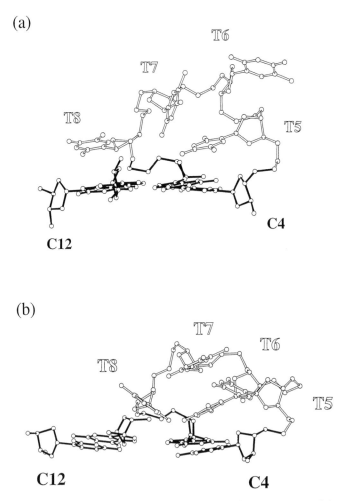

Fig. 13.27. The fold of the T5–T6–T7–T8 hairpin loop in the solution structure of the d($G_3CT_4G_3C$) quadruplex formed through head-to-tail dimerization of a pair of hairpins in: (a) Na$^+$ solution (116) and (b) K$^+$ solution (117).

(a)

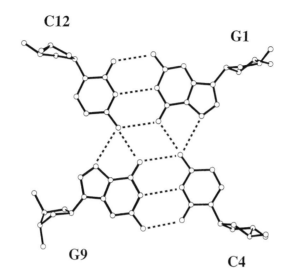

(b)

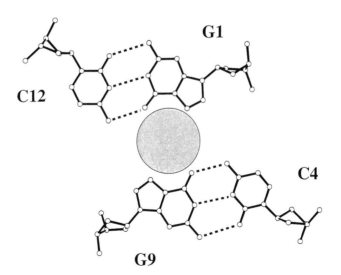

Fig. 13.28. The alignment of opposing Watson–Crick G:C base pairs along their major groove edges in the solution structure of the d($G_3CT_4G_3C$) quadruplex formed through head-to-tail dimerization of a pair of hairpins in: (a) Na^+ solution (116) and (b) K^+ solution (117).

Note the role of the potentially bound K^+ cation in coordinating to the O6 and N7 acceptor atoms of guanines whose Hoogsteen edges are directed towards each other in (b).

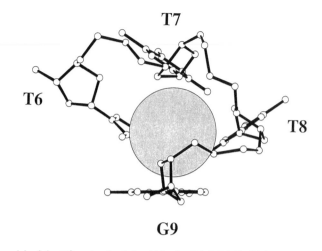

Fig. 13.29. A model of the K⁺ cation buried within the T6–T7–T8–G9 loop segment in the solution structure of the d(G₃CT₄G₃C) quadruplex formed through head-to-tail dimerization of a pair of hairpins in K⁺ solution (117).

the major groove (Fig. 13.28a,b for Na⁺ and K⁺ conformations, respectively), and the total number of potential monovalent cation-binding sites (116,117). Single K⁺-binding cavities were proposed within each of the symmetry-related T₃G loop-spanning segments (Fig. 13.29) resulting in two additional potential monovalent cation-binding sites in the K⁺-stabilized d(G₃CT₄G₃C) quadruplex (Plate XVIb) relative to its Na⁺-stabilized counterpart (Plate XVIa). The major groove edges of opposing guanines from Watson–Crick G:C base pairs are bridged by potential coordinated K⁺ cations in the d(G₃CT₄G₃C) quadruplex conformation in K⁺ solution (Fig. 13.28b) (117), in contrast to the G:C:G:C tetrad formation in Na⁺ solution (Fig. 13.28a) (116).

The solution structure of the K⁺-stabilized d(G₃CT₄G₃C) quadruplex defines the principles involved in potential K⁺ coordination within a T₃G segment, resulting in a defined loop architecture whose outwardly pointing functional groups can provide a unique folded topology that can target potential receptor sites (117). Indeed, the biological significance of this result is likely to be related to the independent demonstration of K⁺-selective folding of loop domains within intramolecular G quadruplexes, with these uniquely folded loops responsible for the potent oligonucleotide inhibitory activity against HIV integrase (86–88).

3.2 Quadruplexes containing G:C:G:C tetrads formed through alignment of minor groove edges of Watson–Crick G:C pairs

The examples above defined the alignment associated with the pairing of two Watson–Crick G:C base pairs through their major groove edges to form G:C:G:C tetrads (Fig. 13.24a) which are stabilized through their participation with G:G:G:G tetrads in quadruplex formation (108,116). Such G:C:G:C tetrad formation (Fig. 13.24a) is facilitated by the glycosidic bonds being directed towards four corners of the tetrad, as they do for G:G:G:G tetrad formation (Fig. 13.1). An interesting issue relates to whether G:C:G:C tetrads can also form through alignment of the minor groove edges of two Watson–Crick G:C base pairs. In this case, pairs of glycosidic bonds would be directed towards each other and steric constraints may require departures from base

(a)

(b)

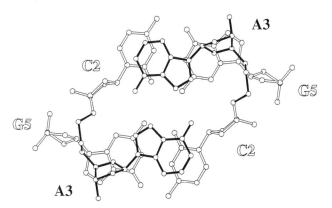

Fig. 13.30. (a) Schematic of the X-ray crystallographic structure-based folding topology adopted by the quadruplex formed through head-to-head dimerization of the d(GCATGCT) sequence (118). This topology contains a pair of G:C:G:C tetrads flanked on one side by a reversed A:A mismatch. The G:C:G:C tetrads involve alignment across the minor groove edges of Watson–Crick G:C base pairs. The A–T loops are of the lateral type. Reproduced with permission of *Structure*. (b) A view down the helix axis showing the stacking between the adjacent reversed A:A mismatch (filled bonds) and the G:C:G:C tetrad (open bonds) in the X-ray structure of the d(GCATGCT) quadruplex (118).

planarity around this alternative G:C:G:C tetrad alignment. Recent X-ray structures of specific G:C- (118) and A:T- (119) containing sequences, which are described below, have provided molecular views defining the alignment of Watson–Crick G:C pairs (and Watson–Crick A:T pairs) along their minor groove edges.

3.2.1   Crystal structure of the d(GCATGCT) quadruplex

The 1.8 Å X-ray structure of the d(GCATGCT) sequence has defined a new quadruplex architecture (118). The structure involves head-to-head dimerization of a pair of hairpins, as shown schematically in Fig. 13.30a, with the structure shown in Fig. 13.31a. The quadruplex structure contains two stacked G:C:G:C tetrads and one A:A mismatch. The quadruplex fold contains a twofold element of symmetry with adjacent strands running antiparallel to each other, all glycosidic torsion angles in the *anti* range, and all sugar puckers in the C2′-*endo* range.

Formation of Watson–Crick G:C pairs involve cross-strand alignment of guanines and cytosines, with further pairing of the minor groove edges of the G:C pairs through two hydrogen bonds to form the G:C:G:C tetrads shown schematically in Fig. 13.24b (118). The bases in the G:C:G:C tetrad are not coplanar but are tilted by *c*. 30°. There is extensive stacking between adjacent G:C:G:C tetrads in the core of the quadruplex through overlap of the cytosine pyrimidine rings and the guanine six-membered rings, as shown in Fig. 13.31b.

The adenines form an A:A mismatch through cross-strand alignment involving a pair of hydrogen bonds along their major groove Hoogsteen edges. This A:A mismatch, involving A residues in the TA loops, anchor the quadruplex achitecture. The stacking between the A:A mismatch and the G:C:G:C tetrad is shown in Fig. 13.30b (118). In addition, the purine ring of G5 stacks over the sugar ring of A3 in a van der Waals interaction similar to that which has been observed previously in the crystal structure of Z-DNA.

The sugar–phosphate backbones of these G:C:G:C tetrads formed through minor groove alignment in the d(GCATGCT) quadruplex (118) are distinct from those observed for the G:C:G:C tetrads formed through major groove alignment in the d($GCG_2T_3GCG_2$) (108) and d($GC_3T_4GC_3$) (116) quadruplexes presented earlier in this chapter. There is a close juxtaposition of the backbone phosphates of C2 and C6 which are coordinated to a cation in the structure of the d(GCATGCT) quadruplex. Furthermore, two molecules of the d(GCATGCT) quadruplex are aligned in the crystallographic lattice through T:T mismatch formation involving the looped out T residue of the AT loop (118).

3.2.2   Crystal structure of the d<pAT_2CAT_2C> quadruplex

The novel architecture that defines the structure of the d(GCATGCT) quadruplex (118) presented above has recently been observed in the d<pAT_2CAT_2C> DNA oligomer as well (cyclized in this case). A key feature common to both sequences is the separation of complementary 5′-purine–pyrimidine dinucleotide steps within the d(··RYNYRYN··) sequence context, where R is a purine, Y is a pyrimidine, and N is any nucleotide.

The high resolution X-ray structure of the cyclic octanucleotide d<pAT_2CAT_2C> establishes quadruplex formation through dimerization (119). This structure involves

(a)

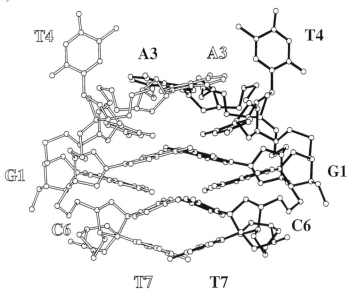

(b)

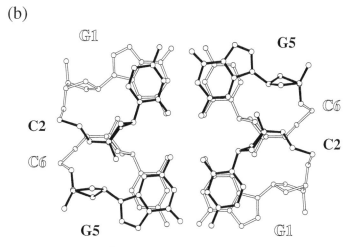

Fig. 13.31. (a) A view of the 1.8 Å X-ray crystallographic structure of the d(GCATGCT) quadruplex formed through head-to-head dimerization of a pair of hairpins (118). One strand is shown with darkened bonds while the other strand is shown with open bonds. The bases in the G:C:G:C tetrads depart significantly from planarity. (b) A view down the helix axis showing the stacking between the adjacent G:C:G:C tetrads (filled and open bonds, respectively) in the X-ray structure of the d(GCATGCT) quadruplex (118).

cross-strand formation of Watson–Crick A:T base pairs involving the A–T steps, with the minor groove edges of these A:T base pairs directed towards each other. The A:T base pairs are inclined by *c.* 32° within each layer of the quadruplex. Since A:T pairs

(a)

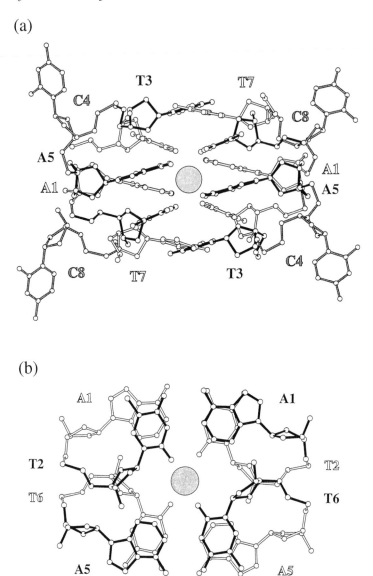

Fig. 13.32. (a) A view of the X-ray crystallographic structure of the d<pAT₂CAT₂C> quadruplex (119). The bases in the A:T:A:T tetrads depart significantly from planarity. (b) A view down the helix axis showing the stacking between the adjacent A:T:A:T tetrads (filled and open bonds, respectively) in the X-ray structure of the d<pAT₂CAT₂C> quadruplex (119).

cannot dimerize through hydrogen bond alignments involving their minor groove edges (contain only acceptor atoms), a sodium ion occupies the centre of the quadruplex and is coordinated to the thymine O2 oxygens of four A:T paired thymines, as shown in Fig. 13.32a (119). The stacking in the central core of the d<pAT₂CAT₂C> quadruplex is shown in Fig. 13.32b.

There is a striking similarity in the crystallographic structures of the central core of the d(GCATGCT) quadruplex (118) in Plate XVIIa and the central core of the d<pAT$_2$CAT$_2$C> quadruplex (119) in Plate XVIIb. It has been proposed that this quadruplex architecture containing common structural elements, called a bi-loop motif, could play a role in biological processes involved in strand exchange (119).

3.3 Other potential purine-containing tetrads

The demonstration of major groove-aligned G:C:G:C tetrad (108,116) and minor groove-aligned G:C:G:C tetrad (118) formation, in addition to the long-established formation of G:G:G:G terads (8), suggests that other purine-containing tetrad alignments may also stabilize quadruplex formation. Possibly the most interesting of these are tetrads containing G and A purine residues which have the potential to align through the major groove edges of either G(*anti*):A(*anti*) or G(*anti*):A(*syn*) mismatch pairs to form G:A:G:A tetrads (see models proposed in ref. 68). The identification of G:A:G:A tetrads and determination of their alignment geometry represents a future challenge. This goal may be approachable based on the reported equilibrium between duplex and quadruplex states for d(AG)$_{10}$ at neutral pH (120).

3.4 Biological relevance of quadruplexes containing G:C:G:C tetrads

The phase of the d(CGG)$_n$ fragile X syndrome triplet repeat can be either CGG, GGC, or GCG. The ability of d(GCG$_2$T$_3$GCG$_2$), which contains GCG and CG$_2$ repeats (108), and d(G$_3$CT$_4$G$_3$C), which contains G$_2$C repeats (116), to form G quadruplexes stabilized by G:C:G:C and G:G:G:G tetrads suggests a potential biological role for G:C:G:C tetrads. Such G:C:G:C tetrad-containing G quadruplex structures could serve as potential blockage sites for the progress of replication forks (121) and might account for the blockage of the fragile X locus observed experimentally (122).

The key demonstration establishing formation of G:C:G:C tetrads through alignment of Watson–Crick G:C base pairs along either their major groove (108,116) or minor groove (118) edges has potential implications in genetic recombination. Homologous DNA segments could be brought into register through G:C:G:C (and A:T:A:T) tetrad formation as a first step prior to the onset of strand exchange mediated through a pair of Holliday junction cross-over sites.

4 i-motif quadruplexes containing intercalated C:CH$^+$ mismatch pairs

4.1 Four-stranded i-motif quadruplexes

The formation of C:CH$^+$ mismatch pairs for poly C at acidic pH was proposed over three decades ago. Furthermore, the the X-ray fibre diffraction pattern of poly C was interpreted in terms of a parallel-stranded C:CH$^+$ mismatch-containing duplex (123,124). Direct evidence for formation of C:CH$^+$ pairs (Fig. 13.33a) in parallel-stranded DNA duplexes emerged following the structural characterization of the par-

(a)

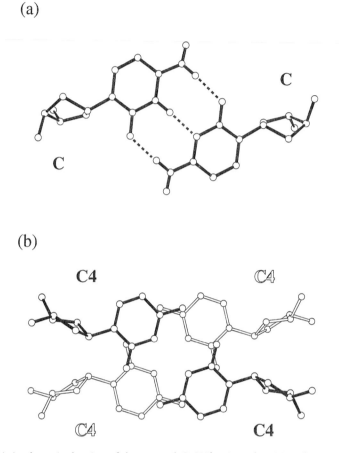

(b)

Fig. 13.33. (a) A schematic drawing of the reversed C:CH⁺ mismatch pairing alignment. (b) A view down the helix axis showing the stacking between the adjacent C:CH⁺ mismatch pairs in the NMR-based solution structure of the d(TC$_5$) i-motif quadruplex at acidic pH (15). One C:CH⁺ mismatch is shown with filled bonds and the other with open bonds.

allel-stranded d(CA) duplex at acidic pH in the crystalline state (125), and the d(TCGA) duplex at acidic pH in solution (126,127). Subsequently, solution NMR studies have identified a higher order quadruplex structure involving C:CH⁺ pairs as the basic repeat unit (15). This quadruplex architecture, called the i-motif (15), has added a new dimension to our understanding of multistranded nucleic acid structures.

4.1.1 Solution structure of d(TC$_5$) i-motif quadruplex

The NMR spectra of d(TC$_5$) at acidic pH exhibit an unusual set of chemical shifts and NOE patterns consistent with the formation of a folded higher order solution structure (15). A concentration-dependent study of d(TC$_5$) at acidic pH by gel electrophoresis established that this sequence forms a four-stranded quadruplex at mM concentrations (15). A single set of resonances were observed for d(TC$_5$) at acidic pH consistent with formation of a four-stranded quadruplex with a fourfold element of symmetry.

The observation of imino proton resonances between 15 and 16 ppm established the formation of C:CH$^+$ mismatch pairs. A set of diagnostic NOEs were observed between sugar H1′ protons on partner strands for the d(TC$_5$) quadruplex (15), a feature not observed for right-handed antiparallel double helical DNA. Furthermore, a set of critical NOEs of a non-sequential nature were identified, which reflected the order of base pair stacking within the quadruplex. These NOEs exhibited the sequential pattern T1–C6–C2–C5–C3–C4 in the d(T1–C2–C3–C4–C5–C6) quadruplex and provided critical restraints for structure determination.

The folding topology of the four-stranded (TC$_5$) quadruplex (15) is shown in a schematic view in Fig. 13.34. It consists of two parallel-stranded C:CH$^+$ mismatch paired duplexes that are interlocked through interdigitation of C:CH$^+$ pairs from individual duplexes that are aligned antiparallel to each other. Individual strands are aligned antiparallel to their neighbours and adjacent C:CH$^+$ mismatch pairs are approximately orthogonal to each other (Fig. 13.34).

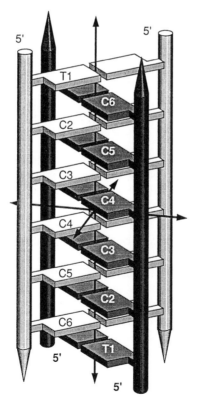

Fig. 13.34. A schematic of the NMR solution structure-based folding topology adopted by the i-motif quadruplex formed by four strands of d(TC$_5$) in acidic pH solution (15). Two parallel-stranded C:CH$^+$ paired duplexes interdigitate into each other in an antiparallel orientation. (Reproduced with permission of *Nature*).

T1   **T1**

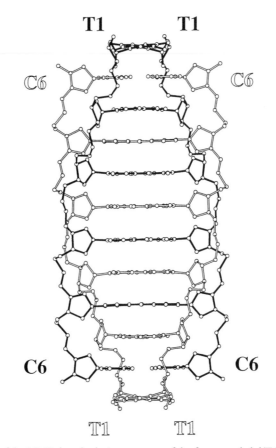

T1   **T1**

Fig. 13.35. A view of the NMR-based solution structure of the four-stranded d(TC₅) i-motif quadruplex in acidic pH solution (15). One parallel-stranded duplex is shown using filled bonds while the other, aligned antiparallel to the first, is shown using open bonds.

The solution structure of the four-stranded (TC₅) i-motif quadruplex is shown in Fig. 13.35 (15). The quadruplex is right-handed with an *c.* 16° twist between mismatch pairs. The i-motif quadruplex contains a pair of opposing wide grooves and a pair of opposing narrow grooves. There is a pairwise association of sugar–phosphate backbones, which results in close van der Waals contacts between sugar rings spanning the minor groove. This architecture explains the strong sugar H1′–sugar H1′ NOEs across the minor groove that are characteristic of the i-motif quadruplex (15).

The base overlap alignments between adjacent face-to-face stacked C:CH⁺ mismatch pairs is shown in Fig. 13.33b (15). There is no overlap between the cytosine rings themselves but, rather, there is overlap between the exocyclic amino groups and between the exocyclic carbonyl groups (Fig. 13.33b). There is a reversed orientation of the amino and carbonyl dipoles between adjacent stacked C:CH⁺ pairs and a maximal separation of the cytosine N3 nitrogens in this overlap pattern. The C:CH⁺ mismatch is of the reversed type with one hydrogen bonded and one exposed amino proton for individual cytosines in the pair. The exchange characteristics of the cytosine imino and amino protons in the i-motif quadruplex imply imino proton hopping between cytosines within the C:CH$^+$ pair of $\geq 80\ 000$ s^{-1} (15,128).

The seminal discovery of the i-motif C:CH$^+$ quadruplex was unanticipated and emerged from an in-depth and long-standing attempt at understanding the hydrogen exchange properties of dC$_n$-containing sequences at acidic pH (15). It became quickly apparent that other dC$_n$-rich sequences, in addition to d(TC$_5$), also adopt this quadruplex architecture at acidic pH (128).

4.1.2 Solution structure of d(TC$_2$) and d(m^{5}CCT) i-motif quadruplexes

The solution structure determination of the d(TC$_5$) i-motif quadruplex (15) was followed by a higher resolution structure determination of the simpler d(TC$_2$) i-motif quadruplex in acidic pH solution (129). The latter NMR studies identified additional NOEs characteristic of the i-motif quadruplexes that were in addition to the previously identified strong sugar H1′–H1′ cross-peaks between adjacent strands across the narrow groove (15). The most critical newly identified restraints included strong NOEs observed between the cytosine amino protons and sugar H2′,2″ protons on adjacent strands across the wide groove (129). The solution structure of the d(TC$_2$) i-motif quadruplex established that sequences containing as few as two successive cytosines are sufficient for formation of an i-motif quadruplex (129).

The d(m^{5}CCT) sequence forms two i-motif quadruplexes of comparable proportions in equilibrium under acidic pH conditions (129). The analysis of the NMR data established that one of these conformers was the maximally intercalated i-motif quadruplex similar to its d(TC$_2$) counterpart, while the other involved a shifting in registry of the intercalated C:CH$^+$ mismatch pairs, resulting in a partial loss of intercalation contributions. The latter conformer presumably reflects relief of methyl group steric clashes in the fully intercalated d(m^{5}CCT) i-motif quadruplex.

4.1.3 Solution structure of the d(m^{5}CCTC$_2$) i-motif quadruplex

A more recent solution structural study has addressed the issue related to whether intervening residues such as T:T mismatches can be accommodated within an i-motif quadruplex containing intercalated C:CH$^+$ mismatch pairs. The solution structure determination of the d(m^{5}CCTC$_2$) i-motif quadruplex has definitively addressed this issue and come up with an unanticipated answer (130).

The solution structure of the d(m^{5}CCTC$_2$) i-motif quadruplex shown in Fig. 13.36 establishes that the thymine bases of one parallel-stranded duplex component intercalate as a symmetrical T:T mismatch pair between C:CH$^+$ mismatch pairs, while those on the other parallel-stranded duplex component are unpaired and loop out into solution (130). Furthermore, the interconversion between paired and looped out thymine bases can be monitored by NMR and occurs at a rate of 1.4 s^{-1} at 0°C with an activation energy of 94 kJ mol^{-1} (130). This opening–closing process is concerted and occurs without disruption of the entire i-motif quadruplex. Interestingly, the interconversion rate increases to 40 s^{-1} at 0°C with a reduced barrier of 55 kJ mol^{-1} for the d(m^{5}CCUC$_2$) i-motif quadruplex where a U has replaced the internal T residue. Thus the swinging of the pyrimidine residue associated with the opening–closing process is impeded by the methyl group. These studies represents an elegant example of a bistable DNA motif with broken symmetry which has been

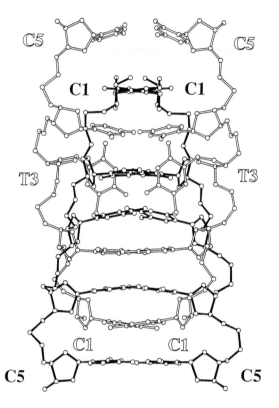

Fig. 13.36. A view of the NMR-based solution structure of the four-stranded d(m⁵CCTC₂) i-motif quadruplex in acidic pH solution (130). One parallel-stranded duplex is shown using filled bonds while the other, aligned antiparallel to the first, is shown using open bonds. The pair of looped out thymine residues can be clearly seen positioned in the grooves.

characterized both structurally and using hydrogen exchange measurements (130). It is also important to emphasize that the main features of the i-motif quadruplex are mantained despite incorporation of the T:T mismatch pair between C:CH⁺ mismatch pairs.

4.1.4 Base pair opening in the i-motif quadruplex

Hydrogen exchange of imino protons in dC$_n$ sequences at acidic pH is limited by base mismatch opening of the i-motif quadruplex (128). The lack of an effect of added catalysts on the hydrogen exchange of the imino protons of C:CH⁺ mismatch pairs (128) must reflect the predominant contribution of intrinsic catalysis across the C–N3H⁺···C–N3 pair in the i-motif quadruplex. The measured C:CH⁺ mismatch pair lifetimes are two orders of magnitude longer than the corresponding values for Watson–Crick pairs in B-form DNA (128). This could reflect the intercalation of the C:CH⁺ mismatch pairs within the structure of the i-motif quadruplex. A free energy value of −8.5 kJ mol⁻¹ per C:C⁺ mismatch pair was deduced for formation of the d(TC₅) i-motif quadruplex from single strands (128).

4.1.5 *Crystal structures of d(C_4) and d(C_3T) i-motif quadruplexes*

The publication of the NMR-based solution structure of the d(TC$_5$) i-motif quadruplex (15) has, in turn, stimulated efforts to elucidate the structure of the i-motif quadruplex in the crystalline state. These efforts have been quite successful, starting with the 2.3 Å crystal structure of d(C_4) (131) and the 1.4 Å crystal structure of d(C_3T) shown in Fig. 13.36 (132). There is good agreement between the helical features of the i-motif quadruplex architecture of the NMR-based solution structures (15,129,130) and X-ray-based crystal structures (131,132).

Thus, the average right-handed helical twists of 12.4° and 17.1° observed in the crystal structures of d(C_4) (131) and d(C_3T) (132) i-motif quadruplexes, respectively, compare favourably with the helical twist of 16° reported in the solution structure of the d(TC$_5$) i-motif quadruplex (15). In addition, the overlap geometries between adjacent C:CH$^+$ tetrads are very similar between the solution (15) and crystal (131,132) structures of the i-motif quadruplex. The base stacking distance between successive C:CH$^+$ base mismatches is 3.1 Å in the two crystal structures of the i-motif quadruplexes (131,132), which is consistent with the same meridional spacing in the X-ray fibre diffraction pattern of polycytidylic acid (123).

It has been pointed out from the crystal structures that the i-motif quadruplex has a flat and ribbon-shaped architecture with very wide grooves along two sides and very narrow grooves at the ends. Furthermore, there is a complementarity in the fit between the zigzag pathway of the sugar–phosphate backbones of adjacent antiparallel strands at the narrow end of the twisted ribbon (131,132). This close packing is reflected in the strong NOE between the sugar H1′ protons on adjacent strands observed in the solution structures of i-motif quadruplexes (15), which is readily explained by the observed separation of *c.* 3.1 Å between these proton pairs in the crystal structures (131,132).

The glycosidic torsion angles are in the high *anti* range while there is considerable variation in the sugar puckers within the crystal structures of the i-motif quadruplexes (131,132). There is also considerable asymmetry in the phosphate positions, as reflected by the spread in phosphorus–phosphorus separations across the wide and narrow grooves in the crystal structures of the i-motif quadruplexes. Several opposing phosphate groups on one strand in the wide groove extend away from the centre of the molecule, while those on the opposing strand in this groove bend over towards each other.

The C:CH$^+$ mismatch pairs are well defined in the 1.4 Å crystal structure of the d(C_3T) i-motif quadruplex (Fig. 13.37) (132), with central N–H···N heteroatom distances of 2.74 Å and N–H···O heteroatom distances of 2.77 Å. Furthermore, the exocyclic amino group of each cytosine is hydrogen bonded to a water molecule with N–H···O heteroatom distances of 3.00 Å. There are 59 solvent molecules in the asymmetric unit of the d(C_3T) i-motif quadruplex, with a small subset bridging cytosine exocyclic amino groups and phosphate oxygens on adjacent strands. Several sodium cations have been identified in the high resolution crystal structure of d(C_3T), with their octahedral coordination spheres containing water molecules and phosphate oxygens (132).

In summary, very similar internal i-motif, intercalated C:CH$^+$ architectures have been determined for the d(TC$_5$) solution structure (see Plate XVIIIa) (15) and d(C_3T) crystal structure (Plate XVIIIb) (132).

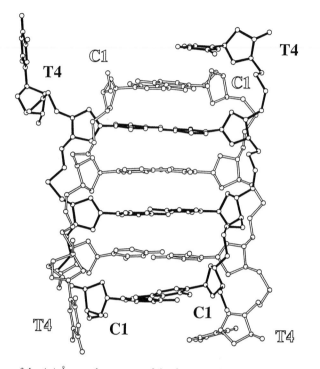

Fig. 13.37. A view of the 1.4 Å crystal structure of the four-stranded d(C₃T) i-motif quadruplex (132). One parallel-stranded duplex is shown using filled bonds while the other, aligned antiparallel to the first, is shown using open bonds.

4.1.6 Crystal structures of d(C₃A₂T) and the human telomere d(TA₂C₃) i-motif quadruplexes

The duplex segment of human telomeres contains d(C₃TA₂)ₙ cytosine-rich and d(T₂AG₃)ₙ guanine-rich repeats on complementary strands. It was therefore of great interest to determine whether the cytosine-rich segments of d(C₃TA₂)ₙ repeats [or d(TA₂C₃)ₙ repeats depending on the phase] can form i-motif quadruplexes and, in addition, elucidate the folding topology of the TA₂ segment. Considerable progress has been made towards these goals with the publication of the crystal structure of the single repeat human telomere d(TA₂C₃) i-motif quadruplex at 1.9 Å resolution (133) and the crystal structure of a sequence variant, d(C₃A₂T) i-motif quadruplex, at 2.0 Å resolution (134). The cytosine segments form i-motif quadruplexes in both structures (133,134) with helical parameters similar to those reported for the earlier crystal structure of the d(C₃T) i-motif quadruplex solved to very high resolution (132). The A:T-rich segments in the d(TA₂C₃) and d(C₃A₂T) i-motif quadruplexes adopt novel folding topologies and these are discussed below.

The crystal structure of the terminal segments of the four-stranded d(TA₂C₃) i-motif quadruplex is shown in Fig. 13.38 (133). The 5′-TA₂ segments exhibit different conformations, with one of them adopting a novel tight loop fold in which the 5′- and 3′-ends of adjacent strands are brought into close proximity. This folded segment

(a)

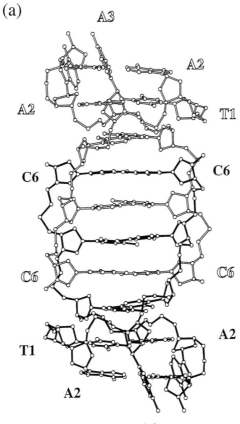

(b)

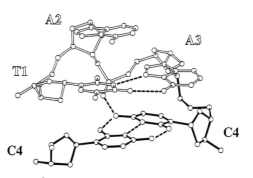

Fig. 13.38. (a) A view of the 1.9 Å crystal structure of the four-stranded d(TA$_2$C$_3$) i-motif quadruplex (133). One parallel-stranded duplex is shown using filled bonds while the other, aligned antiparallel to the first, is shown using open bonds. (b) A view emphasizing the stacking of A$_2$ on the T1:A3 Hoogsteen pair which is in turn stacked on the C:CH$^+$ mismatch pair in the crystal structure of the d(TA$_2$C$_3$) i-motif quadruplex (133).

is stabilized by formation of a Hoogsteen T:A base pair between the thymine and the 3′-adenine, which in turn stacks over the terminal C:C$^+$ mismatch pair. The central adenine of this TA$_2$ segment stacks on the other side over the Hoogsteen T:A base pair and caps the end of the i-motif quadruplex (Fig. 13.38). This structure, with its novel TA$_2$-folded segment, provides insight into the potential folding topologies of d(TA$_2$C$_3$)$_n$ (n =2 and 4) i-motif quadruplexes. Most importantly, isomorphous crystals of d(TA$_2$C$_3$) can be grown between pH 5.5 and 7.5, suggesting that the stability of the crystal lattice has raised the apparent pK_a for hemiprotonation of the C:C$^+$ mismatch pair (133).

   The crystal structure of the terminal segments of the four-stranded d(C$_3$A$_2$T) i-motif quadruplex is shown in Fig. 13.39 (134). An asymmetric A(*anti*):A(*clinal*) mismatch pair stacks over the terminal C:CH$^+$ mismatch pair (Fig. 13.40b) and extends the i-motif architecture by one step in either direction. This asymmetric A:A mismatch, which involves pairing through the Watson–Crick and Hoogsteen edges of the adenines, stacks in turn over a symmetrical A(*anti*):A(*syn*) mismatch (Fig. 13.40a), which involves pairing along the Watson–Crick edges of both adenines. Each of these two distinct A:A mismatches participates in an A:A:T base triple with a thymine from

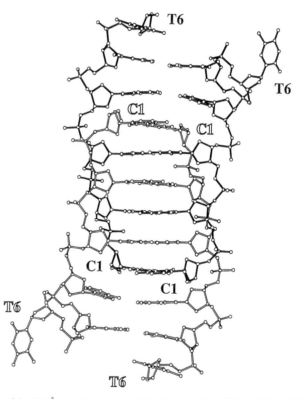

Fig. 13.39. A view of the 2.0 Å crystal structure of the four-stranded d(C$_3$A$_2$T) i-motif quadruplex (134). One parallel-stranded duplex is shown using filled bonds while the other, aligned antiparallel to the first, is shown using open bonds.

(a)

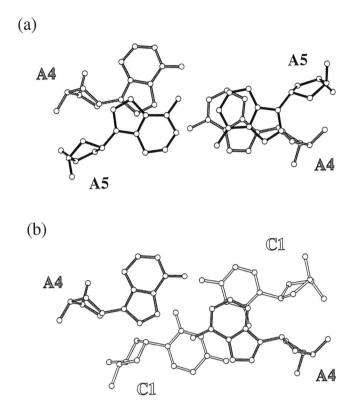

(b)

Fig. 13.40. Views down the helix axis showing the stacking between adjacent (a) A4:A4 and A5:A5 mismatch pairs and (b) A4:A4 and C1:C1H$^+$ mismatch pairs in the crystal structure of the d(C$_3$A$_2$T) i-motif quadruplex (134).

a symmetry-related i-motif in the crystallographic asymmetric unit (134). Isomorphous crystals of d(C$_3$A$_2$T) could also be grown over the pH range 5.0 to 7.5.

4.2 i-Motif quadruplexes formed through dimerization of loop containing segments

Several groups have investigated the folding topologies of d(C$_n$N$_m$C$_n$) sequences at acidic pH with the understanding that such sequences can fold back to form C:CH$^+$ mismatch pairs, which in turn can dimerize to form i-motif quadruplexes (135,136). High resolution NMR has been used more recently to determine the solution structures of i-motif quadruplexes formed through dimerization of d(C$_n$N$_m$C$_n$) sequences at acidic pH and the available results are outlined below.

4.2.1 Solution structure and opening kinetics of the d(m^{5}CCT$_3$AC$_2$) i-motif quadruplex

The d(m^{5}CCT$_3$AC$_2$) sequence gives well-resolved proton NMR spectra and NOE patterns characteristic of i-motif quadruplex formation (137). The concentration dependence of the equilibrium between multimer and single strand conformers

$$[d(5mC\ C\ T\ T\ T\ X\ C\ C)]_2$$

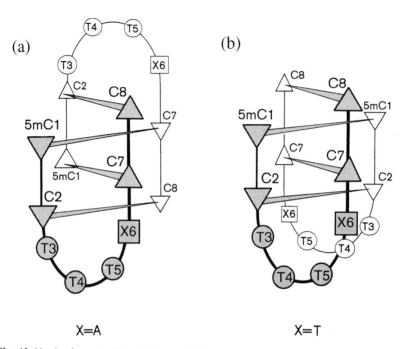

X=A   X=T

Fig. 13.41. A schematic of the NMR-based solution structure-based folding topology adopted by the i-motif quadruplex formed by: (a) the head-to-tail dimerization of a pair of d(m⁵CCT₃AC₂) hairpins in acidic pH solution (137); and (b) the head-to-head dimerization of a pair of d(m⁵CCT₄C₂) hairpins in acidic pH solution (137). Reproduced with permission of *Structure*.

established i-motif quadruplex formation through dimerization. An apparent pK_a of *c.* 6.5 was estimated for d(m⁵CCT₃AC₂) i-motif quadruplex formation. The solution structure was solved by a combined NMR and molecular dynamics structural characterization including intensity refinement. The folding topology consists of an i-motif quadruplex core containing intercalated C:CH⁺ outer pairs and m⁵C:CH⁺ inner pairs linked at opposite ends by T₃A loops that span the wide groove, as shown schematically in Fig. 13.41a (137). The adenine residue in both loops stack on the outer C:CH⁺ pairs thus extending the stacking beyond the central i-motif core. The solution structure of the d(m⁵CCT₃AC₂) i-motif quadruplex is shown in Fig. 13.42.

Hydrogen exchange kinetics of base mismatch opening establish that the lifetime is 1 ms at 15°C, with an activation energy of 60 kJ mol⁻¹ for the outer C:CH⁺ mismatch pairs in the d(m⁵CCT₃AC₂) i-motif quadruplex (137). This number is one order of magnitude longer than the corresponding mismatch lifetimes of terminal C:CH⁺ pairs in the d(TC₂) i-motif quadruplex (129). This could reflect the contributions of the T₃A loop to the stability of this outer C:CH⁺ pair in the d(m⁵CCT₃AC₂) i-motif quadruplex. By contrast, the mismatch lifetime is three orders of magnitude longer at

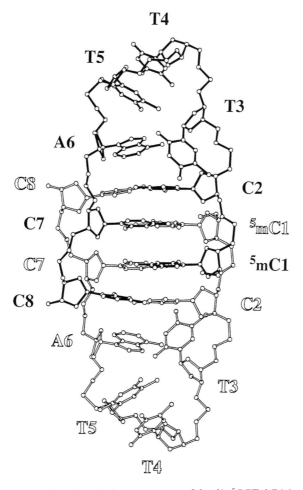

Fig. 13.42. A view of the NMR-based solution structure of the d(m^{5}CCT$_3$AC$_2$) i-motif quadruplex in acidic pH solution (137). One strand is shown using filled bonds and the other strand is shown with open bonds.

1 s at 15°C, with an activation energy of 100 kJ mol^{-1} for the inner m^{5}C:CH$^+$ pairs in the d(m^{5}CCT$_3$AC$_2$) i-motif quadruplex (137). This number is comparable with those determined for the internal C:CH$^+$ mismatch lifetimes in the d(TC$_2$) i-motif quadruplex (129). The exchange characteristics of the thymine imino protons also suggest that the loop is closed by a Hoogsteen-like alignment involving the loop-closing T and A residues bridged by a bound water molecule.

The d(m^{5}CCT$_4$C$_2$) sequence, where the A residue is replaced by T, also forms an i-motif quadruplex through dimerization, except that the loops are positioned on the same side of the i-motif (Fig. 13.41b) (137). This result emphasizes the striking change in folding topology of the i-motif quadruplexes associated with a switch in a single loop residue.

4.2.2 Solution structure of the insulin minisatellite repeat d(C₄TGTC₄) i-motif quadruplex

The insulin minisatellite sequence located upstream of the human insulin gene (138) exhibits polymorphism in both repeat length and sequence. The pyrimidine-rich $d(C_4ACAC_4TGT)_n$ strand contains C_4 segments in the repeat element. A combined NMR and molecular dynamics study has been undertaken to define the solution structure of the $d(C_4TGTC_4)$ domain at acidic pH (139). The folding topology reflects formation of an i-motif quadruplex through dimerization of fold-back segments, with the TGT turns positioned at opposite ends of the twofold symmetric quadruplex. The pH dependence of i-motif quadruplex formation exhibits an apparent pK_a of 6.5. There is some concern about the robustness of the refinements based on the listed statistics for the refined structures of the $d(C_4TGTC_4)$ i-motif quadruplex. Thus, the five refined structures exhibit an unusually large number of NOE violations (42 violations, > 0.5 Å and < 1.0 Å) (139) and this discrepancy needs further clarification.

4.2.3 Solution structure of the centromeric α satellite repeat d(TC₃GT₃C₂A) i-motif quadruplex

The centromeric CENP-B protein is known to target the $d(TC_3GT_3C_2A_2CGA_2G)_n$ box repeat of α satellite DNA located at the centromeric regions of human chromosomes (140). The NMR-based solution structure of the $d(TC_3GT_3C_2A)$ sequence at acidic pH has been determined to high resolution and shown to form an i-motif quadruplex through dimerization of a pair of fold-back segments, with the GT_3 turns positioned on the same side of the twofold symmetric quadruplex (141). The two hairpin turns positioned at one end of the i-motif quadruplex interact with each other through formation of a novel T:G:G:T tetrad. This T:G:G:T tetrad alignment involves the dimerization of two wobble G:T pairs through pairing of their guanine minor groove edges, as shown schematically in Fig. 13.43. This structure exhibits excellent

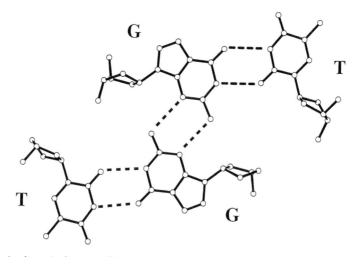

Fig. 13.43. A schematic drawing of the T:G:G:T tetrad pairing alignment observed in the NMR-based solution structure of the $d(TC_3GT_3C_2A)$ i-motif quadruplex at acidic pH (141).

refinement statistics with both low pairwise rmsd values (0.44±0.14 Å) and a low number of NOE violations (two violations, > 0.2 Å) (141).

4.3 Intramolecularly folded i-motif quadruplexes

Several groups have attempted to generate intramolecularly folded i-motif quadruplexes from DNA sequences containing four C_n repeats under acidic pH conditions (139,141–145). The structural characterization of such an intramolecularly folded i-motif represents a challenge because of complications from conformational heterogenity. Initially, some progress was made on the human telomere $d[(C_3TA_2)_3C_3]$ i-motif quadruplex system (142). More recently, a high resolution structure of the intramolecularly folded $d(m^5CCT_3C_2T_3AC_2T_3C_2)$ i-motif quadruplex has been solved (146). These results are summarized briefly below.

4.3.1 Human telomere d[(C₃TA₂)₃C₃] i-motif quadruplex

Two groups have recently investigated the potential formation of intramolecularly folded i-motif quadruplexes by the human telomere $d(C_3TA_2)_4$ sequence and its variants under acidic pH conditions (142,143). One of these groups used UV absorbance melting curves, chemical modification, and non-denaturing gel electrophoresis to monitor the folded state of $d(C_3TA_2)_4$ at acidic pH (143). The other group used UV absorbance and gel filtration, and, in addition, monitored the characteristic NOE patterns to establish intramolecular i-motif quadruplex formation for $d[(C_3TA_2)_3C_3]$ at acidic pH (142). The NMR resonances were marginally resolved and appear to contain more than one folded conformer for $d[(C_3TA_2)_3C_3]$ at acidic pH (142). Hence, current efforts are focused on designing variants of the human telomere $d(C_3TA_2)_4$ sequence, with the aim of obtaining improved NMR spectra corresponding to a single conformation necessary for a high resolution structure determination of an intramolecularly folded i-motif quadruplex.

4.3.2 Solution structure of the d(m⁵CCT₃C₂T₃AC₂T₃C₂) i-motif quadruplex

Thermal denaturation, gel filtration, and NMR studies have also been used to demonstrate formation of an intramolecularly folded i-motif quadruplex by the $d(C_2T_3C_2T_4C_2T_3C_2)$ sequence at acidic pH (145). Thus, as few as eight cytosines and a total of four intercalated C:CH$^+$ mismatch pairs are sufficient to form an intramolecular i-motif quadruplex.

A significant step forward in our understanding of the i-motif quadruplex has resulted from recent NMR studies of the $d(m^5CCT_3C_2T_3AC_2T_3C_2)$ sequence at neutral pH (146). The NMR parameters are consistent with formation of an intramolecularly folded i-motif quadruplex, with the folding topology shown in Fig. 13.44. A view of the solution structure of the $d(m^5CCT_3C_2T_3AC_2T_3C_2)$ i-motif quadruplex is shown in Fig. 13.45. This structure is formed at neutral pH with a pK_a of 7.45 for the midpoint of the transition.

This i-motif quadruplex structure contains four contiguously stacked C:CH$^+$ mismatch pairs capped at one end by a T_3A loop that spans the wide groove and at the other end by two spatially proximal T_3 loops that span the two narrow grooves (Fig. 13.44). The stacking within the C:CH$^+$ i-motif is extended in one direction by

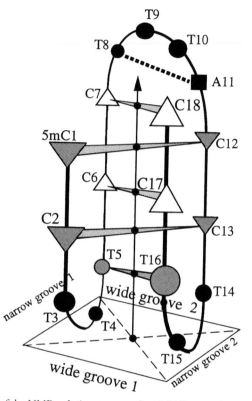

Fig. 13.44. A schematic of the NMR solution structure-based folding topology adopted by the intramolecular i-motif quadruplex formed by the d(m^{5}CCT$_3$C$_2$T$_3$AC$_2$T$_3$C$_2$) sequence (146). Reproduced with permission of *J. Mol. Biol.*

a propeller-twisted reverse Hoogsteen T:A mismatch pair and in the other direction by a T:T mismatch pair involving thymines from the spatially proximal T$_3$ loops (Fig. 13.44) (146).

4.4 Potential biological relevance of the i-motif quadruplex

To date there is no direct evidence to support a biological role for the intercalated C:CH$^+$ mismatch paired i-motif quadruplex. A primary concern is the requirement for acidic pH to favour i-motif quadruplex formation. The pK_a for cytosine N3 protonation is 4.3 at the monomer level, but this pK_a increases to 6.5 for several of the i-motif quadruplexes studied to date (137,139). However, the intramolecularly folded d(m^{5}CCT$_3$C$_2$T$_3$AC$_2$T$_3$C$_2$) i-motif quadruplex exhibits a pK_a of 7.45 consistent with i-motif formation at neutral pH (146). Indeed, crystals of the four-stranded i-motif quadruplex can be grown from solutions at pH values up to 7.5 (133,134). These results suggest that the requirement for slightly acidic pH conditions may not be an issue for intramolecularly folded i-motif quadruplexes (146) and could also be overcome by other factors such as superhelical stress or complex formation with potential proteins that target the i-motif quadruplex.

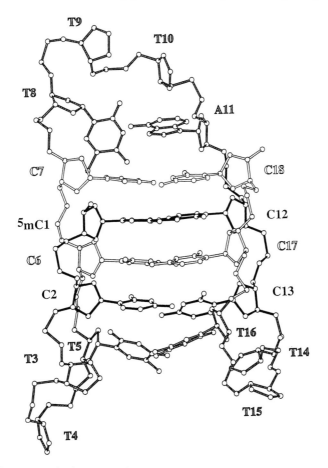

Fig. 13.45. A view of the NMR-based solution structure of the i-motif quadruplex d($m^5CCT_3C_2T_3AC_2T_3C_2$) (137). Alternate C:CH$^+$ pairs are shown by filled and open bonds. The loop segments are shown by hatched bonds. The bases of residues T3, T4, T9, T10, T14, and T15 have been deleted in the interests of clarity.

The i-motif could have potential therapeutic efficacy based on the ability of phosphodithioate dC_n to inhibit HIV-1 integrase (147).

A protein has been identified that binds to the vertebrate cytosine-rich telomeric d($C_3TA_2)_n$ sequence (148). However, no protein that binds with high specificity and affinity to the i-motif quadruplex has been isolated to date, and perhaps more time is needed to pursue this goal given that the i-motif (15) was only discovered five years ago.

5 Future directions

DNA quadruplexes have the potential to play a critical role in self-recognition involving systems ranging from chromosomal pairing to recombination. The repertoire of tetrad alignments is currently limited, with the emphasis on G:G:G:G and G:C:G:C alignments. Future efforts should be directed towards the identification and characterization of potential A:T:A:T and G:A:G:A tetrad alignments and the identification of sequence contexts that favour such pairing alignments. The success associated with the interdigitated, reversed protonated C:C mismatch pair-stabilized i-motif quadruplex formation could possibly be extended to the identification and characterization of potential i-motifs containing reversed protonated A:C and reversed A:A mismatch pairs. It should also be possible to extend the limited repertoire of base triad alignments by designing sequences where potential base triad alignments are stabilized through stacking with adjacent G:G:G:G tetrads.

There is a critical need to characterize structurally G quadruplexes and i-motif quadruplexes complexed with ligands ranging from small molecules to saccharides, peptides, and proteins. The diversity associated with the four grooves of different dimensions, together with novel loop folding topologies in the case of intramolecularly folded quadruplexes, makes these higher order nucleic acid architectures attractive targets for therapeutic intervention.

Coordinates deposition

We have prepared tables listing the structures discussed in this chapter with currently available PDB (Protein Database) accession numbers for deposited coordinates. The accession numbers for guanine-rich G:G:G:G-containing G quadruplex-forming sequences, for guanine-rich G:C:G:C-containing (and related) quadruplex-forming sequences, and for cytosine-rich interdigitated C:CH$^+$ mismatch-containing i-motif quadruplex-forming sequences are listed in Tables 13.1–13.3.

Table 13.1. A listing of NMR and X-ray based structures of guanine-rich G:G:G:G-containing G quadruplex-forming sequences along with PDB accession number for deposited coordinates

Section	Sequence	Conditions	Ref	Accession no.
2.1.1	d(T$_2$G$_4$T)	Na$^+$, solution	35	139d
2.1.3	d(TG$_4$T)	Na$^+$, crystal, 0.95 Å	40	352d
2.1.4	r(UG$_4$U)	K$^+$, solution	42	1rau
2.2.1	d(G$_4$T$_4$G$_4$)	K$^+$, crystal, 2.5 Å	45	1d59
2.2.2	d(G$_2$T$_2$G$_2$TGTG$_2$T$_2$G$_2$)	K$^+$, solution	51	148d
	d(G$_2$T$_2$G$_2$TGTG$_2$T$_2$G$_2$)	Na$^+$, K$^+$, solution	52	1qdf
2.2.3	d(G$_2$T$_2$G$_2$TGTG$_2$T$_2$G$_2$) plus thrombin	Na$^+$, crystal, 2.9 Å	54	1hut
2.3.2	d[AG$_3$(T$_2$AG$_3$)$_3$]	Na$^+$, solution	59	143d
2.3.4	d(G$_4$T$_4$G$_4$)	Na$^+$, solution	47	156d
2.3.5	d[G$_4$(T$_4$G$_4$)$_3$]	Na$^+$, solution	61	201d
	d[G$_4$(T$_4$G$_4$)$_3$]	Na$^+$, solution	62	230d
2.3.8	d(G$_3$T$_4$G$_3$)	Na$^+$, solution	76	1fqp
2.4.1	d(T$_2$G$_4$)$_4$	Na$^+$, solution	77	186d
2.5.1	d(TAG$_2$)	Na$^+$, solution	78	1aff

Table 13.2. A listing of NMR and X-ray based structures of guanine-rich G:C:G:C-containing (and related) quadruplex-forming sequences along with PDB accession numbers for deposited coordinates

Section	Sequence	Conditions	Ref	Accession no.
3.1.1	d(GCG$_2$T$_3$GCG$_2$)	Na$^+$, solution	108	1a6h
3.1.2	d(G$_3$CT$_4$G$_3$C)	Na$^+$, solution	116	1a8n
3.1.3	d(G$_3$CT$_4$G$_3$C)	K$^+$, solution	117	1a8w
3.2.1	d(GCATGCT)	Li$^+$, Mg^{2+}, X-ray, 1.8 Å	118	184d
3.2.2	d<AT$_2$CAT$_2$C>	Na$^+$, Ba^{2+}, X-ray	119	284d

Table 13.3. A listing of NMR and X-ray based structures of cytosine-rich interdigitated C:CH$^+$ mismatch-containing i-motif quadruplex-forming sequences along with PDB accession numbers for deposited coordinates

Section	Sequence	Conditions	Ref	Accession no.
4.1.1	d(TC$_5$)	Solution, low pH	15	225d
4.1.2	d(TC$_2$)	Solution, low pH	129	105d
	d(m^{5}CCT)	Solution, low pH	129	106d
4.1.3	d(m^{5}CCTC$_2$)	Solution, low pH	130	1rme
4.1.5	d(C$_4$)	Crystal, 2.3 Å	131	190d
	d(C$_3$T)	Crystal, 1.4 Å	132	191d
4.1.6	d(TA$_2$C$_3$)	Crystal, 1.9 Å, pH 5.5–7.5	133	200d
	d(C$_3$A$_2$T)	Crystal, 2.0 Å, pH 5.0–7.5	134	241d
4.2.1	d(m^{5}CCT$_3$AC$_2$)	Solution, low pH	137	1bae
4.3.2	d(m^{5}CCT$_3$C$_2$T$_3$AC$_2$T$_3$C$_2$)	Solution, neutral pH	146	1a83

Acknowledgements

The DNA quadruplex research in our laboratory is funded by NIH grant GM 34504. We thank Drs. R. Ajay Kumar and Andrey Gorin for helpful discussions. We thank Drs Jean-Louis Leroy and Maurice Gueron of the Ecole Polytechnique, Palaiseau, France, for providing a preprint and the coordinates of their solution structure of the intramolecularly folded i-motif quadruplex (146) prior to publication.

References

1. Sun, J.S. and Helene, C. (1993) *Curr. Opin. Struct. Biol.* **3**, 345.
2. Radhakrishnan, I. and Patel, D.J. (1994) *Biochemistry* **33**, 11405.
3. Plum, G.E., Pilch, D.S., Singleton, S.F. and Breslauer, K.J. (1995) *Annu. Rev. Biophys. Biomol. Struct.* **24**, 319.
4. Rhodes, D. and Giraldo, R. (1995) *Curr. Opin. Struct. Biol.* **5**, 311.
5. Pilch, D.S., Plum, G.E. and Breslauer, K.J. (1995) *Curr. Opin. Struct. Biol.* **5**, 334.
6. Lilley, D.M.J. and Clegg, R.M. (1993) *Annu. Rev. Biophys. Biomol. Struct.* **22**, 299.
7. Altona, C., Pikkemaat, J.A. and Overmars, F.J.J. (1996) *Curr. Opin. Struct. Biol.* **6**, 305.
8. Gellert, M., Lipsett, M.N. and Davies, D.R. (1962) *Proc. Natl. Acad. Sci. USA* **48**, 2013.

9. Henderson, E.R., Moore, M. and Malcolm, B.A. (1990) *Biochemistry* **29**, 732.

10. Sen, D. and Gilbert, W. (1988) *Nature* **334**, 364.

11. Sundquist, W.I. and Klug, A. (1989) *Nature* **342**, 825.

12. Williamson, J.R., Raghuraman, M.K. and Cech, T.R. (1989) *Cell* **59**, 871.

13. Guschlbauer, W., Chantot, J.F. and Thiele, D. (1990) *J. Biomol. Struct. Dynamics* **8**, 491.

14. Williamson, J.R. (1994) *Annu. Rev. Biophys. Biomol. Struct.* **23**, 703.

15. Gehring, K., Leroy, J.-L. and Gueron, M. (1993) *Nature* **363**, 561.

16. Blackburn, E.H. and Szostak, J.W. (1984) *Annu. Rev. Biochem.* **53**, 163.

17. Yu, G.L., Bradley, J.D., Attardi, L.D. and Blackburn, E.H. (1990) *Nature* **344**, 126.

18. Arnott, S., Chandrasekaran, R. and Marttila, C.M. (1974) *Biochem. J.* **141**, 537.

19. Zimmerman, S.B., Cohen, G.H. and Davies, D.R. (1975) *J. Mol. Biol.* **92**, 181.

20. Sasisekharan, V., Zimmermann, S.B. and Davies, D.R. (1975) *J. Mol. Biol.* **92**, 171.

21. Pinnavaia, T.J., Marshall, C.L., Mettler, C.M., Fisk, C.L., Miles, H.T. and Becker, E.D. (1978) *J. Am. Chem. Soc.* **100**, 3625.

22. Howard, F.B. and Miles, H.T. (1982) *Biochemistry* **21**, 6736.

23. Hardin, C.C., Henderson, E., Watson, T. and Prosser, J.K. (1991) *Biochemistry* **30**, 4460.

24. Xu, Q., Deng, H. and Braunlin, W.H. (1993) *Biochemistry* **32**, 13130.

25. Sen, D. and Gilbert, W. (1990) *Nature* **344**, 410.

26. Hardin, C.C., Watson, T., Corregan, M. and Bailey, C. (1992) *Biochemistry* **32**, 833.

27. Miura, T., Benevides, J.M. and Thomas, G.J., Jr (1995) *J. Mol. Biol.* **248**, 233.

28. Williamson, J.R. (1993) *Curr. Opin. Struct. Biol.* **3**, 357.

29. Henderson, E., Hardin, C.C., Walk, S.K., Tinoco, I., Jr and Blackburn, E.H. (1987) *Cell* **51**, 899.

30. Wang, Y. and Patel, D.J. (1992) *Biochemistry* **31**, 8112.

31. Aboul-ela, F., Murchie, A.I.H. and Lilley, D.M. (1992) *Nature* **360**, 280.

32. Jin, R., Gaffney, B.L., Wang, C., Jones, R.A. and Breslauer, K.J. (1992) *Proc. Natl. Acad. Sci. USA* **89**, 8832.

33. Sen, D. and Gilbert, W. (1992) *Biochemistry* **31**, 65.

34. Sen, D. and Gilbert, W. (1991) *Curr. Opin. Struct. Biol.* **1**, 435.

35. Wang, Y. and Patel, D.J. (1993) *J. Mol. Biol.* **234**, 1171.

36. Gupta, G., Garcia, A.E., Guo, Q., Lu, M. and Kallenbach, N.R. (1993) *Biochemistry* **32**, 7098.

37. Aboul-ela, F., Murchie, A.I.H., Norman, D.G. and Lilley, D.M. (1994) *J. Mol. Biol.* **243**, 458.

38. Guo, Q., Lu, M. and Kallenbach, N.R. (1993) *Biochemistry* **32**, 3596.

39. Laughlan, G., Murchie, A.I., Norman, D.G., Moore, M.H., Moody, P.C., Lilley, D.M. and Luisi, B. (1994) *Science* **265**, 520.

40. Phillips, K., Dauter, Z., Murchie, A.I.H., Lilley, D.M.J. and Luisi, B. (1997) *J. Mol. Biol.* **273**, 171.

41. Kim, J., Cheong, C. and Moore, P.B. (1991) *Nature* **351**, 331.

42. Cheong, C. and Moore, P.B. (1992) *Biochemistry* **31**, 8406.

43. Marsh, T.C. and Henderson, E. (1994) *Biochemistry* **33**, 10718.

44. Marsh, T.C., Vesenka, J. and Henderson, E. (1995) *Nucl. Acids Res.* **23**, 696.

45. Kang, C., Zhang, X., Ratliff, R., Moyzis, R. and Rich, A. (1992) *Nature* **356**, 126.

46. Smith, F.W. and Feigon, J. (1992) *Nature* **356**, 164.

47. Schultze, P., Smith, F.W. and Feigon, J. (1994) *Structure* **2**, 221.

48. Bock, L.C., Griffin, L.C., Lantham, J.A., Vermaas, E.H. and Toole, J.J. (1992) *Nature* **355**, 564.

49. Macaya, R.F., Schultze, P., Smith, F.W., Roe, J.A. and Feigon, J. (1993) *Proc. Natl. Acad. Sci. USA* **90**, 3745.

50. Wang, K.Y., McCurdy, S., Shea, R.G., Swaminathan, S. and Bolton, P.H. (1993) *Biochemistry* **32**, 1899.
51. Schultze, P., Macaya, R.F. and Feigon, J. (1994) *J. Mol. Biol.* **235**, 1532.
52. Wang, K.Y., Krawczyk, S.H., Bischofberger, N., Swaminathan, S. and Bolton, P.H. (1993) *Biochemistry* **32**, 11285.
53. Marathias, V.M., Wang, K.Y., Kumar, S., Pham, T.Q., Swaminathan, S. and Bolton, P.H. (1996) *J. Mol. Biol.* **260**, 378.
54. Padmanabhan, K., Padmanabhan, K.P., Ferrara, J.D., Sadler, J.E. and Tulinsky, A. (1993) *J. Biol. Chem.* **268**, 17651.
55. Kelly, J.A., Feigon, J. and Yeates, T.O. (1996) *J. Mol. Biol.* **256**, 417.
56. Hammond-Kosack, M.C., Dobrinski, B., Lurz, R., Dochert, K. and Kilpatrick, M.W. (1992) *Nucl. Acids Res.* **20**, 231.
57. Catasti, P., Chen, X., Moyzis, R.K., Bradbury, E.M. and Gupta, G. (1996) *J. Mol. Biol.* **264**, 534.
58. Wang, Y., de los Santos, C., Gao, X., Greene, K., Live, D. and Patel, D.J. (1991) *J. Mol. Biol.* **222**, 819.
59. Wang, Y. and Patel, D.J. (1993) *Structure* **1**, 263.
60. Wang, K.Y., Swaminathan, S. and Bolton, P.H. (1994) *Biochemistry* **33**, 7517.
61. Wang, Y. and Patel, D.J. (1995) *J. Mol. Biol.* **251**, 76.
62. Smith, F.W., Schultze, P. and Feigon, J. (1995) *Structure* **3**, 997.
63. Sundquist, W.I. and Heaphy, S. (1993) *Proc. Natl. Acad. Sci. USA* **90**, 3393.
64. Skripkin, E., Paillart, J.-C., Marquet, R., Ehresmann, B. and Ehresmann, C. (1994) *Proc. Natl. Acad. Sci. USA* **91**, 4945.
65. Paillart, J.-C., Skripkin, E., Ehresman, B., Ehresman, C. and Marquet, R. (1996) *Proc. Natl. Acad. Sci. USA* **93**, 5572.
66. Christiansen, J., Kofod, M. and Nielsen, F.C. (1994) *Nucl. Acids Res.* **22**, 5709.
67. Smith, F.W. and Feigon, J. (1993) *Biochemistry* **32**, 8682.
68. Murchie, A.I. and Lilley, D.M. (1994) *EMBO J.* **13**, 993.
69. Balagurumoorthy, P. and Brahmachari, S.K. (1994) *J. Biol. Chem.* **269**, 21858.
70. Jin, R., Breslauer, K.J., Jones, R.A. and Gaffney, B.L. (1990) *Science* **250**, 543.
71. Hud, N.V., Smith, F.W., Anet, F.A.L. and Feigon, J. (1996) *Biochemistry* **35**, 15383.
72. Bouaziz, S. and Patel, D.J. (1998) submitted.
73. Scaria, P.V., Shire, S.J. and Shafer, R.H. (1992) *Proc. Natl. Acad. Sci. USA* **89**, 10336.
74. Strahan, G.D., Shafer, R.H. and Keniry, M.A. (1994) *Nucl. Acids Res.* **22**, 5447.
75. Smith, F.W., Lau, F.W. and Feigon, J. (1994) *Proc. Natl. Acad. Sci. USA* **91**, 10546.
76. Keniry, M.A., Strahan, G.D., Owen, E.A. and Shafer, R.H. (1995) *Eur. J. Biochem.* **233**, 631.
77. Wang, Y. and Patel, D.J. (1994) *Structure* **2**, 1141.
78. Kettani, A., Bouaziz, S., Wang, W., Jones, R.A. and Patel, D.J. (1997) *Nature Struct. Biol.* **4**, 382.
79. Kuryavyi, V.V. and Jovin, T.M. (1995) *Nature Genetics* **9**, 339.
80. Cate, J.H., Gooding, A.R., Podell, E., Zhou, K., Golden, B.L., Szewczak, A.A., Kundrot, C.E., Cech, T.R. and Doudna, J.A. (1996) *Science* **273**, 1696.
81. Guo, Q., Garcia, A.E., Guo, Q., Lu, M. and Kallenbach, N.R. (1993) *Biochemistry* **31**, 2451.
82. Chen, Q., Kuntz, I.D. and Shafer, R.H. (1996) *Proc. Natl. Acad. Sci. USA* **93**, 2635.
83. Li, Y., Geyer, C.R. and Sen, D. (1996) *Biochemistry* **35**, 6911.
84. Li, Y. and Sen, D. (1997) *Biochemistry* **36**, 5589.
85. Wyatt, J.R., Vickers, T.A., Roberson, J.L., Buckheit, R.W., Jr, Klimkait, T., DeBaets, E., Davis, P.W., Rayner, B., Imbach, J.L. and Ecker, D.J. (1994) *Proc. Natl. Acad. Sci. USA* **91**, 1356.

86. Rando, R.F., Ojwang, J., Elbaggari, A., Reyes, G.R., Tinder, R., McGarth, M.S. and Hogan, M.E. (1995) *J. Biol. Chem.* **270**, 1754.

87. Bishop, J.S., Guy-Caffey, J.K., Ojwang, J.O., Smith, S.R., Hogan, M.E., Cossum, P.A., Rando, R.F. and Chaudhary, M. (1996) *J. Biol. Chem.* **271**, 5698.

88. Mazumdar, A.D., Neamati, N., Ojwang, J.O., Sunder, S., Rando, R.F. and Pommier, Y. (1996) *Biochemistry* **5**, 13762.

89. Jing, N., Gao, X., Rando, R.F. and Hogan, M.E. (1997) *J. Biomol. Struct. Dynamics* **15**, 573.

90. Jing, N., Rando, R.F., Pommier, Y. and Hogan, M.E. (1997) *Biochemistry* **36**, 12498.

91. Fang, G. and Cech, T.R. (1993) *Biochemistry* **32**, 11646.

92. Fang, G. and Cech, T.R. (1993) *Cell* **4**, 875.

93. Giraldo, R. and Rhodes, D. (1994) *EMBO J.* **13**, 2411.

94. Walsh, K. and Gualberto, A. (1992) *J. Biol. Chem.* **267**, 13714.

95. Weisman-Shomer, P. and Fry, M. (1993) *J. Biol. Chem.* **268**, 3306.

96. Schierer, T. and Henderson, E. (1994) *Biochemistry* **33**, 2240.

97. Liu, Z. and Gilbert, W. (1994) *Cell* **77**, 1083.

98. Frantz, J.D. and Gilbert, W. (1995) *J. Biol. Chem.* **270**, 9413.

99. Bashkirov, V.I., Scherthan, H., Solinger, J.A., Buerstedde, J.-M. and Heyer, W.-D. (1997) *J. Cell Biol.* **136**, 761.

100. Murchie, A.I. and Lilley, D.M. (1992) *Nucl. Acids Res.* **20**, 49.

101. Simonsson, T., Pechinka, P. and Kubista, M. (1998) *Nucl. Acids Res.* **26**, 1167.

102. Zahler, A.M., Willaimson, J.R., Cech, T.R. and Prescott, D.M. (1991) *Nature* **350**, 718.

103. Caskey, C.T., Pizzuti, A., Fu, Y.H., Fenwick, R.G. and Nelson, D.L. (1992) *Science* **256**, 784.

104. Sinden, R.R. and Wells, R.D. (1992) *Curr. Opin. Biotech.* **3**, 612.

105. Nelson, D.L. (1995) *Sem. Cell. Biol.* **6**, 5.

106. Sutherland, G.R. and Richards, R.I. (1995) *Proc. Natl. Acad. Sci. USA* **92**, 3636.

107. Fry, M. and Loeb, L.A. (1994) *Proc. Natl. Acad. Sci. USA* **91**, 4950.

108. Kettani, A., Kumar, R.A. and Patel, D.J. (1995) *J. Mol. Biol.* **254**, 638.

109. Brunger, A. (1992) *X-PLOR. A System for X-ray Crystallography and NMR.* Yale University Press, New Haven.

110. Nilges, M., Habazettl, J., Brunger, A.T. and Holak, T.A. (1991) *J. Mol. Biol.* **219**, 499.

111. Nilges, M. (1995) *J. Mol. Biol.* **245**, 645.

112. O'Brien, E.J. (1967) *Acta Cryst.* **23**, 92.

113. McGavin, S. (1971) *J. Mol. Biol.* **55**, 293.

114. Mitas, M., Yu, A., Dill, J. and Haworth, I.S. (1995) *Biochemistry* **34**, 12803.

115. Berns, K.I. and Linden, R.M. (1995) *Bioessays* **17**, 237.

116. Kettani, A., Bouaziz, S., Gorin, A., Zhao, H., Jones, R. and Patel, D.J. (1998) *J. Mol. Biol.* **282**, 619.

117. Bouaziz, S., Kettani, A. and Patel, D.J. (1998) *J. Mol. Biol.* **282**, 637.

118. Leonard, G.A., Zhang, S., Peterson, M.R., Harrop, S.J., Helliwell, J.R., Cruse, W.B., d'Estaintot, B.L., Kennard, O., Brown, T. and Hunter, W.N. (1995) *Structure* **3**, 335.

119. Salisbury, S.A., Wilson, S.E., Powell, H.R., Kennard, O., Lubini, P., Sheldrick, G.M., Escaja, N., Alazzouzi, E., Granada, A. and Pedroso, E. (1997) *Proc. Natl. Acad. Sci. USA* **94**, 5515.

120. Shiber, M.C., Braswell, E.H., Klump, H. and Fresco, J.R. (1996) *Nucl. Acids Res.* **24**, 5004.

121. Hansen, R.S., Gartler, S.M., Scott, C.R., Chen, S.H. and Laird, C.D. (1992) *Hum. Mol. Genet.* **1**, 571.

122. Hansen, R.S., Canfield, T.K., Lamb, M.M., Gartler, S.M. and Laird, C.D. (1993) *Cell* **73**, 1403.

123. Langridge, R. and Rich, A. (1963) *Nature* **298**, 725.

124. Hartman, K.A. and Rich, A. (1965) *J. Am. Chem. Soc.* **87**, 2033.

125. Cruse, W.B., Egert, E., Kennard, O., Sala, G.B., Salisbury, S.A. and Viswamitra, M.A. (1983) *Biochemistry* **12**, 1833.

126. Robinson, H., van der Marel, G., van Boom, J.H. and Wang, A.H. (1992) *Biochemistry* **31**, 10510.

127. Wang, Y. and Patel, D.J. (1994) *J. Mol. Biol.* **242**, 508.

128. Leroy, J.-L., Gehring, K., Kettani, A. and Gueron, M. (1993) *Biochemistry* **32**, 6019.

129. Leroy, J.-L. and Gueron, M. (1995) *Structure* **3**, 101.

130. Nonin, S. and Leroy, J.-L. (1996) *J. Mol. Biol.* **261**, 399.

131. Chen, L., Cai, L., Zhang, X. and Rich, A. (1994) *Biochemistry* **33**, 13540.

132. Kang, C.-H., Berger, I., Lockshin, C., Ratliff, R., Moyzis, R. and Rich, A. (1994) *Proc. Natl. Acad. Sci. USA* **91**, 11636.

133. Kang, C.-H., Berger, I., Lockshin, C., Ratliff, R., Moyzis, R. and Rich, A. (1995) *Proc. Natl. Acad. Sci. USA* **92**, 3874.

134. Berger, I., Kang, C.-H., Fredian, A., Ratliff, R., Moyzis, R. and Rich, A. (1995) *Nature Struct. Biol.* **2**, 416.

135. Rohozinski, J., Hancock, J.M. and Keniry, M.A. (1994) *Nucl. Acids Res.* **22**, 4653.

136. Ahmed, S. and Henderson, E. (1992) *Nucl. Acids Res.* **20**, 507.

137. Nonin, S., Phan, A.T. and Leroy, J.-L. (1997) *Structure* **5**, 1231.

138. Bell, G.I., Karam, J.H. and Rutter, W.J. (1981) *Proc. Natl. Acad. Sci. USA* **78**, 5759.

139. Catasti, P., Chen, X., Deaven, L.L., Moyzis, R.K., Bradbury, E.M. and Gupta, G. (1997) *J. Mol. Biol.* **272**, 369.

140. Masumoto, H., Masukata, H., Muro, Y., Nozaki, N. and Okazaki, T. (1989) *J. Cell. Biol.* **109**, 1963.

141. Gallego, J., Chou, S.-H. and Reid, B.R. (1997) *J. Mol. Biol.* **273**, 840.

142. Leroy, J.-L., Gueron, M., Mergny, J.-L. and Helene, C. (1994) *Nucl. Acids Res.* **22**, 1600.

143. Ahmed, S., Kintanar, A. and Henderson, E.(1994) *Nature Struct. Biol.* **1**, 83.

144. Manzini, G., Yathindra, N. and Xodo, L.E. (1994) *Nucl. Acids Res.* **22**, 4634.

145. Mergny, J.-L., Lacroix, L., Han, X., Leroy, J.-L. and Helene, C. (1995) *J. Am. Chem. Soc.* **117**, 8887.

146. Han, X., Leroy, J.-L. and Gueron, M. (1998) *J. Mol. Biol.* **278**, 949.

147. Marshall, W.S., Beaton, G., Stein, C. A., Matsukura, M. and Caruthers, M.H. (1992) *Proc. Natl. Acad. Sci. USA* **89**, 6265.

148. Marsich, E., Piccini, A., Xodo, L.E. and Manzini, G. (1996) *Nucl. Acids Res.* **24**, 4029.

14

DNA bending by adenine–thymine tracts

Donald M. Crothers[1] and Zippora Shakked[2]
[1]*Department of Chemistry, Yale University, New-Haven, CT, 06520, USA*
[2]*Department of Structural Biology, Weizmann Institute of Science, Rehovot, Israel*

1. Global and spectroscopic properties of DNA curvature induced by A-tracts

1.1 Identification of A-tracts as the primary source of DNA curvature

Fifteen years have elapsed since the observations of Marini *et al.* (1) which associated DNA bending or curvature with the anomalously slow electrophoretic mobility and fast overall rotational relaxation observed for DNA restriction fragments from the kinetoplast body of *Leishmania tarentolae.* Confirmation of increased curvature soon followed, using techniques such as electric birefringence decay (2) and electron microscopy (3). The sequences responsible for bending were identified as tracts of oligo (dA):oligo (dT), each about half a helical turn long, repeated in phase with the DNA helical screw (4). The experiment on which this conclusion was based relied on the slow electrophoretic mobility of a molecule containing a bend at its centre, compared with a circularly permuted sequence variant in which the bend is at the end. A simple rule of thumb in interpreting such experiments is that the shorter the end-to-end distance in molecules of equal contour length, the slower the electrophoretic mobility (5). Gel electrophoretic methods for characterizing DNA bending have been reviewed by Crothers and Drak (6).

The importance of the observed phasing of the A-tracts was confirmed by experiments that compared the mobilities of DNA ligation ladders containing A-tracts at variable phasings (7,8). Repetition of A-tracts in phase with the helical repeat of DNA causes their effects to be additive, leading to a circular shape; if the phase match is only approximate, a left- or right-handed superhelix results, which is of higher mobility than the planar circle. Since repetition of A-tracts at 1.5 helical turn phasing results in no observed curvature, hyperflexibility in a plane associated with A-tracts can be ruled out as a source of the electrophoretic anomaly and fast rotational relaxation (8).

The dominance of A-tracts as the primary source of DNA curvature is indicated by experiments such as two-dimensional gel electrophoresis (9) and selection amplification (10), both of which yielded a number of molecules containing phased A-tracts. The latter experiments also assigned a role to C–A (T–G) dinucleotide steps in conferring reduced electrophoretic mobility. Recent amplification experiments starting with genomic DNA and selecting for molecules easily bent to form nucleosome core particles also revealed a role for repeated C–A steps, and for short ($n = 3$–4) A-tracts (11).

Given that systematic DNA bending or curvature is associated with A-tracts that are phased with the DNA helical repeat, the questions that remain can be divided into two categories: what are the global properties of the bend, specifically, its direction and magnitude; and what is the structural basis for curvature at the molecular level? Earlier reviews of this general subject have been provided by Hagerman (12) and Crothers *et al.* (13).

1.2 Direction of A-tract bends

The first indication of the direction of the DNA bend induced by A-tracts was provided by the experiments of Koo *et al.* (8) who measured the mobility of molecules in which A-tracts alternated with T-tracts and compared them with the values observed when all of the A-tracts were on the same strand. Since the mobilities were nearly the same, one can conclude that the overall direction of curvature of an A-tract is little affected by rotation about the pseudo-dyad axis that runs through the centre of the tract, thus interchanging the A and T strands. In other words, the vector that bisects the bend angle is parallel to the pseudo-dyad axis running between major and minor grooves at the centre of the A-tract. This result allowed Koo *et al.* (8) to conclude that the bend is towards either the major or the minor groove at the centre of the A-tract. Based on fibre diffraction studies of poly (dA):poly (dT) (see below), they proposed a model in which the bend is towards the minor groove at a locus at or near the centre of the A-tract. The structural basis for the bend cannot be established by these experiments, but they do allow exclusion of specific models, for example, that the bend in solution is due to the large roll angle immediately adjacent to the A-tract, as observed in crystals and NMR structures (see below). However, models with positive roll distributed over the adjacent base pairs, or negative roll in the A-tract, or tilt of appropriate sign at the junctions are consistent with these results.

Gel electrophoresis methods can be used to determine the direction of the A-tract bend by comparing the mobility of constructs in which the A-tracts are at variable phasings relative to a bend of known direction. When the two bends are in the same direction, the curvature is maximal, the end-to-end distance is minimized, and the mobility reaches a minimum (14,15). (There are some exceptions at high gel percentage; see ref. 16.) Zinkel and Crothers (14) used the DNA bend induced by *E. coli* CAP protein as a standard, and concluded that the A-tract bend is towards the minor, not the major, groove, at or near the centre of the A-tract.

1.3 Polarity and imperfect dyad symmetry of A-tracts

Many solution experiments indicate that the structure of A-tracts varies from the 5' to the 3'-end, implying that the dyad symmetry deduced by Koo *et al.* (8) from A- and T-tract interchange is imperfect. The experiments of Koo *et al.* suggested imperfect symmetry, but the observations of Hagerman (17) were decisive: multimers of the form $(A_4T_4N_2)_n$ are highly curved, whereas those of the form $(T_4A_4N_2)_n$ are straight. This is not consistent with a fully dyad symmetric structure for an A-tract; the structural basis for the difference in curvature might reside in the different characteristics of the central base pair steps, A–T versus T–A, (see below).

Other solution experiments supporting an imperfect dyad include the hydroxyl radical footprinting results of Burkhoff and Tullius (18), which showed a progressive narrowing of the minor groove in the 5′ to 3′ direction. NMR experiments (19,20) revealed a steady shift towards lower field, totaling about one ppm, of the imino proton resonances in A-tracts, as one moves from the 5′ to the 3′-end of the tract. NOE measurements also provided evidence for narrowing of the A-tract minor groove over the first three base steps, followed by a region of approximately constant width in longer A-tracts (see below). The structural basis that gives rise to these observations remains a matter for conjecture, as discussed below.

1.4 Temperature dependence of A-tract structure and curvature

Early studies of A-tracts by electrophoretic methods revealed that the mobility anomaly is strongly reduced at elevated temperatures (21,22); reviewed by Breslauer (23). A premelting structural change in poly (dA): poly (dT) can be detected by UV absorbance (24) and CD spectroscopy (25). From the width of the transition curve, centred around 30–40°C, both groups estimated an apparent or van't Hoff enthalpy of about 20 kcal/mol. Chan *et al.* (26) used CD and scanning calorimetry to characterize the transition in a molecule containing phased A-tracts. They suggest that the transition follows a two-state model, since isoelliptic points are observed. The calorimetric result, about 4. 4 kcal/mol per A–A dinucleotide step, together with their estimate of 16 kcal/mol for the van't Hoff enthalpy from the width of the CD transition curve, can be used to estimate a length of about 5 bp for the cooperative unit in the premelting transition. This result, together with data such as those reported by Haran and Crothers (27), shows unambiguously that formation of the aberrant A-tract structure is cooperative, with an entire A-tract of 5 bp undergoing the transition as an effective cooperative unit.

Recent temperature-dependent resonance Raman studies of the premelting structural transition in poly (dA):poly (dT) by Chan *et al.* (28) provide important evidence concerning the underlying physical phenomenon. They concluded from deconvolution of the resonance Raman spectrum that a thymine C4=O carbonyl stretching frequency, normally observed at 1684–1686 cm^{-1} in poly (dA–dT) at 5 and 55°C, and in poly (dA):poly (dT) at 55°C, is anomalously red-shifted to about 1679 cm^{-1} in poly (dA):poly (dT) at 5°C. A similar anomaly, although smaller in scale, is also observed for the temperature dependence of a vibrational mode assigned to the adenine amino group.

These observations strongly favour associating the low temperature form of poly (dA):poly (dT) with the A-tract structure having propeller twisted base pairs with bifurcated hydrogen bonds, which has been observed in crystals of molecules containing A-tracts (see below). Formation of an extra (bifurcated) hydrogen bond to the thymine carbonyl is consistent with the observed reduction in the force constant for its stretching vibration. Thus one can now, with considerably increased confidence, associate disappearance of this structural feature at elevated temperature with the premelting transition of poly (dA):poly (dT) and accompanying loss of DNA bending in solution.

Another temperature-dependent feature of A-tract structure is the downfield-shifted position of the thymine imino proton resonances, particularly for those base pairs near the 3'-end of the A-tract (the 5'-end of the T-tract) (19,20). This could arise from strengthening the N–H···N hydrogen bond in the propeller twisted state. The imino proton chemical shifts move progressively to higher field as temperature is increased in the range of the premelting transition. The observed narrow minor groove, particularly towards the 3'-end of the A-tract, is also a reasonable consequence of this structural feature.

Thus there is now persuasive solution spectroscopic evidence for associating the low-temperature, bent state of DNA containing A-tracts with the structure having propeller twisted base pairs in the A-tract. The cooperative and two-state character of the thermal transition means that the A-tracts tends to convert as a unit into a structure lacking propeller twisting, presumably one that more closely resembles B-DNA. In order to yield a single imino proton resonance position for each base pair, the structures must equilibrate on a time-scale faster than 100 μs. However, the average extent of propeller twisting along the A-tract does not seem to be uniform in solution, but apparently increases from the 5' to the 3'-end. Since the A-tract tends to act as a cooperative unit, this cannot be explained by a higher occupancy of the high-temperature state by base pairs at the 5'-end of the A-tract. It is more likely that the extent of propeller twisting in the low-temperature state in solution is greater for base pairs near the 3'-end of the A-tract than for those near the 5'-end.

1.5 Bend magnitude

Estimates of the extent of bending produced by phased A-tracts have varied from about 11° per tract using gel electrophoresis (29) to about 28° from the rate of cyclization in ligation ladder experiments (30). Measurement of rotational relaxation gave 18° (31), as did computer simulation of the experiments on the relative rate of cyclization versus dimerization of DNA fragments containing phased A-tracts at 25°C (32). The uncertainty in this angle is estimated to be about 10%. Comparative electrophoresis experiments reveal that the curvature is modulated by only about ±10% by changes in the nature of the DNA sequence between the A-tracts (33). The temperature dependence of bending, as well as the effects of ionic conditions, should also be taken into account when estimating the bend angle.

1.6 Structural evidence from NMR spectroscopy

Intrinsic curvature of DNA of the observed magnitude requires only small deviations from the normal B-DNA structure. For example, the observations could result from systematic roll of about −6° in the A-tracts or +6° in the DNA segment between the A-tracts. This amount of roll is approximately equal to the rms fluctuations in the angle between adjacent base planes that results from thermal motion. NMR structures reflect information contained in a set of proton–proton vectors, as well as scalar coupling constants. So far, the structural resolution that has been achieved for nucleic acids has not been sufficient to yield a definitive solution structure that explains the global curvature. However, the NMR data contain a number of interesting features that must ultimately be explained by a definitive structural model.

A series of NMR studies of poly (dA):poly (dT) and oligonucleotides containing A-tracts have been reported(19,20,34–43). It is generally agreed that the structure is a member of the B family, but with some of the deoxyribose rings showing deviations from the standard C2′-*endo* conformation (pseudo-rotation angle, $P = 150$–$180°$). Inferred P values of some of the residues, particularly dT, fall in the range 90–130° range (38,40). It is also generally agreed that there is an unusually strong cross-strand NOE between adenine H2 and deoxyribose H1′, which reflects a narrowed minor groove and is probably associated with propeller twisting of the base pairs. Minor groove width seems to decrease along the A-tract from 5′ to 3′. Imino protons in the A-tracts have unusually long lifetimes, with the shortest lifetime corresponding to the residue at the 5′-end of the tract (19,39). As discussed above, the imino protons vary in chemical shift depending on the length of the tract and position in it.

Particular attention has been paid to conformational features at the junctions between the A-tracts and the adjacent DNA (36,40,43). However, these findings are not consistent with A-tract-induced bending, since the direction of the proposed bending in these molecules, namely, helix axis deflections corresponding to roll at the A-tract junctions, is not in agreement with the bend direction deduced from the electrophoresis experiments (14). Indeed, bending by roll at the junction does not satisfy the requirement for being unaltered in direction when the A-tract is rotated about its central dyad axis to interchange the A- and T-tracts.

2. X-ray crystallographic studies

2.1 Fibre diffraction studies of poly (dA):poly (dT)

Fibres of poly (dA):poly (dT) were first analysed by Arnott and Selsing (44) who obtained two X-ray patterns different from the classical A and B patterns of general sequence DNA. One pattern (α) obtained at above 85% relative humidity, indicated tenfold symmetry with a rise per base pair of 3.29 Å and a second pattern (β), obtained below 77% relative humidity, indicated a tenfold symmetry with a rise of 3.24 Å per base pair. Analysis of the polycrystalline β pattern yielded a heteronomous structure where each chain has a different conformation, the adenine chain adopting a conformation with C3′-*endo*-puckered sugar rings characteristic of the A-DNA family, and the thymine chain adopting a conformation with C2′-*endo* puckered rings characteristic of the B-DNA family (45).

Diffraction patterns similar to those of poly (dA):poly (dT) have also been observed with poly (dI):poly (dC) and with poly (dA–dI):poly (dT–dC) (46).

An X-ray analysis of fibres of the Ca^{2+} salt of poly (dA):poly (dT) indicated a symmetric structure in which the two chains are conformationally identical with a B-DNA-type backbone (47). The revised analysis of the Na^+ salt of the polymer studied previously (see above) yielded a structure that is only slightly heteronomous and fairly similar to the Ca^{2+} structure (47).

In a more recent analysis of the sodium salt of the homopolymer (48), several constraints were introduced in the refinement of the model in order to maintain conformational parameters close to those observed in the crystal structure of the A_3T_3-containing dodecamer (49).

The unique and common features of the fibre-based structures of poly (dA):poly (dT), which distinguish them most from general sequence B-DNA, are negative inclination of the base pairs with respect to the helix axis (average −6°), high propeller twisting of the base pairs (average −26°) and a very narrow minor groove (average 3.4 Å). In the model of Aymami *et al.* (48), the large propeller twisting is associated with bifurcated hydrogen bonds across the major groove. These features and the exceptionally narrow minor groove are also characteristic of short A-tracts studied by single-crystal X-ray crystallography (see below).

2.2 Crystal structures of A-tracts and related sequences

2.2.1 Helical conformations

Crystallographic studies of short DNA oligomers have been carried out over the past two decades, demonstrating that the structure of the DNA double helix is dependent on both the base sequence and the environment (50; Chapter 6 and references cited therein). A special effort has been directed towards the elucidation of A-tract-containing duplexes in an attempt to reveal the structural basis of A-tract-induced curvature (49,51–54). These studies have shown that A-tract DNA assumes a conformation in which the helix axis is straight, the base pairs are perpendicular to the helix axis and the helical periodicity is close to 10 base pairs per turn (Table 14.1). The sugar pucker within the A-tract regions reflects a broad range of conformations, as for the other B-type structures. However, the resolution of the diffraction data of the A-tract structures (1.9–2.6 Å) is not sufficient to allow accurate determination of the sugar conformations.

Two structural features specific to A-tracts were observed in the various crystal structures: an exceptionally narrow minor groove and highly propeller-twisted base pairs (Table 14.1). The extent of propeller twisting of the A:T base pairs was found to

Table 14.1. Average helical parameters of A-tracts and related sequences[a]

Sequence[b]	A-tract/ I-tract	Helix twist (°)	Roll angle (°)	Propeller twist (°)	Minor groove width (Å)	Reference
CGCGAATTCGCG	AATT	34.9	−1.6	−17.0	4.1	55
CGTGAATTCACG	AATT	34.7	−2.7	−15.9	3.7	56,57
CGCAAAAAAGCG	AAAAAA	36.2	−0.7	−19.8	3.7	51
CGCAAAAATGCG	AAAAAT	35.7	−0.3	−18.2	3.4	52
CGCGAAAAAACG	AAAAAA	35.1	0.5	−21.5	3.7	53
CGCAAATTTGCG	AAATTT	36.1	0.8	−16.6	4.7	54
CGCIAATTCGCG	IAATTC	36.8	−0.6	−17.1	4.1	58
CGCAIATMTGCG	AIATMT	36.2	−1.0	−19.2	3.5	59
CCIIICCCGG	IIICCC	35.8	−0.2	−12.6	3.7	59

[a] Adapted from reference 60. In cases of multiple sites or different studies of the same sequence, the values correspond to the average of individual averages.
[b] I = inosine, M = 5-methylcytosine.

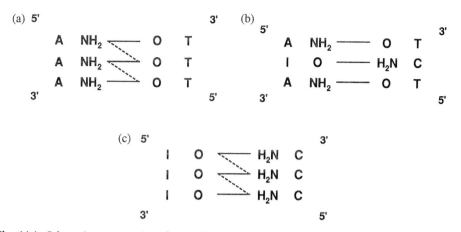

Fig. 14.1. Schematic representation of potential cross-strand interactions in (a) AAA, (b) AIA, and (c) III. Watson–Crick hydrogen bonds are shown as heavy lines and cross-strand bifurcated hydrogen bonds as broken lines.

be sufficiently large to result in interstrand bifurcated hydrogen bonds between adjacent A and T bases across the major groove, as illustrated schematically in Fig. 14.1a.

It has been proposed that the high propeller twisting associated with bifurcated hydrogen bonds observed in the crystal structures might be important for the distinctive ability of A-tracts to induce DNA curvature (49,51). However, the roles of both propeller twist and cross-strand interactions were subsequently challenged on the basis of the observation that curvature is only weakly affected by substituting some of the A:T base pairs for I:C or I:M (M = 5-methylcytosine) base pairs (e.g. AAIAA and AIAIA), whereas curvature decreases abruptly for pure inosine tracts (I-tracts) (61,62). Since a bifurcated hydrogen bond does not seem to be supported by an I:C base pair that is flanked by A:T pairs (Fig. 14.1b), it appeared unlikely that the proposed hydrogen bond could be the principal component stabilizing the A-tract structure.

In an attempt to identify the structural features of A-tract and A-tract-like regions, and to distinguish how they differ from other AT-rich sequences, X-ray crystallography and gel electrophoresis studies of several oligomers incorporating A:T, I:C, or I:M (M = 5-methylcytosine) base pairs have been performed recently (59). The X-ray crystallographic analysis demonstrated that an alternating purine region of the type −AIA− is structurally similar to a pure A-tract in that both are characterized by a remarkably uniform stacking geometry associated with high propeller twisting of the base pairs (Table 14.1). Close interstrand contacts at the major groove between amino groups across A–I base pair steps were observed. This interaction appears to stabilize the geometry of such steps and makes them compatible with A–A steps (see below). In contrast to A-tract and A-tract-like regions, I-tract regions of the type IIICCC display a variable pattern of base stacking geometry and significantly lower propeller twisting (Table 14.1) with no indication of close interstrand interactions across I–I steps (i.e. between inosine carbonyl groups and cytosine amino groups). The inosine runs, however, share two features in common with A-tracts: a helical repeat of 10 bp/turn and a narrow minor groove occupied by a spine of hydration (Table 14.1).

The majority of the duplexes incorporating A-tracts (4–6 bp long) display an overall asymmetric bend as a result of crystal packing interactions, the extent of bending (10–22°) depending on the temperature and crystallization conditions used (63). The direction of the bending, found to be localized at the GC-rich region or at the junction between the A-tract and the flanking GC-rich segment, is about 90° away from that deduced for phased A-tracts by gel electrophoresis (14). It should be emphasized that the bending observed in crystallized oligomers is not necessarily related to that observed in solution. This bending appears to be induced at flexible sites by crystal packing effects (50). However, the short A-tracts flanked by G:C base pairs are unbent and rather resistant to deformations that might be caused by crystal forces. Unlike the structural uniformity of A-tracts, regions of alternating A and T bases of the type $(AT)_n$ ($n = 2$–3) are conformationally polymorphic (59,64,65).

The base-stacking patterns displayed by the homopurine steps of the type A–A, A–I, and I–A are very similar (59). The propeller-twisted conformation observed in such steps should be supported by the various components of base-stacking interac-

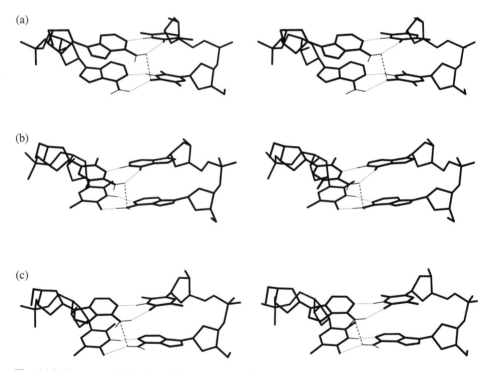

Fig. 14.2. Stereoscopic drawings of base pair steps showing propeller-twisted base pairs with bifurcated hydrogen bonds at the major groove. Watson–Crick hydrogen bonds are shown as dotted lines. Bifurcated hydrogen bonds are shown as broken lines and are between bases at the 5′-ends of the two strands. (a) A–A/T–T step: the bifurcated hydrogen bond is between the amino hydrogen of an adenine base and the carbonyl oxygen of a thymine base (taken from the crystal structure of CGCAAAAAAGCG, ref. 51). (b) A–I/M–T step: the bifurcated hydrogen bond is between the amino hydrogen of a 5-methylcytosine base and the amino nitrogen of an adenine base. (c) A–T step: the bifurcated hydrogen bond is between the amino nitrogen of one adenine base and the amino hydrogen of another one. (b) and (c) were taken from the crystal structure of CGCAIATMTGCG (59).

tions, but other factors may also contribute to the stability of this conformation. Interstrand electrostatic interactions between amino groups and carbonyl oxygens across A–A steps (Fig. 14.2a) or between amino groups across A–I steps (Fig. 14.2b) could stabilize the high propeller twist and thus confer structural invariance to such regions.

Attractive interactions between functional groups across the major groove are also likely to occur in other steps, such as A–T (59,66). In this case, the interactions are between the amino groups of the adenine bases across the groove, as seen in several crystal structures and illustrated in Fig. 14.2c. This interaction might explain, in part, the relative structural uniformity of such steps, in contrast to the large variability of T–A steps observed in the crystal structures (compiled in 67–70 and Chapter 6). Crystal structure data on A–T steps show that they adopt a small roll angle and considerable propeller twisting when adjacent to short A-tracts. These features are compatible with those observed for A–A steps. Hence, the introduction of an A–T step within an A-tract (49,54–59) does not disrupt the conformational uniformity and stability of such regions. In contrast to A–T steps, T–A steps separating short A-tracts are characterized by a positive roll (i.e. bending into the major groove) and modest propeller twisting (71,72). Thus, the insertion of an incompatible hinge like a T–A step into an A-tract can disrupt structural uniformity and optimal base stacking, unlike the effect of an A–T step insertion. The different structural effects of the two insertions correspond with markedly distinct melting behaviours (73) and gel migration data (17,27).

X-ray and gel migration studies have shown that single substitutions of I:C or I:M base pairs within A-tracts have little consequence for either local or global structural properties. However, there are clear differences in the behaviour of I:M or I:C versus A:T base pairs, which become more pronounced as additional substitutions are made. Phased runs of I:M or I:C base pairs display only a small fraction of the curvature seen for A:T pairs (59 and references therein).

The X-ray study of CCIIICCCGG has shown that inosine stretches display low propeller twisting (Table 14.1). As a result, interstrand distances at the major groove between opposing amino and carbonyl groups are relatively long (average 3.6 Å) compared with the equivalent N···O contacts within A-tracts (average 3.2 Å). It therefore appears that this region is not stabilized by a network of bifurcated hydrogen bonds. The low propeller twisting and lack of interstrand interactions between inosine and cytosine bases across I–I steps have been suggested as underlying causes for the variable pattern of their base-stacking geometries in contrast to the relatively uniform base-stacking geometry of A-tracts (59). Therefore, it is likely that I-tracts in solution adopt a structure that is more variable than A-tracts and closer to general sequence B-DNA. This may explain the large reduction in macroscopic curvature for I-tracts with respect to A-tracts.

Gel migration studies have shown that methylation of cytosines has a weak effect on curvature in cases where inosine bases are adjacent to adenine base pairs. However, there appears to be a cooperative effect of the methyl group in the case of I:M base pairs, since the curvature increases significantly for pure I-tract as a result of such a modification (59). No structural data are available on methylated I-tracts to explain this observation at the molecular level.

2.2.2 Hydration patterns

In several of the A-tract-containing helices, a single spine of hydration was observed in the minor groove, spanning the 4–6 A:T base pairs where the groove width is remarkably narrow (3–4 Å). The spine consists of first and second shell hydration molecules, as illustrated schematically in Fig. 14.3a. The first shell molecules link the cross-strand minor groove acceptor atoms, N3 of purine bases and O2 of pyrimidine bases, which are positioned at nearly identical sites in the minor groove. The second shell molecules interact with the first water shell to form a zigzag structure. This characteristic hydration was first observed in the central region of the B-DNA dodecamer CGC-GAATTGCGC (74).

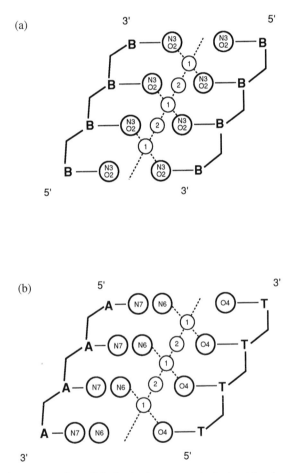

Fig. 14.3. Schematic representation of idealized minor groove hydration (a) and major groove hydration (b) where **B** denotes any base. Acceptor and donor atoms of the bases are shown as big circles with the corresponding atom names. Water molecules are shown as small circles where first and second hydration shell molecules are denoted by 1 and 2, respectively. Hydrogen bonds are shown as broken lines.

A spine of hydration has been also observed in the CCIIICCCGG decamer and in several other B-DNA helices in regions where the minor groove is narrow (59; Chapter 9).

Unlike the minor groove, where common hydration patterns were observed [a single spine for a narrow groove and a double ribbon for a wide groove, reviewed by Berman (75,76; Chapter 9)], the B-DNA major groove has not revealed any common hydration motif. The possibility of a unique hydration motif that is specific to A-tracts and related sequences was demonstrated by the crystal structure of CGCA-IATMTGCG (59). The major groove hydration of this dodecamer indicated the existence of a continuous chain formed by first and second shell molecules along the major groove, as illustrated schematically in Fig. 14.3b.

The cross-strand water-mediated interactions in the minor groove link bases that are neighbouring in the 3' direction, whereas the water-mediated major groove contacts link cross-strand bases neighbouring in the 5' direction (Fig. 14.3). These hydration patterns can therefore stabilize the propeller-twisted conformation. In this manner the specific hydration contributes to the unique stability and structural uni-formity of A-tract regions.

Based on the crystal structure data and recent observations of the effect of MPD (2-methyl-2,4-pentanediol) on gel mobility of DNA fragments incorporating phased A-tracts (77,78), it has been suggested that any disruption of the minor and major groove hydration by dehydrating agents such as MPD would lead to a more flexible structure that is similar to that of general DNA sequence, and thus would reduce A-tract-dependent curvature (59).

2.3 A-tracts in protein–DNA complexes

Several crystal structures of protein–DNA complexes have been determined where the DNA target incorporates short A-tracts. These A-tracts are of the kind A_n and A_nT_m. In several of these complexes the DNA target is severely deformed and the A-tract is bent at the minor groove (79–86). The A-tract bending is achieved by a combination of local roll and tilt angles, resulting in negatively inclined base pairs with respect to the helix axis. The contribution of the short A-tracts (4–6 base pairs) to the overall DNA curvature is modest, ranging from 4 to 13°, whereas the major contribution is achieved by major groove compression at the flanking regions (Table 14.2 and refs 79–86). Like the unbound A-tracts, the minor grooves of the complexed A-tracts are narrow, the average helix twist is close to 36° or slightly overwound, and the base pairs display a large propeller twisting (Table 14.2).

The structural similarity between the various A-tract-containing helices is illustrated in Plate XIX. Two representatives of the complexed A-tract-containing fragments (A_6, with the IHF protein, and A_5, with the DNA-binding domain of the 434 repressor protein, Table 14.2) are displayed together with the fibre structure of poly (dA):poly (dT) derived by Alexeev *et al.* (47) and the crystal structure of an A_6-containing dodecamer (51). Also shown for comparison is the fibre-based structure of the general sequence B-DNA helix (87). The minor grooves of the A-tract regions are narrow and the base pairs are highly propeller twisted. In contrast, the minor groove of the general sequence B-DNA is wide and the base pairs are essentially flat. The

Table 14.2. Average helical parameters of A-tracts complexed to proteins[a]

Protein	A-tract	Helix twist (°)	Roll angle (°)	Propeller twist (°)	Minor groove width (Å)	A-tract bending (°)	Reference
CAP	AAAA	37.5	−0.1	−21.8	3.7	9.4	79,80
434 repressor	AAAAA	36.5	−1.5	−16.6	4.0	9.8	81
NF-κB	AATT	38.0	−3.0	−16.6	3.5	6.4	82
SRF	AATT	37.8	−5.0	−14.7	3.4	8.8	83
MATa1/MATα2	AATTT	34.7	−2.6	−15.0	3.3	13.1	84
Oct-1POU	AAAT	36.2	−2.7	−16.2	4.5	8.9	85
IHF	AAAAAA	35.5	−2.0	−15.7	3.4	4.2	86

[a] Adapted from reference 60 (as for Table 14.1).

four A-tract helices differ in the degree of inclination of the base pairs, which display gradual change from −6° in poly (dA):poly (dT) through −4 and −2° in the complexed A_5 and A_6 regions, to nearly 0° in the A_6 region of the unbound oligomer.

3. The stereochemical basis of A-tract-dependent curvature

The key question is what kind of mechanism, at the molecular level, is causing the observed macroscopic curvature of phased, 4–6 base pair-long A-tracts. Since no single structure explains the whole phenomenon, it is necessary at present to rely on models, several of which have been proposed (8,13,88–94). They generally conform to the gel migration data, which suggest that the centre of curvature is towards the minor groove of the A-tracts and/or towards the major groove of the intervening general sequences (14). However, they differ substantially in the details of the stereochemical origin of curvature.

The crystal structure, spectroscopic, and gel migration data support a model where macroscopic curvature of A-tract-containing DNA and related stretches is correlated with a unique structure conferred by a narrow minor groove, propeller-twisted base pairs, cross-strand bifurcated hydrogen bonds, and characteristic hydration.

The various oligonucleotide crystal structures show that A-tract DNA is straight. Here, we use the term 'straight DNA' to mean a structure where the base pairs are perpendicular to the straight helix axis so that the roll and tilt angles between adjacent base pairs are essentially zero. This should be clearly distinguished from other structures, where the helix axis is straight but the base pairs are uniformly inclined with respect to a plane that is perpendicular to this axis and the base steps have essentially no variation in roll and tilt angles. In general, positive inclination of the base pairs is associated with a positive roll and a narrow major groove, whereas negative inclination is accompanied by a negative roll and a narrow minor groove. Examples are the fibre structure of poly (dA):poly (dT), where the base pairs are negatively inclined (see above) and the fibre and crystal structures of A-DNA where the base pairs are positively inclined to the helix axis (87,95; Chapter 5 and references therein). If the latter type of DNA segment were joined with a 'straight' one, a change in DNA trajectory would result at the junction between the two, as demonstrated recently in the DNA complexed to the TATA-box binding protein (96). Since the crystal structures of short A-tracts are straight in the above sense and exhibit little structural variation amongst a number of crystal structures, it has been proposed that bending must occur outside such regions (51,63).

Related variants of the straight A-tract model suggest that a gentle roll-induced writhe is a property of all B-DNA sequences to a varying degree, with A–A steps exhibiting an average roll close to zero (88,89). In such a manner, the global curvature of DNA containing short A-tracts in phase with the helical repeat is a consequence of a net positive roll (i.e. major groove compression) accumulated in the intervening sequences. Several studies of B-DNA crystal structures have shown that GC-rich and general sequences can bend into the major groove (63,97). A recent crystal structure has demonstrated that a general B-DNA sequence can adopt a roll-induced writhe so that the base pairs are uniformly and positively inclined to the helix axis by nearly 7° (98). It should be noted, however, that the average roll angle determined from B-DNA crystal structures has been estimated to be near zero (69).

On the other hand, the fibre diffraction studies of the homopolymer dA:dT and the crystal structure studies of A-tracts bound to proteins indicate the possibility of a 'bent' A-tract structure; i.e. the base pairs are negatively inclined to the helix axis. In this manner, the global curvature of phased A-tracts separated by 'straight' B-DNA segments is a consequence of a net negative roll (i.e. minor groove compression) at the A-tract regions. This model, originally called the 'junction model', was proposed by Koo *et al.* (8).

The straight A-tract model and the bent A-tract model present the two extreme views of a scientific controversy lasting for more than a decade. The true stereochemical mechanism probably lies somewhere between the two extremes. Indeed, we have become increasingly of the opinion that A-tract curvature may be delocalized, in the sense that there are contributions from negative roll in the A-tracts and positive roll in the adjacent DNA segments, and there may even be small tilt contributions at the junctions. It should also be kept in mind that overall curvature may be the result of anisotropic bendability. Both A-tracts and B-DNA segments may be essentially straight in their lowest energy state, but if bending excursions that compress the minor groove in the A-tracts and the major groove in B-DNA are less costly energetically than motions in the opposite directions, the average result will be curvature of the molecule in solution. If the effect is operative in both sequences, the average excess roll in the preferred direction need be only about 3° to explain the magnitude of the observed global curvature. Experimental verification of such a small effect is an imposing challenge. Further studies are needed to establish the mechanism and relative contributions of A-tracts and the adjacent sequences to the observed macroscopic curvature.

Acknowledgements

This work was supported by grants from the National Institutes of Health (GM-21966 to D.M.C.) and the Israel Science Foundation administered by the Israel Academy of Sciences and Humanities (to Z.S.).

References

1. Marini, J.C., Levene, S.D., Crothers, D.M. and Englund, P.T. (1982) *Proc. Natl. Acad.* Sci. *USA* **79**, 7664.
2. Hagerman, P.J. (1984) *Proc. Natl. Acad. Sci.USA* **81**, 1763.
3. Griffith, J., BleymanM., Rauch, C.A., Pitchin, P.A. and Englund, P.T. (1986) *Cell* **46**, 717.
4. Wu, H.-M. and Crothers, D.M. (1984) *Nature* **308**, 509.
5. Lumpkin, O.J. and Zimm, B.H. (1982) *Biopolymers* **21**, 2315.
6. Crothers, D.M., Drak, J., Kahn, J.D.and Levene, S.D. (1992) *Meth. Enzymol.* **212B**, 3.
7. Hagerman, P.J. (1985) *Biochemistry* **24**, 7033.
8. Koo, H.-S., Wu, H.-M. and Crothers, D.M. (1986) *Nature* **320**, 501.
9. Anderson, J.N. (1986) *Nucl. Acids Res.* **14**, 8513.
10. Beutel, B.A. and Gold, L. (1992) *J. Mol. Biol.* **228**, 803.
11. Widlund, H.R., Cao, H., Simonsson, S., Magnusson, E., Simonsson, T., Nielsen, P.E., Kahn, J.D., Crothers, D.M. and Kubista, M. (1997) *J. Mol. Biol.* **267**, 807.
12. Hagerman, P.J. (1990) *Annu. Rev. Biochem.* **59**, 755.
13. Crothers, D.M., Haran, T.E. and Nadeau, J.G. (1990) *J. Biol. Chem.* **265**, 7093.
14. Zinkel, S.S. and Crothers, D.M. (1987) *Nature* **328**, 178.
15. Salvo, J.J. and Grindley, N.D. F. (1988) *EMBO J.* **7**, 3609.
16. Drak, J. and Crothers, D.M. (1991) *Proc. Natl. Acad. Sci. USA* **88**, 3074.
17. Hagerman, P.J. (1986) *Nature* **321**, 449.
18. Burkoff, A.M. and Tullius, T.D. (1987) *Cell* **48**, 935.
19. Leroy, J.L., Cherretier, E., Kochoyan, M. and Guéron, M. (1988) *Biochemistry* **27**,8894.
20. Nadeau, J.G. and Crothers, D.M. (1989) *Proc. Natl. Acad. Sci. USA* **86**, 2622.
21. Marini, J.C., Effron, P.N., Goodman, T.C., Singleton, C.K., Wells, R.D., Wartell, R.M. and Englund, P.T. (1984) *J. Biol. Chem.* **259**, 8974.
22. Dieckmann, S. (1987) *Nucl. Acids Res.* **15**, 247.
23. Breslauer, K.J. (1991) *Curr. Opin. Struct. Biol.* **1**, 416.
24. Herrera, J.E. and Chaires, J.B. (1989) *Biochemistry* **28**, 1993.
25. Chan, S.S., Breslauer, K.J., Hogan, M.E., Kessler, D.J., Austin, R.H., Ojemann, J., Passner, J.M. and Wiles, N.C. (1990) *Biochemistry* **29**, 6161.
26. Chan, S.S., Breslauer, K.J., Austin, R.H. and Hogan, M.E. (1993) *Biochemistry* **32**, 11776.
27. Haran, T.E. and Crothers, D.M. (1989) *Biochemistry* **28**, 2763.
28. Chan, S.S., Austin, R.H., Mukerji, I. and Spiro, T.G. (1997) *Biophys. J.* **72**, 1512.
29. Calladine, C.R., Drew, H.R. and McCall, M.J. (1988) *J. Mol. Biol.* **201**, 127.
30. Ulanovsky, L.E., Bodner, M., Trifonov, E.N. and Choder, M. (1986) *Proc. Natl. Acad. Sci. USA* **83**, 862.
31. Levene, S.D., Wu, H.-M. and Crothers, D.M. (1986) *Biochemistry* **25**, 3988.
32. Koo, H.-S., Drak, J., Rice, J.A. and Crothers, D.M. (1990) *Biochemistry* **29**, 4227.
33. Haran, T.E., Kahn, J.D. and Crothers, D.M. (1994) *J. Mol. Biol.* **244**, 135.
34. Behling, R.W. and Kearns, D.R. (1986) *Biochemistry* **25**, 3335.
35. Behling, R.W., Rao, S.N., Kollman, P. and Kearns, D.R. (1987) *Biochemistry* **26**, 4674.
36. Katahira, M., Sugeta, H., Kyogoku, Y., Fujii, S., Fujisawa, R. and Tomita, K. (1988) *Nucl. Acids Res.* **16**, 8619.
37. Gupta, G., Sarma, M.H. and Sarma, R.H. (1988) *Biochemistry* **27**, 7909.
38. Celda, G., Widmer, H., Leupin, W., Chazin, W.J., Denny, W.A. and Wutrich, K. (1989) *Biochemistry* **28**, 1462.
39. Moe, J.G. and Russu, I.M. (1990) *Nucl. Acids Res.* **18**, 821.
40. Searle, M.S. and Wakelin, L.P. (1990) *Biochim. Biophys. Acta* **1049** 69.

41. Karahira, M., Sugeta, H. and Kyogoku, Y. (1990) *Nucl. Acids Res.* **18**, 613.

42. Chen, S.M., Leupin, W. and Chazin, W.J. (1992) *Int. J. Biol. Macromol.* **14**, 57.

43. Young, M.A., Srinivasan, J., Goljer, I., Kumar, S., Beveridge, D.L. and Bolton, P.H. (1995) *Meth. Enzymol.* **261**, 121.

44. Arnott, S. and Selsing, E. (1974) *J. Mol. Biol.* **88**, 509.

45. Arnott, S., Chandrasekaran, R. Hall, I.H. and Puigjaner, L.C. (1983) *Nucl. Acids Res.* **11**, 4141.

46. Leslie, A.G.W., Arnott, S., Chandrasekaran, R. and Ratliff, R.L. (1980) *J. Mol. Biol.* **143**, 49.

47. Alexeev, D.G., Lipanov, A.A. and Skuratovskii, I.Y. (1987) *Nature* **325**, 821.

48. Aymami, J., Coll, M.,Frederick, C.A., Wang, A.H.-J. and Rich, A. (1989) *Nucl. Acids Res.* **17**, 3229.

49. Coll, M., Frederick, C.A., Wang, A.H-J. and Rich, A. (1987) *Proc. Natl. Acad. Sci. USA* **84**, 8385.

50. Shakked, Z. (1991) *Curr. Opin. Struct. Biol.* **1**, 446.

51. Nelson, H.C.M., Finch, J.T., Luisi, B.F. and Klug, A. (1987) *Nature* **330**, 221.

52. DiGabriele, A.D., Sanderson, M.R. and Steitz, T.A. (1989) *Proc. Natl. Acad. Sci. USA* **86**, 1816.

53. DiGabriele, A.D. and Steitz, T.A. (1993) *J. Mol. Biol.* **321**, 1024.

54. Edwards, K.J., Brown, D.G., Spink, N., Skelly, J.V. and Neidle, S. (1992) *J. Mol. Biol.* **226**, 1161.

55. Dickerson, R.E. and Drew, H.R. (1981) *J. Mol. Biol.* **149**, 761.

56. Larsen, T.A., Kopka, M.L. and Dickerson, R.E. (1991) *Biochemistry* **30**, 4443.

57. Narayana, N., Ginell, S.L., Russu, I.M. and Berman, H.M. (1991) *Biochemistry* **30**, 4449.

58. Xuan, J.-C. and Weber, I.T. (1992) *Nucl. Acids Res.* **20**, 5457.

59. Shatzky-Schwartz, M., Arbuckle, N.D., Eisenstein, M., Rabinovich, D., Bareket-Samish, A., Haran, T.E., Luisi, B.F. and Shakked, Z. (1997) *J. Mol. Biol.* **267**, 595.

60. Shatzky-Schwartz, M. (1997) PhD Thesis. Weizmann Institute of Science, Israel.

61. Koo, H.-S. and Crothers, D.M. (1987) *Biochemistry* **26**, 3745.

62. Diekmann, S., Mazzarelli, J.M., McLaughlin, L.W., von Kitzing, E., and Travers, A.A. (1992) *J. Mol. Biol.* **225**, 729.

63. Dickerson, R.E., Goodsell, D. and Kopka, M.L. (1996) *J. Mol. Biol.* **256**, 108.

64. Yoon, C., Privé, G.G., Goodsell, D.S. and Dickerson, R.E. (1988) *Proc. Natl. Acad. Sci. USA* **85**, 6332.

65. Yuan, H., Quintana, J.R. and Dickerson, R.E. (1992) *Biochemistry* **31**, 8009.

66. Sponer, J. and Kypr, J. (1994) *Int. J. Biol. Macromol.* **16**, 3.

67. Shakked, Z., Guzikevich-Guerstein, G., Frolow, F., Rabinovich, D., Joachimiak, A. and Sigler, P.B. (1994) in *Structural Biology: the State of the Art*, (Sarma, R.H. and Sarma, M.H., eds), Vol. 1, pp. 199–216. Adenine Press, New York.

68. Suzuki, M. and Yagi, N. (1995) *Nucl. Acids Res.* **23**, 2083.

69. Gorin, A.A., Zhurkin, V.B. and Olson, W.K. (1995) *J. Mol. Biol.* **247**, 34.

70. El Hassan, M.A. and Calladine C.R. (1997) *Phil. Trans. R. Soc. Lond.* **A355**, 43.

71. Goodsell, D.S., Kaczor-Grzeskowiak, M. and Dickerson, R.E. (1994) *J. Mol. Biol.* **239**, 79.

72. Balendiran, K., Rao, S.T., Sekharudu, C.Y., Zon, G. and Sundaralingam, M. (1995) *Acta Cryst.* **D51**, 190.

73. Park, Y.W. and Breslauer, K.J. (1991) *Proc. Natl. Acad. Sci. USA* **88**, 1551.

74. Drew, H.R. and Dickerson, R.E. (1981) *J. Mol. Biol.* **151**, 535.

75. Berman, H.M. (1991) *Curr. Opin. Struct. Biol.* **1**, 423.

76. Berman, H.M. (1994) *Curr. Opin. Struct. Biol.* **4**, 345.

77. Sprous, D., Zacharias, W., Wood, Z.A. and Harvey, S.C. (1995) *Nucl. Acids Res.* **23**, 1816.

78. Dlakic, M., Park, K., Griffith, J.D., Harvey, S.C. and Harrington, R.E. (1996) *J. Biol. Chem.* **271**, 17911.

79. Schultz, S.C, Shields, G.C. and Steitz, T.A. (1991) *Science* **253**, 1001.

80. Parkinson, G., Wilson, C., Gunasekera, A., Ebright, Y.W., Ebright, R.H. and Berman, H.M. (1996) *J. Mol. Biol.* **260**,395.

81. Rodgers, D.W. and Harrison, S.C. (1993) *Structure* **1**, 227.

82. Ghosh, G., Van Duyne, G., Ghosh, S. and Sigler, P.B. (1995) *Nature* **373**, 303.

83. Pellegrini, L., Tan, S. and Richmond, T.J. (1995) *Nature* **376**, 490.

84. Li, T, Stark, M.R., Johnson, A.D. and Wolberger, C. (1995) *Science* **270**, 262.

85. Klemm, J.D., Rould, M.A., Aurora, R., Herr, W. and Pabo, C.O. (1994) *Cell* **77**, 21.

86. Rice, P.A., Yang, S.-W., Mizuuchi, K. and Nash, H.A. Cell, **87**, 1295 (1996).

87. Chandrasekaran, R. and Arnott, S. in *Landolt–Bornstein, New Series, Group VII (Biophysics)*, (Saenger, W., ed.), Vol. 1b, pp. 31–170. Springer-Verlag, Berlin.

88. Calladine, C.R., Drew, H.R., and McCall, M.J. (1988) *J. Mol. Biol.* **201**, 127.

89. Maroun, R.C. and Olson, W.K. (1988) *Biopolymers* **27**, 585.

90. De Santis, P. Palleschi, A. Savino, M and Scipioni, A. (1990) *Biochemistry* **29**, 9269.

91. Bolshoy, A., McNamara, P., Harrington, R.E., and Trifonov, E.N. (1991) *Proc. Natl. Acad. Sci. USA* **88**, 2312.

92. Zhurkin, V.B., Ulyanov, N.B., Gorin, A.A. and Jernigan, R.L. (1991) *Proc. Natl. Acad. Sci. USA* **88**, 7046.

93. Olson, W.K., Marky, N.L., Jernigan, R.L. and Zhurkin, V.B. (1993) *J. Mol. Biol.* **232**, 530.

94. Goodsell, D.S. and Dickerson, R.E. (1994) *Nucl. Acids Res.* **22**, 5497.

95. Haran, T.E. and Shakked, Z. (1988) *J. Mol. Struct. (Theochem.)* **179**, 367.

96. Guzikevich-Guerstein, G. and Shakked, Z. (1996) *Nature Struct. Biol.* **4**, 32.

97. Goodsell, D.S., Kopka, M.L., Cascio, D. and Dickerson, R.E. (1993) *Proc. Natl. Acad. Sci. USA* **90**, 2930.

98. Rozenberg, H., Rabinovich, D., Frolow, F., Hegde, R.S. and Shakked, Z. (1998) *Proc. Natl. Acad. Sci. USA*, in press.

15

Structures and interactions of helical junctions in nucleic acids

David M. J. Lilley

CRC Nucleic Acid Structure Research Group, Department of Biochemistry, The University, Dundee DD1 4HN, UK

1. The occurrence of helical junctions in biology

Helical junctions in nucleic acids are branch points where double helical segments intersect with axial discontinuities, such that strands are exchanged between the different helical sections. While bulges can be brought into this definition, we will restrict our attention to helical junctions in this chapter, of which the most common are three- or four-way junctions (Fig. 15.1). These can be perfect junctions, where every base is paired with its Watson–Crick complement, or they can contain mismatches or unpaired bases; the latter can have significant effects on the folding of the structures in some cases. A systematic nomenclature exists for the unambiguous description of different junctions (1).

Helical junctions are quite common in RNA species. For example, if we look at the secondary structure of a rRNA species we will find examples of three- and four-way junctions. They are seldom perfect however, and one or more single-stranded bases are often present at the point of strand exchange. A number of functional catalytic RNA molecules are based around helical junctions, such as the hairpin ribozyme (2), which is a four-way junction in the tobacco ringspot viral RNA, and the hammerhead ribozyme, which can be regarded as an imperfect three-way junction (3,4) (see also Chapter 17).

In the case of DNA, the main biological significance of branched helical species is as intermediates in DNA rearrangements of various kinds, notably in recombination events. The four-way junction has been proposed to be the central intermediate in

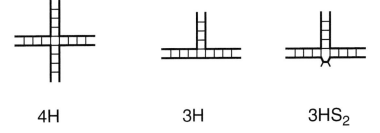

Fig. 15.1. Helical junctions in nucleic acids. The junctions of biological significance are three- and four-way branch points. Junctions can be perfectly base paired, or they can be modified by the addition of unpaired bases. The nomenclature used is the IUPAB scheme explained in ref. 1.

homologous genetic recombination (5–11), created by strand invasion between two homologous DNA molecules. In the integrase class of site-specific recombination events there is good evidence for a four-way junction intermediate (12–16). DNA junctions can also arise in other ways, including the replication of DNA, as exemplified by bacteriophage T4 (17).

DNA junctions are substrates for proteins involved in the later stages of genetic recombination. Proteins accelerate the process of branch migration, and ultimately resolve the branch point to recreate two independent duplex species. Such proteins recognize their DNA substrates at the level of tertiary structure, a process that should reflect molecular recognition of DNA structure on a relatively large scale. More recently it has become apparent that, as well as recognizing branched DNA structure, such proteins also alter the very structure that they recognize in many cases.

One question that we might usefully pose in this review is whether we can establish some general folding principles for helical branch points in nucleic acids. Two candidate principles offer themselves at this stage, and we will return to these at the end of the chapter to see how well they bear up.

- *Coaxial helical stacking.* The formation of branch points potentially involves unstacking and exposure of base pairs to solvent. Coaxial stacking of helical arms maximizes base stacking interactions, and thus folding based on coaxial stacking might be expected. An early example of this can be seen in the tertiary structure of tRNA (18,19). Coaxial stacking can create alternative conformers, the relative stabilities of which are usually dependent on local sequence.

- *Ion-dependent folding.* Nucleic acids are highly charged polyelectrolytes. Thus, their folding is going to be quite different in principle from that of proteins. Phosphate–phosphate repulsion will tend to keep the structure extended in the absence of charge neutralization, and thus metal ions will play an important role in the folding. The folding may, in turn, create specific ion-binding pockets, and such site-bound ions can themselves be very important in the function of the nucleic acid, notably in ribozyme catalysis.

2. Approaches to the study of branched nucleic acids

Helical junctions are extended species, and the analysis of their structure generally requires the description of conformation on a relatively large scale. Initially it is the global structure that is analysed, and information about the relative configurations of helical arms in space and the angles between the helical axes is sought. At such early stages of the investigation high resolution methods such as NMR spectroscopy are not appropriate, and the complexity and size of the structures makes their application difficult. Techniques are required that are sensitive to distances over a relatively long range (e.g. 20–100 Å), and that can report on the relative disposition of entire helical arms. Two approaches have been particularly valuable, namely comparative gel electrophoresis and fluorescence resonance energy transfer.

2.1 Comparative gel electrophoresis

Gel electrophoresis has been extensively applied to the study of nucleic acid structure, and has provided a large body of valuable data despite the relative simplicity of the approach. For example, electrophoresis provided many of the key observations in the analysis of sequence-directed DNA curvature, and has continued to provide great insight into DNA structures of various kinds. The problem inherent in the technique is the lack of a detailed physical understanding lying behind the method. Yet, despite this drawback, valuable contributions have been made towards our understanding of important structures in both DNA and RNA. The essential observation made in many systems is that deviations from linearity in double-stranded nucleic acids results in anomalously slow migration in polyacrylamide gels (20–23), and that the fragments migrate most slowly when the sequence causing the axial deformation is centrally located (24). Various theories can provide at least qualitative agreement with experimental data (25–27). Most are based upon the idea of the nucleic acid reptation (28), in which the nucleic acid is considered to move through the gel in a tube created by the matrix, under the influence of the electric field. Lumpkin and Zimm (26) derived a relationship between the rate of migration (μ) and the end-to-end distance of the molecule:

$$\mu = \frac{Q}{\zeta} \cdot \left\langle \frac{h_x^2}{L^2} \right\rangle \tag{15.1}$$

where Q is the charge on the molecule, ζ is the frictional coefficient for translation along the tube, L is the contour length of the molecule, and h_x is the component of the end-to-end vector $\boldsymbol{h}$ in the direction of the electric field. The brackets indicate an average over an ensemble of configurations. The dependence on end-to-end distance can explain the sensitivity to shape, since this will be reduced by curvature or kinking, and thus such fragments will migrate more slowly. Using Monte Carlo methods to generate an ensemble of chain trajectories, Levene and Zimm (27) calculated the behaviour of curved DNA fragments under electrophoresis in polyacrylamide. They found it necessary to include cross-interaction between the bendability of the DNA and the elastic properties of the gel matrix to obtain a good fit with experimental data. Calladine and coworkers (29,30) have taken a different approach to explain the reduced mobility of curved DNA, calculating the probability of the cylindrical envelope of a superhelix intersecting randomly located gel fibres. The cylindrical radius expands with the curvature, increasing the probability of obstruction to forward motion.

Gel electrophoresis is very powerful in the analysis of the global structures of DNA junctions. It was demonstrated over a decade ago that such species exhibited anomalously slow migration in polyacrylamide (31), and that the mobility depended on the metal ions present (32). In the application of comparative gel electrophoresis to branched DNA, a set of subspecies are created having two arms that are significantly longer than the remaining arm(s). This can be done by ligating reporter arms on to a junction core (33), or perhaps more easily by shortening the arms (typically from 40 to 12 bp) by restriction cleavage (34). In the case of four-way junctions, there are six different species with two long arms, while in the case of three-way junctions there are three. The electrophoretic mobility of the two-long-arm species in polyacrylamide are compared, and the results analysed on the assumption that faster mobility reflects a longer angle between the long arms. In this way we can derive an overall shape for the branched molecule. This comes from comparison of the mobilities of a set of similar species, and relies on symmetry and shape arguments; thus the lack of a fully developed physical basis for electrophoresis need not prevent a qualitative picture of the global structure from emerging. Indeed, our experience using this approach for the study of a number of different branched species indicates that it is very powerful if used carefully, and comparisons with independent techniques have always confirmed the conclusions from the electrophoresis.

2.2 Fluorescence resonance energy transfer

Fluorescence methods can contribute significantly to our understanding of the structure and dynamics of macromolecules (35–40). In conjunction with modern solid phase synthetic methods for both DNA and RNA, and the variety of fluorophores now available (41), it has become a powerful method for obtaining distance information in folded nucleic acids.

In fluorescence resonance energy transfer (FRET) experiments, two different fluorophores (e.g. fluorescein and tetramethyl rhodamine) are coupled to known positions in the macromolecule. In the case of nucleic acids, the 5′-termini of individual strands provide a convenient location in many applications. Upon excitation of the donor (fluorescein in the above example), dipolar coupling between the transition moments of the fluorophores leads to a transfer of excitation from the donor to the acceptor, reducing the fluorescent quantum yield and lifetime of the donor and increasing the fluorescent emission from the acceptor. Because of the dipolar coupling, the efficiency of the energy transfer depends on the inverse sixth power of the distance between the dyes, and thus the efficiency of energy transfer (E) is greater for short separations and falls off as the distance is increased, i.e.

$$E = \frac{1}{\left[1 + \left(R/R^0\right)^6\right]} \tag{15.2}$$

where R is the distance and R^0 is the distance at which energy transfer is 50% efficient.

The most sensitive way to observe energy transfer is to measure the enhanced emission from the acceptor. Since the emission from the donor also contains a component from direct excitation, this must be normalized (40), and this allows the efficiency of the transfer to be calculated. The most reliable results derived from FRET have been acquired by synthesizing a series of DNA molecules that differ only in the positions where the donor and acceptor molecules are attached to the DNA molecules (42,43). In this way we can map relative distances within an ensemble of DNA molecules that have the same global structure except at the local positions of the dye molecules. The conclusions are therefore drawn from comparisons between the energy transfer efficiencies measured from a series of isomeric or very similar molecules, rather than the determination of absolute distances. This removes many uncertainties that might be present, such as an exact knowledge of the orientation parameter κ^2 and R^0. We have applied the FRET method to the study of a series of DNA duplexes of length varying between 8 and 20 bp (44). Overall the FRET efficiency reduced with increasing length of the helix as expected, but, in addition, we observed the cylindrical geometry of the DNA as a sinusoidal modulation of the efficiency. Good agreement was found between the experimental data and the calculated values based on dipolar energy transfer and a knowledge of the geometry of double-stranded DNA. In another study, we observed an increasing kinking of DNA and RNA duplexes as bulges of different sizes were introduced into the centre of the molecule (45); the efficiency of FRET between fluorophores attached to the two 5′-termini increased as the end-to-end distance shortened as a result of kinking.

As applied to branched nucleic acids, the FRET approach requires the attachment of fluorophore donor–acceptor pairs to the 5′-termini of pairs of arms of a junction with arms of equal length. Thus, for a four-way junction, six different species of pairwise-labelled species are prepared, and the efficiencies of energy transfer are measured under a given set of conditions. This then provides a measure of the relative end-to-end distances between the different arms, and from this the global structure may be deduced.

3. The four-way DNA junction

The structure of the four-way DNA junction has been extensively studied in the last decade.

3.1 The global structure of the four-way DNA junction

The four-way (4H) junction can exist in a number of different structures, and undergoes ion-dependent folding transitions (Fig. 15.2). In the absence of added cations the

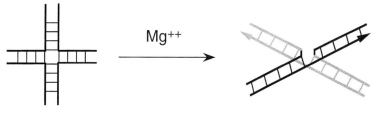

Mg^{++}

extended   stacked X-structure

Fig. 15.2. Ion-dependent folding of the four-way DNA junction into the stacked X-structure. The four-way junction in DNA exists as an open extended structure in the absence of added metal ions. Upon addition of ions (e.g. 100 μM magnesium ions) the junction undergoes a folding transition based on the coaxial stacking of helical arms in pairs. There are two alternative conformers of this structure, which differ in the choice of stacking partners. The folding creates two different kinds of strand. The continuous strands turn about the helical axis of the stacked helices, while the exchanging strands pass from a helix in one coaxial stack to the other at the exchange point. In the antiparallel structure the continuous strands run in opposite directions (their chemical polarity is indicated by the arrow heads).

structure is unfolded; the arms remain unstacked and fully extended in a square configuration (46). Upon addition of sufficient metal ions (such as $\geq$ 100 μM magnesium ions) the four-way DNA junction undergoes a precise folding via the coaxial stacking of pairs of helical arms, to generate the stacked X structure. The essential features of this structure are as follows.

- The arms of the junction associate in pairs by helix–helix stacking. Two stereochemically equivalent conformers are possible (34), depending upon the choice of stacking partners. The relative stability of stacking conformers depends on local sequence.

- The two pairs of stacked helices are rotated, rather like opening a pair of scissors. This minimizes electrostatic repulsion without disturbing the helix–helix stacking.

- The twofold symmetry of the structure generates two sets of inequivalent strands in the structure. The members of one pair (the continuous strands) are related by a helix axis that passes continuously through the point of strand exchange. The other pair (the exchanging strands) pass between the two coaxial stacks at the point of strand exchange.

- The exchanging strands are disposed about the smaller angle of the X structure, and do not cross. This generates an approximately antiparallel alignment of the continuous strands of the DNA helices (34,42,47). The two coaxial helical stacks lie across each other with a right-handed sense (42), allowing a favourable juxtaposition between DNA strands and grooves (see Fig. 15.3); the alignment is best for a small angle of about 60°. Similar strand–groove alignment has been observed between DNA duplexes packed into crystal lattices (48,49). If the backbone of one of the exchanging strands of the four-way junction is interrupted by a covalent discontinuity (nick), the helical pairs appear to disengage (while remaining stacked) and take up a new angle of crossing of about 90° (50).

- The structure presents two sides of different character. This arises because the four base pairs at the point of strand exchange are oriented in the same direction. On one side of the junction (the major groove side) the point of strand exchange has major groove characteristics, while the other side (the minor groove side) has minor groove characteristics.

- The structure can accommodate single base mismatches without extensive disruption to the global structure (51). Some mismatches do not appear to destabilize the structure significantly, while others elevate the concentration of ions required to permit folding into the stacked X structure.

The global structure is consistent with all available experimental evidence. The first indication of the stacked X structure came from the analysis of the overall shape by means of comparative gel electrophoretic experiments (34). Data for one example are shown in Fig. 15.4. Three pairs of mobilities are observed, i.e. slow, intermediate, and fast, consistent with a twofold symmetrical X-shaped structure. The fast mobility of the BX and HR species indicates that for this junction folding occurs by pairwise stacking

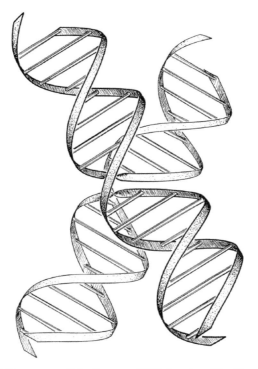

Fig. 15.3. The stacked X-structure of the four-way DNA junction. The illustration uses a ribbon to indicate the path of the backbones in the right-handed, antiparallel stacked X-structure (42). The two sides of the structure are not equivalent. The right side of the junction presents major groove edges of the base pairs at the point of strand exchange, while at the left side the minor groove edges are presented.

of B on X and H on R arms. However, when the central sequence was altered, results indicating the formation of the alternative stacking conformer were obtained (34). The slow mobility of the BH and RX species indicates that the B–centre–H and R–centre–X angles were small; this tells us that the b and r strands turn about the small angle of X, i.e. the relative polarity of the h and x strands is antiparallel. These conclusions were supported and extended by FRET studies (42,43), which found the largest efficiency of energy transfer for the vectors BH and RX in junction 3. This confirmed the antiparallel structure, and studies of other junctions confirmed the formation of alternative stacking isomers for different sequences. Further experiments in which one of the fluorophores was moved around the arms to map the juxtaposition of helical faces indicated that the stacked X structure was right-handed (42). The structure is consistent with other experiments. Seeman and coworkers (52) studied the accessibility of the ribose–phosphate backbone of a four-way junction (of different base sequence from those above) to attack by hydroxyl radicals, and concluded that the structure was twofold symmetrical. Using the same junction sequence, Cooper and Hagerman (53) compared the rotational dynamics of species with pairwise extended arms by means of transient electric birefringence. Their results were consistent with an antiparallel X-shaped structure. Time-resolved fluorescence measurements indicate that there is some scissoring motion of the arms of the junction (54).

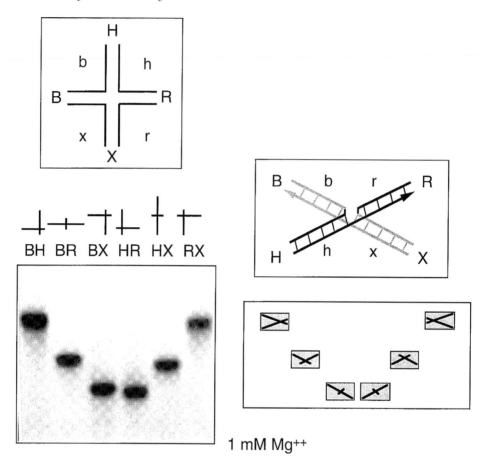

1 mM Mg++

Fig. 15.4. Analysis of the global structure of the four-way DNA junction in the presence of magnesium ions by comparative gel electrophoresis. The junction comprises four arms (each of length 40 bp) labelled B, H, R, and X, generated by the association of the strands b, h, r, and x (each of length 80 nt). By means of selective restriction enzyme cleavage, the six possible species with two shortened arms (reduced to 15 bp) are generated, and their electrophoretic mobility in polyacrylamide compared. The species are named by their two long arms, e.g. the species BH has shortened R and X arms. The pattern of mobilities generated in 1 mM magnesium ions can be described by slow, intermediate, fast, fast, intermediate, slow, and may be explained by the stacked X-structure. Thus the angles subtended between the long arms are acute, obtuse, linear, linear, obtuse, acute, in good agreement with the pattern of electrophoretic mobilities.

3.2 The role of metal ions in the structure of the four-way DNA junction

Metal ions play a critical role in the structure of the four-way DNA junction. In the absence of added cations the junction is unable to undergo folding to form the stacked X structure, but remains in an extended conformation with no coaxial stacking of helical arms. This is indicated by many different experiments. Comparative gel electrophoretic experiments show that the junction adopts a structure with approximately square symmetry in the absence of metal ions (34,46) (Fig. 15.5) and this is confirmed by FRET experiments (55). Thymine bases are reactive to addition by osmium tetroxide in the extended structure of the junction under low salt conditions (34). A variety of ions are able to bring about the folding (46). Group II metals (e.g. magnesium and calcium) fold the junction at concentrations greater than about 100 μM, while complex ions and polyamines are more efficient; 2 μM [Co(NH$_3$)$_6$](III) or 25 μM spermine are sufficient to promote folding. Group I metal ions, such as sodium or potassium, bring about at least a partial folding of the junction (43), but very high concentrations are required and the junction-proximal helical termini remain accessible to addition by osmium tetroxide (46). The ability of monovalent ions to achieve something like the correct folded geometry overall suggests that site-specific binding is not required for these processes. However, uranyl-induced photo-cleavage experiments indicate the presence of a specific ion-binding site near the point of strand exchange in the folded junction (56) (Fig. 15.6). Experiments in which selected phosphate groups were electrically neutralized by replacement with methyl phosphonates (46) revealed that repulsion between phosphates at the point of strand exchange was very significant, as might be expected. Folding the junction probably generates an electronegative cleft that binds divalent ions with increased affinity, whereupon the central bases become inaccessible to osmium tetroxide.

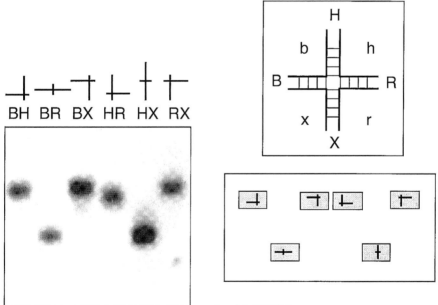

1mM EDTA

Fig. 15.5. Analysis of the global structure of the four-way DNA junction in the absence of added ions by comparative gel electrophoresis. An equivalent set of six species with two long and two short arms used for the analysis in Fig. 15.4 was electrophoresed in a polyacrylamide gel in the presence of 1 mM EDTA. In marked contrast to the pattern of mobilities observed in the presence of magnesium ions, the pattern in the absence of added ions can be described by slow, fast, slow, slow, fast, slow. This is in good agreement with the extended, square geometry of the junction under these conditions, giving the angles between the long arms of 90, 180, 90, 90, 180, and 90°.

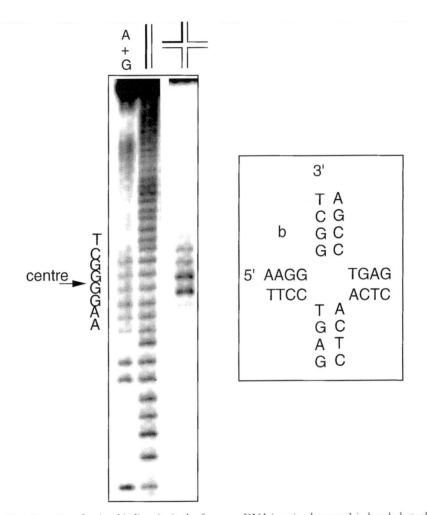

Fig. 15.6. Location of an ion-binding site in the four-way DNA junction by uranyl-induced photocleavage. In this method, a nucleic acid is irradiated with light at 420 nm in the presence of uranyl ion (UO_2^{2+}), whereupon the (deoxy)ribose–phosphate backbone can be broken in the vicinity of the binding site (126). Identification of cleavage sites thus localizes any specific ion-binding sites. The selectivity of the probing can be increased by inclusion of citrate ion, which suppresses non-specific reaction. A four-way junction with the central sequence shown was assembled from four strands, one of which (strand b) was radioactively [5'-^{32}P] labelled (56). The same strand was also hybridized to its complement, to give a perfect duplex species for comparison. The radioactive b strand was subjected to a formate (A+G) sequencing reaction (left track). The duplex species was irradiated in 10 mM Tris-HCl, pH 7.2 (middle track), giving an even level of cleavage along the length of the duplex. The junction species was photoreacted in 50 mM Tris-HCl, pH 7.2, 0.75 mM citrate (right track). The sequence at the centre of strand b is indicated on the left, and the arrows indicate the point of strand exchange. Note the pronounced photocleavage observed around the point of strand exchange in the four-way junction.

3.3 The local stereochemistry of the point of strand exchange in the four-way DNA junction

There have been a number of attempts to model the stereochemistry of the exchange point of the four-way junction (47,49,57), but, experimentally, this must be approached by NMR or crystallography. The latter has been hampered by the lack of suitable quality crystals to date, but despite the almost heroic scale of the problem, significant progress has been made in solution by ¹H NMR in the laboratories of Chazin (58–60) and Altona (61). While full structural determination has not yet been achieved, clear evidence has been obtained for a number of aspects of the structure. Thus, the overall DNA geometry is essentially B-like, with no evidence of broken base pairing at the point of strand exchange. Critically, evidence for base–base stacking across the exchange point has been obtained for several junctions (59,61), and a sequence-dependent stacking conformer bias has been observed (60).

4. The three-way DNA junction

The three-way junction provides a test of the generality of the stereochemical principles established with the four-way junction.

4.1 The perfectly base paired three-way junction

The first three-way junctions studied in DNA were constructed analogously to the usual four-way junctions, such that three helices were connected without the intervention of unpaired bases (3H junctions, see Fig. 15.1). Comparative gel electrophoretic experiments (62) indicated that the three angles between the arms of such perfect three-way DNA junction were much closer to being equal than were the six angles relating the arms of the four-way junction. This was later supported by FRET experiments, where the three end-to-end distances of a three-way junction were found to be closely similar (63). This suggested that the arms fail to undergo the kind of pairwise stacking exhibited by four-way junctions, which was consistent with the permanent reactivity of thymine bases even at high magnesium concentrations (62). Simple model building leads one to expect this result; if we attempt to construct a three-way junction by fusing an additional helix to a broken phosphodiester linkage in one strand of a duplex, we must insert at least the width of the minor groove into the space previously occupied by just a single phosphate group. This is not normally possible, at least if full base pairing is maintained. This conclusion has been partially questioned for other sequences (64), and the structure is probably not fully symmetrical. But while the angles are probably not exactly 120° between each pair of arms, the differences still appear to be smaller than the vari-ation observed between the angles of the four-way junction, and the extended unstacked structure is likely to be broadly correct for most sequences.

4.2 The effect of unpaired bases

The perfect 3H three-way junction is unable to satisfy the principles outlined at the start, namely, that helical junctions tend to undergo coaxial helical stacking and

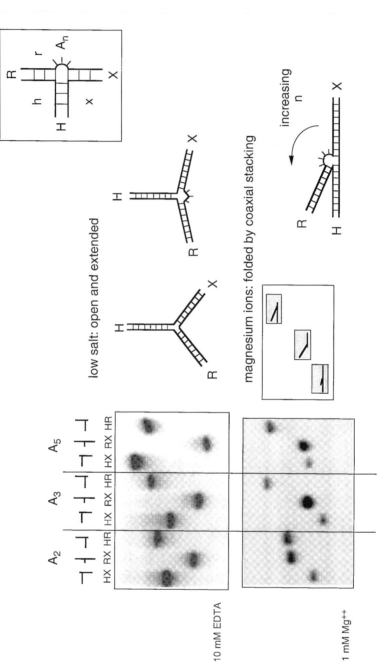

Fig. 15.7. Analysis of the global structure of three-way DNA junctions by comparative gel electrophoresis. In order to analyse the structure of three-way junctions we compare the three species with one shortened arm. The $3HS_n$ junctions are based on a sequence with three arms, H, R, and X, and n unpaired adenine bases (n = 2, 3, or 5) on the r strand, lying opposite the H arm as shown. For the perfectly paired 3H junction (n = 0) the mobilities of the three two-long-arm species are closely similar under all conditions (not shown). However, this is clearly not the case for the $3HS_n$ junctions (66). In the absence of added ions (upper), electrophoretic mobility patterns described by slow, fast, slow are obtained, where the difference between fast and slow becomes greater as n increases. This is simply interpreted in terms of a widening of the angle containing the unpaired bases, i.e. between the R and X arms. In the presence of 1 mM magnesium ions (lower) the behaviour is more complex, and is consistent with a model where there is coaxial stacking between the H and X arms (not possible in the perfect 3H junction), and a reduction of the angle between the H and R arms as n increases.

ion-dependent folding. The rigid framework of the fully paired three-way junction effectively removes the possibility of such folding. However, this stereochemical restraint could be relaxed if some additional conformational flexibility were provided by the addition of a single-stranded region between the helical arms, creating a $3HS_n$ junction (see Fig. 15.1 for an example of a $3HS_2$ junction). It had been shown that such bulged three-way junctions had increased stability in gel electrophoresis (65), and, using electrophoresis (66) and FRET (63), we have demonstrated that such junctions undergo a magnesium-dependent conformational change in which the angles between arms become markedly different (Fig. 15.7). These results can be interpreted in terms of the formation of a structure in which two arms are now coaxially stacked, while the third subtends an angle that is set by the number of unpaired bases. This global structure is also consistent with recent FRET studies (63), in which the distance between the ends of the two helices became increasingly shorter as the number of unpaired bases is increased. Changes in helix–helix lengths in three-way junctions with the introduction of unpaired bases were also observed by time-resolved FRET measurements (67). The distinct conformation of bulged junctions can also explain the lowered rates of cyclization of DNA containing a bulged junction, compared with those carrying a perfectly paired junction (68). Thus we find that once the structural restraints imposed by the perfect three-way junction are removed, three- and four-way junctions exhibit the same general principles of folding. If electrostatic repulsion and steric factors can be reduced sufficiently, then coaxial helix–helix stacking will drive the folding process, resulting in a stacked conformation.

Three-way DNA junctions containing two unpaired bases ($3HS_2$ junctions) have been the subject of two studies by nuclear magnetic resonance (NMR). Junctions of different sequence were studied independently by two groups (69–72). Both studies found structures based upon coaxial stacking of two helices, with the third helix unstacked and extended away from the point of strand exchange (Plate XX). Closer examination of the two NMR structures reveals that they are very different. Like the four-way junction, there are two conformers possible for the three-way junction, which differ in the choice of stacking partners. However, in marked contrast to the four-way junction, these are not stereochemically equivalent structures, and are therefore unlikely to be equally stable. In one structure the polarity of the bulge sequence is 3′ to 5′ as it leaves the stacked helices (conformer I), while in the other it is 5′ to 3′ (conformer II) (Fig. 15.8). The structure solved by the Leontis laboratory is an example of conformer I, while that solved by Rosen and Patel is conformer II. A more recent NMR study of two further $3HS_2$ junctions by Altona and coworkers (73) revealed additional examples of conformer II structures. We have studied a number of different sequences by comparative gel electrophoresis and FRET, and have found that they fold into conformer I on addition of magnesium ions. Nevertheless, when we studied the same sequence as that investigated by Rosen and Patel we found that this adopted the alternative stacking conformer (74), in complete agreement with the NMR analysis. Thus, despite the stereochemical differences between the two structures, both can be adopted, and the relative stability is clearly governed by local DNA sequence. In our experience the formation of conformer II is relatively rare, yet thermal stability measurements indicate that the Rosen–Patel sequence is the most

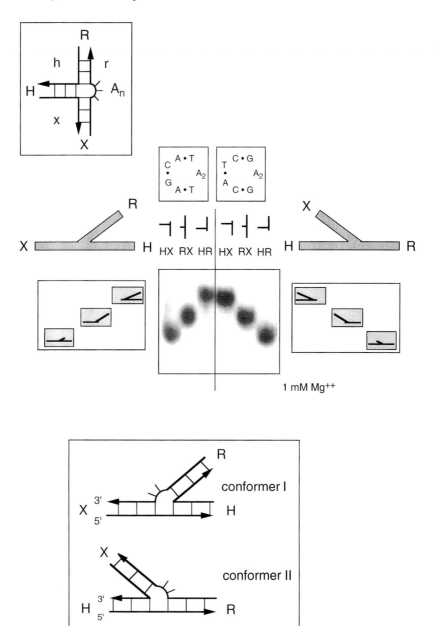

Fig. 15.8. Alternative stacking conformers formed by bulged three-way junctions. Comparative gel electrophoretic analysis of two different 3HS$_2$ junctions (74). The junction on the left is based on the same central sequence as those analysed in Fig. 15.7, while the junction on the right is based on a sequence studied by Rosen and Patel (70)—the central sequences are presented above the autoradiographs. Note that the electrophoretic patterns are virtually mirror images, indicating that the change in central sequence has provoked a change in structure. The left mobility pattern indicates a stacking of H and X arms, while that on the right requires a different model, i.e. H on H stacking. These alternative conformers are not stereochemically equivalent. Note that the polarity of the strand running through the A$_2$ bulge is opposite in the two structures. The structure deduced from this experiment for the Rosen–Patel sequence is completely consistent with the NMR study (see Plate XX).

stable three-way junction that we have examined; this is reflected in both a higher melting temperature compared with other sequences and folding in the presence of just 30 mM sodium ions.

As we would expect, given the formal stereochemical difference between the conformers, the two NMR structures contain significant differences. The path of the backbone of the bulged section of the Leontis structure (isomer I) (69) is relatively looped compared with that of the Rosen–Patel structure (isomer II) (70), where the backbone passes quite smoothly from the stacked helices to the unstacked arm. The unstacked helix of the Leontis structure is largely coplanar, and lies at approximately 90° to the stacked helices, although the angle is probably not well determined by the available NMR data in any of the structures. In the Rosen–Patel structure, the third arm is less coplanar and, in addition, it is bent back at an acute angle, just as our electrophoretic data would indicate. Interestingly, the overall folding of this junction is remarkably similar to that which would be derived by the removal of one helical arm from the right-handed stacked X structure of the four-way DNA junction (42,47). As discussed above, the four-way DNA junction appears to be stabilized by the juxtaposition of the backbone of one stacked helix in the major groove of the other, and a similar feature may be observed in the Rosen–Patel structure (70), where the backbone of stem II is located in the major groove of stem III.

Thus we find that the three-way junction can exhibit many of the same folding properties exhibited by the four-way junction, provided a little extra conformational flexibility is added. Three-way junctions undergo ion-dependent folding by pairwise coaxial stacking of helices, into one of two alternative conformers determined by local sequence.

5. The four-way RNA junction

Given the importance of backbone–groove interactions in the folding of the four-way DNA junction, it might be expected that four-way RNA junctions might fold differently, since RNA adopts an A-form helix with substantially different geometry from the B-form helix of DNA. We have recently examined the global structure of a number of 4H RNA junctions of different central sequence, using the comparative gel electrophoresis technique technically modified for the analysis of RNA.

We initially examined two different RNA junctions with sequences equivalent to junctions that we had studied extensively in DNA (75). From the electrophoretic analysis it was quickly apparent that there were both similarities and differences compared with the DNA equivalents (Fig. 15.9). The RNA junctions apparently fold by coaxial helical stacking, and even seem to exhibit the same choice of stacking partners as the same sequences in DNA. However, the global structure is different, and responds to changes in ionic conditions in a very different way. The general structure of the RNA junction in the presence of moderate (e.g. 1 mM) magnesium ion concentrations is a 90° cross of helical stacks, i.e. a structure that is neither parallel nor antiparallel. One of the biggest surprises came when we performed the analysis of global structure in the absence of added metal ions. In marked contrast to DNA junctions, the RNA species did not suffer loss of coaxial stacking but tended to rotate into a parallel-stranded form. The parallel distortion was rather sequence-dependent, but

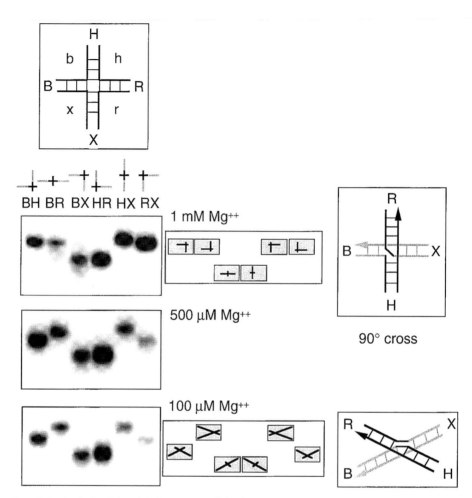

Fig. 15.9. Analysis of the global structure of the four-way RNA junction by comparative gel electrophoresis. The two-long, two-short arms must be prepared differently from the corresponding DNA species because of the difficulty in synthesizing very long RNA molecules, and the impossibility of shortening arms by restriction cleavage. The molecules analysed therefore had RNA cores of 10 bp in each arm, and the remaining portion of each arm comprised DNA. The six species were prepared by synthesis of each of the component strands. The electrophoretic analysis was performed analogously to that on DNA four-way junctions (e.g. Fig. 15.4), in the presence of 1 mM (upper), 500 μM (middle), or 100 μM (lower) magnesium ions. In the presence of 1 mM magnesium ions the electrophoretic mobility pattern can be described as slow, slow, fast, fast, slow, slow, and is explained by a model based on coaxial stacking of B on X and H on R arms, where the angle between the two axes is 90° (75). This gives angles between the long arms of 90, 90, 180, 180, 90, and 90°, as shown. On reduction of the magnesium ion concentration the electrophoretic mobility pattern changes to intermediate, slow, fast, fast, slow, intermediate, and can be interpreted in terms of a rotation of the two axes to give a structure in which the continuous strands are parallel.

was the first time a parallel orientation had been observed for any nucleic acid four-way junction. By contrast, when the junction was placed in 0.5 mM calcium ions, or elevated concentrations (5 mM or higher) of magnesium ions, the junctions rotated in the opposite direction to adopt an antiparallel structure.

Thus, the conversion from DNA to RNA has significant consequences for the global folding of the four-way junction. Some of the differences are likely to derive from the formation of an A-form helix by RNA, where the similarity in the widths of the major and minor grooves suggests that this backbone–groove juxtaposition will be less favourable. If the thermodynamic advantage of strand–groove alignment is denied the junction, then the balance of other steric and electrostatic factors may result in a new global conformational minimum free energy. This appears to be the case for the RNA junction in the presence of 1 mM magnesium ions. The absence of a transition to an unstacked extended structure in the absence of added ions contrasts strongly with the behaviour of DNA junctions, and suggests that overall electrostatic repulsion in the RNA junction is lower.

There are a number of cases of four-way junctions occurring in places that suggest an important biological role. A good example is found in the U1A snRNA, that is involved in splicing of mRNA. The central sequence of this junction is shown in Fig. 15.10. The junction sequence is conserved in mammalian, avian, and amphibian sequences (76,77), and is perfectly base paired for at least three base pairs in each arm, except for the single G:A mismatch located at the point of strand exchange. We analysed the global structure of a junction in which the central RNA core was based upon the U1 sequence, including the G:A mismatch. We found that this adopted a folded structure based on coaxial helical stacking in the conformer in which the adenine base of the G:A mismatch was located on the continuous strand (A_c stacking conformer). This was in good agreement with the results of Krol *et al.* (78), based on differences in sensitivity to ribonuclease V1. We found that the two stacks subtended 90° under all ionic conditions tested. Interestingly, the G:A mismatch did not appear to destabilize the structure, nor did it influence the global structure adopted, since its 'repair' to either G:C or T:A did not alter the overall conformation. While the G:A mismatch is conserved in the sequences of many U1 snRNA species, it is replaced by an A:U base pair in the U1 snRNA of *Drosophila melanogaster*. We analysed the global structure of a junction in which the RNA sequence flanking the point of strand exchange was based on the *Drosophila* sequence, and found that the junction folded in the same way as the mammalian sequence. Once again the structure was based on coaxial stacking of arms in the A_c stacking conformer, with perhaps a little extra rotation in the antiparallel direction. This suggests that there is conservation of three-dimensional structure by the different U1 snRNA species that transcends changes in sequence.

Another biological example of a four-way RNA junction can be found in the hairpin ribozyme of the tobacco ringspot virus (2,79). This ribozyme is usually studied in the form of a nicked duplex containing two bulged regions, one of which contains the scissile phosphodiester bond. The essential sequences are largely located in the two bulges, and evidence suggests that these two regions associate to generate the active site for self-cleavage. In the natural viral sequence, the proposed secondary structure places the two bulges on successive arms of the four-way junction, and thus it would seem probable that the junction should fold in such a way that the two bulges

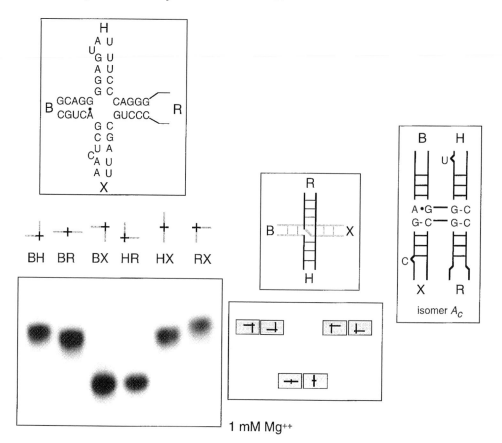

Fig. 15.10. The four-way junction in U1A snRNA. U1A snRNA contains the four-way junction shown (76), which is perfect apart from a G:A mismatch. The sequence is well conserved in mammals, birds, and amphibians (77). Comparative gel electrophoresis in the presence of 1 mM magnesium ions gives a slow, slow, fast, fast, slow, slow pattern of mobilities that is consistent with the stacked geometry in the conformer illustrated to the right (75).

would be brought together. We have analysed the global structure of the tobacco ringspot virus junction (in the absence of the bulges themselves), and found that it naturally adopts the stacking isomer that places the would-be bulge-containing arms on opposite stacks (126). Moreover, as the concentration of magnesium ions was raised, the junction adopted a progressively more antiparallel conformation, whereby the potential bulges would be brought close together. Thus the tobacco ringspot virus junction has exactly the propensity required if the bulges are to be associated to generate the active ribozyme.

6. Interaction between DNA junctions and proteins

Four-way DNA junctions are the substrates for an important class of proteins that exhibit fundamentally structure-selective binding. It is an exciting challenge to understand the manner of the recognition of DNA structure by proteins.

6.1 A class of structure-selective proteins

Enzymes that exhibit selectivity for DNA junctions are probably a ubiquitous class of proteins. These can be junction-specific nucleases (i.e. resolving enzymes) or proteins involved in other processes such as the acceleration of branch migration. They have been isolated from a wide variety of sources, from bacteriophage to mammals, and these are summarized in Table 15.1. In *Escherichia coli* the resolution of four-way junctions is carried out by RuvC (80–82), an enzyme of 172 amino acids. This has been extensively studied (83–85) and the structure is known (86). The gene encoding another resolving enzyme (RusA) has also been found in *E. coli* (87,88); however, this is carried by a prophage and is constitutively repressed. The RuvAB complex facilitates branch migration in *E. coli* (89,90). The junction-selective component of this is a tetramer of RuvA, the crystal structure of which has recently been solved (91). RecG is another *E. coli* protein that exhibits branch migration-facilitating activity (92).

Some bacteriophages encode junction-resolving enzymes whose physiological role appears to be the resolution of branches that are left following replication of DNA. The best characterized is endonuclease VII from phage T4 (93), the product of gene *49*. The enzyme cleaves isolated four-way junctions of various sequence *in vitro* (34,94), as well as supercoil-stabilized cruciform structures (95,96). We have expressed endonuclease VII from a synthetic gene, and constructed a number of site-directed mutants (97). The protein appears to have a modular construction. The N-terminal section contains four cysteine residues that coordinate a single zinc ion (97). In the centre of this 39 amino acid, autonomously folding region lies a cluster of histidine and acidic residues, a number of which appear to be required for the catalysis of DNA cleavage (98). At the C-terminus is a section that is 47% identical to a region of the

Table 15.1. Junction-resolvase and binding activities isolated from various sources. See text for references.

Source	Enzyme	Cleavage specificity	Gene	Size (amino acids)	Structure determined
Junction-resolving enzymes					
Bacteriophage T4	Endonuclease VII		*49*	157	
Bacteriophage T7	Endonuclease I		*3*	149	
Lambdoid prophage	RusA		*rusA*	120	
E. coli	RuvC	TT$^{\Downarrow}$	*ruvC*	172	Yes
Yeast	Endonuclease X1				
Yeast	CCE1	CT$^{\Downarrow}$	*CCEI*	353	
Calf thymus					
CHO cells					
Vaccinia					
Branch migration proteins					
E. coli	RuvA	N/A	*ruvA*		Yes
E. coli	RecG	N/A	*recG*		

T4 repair enzyme endonuclease V. The structure of the latter enzyme is known, and the region of similarity is a helix and turn (99); interestingly, when the sequence from endonuclease V was used to replace the corresponding section of endonuclease VII, the resulting chimeric enzyme had suffered no detectable loss in its selectivity for the cleavage of DNA junctions (97). Lying between the N- and C-terminal sections is a section with weak similarity to T7 endonuclease I, and we have isolated one mutant in this region that lacks catalytic activity but retains the full selectivity for binding to DNA junctions (100). Phage T7 possesses a similar resolvase activity, called endonuclease I (101–103), that is the product of gene 3. We have isolated a number of catalytically deficient mutants of endonuclease I that retain their structural selectivity for binding to DNA junctions (104).

At least two different resolving activities have been isolated from *Saccharomyces cerevisiae*. An as yet poorly characterized activity called endonuclease X1 was isolated (105), which cleaved isolated four-way junctions (106). A different activity (variously called CCE1, MGT1, or endonuclease X2) cleaved the four-way junctions of supercoil-stabilized cruciform structures and figure-eight molecules (107). This has recently been cloned and expressed and studied in greater detail (108). Although encoded by the nuclear *CCE1* gene, CCE1 enzyme is targeted to the mitochondrion (109). It is believed to play an important role in resolving junctions left in mitochondrial DNA, without which segregation is hindered; *cce1* mutants display a raised incidence of petite cells and an increased frequency of junctions in mtDNA (110).

Junction-resolving enzyme activity has also been isolated from higher eukaryotic cells. West and coworkers (111,112) have isolated proteins that cleave synthetic DNA junctions with a specificity comparable to that of the phage enzymes. An activity has also been reported to be encoded by vaccinia virus (113).

6.2 Structure-selective recognition of DNA junctions

The resolving enzymes cleave DNA junctions in a very precise manner. Thus T4 endonuclease VII will, in general, cleave at just two phosphodiester bonds within a given four-way junction (Fig. 15.11). These enzymes bind DNA junctions in dimeric form (100,108) and the complexes migrate as discrete retarded species in polyacrylamide electrophoresis. A number of nuclease-defective mutants of T7 endonuclease I (104), T4 endonuclease VII (100), and yeast CCE1 (M.F. White and D.M.J. Lilley, unpublished data) retain their selectivity for binding to DNA junctions, showing that the binding and catalytic functions are divisible. In general the junction-interacting proteins exhibit a substantial selectivity for the structure of branched species. Thus the protein–junction complexes cannot be displaced by 1000-fold excesses of duplex DNA of the same sequence (83,100,104,108). In another experiment, tethering was used to constrain the structure of a junction of constant sequence into alternate forms (114), which were cleaved by T4 endonuclease VII. It was found that the cleavage pattern depended on the structure of the junction (115), showing that structure rather than sequence was the important element.

In every case studied, it has been found that binding of resolving enzymes to DNA is totally dependent on structure, and independent of base sequence. However, the subsequent cleavage of the junctions can exhibit sequence selectivity for some of the

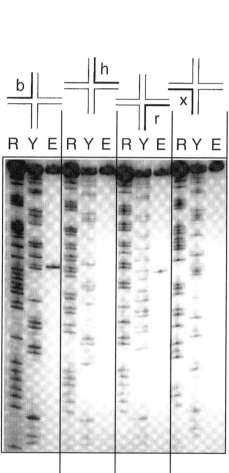

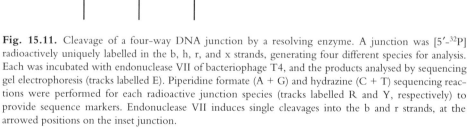

Fig. 15.11. Cleavage of a four-way DNA junction by a resolving enzyme. A junction was [5′-³²P] radioactively uniquely labelled in the b, h, r, and x strands, generating four different species for analysis. Each was incubated with endonuclease VII of bacteriophage T4, and the products analysed by sequencing gel electrophoresis (tracks labelled E). Piperidine formate (A + G) and hydrazine (C + T) sequencing reactions were performed for each radioactive junction species (tracks labelled R and Y, respectively) to provide sequence markers. Endonuclease VII induces single cleavages into the b and r strands, at the arrowed positions on the inset junction.

enzymes. While this is a relatively weak preference in the case of the phage enzymes, the sequence selectivity is considerably stronger for RuvC of *E. coli* (116) (cleavage 3′ to TT) and for CCE1 of yeast (108) (cleavage 3′ to CT). RuvC has been stated to require DNA junctions with a degree of homology (117), such that they can branch migrate. However, this is probably a consequence of the sequence-selectivity cleavage filter, such that a junction that can branch migrate provides more chances of displaying the preferred sequence in the required place relative to the point of strand exchange.

6.3 Manipulation of junction structure by proteins

Quite recently it has emerged that as well as recognizing the structure of DNA junctions, the resolving enzymes also distort that structure in general. This has been shown for T7 endonuclease I (104), T4 endonuclease VII (100), RuvC (85), and CCE1 (118). While each of these enzymes distorts the global structure of the junction, the resulting structure is different in every case. Perhaps the most extreme is that of CCE1, where the resulting structure imposed on the DNA is very close to that of the extended square conformation, just like that of the free junction in the absence of added ions. However, the CCE1–junction complex exists in this extended structure with or without added metal ions. The open centre of the CCE1–junction complex can be demonstrated by the accessibility of thymine bases at the point of strand exchange to attack by potassium permanganate (118).

Distortion of junction structure is not restricted to the resolving enzymes. RuvA also distorts the structure into something very close to an extended square structure (119), and this can be readily rationalized in terms of the recently determined crystal structure of the protein (91). RuvA is a tetrameric junction-selective protein that acts in concert with two hexameric rings of RuvB to facilitate branch migration of junctions. The compact folded structure of the junction suggests that branch migration might require significant disruption of the structure, and recent measurements of the rates of branch migration under conditions where the junction is expected to be folded into the stacked X structure indicate that the process is indeed slow. Panyutin and Hsieh have observed that the rate of branch migration is slower by a factor of 1000 in magnesium, compared with that found in sodium (120). If the structure could therefore be opened, the rate of the exchange of base pairing should be increased, and thus the distortion imposed by RuvA would be expected to facilitate the process.

7. Some final conclusions

Branched nucleic acids undergo folding transitions to generate folded conformations. At the outset we proposed two general features of these folding processes: that metal ions would be an important effector in the conformational transitions, and that coaxial helix–helix stacking would be a common feature of the folded states. We can now look back over the available data to see how well these principles bear up.

In general, branched nucleic acids undergo metal ion-induced folding transitions, driven by the reduction in electrostatic repulsion. The importance of electrostatic interactions is clearly seen in the four-way DNA junction (4H), where selective phosphate neutralization can switch the folding between alternative conformations. In the absence of added metal ions the four-way DNA junction is completely unfolded. Surprisingly, however, this is not true for the corresponding RNA junction, which remains folded even under very low salt conditions. Nevertheless, the global conformation of the four-way RNA junction is responsive to the nature of the metal ions present, and can change between parallel and antiparallel forms. The nature of the ion binding that leads to conformational change is not yet fully resolved. In general, divalent ions like magnesium are much more efficient than monovalent ions like sodium, and specific ion-binding sites have been revealed in the four-way DNA junction by uranyl-induced photocleavage reactions. Yet, in some circumstances at least, partial folding can be induced by monovalent ions, for which site binding can probably be excluded in these systems. Probably a combination of site binding and more general overall charge neutralization is important in general.

Coaxial stacking of pairs of helices is seen to be a very common feature of the folding of branched nucleic acids. Folding of four-way junctions in both DNA and RNA is based on pairwise coaxial stacking, and in each case this generates alternative conformers based on the two possible choices of stacking partners. The choice seems largely determined by the bases flanking the point of strand exchange. Usually one form is thermodynamically favoured over the other, although there are examples of junctions that exhibit no strong isomer bias.

The three-way DNA junction is an interesting case that challenges, but ultimately obeys, these general folding principles. The perfect three-way junction (3H) does not appear to change conformation with addition of metal ions, nor does it undergo coaxial helical stacking. This is a result of the rigid framework of the backbone, which would require loss of base pairing to permit helix stacking. However, when extra unpaired bases are added ($3HS_n$ junctions) the situation is completely changed. The extra conformational freedom allows the junctions to undergo metal ion-induced folding via pairwise coaxial stacking, and once again two (now stereochemically inequivalent) stacking isomers are possible. It really is a case of the exception that proves the rule.

In principle the general folding characteristics established for the model junctions should be applicable to natural helical junctions. The hammerhead ribozyme provides an interesting example of ion-induced folding in a slightly more complex three-way RNA junction. The core of this self-cleaving RNA species is a $HS_1HS_7HS_3$ junction, and the folded structure has been determined in two crystallographic studies (121,122). We have found that in the absence of added metal ions the hammerhead core is unfolded and extended, and upon addition of divalent metal ions it undergoes a two-stage folding process (123,124). The first step (occurring at about 1 mM magnesium ions) involves the coaxial alignment of two of the helical arms, leaving the rest of the core relatively unstructured. In the second stage (occurring at about 1 mM magnesium ions), the probable catalytic core folds, causing a rotation of the remaining helical arm in space. This happens over the same range of magnesium ion concentration that leads to the activation of ribozyme activity, and must generate a conformation that facilitates the trajectory into the transition state of the S_N2 cleavage reaction; this would require colinear alignment of the attacking 2' oxygen atom, the phosphorus atom, and the leaving oxygen atom. There is metal ion participation in the cleavage reaction (125), and the folding would be expected to generate some kind of electronegative binding site for one or more metal ions; using uranyl-induced photocleavage we have detected a high affinity metal ion-binding site within the proposed catalytic core (123).

Thus, many of the folding principles established in DNA and RNA junctions do appear to have general validity, and can be usefully applied to natural and functional nucleic acids.

Acknowledgements

It is a pleasure to thank many of my past and present colleagues for collaborations on the structures of branched nucleic acids, especially Derek Duckett, Alastair Murchie, Robert Clegg, Gurminder Bassi, Richard Pöhler, Jon Welch, Marie-Jo Giraud-Panis, Malcolm White, Niels-Erik Møllegaard, and Eberhard von Kitzing. I thank Dr D. Patel for providing coordinates and the Cancer Research Campaign for financial support.

Note added in proofs 23 August 1998

DNA junctions

A major topic of interest in DNA junctions in the last 18 months has been the demonstration of exchange between stacking conformers. Miick et al. (127) used a combination of NMR and time-resolved FRET measurements to demonstrate the presence of an exchanging population in four-way junctions, depending on central base sequence. While most junction sequences are strongly biased towards one particular stacking conformer, we found a new junction sequence that adopted both conformers in about equal population, with interconversion between them (128). To our surprise, we observed relatively long-range influences of sequence on the relative conformer population.

DNA junction–protein interaction

This remains a very active area that has seen considerable progress in the last 18 months. Some of this has been reviewed by us in White et al. (129). A new junction-resolving enzyme has been discovered in Schizosaccharomyces pombe (130–132), with properties closely similar to CCE1 of S. cerevisiae. The sequence specificity for cleavage of DNA junctions has been studied in depth for CCE1 (133). A tetranucleotide consensus cleavage sequence 5'-ACT⇓A has been identified, although specificity is determined mainly by the central CT dinucleotide. All the junction-resolving enzymes studied to date bind in dimeric form to DNA junctions, consistent with the bilateral resolution reaction. However, subunit exchange reactions in free solution vary widely, and we have recently found that in contrast to most of these enzymes, the exchange rate for endonuclease I of phage T7 is extremely slow (134). Using heterodimeric mutant forms of T4 endonuclease VII, we showed that the two subunits act independently in their cleavage reactions (135). However, both cleavages normally occur within the lifetime of the enzyme–junction complex, leading to the bilateral cleavage required for productive resolution of the junction. While recognising the structure of the four-way junction, all the resolving enzymes appear to distort the global geometry of the junction, and this has been recently extended to the lambdoid enzyme RusA (136,137). In addition to the junction-resolving enzymes, a number of other proteins interact with four-way DNA junctions with some degree of selectivity. In some cases at least, the biological relevance of this interaction is questionable. In general, the HMG-box proteins exhibit selective interaction with DNA junctions. We have recently shown that HMG boxes of diverse origin bind to junctions in the open-square conformation (138), and have suggested that the primary site of interaction is the widened minor groove at the point of strand exchange.

RNA junctions and ribozymes

The global conformation of the four-way RNA junction has been studied using FRET (139). This has confirmed the general folding principles, including the stacking conformers adopted by junction 1 and the U1 snRNA junction. The hairpin ribozyme has been studied in its natural conformation as a four-way junction (126). The ribozyme was found to be active in this form, and the level of activity could be modulated by altering the structure of the junction. FRET studies showed that the ribozyme adopts the stacking conformer that places the unpaired loops (the A and B loops) on opposite stacked helical pairs. Addition of magnesium, calcium or strontium ions induces a change of conformation, in which the helices rotate in an antiparallel direction, leading to a close association between the arms carrying the unpaired loops (1,140). This is presumed to generate the active site that leads to the cleavage reaction. The ion-induced two-stage folding of the hammerhead ribozyme has also been extensively studied using FRET (141). The results indicate two sequential single-ion-induced folding events that mostly likely correspond to the formation of domain II and domain I respectively (see Chapter 17 by Masquida and Westhof in this Volume).

References

1. Lilley, D.M.J., Clegg, R.M., Diekmann, S., Seeman, N.C., von Kitzing, E. and Hagerman, P. (1995) *Eur.J. Biochem.* **230**, 1.
2. Hampel, A. and Tritz, R. (1989) *Biochemistry* **28**, 4929.
3. Forster, A.C. and Symons, R.H. (1987) *Cell* **49**, 211.
4. Hazeloff, J.P. and Gerlach, W.L. (1988) *Nature* **334**, 585.
5. Holliday, R. (1964) *Genet. Res.* **5**, 282.
6. Broker, T.R. and Lehman, I.R. (1971) *J. Mol. Biol.* **60**, 131.
7. Orr-Weaver, T.L., Szostak, J.W. and Rothstein, R.J. (1981) *Proc. Natl. Acad. Sci. USA* **78**, 6354.
8. Potter, H. and Dressler, D. (1976) *Proc. Natl. Acad. Sci. USA* **73**, 3000.
9. Potter, H. and Dressler, D. (1978) *Proc. Natl. Acad. Sci. USA* **75**, 3698.
10. Sigal, N. and Alberts, B. (1972) *J. Mol. Biol.* **71**, 789.
11. Sobell, H.M. (1972) *Proc. Natl. Acad. Sci. USA* **69**, 2483.
12. Kitts, P.A. and Nash, H.A. (1987) *Nature* **329**, 346.
13. Nunes-Düby, S.E., Matsomoto, L. and Landy, A. (1987) *Cell* **50**, 779.
14. Hoess, R., Wierzbicki, A. and Abremski, K. (1987) *Proc. Natl. Acad. Sci. USA* **84**, 6840.
15. Jayaram, M., Crain, K.L., Parsons, R.L. and Harshey, R.M. (1988) *Proc. Natl. Acad. Sci. USA* **85**, 7902.
16. McCulloch, R., Coggins, L.W., Colloms, S.D. and Sherratt, D.J. (1994) *EMBO J.* **13**, 1844.
17. Kemper, B. and Janz, E. (1976) *J. Virol.* **18**, 992.
18. Kim, S.-H., Quigley, G.J., Suddath, F.L., McPherson, A., Sneden, D., Kim, J.J., Weinzierl, J. and Rich, A. (1973) *Science* **179**, 285.
19. Jack, A., Ladner, J.E. and Klug, A. (1976) *J. Mol. Biol.* **108**, 619.
20. Marini, J.C., Levene, S.D., Crothers, D.M. and Englund, P.T. (1982) *Proc. Natl. Acad. Sci. USA* **79**, 7664.
21. Diekmann, S. and Wang, J.C. (1985) *J. Mol. Biol.* **186**, 1.
22. Hagerman, P.J. (1985) *Biochemistry* **24**, 7033.
23. Koo, H.-S., Wu, H.-M. and Crothers, D.M. (1986) *Nature* **320**, 501.

24. Wu, H.-M. and Crothers, D.M. (1984) *Nature* **308**, 509.

25. Lerman, L.S. and Frisch, H.L. (1982) *Biopolymers* **21**, 995.

26. Lumpkin, O.J. and Zimm, B.H. (1982) *Biopolymers* **21**, 2315.

27. Levene, S.D. and Zimm, B.H. (1989) *Science* **245**, 396.

28. de Gennes, P.G. (1971) *J. Chem. Phys.* **55**, 572.

29. Calladine, C.R., Drew, H.R. and McCall, M.J. (1988) *J. Mol. Biol.* **201**, 127.

30. Calladine, C.R., Collis, C.M., Drew, H.R. and Mott, M.R. (1991) *J. Mol. Biol.* **221**, 981.

31. Gough, G.W. and Lilley, D.M.J. (1985) *Nature* **313**, 154.

32. Diekmann, S. and Lilley, D.M.J. (1987) *Nucl. Acids Res.* **14**, 5765.

33. Cooper, J.P. and Hagerman, P.J. (1987) *J. Mol. Biol.* **198**, 711.

34. Duckett, D.R., Murchie, A.I. H., Diekmann, S., von Kitzing, E., Kemper, B. and Lilley, D.M.J. (1988) *Cell* **55**, 79.

35. Weber, G. (1953) *Adv. Protein Chem.* **8**, 415.

36. Steiner, R.F. (ed.) (1983) *Excited States in Biopolymers*. Plenum Press, New York.

37. Lakowicz, J.R. (1983) *Principles of Fluorescence Spectroscopy*. Plenum Press, New York.

38. Jameson, D.M. and Reinhart, G.D. (eds) (1989) *Fluorescent Biomolecules: Methodologies and Applications*. Plenum Press, New York.

39. Lakowicz, J.R. (ed.) (1991) *Topics in Fluorescence Spectroscopy*: Vol. 3, *Biochemical Applications*. Plenum Press, New York.

40. Clegg, R.M. (1992) *Meth. Enzymol.* **211**, 353.

41. Haugland, R.P. (1996) *Molecular Probes: Handbook of Fluorescent Probes and Research Chemicals*. Molecular Probes. Eugene.

42. Murchie, A.I. H., Clegg, R.M., von Kitzing, E., Duckett, D.R., Diekmann, S. and Lilley, D.M.J. (1989) *Nature* **341**, 763.

43. Clegg, R.M., Murchie, A.I.H., Zechel, A., Carlberg, C., Diekmann, S. and Lilley, D.M.J. (1992) *Biochemistry* **31**, 4846.

44. Clegg, R.M., Murchie, A.I.H., Zechel, A. and Lilley, D.M.J. (1993) *Proc. Natl. Acad. Sci. USA* **90**, 2994.

45. Gohlke, C., Murchie, A.I.H., Lilley, D.M.J. and Clegg, R.M. (1994) *Proc. Natl. Acad. Sci. USA* **91**, 11660.

46. Duckett, D.R., Murchie, A.I.H. and Lilley, D.M. J. (1990) *EMBO J.* **9**, 583.

47. von Kitzing, E., Lilley, D.M.J. and Diekmann, S. (1990) *Nucl. Acids Res.* **18**, 2671.

48. Timsit, Y., Westhof, E., Fuchs, R.P. P. and Moras, D. (1989) *Nature* **341**, 459.

49. Goodsell, D.S., Grzeskowiak, K. and Dickerson, R.E. (1995) *Biochemistry* **34**, 1022.

50. Pöhler, J.R.G., Duckett, D.R. and Lilley, D.M.J. (1994) *J. Mol. Biol.* **238**, 62.

51. Duckett, D.R. and Lilley, D.M.J. (1991) *J. Mol. Biol.* **221**, 147.

52. Churchill, M.E., Tullius, T.D., Kallenbach, N.R. and Seeman, N.C. (1988) *Proc. Natl. Acad. Sci. USA* **85**, 4653.

53. Cooper, J.P. and Hagerman, P.J. (1989) *Proc. Natl. Acad. Sci. USA* **86**, 7336.

54. Eis, P.S. and Millar, D.P. (1993) *Biochemistry* **32**, 13852.

55. Clegg, R.M., Murchie, A.I.H., Zechel, A. and Lilley, D.M.J. (1994) *Biophys.J.* **66**, 99.

56. Møllegaard, N.E., Murchie, A.I.H., Lilley, D.M.J. and Nielsen, P.E. (1994) *EMBO J.* **13**, 1508.

57. Srinivasan, A.R. and Olson, W.K. (1994) *Biochemistry* **33**, 9389.

58. Chen, S.M., Heffron, F. and Chazin, W.J. (1993) *Biochemistry* **32**, 319.

59. Chen, S.M. and Chazin, W.J. (1994) *Biochemistry* **33**, 11453.

60. Carlstrom, G. and Chazin, W.J. (1996) *Biochemistry* **35**, 3534.

61. Pikkemaat, J.A., van den Elst, H., van Boom, J.H. and Altona, C. (1994) *Biochemistry* **33**, 14896.

62. Duckett, D.R. and Lilley, D.M.J. (1990) *EMBO J.* **9**, 1659.

63. Stühmeier, F., Welch, J.B., Murchie, A.I.H., Lilley, D.M.J. and Clegg, R.M. (1997) *Biochemistry* **36**, 13530.

64. Lu, M., Guo, Q. and Kallenbach, N.R. (1991) *Biochemistry* **30**, 5815.

65. Leontis, N.B., Kwok, W. and Newman, J.S. (1991) *Nucl. Acids Res.* **19**, 759.

66. Welch, J.B., Duckett, D.R. and Lilley, D.M.J. (1993) *Nucl. Acids Res.* **21**, 4548.

67. Yang, M.S. and Millar, D.P. (1996) *Biochemistry* **35**, 7959.

68. Shlyakhtenko, L.S., Appella, E., Harrington, R.E., Kutyavin, I. and Lyubchenko, Y.L. (1994) *J. Biomol. Struct. Dynamics* **12**, 131.

69. Leontis, N.B., Hills, M.T., Piotto, M., Malhotra, A., Nussbaum, J. and Gorenstein, D.G. (1993) *J. Biomol. Struct. Dynamics* **11**, 215.

70. Rosen, M.A. and Patel, D.J. (1993) *Biochemistry* **32**, 6576.

71. Rosen, M.A. and Patel, D.J. (1993) *Biochemistry* **32**, 6563.

72. Ouporov, I.V. and Leontis, N.B. (1995) *Biophys. J.* **68**, 266.

73. Overmars, F.J. J., Pikkemaat, J.A., Van den Elst, H., Van Boom, J.H. and Altona, C. (1996) *J. Mol. Biol.* **255**, 702.

74. Welch, J.B., Walter, F. and Lilley, D.M.J. (1995) *J. Mol. Biol.* **251**, 507.

75. Duckett, D.R., Murchie, A.I.H. and Lilley, D.M.J. (1995) *Cell* **83**, 1027.

76. Branlant, C., Krol, A. and Ebel, J.-P. (1981) *Nucl. Acids Res.* **9**, 841.

77. Guthrie, C. and Patterson, B. (1988) *Annu. Rev. Genet.* **22**, 387.

78. Krol, A., Westhof, E., Bach, M., Lührmann, R., Ebel, J.-P. and Carbon, P. (1990) *Nucl. Acids Res.* **18**, 3803.

79. Feldstein, P.A., Buzayan, J.M. and Bruening, G. (1989) *Gene* **82**, 53.

80. Connolly, B. and West, S.C. (1990) *Proc. Natl. Acad. Sci. USA* **87**, 8476.

81. Connolly, B., Parsons, C.A., Benson, F.E., Dunderdale, H.J., Sharples, G.J., Lloyd, R.G. and West, S.C. (1991) *Proc. Natl. Acad. Sci. USA* **88**, 6063.

82. Iwasaki, H., Takahagi, M., Shiba, T., Nakata, A. and Shinagawa, H. (1991) *EMBO J.* **10**, 4381.

83. Bennett, R.J., Dunderdale, H.J. and West, S.C. (1993) *Cell* **74**, 1021.

84. Bennett, R.J. and West, S.C. (1995) *Proc. Natl. Acad. Sci. USA* **92**, 5635.

85. Bennett, R.J. and West, S.C. (1995) *J. Mol. Biol.* **252**, 213.

86. Ariyoshi, M., Vassylyev, D.G., Iwasaki, H., Nakamura, H., Shinagawa, H. and Morikawa, K. (1994) *Cell* **78**, 1063.

87. Sharples, G.J., Chan, S.N., Mahdi, A.A., Whitby, M.C. and Lloyd, R.G. (1994) *EMBO J.* **13**, 6133.

88. Mahdi, A.A., Sharples, G.J., Mandal, T.N. and Lloyd, R.G. (1996) *J. Mol. Biol.* **257**, 561.
89. Iwasaki, H., Takahagi, M., Nakata, A. and Shinagawa, H. (1992) *Genes Dev.* **6**, 2214.

90. Muller, B., Tsaneva, I.R. and West, S.C. (1993) *J. Biol. Chem.* **268**, 17179.

91. Rafferty, J.B., Sedelnikova, S.E., Hargreaves, D., Artymiuk, P.J., Baker, P.J., Sharples, G.J., Mahdi, A.A., Lloyd, R.G. and Rice, D.W. (1996) *Science* **274**, 415.

92. Lloyd, R.G. and Sharples, G.J. (1993) *EMBO J.* **12**, 17.

93. Kemper, B. and Garabett, M. (1981) *Eur. J. Biochem.* **115**, 123.

94. Mueller, J.E., Kemper, B., Cunningham, R.P., Kallenbach, N.R. and Seeman, N.C. (1988) *Proc. Natl. Acad. Sci. USA* **85**, 9441.

95. Mizuuchi, K., Kemper, B., Hays, J. and Weisberg, R.A. (1982) *Cell* **29**, 357.

96. Lilley, D.M.J. and Kemper, B. (1984) *Cell* **36**, 413.

97. Giraud-Panis, M.-J.E., Duckett, D.R. and Lilley, D.M.J. (1995) *J. Mol. Biol.* **252**, 596.

98. Giraud-Panis, M.-J.E. and Lilley, D.M.J. (1996) *J. Biol. Chem.* **271**, 33148.

99. Morikawa, K., Matsumoto, O., Tsujimoto, M., Katayanagi, K., Ariyoshi, M., Doi, T., Ikehara, M., Inaoka, T. and Ohtsuka, E. (1992) *Science* **256**, 523.

100. Pöhler, J.R.G., Giraud-Panis, M.-J.E. and Lilley, D.M.J. (1996) *J. Mol. Biol.* **260**, 678.

101. Center, M.S. and Richardson, C.C. (1970) *J. Biol. Chem.* **245**, 6285.
102. Sadowski, P.D. (1971) *J. Biol. Chem.* **246**, 209.
103. de Massey, B., Studier, F.W., Dorgai, L., Appelbaum, F. and Weisberg, R.A. (1984) *Cold Spring Harbor Symp. Quant. Biol.* **49**, 715.
104. Duckett, D.R., Giraud-Panis, M.-E. and Lilley, D.M.J. (1995) *J. Mol. Biol.* **246**, 95.
105. West, S.C. and Korner, A. (1985) *Proc. Natl. Acad. Sci. USA* **82**, 6445.
106. West, S.C., Parsons, C.A. and Picksley, S.M. (1987) *J. Biol. Chem.* **262**, 12752.
107. Symington, L. and Kolodner, R. (1985) *Proc. Natl. Acad. Sci. USA* **82**, 7247.
108. White, M.F. and Lilley, D.M.J. (1996) *J. Mol. Biol.* **257**, 330.
109. Kleff, S., Kemper, B. and Sternglanz, R. (1992) *EMBO J.* **11**, 699.
110. Lockshon, D., Zweifel, S.G., Freeman-Cook, L.L., Lorimer, H.E., Brewer, B.J. and Fangman, W.L. (1995) *Cell* **81**, 947.
111. Elborough, K.M. and West, S.C. (1990) *EMBO J.* **9**, 2931.
112. Hyde, H., Davies, A.A., Benson, F.E. and West, S.C. (1994) *J. Biol. Chem.* **269**, 5202.
113. Stuart, D., Ellison, K., Graham, K. and McFadden, G. (1992) *J. Virol.* **66**, 1551.
114. Kimball, A., Guo, Q., Lu, M., Cunningham, R.P., Kallenbach, N.R., Seeman, N.C. and Tullius, T.D. (1990) *J. Biol. Chem.* **265**, 6544.
115. Bhattacharyya, A., Murchie, A.I. H., von Kitzing, E., Diekmann, S., Kemper, B. and Lilley, D.M.J. (1991) *J. Mol. Biol.* **221**, 1191.
116. Shah, R., Bennett, R.J. and West, S.C. (1994) *Cell* **79**, 853.
117. Benson, F.E. and West, S.C. (1994) *J. Biol. Chem.* **269**, 5195.
118. White, M.F. and Lilley, D.M.J. (1997) *J. Mol. Biol.* **266**, 122.
119. Parsons, C.A., Stasiak, A., Bennett, R.J. and West, S.C. (1995) *Nature* **374**, 375.
120. Panyutin, I.G. and Hsieh, P. (1994) *Proc. Natl. Acad. Sci. USA* **91**, 2021.
121. Pley, H.W., Flaherty, K.M. and McKay, D.B. (1994) *Nature* **372**, 68.
122. Scott, W.G., Finch, J.T. and Klug, A. (1995) *Cell* **81**, 991.
123. Bassi, G., Møllegaard, N.E., Murchie, A.I.H., von Kitzing, E. and Lilley, D.M.J. (1995) *Nature Struct. Biol.* **2**, 45.
124. Bassi, G.S., Murchie, A.I.H. and Lilley, D.M.J. (1996) *RNA* **2**, 756.
125. Dahm, S.C. and Uhlenbeck, O.C. (1991) *Biochemistry* **30**, 9464.
126. Murchie, A.I.H., Thomson, J.B., Walter, F. and Lilley, D.M.J. (1998) *Molecular Cell* **1**, 873.
127. Miick, S.M., Fee, R.S., Millar, D.P. and Chazin, W.J. (1997) *Proc. Natl. Acad. Sci. USA* **94**, 9080.
128. Grainger, R.J., Murchie, A.I.H. and Lilley, D.M.J. (1998) *Biochemistry* **37**, 23.
129. White, M.F., Giraud-Panis, M.-J.E., Pöhler, J.R.G. and Lilley, D.M.J. (1997) *J.Molec. Biol.* **269**, 647.
130. White, M.F. and Lilley, D.M.J. (1997) *Mol. Cell Biol.* **17**, 6465.
131. Whitby, M.C. and Dixon, J. (1997) *J. Molec. Biol.* **272** 509.
132. Oram, M., Keeley, A. and Tsaneva, I. (1998) *Nucleic Acids Res.* **26**, 594.
133. Schofield, M.J., Lilley, D.M.J. and White, M.F. (1998) *Biochemistry* **37**, 7733.
134. Parkinson, M.J. and Lilley, D.M.J. (1997) *Molec. Biol.* **270**, 169.
135. Giraud-Panis, M.-J.E. and Lilley, D.M.J. (1997) *EMBO J.* **16**, 2528.
136. Giraud-Panis, M.-J.E. and Lilley, D.M.J. (1998) *J. Molec. Biol.* **278**, 117.
137. Chan, S.N., Vincent, S.D. and Lloyd, R.G. (1998) *Nucleic Acids Res.* **26**, 1560.
138. Pöhler, J.R.G., Norman, D.G., Bramham, J., Bianchi, M.E. and Lilley, D.M.J. (1998) *EMBO J.* **17**, 817.
139. Walter, F., Murchie, A.I.H., Duckett, D.R. and Lilley, D.M.J. (1998) *RNA* **4**, 719.
140. Walter, F., Murchie, A.I.H., Thomson, J.B. and Lilley, D.M.J. (1998) *Biochemistry*, in press.
141. Bassi, G.S., Murchie, A.I.H., Walter, F., Clegg, R.M. and Lilley, D.M.J. (1997) *EMBO J.* **16**, 7481.

16

DNA higher-order structures

Wilma K. Olson

Department of Chemistry, Rutgers, State University of New Jersey, New Brunswick, NJ 08903, USA

1. Overview

The packaging of DNA within the close confines of the cell imposes a higher order structure on the long, thread-like molecule. The chain must fold within a highly crowded environment as well as adopt arrangements that allow for correct recognition and processing of the genetic message. This organizational structure, which is too unwieldy for direct molecular characterization, can only be inferred from the physical properties of relevant model systems. Isolated DNA supercoils with intertwined double helical strands constitute one such useful model. The well-known interplay between long-range structure and local twisting of the supercoil can be used to drive the folding of DNA around proteins and other packaging agents. The long-range association between interwound strands is relevant to the close packing of DNA, while the local structural changes provide insight into the transient opening of the double helix during biological processes.

This chapter starts with a general discussion of DNA supercoiling, including the topological constraints on the chain molecule and the known biological significance of the supercoiled state. Following a brief review of the intrinsic flexibility of the double helix, and the combined elastic rod/polyelectrolyte character of the chain, we then turn to the models and computational approaches used to deduce the structure of supercoiled DNA. The survey covers novel mathematical representations of the double helical axis, classic parameterization of DNA as an elastic rod, typical energy minimization and dynamics protocols, and efficient numerical solutions of the equations of equilibrium. Section 4 details the equilibrium structures and general structural principles gleaned from a variety of systems, starting with the uncharged, naturally straight, isotropic rod as a point of reference. The examples point to the role of the ionic environment, as measured by different non-bonded energy terms, and the effects of bound proteins on the configuration of the idealized rod. The final section illustrates how it is becoming possible to study the influence of realistic chemical features, such as anisotropic bending, natural curvature, and enhanced bending flexibility, on DNA supercoiling. The chapter concludes with a discussion of the large-scale structural changes observed in dynamical studies and a brief commentary on various perspectives of supercoiled structure.

2. DNA supercoiling

Closed loops of double-stranded DNA are ubiquitous in nature, occurring in systems ranging from plasmids, bacterial chromosomes, and viral genomes, which form single closed loops (1,2), to eukaryotic chromosomes and other linear DNAs, which appear to be organized into topologically constrained domains by DNA-binding proteins or other cellular attachments (3,4). The topological constraints in the latter systems are determined by the spacing of the bound residues along the contour of the chain and the imposed turns and twists of DNA in the intermolecular complexes (5–8). As long as the ends of the DNA stay in place and the duplex remains unbroken, the linking number, *Lk*, or number of times the two strands of the double helix wrap around one another, is conserved. [While the linking number is conventionally associated with a closed duplex (9), a conserved quantity similar to *Lk* can also be defined (I. Tobias, unpublished data) for a spatially anchored linear DNA.] These constraints in *Lk* underlie the well-known supercoiling of DNA, i.e. the deformation of native three-dimensional structure manifested by a higher-order folding of the chain axis and compensatory coiling of the complementary strands. In other words, the stress induced by positioning the ends of the polymer in locations other than the natural (relaxed) state perturbs the overall shape and/or local twisting of the intervening parts of the chain. These structural distortions are the nucleic acid counterparts of the tertiary folding of helical segments in proteins (e.g. coiled coils, twisted sheets), but the changes in structure are spread over a much larger molecular scale in DNA.

2.1 Topological constraints

The interdependence of secondary and tertiary structure in supercoiled DNA is expressed in mathematical terms using White's equation (10),

$$Lk = Wr + Tw. \tag{16.1}$$

In the absence of strand breaks *Lk* has a fixed value which can be decomposed into a contribution *Wr* called the writhing number, which describes the folding of the helix axis, and the total twisting of the two strands, *Tw*. These two parameters are differential geometric quantities that vary continuously with the shape of the duplex, so that when chain ends are spatially constrained, *Lk*, a topological property, is constant. The writhing number, an accounting of pairwise spatial interactions along the helical axis (9), is zero for planar configurations and for out-of-plane symmetric arrangements. Non-zero values are obtained only when the DNA axis is distorted to a non-planar asymmetric arrangement. The writhing number, however, is not a unique characterization of tertiary structure and may be the same for very different spatial arrangements, such as the nicked, circular DNA shown in Fig. 16.1 with a short fragment wrapped in a superhelical pathway around a cylindrical 'phantom' protein or the unrestrained interwound structure that results when the chain is ligated and the protein is removed. The local disposition of chemical residues in different structures with the same writhing number is also quite different.

2.2 Biological importance

The linking number constraint in supercoiled DNA provides a structural basis for comprehending the helical unwinding implicated in significant biological processes

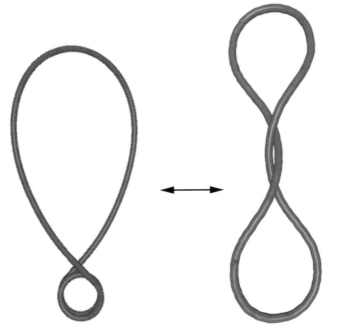

Fig. 16.1. Nicked solenoidal (116) and unnicked interwound (107) configurations of supercoiled DNA with the same magnitude of the writhing number (~1.7).

such as replication and transcription. For example, the binding of different polymerase enzymes to DNA at the starting point of replication is enhanced in negatively super-coiled chains (11), where the linking number is less than that in the native molecule (Lk_0) and the double helix is subjected to a persistent internal strain that tends to unwind regions of local structure. Conversely, the opening of DNA generated upon its complexation with RNA polymerase creates topological subdomains on either side of the moving enzyme, the nucleic acid segments behind the protein assembly adopting a negative ΔLk, where $\Delta Lk = Lk - Lk_0$, and those ahead of it having a positive value (12) (see Fig. 16.2 for a computer-generated representation of the base pair structure in a looped segment of such a DNA). A global response to these locally induced changes in ΔLk, where the unwound residues behind the polymerase convert into configurations with negative writhing number and those ahead of it fold into arrangements of positive Wr (13), helps to account for the uptake of other proteins on the DNA. Specifically, the negatively writhed structures are expected to facilitate the reassembly of DNA on nucleosomes behind the polymerase, while the positively writhed forms may enhance their disassembly. It is well-known that the association of the histone proteins with DNA on the nucleosome forces ~140 bp of the double helix into a left-handed superhelix ($Wr < 0$) (14,15), that nucleosome formation occurs preferentially on negatively rather than positively supercoiled DNA (16), and that positive supercoiling alters nucleosome structure compared to negative coiling (17,18). The positive supercoiling ahead of a moving polymerase may similarly facilitate the

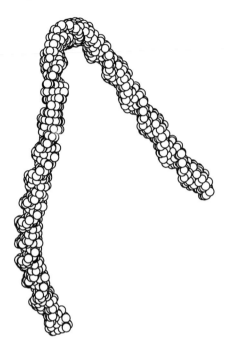

Fig. 16.2. Computer-generated illustration at the base pair level of the topological subdomains created in an anchored DNA loop by the action of enzymes such as RNA polymerase and certain topoisomerases. Underwound segments ($\Delta Lk = -1$) behind the phantom protein on the right end of the loop and overwound segments ($\Delta Lk = 1$) on the left end. Image based on unpublished computer simulations by S.C. Pedersen.

uptake of topoisomerases like *E. coli* DNA gyrase, which removes added superhelical stress and wraps 120–150 bp in a right-handed pathway ($Wr > 0$) around an aggregate of proteins (19–22).

2.3 DNA conformation and flexibility

The manner in which a DNA fragment responds to superhelical stress depends on the native structure and intrinsic flexibility of the chain sequence. A structural code embedded in the DNA base pair sequence helps to organize the folding and determines the flexibility of the long polymer molecule. Some DNAs, for example, form natural superhelices that help to organize the folding of supercoiled states (23–25), while other sequences appear to resist folding deformation (26–29). As pointed out in preceding chapters in this volume, the double helix bends anisotropically at the dinucleotide level. Neighbouring base pairs preferentially roll about their long axes, and hence into the major and minor grooves of the structure, rather than tilt about their short (dyad) axes (30–32). Moreover, the growing database of X-ray crystal structures (33) shows that the bending and twisting of individual base pair steps depend on sequence, with some dimers acting as natural wedges that change the direction of the helical axis and other sequences acting as sites of under- or over-winding (34). The degree of twisting observed in the X-ray structures is further tied to the degree of bending and the base pair displacement with the unwinding of adjacent residues inducing deformations into the major groove and the lateral displacement of base pairs along their long axes (34–38), i.e. an increase in roll and a decrease in slide. Furthermore, the local chain stiffness is sequence dependent with certain residues adapting more easily to imposed stress. For example, severe protein-induced bends of DNA occur predominantly at pyrimidine–purine steps (39–41), the dimers expected to be the most deformable on the basis of steric (34,42) and energetic (30–32,43) arguments.

2.4 DNA as an elastic rod

The influence of fixed ends and enzymatic activity on the overall folding of DNA is analogous to the changes in topology seen in the manipulation of physical models, such as the looping and self-interwinding that results when the free end is rotated and/or translated with respect to the anchored end of a stiff rubber cord or guitar string. Mechanical methods commonly used to analyse these elastic materials are at once applicable to the study of spatially constrained DNA. Importantly, the double helix shares critical material features with the thin, circular elastic rods treated in classical 19th century models (44–47). Supercoiled DNA is clearly longer than it is wide (~20 Å diameter). Furthermore, because of the strong hydrogen bonding and stacking interactions of the constituent base pairs, the DNA molecule is naturally very stiff. The bending, twisting, and stretching of adjacent residues are so limited that chains of 150 bp are almost fully extended, with the computed root-mean-square end-to-end distance equal to roughly 85% the total contour length (48). The deformations of DNA can thus be described in terms of Kirchhoff's rod model with two bending contributions (κ_1 and κ_2), the twist density (κ_3), and the axial extension (ε) at all points s along the chain contour L. The elastic energy of such a system is given by:

$$E_{elastic} = \frac{A_1}{2}\int_0^L (\kappa_1 - \kappa_1^0)^2 \, ds + \frac{A_2}{2}\int_0^L (\kappa_2 - \kappa_2^0)^2 \, ds + \frac{C}{2}\int_0^L (\kappa_3 - \kappa_3^0)^2 \, ds + \frac{Ea}{2}\int_0^L \varepsilon^2 \, ds, \quad (16.2)$$

where the angles are components of the vector, $\kappa(s) = [\kappa_1, \kappa_2, \kappa_3]$ describing the angular rotation of local coordinate frames embedded in cross-sections of the rod at s and $s + ds$, and ε reflects the displacement of adjacent frames along the axis of the rod. The parameters E and a comprising the stretching constant are the Young's modulus and cross-sectional area, respectively. If the rod is divided into a set of discrete elements and the spacing between planar slabs, Δs, is taken as equal to the typical 3.4 Å distance between residues in B-DNA, $\kappa_1 \Delta s$ and $\kappa_2 \Delta s$ approximate the so-called roll and tilt angles, $\kappa_3 \Delta s$ the base pair twist angle, and $\varepsilon \Delta s$ the per residue axial rise (8,49). Lateral/shear displacements in the base pair plane (i.e. slide and shift) are not treated in this scheme. As evident from eqn 16.2, the interdependence of angular and translational variables is also omitted in the model.

Until very recently, supercoiled DNA was always approximated as a naturally straight, inextensible rod that bends with equal likelihood in all directions, i.e.

$A_1 = A_2 = A$, $\kappa_1^0 = \kappa_2^0 = 0$ in eqn 16.2. At this level of simplification, the bending energy reduces to a function of the curvature of the helical axis, $\kappa = (\kappa_1^2 + \kappa_2^2)^{1/2}$, and the twisting contribution simplifies to a function of the writhing number, the imposed value of ΔLk, and the total contour length (50):

$$E_{\text{elastic}} = \frac{A}{2}\int_0^L \kappa(s)^2 ds + \frac{2\pi^2 C}{L}(\Delta Lk - Wr)^2. \tag{16.3}$$

This formulation takes advantage of the fact that the twist density is uniform in the equilibrium configurations of a naturally straight rod. The computational advantages of omitting individual base pairs in this treatment are obvious (i.e. the energetic profile is a function of the duplex axis alone and there is no need to locate individual base pairs), but do not necessarily justify the erroneous representation of base pair structure. The simplification is, of course, necessary in most analytical schemes. The exact results provided by the latter studies serve as critical bench marks for numerical methods aimed at modelling the double helix at a more realistic level.

2.5 Polyelectrolyte character of DNA

As a polyelectrolyte with a net negative charge at every nucleotide residue, double helical DNA is profoundly affected by its ionic environment. Salt effects are particularly important in supercoiled DNA where parts of the chain that are distant in linear sequence may come into close contact. Explicit atomic level treatment of supercoiled DNA, however, is still beyond the capabilities of even the most sophisticated computers. The size limitation problems confronting simulations of supercoiled molecules necessitate the use of primitive models where the chain backbone is reduced to an approximate atomic representation. At the simplest level, the DNA is modelled by hard sphere excluded volume terms that only crudely mimic the electrostatic repulsions of contacted segments (51), while the most detailed models to date (52,53) assign a point charge to each nucleotide residue and use an implicit representation of solvent (Debye charge screening). A number of intermediate schemes (54–57) avoid explicit counting of charged residues by dividing the chain into longer segments of uniform charge density. Recent theoretical work (58,59) points to potential attractive forces stabilizing the association of closely spaced charged rods. These interactions, which are thought to reflect the shared counterion atmosphere of the rods, help to account for the spontaneous aggregation at high salt concentrations of short DNA fragments with increased concentration of polymer and may be relevant to both the long-range contacts brought about by supercoiling and the cholesteric liquid crystal organization of DNA in some organisms (e.g. bacteria, dinoflagellates, mitochondria) (60–62).

3. Computational issues

3.1 Equilibrium vs. dynamic structures

The logical first step in understanding the global folding of supercoiled DNA is to identify the configurations of minimum energy (i.e. equilibrium states). These states must compromise the natural twisting and bending of the chain in order to keep the ends in place and to avoid long-range self-contacts. The forces that satisfy the boundary conditions are initially unknown in rod models but can be determined along with the complete set of structural variables (typically Euler parameters tied to the bending and twisting components of individual reference frames) that minimize the energy of the constrained DNA. External forces, such as those that might be associated with binding proteins or an electric field, can also be included in the total energy. The equilibrium configurations of the system are then obtained by numerical solution of a set of non-linear algebraic equations (8,49,63,64). Other treatments of supercoiled DNA, by contrast, add explicit terms to the potential energy (65–68) or include clever representations of the chain axis (69–72) to satisfy the structural constraints on chain ends. The minimum energy states in such studies are identified by a guided search of configuration space, typically via simulated annealing or other acceleration procedures in Monte Carlo and molecular dynamics studies (70,73–75) or with derivatives of the energy in direct minimizations (71,72).

In general, the thermal fluctuations of the double helix as a whole must be considered alongside the equilibrium structures (76). These entropic effects become especially significant when the DNA is large compared with the persistence length, a classical measure of the distance over which the direction of the chain is maintained (77). In B-DNA this distance is about 500 Å (78), assuming that the measured chain dimensions can be interpreted in terms of the isotropic rod model. The equivalence of this value with the contour length of a ~150 bp duplex is thus a rough indicator of the chain length at which global flexibility starts to become important. The important issue in sufficiently long DNA is how the energy differences between local minima, and the barriers between them, compare with the thermal energy, kT. Part of this problem can be addressed with techniques like Monte Carlo sampling (54,66,67,79) or by Brownian (80–84), Langevin (71,85), and molecular (73–75) dynamics simulations, as well as with analytical theory (86,87). The Monte Carlo method, if care is taken to generate a representative sample of configuration space, will uncover fine details of the global states accessible through thermal fluctuations, while the dynamical studies, because they are based on numerical integration of the equations of motion, will give additional insight into the pathways of overall structural change. The classical rod models, while employed to date almost exclusively in studies of DNA statics, are routinely applied to a great variety of dynamical problems in engineering mechanics. Applications of classical rod dynamics to DNA are just beginning to appear (87–91).

A variety of theoretical and computational approaches are therefore required to deal with the various aspects of DNA supercoiling. For quantitative and qualitative predictions of the effect of local structural changes on the global features of DNA that is long compared with the persistence length, the inclusion of thermal fluctuations is essential, and Monte Carlo or dynamical methods must be used. There are situations, however, when the inclusion of thermal fluctuations is not essential, such as in short stretches of a long molecule (e.g. loops of DNA anchored at their ends by proteins). If one expects the most important information to be contained in the structural details of the equilibrium states (e.g. the path of DNA on the nucleosome or the rotational positioning of bent DNA sequences), one should turn to minimization methods or to one of the numerical or analytical approaches recently developed on the basis of classical rod theory.

3.2 Chain representations

The representation of closed chain molecules with ends confined to a fixed separation and orientation is a long-standing problem in polymer physical chemistry that can be attacked from several points of view. In one approach the configurations of unconstrained linear molecules that meet certain spatial criteria (normally a set of distances and angles between chain ends) are collected through exhaustive simulation studies (92). This method, however, is not practical for studies of the preferred geometry and intrinsic flexibility of supercoiled DNA. The probability of identifying specific configurations from random sampling of the unconstrained chain is so low that it is difficult to accumulate a meaningful set of appropriate states. The method, however, is very useful for simulations of the kinetics of chain cyclization (i.e. ring closure) (93–95) or the formation of closed loops (79,96,97). A second way to study supercoiled DNA is to start with a configuration that meets the desired structural criteria and allow the system to deform subject to some potential function. The major difficulty in such simulations is the preservation of the constraints on chain ends. Individual Cartesian coordinates must be moved in small concerted steps, or internal torsions and valence angles varied in a highly correlated fashion, to maintain the fixed configuration of chain ends (66,98–100). Alternatively, one can introduce explicit energy terms that force the chain ends to a given position. Elastic potentials with no physical significance are typically employed in Cartesian simulations to keep one or more interatomic distances within a desired range (65,73–75,101). Both approaches are computationally intensive.

Two less computationally demanding methods, one using curve fitting techniques and the other involving Euler parameters, can also be taken to identify the preferred configurations of supercoiled DNA. The former method employs simple mathematical formulations, i.e. piecewise B-spline curves or finite Fourier series representations, that automatically satisfy the end-to-end limitations on the constrained DNA axis. These expressions, with a small number of independent variables (the vertices of a polygonal representation of the smoothly folding chain in the case of the B-spline and a set of coefficients for the Fourier series), have been used in numerous simulations of DNA modelled as a naturally straight isotropic rod (5,52,55,72,102–105). The Euler parameters are unknowns determined in the elastic rod treatment of supercoiled DNA (8,49,63,106,107). Both representations aid rapid optimization of chain configuration.

The degree of computed chain movement depends on both the length of achievable simulations and the finest level of chain representation. Large-scale polymer motions become apparent if the DNA is simplified and the number of independent variables is thereby reduced. The treatment of supercoiled molecules frequently entails reduction of the polymer to a sequence of virtual bonds, each of which may sometimes span several helical turns (66,108). The use of such rigid units is justified in short, stiff fragments up to a few helical turns and in very long chains, i.e. of 2000 bp or more according to direct computations of the Gaussian limit for idealized B-DNA duplexes (48), where the extended bonds correspond to hypothetical Kuhn segments (109,110). The representation of intermediate length [$O(10^2)$ bp] DNA as rigid repeating units can be misleading in that chains of this length are flexible enough on the global scale that the mean end-to-end distances differ by 15% or more from the static rod approximation (48,111,112). Furthermore, the bending 'corrections' needed to relate such long segments to the observed persistence length of DNA are exaggerated (112,113) and beyond the limited angular range over which the elastic rod approximation is valid. The global folding is also quite irregular in simplified models generated from extended polymer links (66,67,82).

3.3 Curve fitting techniques: B-splines and finite Fourier series

The main advantage of B-spline parameterization of a closed curve, $r = \Sigma r_i(u)$, is the direct control of the chain pathway provided by the choice of independent parameters, p, called controlling points (114). The order-four (cubic) curves with regional segments,

$$r_i(u) = \frac{-u^3 + 3u^2 - 3u + 1}{6} p_{i-1} + \frac{3u^3 - 6u^2 + 4}{6} p_i +$$
$$\frac{-3u^3 + 3u^2 + 3u + 1}{6} p_{i+1} + \frac{u^3}{6} p_{i+2}, \qquad (16.4)$$

are sufficient for the calculation of topological and energetic parameters of a naturally straight, isotropic rod. The coefficients in this expression assure the smooth connection between successive curve segments and the continuity in first and second derivatives needed to evaluate eqn 16.3. The increments of the mesh parameter, u, which varies between zero and unity, determine the level of structural representation (i.e. virtual bond lengths). In other words, the location of individual residues is implicitly determined by the equations of the closed curve, with the number of computational variables sharply reduced compared with that necessary for explicit specification of all chain units. A subset of controlling points can be fixed during the course of computation to simulate effects of local rigidity within the DNA (103,115,116). In some of the simulations reported below, a set of points describing a superhelix of appropriate proportions is used to model the presence of a protein rigidly bound to DNA. The B-spline procedure, however, has two drawbacks. The complexity of the curve is limited by the number of controlling points: more variables are needed to represent more convoluted pathways. In addition, the controlling points simply guide, but do not lie on, the curve that they define. Only in the limit of an infinite number of controlling points is it possible to represent specific spatial features.

Fourier analysis corrects for the deficiency in B-spline configurational control and provides a direct connection between experimental measurement and computer simulation. Virtually any target function or set of coordinates (e.g. an electron micrographic tracing) can be transformed into a finite Fourier series, the simplicity of which can be exploited for structural manipulation and analysis (5,72,104,105). An expression of the form,

$$\Delta r(u) = \sum_{m=1}^{M} \{a_m \sin(2\pi mu) + b_m [\cos(2\pi mu) - 1]\} + a_{M+1} \sin[2\pi(M+1)u], \qquad (16.5)$$

corresponds to the difference between a given starting structure and an arbitrary chain configuration. The vectorial coefficients, a_m and b_m, are the independent variables that determine the folding of the helix axis, while the increments of the contour parameter, $0 \leq u \leq 1$, determine the level of structural detail. The choice of sines and cosines as basis functions assures that the ends of the DNA coincide with those of the starting state, while the definition of the coefficient a_{M+1},

$$a_{M+1} = \frac{-1}{M+1} \sum_{m=1}^{M} m a_m, \tag{16.6}$$

guarantees that the tangents at the ends of the chain coincide with those of the starting state (5). The latter capability can be used to model specific DNA-binding proteins.

3.4 Euler parameters

Treatment of DNA as a closed elastic rod requires knowledge of the locations of the constituent elements (base pair slabs or extended virtual rod segments). This is achieved with a transformation, $d(s) = T(s)e$, which expresses a local reference frame at position s, $d(s) = [d_1(s), d_2(s), d_3(s)]$, in terms of the global coordinate system, $e = [e_1, e_2, e_3]$. The elements of $T(s)$ may be constructed from a set of Euler angles or with so-called Euler symmetric parameters, a four-variable description of the rotation ϕ about axis u that converts one coordinate frame into another (117). The latter formulation,

$$T(s) = \begin{bmatrix} q_1^2 - q_2^2 - q_3^2 + q_4^2 & 2(q_1 q_2 + q_3 q_4) & 2(q_1 q_3 - q_2 q_4) \\ 2(q_1 q_2 - q_3 q_4) & -q_1^2 + q_2^2 - q_3^2 + q_4^2 & 2(q_2 q_3 + q_1 q_4) \\ 2(q_1 q_3 + q_2 q_4) & 2(q_2 q_3 - q_1 q_4) & -q_1^2 - q_2^2 + q_3^2 + q_4^2 \end{bmatrix}, \tag{16.7}$$

where $q_i = u_i \sin[\phi(s)/2]$, $i = 1, 3$, $q_4 = \cos[\phi(s)/2]$ and $u = [u_1(s), u_2(s), u_3(s)]$, avoids singularities and the computational costs associated with the trigonometric parameterization (49). The elements of $T(s)$ are further related to the components of $\kappa(s)$, the parameters used in eqn 16.2 to monitor the bending and twisting of the rod. The $\kappa(s)$ values are given by the scalar products of the d_i with their derivatives with respect to arc length, $dd_i/ds = \kappa(s) \times d_i$, e.g. $\kappa_1 = d_3 \cdot dd2/ds = 2(q_1' q_4 + q_2' q_3 - q_3' q_2 - q_4' q_1)$. Thus, optimization of the energy of the constrained DNA ultimately yields the base pair axes.

3.5 Energy minimization procedures

Minimum energy forms of supercoiled DNAs can be identified using stochastic (e.g. Monte Carlo), deterministic (e.g. direct minimization), and iterative methods. The Monte Carlo schemes entail random variation of independent chain parameters (e.g. polygonal vertices, B-spline controlling points, Fourier coefficients) with configurational acceptance based on the standard Metropolis criterion (118). The simplicity of the algorithm and the ease of programming are counterbalanced by the long times required to identify the global energy minimum. Monte Carlo simulations (5,103,119) can be carried out at a fixed (high) temperature with the repetition of successful downhill moves to accelerate convergence (70), or gradually over a series of temperatures in a simulated annealing scheme (120).

Direct optimization methods entail computation of the energy and its first and second derivatives with respect to the three components of the independent variables (55,72,102,104,105). The requisite programming is more demanding than the Monte Carlo method, but the computational time is significantly enhanced; see Table 5.1 in ref. 116 for timings.

3.6 Elastic equilibrium conditions

Iterative procedures are used in solving the set of non-linear differential equations of equilibrium for a spatially constrained DNA rod (8,49,63,106). The equations follow from the equilibrium conditions, $dF/ds + f = 0$ and $dM/ds - (1 + \varepsilon)F \times d_3 = 0$, which detail the balance of internal and applied forces and the net moment at each cross-section (i.e. base pair) of the rod. The unknowns in these expressions are the components of the internal force, $F = F_1 e_1 + F_2 e_2 + F_3 e_3$, and the Euler parameters within $M = A_1(\kappa_1 - \kappa_1^0)d_1 + A_2(\kappa_2 - \kappa_2^0)d_2 + C(\kappa_3 - \kappa_3^0)d_3$ (recall the dependence of the d_j and κ_j, $j = 1, 3$, on the q_i, $i = 1, 4$ noted in Section 3.4). The geometric constraints on the ends of the rod can be introduced through a Lagrange multiplier, Λ, another unknown that acts as the force keeping the chain ends in place. The external force f has been treated as an unknown in some calculations (63,106,121), set to zero in other studies (8), and expressed as a sum of screened Coulombic forces acting on each residue (49,107),

$$f = \sum_{m \neq N} \frac{\delta_m \delta_N e^{-\kappa_D r}}{4 \pi \varepsilon_0 \varepsilon_w r^2} \left(\frac{1}{r} + \kappa_D \right) (r_m - r_N). \tag{16.8}$$

Here, δ_N is the charge on the residue of interest, r the magnitude of the vector $r_m - r_N$ radiating from residue N to m, κ_D the Debye length (with standard dependence on ionic strength), ε_0 the permittivity of free space (not to be confused with the axial extension ε which lacks a subscript), and ε_w the dielectric constant of water. The repulsive nature of f prevents long-range self-contacts at low salt conditions. A stronger term, however, must be introduced to prevent self-intersection at high salt (107).

The equations of equilibrium of supercoiled DNA were first solved using a finite element analysis in which the elastic rod was divided into a small number of elements (63,106,121). The equilibrium configurations and associated energies of the system were obtained by solving for the nodal variables at the boundaries between elements, and the variables within each element were interpolated from the values at adjoining nodal points. The numerical methods have since been improved. One solution re-expresses the non-linear equations of equilibrium as a set of linear equations using the Newton–Raphson method and finds corrections to a set of trial values for the unknowns (8,49,107). The corrections are added to the initial guess and the process is reiterated until the corrections are less than a selected level of accuracy. The chain axis can also be perturbed directly using a set of correlated moves that preserve the positioning of polymer ends and provide a subset of the Euler angles at nodal points (S.C. Pedersen, unpublished data). The remaining geometric variables are found by optimizing the elastic energy with added knowledge (from $\Delta Lk - Wr$) of the total twist, sampling both equilibrium and non-equilibrium states. The approach is equivalent to finding the edges of a ribbon that lies along a specified curve. The iterative solution of the equations of equilibrium is practical for chains with $O(10^3)$ elements, while the sampling of nearby configuration space via correlated configurational changes limits analyses to $O(10^2)$ bp. A third very rapid method takes advantage of the mathematical symmetries of the problem (64). All three approaches lend themselves to studies of DNA at the base pair level with each element of the rod assigned intrinsic angular parameters (κ_i^0, $i = 1, 3$) and elastic constants (A_1, A_2, C, Ea). Thus, it is quite easy to study DNAs with special features, such as curvature or bound proteins (8,63,90,106,107,116,122).

3.7 Langevin dynamics

Simulating the large-scale motions of supercoiled DNA requires more than a simplified chain model. The range of configurations sampled in molecular simulations can be extended using the Langevin formalism (71,85,104,123) or Brownian dynamics methods (80–84,124) with phenomenological terms that mimic forces in the local chemical environment. Molecular dynamics studies (73–75), by contrast, focus on the intrinsic molecular motions of supercoiled DNA in the absence of solvent. With an appropriate choice of parameters to specify the strength of the imposed forces, it is possible to sample a wide range of configuration space and to mimic physically relevant time-scales. The equations of motion are solved numerically in terms of the independent chain parameters. The current implementation of dynamics for B-spline and Fourier series-represented chains (71,85,104,123), however, is oversimplified in that the time-dependent fluctuations of twist, known to be important even for the isotropic rod (87,91), cannot be determined on the basis of a single curve. The next step to take in improving the numerical studies is to employ a ribbon-like representation of DNA that keeps track of both smooth bending and local helical twist. Such features have been incorporated in recent Monte Carlo and Brownian dynamics simulations of DNA modelled in terms of extended rod segments (68,125,126). Solution of the equations of motion in terms of spline points or Fourier coefficients also renders the absolute time-scale of the simulations ambiguous. This uncertainty can be avoided by basing the dynamics on the time evolution of points on the curve mesh (127).

4. Equilibrium structures

4.1 Isotropic elastic rod

Studies of the naturally straight, isotropic rod serve as a logical starting point for simulations of supercoiled DNA. As noted above, the analytical results derived from this idealized chain representation provide important standards against which numerical treatments can be compared, such as the magnitude of ΔLk at which the closed rod of given length and elastic profile (i.e. A, C in eqn 16.3) suddenly converts from a circle to a figure-eight (128). The exact mathematical solutions of the equations of elasticity not only check the accuracy of the numerical work but also give a point of reference for interpreting the effect of added variables in more realistic molecular models. The comparability of different numerical approaches strengthens predictions made in systems beyond the scope of present day theory, such as the interwound states found at higher levels of supercoiling which are subject to long-range self-contacts. The theory is therefore routinely combined with a variety of chain representations and computational schemes to reach structural conclusions.

The variation of writhing number with degree of supercoiling in Fig. 16.3 is typical of the configurational profiles obtained numerically for the isotropic rod (49,63,72,102,107,121). The influence of ΔLk on chain folding is manifested in a wide variety of supercoiled structures. Of the different equilibrium states found at a given ΔLk, the energetically preferred form is either a circle ($Wr = 0$) or one of the interwound shapes that evolve from the figure-eight ($Wr = \pm 1$). Three minimum energy states, identified by direct optimization of a 1000-bp Fourier series-represented DNA rod with $\Delta Lk = 4$ (72), are illustrated in Fig. 16.4. The observed states reflect a tug of war between bending and twisting deformations and van der Waals forces. The bending energy dominates at this chain length (see figure legend), accounting for the significant preference for the linear form over the branched states. The difference in energy would be much smaller in a longer DNA of the same shape since the bending energy scales inversely with the chain length. As evident from the slower increase in Wr with ΔLk in Fig. 16.3, the branched forms (noted by Roman numerals) resist the onset of interwinding more strongly than the linear structures, following theoretical expectations for the small hairpin loops in these structures (6). The influence of axial extension is a secondary effect in the elastic equations (49,107), so that the penalty terms routinely used to constrain total chain length in different minimization procedures (69,71,72) have no significant effect.

While all numerical models show the above trends, there are notable differences associated with the choice of chain representation and treatment of long-range contacts. For example, higher values of Wr and lower energies are consistently found upon minimization of Fourier series-represented chains compared with B-spline curves under the same computational conditions (72,102). The observed differences reflect the global nature of the Fourier series pathway and its different control compared with the B-spline curves. The variation of a Fourier coefficient alters the shape of the whole curve, whereas a given spline point affects the configuration locally (cf. eqns 16.4 and 16.5). Furthermore, because the B-spline points guide, but do not fall along, the curves that they determine, the fine tuning of overall shape is difficult. Because of the built-in connection between a Fourier series and a known curve, energy minimization in terms of Fourier coefficients is equivalent to minimization against a set of controlling points on the curve (see ref. 72 for further discussion).

The relatively 'smooth' changes in Wr with ΔLk in Fig. 16.3 are characteristic of the van der Waals term used to prevent self-contacts in this example. As noted above, the small attractive contribution assigned to points in close contact competes with the bending and twisting terms in determining chain configuration (72,102). The attrac-

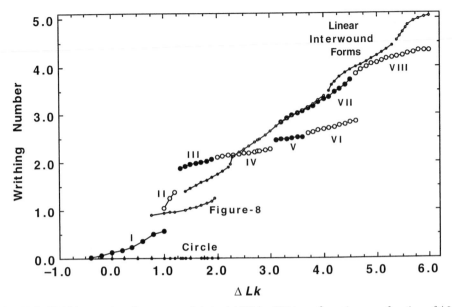

Fig. 16.3. Writhing numbers for energy-minimized 1000 bp DNA configurations as a function of ΔLk. Data based on direct energy minimization of a Fourier series-represented helical axis subject to bending, twisting, and van der Waals energy contributions with $A = C = 2.32 \times 10^{-19}$ erg-cm (72). Roman numerals I–VIII distinguish different families of branched configurations noted by the circular symbols. Global energy minima (circle, figure-eight, and interwound energy minima) found over the same range of ΔLk are designated by small dots.

tive term enhances polymer interwinding compared with that found in chains subject to hard sphere contacts and leads to smooth interconversion of supercoiled states at higher ΔLk (72,102). By contrast, systems modelled with a hard sphere potential (63,121) exhibit discontinuous transitions in Wr at higher ΔLk, much like the sharp configurational jump observed in Fig. 16.3 between the circle and figure-eight. Stepwise changes in Wr also appear in minimization studies of 'flattened' rods, where a penalty term introduced to keep the molecule in a plane cancels the van der Waals attraction (105). Importantly, the twist energy drops sharply in both the smooth and sudden configurational transitions, a finding that suggests how nicked DNA, under suitable conditions, might relax to (or become trapped in) forms other than a circle.

4.2 'Electrostatic' effects

The structures generated under the constraints of an electrostatic potential are less compact than those subjected to hard sphere or van der Waals interactions. Furthermore, the configurational differences are more pronounced at low than at high salt concentrations. The electrostatic contributions influence the close contacts of interwound configurations and shift the critical value of ΔLk at which the circle collapses into a figure-eight (49,52,55,107). The configurational preferences stem from a three-way contest between bending, twisting, and electrostatic components, with the

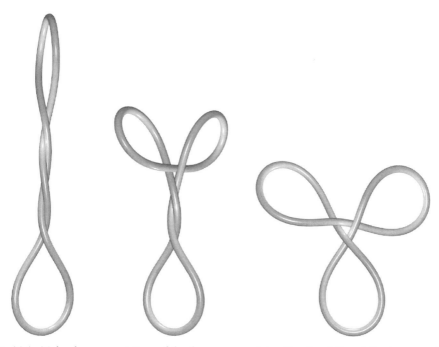

Fig. 16.4. Molecular representations of the three energy minima found at $\Delta Lk = 4$ for the supercoiled DNA described in Fig. 16.3 at 298 K (72): (left) linear interwound configuration ($Wr = 3.4$, $E_{\text{bend}} = 18.2kT$, $E_{\text{twist}} = 1.2kT$, $E_{\text{vdW}} = -13.3kT$); (centre) twofold symmetric branched state (family VII: $Wr = 3.3$, $E_{\text{bend}} = 27.9kT$, $E_{\text{twist}} = 1.5kT$, $E_{\text{vdW}} = -11.9kT$); (right) symmetric three-lobed structure (family VI: $Wr = 2.6$, $E_{\text{bend}} = 23.6kT$, $E_{\text{twist}} = 5.6kT$, $E_{\text{vdW}} = -10.3kT$).

repulsive forces between charged residues tending to open the structure. As a result, the chain resists folding into tightly associated interwound forms, and, when supercoiled, adopts a different pattern of long-range interactions from 'null' (uncharged) DNA: the zone of non-bonded contacts found in the centre of the neutral molecule opens to a few isolated points of contact in the polyelectrolyte model; see the images of representative 1000 bp charged elastic DNA rods at $\Delta Lk = -6$ in Fig. 16.5. Significantly, the predicted opening of the chain at low salt is consistent with the observed transition from tightly to loosely interwound supercoils observed upon decrease of salt concentration in cryoelectron and atomic force microscopy studies (129,130). Another notable result is the opening of the charged supercoil with increase in chain length at both low and high salt (52,55,107). The additive contributions of the electrostatic energy overwhelm the reduction in bending energy that enhances the folding of long uncharged chains. Computational limitations on direct enumeration of the electrostatic energy, however, preclude routine optimization of chains longer than 1000 bp (49,107) or Monte Carlo simulations of chains with more than 300 point charges (52). New speed-up algorithms that reduce the time to evaluate all pairwise contributions to the electrostatic energy in model chains (53) promise to enhance studies of long supercoiled molecules.

0.02 *M*   0.1 *M*   0.5 *M*

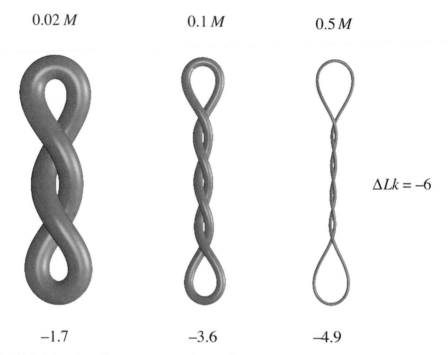

$\Delta Lk = -6$

−1.7   −3.6   −4.9

Fig. 16.5. Selected equilibrium structures of a 1000 bp DNA elastic rod at $\Delta Lk = -6$ simulated under different monovalent salt conditions (107). The collapse of the molecule upon addition of salt is represented by the decrease in effective radius of the rod. Salt concentration is noted above and writhing number below the images.

The preceding picture persists when a Debye–Hückel model of interacting rods is used in place of explicit evaluation of the energy between point charges (55). The DNA is divided into rigid segments of uniform charge density and the solvent is modelled in terms of coefficients obtained from the Poisson–Boltzmann solution to the electrostatic interactions of two infinitely long, charged cylinders (131). Because the Debye–Hückel approximation is valid only in dilute solution, there is some question as to the validity of the potential for treating the closely spaced interwound regions of supercoiled DNA. This simple treatment yields a branched minimum at high salt not yet detected with the detailed approach. The structure, shown on the left in Fig. 16.6, resembles a figure-eight with extra wrapping in one of the two loops. With added supercoiling and decrease of salt, the intertwined loop gradually opens like a clam shell and two distinct lobes emerge (Fig. 16.6, middle and right). The lobes reorient and eventually form a three-fold symmetric, roughly planar arrangement. Because the superhelical wrapping resembles the folding of DNA around the 'phantom' protein illustrated in Fig. 16.1, the sequence of opening suggests a potential supercoil-induced pathway for disassembly of protein–DNA complexes. The gradual opening of structure with decrease of salt is also qualitatively consistent with observed changes in low-angle X-ray scattering of DNA (55,132,133).

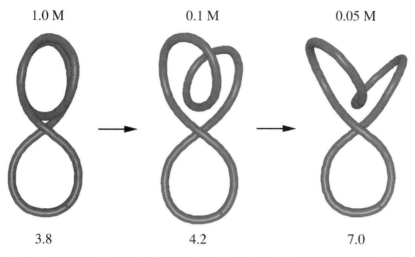

1.0 M · 0.1 M · 0.05 M

3.8 · 4.2 · 7.0

Fig. 16.6. Unwrapping of a pseudo-figure-eight with a solenoidal loop into a three-lobed branched configuration with added supercoiling and decrease of salt (55). Salt concentration is noted above and ΔLk below each image.

4.3 Protein-induced structural changes

To model the effects of protein binding on overall structure (5,103,116), a DNA rod must be divided into two parts. The part associated with protein is typically represented by a rigid superhelix of appropriate proportions (e.g. 45 Å radius, 30 Å pitch for the nucleosome assembly). The rest of the chain, the protein-free double helical loop, is allowed to relax to a configuration that minimizes the energy and satisfies the boundary conditions at the junction between the two sections. If the DNA is modelled as a twist-free, naturally straight, thin, isotropic rod, only the bending energy is included in the optimization. The resulting global minimum corresponds to the equilibrium rest state of a nicked DNA loop. Each such configuration has a writhing number that depends on both the shape of the loop and that of the rigid, protein-bound segment. If the nicked strand is sealed, the DNA will be characterized, following eqns 16.1 and 16.3, by a linking number difference numerically equal to the writhing number (since $\Delta Lk - Wr = 0$ in the torsionally relaxed state). Upon release of the protein, the molecule will assume a new shape, consistent with the value of ΔLk so determined. At this point the DNA is no longer twist-free, but converted to a new protein-free equilibrium state with a new writhing number, Wr', and twist energy proportional to $(\Delta Lk - Wr')^2$.

Figure 16.7 illustrates the computed change in writhing number of a nicked DNA with varying degrees of superhelical wrapping around an idealized nucleosome core (5,116). Notable are the series of sudden drops in Wr at roughly half-integral superhelical turns. The unbound loop snaps at these points from an open, contact-free state to a collapsed configuration with a point of long-range self-association. The intriguing possibility that this large-scale transition, brought about by a small change in local structure, may be related to the delicate control of three-dimensional gene structure in the cell and the resemblance of this collapse to the circle–figure-eight transition (cf. Figs 16.3 and 16.7), has prompted analytical studies of the explicit expressions, exact within the framework of Kirchhoff's theory, for the equilibrium configurations of elastic rods with fixed ends (6). The theory shows that the equilibrium configuration of such a rod depends on three factors: the extent of bending (related to the wrapping in Fig. 16.7), the degree of non-planarity of the bend (determined in the example by the proportions of the DNA superhelix wrapped around the protein), and the ratio of the end-to-end and contour lengths of the unbound fragment. The variation in *Wr* will thus reflect the choice of superhelical pitch and chain length, as is observed in numerical studies (5). Computer simulations further reveal an intriguing structural asymmetry when the DNA–protein complex is subjected to added supercoiling. The local arrangement of the DNA loop in under- and over-wound minichromosome models is noticeably different (5), suggesting ways in which a protein bound to one part of the chain might regulate the structure and interactions in distant parts.

The complete configurational profile, including the precise points and angles of long-range contact in the collapsed DNA, follows from the theory (7). In other words, the degree of superhelical wrapping determines which residues of the unbound DNA are brought into direct contact, as well as the orientation and precise distance between interacting residues. The analytical results have not only provided an important check of the numerical studies but also prompted computations (8) of the looping induced by proteins like the lac repressor, which interact with sequentially distant sites on DNA (134). Figure 16.8 illustrates the large-scale configurational collapse brought about by minor reorientation of the ends of a nicked, spatially anchored elastic loop.

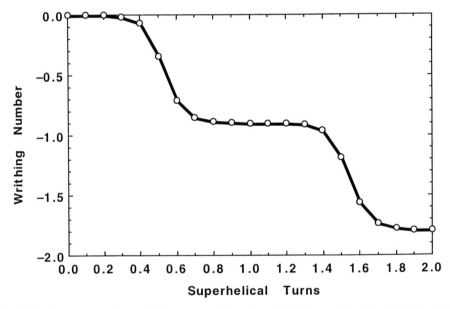

Fig. 16.7. Comparative change of *Wr* versus number of superhelical turns of a nicked, 1000 bp DNA constrained to wrap in a superhelical pathway (radius = 45 Å; pitch = 30 Å) around an ideal nucleosome core. Curves generated with data obtained from numerical solution of the Euler–Lagrange equations at 0.0125 turn increments of wrapping (116).

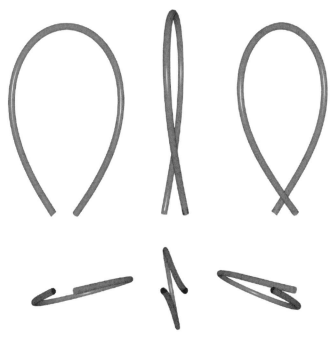

Fig. 16.8. Large-scale configurational collapse induced by reorientation of anchored ends of a 1000 bp, naturally straight, nicked DNA loop. Pseudo-valence angles between end tangents and end-to-end vector increase from (left) 53° to (centre) 90° and (right) 127°. End-to-end separation held at 170 Å and virtual torsion at 10°. See ref. 8 for further details.

The top views show the ±37° repositioning of terminal residues in the open and closed forms, and the bottom views the associated out-of-plane rotation. The numerical work has further revealed a mechanism of configurational collapse beyond the scope of the theory: the shape of the loop depends on whether the DNA is nicked or unnicked, and a transition between open and closed forms can be induced by nicking (8).

4.4 Multinucleosomal DNA

Interest in modelling DNA with several bound nucleosomes (103) was stimulated by the so-called linking number paradox, the apparent discrepancy between the roughly −1.7 change in ΔLk expected for a DNA rod wrapped the ~1.75 superhelical turns found in the nucleosome core particle (14,15) (cf. Fig. 16.7), and the value of −1.0±0.1 per nucleosome deduced from reconstitution experiments (135,136). A new resolution of the discrepancy (103) between the folding exhibited by model curves and natural minicircles in terms of the number of nucleosomes present on the DNA adds to the many solutions offered to resolve the paradox: such as altered duplex twist on the nucleosome (137–139), a non-cylindrically shaped protein core (5), or non-tangential departure of DNA from the histone assembly (140). The construction and manipulation of multinucleosomal structures on the computer is difficult in that the loop closure problem must be solved between all successive pairs of bound proteins.

The equilibrium structure of a 500 bp DNA ring bearing two evenly spaced ~1.75 turn superhelical constraints (103) has a writhing number per nucleosome of −1.15 that closely approximates the experimentally derived value of $\Delta Lk = -1$ per nucleosome. In other words, the contribution that a nucleosome makes to the writhing number is not a multiple of the −1.7 value of ΔLk found for a single superhelical fragment. Rather, Wr is a global property of the entire structure and not an additive quantity obtained by assigning a unit value per bound protein. Therefore, as system conditions change in terms of the number of bound proteins, linker and total chain lengths, and precise amount of DNA per core, the writhing number per core and its contribution to the change in ΔLk upon ligation of nicked chains also changes (141). Various experimental measurements [i.e. nuclease digestion (137) and hydroxyl radical cleavage (142)] appear, at first glance, to support a change in the twist of DNA in the core particle compared to its unbound states. The interpretation of these results, however, is not as straightforward as once believed. When the degree of experimental error of the footprinting techniques is taken into account, in the absence of firm knowledge of the shape of the underlying protein and the path of the DNA on that surface, it is generally not possible to conclude with certainty whether the bound DNA is over- or under-wound (5).

While it is relatively straightforward to simulate the structures of open DNA chains covered by large numbers of nucleosomes (143,144) and simple dinucleosomal DNA circles (103), it is far more difficult to model spatially constrained systems like the 5243 bp SV40 minichromosome with 21–26 bound nucleosomes studied experimentally (135). As with any closed polymer, the influence of local structural changes on global chain folding is unpredictable. One of the biggest stumbling blocks is finding a reasonable starting structure for optimization or dynamics. The ~13000 bp model in Fig. 16.9 with 64 evenly spaced nucleosomes, each wrapped 1.75 turns around a phantom histone core, is a regular fourth-order helix of zero pitch that satisfies a simple relationship between the number of bound nucleosomes (N), the radius of the assembly (R), the linker length (L), and the degree of wrapping on the nucleosome measured by φ, a virtual valence angle between linker and bound DNA segments (141):

$$\cos\left(\frac{\pi}{N}\right) = \sin\left(\frac{\varphi}{2}\right), \quad 2\left[1 - \cos\left(\frac{2\pi}{N}\right)\right] = \left(\frac{L}{R}\right)^2. \tag{16.9}$$

These expressions arise from equations commonly used to classify regular structures of synthetic polymers (145). Because the linker lengths are straight, the example illustrated in the figure is the lowest energy state that a naturally straight DNA with bound nucleosomes of this number and proportion can attain. Models like this one, which incorporate the chain lengths and nucleosome numbers found in naturally occurring minichromosomes, provide a good starting point for the study of topological issues tied to the linking number paradox, as well as for dynamic simulations of the chromatin machinery.

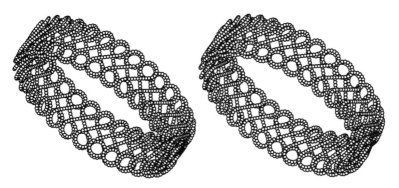

Fig. 16.9. Relaxed DNA minichromosome model with 64 nucleosomes, each bound by ~140 bp of DNA (141).

4.5 Sequence effects

The preferred configurations of closed DNAs with realistic chemical features are strikingly different from those of homogeneous, naturally straight rods. Intrinsically curved DNA fragments not only adopt altered three-dimensional arrangements from chains that obey the isotropic rod model, but also show pronounced variations in the local twist of successive base pairs. A given supercoiled structure is thus characterized by regions that are underwound or overwound with respect to the natural twist density. The mere presence of an intrinsically curved sequence can therefore bring about a situation in which portions of the molecule are more or less susceptible to interaction with other agents. Bending anisotropy (i.e. the tendency of DNA to deform through rolling rather than tilting motions) similarly affects the optimum shape and the distribution of twist. While determination of the equilibrium solutions for a closed, naturally curved rod is beyond the scope of analytical work, a few states have been identified and an expression for the rate of change of local twist along a naturally curved circle has been derived (146). This latter equation is useful in checking the computed twist density of supercoiled 'O-rings' that bend naturally into a smooth circle. Such chains deform gradually with change of ΔLk into collapsed figure-eight structures (107,122), and subsequently open into symmetric figure-eights and linear interwound structures (107).

The sequential variation in twist in curved DNA is illustrated in Fig. 16.10 for two elastic models closed into 150 bp circles with low levels of imposed supercoiling (ΔLk = −0.2). In one case, the O-ring, the intrinsic bending of the chain arises from sinusoidal changes in the roll and tilt angles of successive base pairs, with successive residues twisted in the relaxed state by 36°. The second case more closely resembles naturally curved DNA sequences, such as the $A_m X_{10-m}$ (m = 4–6) repeating motif found in the kinetoplast DNA of *Crithidia fasiculata* (147,148), which are a composite of parallel and slightly wedged base pair steps. Here the chain sequence contains five naturally straight dimer steps with an intrinsic twist of 36° regularly interspersed between five steps with an intrinsic roll angle of 7.4° and a natural twist of 35.3°. In both step types the intrinsic tilt angle is zero and the bending is assumed to be

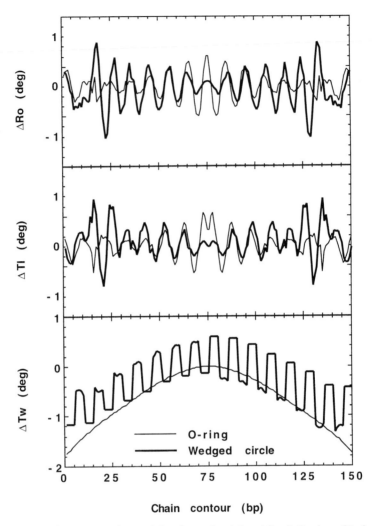

Fig. 16.10. Sequential variations in base pair bending and twisting, $\Delta\theta = \theta - \theta_0$ where θ is the roll (Ro), tilt (Tl), or twist (Tw) angle and θ_0 is the intrinsic value, in curved DNA models at low levels of imposed supercoiling ($\Delta Lk = -0.2$) (S. C. Pedersen, unpublished data). Chain closed naturally into a 150 bp circle by smooth bending (i.e. O-ring) or with local wedges phased with the helical repeat. See text for further details.

isotropic with force constants, $A_1 = A_2 = A = 2.1 \times 10^{-19}$ erg-cm, matching the persistence length of duplex DNA, i.e. $a = A/kT = 500$ Å. The choice of twisting constant, $C = A$, mimics the recently observed torsional deformations of linear DNA (149). The linking number deficit of -0.2 corresponds to a change of $-0.5°$ in the per residue twist of the naturally straight isotropic DNA of the same chain length, i.e. $\Delta\omega = 360°\Delta Tw/150$ bp $= 360°\Delta Lk/150$ bp.

The energy minimized states in Fig. 16.10 are identified through numerical solutions of the equations of equilibrium (S.C. Pedersen, unpublished work). The variation in twist along the curved DNAs, while fairly limited under the conditions of these calculations (fluctuating by less than a degree about the intrinsic twist values in both the O-ring and the wedged model) illustrates the potential specificity of ligand interactions in moderately supercoiled, curved DNAs. Segments of minimum twist and high bending in the more realistic wedged supercoil, for example, could direct the binding of drugs and proteins that induce these local structural deformations [e.g. the diamino steroid irehdiamine A (150), the antitumour drug actinomycin D (151), or the TATA box-binding protein (152–154)]. By contrast, the roll and tilt deformations are slightly more pronounced in the segments of larger twist in the supercoiled O-ring.

4.6 Intrinsic curvature and deformability

A naturally straight DNA rod that is allowed to bend in an anisotropic manner $(A_1 \neq A_2$ in eqn 16.2) adopts a polygonal rather than a circular twist-free structure (122). Such molecules bend sharply and preferentially along the roll axes of selected base pairs. If the contour length of the DNA is a multiple of the double helical repeat, the chain relieves the elastic stress associated with ring closure through large changes in the roll angle at points where the base pairs can bend into either the major or minor groove, leading to a pattern of structural change closely resembling that illustrated for the wedged DNA model in Fig. 16.10 (122). By contrast, the tilt angle shows limited variation along the contour of the rod, as predicted almost 20 years ago from all-atom energy studies (43). The numerical calculations additionally find small changes in twist at the sites of significant base pair roll.

The presence of a curved insert (106,155) or a fragment with different elastic constants (63) introduces asymmetry in the global equilibrium structures of the naturally straight, isotropic rod. The closed 'circular' forms of such models are irregular ovoid shapes (106), while the interwound structures contain terminal loops of uneven sizes. The differences are most pronounced at low superhelical density, where the inserts not only distort the overall shape but also rearrange the points of long-range contact (see, for example, the change in branching effected by a small curved insert in the naturally straight DNA supercoil in Fig. 16.11). Such configurational changes make it easy to understand how intrinsic sequence or certain proteins that deform the local structure of DNA can govern 'action at a distance' (156,157) along the chain. The curved insert, however, does more than change the shape of the supercoiled molecule. Accompanying changes in twist (Fig. 16.12) introduce a subtle level of added structural control, bringing parts that lie on the inside of the closed duplex to the outside of the DNA with a curved insert and vice versa. The twist is non-uniform in the curved insert, in agreement with elasticity theory (146), and undergoes sharp oscillations within this fragment, but is uniform elsewhere. Because the curved segment is somewhat straightened and smoothly distorted out of the plane into a superhelical shape, the average increment in twist is lower in the curved segment (0.047°/bp) than in the remainder of the chain (0.060°/bp). The incremental twist in the naturally straight DNA, by contrast, is uniform and intermediate in value, i.e. (360°/6441 bp) ΔTw = (360°/6441 bp) $(\Delta Lk - Wr)$ = 0.057°/bp.

Fig. 16.11. Refolding of branched equilibrium configurations effected by the presence of a 215 bp, smoothly curved segment in a naturally straight 6441 bp DNA at low superhelical density, $\sigma = \Delta Lk/Lk_0 =$ 0.007 (106). (Left) naturally straight chain; (right) chain with insert intrinsically deformed into a circular arc of 209°.

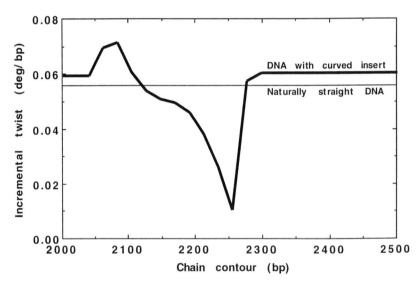

Fig. 16.12. Change in incremental twist, in °/bp, versus chain contour in the branched supercoiled DNAs illustrated in Fig. 16.11 (106).

4.7 Global structural organization

Curved segments do not concentrate solely in the hairpin loops of DNA at higher superhelical density, in apparent contradiction with the notion that curved DNA fixes the topology of a supercoil (106). If there is no difference in intrinsic flexibility, the curved insert can also appear in the central interwound contact zone, yielding the equivalent of a bent linear configuration or a Y-shaped branched structure missing the arm that contains the curved fragment. Moreover, the two arms of this bent minimum can vary in length so that the deformed curved insert can appear virtually anywhere along the chain contour. When two or more naturally curved segments of the same size and intrinsic proportions are introduced in a closed naturally straight DNA, their relative spacing becomes an important factor (106,155). Adjusting the intrinsic curva-

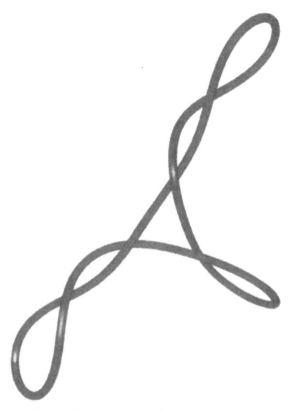

Fig. 16.13. Three naturally curved planar segments of 200 bp used to organize the folding of a 5000 bp, supercoiled chain (155).

ture and relative positioning of curved segments can therefore facilitate or inhibit distant residues from coming into close contact (Fig. 16.13). Such changes suggest structural mechanisms whereby specific base pair sequences can regulate intramolecular interactions and related biochemical processes.

The folding patterns at low levels of supercoiling follow from theoretical understanding of constrained loops (6). Chains where the end-to-end distance of straight fragments is comparable with their contour length tend to adopt open forms, while those where the through space distance between chain ends is smaller tend to be collapsed (103,106,155). Numerical studies reveal that the magnitude of the curved insert determines the ease of structural collapse (8). Curved inserts divide straight DNA into regions of uniform, although different, local twist (see Fig. 16.12). These subtle distinctions make parts of the chain potentially more susceptible to melting than others. Furthermore, upon supercoiling, two inserts may show completely different patterns in local twisting (106). Since the change in twist at the base pair level can substantially affect the binding of ligands to DNA, the affinity of a drug or protein for a naturally curved segment can therefore vary at different sites along a supercoiled duplex, even if the curved fragments are identical to one other in the relaxed stress-free state. Thus, one can begin to see how the combination of intrinsic curvature and supercoiling can facilitate the recognition of specific parts of a DNA.

4.8 DNA dynamics

The ability to treat the dynamics of supercoiled DNA lags behind understanding of their equilibrium structures. The limitations of methodology notwithstanding, studies to date based on the Langevin dynamics formalism (55,102,104,105) reveal the potential to simulate massive changes of macromolecular shape while keeping close track of local chemical structure. The representation of the double helical axis in terms of B-spline curves or finite Fourier series expressions makes it possible to simulate large-scale

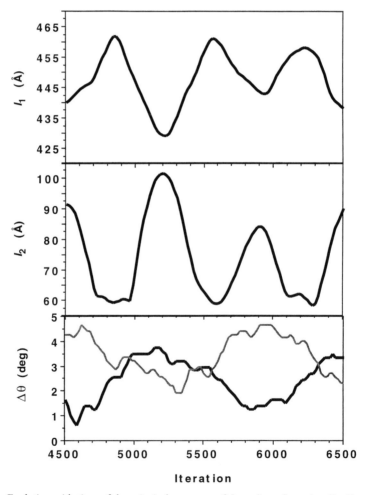

Fig. 16.14. Evolution with time of the principal moments of the radius of gyration (I_1, I_2) and bending angles $\Delta\theta$ at two sites in the interwound core of a 1000 bp DNA supercoil. Langevin dynamics simulations (104,105) started from the interwound global equilibrium configuration illustrated in Fig. 16.4.

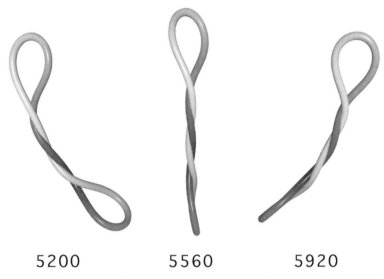

5200   5560   5920

Fig. 16.15. Molecular images of three supercoiled DNA configurations sampled at iteration times given in Fig. 16.14. Each 180° turn of the long axis occurs in concert with a planar wagging movement that bends the molecule in opposing directions. Roughly 45° of the end-over-end tumbling is described in the figure. The shading along the chain contour reveals limited slithering motions over the time-scale of the tumbling.

configurational changes while following the fine structure of the constituent residues. The details of local fine structure are incorporated in the mesh parameters of the curve, and can, in principle, be used to account for the sequence-dependent secondary structure of the double helix. These formulations are alternatives to the traditional polygonal models (66,81) used to describe the motions of DNA in terms of a limited number of independent variables and a phenomenological description of chain structure.

Figure 16.14 illustrates changes in local angular parameters tied to the end-over-end tumbling and global flexing of a supercoiled chain generated with the Langevin protocol (104,105). Regular deformations of bending angles in the interwound core of the model, $\Delta\theta$, generate a planar wagging movement detected by cyclic changes in principal moments I_1 and I_2. One sweep of this wagging (a half cycle) is depicted by the molecular images in Fig. 16.15 along with the associated tumbling of the molecule as a whole. The period of local angular variation in the $\Delta\theta$, bending angles at points separated by 100 bp of the DNA contour, corresponds to a complete cycle of wagging.

With an appropriate choice of parameters to simulate the solvent forces, the simulations span a broad range of configuration space and, at selected levels of supercoiling, interconvert between competing minima (e.g. circle and figure-eight states) (104). The common pathways covered in independent simulations demonstrate the utility of the approach in unravelling the physicochemical mechanisms behind large-scale configurational transitions. The introduction of a long-range electrostatic term further enhances the global motions of the model (55), whereas the adoption of a van der Waals term tends to stabilize the interwound segments of the polymer (102). The decrease of salt in the electrostatic term enhances the opening and closing of the chain as a whole and promotes the rapid slithering of individual residues past one another (55).

Current investigations of DNA dynamics are also exploiting new solutions of the dynamical equations of elastic rods. Two types of motion in the vicinity of the closed circle have been reported recently: (a) a class of global twisting motions in which the axial curve of the rod remains stationary while the cross-sections rotate in concert about their centres (87); and (b) the normal modes of vibration of the circle, involving both small amplitude distortions of the circular shape and position-dependent variations of the twist density (91). The global rotation of the naturally straight rod is found to vary linearly with time so that the complete surface of the rod is uniformly exposed (i.e. the DNA model rotates freely about the helical axis). By contrast, the naturally curved rod resists deformation from the circular state and moves in an oscillatory fashion, exposing some parts of the model more than others. There have been suggestions that such global motions might play a role in transcription (158). To treat the thermal fluctuations associated with global rotation, classical statistical thermodynamic methods have been applied to an ensemble of closed rings and selected parameters (e.g. the root-mean-square angle of rotation of the cross-sections as a function of the degree of intrinsic curvature of the rod) have been computed (87). The global rocking of the circle about an axis passing through its centre is linked to time- and position-dependent changes in twist density, even in the naturally straight rod (91). The latter result provides an important check of numerical work.

5. Summary

Current understanding of supercoiled DNA combines new computational approaches with developments in elastic rod theory to examine the configurations and properties of supercoiled DNA. No single approach can deal with all aspects of supercoiled structure. The global equilibrium configuration of a long circular DNA, while only one of the multitude of energetically possible low energy states, serves as an important reference point for monitoring chain deformations. The thermal fluctuations and pathways monitored in Monte Carlo and molecular dynamics studies are critical for quantitative and qualitative understanding of polymers that are long compared with the persistence length.

Research is rapidly progressing beyond simplified treatments of DNA as an ideal elastic rod to models that incorporate base sequence-dependent features and the polyelectrolyte character of the double helix. Previous work on closed circular duplexes has been generalized to include loops with arbitrarily fixed end-points so that it is now possible to study systems closely resembling the fragments of supercoiled DNA anchored by proteins in the cell. A variety of methods are at hand for rapid identification of equilibrium structures, for treating the dynamics of the long thread-like polymer, and for increased understanding of the principles governing the local and global features of the chain.

The new capability for treating base sequence-dependent features and bound proteins in the identification of the equilibrium structures of supercoiled models makes it possible to analyse naturally occurring DNA sequences and to address questions of biological interest, such as how the assembly of nucleosomes influences the overall folding of minichromosomes. Further improvements to the dynamics protocols will make it possible, in time, to follow the pathways of configurational change in realistic models of supercoiled DNA.

Acknowledgements

I am grateful to Helen Berman, Bernard Coleman, Gerald Manning, Tamar Schlick, Irwin Tobias, and Victor Zhurkin for valuable discussions related to this work. I am also indebted to current and former members of my research group, especially Guohua Liu, Jennifer Martino, Scott Pedersen, Timothy Westcott, and Yang Yang, for sharing data and ideas related to the topics in this chapter and to Andrew Olson and A.R. Srinivasan for original computer graphics. This research has been generously supported by the US Public Health Service (grant GM34809). Calculations were performed at the Rutgers Center for Computational Chemistry.

References

1. Vinograd, J., Lebowitz, J., Radloff, R., Watson, R. and Laipis, P. (1965) *Proc. Natl. Acad. Sci. USA* **53**, 1104.
2. Bauer, W.R. (1978) *Annu. Rev. Biophys. Bioeng.* **7**, 287.
3. Bates, A.D. and Maxwell, A. (1993) *DNA Topology*. IRL Press, Oxford.
4. Travers, A. (1993) *DNA–Protein Interactions*. Chapman & Hall, London.
5. Zhang, P., Tobias, I. and Olson, W.K. (1994) *J. Mol. Biol.* **242**, 271.
6. Tobias, I., Coleman, B. and Olson, W.K. (1994) *J. Chem. Phys.* **101**, 10990.
7. Coleman, B.D., Tobias, I. and Swigon, D. (1995) *J. Chem. Phys.* **103**, 9101.
8. Westcott, T.P., Tobias, I. and Olson, W.K. (1995) *J. Phys. Chem.* **99**, 17926.
9. White, J.H. (1989) in *Mathematical Methods for DNA Sequences*, (Waterman, M.S., ed.), p. 225. CRC Press, Boca Raton.
10. White, J.H. (1969) *Am. J. Math.* **91**, 693.
11. Wang, J.C. and Liu, L.F. (1990) in *DNA Topology and Its Biological Effects*, (Cozzarelli, N.R. and Wang, J.C., eds), p. 321. Cold Spring Harbor Laboratory Press, Cold Spring Harbor, NY.
12. Wu, H.-Y., Shyy, S.H., Wang, J.C. and Liu, L.F. (1988) *Cell* **53**, 433.
13. Liu, L.F. andWang, J.C. (1987) *Proc. Natl. Acad. Sci. USA* **84**, 7024.
14. Richmond, T.J., Finch, J.T., Rushton, B., Rhodes, D. and Klug, A. (1984) *Nature* **311**, 532.
15. Klug, A., Finch, J.T. and Richmond, T.J. (1985) *Science* **229**, 1109.
16. Clark, D.J. and Felsenfeld, G. (1991) *EMBO J.* **10**, 387.
17. Clark, D.J., Ghirlando, R., Felsenfeld, G. and Eisenberg, H. (1993) *J. Mol. Biol.* **234**, 297.
18. Jackson, V. (1993) *Biochemistry* **32**, 5901.
19. Liu, L.F. and Wang, J.C. (1978) *Cell* **15**, 979.
20. Morrison, A. and Cozzarelli, N.R. (1981) *Proc. Natl. Acad. Sci. USA* **78**, 1416.
21. Kirchausen, T., Wang, J.C. and Harrison, S. (1985) *Cell* **41**, 933.
22. Bates, A.D. and Maxwell, A. (1989) *EMBO J.* **8**, 1861.
23. Hsieh, C.-H. and Griffith, J.D. (1988) *Cell* **52**, 535.
24. Shrader, T.E. and Crothers, D.M. (1990) *J. Mol. Biol.* **216**, 69.
25. Laundon, C.H. and Griffith, J.D. (1988) *Cell* **52**, 545.

26. Rhodes, D. (1979) *Nucl. Acids Res.* **6**, 1805.
27. Prunell, A. (1981) *EMBO J.* **1**, 173.
28. Hogan, M.E. and Austin, R.H. (1987) *Nature* **329**, 263.
29. Puhl, H.L. and Behe, M.J. (1995) *J. Mol. Biol.* **245**, 559.
30. Ulyanov, N.B. and Zhurkin, V.B. (1985) *Mol. Biol. USSR (Engl. Ed.)* **18**, 1366.
31. Srinivasan, A.R., Torres, R., Clark, W. and Olson, W.K. (1987) *J. Biomol. Struct. Dynamics* **5**, 459.
32. Sarai, A., Mazur, J., Nussinov, R. and Jernigan, R.L. (1989) *Biochemistry* **28**, 7842.
33. Berman, H.M., Olson, W.K., Beveridge, D.L., Westbrook, J., Gelbin, A., Demeny, T., Hsieh, S.-H., Srinivasan, A.R. and Schneider, B. (1992) *Biophys. J.* **63**, 751.
34. Gorin, A.A., Zhurkin, V.B. and Olson, W.K. (1995) *J. Mol. Biol.* **247**, 34.
35. Bhattacharyya, D. and Bansal, M. (1990) *J. Biomol. Struct. Dynamics* **8**, 539.
36. Yanagi, K., Privé, G.C. and Dickerson, R.E. (1991) *J. Mol. Biol.* **217**, 201.
37. Babcock, M.S. and Olson, W.K. (1994) *J. Mol. Biol.* **237**, 98.
38. Heinemann, U., Alings, C. and Hahn, M. (1994) *Biophys. Chem.* **50**, 157.
39. Gorin, A.A., Zhurkin, V.B. and Olson, W.K. (1995) *J. Biomol. Struct. Dynamics* **12**, a74.
40. Suzuki, M. and Yagi, N. (1995) *Nucl. Acids Res.* **23**, 2083.
41. Young, M.A., Ravishankar, G., Beveridge, D.L. and Berman, H.M. (1995) *Biophys. J.* **68**, 2454.
42. Calladine, C.R. (1982) *J. Mol. Biol.* **161**, 343.
43. Zhurkin, V.B., Lysov, Y.P. and Ivanov, V. (1979) *Nucl. Acids Res.* **6**, 1081.
44. Kirchhoff, G.J. (1859) *J. f. Reine. Angew. Math. (Crelle)* **56**, 285.
45. Clebsch, A. (1862) *Theorie der Elasticität fester Körper.* B.G. Tuebner, Leipzig.
46. Kirchhoff, G. (1876) *Vorlesungen über mathematische Physik, Mechanik.* B.G. Tuebner, Leipzig.
47. Clebsch, A. (1883) Théorie de l'Elasticité des Corps Solides. Dunod, Paris.
48. Marky, N.L. and Olson, W.K. (1994) *Biopolymers* **34**, 121.
49. Westcott, T.P., Tobias, I. and Olson, W.K. (1997) *J. Chem. Phys.* **107**, 3967
50. Fuller, F.B. (1971) *Proc. Natl. Acad. Sci. USA* **68**, 815.
51. Hao, M.-H. and Olson, W.K. (1989) *Macromolecules* **22**, 3292.
52. Fenley, M.O., Olson, W.K., Tobias, I. and Manning, G.S. (1994) *Biophys. Chem.* **50**, 255.
53. Fenley, M.O., Chua, K., Boschitsch, A.H. and Olson, W.K. (1996) *J. Comp. Chem.* **17**, 976.
54. Tesi, M.C., Janse van Rensburg, E.J., Orlandini, E., Sumners, D.W. and Whittington, S.C. (1993) *Phys. Rev.* **E49**, 868.
55. Schlick, T., Li, B. and Olson, W.K. (1994) *Biophys. J.* **67**, 2146.
56. Vologodskii, A.V. and Cozzarelli, N.R. (1995) *Biopolymers* **35**, 289.
57. Gebe, J.A., Delrow, J.J., Heath, P.J., Stewart, D.W. and Schurr, J.M. (1996) *J. Mol. Biol.* **262**, 105.
58. Odijk, T. (1994) *Macromolecules* **27**, 4998.
59. Ray, J. and Manning, G.S. (1994) *Langmuir* **10**, 2450.
60. Livolant, F. (1984) *Eur. J. Cell Biol.* **33**, 300.
61. Livolant, F. (1987) *J. Phys. (Les Ulis)* **48**, 1051.
62. Livolant, F. and Maestre, M.F. (1988) *Biochemistry* **27**, 3056.
63. Yang, Y., Tobias, I. and Olson, W.K. (1993) *J. Chem. Phys.* **98**, 1673.
64. Dichmann, D.J., Li, Y. and Maddocks, J.H. (1996) in *Mathematical Approaches to Biomolecular Structure and Dynamics*, (Mesirov, J.P., Schulten, K. and Sumners, D., eds), p. 71. Springer-Verlag, New York.
65. Tan, R.K.-Z. and Harvey, S.C. (1989) *J. Mol. Biol.* **205**, 573.

66. Klenin, K.V., Vologodskii, A.V., Anshelevich, V.V., Dykhne, A.M. and Frank-Kamenetskii, M.D. (1991) *J. Mol. Biol.* **217**, 413.
67. Gebe, J.A., Allison, S.A., Clendenning, J.B. and Schurr, J.M. (1995) *Biophys. J.* **68**, 619.
68. Klenin, K.V., Frank-Kamenetskii, M.D. and Langowski, J. (1995) *Biophys. J.* **68**, 81.
69. Hao, M.-H. and Olson, W.K. (1989) *Biopolymers* **28**, 873.
70. Zhang, P., Olson, W.K. and Tobias, I. (1991) *Comp. Polymer Sci.* **1**, 3.
71. Schlick, T. and Olson, W.K. (1992) *J. Mol. Biol.* **223**, 1089.
72. Liu, G., Olson, W.K. and Schlick, T. (1995) *Comp. Polymer Sci.* **5**, 7.
73. Tan, R.K.-Z., Sprous, D. and Harvey, S.C. (1996) *Biopolymers* **38**, 259.
74. Sprous, D., Tan, R.K.-Z. and Harvey, S.C. (1996) *Biopolymers* **38**, 243.
75. Sprous, D. and Harvey, S.C. (1996) *Biophys. J.* **69**, 1903.
76. Langowski, J., Olson, W.K., Pedersen, S.C., Tobias, I., Westcott, T.P. and Yang, Y. (1996) *TIBS* **21**, 50.
77. Kratky, O. and Porod, G. (1949) *Rec. Trav. Chim. Pays-Bas* **68**, 1106.
78. Hagerman, P.J. (1988) *Annu. Rev. Biophys. Biophys. Chem.* **17**, 265.
79. Vologodskii, A.V., Levene, S.D., Frank-Kamenetskii, M.D. and Cozzarelli, N.R. (1992) *J. Mol. Biol.* **227**, 1224.
80. van Waveren, M., Bishop, M. and Michels, J.P. J. (1988) *J. Chem. Phys.* **88**, 1326.
81. Chirico, G. and Langowski, J. (1991) *J. Chem. Phys.* **88**, 2561.
82. Chirico, G. and Langowski, J. (1992) *Macromolecules* **25**, 769.
83. Dwyer, J.D. and Bloomfield, V.A. (1993) *Biophys. J.* **65**, 1810.
84. Chirico, G. and Langowski, J. (1994) *Biopolymers* **34**, 415.
85. Ramachandran, G. and Schlick, T. (1995) *Phys. Rev. E* **51**, 6188.
86. Marko, J.F. and Siggia, E.D. (1995) *Phys. Rev. E.* **52**, 2912.
87. Tobias, I., Coleman, B.D. and Lembo, M. (1996) *J. Chem. Phys.* **105**, 2517.
88. Klapper, J. and Tabor, M. (1994) *J. Phys. A: Math. Gen.* **27**, 4919.
89. Klapper, J. (1996) *J. Comp. Phys.* **125**, 325.
90. Manning, R.S. and Maddocks, J.H. (1996) *J. Chem. Phys.* **105**, 5626.
91. Coleman, B.D., Lembo, M. and Tobias, I. (1996) *Meccanica* **31**, 565.
92. Yoon, D.Y. and Flory, P.J. (1974) *J. Chem. Phys.* **61**, 5366.
93. Suter, U.W., Mutter, M. and Flory, P.J. (1976) *J. Am. Chem. Soc.* **98**, 5733.
94. Hagerman, P.J. (1985) *Biopolymers* **24**, 1881.
95. Levene, S.D. and Crothers, D.M. (1986) *J. Mol. Biol.* **189**, 61.
96. Marky, N.L. and Olson, W.K. (1982) *Biopolymers* **21**, 2329.
97. Rippe, K., von Hippel, P.H. and Langowski, J. (1995) *TIBS* **20**, 500.
98. Frank-Kamenetskii, M.D., Lukashin, A.V. and Vologodskii, A.V. (1975) *Nature (London)* **258**, 398.
99. Vologodskii, A.V., Lukashin, A.V., Anshelevich, V.V. and Frank-Kamenetskii, M.D. (1979) *Nucleic Acids Res.* **6**, 967.
100. Frank-Kamenetskii, M.D. and Vologodskii, A.V. (1981) *Sov. Phys. Usp. (Eng. Ed.)* **24**, 679.
101. Malhotra, A., Tan, R.K.-Z. and Harvey, S.C. (1994) *Biophys. J.* **66**, 1777.
102. Schlick, T., Olson, W.K., Westcott, T. and Greenberg, J.P. (1994) *Biopolymers* **34**, 565.
103. Martino, J.A. and Olson, W.K. (1997) *Biopolymers* **41**, 419.
104. Liu, G.H., Olson, W.K., Olson, A.J. and Schlick, T. (1997) *Biophys. J.* **73**, 1742.
105. Liu, G. (1996) *PhD Thesis*, Rutgers, The State University of New Jersey, New Brunswick, NJ.
106. Yang, Y., Westcott, T.P., Pedersen, S.C., Tobias, I. and Olson, W.K. (1995) *TIBS* **20**, 313.
107. Westcott, T.P. (1996) *PhD Thesis*, Rutgers, the State University of New Jersey, New Brunswick, NJ.

108. Olson, W.K. (1979) *Biopolymers* **18**, 1235.
109. Kuhn, W. (1936) *Kolloid-Z.* **76**, 258.
110. Kuhn, W. (1939) *Kolloid-Z.* **87**, 3.
111. Zhurkin, V.B., Ulyanov, N.B., Gorin, A.A. and Jernigan, R.L. (1991) *Proc. Natl. Acad. Sci. USA* **88**, 7046.
112. Olson, W.K., Marky, N.L., Jernigan, R.L. and Zhurkin, V.B. (1993) *J. Mol. Biol.* **232**, 530.
113. Schellman, J.A. and Harvey, S.C. (1995) *Biophys. Chem.* **55**, 95.
114. Gordon, W.J. and Riesenfeld, R.E. (1974) in *Computer Aided Geometric Design*, (Barnhill, R.E. and Riesenfeld, R.E., eds), p. 95. Academic Press, New York.
115. Olson, W.K. and Zhang, P. (1991) in *Molecular Design and Modeling: Concepts and Applications, Part B, Antibodies and Antigens, Nucleic Acids, Polysaccharides, and Drugs* (Langone, J.J., ed.), p. 403. Academic Press, Orlando.
116. Olson, W.K., Westcott, T.P., Martino, J., A. and Liu, G.-H. (1996) in *Mathematical Approaches to Biomolecular Structure and Dynamics*, (Mesirov, J.P., Schulten, K. and Sumners, D., eds), p. 195. Springer-Verlag, New York.
117. Wertz, J.R. (ed.) (1980) *Spacecraft Attitude Determination and Control*. D. Reidel Publishing Co., Dordrecht, Holland.
118. Metropolis, N., Rosenbluth, A.W., Rosenbluth, M.N., Teller, A. and Teller, E. (1953) *J. Chem. Phys.* **21**, 1087.
119. Olson, W.K. (1992) in *Computation of Biomolecular Structures: Achievements, Problems, and Perspectives*, (Soumpasis, D.M. and Jovin, T.M., eds), p. 55. Springer Verlag, Berlin.
120. Kirkpatrick, S., Gelatt, C.D. and Vecchi, M.P. (1983) *Science* **220**, 671.
121. Yang, Y., Tobias, I. and Olson, W.K. (1992) *Adv. Bioenging, ASME, Bioengng Div.* **22**, 5.
122. Olson, W.K., Babcock, M.S., Gorin, A., Liu, G.-H., Marky, N.L., Martino, J.A., Pedersen, S.C., Srinivasan, A.R., Tobias, I., Westcott, T.P. and Zhang, P. (1995) *Biophys. Chem.* **55**, 7.
123. Ramachandran, G. and Schlick, T. (1996) in *DIMACS Series in Discrete Mathematics and Theoretical Computer Science*, (Pardalos, P. and Shalloway, D., eds), p. 215. American Math. Society, Providence.
124. Allison, S.A. and McCammon, J.A. (1984) *Biopolymers* **23**, 167.
125. Chirico, G. and Langowski, J. (1996) *Biophys. J.* **71**, 955.
126. Katritch, V. and Vologodskii, A.V. (1997) *Biophys. J.* **72**, 1070.
127. Jian, H., Vologodskii, A.V. and Schlick, T. (1997) *J. Comp. Phys.* **136**, 168.
128. Le Bret, M. (1979) *Biopolymers* **18**, 1709.
129. Bednar, J., Furrer, P., Stasiak, A., Dubochet, J., Egelman, E.H. and Bates, A.D. (1994) *J. Mol. Biol.* **235**, 825.
130. Lyubchenko, Y.L. and Shlyaktenko, L.S. (1997) *Proc. Natl. Acad. Sci. USA* **94**, 496.
131. Stigter, D. (1977) *Biopolymers* **16**, 1435.
132. Brady, G.W., Fein, D.B., Lambertson, J., Grassian, V., Foos, D. and Benham, C.J. (1983) *Proc. Natl. Acad. Sci. USA* **80**, 741.
133. Brady, G.W., Foos, D., Satkowski, M. and Benham, C.J. (1987) *J. Mol. Biol.* **195**, 185.
134. Lewis, M., Chang, G., Horton, N.C., Kercher, M.A., Pace, H.C., Schumacher, M.A., Brennan, R.G. and Lu, P. (1996) *Science* **271**, 1247.
135. Germond, J.E., Hirt, B., Oudet, P., Gross-Bellard, M. and Chambon, P. (1975) *Proc. Natl. Acad. Sci. USA* **72**, 1843.
136. Simpson, R.T., Thoma, F. and Brubaker, J.M. (1985) *Cell* **42**, 799.
137. Rhodes, D. and Klug, A. (1981) *Nature* **292**, 378.
138. Klug, A. and Travers, A.A. (1989) *Cell* **56**, 10.
139. White, J.H., Gallo, R. and Bauer, W.R. (1989) *J. Mol. Biol.* **207**, 193.

140. Furrer, P., Bednar, J. and Dubochet, J. (1995) *J. Struct. Biol.* **114**, 177.

141. Martino, J.A. (1997) *PhD Thesis*, Rutgers, The State University of New Jersey, New Brunswick, NJ.

142. Hayes, J.J., Tullius, T.D. and Wolffe, A.P. (1990) *Proc. Natl. Acad. Sci. USA* **87**, 7405.

143. Woodcock, C.L., Grigoryev, S.A., Horowitz, R.A. and Whitaker, N. (1993) *Proc. Natl. Acad. Sci. USA* **90**, 9021.

144. Yang, G., Leuba, S.H., Bustamante, C., Zlatanova, J. and van Holde, K. (1994) *Struct. Biol.* **1**, 761.

145. Miyazawa, T. (1961) *J. Polymer Sci.* **55**, 215.

146. Tobias, I. and Olson, W.K. (1993) *Biopolymers* **33**, 639.

147. Marini, J.C., Levene, S.D., Crothers, D.M. and Englund, P.T. (1982) *Proc. Natl. Acad. Sci. USA* **79**, 7664.

148. Griffith, J.D., Bleyman, M., Rauch, C.A., Kitchin, P.A. and Englund, P.T. (1986) *Cell* **40**, 717.

149. Heath, P.J., Clendenning, J.B., Fujimoto, B.S. and Schurr, J.M. (1996) *J. Mol. Biol.* **260**, 718.

150. Dattagupta, N., Hogan, M. and Crothers, D.M. (1978) *Proc. Natl. Acad. Sci. USA* **75**, 4286.

151. Shinomiya, M., Chu, W., Carlson, R.G., Weave, R.F. and Takusagawa, F. (1995) *Biochemistry* **34**, 8481.

152. Kim, J.L., Nikolov, D.B. and Burley, S.K. (1993) *Nature* **365**, 520.

153. Kim, Y., Geider, H.J., Hahn, S. and Sigler, P.B. (1993) *Nature* **365**, 512.

154. Parvin, J.D., McCormick, R.J., Sharp, P.A. and Fisher, D.E. (1995) *Nature* **35**, 724.

155. Olson, W.K. (1996) *Curr. Opin. Struct. Biol.* **6**, 242.

156. Burd, J.F., Wartell, R.M., Dodgson, J.B. and Wells, R.D. (1975) *J. Biol. Chem.* **250**, 5109.

157. Lobell, R.B. and Schleif, R.F. (1991) *J. Mol. Biol.* **218**, 45.

158. Ten Heggeler-Borider, B., Wahli, W., Adrian, M., Stasiak, A. and Dubochet, J. (1992) *EMBO J.* **11**, 667.

17
Crystallographic structures of RNA oligoribonucleotides and ribozymes

B. Masquida and E. Westhof

Institut de Biologie Moléculaire et Cellulaire, Centre National de la Recherche Scientifique, UPR 9002, 15, rue R. Descartes, F-67084 Strasbourg, France

1. Introduction

Since the resolution of tRNA crystal structures in the late 1970s, X-ray analysis of RNA has made a considerable leap, albeit only recently. RNA structures are, since 1988, solved at an average rate of two structures per year, and this number is rapidly increasing. This cooperative phenomenon mainly results from the overcoming of experimental limitations such as the obtaining of 'heaps' of crystal-pure RNA molecules of any size or the development of efficient screening methods for crystallization conditions. The recent appearance of the structures of 13 RNA duplexes, three hammerhead ribozymes, and the P4–P6 domain of the group I self-splicing introns, considerably broadens our views of the fascinating RNA structural landscapes.

2. Crystallization

2.1 Synthesis and purification

Although it is common to purify naturally occurring RNA molecules from cell extracts, the only way to produce the original systems is by the use of *in vitro* polymerization of nucleotides. This can be performed either by enzymatic or chemical synthesis. In the enzymatic procedure, any DNA matrix, single- or double-stranded, of whatever sequence, is transcribed by the T7 DNA-dependent RNA polymerase (1,2). This method can yield around a hundred RNA copies per DNA molecule. This ratio depends on the magnesium concentration as well as on the presence of polyamines such as spermine or spermidine. A drawback of the method is that the T7 enzyme has a tendency to polymerize, randomly, additional nucleotides at the 3'-extremities of the RNA copies, thus generating heteromers, which may be difficult to separate and requires G residues at the 5'-end. At this stage, ribozyme technology is of help. By flanking the RNA-coding DNA matrix with sequences expressing two hammerhead ribozymes, or a 5'-hammerhead and a 3'-hairpin ribozyme, one can obtain, after hydrolysis, the desired RNA (3). Although *in vitro* transcription allows for the design of original sequences, the incorporation of exotic nucleotides at chosen sites of the RNA chain can only be performed by stepwise solid-phase chemical synthesis of RNA using specially devised silylated synthons (4,5) This, now well-established, chemistry not only enables the design of RNA molecules containing modified nucleotides (of great interest to crystallographers, especially in the case of heavy atom-substituted residues in order to solve the phase problem), but is considerably simpler for handling on automated systems than litre-scale transcription. However, the method does not treat the polymerization of hundreds of nucleotides, a feat easily achieved by *in vitro* transcription. The chemical synthesis of an active f-Met-tRNA has been reported (6) as well as that of a fully modified tRNA (7).

Various methods can be used to purify the RNA before attempts at crystallization. The most common method is polycrylamide gel electrophoresis (PAGE), which enables the purification of short oligonucleotides as well as of hundred-mers. None the less, running several five hour long electrophoreses, followed by overnight elution is time consuming. Otherwise, for oligomers of lengths not exceeding about 40 bases, chromatography is a more accurate method and is thus more often undertaken (Table 17.1). Ion-exchange or reverse-phase high performance liquid chromatography (HPLC) enables the purification of large amounts of RNA (depending on the loading capacity of the column) within 20 minutes for a salt or acetonitrile gradient, respectively. Afterwards, the fraction containing the desired product is evaporated to dryness, preceded by a quick desalting step in the case of ion-exchange HPLC (27–29). The coupling of chemical synthesis to accurate preparative-scale HPLC constitutes a prerequisite for the production of 'crystal pure' RNAs.

2.2 Screening, heavy atom derivatives, and resolution

Resolution of these technical limitations has led to further investigations of the factors modulating crystal growth. In the vapour diffusion method, these are mainly the type and concentration of the precipitating agent, the chemical nature of the additives such as salts and organic molecules (polyamines, detergents, etc.), the buffer used to control the pH, and the temperature. Based on statistical approaches, several authors have thus developed ready-to-use sparse matrices for representative subsets of all possible conditions (30,31). In addition, the rapid handling of large quantities of RNA allows one to screen for the molecule yielding the best crystals by changing key residues. This double screening procedure was successfully employed for obtaining the first crystals of an all-RNA hammerhead ribozyme (24). More surprising were the recent observations that slight sequence modifications could often facilitate crystallization even when testing a narrow set of crystallization conditions (32). This stresses how small structural changes can modify the intimate intermolecular contacts of the crystal packing. As a consequence, sequence screening appears to be a very relevant strategy for obtaining high quality RNA crystals.

Most conditions use the vapour diffusion method in wells containing a 15–30% range of an alcoholic derivative [2-methyl-2,4-pentane-diol (MPD), polyethyleneglycol (PEG 400, 4000, 6000), 2-propanol] or ammonium sulfate as precipitant, buffered in aqueous sodium cacodylate (average pH 6.5). Thus, the specificity of a given solution is given by the nature of the additives, the role of which is either to compensate for the negative charges of the phosphate groups (polyamines, magnesium salts) or to look for heavy atom derivatives (cobalt, iridium, or rhodium complexes) to solve the phase problem. Table 17.1 presents the crystallization conditions of all RNA structures reviewed here (up to the beginning of March 1997). The crystallographic parameters

Table 17.1. Crystallization conditions for the 13 oligoribonucleotides, the hammerhead structures, and the P4–P6 domain reviewed

	RNA type	Synthesis and purification	Crystallization method	Droplet	Reservoir	T (°C)	Reference
3.1.1	[r(U(UA)₆A)]₂	Solid-phase chemical synthesis Ion-exchange HPLC	Vapour diffusion (hanging drop)	4 mM RNA 40 mM Na cacodylate (pH 6.5) 35% MPD 400 mM MgCl₂		35	8
3.1.2	[r(AUCCCCCGUGCC)·r(GGUGCGGGGGAU)]	Solid-phase chemical synthesis	Vapour diffusion (hanging drop) and seeding	5 mg/ml RNA 20 mM Na cacodylate (pH 6.5) 5% MPD 200 mM MgCl₂	30% MPD 400 mM MgCl₂ 40 mM Na cacodylate (pH 6.5)	45	9
3.1.3	[r(CGCGAAUUAGCG)]₂	Solid-phase chemical synthesis Reverse-phase HPLC	Reverse vapour diffusion (sitting drop)	1 mM RNA 12.5 mM Na cacodylate (pH 6.5) 6.2% MPD 50–60 mM MgCl₂ 0.9 mM spermine(HCl)₄	MPD	20–25	10
3.1.4	[r(UAAGGAGGUGAU)·r(AUCACCUCCUUA)]	Solid-phase chemical synthesis Reverse-phase HPLC	Vapour diffusion (hanging drop)	RNA 5 mg/ml Na cacodylate, pH 6.5 5% MPD 200 mM MgCl₂	40 mM Na cacodylate (pH 6.5) 30% MPD 400 mM MgCl₂	32	11
3.1.5	[r(CCCCGGGG)]₂	Solid-phase chemical synthesis Ion-exchange HPLC	Vapour diffusion (hanging drop)	Rhombohedral 1 mM RNA 50 mM Na acetate (pH 4.6) 1 M ammonium sulfate Hexagonal 1 mM RNA 50 mM Na cacodylate (pH 4.5) 15% MPD 7.8 mM CaCl₂	100 mM Na acetate (pH 4.6) 2 M ammonium sulfate 30% MPD	Room Room	12,13
3.1.6	[r(GUAUAUA)d(C)]₂	Solid-phase chemical synthesis Reverse-phase HPLC	Vapour diffusion (hanging drop)	2 mM complex 6 mM Co(NH₃)₆Cl₃ 100 mM Na cacodylate water (5 μl) MPD (1 μl)	60% MPD		14

Table 17.1. *Continued*

	RNA type	Synthesis and purification	Crystallization method	Droplet	Reservoir	T (°C)	Reference
3.2.1	[r(GGACUUCGGUCC)]₂			0.2 M Na citrate	100 mM Na citrate 50 mM Tris pH 8.5 2.5 mM MgCl₂ 12–32% PEG 400		15
3.2.2	[r(GGACUUUGGUCC)]₂		Vapour diffusion (hanging drop)	2 mM RNA 0.1 mM EDTA 50% (v/v)	0.2 M ammonium acetate 0.1 M Na citrate (pH 5.6) 30% PEG 4000 8 mM spermine(HCl)₄		16
3.2.3	[r(GGCGCUUGCGUG)]₂	*In vitro* transcription PAGE followed by ion-exchange HPLC	Vapour diffusion (hanging drop or sitting drop)	0.63 mg/ml RNA 25 mM Na cacodylate (pH 7.0) 100–300 mM Na cacodylate 2.5 or 5% MPD 12.5 mM MgCl₂	50 mM Na cacodylate (pH 7.0) 5 or 10% MPD 15 mM MgCl₂	37	17
3.2.4	[r(GGCCGAAAGGCC)]₂	*In vitro* transcription	Vapour diffusion (hanging drop)	2 mM RNA	20 mM NaCl 5 mM MnCl₂ 50 mM Tris·HCl pH 7.5 30% PEG 400	Room	18
3.3.1	[r(GCUUCGGC)d^{Br}U]₂	Solid-phase chemical synthesis Ion-exchange and reverse-phase HPLC	Reverse vapour diffusion (sitting drop)	0.1 mg RNA 20 mM Li cacodylate (pH 6.3) 6–10 mM Rh(NH₃)₆Cl₃ or Ir(NH₃)₆Cl₃	40% MPD	4	19
3.3.2	[r(UUCGCG)]₂[a]	Solid-phase chemical synthesis Ion-exchange HPLC	Vapour diffusion (hanging drop)	2-propanol Mg²⁺ spermine		4	20
3.3.3	[r(GCG)d(ATATA)r(CGC)]₂	Solid-phase chemical synthesis Ion-exchange HPLC		Spermine form crystal 0.8 mM complex 15 mM Na cacodylate (pH 6.9) 50 mM spermine(HCl)₄ MgCl₂	40% MPD		21

Table 17.1. *Continued*

	RNA type	Synthesis and purification	Crystallization method	Droplet	Reservoir	T (°C)	Reference
4.1.1	Hammerhead RNA:DNA ribozyme–inhibitor complex	RNA: *in vitro* transcription DNA: solid-phase synthesis	Vapour diffusion (hanging drop)	Mg form crystal 0.5 mM complex 10 mM Na cacodylate (pH 7.4) 33 mM spermine(HCl)$_4$ 2 mM MgCl$_2$ 10 mM Na cacodylate (pH 6.0) 1.9–2.2 M AS or 1.5–2.2 M LiSO$_4$ 0–100 mM spermine	40% MPD	4	22,23
4.1.2	All-RNA hammerhead ribozyme	Solid-phase chemical synthesis	Reverse vapour diffusion (sitting drop)	0.5 mM ribozyme 10 mM NH^{4+} cacodylate (pH 6.5) 50%(v/v)	50 mM ammonium cacodylate (pH 6.5) 100 mM ammonium acetate 10 mM Mg(acetate)$_2$ 1 mM spermine 23% PEG 6000 5% glycerol	20	24
4.3	P4–P6 domain of the *Tetrahymena thermophila* group I intron	*In vitro* transcription	Native crystals Vapour diffusion (hanging drop) Heavy atom derivatives Vapour diffusion (microseeding in hanging drop)[b]	50 mM K cacodylate (pH 6.0) 20% MPD 20 mM MgCl$_2$ 2:8:5 ratio of solutions *a*, *b*, *c*: [a] 250 mM K cacodylate (pH 6.0) 200 mM MgCl$_2$ 2.5 mM spermine [b] 4 mg/ml P4–P6 RNA 63 mM K cacodylate (pH 6.0) 13 mM MgCl$_2$ 160 μM Co(NH$_3$)$_6$Cl$_3$ (2eq RNA) 0.1 mM EDTA (pH 8.0) [c] 20–5% (w/v) MPD	30 17% (w/v) MPD 170 mM NaCl	30	25,26

[a] The duplex is formed between the bold-face nucleotides generating UU-3′ overhangs.

[b] Multiwave length anomalous diffraction was then performed on native crystals or heavy derivatives after soaking in Os^{3+} and/or in Sm^{3+} hexammine solution.

Table 17.2. Crystallographic parameters and refinement data for the RNA structures reviewed

	Number of reflections (% Completeness)	Space group	Unit cell parameters a ; b ; c (Å) / α ; β ; γ (°)	Resolution range (Å)	R-factor (%)	Refinement[a]	Reference
3.1.1	2437(71)	$P2_12_12_1$	34.11 ; 44.61 ; 49.11	10.0–2.25	13.1	NUCLSQ	8
3.1.2	2477(83.5)	$P4_3$	30.10 ; 30.10 ; 86.80	8.0–2.38	18	XPLOR & NUCLSQ	33
3.1.3	3147(49)	$P2_1$	41.69 ; 34.62 ; 32.13 ; $\beta = 127.6°$	7.0–1.8	19	NUCLSQ	10
3.1.4	3050(83.6)	P1	27.69 ; 30.90 ; 41.98 / 90.92 ; 103.63 ; 113.81	8.0–2.6	19.5	XPLOR	11
3.1.5	4382(62)	R32	42.40 ; 42.40 ; 131.70	10.0–1.8	20.1	NUCLSQ & XPLOR	12
	710(20.5)	$P6_122$	39.74 ; 39.74 ; 58.55	10.0–2.9	25	XPLOR	
3.1.6	1775(85.5)	R3	43.07 ; 43.07 ; 59.36	10.0–2.2	15.6	XPLOR	14
3.2.1	2082(83.9)	C2		8.0–2.0	18	XPLOR & NUCLSQ	15
3.2.2	2415(90.1)	$P4_1$	32.44 ; 32.44 ; 85.48	8.0–2.6	19.6	XPLOR	16
3.2.3	5429(95.4)	P1	29.4 ; 28.9 ; 46.5 / 98.9 ; 72.9 ; 96.1	8.0–2.4	19.1	XPLOR	17
3.2.4	1869(92.4)	$P6_522$	37.71 ; 37.71 ; 88.30	8.0–2.27	18.6	XPLOR	18
3.3.1	6093(89.1)	C2	53.80 ; 19.40 ; 50.14 ; $\beta = 109.9°$	10.0–1.5	18	XPLOR	19
3.3.2	2991(86.4)	$C222_1$	30.60 ; 34.00 ; 29.20	8.0–1.4	16.1	XPLOR	20
3.3.3	Spermine form crystal 989(82)	$P6_522$	27.25 ; 27.25 ; 177.40	10.0–2.8	18.7	XPLOR	21
	Mg form crystal 2170(94)	$C222_1$	25.89 ; 33.97 ; 57.56	10.0–1.83	15.9		
4.1.1	24660(95.6)	$P3_221$	89.70 ; 89.70 ; 185.8	8.0–2.6	25.8	XPLOR	22
4.1.2	6529(94.9)	$P3_121$	64.98 ; 64.98 ; 138.14	15.0–3.1	24.7	XPLOR	34
4.3	Native[b] 28567(90/56)	$P2_12_12_1$	74.8 ; 128.7 ; 145.9	16–2.8	22.8	XPLOR	25
	Co^{3+} derivative[b] 34551(91/35)	$P2_12_12_1$	74.8 ; 128.7 ; 145.9	18–2.5	24.2		
	Os^{3+} derivative[b] 26510(90/39)	$P2_12_12_1$	74.6 ; 128.1 ; 145.8	20–2.8	23.0		

[a] NUCLSQ (35), XPLOR (36).
[b] %Completeness: (data to 3.0 Å/last shell).

are summarized in Table 17.2. The resolution ranges of the crystals show the efficiency of these additives in yielding highly diffracting crystals. However, conditions could certainly be found involving chemicals distinct from those described.

3. Oligoribonucleotide crystals

By the end of January 1997, 13 crystal structures of ribo-oligonucleotides had been published. They can be divided into three broad categories, regular RNA helices, helices with unusual base pairs, and helices with bulges or overhangs. They will be described one by one in order of their publication within each of these categories. In each case, we will emphasize their peculiarities with respect to conformation, base pairing, hydration, ion binding, and crystal packing. All figures were drawn using the DRAWNA (37) and MANIP (38) programs.

3.1 Regular RNA helices

3.1.1 The r[U(UA)$_6$A]$_2$ helix

This is the first high resolution (2.25 Å) structure of a RNA helix (39). The RNA was chemically synthesized and its base composition mimics the part of the genomic RNA in alfalfa mosaic virus that binds the amino-terminal domain of the viral protein subunit. Crystals were obtained at high temperature (35°C), as for the Z-DNA structure of d(BrCG)$_6$ (40). The helix is kinked at two places in a sequence-independent manner, with torsion angles about P–O5′ and C4′–C5′ at the kinks in the *trans* domain instead of the usual g^- and g^+ conformations (41). The average helical parameters are very close to those of the fibre model (42): helix twist, 33(3) versus 32.7°; axial rise, 2.8(2) versus 2.81 Å; inclination, 17(4) versus 16.7°; displacement, 3.6(4) versus 4.4 Å. The average propeller twist of the A:U base pairs is large [19(4)°] comparable with the large propeller twist of the AT-rich region in DNA oligomers. Interestingly, the roll angles at the 5′U–3′A steps are much larger (14°) than those at the 5′A–3′U steps (6°). Similar trends are seen in DNA oligomers; however, in the RNA helix there is no concomitant twist variation (43).

The crystal packing presents two interhelical contacts. In one, a terminal base pair stacks against the shallow groove of a symmetry-related molecule (recently, this motif has been called 'ribose zipper'; see structure 4.2 and Fig. 17.13 in Section 4.3.3). This contact is stabilized by hydrogen bonds involving the hydroxyl O2′ atoms, the O2 of pyrimidines, and the hydroxyl O3′ atoms. In the other contact type, the deep grooves of related molecules face each other with hydrogen bonds between the C5–H or C6–H bonds and the hydroxyl O2′ or O3′ atoms. The solvent molecules continuously fill the deep groove, with the adenine Hoogsteen sites more hydrated than the uracil O4 carbonyl groups. The solvation in the shallow groove involves the adenine N3 and uracil O2 atoms equally, as well as the O2′ hydroxyl groups. The O2′ hydroxyl groups interact extensively with water molecules, although 12 of them participate in direct intermolecular contacts. Half of the O2′ hydroxyl groups are involved in one-water bridges with either adenine N3 atoms (or uracil O2 atoms) or with the ring oxygen O4′ of the next residue in the 3′ direction.

3.1.2 The helix of domain A of Thermus flavus 5S rRNA

This decamer helix was solved at 2.4 Å resolution with the inclusion of 159 solvent molecules (33). The axial rise per residue, 2.4 Å, is closer to that seen in tRNA structures than in the duplex discussed in Section 3.1.1. The deep groove of the helix is filled with solvent molecules. The helix contains a tandem of alternating U:G pairs which present an interesting pattern of water-mediated inter-residue hydrogen bonds on the deep groove side. The authors describe the crystal packing 'as a set of crossed cogwheels' as seen in the crystal structure of a DNA dodecamer (44) with the sugar–phosphate backbone of one molecule interacting in the shallow groove of a symmetry-related molecule. In the DNA dodecamer, the contacts involve the sugar–phosphate backbone and bases in the major groove (equivalent to the deep groove of RNA helices) instead.

3.1.3 The r(CGCGAAUUAGCG) helix with two G(anti):A(anti) base pairs

This structure (10) contains, within the regular RNA helix, two non-adjacent G:A pairs of the type found in tRNA crystal structures at the junction between the dihydrouridine (D) and the anticodon (AC) helices (Fig. 17.1a). Both bases adopt the *anti* conformation and the pairs have two hydrogen bonds, O6(G)···N6(A) and N1(G)···N1(A) (Fig. 17.1). The G:A pairs have C1'–C1' distances around 12.5 Å (instead of the usual 10.5 Å found in Watson–Crick pairs) and they do show a high propeller twist [as in the tRNA structures or in a DNA decamer (45)] of around −10°. Interestingly, the central A:U base pairs have very high propeller twists, of −27° and −23°, similar to those seen in AU-rich RNA (8) and AT-rich DNA (46) oligomers. Such a situation is coupled with three-centre cross-strand hydrogen bonding [between N6(A) and O4(U)]. Another noteworthy point is the high propeller twisting of the G:C duplex end base pairs (−20° and −23°). Despite the presence of three interbase hydrogen bonds, such high values for the propeller twist have been observed in an A-form DNA oligomer (47) and in a tRNA crystal structure (48). The crystal packing occurs by end-to-end stacking of duplexes, with a pseudo-helical right-handed rotation of only 9° and an axial rise of 2.9 Å. In addition, each RNA helix is surrounded by four helices. The side-by-side contacts involve numerous van der Waals interactions and 10 hydrogen bonds, all of which implicate O2' hydroxyl groups. No striking hydration pattern is present in the deep groove. Half of the solvent peaks in the shallow groove bridge base edges to the hydroxyl groups, a pattern frequently seen in RNA structures (49).

3.1.4 The dodecamer with the E. coli Shine–Dalgarno sequence

Again, in this study (11), suitable crystals were only obtained at high temperature, 32°C. The crystals contain two independent RNA dodecameric helices which display helical parameters very similar to those derived from fibre work. The crystal packing is interesting on two grounds. First, the helical rotation angle at the pseudo-step between the end-to-end stacked helices is negative, −9° and −6°, respectively. Secondly, the sugar–phosphate backbone of one molecule interacts with the shallow groove of a neighbouring molecule, in a manner reminiscent of the one present in the crystal structure of the DNA dodecamer of Timsit *et al.* (44). The two molecules

(a)

(b)

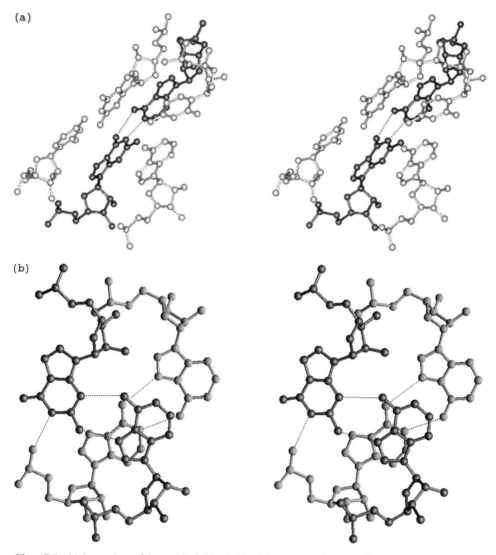

Fig. 17.1. (a) Stereoview of the standard G(*anti*):A(*anti*) base pair within a regular helix in structure 3.1.3 (10) (PDB ID number: 157d). (b) Stereoview of a sheared G:A pair stacked on a *trans* Watson–Crick/Hoogsteen A:A pair in structure 3.2.4 (18) (PDB ID number: 283d).

cross each other at an angle of 57° and the contacts involve the hydroxyl groups of one molecule with the pyrimidine O2 and the purine N3 or N2 atoms of the other (see below and Fig. 17.12). In total 18 direct intermolecular contacts are observed, which can be divided into four ribose–phosphate, five ribose–base, or nine ribose–ribose interactions. In the terminal base pairs, the O3′ hydroxyl groups participate.

3.1.5 The r(C₄G₄) helix in two crystal lattices

The self-complementary RNA octamer, r(C$_4$G$_4$), was studied in two crystal forms, a rhombohedral one (at 1.8 Å) and a hexagonal one (at 2.8 Å) (12). For the rhombohedral form, the data extended to 1.46 Å with synchroton radiation. The helices are very similar in the two forms and both are close to the standard fibre RNA helix. In the middle of the helix, the 5′C–3′G step presents a pronounced interstrand stacking of the purine rings, as is common in RNA helices. In the rhombohedral crystal form, adjacent double helices stack head-to-tail and form infinite columns (the local pseudo-twist angle between duplexes is 4°). In the hexagonal crystal form, the helices stack head-to-head (they are related by a twofold axis), which leads to a pseudo-right-handed superhelix. The packing contacts in the hexagonal form are restricted to stacking interactions between terminal base pairs, while in the rhombohedral crystal the helices are interlocked with insertions of the sugar–phosphate backbone of one helix in the shallow groove of another helix. In the latter case, the contacts are made essentially by hydrogen bonds between hydroxyl and phosphate groups. The different packing contacts might explain the lower resolution of the hexagonal form.

The RNA hydration was studied in detail in a noteworthy article (50). The O1P phosphate oxygen atoms, the *pro*-R$_P$ oxygen atoms, are systematically bridged by water molecules on both strands (51). These bridging water molecules are themselves linked to a string of bridged water molecules binding to hydrophilic atoms in the deep groove (N4, N7). On average, each O2′ hydroxyl group is hydrated by two water molecules (about the same level as the *pro*-R$_P$ oxygen atoms). Except for the terminal base pairs, a water molecule bound to the O2′ atom bridges to the exocyclic O2(Y) or ring N3(R) atoms, as is frequently seen. Around the O2′ hydroxyl groups, water molecules cluster into four regions, indicating that the bound water molecules possess additional contacts to the surrounding polar atoms, like O3′, O4′, and O2P (*pro*-S$_P$) atoms. The average distance between O2′ of residue (i) and O4′ of residue ($i + 1$) is 3.68 Å, longer than that of a typical hydrogen bond. It was therefore concluded that water molecules are better acceptors than the ring ribose O4′ atoms. The water structure in the deep groove is highly organized and displays pentagonal arrangements. In the shallow groove, at the packing contacts, the hydrophilic atoms of one duplex (especially the O2′ hydroxyl group) can replace a water molecule of the hydration network. Interestingly, compared with the same sequence with deoxyribose sugars, the ribo-oligomer is strongly stabilized ($\Delta T_m = 25.5°$) and the stabilization is enthalpy driven.

3.1.6 The alternating purine–pyrimidine r(GUAUAUA)d(C) helix

In this duplex, the 3′-terminal residue contains a deoxyribose sugar and not a ribose. Crystals were never obtained in that case, an observation that is not understood (14). The crystals belong to the rhombohedral space group R3 (one of the previous structures belongs to R32) with head-to-tail packing of helices with a negligible pseudo-twist angle at the junction. Each duplex is surrounded by three other duplexes and possesses three types of environment, two of which present packing contacts whereby the sugar–phosphate of one duplex faces the shallow groove of a neighbouring one.

Within each duplex, the roll angles alternate between the large positive values at UpA steps (13.3°) and the small values at ApU steps (3.6°) with, in both cases, large negative values for the propeller twist [−18(3)°]. Eleven of the 14 hydroxyl groups are hydrated and four of them directly contact the ring O4′ atom of the next residue. In a couple of instances, two-water bridges link the O2′ to the O2(Y)/N3(R) or the O2′ to the O4′.

3.2 Helices with unusual internal base pairs

The observation that three families of tetraloops, the −GNRA−, the −UNCG−, and the −CUUG− tetraloops, are overwhelmingly present in large RNAs like ribosomal RNAs, or self-splicing introns (52,53), encouraged investigators to attempt to crystallize them. Some of these tetraloops have been analysed by NMR methods in solution (54–56; see Chapter 18). However, since such hairpin loops are attached to an RNA duplex with Watson–Crick complementarity, at the high RNA and salt concentrations typical of crystallization conditions, they tend to form intermolecular duplexes with non-canonical base pairs in their middle instead of intramolecular hairpin loops. This led to structural information on non-canonical base pairs, albeit sometimes in somewhat unnatural environments.

3.2.1 The helix with two U:C mismatches between two G:U wobble pairs

The dodecamer GGACUUCGGUCC crystallizes with a twofold axis between the two central U:C base pairs (15). There is an additional twofold axis between adjacent dodecamers so that they stack in a head-to-tail fashion with a pseudo-twist angle of 16.1° and a rise of 2.12 Å. The helical parameters (32.1° and 2.93 Å) are typical of RNA helices with the C1′–C1′ distance at the U:C base pair increased by 1 Å, and the angle between the glycosyl bonds decreased by 15°. The U:C base pair contains only one direct hydrogen bond, between O4(U) and N4(C), with the two ring N3 nitrogen atoms bridged by a water molecule. Interestingly, two-water bridges occur between the N4(C) [or the O4(U)] and the *pro*-R_p anionic phosphate oxygen of the attached 5′-phosphate group (see Fig. 17.2a). Such a two-water bridge occurs also in the G:U pair where it involves the N7(G). In the G:U pair, a water molecule links the N2(G) and the O2′(U), instead of the N2(G) and O2(T) in DNA G:T base pairs (57). The water molecules in the deep groove of the G:U and U:C base pairs have isotropic B factors about twice as high as those in the shallow groove. The width of the deep groove, normally around 4 Å, is almost doubled in the present structure, while the width of the shallow groove, normally around 11 Å, is almost unchanged at 9 Å. In addition, it is worth noting that the hydration pattern of the G:U remains qualitatively unchanged when the nature of the flanking base pairs changes, as seen in recent crystal structures (58,59) where tandems of alternate G:U pairs take place. In the shallow groove, a water molecule is present that contacts the N2 of the G together with the O2′ and O2 atoms of the U. This pattern of hydration is typical of G:U pairs in crystals and in molecular dynamics simulations (see ref. 60 for discussion). The sequence order, 5′-UG-3′ or 5′-GU-3′, mainly affects the twist angle between the tandem G:U pairs by increasing it to 38.1°, or decreasing it to 25.3°, respectively.

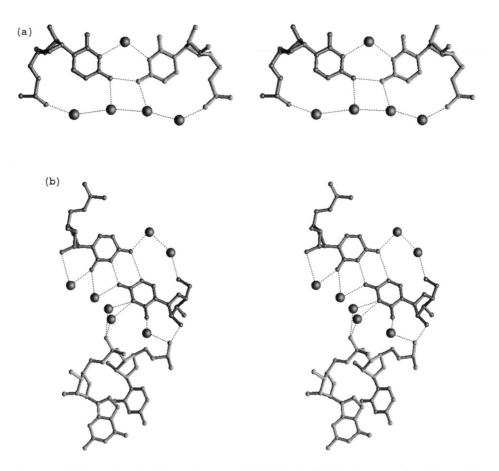

Fig. 17.2. (a) Stereoview of a U:C pair in structure 3.2.1 (U in black) with the surrounding solvent molecules (dark spheres) (15) (PDB ID number: 255d). (b) Stereoview of the Calcutta *trans* U:U pair in structure 3.3.2 with its solvent and crystal packing environment (20) (PDB ID number: 1osu).

3.2.2 The helix with two U:U mismatches between two G:U wobble pairs

Compared with the preceding structure, this structure presents two U:U pairs instead of two U:C pairs (16). Interestingly, the space group is $P4_1$, instead of C2, and the resolution is only 2.6 Å, instead of 2 Å. There are two possibilities for U:U pairs in an asymmetric environment (35), involving either $O2(U)\cdots N3(U')$ and $N3(U)\cdots O4(U')$ or $O4(U)\cdots N3(U')$ and $N3(U)\cdots O2(U')$. Both are observed in this crystal structure, but one base pair is highly twisted (about 45°) and only one hydrogen bond is formed between N3 and O2. A water molecule links the two carbonyl groups, O4, of the adjacent U:U pairs. There is a very large positive twist angle between the two U:U pairs (55°) which might be related to the presence of the double *trans* conformations for the torsion angles about C4'–C5' and P–O5' between the U:U and G:U pairs.

Unlike the situation in the previous structure (Section 3.2.1), the widths of the groove are standard, 10 and 4 Å, respectively.

3.2.3 The helix with two U:U mismatches between two G:C pairs

This structure shows variability in geometry and dependence on the neighbouring base pairs in non-canonical base pairs (17). Between the two G:C pairs, two U:U pairs are formed, both with two hydrogen bonds. The C1′–C1′ distances are around 8.5 Å, with the angle between the glycosyl bonds (normally around 50°) quite disymmetric, with one normal and the other one around 80°. The uridine contributing the carbonyl O4 atom to the hydrogen bonding has a standard value for the angle between the glycosyl bonds, while that contributing the O2 has the abnormal value, since the latter is displaced into the deep groove. It was suggested that the hydrogen bond pattern is dictated by stacking with the neighbouring G:C pairs. Again, the twist angle between the two U:U pairs is very large (55–57°) with good interstrand stacking between uridines. In the crystal, the duplexes organize themselves as stacks of two quasi-continuous helices inclined at 90° with respect to each other. The interhelical contacts occur in the shallow grooves and involve 21 direct hydrogen bonds, 13 of them made by hydroxyl O2′ atoms. Eleven other interactions are mediated by solvent molecules.

3.2.4 The helix with two A:A mismatches between two sheared G:A pairs

In this very unusual structure, the central base pairs are asymmetric A:A pairs flanked by sheared G:A pairs (18) (see Figs 17.1b and 17.5). Asymmetric A:A pairs were previously observed in crystals of nucleotides (61) and sheared G:A pairs occur in the structures of the hammerhead and P4–P6 domain of group I intron (see Section 4.2). The presence of the internal loop leads to an end-to-end bend of 34° of the helical axis. In the internal loop, the deep groove is compressed (2 Å) and the shallow groove expanded (13.5 Å). At the same time, the helix diameter is reduced from 19.6 Å (in the helical region) to 13.3 Å. Two hydrogen bonds occur in the A:A pair (N6···N1 and N7···N6). The central tandem A:A pairs stack with the ribose ring oxygen, O4′, of adenine residues interacting with the pyrimidine ring of the adjacent adenine, but there is a small interstrand overlap of the adenine bases. In the G:A pairs [with N7(A)···N2(G) and N6(A)···N3(G) hydrogen bonds], the guanine imino hydrogen binds to the *pro-R*$_p$ phosphate oxygen of the residue preceding the pairing adenine, and the second hydrogen of the adenine amino group forms a hydrogen bond to the hydroxyl O2′ of the pairing G. Interestingly, a manganese ion links the *pro-R*$_p$ oxygen of the 5′-phosphate of the adenine of the G:A pairs with the N7(G) of the adjacent C:G pair. An identical ion site is seen in the hammerhead crystal structure between a G:A and a C:G pair. The C1′–C1′ distance in the G:A pair is 9.5 Å and that in the A:A pair is 11.1 Å. The twist angle between the G:A and A:A pairs is around 60°.

3.3 Helices with bulges and overhangs

3.3.1 The nonamer r(GCUUCGGC)d(Br^{5}U)

A further attempt to crystallize the UUCG loop again led to a structure for the dimerized oligonucleotide (19). Two U:C mismatches sandwiched between two G:U wobble pairs were observed with water bridging molecules, which is commonly seen in such mispairings (15). In the U:C base pairs, the C1'–C1' distance is increased to about 12 Å and the angle between the glycosyl bonds varies between 31 and 39°. The resulting RNA helix has helical parameters corresponding to a slightly overtwisted A-form duplex with a rise of 2.8 Å and a rotation of 37.4°. The parallel helical stacks of duplexes are stabilized by interactions with solvent and ions in which rhodium hexammine cations $[Rh(NH_3)_6]^{3+}$ play a crucial role. Surprisingly, both 3'-d(Br⁵U) of a given duplex are not symmetry related. The O4' and O3' atoms of the first 3'-d(Br⁵U) forms two hydrogen bonds with the O2 and N2 atoms, respectively, of the preterminal C:G pair of the stacked adjacent duplex. The N3 atom of the same Br⁵U interacts with *pro*-R_P (U3) of a second symmetric duplex belonging to a neighbouring helical column. The pinch of this Br⁵U between two translationally related duplexes is greatly facilitated by additional coordination bonds resulting from the presence in this region of a rhodium complex. The second type of interaction between 3'-d(Br⁵U) occurs within two symmetry-related duplexes that form alternative base pairs, (N3,O4)Br⁵U:(O2,N3)Br⁵U or (O2,N3)Br⁵U:(N3,O4)Br⁵U, leading to disorder in the electron density map in this region.

3.3.2 The Calcutta base pair: the trans U:U

The hexamer 5'-r(UUCGCG)-3' dimerizes to form a helical core consisting of four canonical C:G pairs with peripheral 5'-UU dangling ends (20). In the crystal, these overhangs have the ability to be cohesive. However, their geometry is very unusual and results in the formation of infinite helical stacks of duplexes. The adjacent U:U mismatches adopt a *trans* geometry and the high resolution (1.4 Å) of the structure enables the unambiguous observation of two 'Hoogsteen-like' U:U base pairs involving C–H···O bonds (see Fig. 17.2b). The two bases are in the *anti* conformation and, since the pairing involves Watson–Crick and Hoogsteen sites, the strands are antiparallel (62). Two overhanging bases interact together according to the following scheme: the O4 and C5(H) (respectively, N3 and O4) atoms of the 5'-U (respectively, 3'-U) of the dangling end of a given duplex interact with the N3 and O4 [respectively, O4 and C5(H)] atoms of the 3'-U (respectively, 5'-U) of the symmetry-related duplex. This arrangement is further stabilized by additional water molecules. The helical parameters correspond to those of an A-form duplex although there is a high–low twist alternation from the C–G (37°) to the G–C (31°) step. The same phenomenon occurs from the U(Hoogsteen):U(Watson–Crick) to the U(Watson–Crick): U(Hoogsteen) step (33 and 22°, respectively). The contribution to the packing of a single intermolecular direct contact between O2'(U1) and *pro*-S_P (C5) of a symmetry-related duplex was noticed. This structure is remarkable because it demonstrates that RNA molecules can exploit C–H bonds for forming hydrogen bonded pairs besides the usual hydrophilic atoms. The role and importance of C–H···O/N bonds has also been shown in molecular dynamics simulations of RNA systems (63).

3.3.3 The chimeric RNA:DNA helix with a bulging dA

The structures of two crystal forms of the self-complementary chimeric RNA–DNA undecamer [r(GCG)d(ATATA)r(CGC)]₂ containing single 3'-dA bulges have been reported(21). In both forms, the crystal packing as well as the conformation of the bulging dAs are completely different. In the hexagonal form, the crystal packing forces the bulged dAs from two adjacent duplexes to stack one upon the other and to adopt *anti* conformations, with the plane of the sugars nearly parallel to the plane of the base pairs. The bulged dAs interact in the shallow groove of a column of duplexes at the junction step that exhibits a reduced twist. This architecture is stabilized by a C–H⋯O

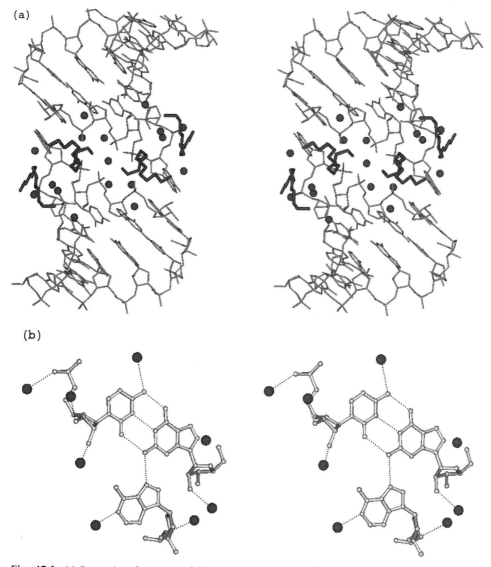

Fig. 17.3. (a) Stereoview down a twofold axis in structure 3.3.3 illustrating how spermine and water molecules stabilize the contacts between two helices (21) (PDB ID numbers: 247d, 248d). (b) Contacts between the bulging A of structure 3.3.3 and a G:C terminal pair.

bond between the C8(H) atom of the bulged dA and the O2′ of the 5′-G of the facing duplex. In the orthorhombic form, the high resolution of 1.8 Å enabled observation of two spermine molecules per asymmetric unit, which form intimate contacts with phosphate groups, base atoms, and water molecules (Fig. 17.3a). The bulge dAs adopt *syn* conformations with the ribose rings perpendicular to the plane of the base pairs. The helical columns of duplexes are largely overtwisted at the junction step, maximizing 5′-G stacking and leaving the 3′-C sufficiently free to stack upon the bulge dAs of symmetry-related duplexes. This arrangement, in addition, provides hydrogen bonding between the N7 atom of the bulge dAs and the N2 atom of the 5′-Gs (Fig. 17.3b).

4. Catalytic RNAs

In the RNA field, the crystallographic achievement of the mid-1990s is the crystallographic resolution of one catalytic RNA molecule, the hammerhead ribozyme. A couple of years later, the crystal structure of the largest RNA molecule ever solved appeared, that of the P4–P6 domain of the archetypal group I intron from *Tetrahymena thermophila*. An enormous amount of literature relating to these structures exists that cannot be considered here. We will restrict our attention to the structural information contained in those structures. Several reviews exist on both systems (64–66).

4.1 Hammerhead ribozymes

Two structures have been published, both designed so as to prevent catalysis occurring. In one, the substrate strand is made of DNA (22) and in the second the substrate strand is all-RNA with a 2′-O-methyl group at the cleavable position (34). Both structures are extremely similar and we will discuss them together, stressing only the differences. The average root mean square (rms) between the common residues in the two structures is either 0.77 or 0.92 Å, depending on the molecule chosen from the asymmetric unit. The secondary structure consists of three base paired stems linked by two single-stranded regions, normally arranged to resemble the form of a hammerhead. Between stems I and II, there is a seven–nucleotide junction ($J_{I–II}$) and between stems II and III there is a three–nucleotide junction ($J_{II–III}$). Both junctions are formed by quasi-invariant residues. The cleavage occurs 3′ of a single residue located between stems III and I. It requires divalent cations (e.g. magnesium or manganese) and the rate of the chemical cleavage step is around 1 min^{-1}. The overall structure looks like a wishbone with almost co-axial stacking of stems II and III and stem I at an angle (Fig. 17.4). The seven–nucleotide junction $J_{I–II}$ forms two structural motifs: the 5′-end adopts the classical U-turn of tRNAs and the 3′-end forms non-canonical sheared G:A pairs and a highly twisted one H-bond A:U pair with the three-nucleotide junction $J_{II–III}$.

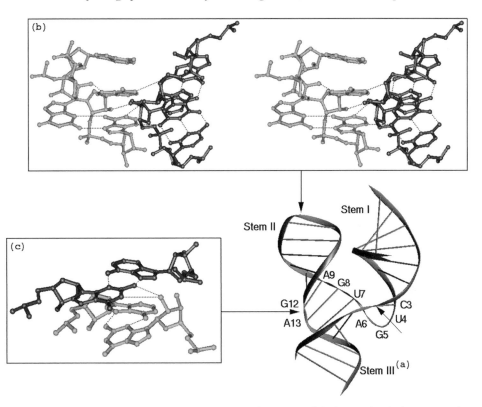

Fig. 17.4. (a) Ribbon drawing of the three-dimensional structure of the hammerhead ribozyme empha-sizing the key positions discussed in the text. Stem numbering is also indicated. (b) Stereoview of the intermolecular contact between a –GAAA– tetraloop (in grey) and two helical G:C pairs of a neighbour-ing molecule (22) (PDB ID number: 1hmh) (34) (PDB ID number: 1mme). (c) The tandem alternating sheared G:A pairs.

The occurrence of the U–turn in the hammerhead structure led to a revival of its importance. The structural similarity with the tRNA U-turns is, indeed, striking: the mean rms difference between the three hammerhead U-turns (three independent molecules in the unit cell) and the two tRNA U-turns (one in the anticodon loop and one in the thymine loop) is only 0.71 Å. The stabilizing contacts of the U-turn in the two structures are similar: a hydrogen bond between N3(U), in the hammerhead N3(U4) and an anionic 5'-phosphate oxygen (of residue $i + 3$); a hydrogen bond between the O2'(U) and position N7(R) [of residue $(i + 2)$]; and stacking between the pyrimidine ring of U with the phosphate of residue $(i + 2)$. Two points are worth mentioning. First, in the tRNA structure, the N3–H(U) atom binds to the *pro*-R_P phosphate oxygen, while in the hammerhead it is to the *pro*-S_P oxygen. This is because in the tRNA structure the loop continues in the helical conformation whereas in the hammerhead structure there is a sharp turn at residue $(i + 3)$, which adopts a conformation similar to that of the phosphate of residue 49 after the variable loop when entering the thymine stem in tRNA structures. In the hammerhead structure, the residue ($i + 3$) (U7) starts stem II. Secondly, the hydrogen bond between the hydroxyl group of U4 and the N7 site of purine A6 favours sequences of the type –UNR– for the U-turn. However, when position ($i + 2$) is a pyrimidine, there is a C–H···O bond between position C5(Y) and the hydroxyl group (63). The hydroxyl group of residue G5, which follows the U-turn, hydrogen bonds to the hydroxyl group of one ribose of the first fully formed Watson–Crick base pair in stem III.

Since the determination of the crystal structure, the structural importance of the second domain, which contains a tandem of sheared G:A pairs, has increased considerably. Indeed, such tandem G:A pairs are found in ribosomal RNAs (67) and they have been shown to play a role in the region specific for selenocysteine insertion in some eukaryotic messengers (68). The C1′–C1′ distances in the two G:A pairs, G8:A13 and A9:G12, are 9.3 and 7.9 Å, respectively, and they have a large positive twist angle between them of about 90°. They do not form an identical network of hydrogen bonds, mainly because G8 adopts a C2′-*endo* sugar pucker. Thus, N6(A13) hydrogen bonds to the ribose ring oxygen O4′ of G8, while it binds the O2′ hydroxyl in A9:G12. At the same time, the imino hydrogen of G8 does not form a hydrogen bond, while N1(G12) forms a hydrogen bond to the hydroxyl of G8 (Fig. 17.5). There is a divalent ion-binding site linking the last A of the tandem G:A (A9) and the first base pair of stem II, in fact to N7(G10.1). In the crystal structure with a DNA strand, the ion is manganese and it links the *pro*-R_P oxygen of A9 to N7 of the following G. In the second crystal structure (69) the ion is a fully hydrated magnesium that binds via water molecules in the same pocket (Fig. 17.6). Tandem G:A pairs would then be stabilized by divalent ions when followed in the 3′ direction by a purine and especially a guanine (5′-GAR-3′). A similar manganese binding site has been seen in structure 3.2.4 at a 5′-AAG-3′ sequence. In that structure, the first base pair is an A:A Watson–Crick/Hoogsteen pair followed by a sheared G:A and the Watson–Crick G:C pair. Thus, the ion-binding site is formed at the junction between a sheared G:A pair and a Watson–Crick G:C pair. However, in order to be inserted in a regular helix, this motif needs to be preceded by a non-canonical base pair with a short C1′–C1′ distance.

The G:A tandem is preceded by a non-Watson–Crick base pair between A14 and U7, which is not without some resemblance to the contact between A21 and U8 in tRNAs (see Fig. 17.13). In both cases, the N1(A) hydrogen bonds to the O2′ hydroxyl of the uridine but, while in tRNA the nucleotides are parallel, in the hammerhead structure they are antiparallel so that the N6 amino group can form a hydrogen bond to the carbonyl O2 of the uridine. In the hammerhead structure the base pair has a large propeller twist so that the imino proton at N3 is able to form a hydrogen bond to the carbonyl O6 of the following first G:A pair.

The hammerhead structure has two other base pairs with only a single hydrogen bond: the first base pair of stem III, N6(A)···O4(Y), and the last base pair of stem I, N4(C)···N3(C). The scissile bond occurs 3′ to the cytosine with its amino group in hydrogen bond contact with the N3 of the cytosine starting the seven-nucleotide junction between stems I and II.

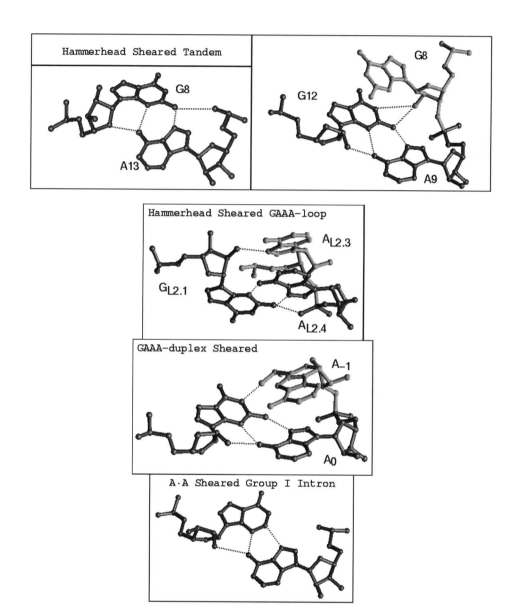

Fig. 17.5. Examples of sheared G:A pairs (from top to bottom). The hydrogen bonding network of each sheared G:A pair in the tandem alternating domain of the hammerhead structure (22) (PDB ID number: 1hmh). The sheared G:A pair in the tetraloop –GAAA–. The same sheared G:A pair as in Fig. 17.1b. An example of a sheared A:A pair from structure 4.2 (64) (PDB ID number: 1gid) to illustrate the isostericity between a sheared A:A pair and a G:A pair.

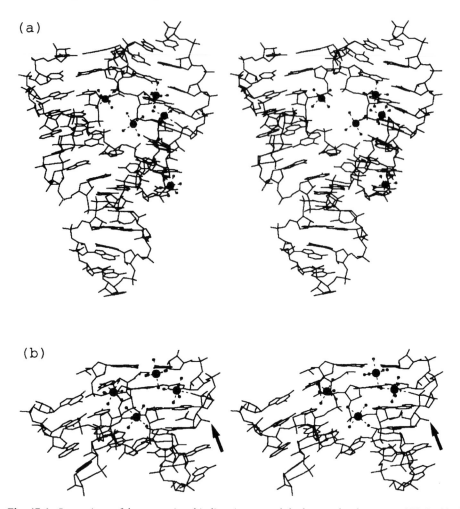

Fig. 17.6. Stereoviews of the magnesium binding sites around the hammerhead structure (69). In (a), the full structure is represented. In (b), only the core structure is shown in the same orientation as that in (a). The nucleophilic 2′-hydroxyl hydrolysing the phosphate group is marked by a black arrow. The waters of hydration are those observed after equilibration of molecular dynamics simulations of the crystallographic structure of the RNA and the magnesium positions identified by crystallography are fixed as described in ref. 70.

4.2 The P4–P6 domain of group I introns

The largest RNA structure ever solved contains, per asymmetric unit, two molecules of the P4–P6 domain (154 nucleotides), of a group I intron within a gene of the large ribosomal subunit of *Tetrahymena thermophyla*, and was solved at 2.5–2.8 Å resolution (25). The secondary structure consists of eight helices called P4, P5, P5a, P5b, P5c, P6, P6a, and P6b separated by junctions J4/5, J5/5a, J5a/5b, J5b/5c, J6/6a, and

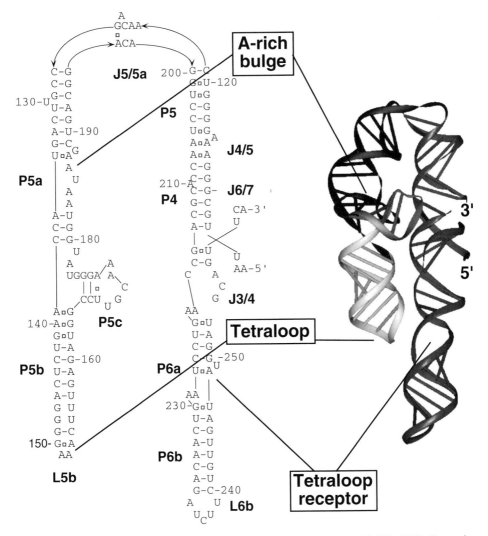

Fig. 17.7. The secondary (left) and tertiary (right) structures of structure 4.2 (25) (PDB ID number: 1gid), the P4–P6 domain of the *Tetrahymena thermophilus* group I intron. Empty squares between bases indicate non-canonical base pairings. Some important three-dimensional motifs are indicated.

J6a/6b (Fig. 17.7). The P4–P6 domain is connected to the ribozyme core by two junctions: at the 5′-end J3/4 and at the 3′-end J6/7 (Fig. 17.7). This remarkable crystallographic achievement brings a wealth of stunning interactions, contacts, and new motifs (26,71). Besides, since the *Tetrahymena* ribozyme has been exstensively studied in solution, comparisons between the crystal and solution data can be made. Furthermore, sequence comparisons have led to a model structure of the catalytic core (72), and, more recently, of the full intron (73).

Helices P6b, P6a, P6, P4, and P5 form one helical domain and helices P5a and P5b form another stack. Between the two stacked columns, there is a 150° turn made by the internal loop J5/5a so that the two helical domains are packed side by side (overall length 110 Å, width 50 Å, and thickness 25 Å). Helices P4 and P6 stack on top of each other with the 5′-entering strand J3/4 binding into the shallow groove of P6 and the 3′-leaving strand J6/7 binding into the deep groove of P4, as predicted (72,74). However, the secondary structure of helix P6 is not as expected (either because there is a crystal contact just below it or because of the presence of an additional and unnatural G at the 5′-end which forms a non-native additional base pair) since it is the 5′-end stretch that base pairs to P6(5′) [and not P6(3′)] with the 3′ dangling part forming triples in the deep groove of the unnatural helix. The internal loop J4/5 (Fig. 17.8a; see also Fig. 17.5) is important because of its predicted role in recognition of the G:U base pair in the substrate helix (not present in the crystal structure, where, instead, two symmetrically related J4/5 loops interact with each other). It consists of a tandem of

(a)

(b)

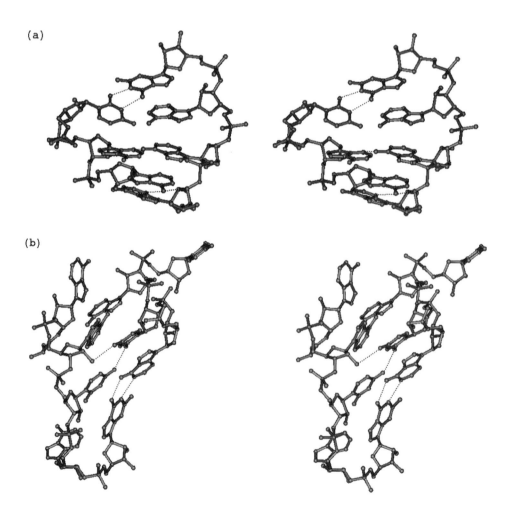

sheared A:A pairs with a third adenine stacked between the last A:A pair and a G:U pair. It is important to note that a sheared A:A pair is isosteric with a sheared G:A pair [with a C2–H···N7 hydrogen bond, instead of the more classical N2(G)···N7(A) hydrogen bond (see Fig. 17.5)].

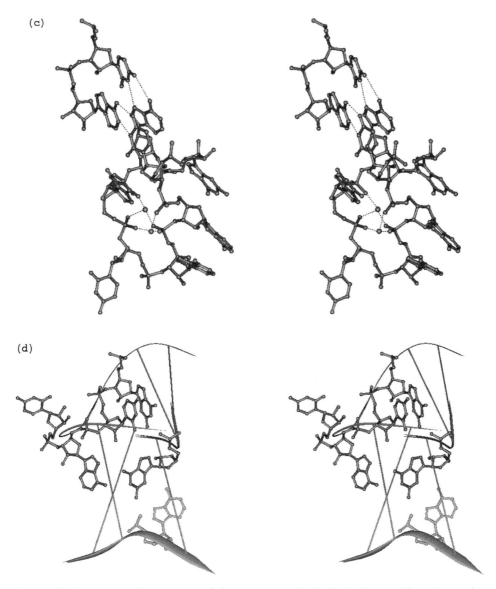

Fig. 17.8. Stereoviews of important motifs from structure 4.2. (a) The J4/5 internal loop. Notice the internally stacked single adenine between the G:U pair and a sheared A:A pair. (b) The J5/5a junction which forms the bend between the two helical domains. Notice the central one-hydrogen bond C:C pair. (c) The A-rich bulge. (d) The three-way junction between P5a, P5b, and P5c.

The turn between the two stacked column of helices is surprising (Fig. 17.8b). The last base pair of P5 is a G:C pair since the following expected A:U is not formed. Instead, the U bulges out and the A stacks in. On the other side, the first base pair in P5a is a *cis* G:A base pair, as in tRNAs, followed by a G:C pair, with the adenine preceding the G:A pair bulging out. Thus, on each strand, there is a bulging base. In this way, two cytosine residues face each other (forming possibly a *trans* C:C pair similar to the Calcutta base pair of structure 3.3.2 with $O2\cdots H-N4$ and $N3\cdots H-C5$ hydrogen bonds) and three adenine residues come close with one base stacked on a possible *trans* Watson–Crick/Hoogsteen A:A pair ($N6\cdots N7$ and $N1\cdots N6$). In P5a, there is an asymmetric bulge, the A-rich bulge (Fig. 17.8c), which is also important for the assembly of the two helical domains. Before the A-rich bulge, the last base pair is a G:C pair followed by a bulging G and, surprisingly, a *cis* Hoogsteen A:U pair. After the adenine residue of the Hoogsteen pair, an adenine stacks below the sugar of the first base pair occurring after the A-rich bulge. The A-rich bulge continues with an outside bulging U and two stacked adenine residues, the last one of which stacks below the bulging G preceding the *cis* Hoogsteen U(*anti*):A(*syn*) pair, the latter constituting the single occurrence of a *syn* base in published RNA X-ray structures. Within the closed loop formed by the A-rich bulge, two magnesium ions have been identified (5.4 Å apart) in interaction with anionic phosphate oxygens of residues of the loop (in which the phosphates point towards the interior and the bases towards the exterior). The A-rich bulge plays an important role in the contact between the two domains via A183 and A184, which bind in the shallow groove to the riboses of base pairs C109:G212/G110:C211 in helix P4, forming a ribose zipper (see below and Fig. 17.12c).

The three-way junction between P5a, P5b, and P5c (Fig. 17.8d) is of a new kind and unlike the one present in the hammerhead structure. Although the overall impression of the P5abc domain is that of a helical column, the helical axes of P5a and P5b are not colinear at the three-way junction. Helix P5c points clearly to the side. The tandem sheared A:G pairs (non-alternating) are instrumental for the left-handed positioning of helix P5a towards P5b. The junction between P5c and P5a is highly unusual. After helix P5c, two residues point towards the sugar–phosphate of the two guanine bases implicated in the tandem G:A pairs, the following A residue stacks under the last base pair of P5a, with the last U bulging out.

Analysis of the structure has revealed several ion-binding sites. Figure 17.9 shows two examples. Interestingly, the magnesium ions bind in the deep groove of the RNA helices with a preference for guanine N7 and O6 atoms, and especially stacked G:U pairs (non-alternating, see Fig. 17.9a). The alternating G:U pairs bind only cobalt hexammines, and not magnesium ions.

4.3 RNA:RNA interaction motifs

Although all RNA molecules have a well-characterized secondary structure, large RNA molecules with a biological function such as recognition or catalysis require a tertiary structure. It is a puzzle to understand how such large and highly charged molecules are able to fold into compact structures, often by themselves without the help of proteins, as in several autocatalytic RNAs (75,76). The arrangements between

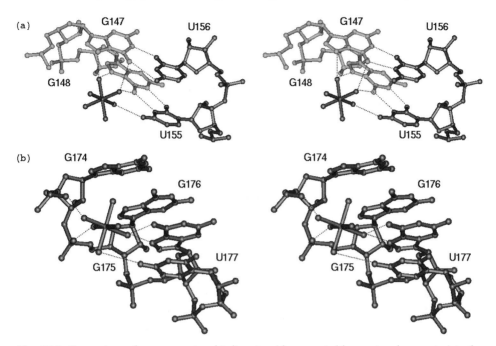

Fig. 17.9. Stereoviews of two magnesium binding sites (also occupied by osmium hexammine) in the P4–P6 domain. Both occur in the deep groove. (a) Binding to two non-alternating G:U pairs. (b) Binding to three adjacent guanine residues.

RNA molecules in the crystal packing reveal possible contacts. The RNA:RNA interaction motifs will be described here with special emphasis on those that have been observed recurrently. Large ribozymes, such as group I introns, rely heavily on two recognition motifs: the loop–loop motif and that between the GNRA family of tetraloops that interact specifically with shallow groove sides of regular or irregular helices (73). The loop–loop motif was first seen in the crystal structure of yeast tRNA[Asp] in which two anticodon loops with almost self-complementarity (–GUC–) form a small helix with a central U:U pair (35,77). The recognition motif involving the GNRA tetraloops were first predicted on the basis of sequence comparisons (72) and proved by a swap with a pseudo-knot motif in a group I intron (78). The more sophisticated recognition motif of the GAAA tetraloop family was discovered by *in vitro* selection experiments (79). Both types of recognition motifs have now been observed by X-ray crystallography: the –GNRA– motif to a shallow groove of G:C pairs in the hammerhead structures, and the GAAA motif to an internal loop in the P4–P6 domain. In crystal packing arrangements, a frequent contact is made between ribose rings of adjacent helices; this motif, termed the ribose zipper motif, has also been seen intramolecularly in the P4–P6 domain. Finally, the crystal structure of the P4–P6 domain led to the discovery of the structurally stunning A:A platforms.

4.3.1 The –GNRA– helix–loop motif

The–GNRA– tetraloops, especially –GNAA– and –GNGA– tetraloops, interact, respectively, with two consecutive C:G pairs (noted 5′-CC:GG) and a C:G stacked on an U:A pair (5′-CU:GA) so that the fourth residue of the loop (always A) binds to the second guanine of the helix (Fig. 17.10; see also Fig. 17.4) and the third residue of the loop, if an A, binds to G and, if a G, binds to A. The recognition occurs in the shallow groove of the helix and the chirality is such that the interacting bases of the loop are parallel to the purine bases of the helix (Fig. 17.10). With the Hoogsteen side of the bases in the loop oriented towards the inside of the loop, the recognition can only occur with the Watson–Crick face of the loop bases. The hydrogen bonding scheme does not involve the N6 amino group of loop adenines but, instead, N1(A) and N3(A), which, for the third base of the loop, bind to N2(G) and O2′(C), and, for the fourth base of the loop, to O2′(G) and N2(G) (see Fig. 17.5). Between the third and fourth adenine of the loop, there is therefore a rotation of about 30°.

The –GAAA– tetraloops also specifically bind an 11-nucleotide internal loop with a complex structure (see Fig. 17.11). It starts with two C:G pairs, followed by a bulged U, a *trans* Hoogsteen A:U pair, and an A:A platform. The three adenine residues of the loop interact with the 11-nucleotide motif and form a stack of four adenine bases with the first A of the A:A platform. Within the 11-nucleotide motif, the bulging U folds back and forms a one hydrogen bond contact with the first A of the A:A platform. The second base of the loop forms a *trans* Watson–Crick A:A symmetrical pair (N6···N1) with the adenine of the 11-nucleotide motif involved in the Hoogsteen pair. The third adenine base of the loop interacts with three hydroxyl groups (one to the G of the loop, one to that of the bulging U, and one to the G of the second G:C pair in the 11-nucleotide motif). Finally, the third adenine of the loop forms the network of hydrogen bonds as the fourth A of the –GNRA– tetraloop interacting with C:G pairs.

4.3.2 The A:A platform motif

This motif is unexpected because two consecutive A residues stay at about the same level and present a pronounced translational shift with the N3 of the 5′-adenine facing the Hoogsteen sites of the following adenine residue (Fig. 17.10). It has been remarked that, following the A:A platform, a G:U pair is generally found with the G 3′ to the As. In the L5c loop, a non-canonical A:U pair with a single hydrogen bond between O4(U) and N6(A) is found instead of the G:U pair. In the latter case, the segment CG of the loop self-pairs between the two molecules of the asymmetric unit, forming a small intermolecular loop–loop helix. It is interesting to remark that, in the present model of the full intron, the same L5c loop forms an intramolecular loop–loop contact with loop L2 (73).

4.3.3 The ribose zipper motif

This motif is dominated by contacts involving the hydroxyl O2′ group of two strands. It is seen in various forms in crystal packing contacts. In the ribose zipper motif, the O2′ of one residue hydrogen bonds with the O2′ and the N3(R) [or the O2(Y)] of an adjacent residue (Fig. 17.12).

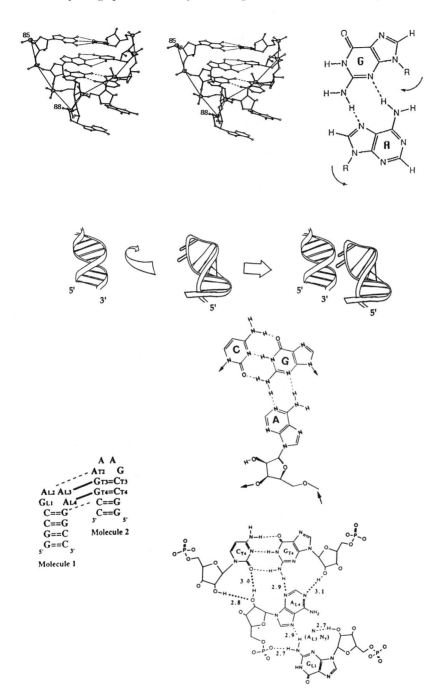

Fig. 17.10. The –GNRA–/helix recognition motif (from top to bottom). Stereoview of a –GNRA–tetraloop as modelled on the basis of chemical probing experiments (80). The rms deviation between the modelled structure and the X-ray structure is 1.54 Å. Left, an idealized sheared G:A pair. Ribbon diagram illustrating the recognition potential of –GNRA– tetraloop with the shallow groove of helices. Below, the idealized binding of a loop adenine residue and a helical G:C pair (78) and the crystallographically observed contact (81) with the intermolecular packing contact at the left.

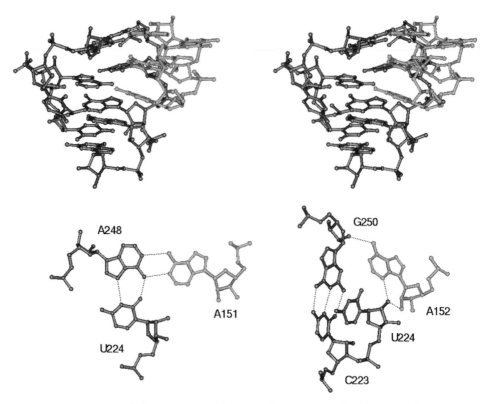

Fig. 17.11. Stereoview of the –GAAA– tetraloop motif and its 11-nucleotide internal loop receptor with two specific triple interactions.

5. Conclusions

Our knowledge of RNA structure and folding has increased considerably in recent years. Besides small RNA fragments, three large crystal structures are now available (tRNAs, hammerhead ribozymes, and the P4–P6 domain). Recurrent three-dimensional motifs, which can be either structural or folding, have been detected (the U-turn, the A:A platform, the ribose zipper, the G:A tandem, and the –GNRA–/shallow groove or the –GAAA–/internal loop contact). Together with the concept of hierarchical folding of large RNAs, the existence of recurrent RNA motifs has led to the RNA tectonics view, according to which large RNA structures can be decomposed into modules and assembled from them (82). At the atomic level, however, the variability in precise contact is subtle. For example, it is worth comparing the variability in the sheared G:A pairs (see Fig. 17.5), where the N1(G) is at times free and at other times engaged in hydrogen bonding (see Fig. 17.13, the tRNA[Ser] structure). In Fig. 17.13, a triple interaction between the deep groove of helix D and residues from the variable loop are shown in three different tRNAs. In tRNA[Phe], residue 46 forms a Watson–Crick/Hoogsteen pair with residue 22 and the phosphate of residue 9 binds to 46 and 13 (which pairs to 22). However, in tRNA[Ser], it is residue

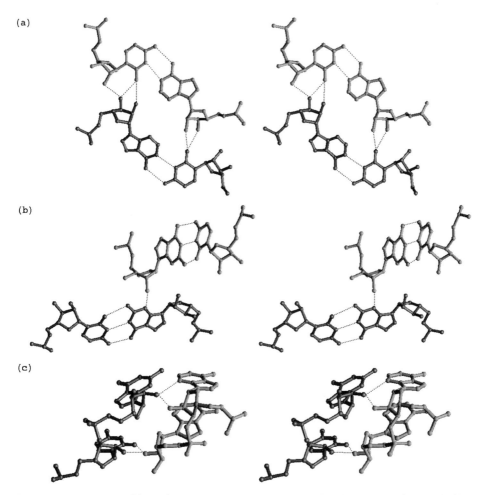

Fig. 17.12. Stereoviews of three ribose zippers. (a) In structure 3.1.1 (39) (PDB ID number: 1rna). (b) In structure 3.1.4 (11) (PDB ID number: 1sdr). (c) In structure 4.2 (25) (PDB ID number: 1gid).

9 that presents its Hoogsteen sites to the Watson–Crick sites of residue 13, which itself forms a sheared G:A pair with residue 22 (at the same time, it is the phosphate of residue 22 that binds to residue 9). Identical overall topological arrangements are, thus, coupled to microheterogeneities in the specific atomic contacts between residues underlying the stability of the global tertiary fold.

The important experimental observation is that topologically distinct molecules share quasi-identical three-dimensional micromotifs. These frequently observed motifs may have been selected during biological evolution because they are able to accommodate, within their folding, variability and heterogeneity. The building and assembly of a three-dimensional database of these motifs could therefore be a considerable help to scientists dealing with RNA for which X-ray or NMR structure models are not available.

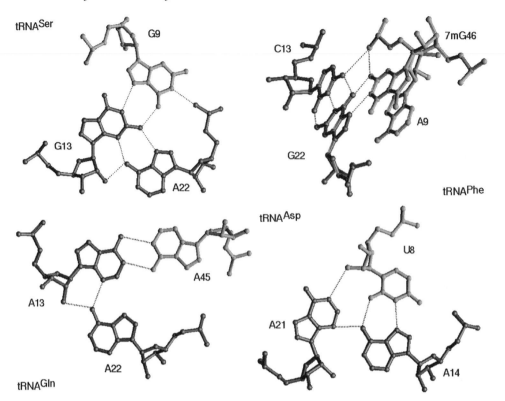

Fig. 17.13. Similarities and differences in triple contacts in four tRNAs. This triple contact occurs in the deep groove of the D helix. The triple of tRNA^Asp is shown to illustrate the similarity of the contacts between A14 and A21 and that between A13 and A22 in tRNA^Gln. There is no hydrogen bond between N1(A14) and O2'(A21) (distance 3.6 Å). The references are for tRNA^ser (83) (PDB ID number: 1ser), for tRNA^phe (84) (PDB ID number: 6tna), for tRNA^gln (85) (PDB ID number: 1gtr), for tRNA^asp (35) (PDB ID number: 3tra).

Acknowledgements

B. M. is supported by a Bourse Docteur-Ingénieur CNRS-Rhône-Poulenc-Rorer. We thank Dr Thomas Hermann for supplying Fig. 17.6 and Quentin Vicens for compiling the tables. E. W. is thankful to the Institut Universitaire de France for support.

References

1. Milligan, J.F. and Uhlenbeck, O.C. (1990) *Meth. Enzymol.* **180**, 51.
2. Chamberlin, M. and Ryan, T. (1982) *Enzymes* **15**, 85.
3. Price, S.R., Ito, N., Oubridge, C., Avis, J.M. and Nagai, K. (1995) *J. Mol . Biol.* **249**, 398.
4. Scaringe, S.A., Francklyn, C. and Usman, N. (1990) *Nucl. Acids Res.* **18**, 5433.
5. Usman, N., Ogilvie, K.K., Jiang, M.-Y. and Cedergren, R.J. (1987) *J. Am. Chem. Soc.* **109**, 7845.
6. Ogilvie, K.K., Usman, N., Nicoghosian, K. and Cedergren, R.J. (1988) *Proc. Natl. Acad. Sci. USA* **85**, 5764.

7. Gasparutto, D.L., T., Bazin, H., Duplaa, A.-M., Guy, A., Khorlin, D., Molko, A., Roget, A. and Teoule, R. (1992) *Nucl. Acids Res.* **20**, 5159.

8. Dock-Bregeon, A.C., Chevrier, B., Podjarny, A., Jonsohn, J., de Bear, J.S., Gough, G.R., Gilham, P.T. and Moras, D. (1989) *J. Mol. Biol.* **209**, 459.

9. Lorenz, S., Fürste, J.P., Bald, R., Zhang, M. and Raderschall, E. (1993) *Acta Cryst.* **D49**, 418.

10. Leonard, G.A., McAuley-Hecht, K.E., Ebel, S., Lough, D.M., Brown, T. and Hunter, W.N. (1994) *Structure* **2**, 483.

11. Schindelin, H., Zhang, M., Bald, R., Fürste, J.P., Erdmann, V.A., ????. (1995) *J. Mol. Biol.* **249**, 595.

12. Portmann, S., Usman, N. and Egli, M. (1995) *Biochemistry* **34**, 7569.

13. Egli, M. and Portmann, S. (1995) *Acta Cryst.* **D51**, 1065.

14. Wahl, M.C., Ban, C., Sekharudu, C., Ramakrishnan, B. and Sundaralingam, M. (1996) *Acta Cryst.* **D52**, 655.

15. Holbrook, S.R., Cheong, C., Tinoco, I. and Kim, S.H. (1991) *Nature* **353**, 579.

16. Baeyens, K.J., De Bondt, H.L. and Holbrook, S.R. (1995) *Nature Struct. Biol.* **2**, 56.

17. Lietzke, S., Barnes, C.L., Berglund, J.A. and Kundrot, C.E. (1996) *Structure* **4**, 917.

18. Baeyens, K.J., Debondt, H.L., Pardi, A. and Holbrook, S.R. (1996) *Proc. Natl. Acad. Sci. USA* **93**, 12851.

19. Cruse, W.B.T., Saludjian, P., Biala, E., Strazewski, P., Prangé, T. and Kennard, O. (1994) *Proc. Natl. Acad. Sci. USA* **91**, 4160.

20. Wahl, M.C., Rao, S.T. and Sundaralingam, M. (1996) *Nature Struct. Biol.* **3**, 24.

21. Portmann, S., Grimm, S., Workman, C., Usman, N. and Egli, M. (1996) *Chem. Biol.* **3**, 173.

22. Pley, H.W., Flaherty, K.M. and McKay, D.B. (1994) *Nature* **372**, 68.

23. Pley, H.W., Lindes, D.S., DeLuca-Flaherty, C. and McKay, D.B. (1993) *J. Biol. Chem.* **268**, 19656.

24. Scott, W.G., Finch, J.T., Grenfell, R., Fogg, J., Smith, T., Gait, M.J. and Klug, A. (1995) *J. Mol. Biol.* **250**, 327.

25. Cate, J.H., Gooding, A.R., Podell, E., Zhou, K., Golden, B.L., Kundrot, C.E., Cech, T.R. and Doudna, J.A. (1996) *Science* **273**, 1678.

26. Cate, J.H. and Doudna, J.A. (1996) *Structure* **4**, 1221.

27. Anderson, A.C., Scaringe, S.A., Earp, B.E. and Frederick, C.A. (1996) *RNA* **2**, 110.

28. Wahl, M.C., Ramakrishnan, B., Ban, C., Chen, X. and Sundaralingam, M. (1996) *Acta Cryst.* **D52**, 668.

29. Wincott, F., DiRenzo, A., Shaffer, C., Grimm, S., Tracz, D., Workmann, C., Sweedler, D., Gonzalez, C., Scaringe, S. and Usman, N. (1995) *Nucl. Acids Res.* **23**, 2677.

30. Doudna, J.A., Grosshans, C., Gooding, A. and Kundrot, C.E. (1993) *Proc. Natl. Acad. Sci. USA* **90**, 7829.

31. Berger, I., Kang, C.H., Sinha, N., Wolters, M. and Rich, A. (1996) *Acta Cryst.* **D52**, 465.

32. Anderson, A.C., Earp, B.E. and Frederick, C.A. (1996) *J. Mol. Biol.* **259**, 696.

33. Betzel, C., Lorenz, S., Fürste, J.P., Bald, R., Zhang, M., Schneider, Th.R., Wilson, K.S. and Erdmann, V.A. (1994) *FEBS Lett.* **351**, 159.

34. Scott, W.G., Finch, J.T. and Klug, A. (1995) *Cell* **81**, 991.

35. Westhof, E., Dumas, P. and Moras, D. (1985) *J. Mol. Biol.* **184**, 119.

36. Brünger, A.T., Krukowski, A. and Erckson, J.W. (1990) *Acta Cryst.* **A46**, 585.

37. Massire, C., Gaspin, C. and Westhof, E. (1994) *J. Mol. Graph.* **12**, 201.

38. Massire, C. and Westhof, E. in prep.

39. Dock-Bregeon, A.C., Chevrier, B., Podjarny, A., Moras, D., de Bear, J.S., Gough, G.R., Gilham, P.T. and Jonsohn, J. (1988) *Nature* **335**, 375.

40. Chevrier, B., Dock, A.-C., Hartmann, B., Leng, M., Moras, D., Thuong, M.T. and Westhof, E. (1986) *J. Mol. Biol.* **188**, 707.
41. Sundaralingam, M. (1973) *Conformation of Biological Molecules and Polymers.* T.I.A.o.S.a. Humanities, Jerusalem.
42. Arnott, S., Hukins, D.W.L., Dover, S.D., Fuller, W. and Hodgson, A.R. (1973) *J. Mol. Biol.* **81**, 107.
43. Shakked, Z. and Rabinovich, D. (1986) *Progr. Biophys. Mol. Biol.* **47**, 157.
44. Timsit, Y., Westhof, E., Fuchs, R.P.P. and Moras, D. (1989) *Nature* **341**, 459.
45. Privé, C.G., Heinemann, U., Chandrasegaran, S., Kan, L.S., Kopka, M.L. and Dickerson, R.E. (1987) *Science* **38**, 498.
46. Nelson, H.C.M., Finch, J.T., Luisi, B.F. and Klug, A. (1987) *Nature* **330**, 221.
47. Conner, B.N., Yoon, C., Dickerson, J.L. and Dickerson, R.E. (1984) *J. Mol. Biol.* **174**, 663.
48. Westhof, E. and Sundaralingam, M. (1986) *Biochemistry* **25**, 4868.
49. Westhof, E. and Beveridge, D.L. (1990) *Water Sciences Reviews 5: The Molecules of Life,* (Franks, F., ed.), p. 24. Cambridge University Press, Cambridge.
50. Egli, M., Portmann, S. and Usman, N. (1996) *Biochemistry* **35**, 8489.
51. Saenger, W., Hunter, W.N. and Kennard, O. (1986) *Nature* **324**, 385.
52. Tuerk, C., Gauss, P., Thermes, C., Groebe, D.R., Gayle, M., Guild, N., Stormo, G., d'Aubenton-Carafa, Y., Uhlenbeck, O.C., Tinoco, I., Brody, E. and Gold, L. (1988) *Proc. Natl. Acad. Sci. USA* **85**, 1364.
53. Woese, C.R., Winker, S. and Gutell, R.R. (1990) *Proc. Natl. Acad. Sci. USA* **87**, 8467.
54. Cheong, C., Varani, G. and Tinoco, I. (1990) *Nature* **346**, 680.
55. Heus, H.A. and Pardi, A. (1991) *Science* **253**, 191.
56. Nikonowicz, E.P., Sirr, A., Legault, P., Jucker, F.M., Baer, L.M. and Pardi, A. (1992) *Nucl. Acids Res.* **20**, 4507.
57. Hunter, W.N. (1987) *J. Biol. Chem.* **262**, 9962.
58. Biswas, R., Wahl, M.C., Ban, C. and Sundaralingam, M. (1997) *J. Mol. Biol.* **267**, 1149.
59. Biswas, R. and Sundaralingam, M. (1997) *J. Mol. Biol.* **270**, 511.
60. Auffinger, P. and Westhof, E. (1997) *J. Mol. Biol.* **269**, 326.
61. Prusiner, P. and Sundaralingam, M. (1976) *Acta Cryst.* **B32**, 161.
62. Westhof, E. (1992) *Nature* **358**, 459.
63. Auffinger, P., Louise-May, S. and Westhof, E. (1996) *J. Am. Chem. Soc.* **118**, 1181.
64. Jaeger, L., Michel, F. and Westhof, E. (1996) *Nucleic Acids and Molecular Biology,* (Eckstein, F. and Lilley, D.M.J., eds), p. 33. Springer Verlag, Berlin.
65. Cech, T.R. and Herschlag, D. (1996) *Catalytic RNA,* (Eckstein, F. and Lilley, D.M.J., eds), p. 1. Springer-Verlag, Berlin.
66. McKay, D.B. (1996) *RNA* **2**, 395.
67. Gautheret, D., Konings, D. and Gutell, R.R. (1994) *J. Mol. Biol.* **242**, 1.
68. Walczak, R., Westhof, E., Carbon, P. and Krol, A. (1996) *RNA* **2**, 354.
69. Scott, W.G., Murray, J.B., Arnold, J.R.P., Stoddard, B.L. and Klug, A. (1996) *Science* **274**, 2065.
70. Hermann, T., Auffinger, P., Scott, W.G. and Westhof, E. (1997) *Nucl. Acids Res.* **25**, 3421.
71. Cate, J.H., Gooding, A.R., Podell, E., Zhou, K., Golden, B.L., Szewczack, A.A., Kundrot, C.E., Cech, T.R. and Doudna, J.A. (1996) *Science* **273**, 1696.
72. Michel, F. and Westhof, E. (1990) *J. Mol. Biol.* **216**, 585.
73. Lehnert, V., Jaeger, L., Michel, F. and Westhof, E. (1996) *Chem. Biol.* **3**, 993.
74. Michel, F., Ellington, A., Couture, S. and Szostak, J.W. (1990) *Nature* **347**, 578.
75. Cate, J.H., Hanna, R.L. and Doudna, J.A. (1997) *Nature Struct. Biol.* **4**, 553.

76. Brion, P. and Westhof, E. (1997) *Annu. Rev. Biophys. Biomol. Struct.* **26**, 113.
77. Westhof, E., Dumas, P. and Moras, D. (1988) *Acta Cryst.* **A44**, 112.
78. Jaeger, L., Michel, F. and Westhof, E. (1994) *J. Mol. Biol.* **236**, 1271.
79. Costa, M. and Michel, F. (1995) *EMBO J.* **14**, 1276.
80. Westhof, E., Romby, P., Romaniuk, P.J., Ebel, J.P., Ehresmann, C., Ehresmann, C. and Ehresmann, B. (1989) *J. Mol. Biol.* **207**, 417.
81. Pley, H.W., Flaherty, K.M. and McKay, D.B. (1994) *Nature* **372**, 111.
82. Westhof, E., Masquida, B. and Jaeger, L. (1996) *Fold. Des.* **1**, R78.
83. Biou, V., Yaremchuk, A., Tukalo, M. and Cusack, S. (1994) *Science* **263**, 1404.
84. Sussman, J.L., Holbrook, S.R., Warrant, R.W., Church, G.M. and Kim, S.-H. (1978) *J. Mol. Biol.* **123**, 417.
85. Rould, M.A., Perona, J.J. and Steitz, T.A. (1991) *Nature* **352**, 213.

18

RNA structure in solution

Jacek Nowakowski and Ignacio Tinoco, Jr

Department of Chemistry, University of California Berkeley and Structural Biology Division, Lawrence Berkeley National Laboratory, Berkeley, CA 94720-1460, USA

1. Introduction

Knowledge of the structures of the molecules involved in a biological reaction is necessary for deducing a mechanism for the reaction. Structures and mechanisms can lead to rational design of drugs, and effective methods for controlling biological reactions. This chapter summarizes what has been learned about the structures of RNA molecules by solution methods, mainly nuclear magnetic resonance spectroscopy. Specific structures will be described with emphasis on the generalizations that can be deduced from the structures. Coordinates for most of the structures described are available in the Protein Data Bank and the Nucleic Acid Data Base (see Chapter 3). For a thorough discussion of nucleic acid structure with an exhaustive bibliography of structural information on nucleic acids see *Principles of Nucleic Acid Structure* (1984) by Wolfram Saenger. *The RNA World* (1993), edited by R.F. Gesteland and J.F. Atkins, and *RNA Structure and Function* (1998), edited by R.W. Simons and M. Grunberg-Manago, contain chapters on all aspects of RNA structure and function written by different authors.

2. RNA structural elements

The sequence of an RNA, the primary structure, contains all the information needed to direct the folding of the RNA into its three-dimensional structure. The three-dimensional structure will depend on the solvent and the temperature, and it may require proteins (1) or other RNA molecules to attain its active biological form. There may be kinetic barriers that slow or prevent the RNA from reaching its most stable conformation and there may be different structures with very similar stabilities. All this means that the three-dimensional structure of the RNA depends on the equilibrium thermodynamics of the solution, as well as on the kinetics of conformational changes.

An RNA structure can be described in terms of several structural elements that are considered to be the elementary building blocks of the structure. The secondary structure is the sum of these non-interacting motifs. The tertiary structure describes how the secondary structure elements interact with each other to fold into the three-dimensional structure. Secondary structure motifs include (see Fig. 18.1), single strands, double helices, bulges or bulge loops, hairpin loops, internal loops, and junctions. Tertiary structure motifs involve the interactions between the secondary structure motifs. They include (see Fig. 18.1), triple strands, quadruplexes, pseudoknots,

Elements of RNA secondary structure

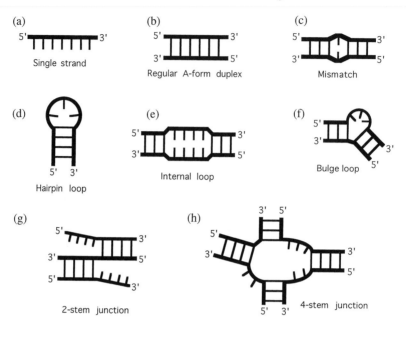

Elements of RNA tertiary structure

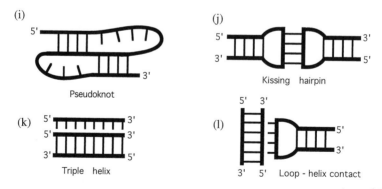

Fig. 18.1. RNA secondary and tertiary structural elements. Thick lines represent the nucleic acid backbone, and thin lines represent the bases.

kissing hairpins, bulge–loop complexes, hairpin loop–internal loop complexes, and so forth. At a more fundamental level, an RNA structure can be analysed in terms of group interactions such as hydrogen bonds, electrostatic interactions, London–van der Waals non-bonded interactions, hydrophobic interactions, etc. We will describe each structural element systematically and discuss the main forces that provide its stability.

3. Secondary structures

3.1 Single strands

The single-stranded state of an RNA is in some ways the most difficult to characterize because it is the least rigid. At low temperature the bases are stacked. As the temperature is raised the stacking decreases, but there is no abrupt transition. The stacking decreases smoothly and continuously. The most convenient way to monitor this process is from the temperature dependence of RNA absorbance at 260 nm which produces a melting curve (2). The UV melting data can be analysed in terms of the fraction of nucleotides stacked at any temperature and the degree of cooperativity of the transition. Analysis of single-stranded polyadenylic acid showed that the stacking process is exothermic and proceeds with a decrease in entropy; the single strand plus solvent is more ordered in the stacked state. There is very little cooperativity, which means that the probability of continuing a stack of bases is not very different from starting a new stack. This is in marked contrast to the highly cooperative polypeptide random coil to α helix transition. From absorption and optical rotatory dispersion measurements on the 16 dinucleoside phosphates (3), the stacking propensity of the bases was found to be, adenine and guanine stack more than cytosine, and uracil stacks the least. As the bases in a single-stranded RNA stack, the ribose rings change conformation. For unstacked single strands as well as mononucleosides the ribose rings are partly C2'-*endo* (S or south conformation) and partly C3'-*endo* (N or north conformation). As the stacking increases the ribose conformations (as measured by NMR) approach pure C3'-*endo* which is the conformation found in double-stranded A-form RNA helices (4).

The forces that determine base stacking are difficult to measure experimentally since many different interactions contribute to the net measured changes of enthalpy and entropy of the transition. The attractive forces between the bases that contribute to the negative enthalpy of stacking are London–van der Waals forces (fluctuation dipole–induced dipole and permanent dipole–induced dipole). These attractive forces are present between all molecules, but planar aromatic molecules have particularly large effects. Permanent dipole interactions can be attractive or repulsive depending on the orientation and type of bases. Two identical bases stacked directly one above the other will have a repulsive permanent dipole interaction, but sliding and rotating the bases can reduce the repulsion or even make the interaction attractive. The decrease in entropy on stacking can be explained by the loss of torsional freedom for rotation about single bonds that occurs during the transition. The solvent contributions are more difficult to assess. When water molecules surrounding each base are released on stacking, the entropy increases but the sign of the enthalpy change is not obvious. Water clearly favours stacking since adding nearly any other solvent tends to unstack the bases. The importance of water is usually described qualitatively as the contribution of the hydrophobic effect to stacking. Stacking decreases the average phosphate–phosphate distance, so increasing salt concentration increases the stacking.

Single-strand extensions at the ends of double helices stabilize the duplexes. The effect of these 'dangling' nucleotides, or single-strand overhangs on double-strand stability has been quantitated (5). The terminal base pair and the identity of the dangling base are crucial to the increase in thermodynamic stability. A single-strand extension on the 3'-end is thermodynamically more stable than an extension on the 5'-end, and a purine usually stabilizes the duplex more than a pyrimidine.

3.2 Double strands

3.2.1 A-form

Double helical RNA with Watson–Crick base pairs has A-form geometry. Crystal and fibre diffraction data indicate a right-handed helix with 11 base pairs per turn (a winding angle of 32.7°) and a 2.73 Å rise per base pair (see Chapters 1 and 2). The base pairs are displaced about 4 Å from the helix axis and have a tilt of 18° with respect to the helix axis (6). The RNA A-form double helix thus differs from B-form DNA where the helix axis passes directly through the 10 base pairs per turn. The displacement of the base pairs in RNA produces a deep and narrow major groove, and a shallow minor groove (Fig. 18.2). Nuclear magnetic resonance (NMR) studies of RNA in solution corroborate the presence of A-form geometry for the double helices (reviewed in refs 7 and 8). The ribose rings are all C3'-endo except at the ends of helices, where mixed puckers occur. There seems to be less sequence variation in A-form RNA helices than there is in B-form DNA helices. For example, continuous sequences of A:U base pairs (A-tracts) do not introduce curvature in double-stranded RNA (9).

The narrow major groove of RNA A-form helices means that protein side chains cannot easily reach into the groove to interact with the bases. This is very different from the accessible DNA major groove. The shallow minor groove also presents a very different binding pocket from that of DNA. Chemical protection studies showed that the major groove of RNA becomes more accessible for ligand binding at the ends of helices (10). Therefore, most RNA-binding sites for ligands are often formed by interrupting the helices by bulges, loops, and non-Watson–Crick pairs.

3.2.2 Z-form

RNA double helices with an alternating rCrG sequence become left-handed Z-form in 6M NaClO$_4$ or other extreme salt concentrations (11). NMR measurements (12) showed that the structure is consistent with that of Z-form DNA (13). In Z-DNA there are 12 base pairs per turn; the repeating unit is a dinucleotide (dCdG) with a winding angle of −60°. The rise per dinucleotide is 7.4 Å. In both Z-RNA and Z-DNA (see Chapter 7) each guanosine (deoxyguanosine) has the base *syn* with a C3'-endo sugar, each cytidine (deoxycytidine) has the base *anti* with a C2'-endo sugar. The transition from A-form to Z-form in RNA and from B-form to Z-form in DNA depends on salt concentration, alcohol concentration, and temperature (6,14). Bromination of guanine at C8 or methylation of cytosine at C5 favours the Z-form.

3.2.3 B-form

No evidence of a B-form RNA has been found. The different forms of RNA (and DNA) helices have characteristic and distinct circular dichroism spectra (6,11) that can be used to follow their transitions. Solvent conditions (such as high water activity)

A-form RNA helix   B-form DNA helix

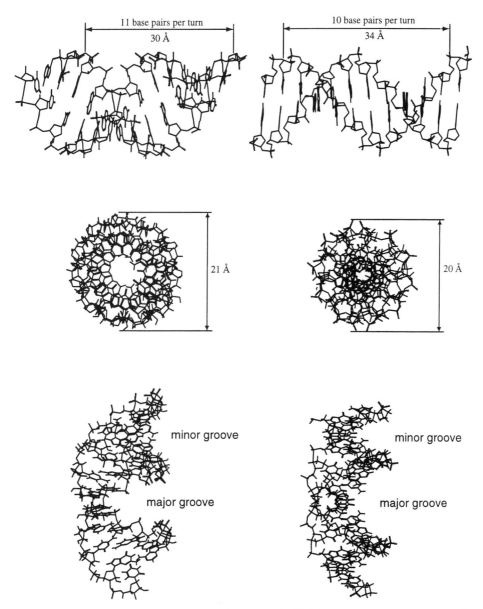

Fig. 18.2. Comparison of A-form RNA and B-form DNA double helices. Top: the side view shows that an RNA helix is shorter than a DNA helix because of tilting of the base pairs. Middle: a view down the helix axis shows distribution of nucleotides within the interior of the helices. The diameter of the helical barrel was measured between van der Waals surfaces of the outermost atoms. Bottom: the different shapes of the grooves in RNA and DNA helices.

that might have been expected to favour a B-form did not cause any apparent A- to B-form transition in RNA. When A-form RNA is destabilized, single strands seem to be more stable than B-form under the conditions tested so far.

3.3 Hairpin loops

An RNA hairpin loop occurs when a single strand folds back on itself to form a Watson–Crick-paired helix leaving some bases unpaired (Fig. 18.1). Secondary structures for RNA molecules such as ribosomal RNAs often show hairpin loops with many unpaired bases (15). These secondary structures (based on Watson–Crick-paired stems found by compensating mutations) can be misleading because the loop bases are usually in non-Watson–Crick pairings. Hairpin loops, as well as other loops, generally have compact structures, with the bases interacting with other bases, riboses, and phosphates by hydrogen bonding and stacking.

3.3.1 Tetraloops

Loops of four nucleotides are the most common hairpin loops in ribosomal RNAs. Statistical analysis of RNA sequences reveals that the most common tetraloop sequences can be grouped into three families, GNRA, UNCG, and CUNG (N is any nucleotide, R is a purine) (16). The structures of tetraloops have been extensively studied by NMR, and their thermodynamic stabilities relative to single strands have been documented (for a review see ref. 17).

   GNRA. Three members of the GNRA tetraloop family, GAGA, GCAA, and GAAA, have been characterized in solution by NMR spectroscopy (18–22). The structures of the loops are very similar to one another although they were determined in different contexts and by different groups (Fig 18.3). They all contain a sheared G:A base pair which actually turns the tetraloop into a biloop. In a sheared G:A pair both bases are *anti* with hydrogen bonding between the amino group of G to N7 of A, and the amino group of A to N3 of G (Fig. 18.4a). The Watson–Crick hydrogen bond donors and acceptors of guanine and adenine are still left available for other interactions. There is a sharp turn between the G and the N. The ribose ring conformations are C3'-*endo* (as in A-form helices) except at the turn, where the N and R nucleotides switch to C2'-*endo*. For the usual backbone torsion angles the distance between two adjacent phosphates on a strand increases by about 1 Å for a C2'-*endo* sugar relative to a C3'-*endo* one (6). The larger distance relative to the A-form that must be spanned at the turn requires the C2'-*endo* conformation. The three bases following the turn (NRA) stack on each other with their Watson–Crick faces extending into the solvent (Fig. 18.3a). The G:A pair in the tetraloop seems to be further stabilized by hydrogen bonds to the phosphate group of the A, and also by hydrogen bonds from the purine above the pair to the 2'-hydroxyl of the G (20). This type of heterogeneous network of hydrogen bonds involving bases, riboses, and phosphates is probably present in all folded RNA structures. The proposed hydrogen bonds are consistent with the effect of functional group substitution on the thermodynamic stability of the GCAA tetraloop (23). For example, changing the G of GNRA to a deoxy-G lowers the melting point by 3°C, and changing it to inosine lowers the melting temperature by 7°C. Of course every change in a nucleotide affects the hydrogen bonding and stacking interactions with its surroundings, so a thermodynamic difference cannot be

RNA tetraloops

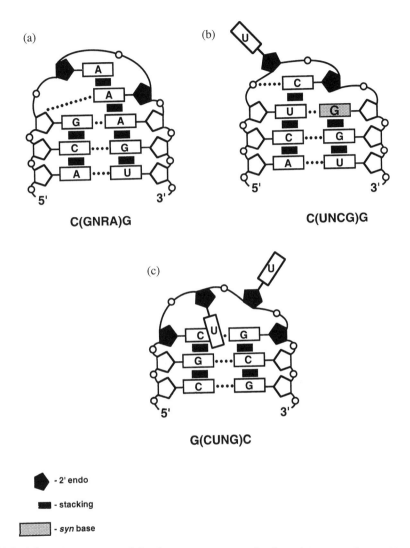

Fig. 18.3. Schematic structures of the three most common families of RNA tetraloops: (a) GAAA tetraloop belonging to the GNRA family (18); (b) UUCG tetraloop of the UNCG family (35,36); and (c) CUUG loop from the CUNG family (19). N = any nucleotide, R = purine, Y = pyrimidine.

assigned to a specific group. X-ray diffraction studies of ribozymes have proved that structures of the GAAA (24,25) and GUAA (26) tetraloops in crystals are essentially identical to solution structures.

The GNRA tetraloops are similar to the U-turn motif found originally in the anticodon loop and the T-loop of transfer RNAs (27), and more recently in the hammerhead ribozyme (24,26). In the anticodon loop of phenylalanine transfer RNA there is a sharp turn after U33 followed by the three bases of the anticodon (Gm34-A35-A36) (see Fig. 18.8d). These bases are stacked and have their Watson–Crick faces positioned for binding to the codon of the messenger RNA. There are hydrogen bonds from the U33 imino proton to the phosphate between A35 and A36, and from the 2′-hydroxyl of U33 to N7 of A35. The loop conformation and the network of heterogeneous hydrogen bonds are very similar to the GNRA motif (19).

In addition to the widespread natural occurrence of the GNRA family, the GAAA loop is found as part of the binding pocket of the RNA molecule (aptamer) that selectively binds adenosine triphosphate (ATP) (28–30). The GAAA motif forms in *trans* with the GAA furnished by the aptamer and the bound AMP pairing with the G to provide the sheared G:A base pair. The bound AMP thus completes the tetraloop; this is discussed further in Section 3.5.

An important function of GNRA loops is to provide binding sites for other RNA molecules or proteins. For instance, the GAGA member of the GNRA family is a substrate for the cytotoxic protein ricin (31) and the GAAA loop in self-splicing group I introns is involved in an intramolecular complex with an RNA internal loop; see Section 4.3 for more details (25,32).

UNCG. The UNCG family of tetraloops are exceptionably thermodynamically stable with respect to the single strand. Hairpins closed by a C:G pair and containing the loop sequence of (UUCG) frequently occur in intercistronic regions of bacteriophage T4 RNA where they efficiently terminate transcription by reverse transcriptase (33). Thermal melting studies of UNCG hairpins provided quantitative data on their stability (34). The tetraloop hairpin GGAC(UUCG)GUCC melts over 10°C higher than the tetraloop with sequence (UUUU). The standard enthalpy of formation relative to the single strand is more favourable (more negative) by 13.8 kcal/mol and the standard free energy of formation at 37°C is more favourable by 2.7 kcal/mol. Switching the closing base pair from C:G to G:C significantly reduces the stability; the difference between G(UUCG)C and G(UUUU)C is only 1.7 kcal/mol in free energy at 37°C.

The structure of the C(UUCG)G tetraloop determined by NMR (35,36) is analogous to that of the GNRA family. The UNCG tetraloop family can even replace GNRA loops functionally (37). There is a U:G pair, the second U of the loop (which can be any base) does not interact specifically with the rest of the loop, and the C stacks on the U:G pair with its C-amino group hydrogen bonded to the phosphate between the two Us. The middle two nucleotides of the tetraloop are C2′-*endo* (as in the GNRA tetraloops) and the other nucleotides are C3′-*endo*. The guanosine of the U:G pair is in a *syn* conformation and the U:G pair was originally characterized as a reverse wobble pair (35). However, the U:G pair has now been found to involve both base–base and base–ribose hydrogen bonding (36). The C(UUCG)G tetraloop structure is shown in Fig. 18.3b. The structure is quite consistent with the thermodynamic data on UNCG tetraloops (34). Clearly the U:G pair is crucial to stability, and the C-amino to phosphate hydrogen bond explains the requirement for cytosine in (UUCG). As expected, the UACG tetraloop has a very similar structure to UUCG and is also thermodynamically extra stable (38). Somewhat surprisingly, UUUG, although less stable, has a very similar structure to UUCG. The third U is moved relative to the position of the C it replaces, but otherwise the structure seems similar (35).

The TTCG or dUdUCG deoxynucleotide loop sequences were not found to be extra stable (34) and the NMR structure of deoxy-GGAC(TTCG)GTCC showed it be a flexible tetraloop with no characteristics (such as a *syn* G) of the compact, extra stable RNA loops (39).

G(CUNG)C. The G(CUUG)C tetraloop is novel in many ways (19). It could also be considered a biloop because the terminal bases of the tetraloop form a normal Watson–Crick pair. The first U in the loop folds into the minor groove and hydrogen bonds to the C:G of the loop and the G:C of the stem (see Fig. 18.3c). The second U in the loop (which can be any base) does not form any specific hydrogen bonds to the rest of the loop.

3.3.2   *Tetraloop summary*

The tetraloop structures described above have compact, relatively rigid, conformations for three constant nucleotides. The most variable nucleotide (N in UNCG, GNRA, and CUNG) is more flexible than the rest of the loop. There is extensive hydrogen bonding and stacking between the constant nucleotides. Their thermodynamic stability and characteristic NMR signatures make them useful as standard motifs in model oligonucleotides. The preferential occurrence of certain tetraloops implies their functional importance. They are clearly involved in protein recognition, in RNA–RNA binding, and in small molecule binding.

3.3.3   *Other hairpin loops*

The thermodynamic stability of a hairpin loop (quantitated by the standard free energy, $\Delta G°$) is the sum of a standard enthalpy, $\Delta H°$, and a standard entropy term, $-T\Delta S°$. Increased hydrogen bonding and stacking in the loop contributes to a favourable (negative) enthalpy change, and steric hindrance and loss of water hydrogen bonding contributes to a positive enthalpy change. The entropy change in loop formation has two main components, an unfavourable loss of strand flexibility (a loss of configurational entropy) and a favourable gain in solvent entropy by release of water in forming the loop. The loss of configurational entropy increases logarithmically with loop size (40,41), so entropy favours small loops. If configurational entropy loss were the dominant effect in loop stability, then all loops of the same size would have equal stability. This is clearly not the case. The specific sequence and interactions of each loop determine its structure and stability.

A growing number of hairpin loops has been studied by NMR; here we can only describe a few. However, thorough reviews appear periodically (17,42). The structure of an RNA hairpin that binds the coat protein of bacteriophage R17 and thus regulates protein translation has been determined (43). An octanucleotide hairpin U(AAUAACUC)G, which regulates translation of its mRNA by binding bacteriophage T4 DNA polymerase, is actually a tetraloop with the stem lengthened by two base pairs (A:C and A:U) (44). The functional relevance of this structure is emphasized by the finding that a mutant with two extra Watson–Crick base pairs binds T4 DNA polymerase with the same binding constant (45). The NMR-determined structures of the native hairpin and the mutant (46) have the same conformations for the four nucleotides in the loop, although the sequences differ significantly. The native sequence UAA(UAAC)UCG and the quadruple mutant UA**G**(CAAC)**CU**G both have a hydrogen bond from the carbonyl of the first loop U or **C** to the amino group of the last loop C. Both have the AAC loop bases stacked in the same orientation. The very similar structures and protein-binding constants suggest that the stacked AAC sequence is the recognition site for T4 DNA polymerase binding.

Not all hairpin loops studied have compact, rigid structures. Not surprisingly, a triloop C(UUU)G was found to have a flexible loop with C2′-*endo* or mixed pucker loop nucleotides (47). Even biologically relevant loops, including the TAR hexaloop from HIV-1 (48) and a hexaloop from the L11 binding region of ribosomal RNA (49,50), can be significantly flexible. Comparison of anticodon loops of the initiator tRNA from yeast tRNA^{fMet} with elongator tRNA^{Met} from *E. coli* reveals significant differences in loop flexibility (51). The loops have identical sequences (in the absence of modifications), but differ in their stem sequences and closing base pairs. The initiator tRNA has three G:C base pairs closing the loop [CAGGG(CUCAUAA)CCCUG], whereas the elongator tRNA has two out of three A:Us closing the loop [CAUCA(CUCAUAA)UGAUG]. The conformations of the two loops are similar to each other, and to the conformations seen in tRNA^{Phe} crystals (6). Both have the characteristic U-turn motif of anticodon loops which stacks the anticodon (here CAU) to position it for interaction with the (AUG) codon. But the imino proton of the A:U closing base pair of the methionine anticodon loop is exchanging too rapidly with water to appear in the NMR spectrum, and the loop riboses all have appreciable C2′-*endo* character. Also, the temperature dependence of the phosphorus spectrum is large. All these suggest greater flexibility for the elongator loop than for the initiator loop. The difference in flexibilities of the two loops is consistent with their S1 nuclease sensitivities. The loop flexibility is not likely to have great biological significance. Probably the difference in stem sequences is the crucial factor for the biological recognition (51).

3.4 Mismatches

A mismatch is defined as two apposed bases that cannot form a Watson–Crick base pair (Fig. 18.1). Mismatches disrupt the A-form helix geometry locally and affect the stability of a duplex. Nucleotide bases involved in non-Watson–Crick pairing are often incorporated into the helix and form hydrogen bonds with each other. Figure 18.4 shows various hydrogen bonding patterns between mismatched bases. Stabilities of mismatches depend on the mismatch sequence and the neighbouring Watson–Crick base pairs. All single mismatches destabilize RNA helices except for the wobble G:U pair, which is as stable as the A:U pair (52). Double mismatches can either stabilize or destabilize the duplex, depending on their sequence. The measured order of stability for double mismatch formation in 1.0 M NaCl at pH 7 is 5′-UG-3′/3′-GU-5′ > GU/UG > GA/AG ≥ AG/GA > UU/UU > CA/AC ≥ CU/UC ~ UC/CU ~ CC/CC ~ AC/CA ~ AA/AA and the closing base pair dependence is 5′-G:C > C:G > U:A > A:U (53). The most stable double mismatch 5′-UG/GU stabilizes a duplex by $\Delta G°$ (37°C) = −4.8 kcal/mol and the least stable AA/AA destabilizes the duplex by $\Delta G°$ (37°C) = +3.0 kcal/mol. There is no evidence of mismatch-induced bending in RNA helices. To illustrate some general features of mismatch geometry, we discuss in detail the structures of G:U, G:A, AH$^+$:C, and G:G pairs.

RNA mismatch pairs

purine - purine

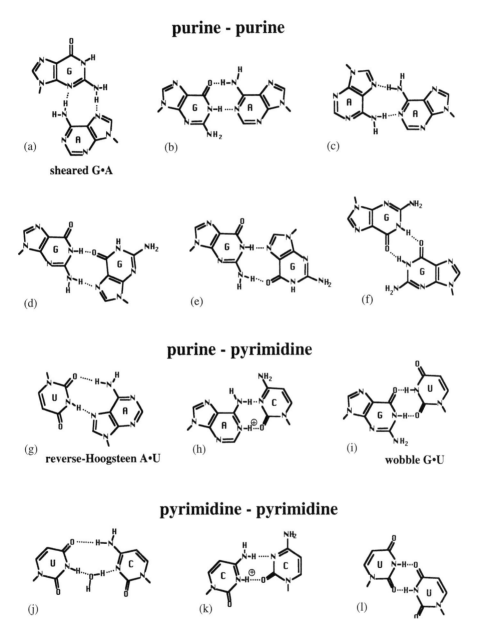

(a) sheared G•A

(b)

(c)

(d)

(e)

(f)

purine - pyrimidine

(g) reverse-Hoogsteen A•U

(h)

(i) wobble G•U

pyrimidine - pyrimidine

(j)

(k)

(l)

Fig. 18.4. Non-Watson–Crick base pairs observed in solution structures of RNA. (a) Sheared G:A pair (18,58); (b) G:A (59); (c) A:A (60,61); (d)–(f) G:G mismatches (29,30,64); (g) reverse Hoogsteen A:U pair (60,61); (h) protonated A+:C pair (62,68); (i) wobble G:U pair (36,72); (j) water-mediated U:C pair (72); (k) protonated C+:C pair (53,75); (l) U:U mismatch (53,75).

3.4.1 G:U wobble pair

G:U mismatches are very common. Replacement of standard Watson–Crick base pairs by 'wobble' G:U pairs (Fig. 18.4i) perturbs the A-form helix only slightly (54). The distance between the C1′ atoms across the minor groove is increased from about 10.6 Å to 12.8 Å (6), and the stacking, twist, and rise of the G:U pair are slightly changed (36). This perturbation can produce sites that facilitate ligand binding. A single G:U pair forms a preferential site for binding Mn^{2+} ions in the major groove of an RNA helix, as indicated by broadening of the NMR resonances caused by the paramagnetic ion (55). Magnesium ions presumably bind in a similar fashion. X-ray diffraction has revealed binding pockets for Co^{3+}–hexammine and Os^{3+}–hexammine in the major groove of an RNA helix with two adjacent G:U base pairs (56).

3.4.2 G:A mismatch

G:A pairs commonly occur in RNAs as a tandem mismatch. The stability and structure of tandem G:A mismatches depend on the closing Watson–Crick base pairs (57). The solution structure of the RNA duplex r(GGC<u>GA</u>GCC)$_2$ shows that the G:A pair is in a 'sheared' conformation with hydrogen bonds between G amino and A N7, and G N3 and A amino (Fig. 18.4a) (58). There is a strong cross-strand G–G and A–A stacking which contributes to the high stability of the motif. When the closing Watson–Crick base pairs are changed, the same tandem mismatch in r(GCG-<u>GA</u>CGC)$_2$ forms an imino-hydrogen bonded structure (Fig. 18.4b) (59) with intrastrand G–A stacking. The two motifs create very different distortions of the A-form helical geometry. The helix is much wider for the G<u>GA</u>C than for the C<u>GA</u>G motif, with the distances between G and A phosphates on opposite strands being 20.4 and 12.5 Å, respectively (the regular A-form distance between opposite strand phosphates is 17.5 Å). The 5′-GA-3′ step is underwound (21°) at the G<u>GA</u>C mismatch and overwound (81°) at the C<u>GA</u>G.

G:A pairs are abundant in biological RNAs. A sheared G:A motif was found in the solution structure of the GCAA hairpin loop (18), the loop E family (see Section 3.5.1) (60,61) and loop A of the hairpin ribozyme (62). The imino-hydrogen bonded G:A pair was observed in a Rev response element (RRE) RNA (63,64) and in the crystal structure of r(CGCGAAUUAGCG) (65). Both G:A hydrogen bonding patterns are present in the structure of a flavin mononucleotide (FMN) aptamer (66).

Functionality of the sheared G:A pairs is attributed to the availability of the Watson–Crick faces of the bases for additional hydrogen bonding. An exposed N7 of G can also form a divalent metal-binding site, as is the case in the highly conserved tandem G:A pair seen in the crystal structure of the hammerhead ribozyme (26,67).

3.4.3 AH$^+$:C pair

The protonated A:C pair is geometrically similar to the G:U wobble pair. It has been observed in a lead-dependent ribozyme (68), loop A of the hairpin ribozyme (62), and in the structure of a small hairpin loop (69). There are hydrogen bonds formed between the A amino and N3 of C, and between the protonated N1 of A and O2 of C (Fig. 18.4h). The evidence for the protonation of adenine N1 comes from the change in chemical shift of the C2 carbon, which can be monitored as a function of pH (62,68). The pH titration curves from these studies indicate that the pK_a of the N1 nitrogen is significantly shifted and has a value of 6.2–6.4 (free adenosine has a pK_a near 4). Similar protonated forms of adenine were observed in a crystal structure of ApA dimers (6). The ribose rings of both mismatched nucleotides are in the usual C3′-*endo* conformation and the pair is incorporated into a helix without significant distortions of the A-form geometry. The AH$^+$:C pair is 2 kcal/mol ($\Delta G°$) less stable than an A:U base pair at 37°C (69).

3.4.4 G:G pairs

Three different hydrogen bonding patterns have been observed for G:G mismatch pairs (Fig. 18.4 d–f). G:G mismatches are common in structures of aptamers identified through *in vitro* selection for binding of various ligands. The structure of an ATP aptamer (29,30) revealed two different G:G pairs bonded as shown in Fig. 18.4d and e. A G:G pair (Fig. 18.4d) was observed in the arginine/citrulline and flavin mononucleotide (FMN) aptamers (66,70). The imino-hydrogen bonded G:G pair (Fig. 18.4f) is present in the RRE RNA internal loop (63,64). G:G mismatches easily dimerize in G-rich RNA sequences forming ultra-stable tetrameric structures (G quartets) consisting of four hydrogen bonded guanine bases (71).

3.4.5 Other mismatches

Many other non-Watson–Crick base pairs have been found in RNA. An A:A pair and reverse Hoogsteen A:U pair (Fig. 18.4c and g) are formed in the eukaryotic 5S rRNA loop E (61) and in the sarcin/ricin loop from 28S rRNA (60). The structure of the U:C mismatch shown in Fig. 18.4j was found in crystals of RNA duplexes containing the internal loop sequence 5′-UUCG (72,73). The U:C pair involves two hydrogen bonds, one directly between the pyrimidine bases and another one mediated through a bridging water molecule. Incorporation of solvent into the hydrogen bond network spreads the bases apart and ensures a good fit of the U:C mismatch to the A-form helical geometry. NMR studies of a duplex r(GGAC$\underline{UC}$GUCC)$_2$ suggest that the structure of the U:C pair in solution is similar to the crystal structure (74). The structures of U:U and C:C$^+$ mismatches shown in Fig. 18.4l and k are strongly supported by one-dimensional NMR data and thermodynamic studies of short duplexes containing these pairs (53,75).

3.4.6 Mismatch summary

RNA bases have the ability to form hydrogen bonded pairs in any combination (for the full list of possible base pairs with two hydrogen bonds see ref. 6, or Appendix I of ref. 76). Mismatches are stabilized by inter- or intra-strand stacking interactions and hydrogen bond networks. Introduction of mismatches into an RNA helix does not change the global A-form geometry to a large extent.

3.5 Internal loops

An internal loop contains nucleotides that cannot form Watson–Crick pairs on both strands of a regular RNA duplex (Fig. 18.1). If the number of unpaired nucleotides is the same on each strand, the internal loop is symmetric. According to this definition, the single and double mismatches discussed in Section 3.4 constitute the smallest symmetrical internal loops. Stabilities and structures of loops vary significantly depending on the loop size and sequence. UV melting studies of internal loops containing unpaired adenines showed that symmetric loops were more stable than asymmetric loops of the same size (77). RNA duplexes containing asymmetric loops A_5,A_n and U_5,U_n ($n \neq 5$) had slower electrophoretic gel mobilities than corresponding symmetric loops, or a regular RNA duplex (78). Slower electrophoretic mobilities can be a consequence of an intrinsically bent conformation or higher flexibility of asymmetric loops, but more detailed structural analysis is needed to distinguish between these effects.

In solution, internal loops can be flexible and disordered, but many have well-defined rigid structures, with non-Watson–Crick base pairs, base–sugar hydrogen bonding, and extended stacking interactions. In order to illustrate the structural complexity of RNA internal loops, we will describe the structures of two classes of internal loops. The first class, represented by loop E from 5S rRNA and the hairpin ribozyme loop A, constitutes loops that are structurally ordered and relatively rigid in solution. NMR spectra of oligoribonucleotides containing these loops normally have well-resolved, sharp proton resonance lines. Internal loops belonging to the second class are disordered and flexible in solution by themselves, but become structured upon binding an external ligand. Examples of the latter class include the ATP aptamer and the 3′-UTR regulatory element of human U1A protein.

3.5.1 Structured loops

Loop E family. Asymmetric internal loops of nine nucleotides with the sequence

$$5'\text{-G A A Y}$$
$$\bullet \quad \bullet \quad \bullet \quad \bullet$$
$$\text{A U A Y-}5'$$
$$\text{N}$$

are found in several biologically important RNAs. This motif is highly conserved in eukaryotic 5S rRNAs (loop E) and 23S/28S rRNAs (sarcin–ricin loop) It also occurs in viroid RNAs and in the hairpin ribozyme. Evidence from UV cross-linking, chemical modification studies, and NMR spectroscopy indicate that the structures of these loops are very similar and thus can be categorized as a single family (Fig. 18.5). Detailed NMR studies of loop E (61) and the sarcin–ricin loop (60) showed that this motif contained several non-Watson–Crick base pairs and a single bulged base. In the structure of loop E, a sheared G:A mismatch (Fig. 18.4a) is stacked on a reverse Hoogsteen A:U pair (Fig. 18.4g) and a non-conserved G residue is bulged out of the helix. The loop is closed by A:A and U:U pairs. All ribose residues are C3′-*endo* (A-form like) except for the bulged G and adjacent A residues. The backbone of the loop is severely distorted at the G:A/U:A step (Fig. 18.5). Electrophoretic gel mobility measurements on RNA duplexes containing a eubacterial loop E indicate that this symmetric (seven nucleotides in each strand) loop introduces a directional bend and an increased helical twist in the A-form geometry (78). In summary, the loop E-like structure is highly ordered and roughly resembles a continuous A-form helix. Its diverse functionalities are most likely accomplished by the accessible sides of non-standard base pairs that are accessible for intermolecular binding (79).

(a)

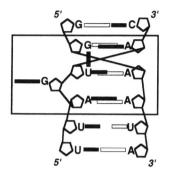

5S rRNA (Loop E)

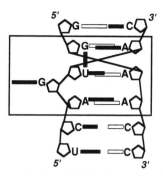

28S rRNA (Sarcin Loop)

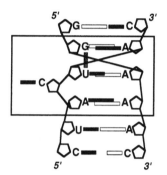

PSTV RNA

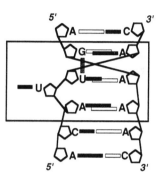

Hairpin Ribozyme (Loop B)

(b)

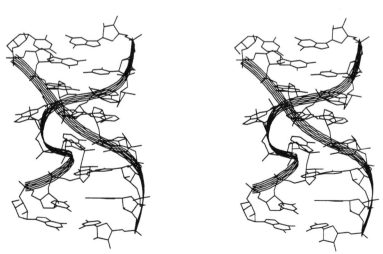

Fig. 18.5. (a) Family of loop E-like sequences from different RNAs. (b) Stereoview of the three-dimensional structure of loop E from 5S RNA (61).

Hairpin ribozyme loop A. Loop A of the hairpin ribozyme is a symmetric internal loop of eight nucleotides that contains the cleavage site (80). The structure of the loop solved by NMR (62) shows that the guanine 3′ to the cleavage site forms a sheared G:A base pair and that the cytosine residue immediately 5′ to the cleavage site is involved in a protonated AH⁺:C base pair. The loop is stabilized by extended A-form stacking between residues adjacent to the cleavage site and by several cross-strand base to sugar hydrogen bonds that are formed by residue G8. They include a hydrogen bond from the G8 carbonyl oxygen to the 2′-OH of G20 and from the G8 amino or imino protons to the O4′ of U21. The structure of the loop has an overall A-form helical shape with a widened major groove.

3.5.2 Flexible loops

ATP aptamer. The technique of *in vitro* selection has been used to isolate RNA aptamers that bind to biological cofactors with high affinity and selectivity (81,82). An aptamer for ATP (or AMP) was found to contain a 12-nucleotide asymmetrical RNA loop flanked by double helical regions (Fig. 18.6a) (28). Two high resolution solution structures are available for this motif, revealing several unusual properties (29,30). In the absence of exogenous AMP, NMR spectra of the aptamer showed that only Watson–Crick base pairs in the flanking helices are formed while the loop itself is largely unstructured. Upon addition of AMP, sharp resonances of *all* imino hydrogens in the loop appeared, indicating formation of a structured core with an extensive network of hydrogen bonds. In the complex, AMP is tightly docked in a binding pocket forming a sheared G:A base pair with the residue G8 (Fig. 18.6a). This base pair, along with residues A9 and A10, forms a GNRA tetraloop fold (see Section 3.3) which is stabilized by stacking on a G11:G7 pair (Fig. 18.4d). Yet another non-Watson–Crick pair, G30:G17 (Fig. 18.4e), forms in the binding pocket providing a stacking platform for residues A12 and U16. The backbone of the 11–base loop forms the shape of a Greek letter ζ, with the middle arch corresponding to the AMP-binding site (29). The entire motif is stabilized by extensive base–base stacking and hydrogen bonding within the loop. The AMP–loop complex introduces a bend of about 100° between the two helical stems according to the NMR structure (30).

3′-UTR RNA. An asymmetrical RNA loop of eight nucleotides (Fig. 18.6b) is involved in regulation of expression of a human U1A protein. The structure of the free loop and the loop bound to a ribonucleoprotein (RNP) domain have been solved by NMR spectroscopy (83,84). The structure of the free RNA indicates that the single-stranded loop region contains local stacking interactions in the context of a generally flexible structure (84). Protein binding orders the internal loop and changes the overall shape of the RNA. In the complex, the RNA is severely bent, with the single-stranded nucleotides positioned across the surface of a four-stranded β sheet. There are no base–base hydrogen bonds formed in the loop, but most of the residues are involved in stacking interactions. The RNA–protein interface is highly structured and consists of extensive intermolecular hydrogen bonds and hydrophobic interactions (83).

Other examples of RNA internal loops that become ordered upon binding to an external ligand include aptamers for flavin mononucleotide (FMN) (66), arginine/citrulline (70), RRE Rev internal loop (64), and an aminoglycoside binding site from *E. coli* 16S rRNA (85).

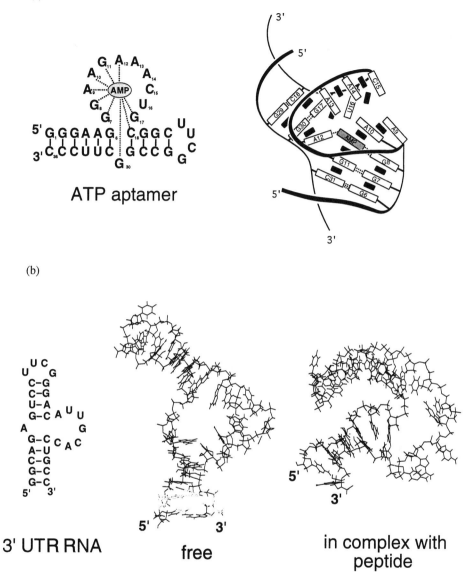

Fig. 18.6. RNA internal loops involved in intermolecular binding. (a) The sequence and overall shape of the ATP aptamer (29,30). Dotted lines represent NOE contacts used in structure determination. The figures are Fig. 1 and Fig. 6 from ref. (29). (b) The structure of the 3'-UTR control element of the human U1A protein, free (left) and bound (right) to its target peptide (83,84).

3.5.3 Internal loop summary

Internal loops are extremely important in the function of RNA molecules. Depending on their size and sequence, internal loops may introduce sites of local flexibility and bending in the RNA double helix. Many internal loops form compact and rigid structures with non–Watson–Crick base pairs, sugar–base, and sugar–sugar interactions. Loops involved in protein recognition or ligand binding are often disordered and flexible in solution but become structured upon binding to ligand.

3.6 Bulge loops

A bulge loop is defined as one or more nucleotides that interrupt one strand of an otherwise continuous Watson–Crick-paired double helix (Fig. 18.1). The stabilities of RNA bulges depend on the size and the sequence of the unpaired region and, additionally, on the sequence of adjacent Watson–Crick base pairs (86). UV melting experiments showed that the stability of bulges containing unpaired adenosines or uridines depended on the sequence of the flanking Watson–Crick base pairs. For instance, a loop of three As was more stable by about 2 kcal/mol of free energy when placed between 5′-**C**–A$_3$–**C** instead of 5′-**G**–A$_3$–**G** adjacent nucleotides (86).

It is well established that bulges bend the A-form double helix (87–89). The extent of bending depends on several factors including the size and sequence of the bulge, the sequences of flanking base pairs, and the presence of divalent metal ions (89–91). Transient electric birefringence measurements on RNA duplexes containing single bulges of a sequence A$_n$ or U$_n$ (where n = 1–6) showed that the magnitude of helix bending increased with increasing size of the bulge. In the absence of Mg^{2+}, for both A$_n$ and U$_n$ series, the angle increment varied from ~20° to ~8° per added nucleotide as n was increased from 1 to 6. The total value of the bend ranged from 7° to 93° (89). In all cases studied, uridine bulges induced smaller bends than adenosine bulges of the same size. The effects of mixed-sequence bulges on helix bending have not yet been studied systematically (91).

3.6.1 Single-nucleotide bulges

An NMR structure of a single adenosine bulge in the stem of a hairpin loop showed that the unpaired base was intercalated into the helix, creating a small kink in an otherwise normal A-form helix (43). The helix axis was bent away from the bulge to allow base stacking on the strand opposite the unpaired A. The intercalated adenine was also stabilized by stacking on adjacent Watson–Crick pairs. The bulge region was more dynamic then the remaining part of the helix as evidenced by a mixed C2′/C3′-*endo* conformation of ribose sugar puckers and by broad imino resonances from flanking Watson–Crick base pairs.

NMR studies of a duplex r(CUGG<u>U</u>GCGG),(CCGCCCAG), which contains a single unpaired uridine residue, provided evidence that the extra U was looped out of the helix (92). Model building studies indicated that an extrahelical residue did not introduce significant bending into the duplex.

A larger number of structural studies exist for DNA single-nucleotide bulges. The equilibrium between the stacked-in and looped-out conformations of single-nucleotide bulges in DNA is dependent on temperature, the unpaired residue, and the sequence of the adjacent base pairs (93).

3.6.2 TAR element from HIV

The TAR element (*trans*-activation response element) from the HIV-1 genome consists of a six-nucleotide hairpin loop and a stem with a three-nucleotide bulge

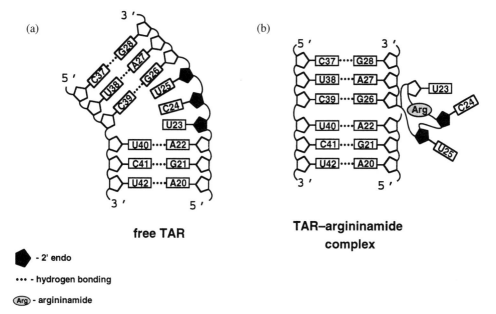

(a)

(b)

free TAR

TAR–argininamide complex

◆ - 2' endo

••• - hydrogen bonding

(Arg) - argininamide

Fig. 18.7. The structure of an HIV-1 TAR three nucleotide bulge (94,95). (a) Bent conformation of the TAR bulge. (b) TAR bulge bound to argininamide.

(Fig. 18.7). The solution structures of HIV-1 TAR elements bound to argininamide or to a 37 amino acid peptide (ADP-1) have been solved by NMR, providing detailed structures of the bulge in a free and complexed RNA (94,95). In the unbound RNA, the nucleotides in the bulge (U23-U25, Fig. 18.7) are flexible but stack, as evidenced by intranucleotide NOEs, consistent with the helical geometry. The stacked structure within the bulge induces bending in the helix axis (90,96). The conformation of the bulge changes significantly upon ligand binding. The stacking between nucleotides A22–U23–C24 is disrupted, and the A-form stems flanking the bulge stack coaxially. In the argininamide–TAR complex, U23 forms a major groove base triple with the U38,A27 pair, and the argininamide is positioned below the triple forming hydrogen bonds with G26 (94). The ADP-1–TAR structure does not provide evidence for the U23–U38–A27 triple formation, but also positions an arginine residue within hydrogen bonding distance from G26–N7. Binding of both the Tat peptide and argininamide straightens the bend introduced by the bulge in the unbound RNA (96).

3.6.3 Bulge loops summary

Unpaired nucleotides in an RNA bulge loop can be positioned inside or outside the helix. The incorporation of unpaired bases into the duplex introduces a directional bend into a regular A-form helix. The amount of bending depends on the size and the sequence of the bulge, and the presence of divalent metal ions. Bulges in which all residues are looped out of the helix allow coaxial stacking of the helical stems and do not bend the helix. Larger bulge loops can form complex binding pockets that serve as RNA–RNA or protein–RNA recognition sites.

3.7 Junctions

RNA junctions are broadly defined as regions where two or more interconnected double helical stems come together (Fig. 18.1). Many types of junctions are possible depending on the number of stems and the size of the branch region (Fig. 18.8). Junctions play an important role in positioning helical domains at specific angles, thus determining global shapes of RNA molecules.

An important force stabilizing multibranched junctions is provided by coaxial stacking between helices. A coaxial stack is formed when the terminal base pairs of two helices are in van der Waals contact forming a straight and quasi-continuous helical domain. The free energy of an end-to-end stacking between two duplexes follows essentially the same sequence dependence as Watson–Crick pairing in a continuous helix, but is usually more favourable. UV melting studies showed that the stacked interfaces can contribute from −0.6 to −1.6 kcal/mol extra stability in free energy than the equivalent nearest neighbour pairs in a continuous helix (97). In more complicated junctions, the stacked stems are often additionally stabilized by hydrogen bonds to unpaired nucleotides in the junction. Based on several structures discussed later in this section, it seems that relatively rigid coaxial stacks are formed at junctions containing an even number of branches. It is difficult to predict which pair of helices will stack coaxially from the nucleotide sequence alone.

RNA and DNA junctions are sites of extra counterion association owing to a high phosphate charge density (98). Junctions from tRNA, group I introns, and the hammerhead ribozyme form specific metal-binding pockets as determined by X-ray diffraction (26,56,67,99). The release of bound counterions upon changes in junction geometry can be an important factor determining stability and the function of RNA junctions (100,101).

3.7.1 Two-way junctions

A highly conserved two-way junction is part of the catalytic core of self-splicing group I introns. The junction consists of two double helical stems (P4 and P6) flanked by single-stranded overhangs at the 3′ and 5′ ends of the branch point (Fig. 18.8a). Comparative sequence analysis and a large amount of biochemical data available on group I introns led to a three-dimensional structure (102). In the Michel–Westhof model, the P4 and P6 stems stack coaxially, forming a continuous helical domain. The right-handed rotation between the stacked helices places the nucleotides of the single strands in opposite RNA grooves where they can form hydrogen bonds with the stems. Two residues from the 5′ single-stranded end bind in the minor groove forming base triples with the P4 stem, and two nucleotides from the 3′ unpaired strand form base triples in the major groove of the P6 stem. The Michel–Westhof model of the P4/P6 region has been shown to be essentially correct by the crystal structure of a 154-nucleotide P4/P6 domain (25).

Oligonucleotide models of the P4/P6 junction have been studied in solution. An NMR structure of a small RNA oligonucleotide containing shortened versions of P4 and P6 stems and the 5′ overhang showed that the stems formed a coaxial stack in solution. The rotation at the junction of the helices was right-handed and almost twice as large as the rotation between two Watson–Crick base pairs in a regular

RNA junctions

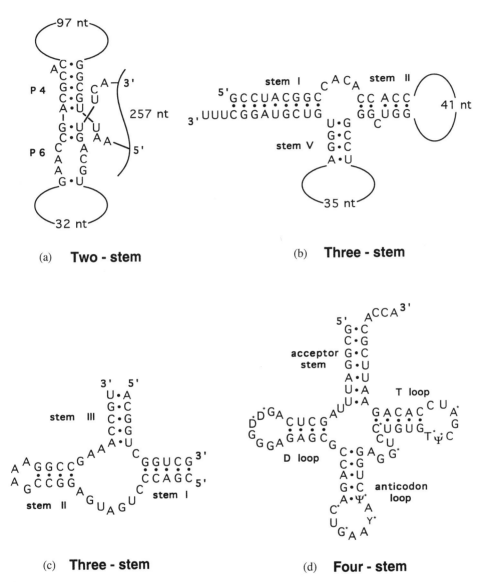

Fig. 18.8. Multibranched RNA junctions. (a) Two-stem P4/P6 junction from *Tetrahymena thermophilus* group I intron. (b) Three-stem junction from 5S rRNA. (c) Three-stem junction of hammerhead ribozyme. (d) Four-stem junction from tRNA[Phe].

A-form duplex (103). As expected from the Michel–Westhof model, nucleotides from the 5′ overhang formed nucleoside triples in the minor groove of P4 (104). The term 'nucleoside triple' is used because the hydrogen bonding involves a ribose as well as the bases. An NMR structure of the junction containing both 3′ and 5′ overhangs showed an entirely different conformation. When the 3′ unpaired nucleotides were included in the model, the stems did not stack coaxially and the nucleoside triples in the minor groove of P4 did not form (105). With the 3′ overhang, the junction was bent with the two helices rotated in a left-handed fashion. Structural analysis of junction mutants with shortened 3′-ends showed that one unpaired nucleotide at the overhang was sufficient to change the conformation of the molecule (105). This study clearly illustrates the sensitivity of global RNA structure to minor changes in the nucleotide sequence.

3.7.2 Four-way junctions

The best structurally characterized RNA junction is the four-way junction from transfer RNAs (Fig. 18.8d). Crystal structures of several tRNAs reveal this junction as a rigid structure of two pairs of coaxially stacked helices; the acceptor stem is stacked coaxially on the T stem, and the D stem is stacked on the anticodon helix (99) (for more details see Chapter 19). The two helical regions are roughly perpendicular to each other and create an overall L-shape for the molecule. Numerous studies have shown that the L-shaped geometry of tRNA is also present in solution. The unpaired nucleotides at the junction form several tertiary contacts with the stems stabilizing the geometry. Several specific metal-binding pockets are also formed in the tRNA junction region.

The L-shaped structure of tRNA is created by separating the two stacked helical domains with unpaired nucleotides at the junction. Another well-characterized example of nucleic acid four-way junctions (DNA Holliday junctions) also consists of two coaxial stacks, but, in the absence of intervening unpaired nucleotides at the branch point, the stems assume a symmetric X shape (106).

3.7.3 Three-way junctions

An odd number of helices at the junction pose several structural questions. Are there coaxial stacks formed between the stems, and if so, what is the spatial relationship of the remaining helix with respect to the stacked domains? Some insights into these questions were provided by low resolution solution studies of two RNA molecules, the 5S ribosomal RNA and the hammerhead ribozyme. Both of these RNAs contain central three-way junctions with several unpaired nucleotides (Fig. 18.8b and c).

The central junction of 5S rRNA, also known as loop A, forms a binding site for the transcription factor IIIA. Transient electric birefringence measurements provided evidence that the 5S rRNA junction from *Sulfolobus acidocaldarius* contains two colinear stems, I and V (Fig. 18.8b). The third stem (helix II) was found to be relatively unconstrained and free to reorient with respect to the I–V axis (107). An entirely different result was obtained from chemical modification data and computer modelling of loop A from *E. coli* and *Xenopus leavis*, which supported a colinear, stacked arrangement of helices II and V (108,109). The two alternative stacking arrangements may not differ greatly in free energy and thus may coexist in solution with different ratios depending on the nucleotide sequence from a particular organism. It has been postulated that interconversion between two forms of 5S rRNA might be of functional significance (110).

A junction between three short helices forms an active site of the hammerhead ribozyme (Fig. 18.8c). In addition to the conserved stem sequences, there are several unpaired nucleotides at the branch point that are necessary for the catalytic activity. In the crystal structures of the ribozyme, the three stems form an overall Y shape with helices I and II forming the upper fork (24,26). Stem II stacks directly on stem III, forming a pseudo-continuous helix. The junction is stabilized by an array of hydrogen bonds from the unpaired nucleotides. The geometry of the hammerhead ribozyme measured in solution by fluorescence resonance energy transfer (FRET) led to the same Y-shaped conformation of the junction found in the crystal structure (111).

3.7.4 Junctions summary

Structures of multibranched junctions often determine the global shapes of biologically functional RNA molecules. The conformations of RNA junctions are difficult to predict and depend on the number of stems and the size of the branch region. An important element of junction structure and stability is provided by coaxial stacking between double helical branches. Coaxial stacks stabilize junctions and join the shorter stems forming quasi-continuous elongated domains. RNA junctions often form complex structures with multiple metal-binding pockets and serve as sites for protein–RNA and RNA–RNA recognition.

4. Tertiary structures, interactions between secondary structures

The secondary structure motifs that have been described can interact (mainly by base pairing) to form tertiary structure. The base pairs formed in secondary structures can be represented by drawing the sequence in a circle with non-crossing lines joining the paired bases (42,126). Lines representing the interactions that characterize tertiary interactions cross the secondary structure lines. This distinction is important in methods to predict structure. Secondary structures can be considered as a sum of structural elements. The non-crossing of the base–base interactions means that the structural elements are independent. Tertiary structures involve base pairs between parts of the secondary structures and thus make them highly dependent. Other definitions of secondary and tertiary structure are also used.

4.1 Base, nucleoside, and nucleotide triples

When a single-stranded nucleotide interacts with nucleotides that are already involved in a base pair, a triple is formed. If the hydrogen bonding involves only the bases it is called a base triple. If base–ribose or ribose–ribose hydrogen bonding is present, we have a nucleoside triple. Phosphate involvement as a hydrogen bond acceptor creates a nucleotide triple. Triple interactions help to orient different regions of secondary structure and stabilize the global three-dimensional folds of large RNAs.

4.1.1 Triple helices

In addition to regular Watson–Crick double helices, some nucleic acid sequences form stable three-stranded complexes (for a review of DNA triple helices see Chapter 12). RNA triple helices consist of two strands forming an A-form Watson–Crick duplex and the third strand bound in either the major or the minor groove of the helix. Although triple helices are stabilized by extensive stacking between repeating base triples, they only form at high ionic strength conditions that overcome the unfavourable electrostatic repulsions between negatively charged phosphates. For example, a regular poly r(A):poly r(U) duplex converts into a stable three-stranded poly r(U):poly r(A):poly r(U) structure upon addition of magnesium, or at higher concentration of monovalent cations (greater than 0.1 M Na$^+$) (112). Fibre diffraction studies on this triplex showed that the extra poly r(U) strand was parallel to poly r(A) and bound in the major groove of a the Watson–Crick duplex forming an array of U:AU base triples (Fig. 18.9a) (113). Other polyribonucleotides also form triple helices in solution. Poly r(C) and poly r(G) have been shown to associate at low pH to form a poly r(C$^+$):poly r(G):poly r(C) triplex (114) with the protonated poly r(C$^+$) strand bound in the major groove (Fig. 18.9b). Evidence for poly r(G):poly r(G):poly r(C) and poly r(A):poly r(A):poly r(U) triple helices was found by using agarose-linked polyribonucleotide affinity columns (115) and the formation of poly r(A):poly r(G):poly r(C) triple helix has been shown by UV mixing curve experiments (116). The formation of poly r(A):poly r(A):poly r(U) and poly r(A):poly r(G):poly r(C) is dependent on the length of the polynucleotide strands participating in the triplex formation. The A:AU triplex forms only when poly r(A) strands are 28–150 nucleotides in length, whereas the size of poly r(U) has no effect on the triplex stability (117). The formation of A:GC triple helices depends on the length of the poly r(C) strand. This triple helix forms readily when the average length of poly r(C) is 100 nucleotides, but does not form when the average length is 500 nucleotides (116).

The two major groove triples, U:AU and C$^+$:GC, have isomorphic structures and therefore can form simultaneously in a mixed pyrimidine–purine–pyrimidine sequences. This triple-stranded motif is also stable when one or two of the participating strands are substituted with deoxyribonucleotides (118,119). A structure of a small unimolecular RNA triple helix containing several alternating U:AU) and C$^+$:GC base triples has been investigated by NMR (120). The sequence of this molecule was based on a DNA oligonucleotide that had been shown previously to form an intramolecular triple helix in solution (121). At pH 4.8, the NMR data showed formation of four U:AU and three C$^+$:GC major groove base triples. Each of the third strand pyrimidines formed two Hoogsteen hydrogen bonds with the Watson–Crick bound purines (Fig. 18.9a and b). Strong evidence for the formation of C$^+$:GC triples was provided by the presence of downfield-shifted imino resonances from protonated N3 of the Hoogsteen-bound cytosines (122). All of the nucleotides involved in base triples had the A-form C3'-*endo* sugar conformation, indicating relative rigidity of the structure. NMR studies of an intramolecular RNA triplex of slightly different sequence also showed formation of alternating major groove U:AU and C$^+$:GC triples (123).

4.1.2 Isolated triples

Isolated triples have been found or predicted in a variety of large RNAs, including tRNAs and group I introns (102,124). Single-nucleotide triples often occur at the interface of coaxially stacked helices within bulges, internal loops, or junctions. If the

RNA triples

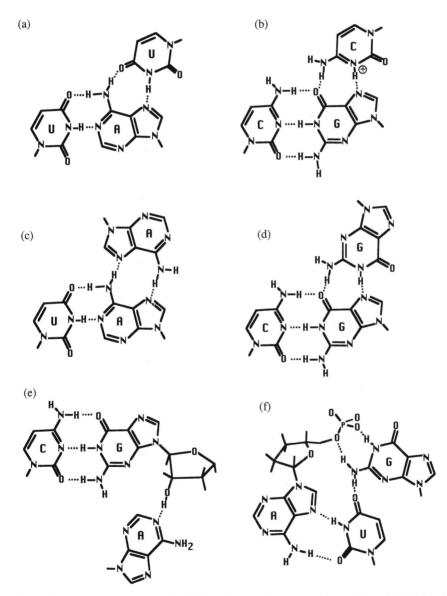

Fig. 18.9. Triple interactions observed in RNA molecules. Major groove base triples: (a) (A:U):U triple (113,120). (b) (G:C):C⁺ triple (120). (c) and (d) (U:A):A and (C:G):G triples from yeast tRNA^Phe. (e) Minor groove nucleoside (C:G):A triple (103). (f) Nucleotide triple (A:U):G from FMN aptamer (66).

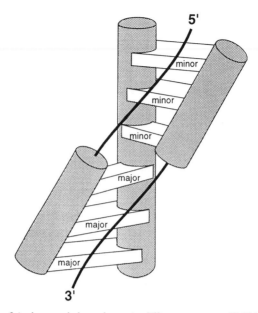

Fig. 18.10. Placement of single-stranded overhangs in different grooves of RNA at the junction of two helices. A right-handed rotation at the junction of the helices positions the 5′-single strand in the minor groove of the helix and the 3′-single strand enters the major groove. This figure is useful in visualizing structures of pseudo-knots, kissing hairpins, and two-stem junctions.

stacked helices continue a right-handed twist at the junction, the unpaired nucleotides at the 5′-end of each duplex enter the minor groove of the other helix. Similarly, each 3′ single-stranded end will be placed in the major groove of the opposite helix (Fig. 18.10). Several base triples observed in different RNA structures are consistent with this simple rule. For example, two G:GC base triples (Fig. 18.9d) observed in crystal structures of tRNA^{Phe} are formed at the junction of two stacked helices. The single-stranded G45 and G46 from the 3′-end of the anticodon stem enter the major groove of the D-stem and form base triples with G10:C25 and G22:C13 Watson–Crick pairs, respectively (Fig. 18.8d). Similarly, two sets of simultaneous major and minor groove triples have been proposed to form at the P4/P6 junction from group I introns (125). An NMR structure of a model of this junction confirmed the formation of two minor groove nucleoside triples, A:GC and U:GU although their structure was different to that proposed by Michel and Westhof (103,104). In the A:GC triple (Fig. 18.9e), the N1 of the single-stranded A forms a hydrogen bond to the 2′-hydroxyl of a Watson–Crick paired G but no base–base contacts were detected.

A single major groove U:AU base triple was formed in the structure of TAR RNA (see Section 3.6) upon binding of the argininamide ligand (94). The geometry of this triple is identical to the major groove triples seen in U:AU triple helices (Fig. 18.9a).

A well-defined G:AU nucleotide triple was identified in the structure of the FMN aptamer solved by NMR (66). The triple is formed upon FMN binding and is involved in generating the intercalation site pocket. A unique feature of this triple is that none of the bases are involved in Watson–Crick pairing. The triple is formed between a reverse Hoogsteen A:U pair and a G residue (Fig. 18.9f). Besides a single hydrogen bond with the Hoogsteen-paired uracil, the external G is in close proximity to phosphate oxygens, possibly forming an additional hydrogen bond (66).

4.2 Pseudoknots

A pseudoknot forms when a single strand pairs to a hairpin loop; two loops and two stems result (see Fig. 18.11a). The name pseudoknot was proposed (126) because if each stem contained more than 11 base pairs, and thus made a complete turn, and if the ends were linked, a topological knot would result. In 1982 experimental evidence was obtained for a pseudoknot structure in turnip yellow mosaic virus (127). Pseudoknots are found in all types of RNA and have a wide variety of biological functions; several reviews describe their importance (128–130).

In Fig. 18.11a a general pseudoknot is shown with two stems and three loops. This figure represents a wide variety of possible pseudoknots if we allow any one of the three loops to have zero length, or if we allow them to fold into further secondary structures, such as hairpins. The simplest pseudoknot has loop 1.5 with zero length; this is the so-called H-type pseudoknot. The two stems can stack coaxially on each other to form a quasi-continuous helix. Because of the right-handed winding of A-form helices, loop 1 crosses the deep major groove of stem 2, whereas loop 2 crosses the shallow minor groove of stem 1. The minimum loop lengths for a given number of base pairs in each stem can be estimated from A-form geometry (Fig. 18.11b) (131). A minimum loop 1 length of one or two nucleotides occurs when stem 2 is seven base pairs long. Loop 2 must be longer. With four base pairs in stem 1 at least three nucleotides are needed in loop 2, and the loop length increases rapidly with stem length. These estimates are based on standard A-form structure, so bending or unusual twisting of the helices can lead to different results. Experimental studies have been done on an H-type pseudoknot with three base pairs in stem 1 and five base pairs in stem 2. The effect of loop lengths (with Us in the loops) on the pseudoknot stability relative to its constituent hairpins was determined (132). Magnesium ion preferentially stabilizes the pseudoknot with respect to its hairpins. In 5 mM Mg^{2+} a minimum of three nucleotides in loop 1 was needed for the five base pairs in stem 2, and a minimum of four nucleotides was needed for the three base pairs in stem 1. These results are consistent with the estimates based on A-form stem geometry. The pseudoknot is only marginally more stable than its constituent hairpins; a decrease in standard free energy of only 1.5 to 2 kcal/mol at 37°C results when the pseudoknot forms.

The structure of the H-type pseudoknot was found to have the two stems coaxially stacked, with only minor distortion in helical stacking at the junction of the two stems (133). Right-handed winding continues at the stem–stem junction with an increase in the winding angle, which helps relieve the crowding of the two loops at the junction. The phosphates from the loops and stems are very close at the stem–stem interface, and may provide the binding site for the Mg^{2+} ions required for pseudoknot formation. Surprisingly, no evidence for base triple formation was seen between the loops and stems. Although model building can place loop 1 in the major groove and loop 2 in the minor groove of the stems, no NMR evidence for loop–stem interaction was seen.

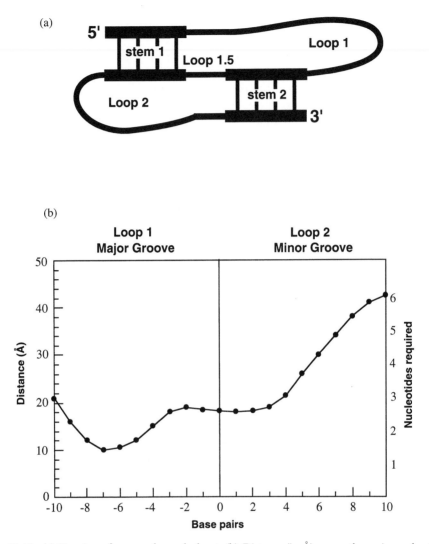

Fig. 18.11. (a) Drawing of a general pseudo-knot. (b) Distance (in Å) across the major and minor grooves of an A-form RNA helix as a function of the number of base pairs (131). The distances were calculated using coordinates from fibre diffraction studies. Indicated on the right-hand side of the graph are the number of nucleotides necessary to cross the indicated distance (assuming that a nucleotide is able to span 7 Å).

Thermodynamic and structural studies have been done on pseudoknots from gene *32* mRNAs from T2, T4, and T6 bacteriophages (134,135). These pseudoknots bind the gene 32 protein to autoregulate the translation of its mRNA. The pseudoknots are stabilized by Mg^{2+} and have coaxially stacked stems. Stem 2 contains seven base pairs and is spanned by a loop of only one nucleotide; the minimum predicted for standard A-form geometry. Stem 1 contains four or five base pairs and is spanned by loops of five or seven nucleotides, respectively. There was a hint from NOEs of loop–stem interactions, but no definite structure could be deduced.

If loop 1.5 (see Fig. 18.11a) is non-zero the direct coaxial stacking of the stems is interrupted. Pseudo-knots with a loop 1.5 of one nucleotide (one nucleotide interrupts the stacking of the stems) are important in the programmed frameshifting used by several retroviruses to synthesize vital enzymes (136). The structure of the pseudo-knot required for frameshifting in mouse mammary tumour virus (137) shows that the adenylate residue between the stems causes a bend in the pseudoknot [Plate XXI (top)]. Removing the intervening nucleotide produces a linear structure with coaxially stacked stems (138). If two nucleotides intervene between the stems (loop 1.5 contains two nucleotides) the stems are not coaxial; instead they are displaced relative to each other (139). It is important to realize that the number of nucleotides in loops 1, 1.5, and 2 can not be deduced simply from the sequence. Whether base pairs form at the ends of the stems, or the bases are part of the loops, must be determined experimentally.

4.3 Loop–loop and loop–helix interactions

These interactions can include any combination of hairpin loops, internal loops, bulges, and helices. These tertiary interactions are important in folding RNA molecules into the specific compact forms required for their biological functions.

4.3.1 Kissing hairpins

Kissing hairpins are formed by base pairing between complementary hairpin loops (Fig. 18.1) (140). They are involved in naturally occurring antisense control of biological function (141). The best-studied example is the control of ColE1 plasmid replication in *E. coli* (142). A kissing hairpin complex forms as the first step in the hybridization of the complementary RNAs. Formation of the loop–loop interaction is faster than the conversion of the complex to the more stable duplex. The latter process is subsequently catalysed by a protein. The thermodynamics (143) and structure (144) of the kissing complex between the RNA I and RNA II stem loops of the ColE1 plasmid have been studied in detail. Imino proton spectra showed that all seven base pairs of the loop–loop helix formed and that the stem base pairs were not disrupted. Two-dimensional NMR NOESY spectra indicated continuous stacking of the base pairs on the 3′-side of each stem. In addition to NMR data, electrophoretic gel mobility experiments showed that the complex was bent. A model consistent with the NMR and electrophoresis results was obtained (144).

The structure of a kissing complex between the HIV TAR hairpin loop and its complement (145) is shown in Plate XXI (bottom). All six nucleotides of each loop form base pairs in the loop–loop helix. As in the ColE1 complex, the two stems plus the loop–loop helix form a quasi-continuous bent helix. The formation of a helix by all the nucleotides that are part of the loop of a stem–loop structure means that a single phosphodiester group must join the bases at the beginning and end of the helix. The shortest distance between the ends of an A-form helix is across the major groove; for 6 or 7 base pairs the distance is about 10 Å (Fig. 18.11b) (131). Although this distance is too long for a phosphate group, bending the helix towards the major groove, and increasing winding angles and propeller twists (145), allows the formation of the complex. Two phosphates (one from each hairpin) bridge the major groove of the loop–loop helix. The phosphate cluster makes a likely Mg^{2+}-binding site. The helix distortions may be part of the recognition mechanism for the Rom (or Rop) protein, which specifically binds kissing hairpins (146)

4.3.2 Loop–helix

The GAAA tetraloop in a hammerhead ribozyme forms an intermolecular contact with the minor groove of stem II of another hammerhead molecule in the crystal structure (24). Only one out of three GAAA tetraloops present in the unit cell is involved in the loop–helix interaction. Remarkably, the structures of the bound and unbound tetraloops are identical, and also closely resemble the structure of the GCAA tetraloop solved in solution by NMR (18). In the complex, the tetraloop stem and the target helix are almost parallel forming a 31° angle between the helix axes. The third and fourth adenines of the tetraloop form minor groove triples with two consecutive C:G base pairs; each A forms four hydrogen bonds with its target C:G pair. In each triple, only one hydrogen bond is formed between the bases; the other three involve 2′-hydroxyl hydrogen bonds.

The P4/P5/P6 domain of the *Tetrahymena thermophila* group I intron contains two loop–helix interactions (25); a GNRA tetraloop binds to its internal loop receptor and an A-rich bulge hydrogen binds to a helix. These interactions hold two helical domains in close and specific contact. The X-ray structure of a group I ribozyme domain (147) shows that the three As of the GAAA tetraloop stack on two adjacent As in the receptor loop and hydrogen bond in the minor groove of the adjacent helix. The hydrogen bonding provides the sequence specificity between the tetraloop receptor and the tetraloop. Their diverse binding capabilities, and the fact that GNRA loops are present in exceptional abundance in natural RNAs (16), suggest that the GNRA tetraloop family may act as a general long-distance docking motif for RNA–RNA recognition (32).

4.4 Prediction of structure

The ultimate goal of methods to predict macromolecular structure is to calculate a high resolution structure from the base sequence, the solvent conditions (salt concentration, pH, etc.), and the temperature. No experiments are done, only calculations. We are far from this goal. Here we will describe methods available for obtaining possible RNA secondary structures, and for modelling their three-dimensional structures. Useful general reviews of this subject are available (148,149).

Secondary and tertiary structure can be obtained from sequence alone by phylogenetic comparison of many RNA molecules with the same function from different species. The sequences are first aligned using invariant and homologous sequence regions as guides. Then covariation of bases is used to establish Watson–Crick base pairs. For example, if an A in one species changes to a C, and a U in the same species changes to a G, they potentially covary. A detailed secondary structure can be constructed if enough sequences are available. Similarly, if there is covariation of a base pair with a third base, tertiary structure interactions can be established. When hundreds of sequences are available, very detailed structures can be determined (150,102). In the following sections we will describe methods that require only one sequence.

4.4.1 Secondary structure

The free energy is a minimum for a system (such as a solution of RNA molecules in a buffer) at equilibrium at constant temperature and pressure. Therefore, if we can calculate the free energies of different RNA secondary structures, we can predict which will actually occur, i.e. the one with the lowest free energy. Algorithms to calculate free energies of RNA secondary structures are based on the nearest-neighbour hypothesis (5). The free energy of a secondary structure is calculated as a sum of, (a) negative (favourable) contributions from adjacent pairs—Watson–Crick and G:U nearest neighbours; and (b) positive (unfavourable) contributions from forming mismatches, loops, and bulges. The free energy values are obtained from experimental data on equilibrium constants for double strand formation, hairpin loop formation, etc. as a function of sequence. The calculated free energy is approximate because of uncertainties in the measured free energies of the structural elements, the need to extrapolate to other loop sequences and loop sizes, the assumption of additivity of the thermodynamic values, and so forth. Thus, algorithms to predict secondary structure must provide not only the optimal structure, but also many possible suboptimal structures (151). A comparison of the thermodynamic predictions of base paired duplexes with those established by extensive phylogenetic comparisons, showed about 90% agreement (152). As more reference thermodynamic data are obtained (153), including junctions and extra stable loop sequences, the thermodynamic prediction of secondary structure should improve.

The effects of solvent (Na^+, K^+, Mg^{2+} concentrations, for example) and temperature need to be explored further. The reference thermodynamic data is mainly available for 1 M Na^+. This was chosen to avoid the hydrolytic effect of Mg^{2+}, but to provide sufficient ionic strength to shield electrostatic repulsion of the phosphates. Free energy values are given for 37°C, but enthalpy and entropy values needed for obtaining free energies at other temperatures are also available.

4.4.2 Tertiary structure

Prediction of tertiary structure from a single sequence is extremely difficult. The strategy is to search for possible base–base interactions among the secondary structure elements. For example, pseudoknots can be predicted by considering further base pairing of the loops and single-stranded regions of the calculated secondary structure. Presumably, eventually, specific RNA structure receptors, such as the tetraloop receptor (25), will be established. At present, however, tertiary structures are nearly completely based on phylogenetic sequence information, chemical reactivity, and spectroscopic measurements.

4.4.3 Three-dimensional structure

Modelling three-dimensional structures for RNA from the sequence is based on building up the structure from measured structures of model RNAs. RNA double-strand helices are essentially A-form, so helices obtained from thermodynamics or phylogenetics can be modelled accurately. The three-dimensional structures of any of the tetraloop families, or loop-E-like sequences described above, can be added. Other sequences can be modelled from a database of possible mononucleotide conformations (154,155). There are restrictions on the seven torsion angles that specify the conformation of each nucleotide, and more constraints are imposed by each particular loop size, or by the requirements of mismatch formation, or a base triple. All these constraints can be used to calculate possible three-dimensional structures for a given sequence. A test of this method for tRNA^{Phe} gave encouraging results (156).

Many other methods for calculating folding of nucleic acids are being actively developed as described in ref. 149.

Acknowledgements

We gratefully acknowledge Dr Kevin Luebke for reading the manuscript and making very useful comments. The work on RNA in our laboratory has been supported by the National Institutes of Health and the Department of Energy. We thank Dr Juli Feigon and Dr Gabriele Varani for providing us with figures.

References

1. Herschlag, D. (1995) *J. Biol. Chem.* **270**, 20871.
2. Sauer, K. (ed.) (1995) *Biochemical Spectroscopy*, Vol. 246, *Methods in Enzymology*. Academic Press, San Diego.
3. Warshaw, M.M. and Tinoco, Jr, I. (1966) *J. Mol. Biol.* **20**, 29.
4. Altona, C. (1982) *Recl. Trav. Chim. Pays-Bas.* **101**, 413.
5. Turner, D.H., Sugimoto, N. and Freier, S.M. (1988) *Annu. Rev. Biophys. Biophys. Chem.* **17**, 167.
6. Saenger, W. (1984) *Principles of Nucleic Acid Structure*. Springer-Verlag, New York.
7. Varani, G. and Tinoco, Jr, I. (1991) *Q. Rev. Biophys.* **24**, 479.
8. Allain, F.H.-T. and Varani, G. (1996) *Progr. Nucl. Magn. Reson. Spectrosc* **29**, 54.
9. Gast, F.U. and Hagerman, P.J. (1991) *Biochemistry* **30**, 4268.
10. Weeks, K.M. and Crothers, D.M. (1993) *Science* **261**, 1574.
11. Hall, K., Cruz, P., Tinoco, Jr, I., Jovin, T.M. and van de Sande, J.H. (1984) *Nature* **311**, 584.
12. Davis, P.W., Adamiak, R.W. and Tinoco, Jr, I. (1990) *Biopolymers* **29**, 109.
13. Wang, A.H., Quigley, G.J., Kolpak, F.J., Crawford, J.L., van Boom, J.H., van der Marel, G. and Rich, A. (1979) *Nature* **282**, 680.
14. Tinoco, Jr, I., Davis, P., Hardin, C.C., Puglisi, J.D., Walker, G.T. and Wyatt, J. (1987) *Cold Spring Harbor Symp. Quant. Biol.* **52**, 135.
15. Noller, H.F. (1984) *Annu. Rev. Biochem.* **53**, 119.
16. Woese, C.R., Winker, S. and Gutell, R.R. (1990) *Proc. Natl. Acad. Sci. USA* **87**, 8467.
17. Varani, G. (1995) *Annu. Rev. Biophys. Biomol. Struct.* **24**, 379.
18. Heus, H.A. and Pardi, A. (1991) *Science* **253**, 191.
19. Jucker, F.M. and Pardi, A. (1995) *Biochemistry* **34**, 14416.
20. Jucker, F.M., Heus, H.A., Yip, P.F., Moors, E.H. M. and Pardi, A. (1996) *J. Mol. Biol.*
21. Orita, M., Nishikawa, F., Shimayama, T., Taira, K., Endo, Y. and Nishikawa, S. (1993) *Nucl. Acids Res.* **21**, 5670.
22. Szewczak, A.A. and Moore, P.B. (1995) *J. Mol. Biol.* **247**, 81.
23. SantaLucia, Jr, J., Kierzek, R. and Turner, D.H. (1992) *Science* **256**, 217.
24. Pley, H.W., Flaherty, K.M. and McKay, D.B. (1994) *Nature* **372**, 111.
25. Cate, J.H., Gooding, A.R., Podell, E., Zhou, K., Golden, B.L., Kundrot, C.E., Cech, T.R. and Doudna, J.A. (1996) *Science* **273**, 1678.

26. Scott, W.G., Finch, J.T. and Klug, A. (1995) *Cell* **81**, 991.

27. Quigley, G.J. and Rich, A. (1976) *Science* **194**, 796.

28. Sassanfar, M. and Szostak, J.W. (1993) *Nature* **364**, 550.

29. Dieckmann, T., Suzuki, E., Nakamura, G.K. and Feigon, J. (1996) *RNA* **2**, 628.

30. Jiang, F., Kumar, R.A., Jones, R.A. and Patel, D.J. (1996) *Nature* **382**, 183.

31. Glück, A., Endo, Y. and Wool, I.G. (1992) *J. Mol. Biol.* **226**, 411.

32. Jaeger, L., Michel, F. and Westhof, E. (1994) *J. Mol. Biol.* **236**, 1271.

33. Tuerk, C., Gauss, P., Thermes, C., Groebe, D.R., Guild, N., Stormo, G., Gayle, M., d'Auberton-Carafa, Y., Uhlenbeck, O.C., Tinoco, Jr, I., Brody, E.N. and Gold, L. (1988) *Proc. Natl. Acad. Sci. USA* **85**, 1364.

34. Antao, V.P., Lai, S.Y. and Tinoco, Jr, I. (1991) *Nucl. Acids Res.* **19**, 5901.

35. Varani, G., Cheong, C. and Tinoco, Jr, I. (1991) *Biochemistry* **30**, 3280.

36. Allain, F.H. -T. and Varani, G. (1995) *J. Mol. Biol.* **250**, 333.

37. Selinger, D., Liao, X. and Wise, J.A. (1993) *Proc. Natl. Acad. Sci. USA* **90**, 5409.

38. Molinaro, M. and Tinoco, Jr, I. (1995) *Nucl. Acids Res.* **23**, 3056.

39. James, J.K. and Tinoco, Jr, I. (1993) *Nucl. Acids Res.* **21**, 3287.

40. Jacobson, H. and Stockmayer, W.H. (1950) *J. Chem. Phys.* **18**, 1600.

41. Gralla, J. and Crothers, D.M. (1973) *J. Mol. Biol.* **73**, 497.

42. Chastain, M. and Tinoco, Jr, I. (1991) *Prog. Nucleic Acid Res. Mol. Biol.* **41**, 131.

43. Borer, P.N., Lin, Y., Wang, S., Roggenbuck, M.W., Gott, J.M., Uhlenbeck, O.C. and Pelczer, I. (1995) *Biochemistry* **34**, 6488.

44. Mirmira, S.R. and Tinoco, Jr, I. (1996) *Biochemistry* **35**, 7664.

45. Tuerk, C. and Gold, L. (1990) *Science* **249**, 505.

46. Mirmira, S.R. and Tinoco, Jr, I. (1996) *Biochemistry* **35**, 7675.

47. Davis, P.W., Thurmes, W. and Tinoco, Jr, I. (1993) *Nucl. Acids Res.* **21**, 537.

48. Jaeger, J.A. and Tinoco, Jr, I. (1993) *Biochemistry* **32**, 12522.

49. Fountain, M.A., Serra, M.J., Krugh, T.R. and Turner, D.H. (1996) *Biochemistry* **35**, 6539.

50. Huang, S., Wang, Y.X. and Draper, D.E. (1996) *J. Mol. Biol.* **258**, 308.

51. Schweisguth, D.C. and Moore, P.B. (1996) *J. Mol. Biol.* **267**, 505.

52. Sugimoto, N., Kierzek, R., Freier, S.M. and Turner, D.H. (1986) *Biochemistry* **25**, 5755.

53. Wu, M., McDowell, J.A. and Turner, D.H. (1995) *Biochemistry* **34**, 3204.

54. Crick, F.H.C. (1966) *J. Mol. Biol.* **19**, 548.

55. Allain, F.H.-T. and Varani, G. (1995) *Nucl. Acids Res.* **23**, 341.

56. Cate, J.H. and Doudna, J.A. (1996) *Structure* **4**, 1221.

57. Walter, A.E., Wu, M. and Turner, D.H. (1994) *Biochemistry* **33**, 11349.

58. SantaLucia, Jr. J. and Turner, D.H. (1993) *Biochemistry* **32**, 12612.

59. Wu, M. and Turner, D.H. (1996) *Biochemistry* **35**, 9677.

60. Szewczak, A.A., Moore, P.B., Chan, Y.-L and Wool, I.G. (1993) *Proc. Natl. Acad. Sci. USA* **90**, 9581.

61. Wimberly, B., Varani, G. and Tinoco, Jr, I. (1993) *Biochemistry* **32**, 1078.

62. Cai, Z. and Tinoco, Jr, I. (1996) *Biochemistry* **35**, 6026.

63. Peterson, R.D., Bartel, D.P., Szostak, J.W., Horvath, S.J. and Feigon, J. (1994) *Biochemistry* **33**, 5357.

64. Battiste, J.L., Mao, H., Rao, N.S., Tan, R., Muhandiram, D.R., Kay, L.E., Frankel, A. and Williamson, J.R. (1996) *Science* **273**, 1547.

65. Leonard, G.A., McAuley-Hecht, K.E., Ebel, S., Lough, D.M., Brown, T. and Hunter, W.N. (1994) *Structure* **2**, 483.

66. Fan, P., Suri, A.K., Fiala, R., Live, D. and Patel, D.J. (1996) *J. Mol. Biol.* **258**, 480.

67. Pley, H.W., Flaherty, K.M. and McKay, D.B. (1994) *Nature* **372**, 68.

68. Legault, P. and Pardi, A. (1994) *J. Am. Chem. Soc.* **116**, 8390.
69. Puglisi, J.D., Wyatt, J.R. and Tinoco, Jr, I. (1990) *Biochemistry* **29**, 4215.
70. Yang, Y., Kochoyan, M., Burgstaller, P., Westhof, E. and Famulok, M. (1996) *Science* **272**, 1343.
71. Cheong, C. and Moore, P.B. (1992) *Biochemistry* **31**, 8406.
72. Holbrook, S.R., Cheong, C., Tinoco, Jr, I. and Kim, S.-H. (1991) *Nature* **353**, 579.
73. Cruse, W.B.T., Saludjian, P., Biala, E., Strazewski, P., Prangé, T. and Kennard, O. (1994) *Proc. Natl. Acad. Sci. USA* **91**, 4160.
74. Lewis, H.A. (1995) *PhD Thesis*. University of California, Berkeley.
75. SantaLucia, Jr, J., Kierzek, R. and Turner, D.H. (1991) *Biochemistry* **30**, 8242.
76. Gesteland, R.F. and Atkins, J.F. (eds) (1993) *The RNA World*. Cold Spring Harbor Laboratory Press, Cold Spring Harbor.
77. Peritz, A.E., Kierzek, R., Sugimoto, N. and Turner, D.H. (1991) *Biochemistry* **30**, 6428.
78. Tang, R.S. and Draper, D.E. (1994) *Biochemistry* **33**, 10089.
79. Wimberly, B. (1994) *Nature Struct. Biol.* **1**, 820.
80. Burke, J.M. (1994) *Nucl. Acids Mol. Biol.* **8**, 105.
81. Burgstaller, P. and Famulok, M. (1994) *Angew. Chem. Int. Ed. Engl.* **33**, 1084.
82. Joyce, G.F. (1994) *Curr. Opin. Struct. Biol.* **4**, 331.
83. Allain, F.H.-T., Gubser, C.C., Howe, P.W., Nagai, K., Neuhaus, D. and Varani, G. (1996) *Nature* **380**, 646.
84. Gubser, C.C. and Varani, G. (1996) *Biochemistry* **35**, 2253.
85. Fourmy, D., Recht, M.I., Blanchard, S.C. and Puglisi, J.D. (1996) *Science* **274**, 1367.
86. Longfellow, C.E., Kierzek, R. and Turner, D.H. (1990) *Biochemistry* **29**, 278.
87. Bhattacharyya, A., Murchie, A.I.H. and Lilley, D.M.J. (1990) *Nature* **343**, 484.
88. Tang, R.S. and Draper, D.E. (1990) *Biochemistry* **29**, 5232.
89. Zacharias, M. and Hagerman, P.J. (1995) *J. Mol. Biol.* **247**, 486.
90. Riordan, F.A., Bhattacharyya, A., McAteer, S. and Lilley, D.M. (1992) *J. Mol. Biol.* **226**, 305.
91. Luebke, K.J. and Tinoco, Jr, I. (1996) *Biochemistry* **35**, 11677.
92. van den Hoogen, Y.T., van Beuzekom, A.A., de Vroom, E., van der Marel, G.A., van Boom, J.H. and Altona, C. (1988) *Nucl. Acids Res.* **16**, 5013.
93. Joshua-Tor, L., Frolov, F., Appella, E., Hope, H., Rabinovich, D. and Sussman, J.L. (1992) *J. Mol. Biol.* **225**, 397.
94. Puglisi, J.D., Tan, R., Calnan, B.J., Frankel, A.D. and Williamson, J.R. (1992) *Science* **257**, 76.
95. Aboul-ela, F., Karn, J. and Varani, G. (1995) *J. Mol. Biol.* **253**, 313; (1996) *Nucl. Acids Res.* **24**, 3974.
96. Zacharias, M. and Hagerman, P.J. (1995) *Proc. Natl. Acad. Sci. USA* **92**, 6052.
97. Walter, A.E., Turner, D.H., Kim, J., Lyttle, M.H., Muller, P., Mathews, D.H. and Zuker, M. (1994) *Proc. Natl. Acad. Sci. USA* **91**, 9218.
98. Olmsted, M.C. and Hagerman, P.J. (1994) *J. Mol. Biol.* **243**, 919.
99. Holbrook, S.R., Sussman, J.L., Warrant, R.W. and Kim, S.-H. (1978) *J. Mol. Biol.* **123**, 631.
100. Weidner, H. and Crothers, D.M. (1977) *Nucl. Acids Res.* **4**, 3401.
101. Dahm, S.C. and Uhlenbeck, O.C. (1991) *Biochemistry* **30**, 9464.
102. Michel, F. and Westhof, E. (1990) *J. Mol. Biol.* **216**, 585.
103. Chastain, M. and Tinoco, Jr, I. (1992) *Biochemistry* **31**, 12733.
104. Chastain, M. and Tinoco, Jr, I. (1993) *Biochemistry* **32**, 14220.
105. Nowakowski, J. and Tinoco, Jr, I. (1996) *Biochemistry* **35**, 2577.

106. Murchie, A.I.H. and Clegg, R.M., von Kitzing, E., Duckett, D.R., Diekmann, S., Lilley, D.M.J. (1989) *Nature* **341**, 763.
107. Shen, Z. and Hagerman, P.J. (1994) *J. Mol. Biol.* **241**, 415.
108. Westhof, E., Romby, P., Romaniuk, P.J., Ebel, J.-P., Ehresmann, C. and Ehresmann, B. (1989) *J. Mol. Biol.* **207**, 417.
109. Brunel, C., Romby, P., Westhof, E., Ehresmann, C. and Ehresmann, B. (1991) *J. Mol. Biol.* **221**(1), 293.
110. Stahl, D.A., Luehrsen, K.R., Woese, C.R. and Pace, N.R. (1981) *Nucl. Acids Res.* **9**, 6129.
111. Tuschl, T., Gohlke, C., Jovin, T.M., Westhof, E. and Eckstein, F. (1994) *Science* **266**, 785.
112. Felsenfeld, G., Davies, D.R. and Rich, A. (1957) *J. Am. Chem. Soc.* **79**, 2023.
113. Arnott, S. and Bond, P.J. (1973) *Nature New Biol.* **244**, 99.
114. Thiele, D. and Guschlbauer, W. (1971) *Biopolymers* **10**, 143.
115. Letai, A.G., Palladino, M.A., Fromm, E., Rizzo, V. and Fresco, J.R. (1988) *Biochemistry* **27**, 9108.
116. Chastain, M. and Tinoco, Jr, I. (1992) *Nucl. Acids Res.* **20**, 315.
117. Broitman, S.L., Im, D.D. and Fresco, J.R. (1987) *Proc. Natl. Acad. Sci. USA* **84**, 5120.
118. Roberts, R.W. and Crothers, D.M. (1992) *Science* **258**, 1463.
119. Han, H. and Dervan, P.B. (1993) *Proc. Natl. Acad. Sci. USA* **90**, 3806.
120. Klinck, R., Liquier, J., Taillandier, E., Gouyette, C. and Tam, H.-D. (1995) *Eur. J. Biochem.* **233**, 544.
121. Sklenar, V. and Feigon, J. (1990) *Nature* **345**, 836.
122. de los Santos, C., Rosen, M. and Patel, D. (1989) *Biochemistry* **28**, 7282.
123. Holland, J.A. and Hoffman, D.W. (1996) *Nucl. Acids Res.* **24**, 2841.
124. Gautheret, D., Damberger, S.H. and Gutell, R.R. (1995) *J. Mol. Biol.* **248**, 27.
125. Michel, F., Ellington, A.D., Couture, S. and Szostak, J.W. (1990) *Nature* **347**, 578.
126. Studnicka, G.M., Rahn, G.M., Cummings, I.W. and Salser, W.A. (1978) *Nucl. Acids Res.* **5**, 3365.
127. Rietveld, K., van Peolgeest, R., Pleij, C.W. A., van Boom, J.H. and Bosch, L. (1982) *Nucl. Acids Res.* **10**, 1929.
128. Pleij, C.W. (1990) *TIBS* **15**, 143.
129. Puglisi, J.D., Wyatt, J.R. and Tinoco, Jr, I. (1991) *Acc. Chem. Res.* **24**, 152.
130. ten Dam, E., Pleij, K. and Draper, D. (1992) *Biochemistry* **31**, 11665.
131. Pleij, C.W., Rietveld, K. and Bosch, L. (1985) *Nucl. Acids Res.* **13**, 1717.
132. Wyatt, J.R., Puglisi, J.D. and Tinoco, Jr, I. (1990) *J. Mol. Biol.* **214**, 455.
133. Puglisi, J.D., Wyatt, J.R. and Tinoco, Jr, I. (1990) *J. Mol. Biol.* **214**, 437.
134. Qiu, H., Kaluarachchi, K., Du, Z., Hoffman, D.W. and Giedroc, D.P. (1996) *Biochemistry* **35**, 4176.
135. Du, Z., Giedroc, D.P. and Hoffman, D.W. (1996) *Biochemistry* **35**, 4187.
136. Brierley, I. (1995) *J. Gen. Virol.* **76**, 1885.
137. Shen, L.X. and Tinoco, Jr, I. (1995) *J. Mol. Biol.* **247**, 963.
138. Chen, X., Kang, H., Shen, L.X., Chamorro, M., Varmus, H.E. and Tinoco, Jr, I. (1996) *J. Mol. Biol.* **260**, 479.
139. Kang, H., Hines, J.V. and Tinoco, Jr, I. (1996) *J. Mol. Biol.* **259**, 135.
140. Eguchi, Y. and Tomizawa, J.I. (1991) *J. Mol. Biol.* **220**, 831.
141. Wagner, E.G.H. and Simons, R.W. (1994) *Annu. Rev. Microbiol.* **48**, 713.
142. Tomizawa, J.I., Eguchi, Y. and Itoh, T. (1991) *Annu. Rev. Biochem.* **60**, 631.
143. Gregorian, R.S. Jr. and Crothers, D.M. (1995) *J. Mol. Biol.* **248**(5), 968.

144. Marino, J.P., Gregorian, Jr, R.S., Csankovszki, G. and Crothers, D.M. (1995) *Science* **268**, 1448.

145. Chang, K. -Y and Tinoco, Jr, I. (1997) *J. Mol. Biol.* **269**, 52

146. Predki, P.F., Nayak, L.M., Gottlieb, M.B. and Regan, L. (1995) *Cell* **80**, 41.

147. Cate, J.H., Gooding, A.R., Podell, E., Zhou, K., Golden, B.L., Szewczak, A.A., Kundrot, C.E., Cech, T.R. and Doudna, J.A. (1996) *Science* **273**, 1696.

148. Jaeger, J., SantaLucia, Jr, J. and Tinoco, Jr, I. (1993) *Annu. Rev. Biochem.* **62**, 255.

149. Louise-May, S., Auffinger, P. and Westhof, E. (1996) *Curr. Opin. Struct. Biol.* **6**, 268.

150. Schnare, M.N., Damberger, S.H., Gray, M.W. and Gutell, R.R. (1996) *J. Mol. Biol.* **256**, 701.

151. Zuker, M. (1989) *Science* **244**, 48.

152. Jaeger, J.A., Turner, D.H. and Zuker, M. (1989) *Proc. Natl. Acad. Sci. USA* **86**, 7706.

153. Turner, D.H. (1996) *Curr. Opin. Struct. Biol.* **6**(3), 299.

154. Gautheret, D., Major, F. and Cedergren, R. (1993) *J. Mol. Biol.* **229**, 1049.

155. Gautheret, D. and Cedergren, R. (1993) *FASEB J.* **7**, 97.

156. Major, F., Gautheret, D. and Cedergren, R. (1993) *Proc. Natl. Acad. Sci. USA* **90**, 9408.

19

Transfer RNA

John G. Arnez and Dino Moras

Laboratoire de Biologie Structurale, Institut de Génétique et de Biologie Moléculaire et Cellulaire,
CNRS/INSERM/ULP, 1, rue L. Fries–BP 163, F-67404 Illkirch, France

1. Introduction

Transfer RNA (tRNA) is the key intermediate in the process of protein synthesis. It is a link between genetic information contained in nucleic acids and its expression in the protein world. The molecule possesses two important ends; one interacts with the codon of the messenger RNA (mRNA) through three specific nucleotides called the anticodon, while the other end serves as the attachment point for the amino acid and is subseqently linked to a growing polypeptide chain during protein synthesis on the ribosome. Thus, the molecule adapts the amino acids to the genetic code.

Its existence as the adaptor molecule was initially postulated at a tie club meeting by Francis Crick after the determination of the three-dimensional structure of DNA. It was discovered in 1957 by Hoagland *et al.* (1). The first tRNA nucleotide sequence, that of yeast tRNA^{Ala}, was determined by Holley *et al.* in 1965 (2), who also first proposed the clover-leaf representation of its secondary structure. Later it was found that all tRNA sequences can be folded in such a structure. Currently, the primary structures of 2700 tRNAs are known (3). In 1966 Crick proposed the wobble hypothesis for the reading of the triplet codons on mRNA by tRNA anticodons (4). The first nucleotide modifications in tRNA were isolated in 1959 (5,6). In 1969, Bernhardt and Darnell (7) reported that tRNAs are transcribed as part of larger precursor RNAs, and in 1971 the first precursor sequence, that of *E. coli* p–tRNA^{Tyr}, was elucidated by Altman and Smith (8). The precursors are then processed to give mature tRNAs; the processing pathways have since been elucidated in many organisms and many of the enzymes involved have been isolated (reviewed in refs 9 and 10). Fraser and Rich (11) and Sprinzl and Cramer (12) noted, in 1975, that amino acids are specifically attached to either the 2'-OH or the 3'-OH of the 3'-terminal adenosine of a tRNA, depending on the aminoacylation system. A functional correlation for this observation was found in 1990 with the partition of aminoacyl–tRNA synthetases into two classes (13). The first three-dimensional structure of a tRNA was that of yeast tRNA^{Phe} and was determined in the early 1970s, by two groups concurrently, one headed by A. Rich at MIT, Cambridge, USA (14) and the other led by A. Klug at the MRC in Cambridge, UK (15), using X-ray crystallography. Subsequently, the crystal structures of a few other tRNAs were determined, that of yeast tRNA^{Asp} (16), *E. coli* initiator tRNA$^{Met}_f$ (17) and yeast initiator tRNA$^{Met}_i$ (18). In addition, X-ray crystal structures have been determined of tRNA complexed with cognate aminoacyl–tRNA synthetases for the *E. coli* glutamine (19), yeast aspartic acid (20), and *T. thermophilus* serine (21) systems, and with the *E. coli* elongation factor

Tu (22). Most recently, tRNA has been observed in the *E. coli* ribosome, using cryo-electron microscopy (23,24).

It was noted very early that tRNA is a substrate for many enzymes. First, tRNA genes are transcribed by RNA polymerase in prokaryotes (25) and by RNA polymerase III in eukaryotes (26); the transcripts are precursor molecules that have 5′ and 3′ extensions in addition to the sequences that correspond to tRNAs and are processed by a series of specific nucleases to give mature tRNAs. The 5′-end is specifically cleaved by ribonuclease P, a ribonucleoprotein that contains a catalytic RNA subunit and a helper protein cofactor (9). The 3′-end is processed by a variety of nucleases, and the integrity of the CCA terminus is maintained by a terminal nucleotidyl transferase (10). Many nucleotides are modified by specific modifying enzymes during and after maturation. Then amino acids are attached to the 3′-adenosine by their cognate aminoacyl–tRNA synthetases. Aminoacyl–tRNAs are bound by the elongation factor Tu (EF-Tu) in prokaryotes and 1α (eEF-1α) in eukaryotes and carried by this factor to the ribosomal A site, where the anticodon interacts with the codon on the messenger RNA and the aminoacylated 3′-CCA end interacts with peptidyl transferase. It is translocated to the P site once the amino acid is incorporated into the growing polypeptide. Initiator tRNAs are different from the majority of tRNAs, called elongators, in that they possess certain features that are specific for initiation factors and against elongation factors; they bind to the P site of the ribosome when these are assembled for protein synthesis. They are usually charged with methionine.

In this chapter we discuss the structure of cytoplasmic tRNA at several stages of its cellular translational 'career.' Although these molecules are seldom free in solution, i.e. uncomplexed to another molecule, be it a protein or a ribonucleoprotein particle, the structures of three tRNAs were determined in their 'free', i.e. uncomplexed, states. The first part of the chapter will thus focus on the three-dimensional (crystal) structures of these tRNAs. The second important milestone in their cellular activity is aminoacylation. Three cases of tRNAs bound to their cognate aminoacyl–tRNA synthetases are known in structural detail and they are described in the second part of this chapter. The third part is devoted to tRNA[Phe] complexed with the elongation factor Tu, which takes it to the ribosome that is already synthesizing a polypeptide chain. Finally, tRNAs have been observed on a ribosome by electron microscopy. The relevant structures are summarized in Table 19.1.

2. The free tRNA

Transfer RNAs are 73–93 nucleotides long and can be folded into a similar clover-leaf secondary structure (2,3,27). There are constant features that are present in all tRNAs (Fig. 19.1) and a number of semi-conserved residues, i.e. constant purines or pyrimidines, that are concentrated in the D (dihydrouridine) and T (thymidine) arms. All base pairs in the stems, with few exceptions, are of the Watson–Crick type. The acceptor stem comprises seven base pairs and four additional residues at the 3′ extremity that are not base paired; the last three of these are CCA. The amino acid is attached to the ribose of the 3′-terminal adenosine. The T arm is the most highly conserved stem–loop structure; the helical stem consists of five base pairs and the loop

Table 19.1. High resolution structures of tRNAs, tRNA-binding proteins, and tRNA-protein complexes

tRNA/protein	Organism	Resolution (Å)	Reference
Transfer RNA			
tRNA^{Phe}	*S. cerevisiae*	2.5	32
			30
tRNA^{Asp}	*S. cerevisiae*	3.0	16
			39
tRNA^{Met_f}	*E. coli*	3.5	17
tRNA^{Met_i}	*S. cerevisiae*	3.0	18
Aminoacyl–tRNA synthetases			
GlnRS:tRNA^{Gln}:ATP	*E. coli*	2.8	19
		2.5	92
GluRS	*T. thermophilus*	2.5	120
TyrRS:TyrAMP	*B. stearothermophilus*	2.3	121
TrpRS:TrpAMP	*B. stearothermophilus*	2.9	122
MetRS	*E. coli*	2.3	118
AspRS:tRNA^{Asp}:ATP	*S. cerevisiae*	2.7	20
			103
AspRS:AspAMP	*T. thermophilus*	2.8	158
LysRS:Lys	*E. coli*	2.8	125
SerRS	*E. coli*	2.5	109
SerRS:ATP; SerRS:SerAMP	*T. thermophilus*	2.5	111
SerRS:tRNA^{Ser}:SerAMP	*T. thermophilus*	2.7	21,112
HisRS:HisAMP	*E. coli*	2.6	128
HisRS:HisOH:ATP	*E. coli*	2.8	159
HisRS:His	*T. thermophilus*	2.7	160
GlyRS	*T. thermophilus*	2.75	127
PheRS	*T. thermophilus*	2.9	129
Elongation factors			
EF-Tu:GDP	*E. coli*	2.5	135
			136
EF-Tu:GppNHp	*T. thermophilus*	1.7	137
EF-Tu:GppNHp	*T. aquaticus*	2.5	138
EF-Tu:Phe-tRNA^{Phe}:GTP	*T. aquaticus*	2.7	22
EF-Tu:EF-Ts:	*E. coli*	2.5	139
EF-G:GDP	*T. thermophilus*	2.7	161
EF-G	*T. thermophilus*	2.85	162
tRNA-modifying enzymes			
tRNA-guanine transglycosylase	*Zymomonas mobilis*	1.85	50
Met-tRNA^{Met_f} formyltransferase	*E. coli*	2.0	132
Ribosomal particle			
Ribosome	*E. coli*	23.	151
Ribosome:tRNA	*E. coli*	25.	23
		20.	24

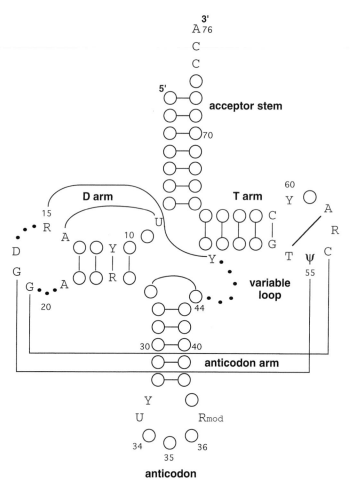

Fig. 19.1. Clover-leaf diagram of the secondary structure of a generalized tRNA. The conserved residues are marked in capital letters; the conserved purines as R and pyrimidines as Y. R$_{mod}$ stands for a heavily modified R. Variant residues in fixed positions are indicated by circles. Elements of variable size are drawn as bold dots. Some conserved tertiary interactions are shown by connecting lines.

contains seven nucleotides whose sequence is TΨCRANY, where Ψ is a pseudouridine, N can be any nucleotide, R is a purine, and Y a pyrimidine. The stem ends with a G:C base pair on the T loop side. The anticodon stem is built of five base pairs and the loop has seven nucleotides. The three central bases comprise the anticodon and thus vary according to the accepting activity of the tRNA. The D arm is more variable; its stem is three or four base pairs long and the D loop may have 7–11 residues. The D loop contains some conserved residues, such as two invariable Gs and an A at the beginning of the loop. There is a conserved U at position 8, between the acceptor and D stems. The variable loop is the most variable element, ranging in length from 4 to 21 bases; however, most of the variable loops are short. The structures of tRNAs were recently reviewed by Dirheimer *et al.* (28).

The two tRNAs whose structures were determined first, tRNA^{Phe} and tRNA^{Asp}, are both elongator tRNAs, i.e. they participate in the elongation cycle of protein synthesis. The structures of two initiator tRNAs have also been determined, one prokaryotic and one eukaryotic. They both possess distinct features that enable them to help initiate protein synthesis on the ribosome. All these structures were solved by X-ray crystallography. The three-dimensional structure of yeast tRNA^{Ser} was deduced from biochemical data using the structural framework of the two elongator tRNAs.

2.1 Yeast tRNA^{Phe}

Yeast tRNA^{Phe} is considered the canonical molecule since it was the first known tRNA structure. The numbering of all tRNA sequences is based on that of tRNA^{Phe}. Its crystal structure was determined in the 1970s by two groups (14,15,29–34). Several reviews were written on its structure in the same period (27,35–37). The clover-leaf secondary structure contains the constant features as described above (Fig. 19.2a). It has a wobble G4:U69 base pair in the acceptor stem. The D stem comprises four base pairs, and the D loop contains eight nucleotides. The variable loop is small, extending over five nucleotides.

The molecule is folded into an L-shaped structure (Fig. 19.2b), with the two limbs nearly perpendicular to each other, and is 20 Å thick. The main structural element is the A-form RNA double helix, which has 11 base pairs per turn (Table 19.2). The principal characteristics of this form are a wide and shallow minor groove, a deep major groove, and base pairs tilted relative to the helix axis (see Chapters 1 and 17). The segments having this structure correspond to the base paired portions of the clover-leaf. The acceptor stem stacks onto the T stem; this combination forms one limb of the L structure. The anticodon and D stems stack to form the other limb of the L.

Table 19.2. Average helical parameters of A-RNA, tRNA^{Phe} and tRNA^{Asp} [a]

Stem	Twist/residue (°)	Rise/residue (Å)	Residues/turn
A-RNA	32.7	2.8	11.0
tRNA^{Phe}			
D	35.8	2.36	10.1
Anticodon	32.6	2.68	11.0
T	31.1	2.71	11.6
Acceptor	33.6	2.51	10.7
tRNA^{Asp}			
D	33.0	2.63	10.9
Anticodon	32.1	2.52	11.2
T	34.3	2.03	10.5
Acceptor	32.5	2.62	11.1

[a] Adapted from ref. 163.

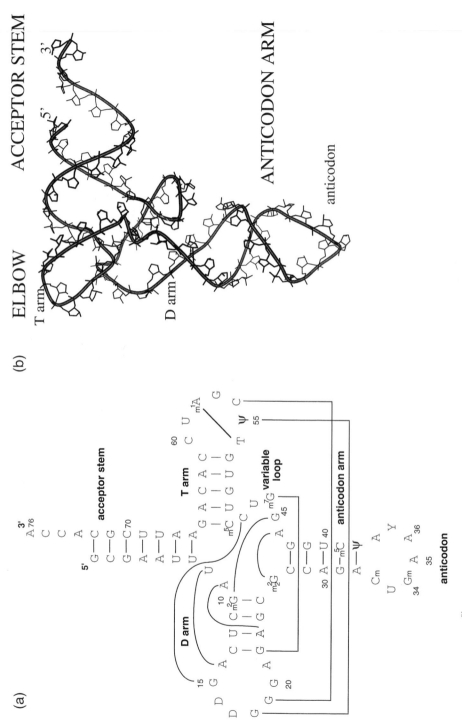

Fig. 19.2. Yeast tRNA^Phe. (a) Clover-leaf representation. The tertiary interactions are shown by connecting lines. (b) Three-dimensional fold. The backbone is shown as a stick rendering and the phosphate atoms are traced as a thick black line. (From ref. 124, by permission of Oxford University Press.)

The two limbs are held together at the elbow, which is structurally the most complex part of the tRNA. It is stabilized by tertiary interactions between the D and T loop and strongly anchored by what is known as the augmented D helix. The latter is formed by the helical portion of the D stem, the two residues between the acceptor and D stems, namely U8 and A9, and the flanking residues of the variable loop. These bases form tertiary base triples with the D helix. Most tertiary interactions involve hydrogen bonds between bases that result in base pairs or base triples that are not of the Watson–Crick type (Fig. 19.3). Starting at the bottom of the acceptor stem, as seen in the clover-leaf diagram, and moving towards the D stem, residue U8 makes a reverse Hoogsteen interaction with A14; the backbones are antiparallel. The following residue, A9, interacts with base A23 in a symmetric fashion, and the backbones are parallel; A23 pairs in the Watson–Crick manner with U12. Proceeding along the strand, G10, which is the first residue in the D stem, forms a standard Watson–Crick base pair with C25 and a tertiary interaction with G45 of the variable loop, whose backbone runs parallel with that of G10. Base pair C11:G24 is not involved in any tertiary interactions with bases, but wedges in between two triples, changing the helix axis as a result. U12, as already mentioned above, is involved in a base triple with A23 and A9. It is followed by C13, which base pairs in the standard Watson–Crick fashion with G22; the latter interacts in a non-standard and asymmetric fashion with G46, which is part of the variable loop and whose backbone runs antiparallel to that of G22. A14 forms the 5′-flank of the D loop and associates with U8. G15 interacts in a reverse Watson–Crick fashion with C48 of the variable loop; their backbones run parallel to each other. This is also known as the Levitt pair, for it had been predicted by Levitt (38) before the crystal structure was determined. Two interactions between the D and T loop follow: the non-standard G18:Ψ55 pair, and the Watson–Crick pair G19:C56. An intra-T loop strut is formed by the reverse Hoogsteen base pair T54:A58. The transition from the D stem to the anticodon stem is marked by a 24° kink in the helical axis between the two, introduced by the hinge formed by the symmetrical heteropurine base pair G26:A44. Furthermore, several bases form hydrogen bonds to the backbone. C11 of the D stem contacts the 2′-OH of A9. A21, which flanks the D loop on the 3′-end, interacts with the 2′-OH of the ribose of U8. Ψ55 contacts the phosphate of residue 58. G57, which follows the TΨC in the T loop, hydrogen bonds with the 2′-OH of the riboses of residues 18 and 55, and with the 4′-O of residue 19. The base of G57 also intercalates between base pairs G18:Ψ55 and G19:C56 and thus enhances the stability of the junction. Most bases of the tRNA are engaged in stacking interactions, which provides additional stabilization of the tertiary structure. Only D16, D17, and G20 of the D loop, and U47 of the variable loop, point into the solvent and do not engage in stacking interactions.

The anticodon loop is similar to the T loop in that both contain seven residues. They also have similar conformations of the backbone, which makes a sharp turn, and a U residue at the bend (U33 and Ψ55) that stabilizes the bend by interacting with a phosphate moiety on the opposite strand of the loop. This turn was dubbed the uridine, or U turn (Fig. 19.4). The U base terminates the hydrophobic stack emanating from the anticodon stem by making a van der Waals contact with a phospate group.

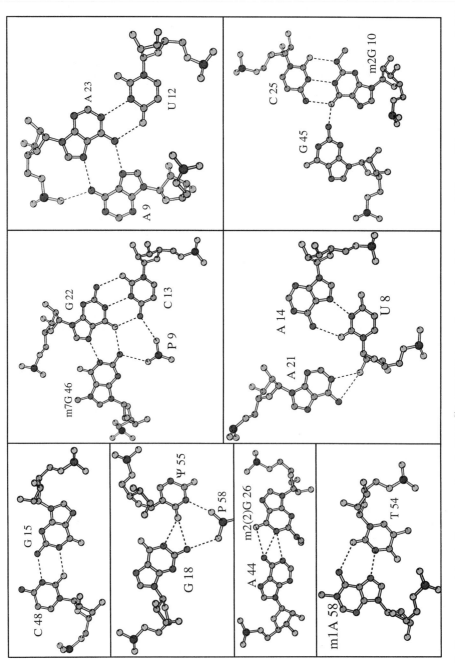

Fig. 19.3. Some representative tertiary interactions in tRNA^{Phe}. G19:C56 is not shown, since it is a standard Watson–Crick base pair. (From ref. 124, by permission of Oxford University Press.)

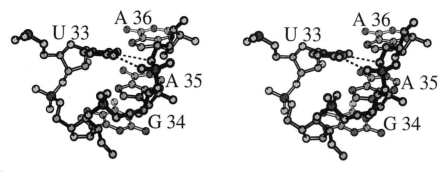

Fig. 19.4. Stereoview of the anticodon U turn in tRNA^Phe.

The acceptor stem contains a wobble G4:U69 base pair, which introduces a series of rotations in the backbone that result in the displacement of phosphate 5 by about 2 Å from what would be its normal position in a standard double helix. The conformation is stabilized by a water molecule.

There are four hexacoordinated Mg^{2+} binding sites in tRNA^Phe (Fig. 19.5), two on the back side of the elbow, one in the augmented helix region, and the fourth in the anticodon loop. While some of the metal coordination sites are filled by phosphate oxygens, most direct binding is carried out by water molecules, which then interact mostly with the phosphates of the tRNA, although some of them are liganded by nucleotide bases.

2.2 Yeast tRNA^Asp

The structure of yeast tRNA^Asp (16,39) is globally similar to that of tRNA^Phe. Its clover-leaf secondary structure shows the constant features as indicated above (Fig. 19.6a). As for the variable features, it possesses a three-base pair D stem, a 10-nucleotide D loop, and its variable loop is shorter than that of tRNA^Phe, consisting of four residues. While the D loop contains no G:C base pairs, both the acceptor and anticodon stems are rich in them.

The overall folding of tRNA^Asp (Fig. 19.6b) is the same as that of tRNA^Phe; both have similar L-shaped structures and are 20 Å thick. However, the conformation of tRNA^Asp is more open, resembling a boomerang. The angle between the helical axes of the acceptor T stem helix and the anticodon D stem limb is by about 10° more obtuse. The double helical segments are based on the RNA-type double helix (Table 19.2). The relative positions of the D and T loops are different as well. Similarly to tRNA^Phe, the transition from the D stem to the anticodon stem is marked by a 25° break in the helical axes between the two, introduced by the hinge formed by the mismatched symmetrical purine–purine base pair G26:A44.

Tertiary interactions (Fig. 19.7) are for the most part similar to those found in tRNA^Phe. Unlike tRNA^Phe, all bases of the variable loop participate in such contacts; the shorter variable loop induces a different interaction of the base of A21, which interacts with the base of A14 and also contacts the ribose of U8. U8 and A14 are

(a)

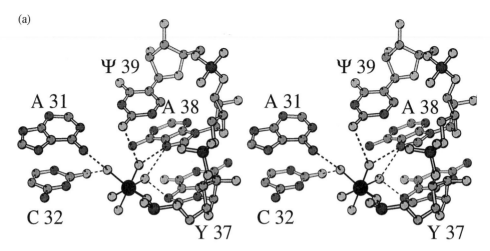

(b)

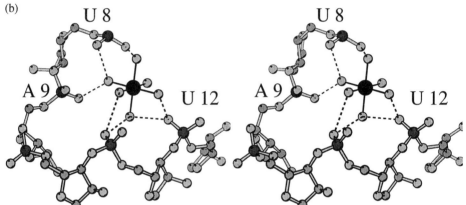

(c)

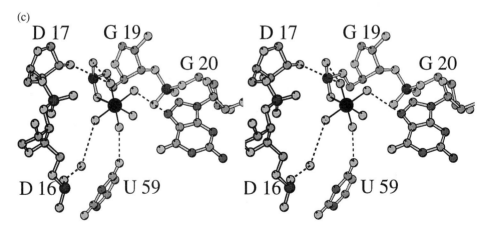

Fig. 19.5. Stereoviews of three magnesium ion-binding sites in tRNA[Phe]: (a) in the anticodon loop, (b) in the augmented D helix, and (c) in the D loop.

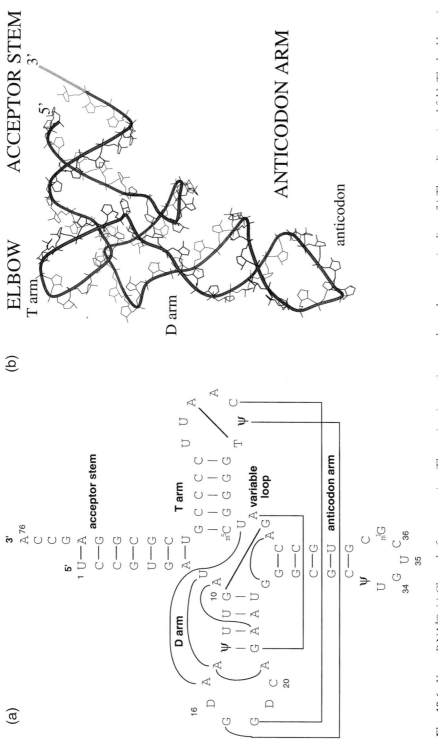

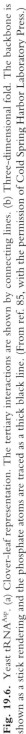

Fig. 19.6. Yeast tRNA^Asp. (a) Clover-leaf representation. The tertiary interactions are shown by connecting lines. (b) Three-dimensional fold. The backbone is shown as a stick rendering and the phosphate atoms are traced as a thick black line. (From ref. 85, with the permission of Cold Spring Harbor Laboratory Press.)

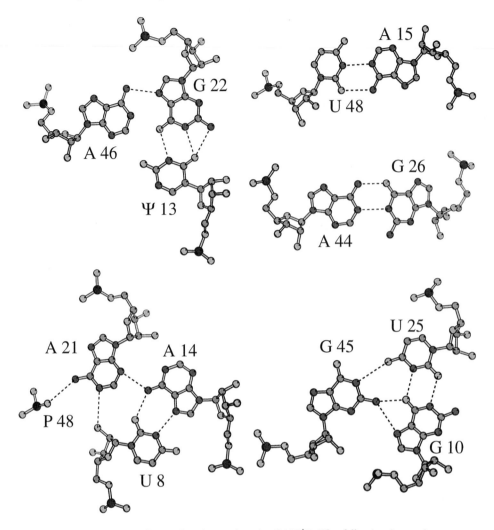

Fig. 19.7. Some representative tertiary interactions in tRNA[Asp]. The following interactions are not shown, since they are very similar to the ones in tRNA[Phe]: A9:A23:U12, G17:Ψ55, G18:C56, and T54:A58.

engaged in a reverse Hoogsteen base pairing. As in tRNA[Phe], A9 interacts in a symmetrical fashion with base A23, which in turn pairs in the Watson–Crick manner with U12; G10 forms a wobble base pair with U25 and a tertiary interaction with G45. Base pair U11:A24 is not involved in any tertiary interactions with bases, but forms a wedge between the two neighbouring triples. Unlike tRNA[Phe], Ψ13 forms a wobble pair with G22; the latter interacts with G46, also in a non-standard way. The Levitt pair in this case is the reverse Watson–Crick A15:U48 base pair; in tRNA[Phe] this is a reverse Watson–Crick G:C pairing. Owing to the different relative positions of the invariant Gs in the D loop, the interactions between the D and T loop are slightly different from those in tRNA[Phe]. G17 interacts with Ψ55 and G18 forms a

Watson–Crick base pair with C56. A57 forms a backbone contact with the ribose of Ψ55, and intercalates between base pairs G17:Ψ55 and G18:56, much as it does in tRNA^Phe. Ψ55 also interacts with the phosphate of A58. An intra-T loop strut is formed by the reverse Hoogsteen base pair T54:A58. Residues D16, D19, and G20 of the D loop project into the solvent and do not participate in stacking interactions.

The anticodon loop of tRNA^Asp has the same fold as that in tRNA^Phe. It has the U turn structure and the same stacking pattern. In the crystal, the anticodons of two tRNA^Asp molecules interact via their self-complementary sequences in a two-fold symmetrical fashion. This duplex formation is most likely responsible for the wider angle between the anticodon and acceptor stems, as suggested by solution studies (40).

2.3 E. coli initiator tRNA^Met_f

The structure of *E. coli* tRNA^Met_f (17) is globally similar to those of elongator tRNA^Phe and tRNA^Asp. In fact, its structure was solved by molecular replacement using yeast tRNA^Phe as the search model. Overall, it is 77 nucleotide residues long, and its secondary structure is a clover-leaf (Fig. 19.8) that shows most of the constant

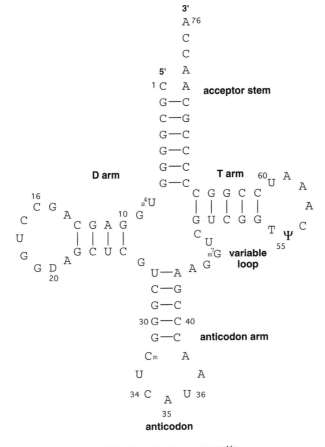

Fig. 19.8. Clover-leaf representation of the *E. coli* initiator tRNA^Met_f.

features indicated above, with the exception that the first base pair of the acceptor stem is a C:A mismatch. As for the variable features, it possesses a four-base pair D stem and a nine-nucleotide D loop. Its variable loop consists of five residues and is thus the same length as that of tRNA^{Phe}. All helical stems are rich in G:C base pairs.

The overall folding of tRNA$^{Met}_f$ is the same L-shaped structure as that of the elongator tRNAs. The conformation of tRNA$^{Met}_f$ is more akin to that of tRNA^{Phe}, since the helical axes of the acceptor T stem helix and the anticodon D stem limb are nearly orthogonal. As in the two elongator tRNAs, the double helical segments are based on the RNA-type double helix. A mismatched terminal pair of bases of the acceptor stem makes the acceptor end more flexible, which is reflected in the more curved conformation.

Most of the tertiary interactions observed in tRNA^{Phe} appear to be present in tRNA$^{Met}_f$, although some differences do exist. The base of residue A57 intercalates between the bases of G18 and G19. The nucleotide does not make as many backbone interactions as its analogue in tRNA^{Phe}, G57. The D loop is one base longer and more tightly organized than that of tRNA^{Phe}; it is also folded towards the core. Residues C17 and U17a do not extend into the solvent but are closer together and the bases stack on each other.

Outside of the core, the main differences lie at the end of the acceptor arm and the anticodon loop. The terminal base pair of the acceptor stem is a mismatch. The anticodon loop is superficially similar to that of tRNA^{Phe} for its nearly similar stacking. However, the orientation of U33 is dramatically different. The base points into the solvent, whereas it is stacked in tRNA^{Phe}. Thus, it cannot hydrogen bond with the phosphate of residue 36, but the ribose does. Hence the phosphate is in a slightly different position, which results in a marked shift away from the loop in the position of the phosphate moeity of nucleotide 35.

2.4 Yeast initiator tRNA$^{Met}_i$

The structure of yeast tRNA$^{Met}_i$ (18,41) is globally similar to those of elongator tRNA^{Phe} and tRNA^{Asp}. Overall, it is 75 nucleotide residues long, and its secondary structure is a clover leaf that exhibits all of the constant features indicated above (Fig. 19.9a). As for the variable features, it possesses a four-base pair D stem and a shorter, seven-nucleotide, D loop. Its variable loop consists of five residues and is thus the same length as that of tRNA^{Phe}.

The overall folding of tRNA$^{Met}_i$ is the same L-shaped structure as that of the elongator tRNAs (Fig. 19.9b). In tRNA$^{Met}_i$, like tRNA$^{Met}_f$, the helical axes of the acceptor T stem helix and the anticodon D stem limb are nearly orthogonal and the double helical segments are of standard RNA-type.

All of the tertiary interactions seen in tRNA^{Phe} are present. The U8:A14:A21 and C13:G22:m^{7}G46 triples, the G15:C48 reverse Watson–Crick, and the m^{22}G26:A44 symmetrical heteropurine interactions are essentially identical in the two tRNAs. In tRNA$^{Met}_i$, G18 interacts with U55 instead of a Ψ at the same position in tRNA^{Phe}. Other interactions are very similar. The G10:C25 pair interacts with U45 instead of a G, and the essentially homologous triple G9:C23:G12 replaces the A9:A23:U12 of tRNA^{Phe}.

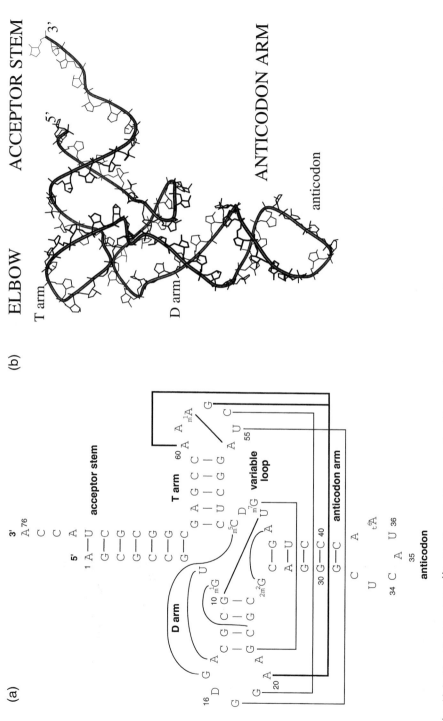

Fig. 19.9. Yeast initiator tRNA^Met_i. (a) Clover-leaf representation. The tertiary interactions are shown by connecting lines. Novel interactions are shown by bold lines. (b) Three-dimensional fold. The backbone is shown as a stick rendering and the phosphate atoms are traced as a thick black line.

The unique features of tRNA$^{Met}_i$ cluster in a region of the core of the three-dimensional structure, giving rise to a unique contiguous surface. They form a substructure specific for eukaryotic initiator tRNAs that is characterized by a shortened D loop, A20 instead of a D, and A54 instead of the T in elongator tRNAs. These lead to some novel tertiary interactions (Fig. 19.10). The A54:A58 strut is analogous to that seen in elongators, although different in the nature of the bases. The asymmetric homopurine pair shifts the position of the backbone at residue 58 slightly. The nearly invariant pyrimidine at position 60 in the T loop is replaced by an A. The substructure is stabilized by a network of hydrogen bonds. Residue A20 of the D loop interacts with G57, A59, and A60 in the T loop; this interaction is sequence specific and forms a strong bridge between the two loops. It seems to fill the role of the Mg^{2+}

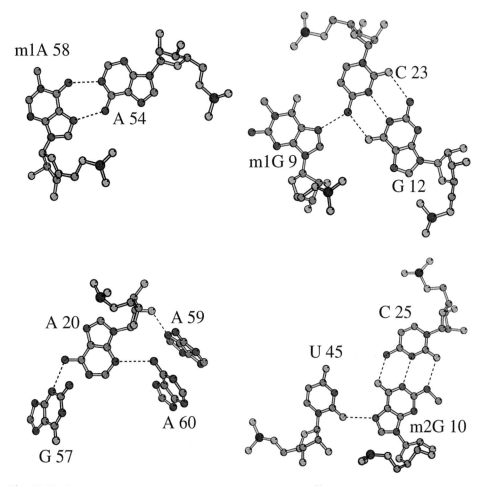

Fig. 19.10. Some representative tertiary interactions in yeast tRNA$^{Met}_i$. The following interactions are not shown, since they are very similar to the ones in tRNA^{Phe}: U8:A14:A21, G22:C13:m^{7}G46, G15:C48, m$^2_2$G26:A44, G18:U55, G19:C56, and T54:A58.

coordinated in the same region of tRNA^{Phe}. These elements represent a functional differentiation within the common tRNA fold.

The tRNA shows yet another unique feature, which is a novel modification, a 5′-phosphoryl group O2′ glycosylated to the ribose of residue A64. It appears on the surface in the minor groove and is accessible to solvent and other molecules. The phosphoryl group interacts with the base of the neighbouring residue 63. Its role seems to be a rejection signal for elongation factors.

The anticodon arm is not defined well enough in the electron density map to furnish detailed structural information (18). The sequence is distinct from that of elongator tRNA$^{Met}_m$; the invariant G:C base pairs of the anticodon stem appear essential for the initiation function and are not found in elongators (42).

2.5 Yeast tRNA^{Ser} in solution

The clover-leaf secondary structure of yeast tRNA^{Ser} (Fig. 19.11a) has the standard constant features, a three-base pair D stem, a 10-nucleotide D loop, and a large variable loop. The latter is built of a four-base pair stem and a three-nucleotide loop, and is flanked by one nucleotide at the anticodon stem and two residues at the T stem. Its structure in solution was probed with a variety of chemical agents (43), along with those of tRNA^{Phe} and tRNA^{Asp}. The sequence and the resulting comparison of protection patterns were combined with the three-dimensional folding of tRNA^{Phe} and tRNA^{Asp} to obtain a model of tRNA^{Ser}. The coordinates of the structure of tRNA^{Asp} were used for the actual model, for it has a more similar D loop. The resulting model (Fig. 19.11b) has the classical tRNA L shape with the extra arm nearly in the plane of the two limbs of the L. There are slight differences in the anticodon loop. The interactions within the T loop are maintained. The variable stem and loop are characterized by tight folding, with a three-nucleotide mini loop capping a four-base pair stem. It is joined to the body of the tRNA in a fashion more akin to that of tRNA^{Asp}. The large variable loop engenders some replacements in the tertiary interactions in the augmented D helix, while some are preserved. Base pair G10:C25 does not seem to interact with base 45, for the latter is engaged in base pairing within the variable loop; the N7 of G10 is accessible to chemical agents. Residue 9 is likely to interact with base pair 12:23 in a different way. The base of residue G47:9, which is analogous to residue 46 in tRNA^{Asp} or tRNA^{Phe}, stacks between bases G9 and A21, which locks the variable stem in its position relative to the body of tRNA^{Ser}.

2.6 Comparison in solution of yeast tRNA^{Phe} and tRNA^{Asp}

The structures of yeast tRNA^{Phe} and tRNA^{Asp} were probed by chemical modification (44). The principal differences were observed in the acceptor stem (namely in purines 4, 71, and 73), in residue A21 of the D loop, and in residue G45 of the variable loop. The N7 of A21 was found to be reactive in tRNA^{Asp} and unreactive in tRNA^{Phe}. The movement of residue A46 towards the interior of the molecule in tRNA^{Asp} and the absence of residue 47 result in a different shape of the variable loop and expose the N7 of A21; the group is protected in tRNA^{Phe} by the modified m^{7}G46. The tertiary

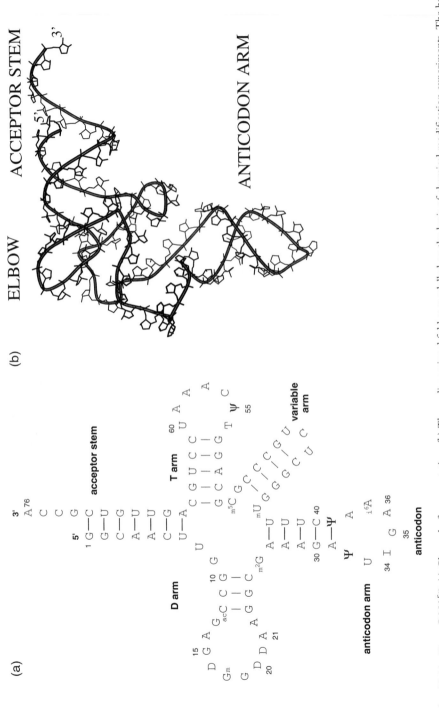

Fig. 19.11. Yeast tRNA^Ser. (a) Clover-leaf representation. (b) Three-dimensional fold as modelled on the basis of chemical modification experiments. The back-bone is shown as a stick rendering and the phosphate atoms are traced as a thick black line.

interaction U8:A14:A21 is different in the two tRNAs. The N7 of G45 is reactive in tRNA^{Phe} and protected in tRNA^{Asp} because of the different stacking of residue 9 between bases 45 and 46. In the acceptor stem, G4 is reactive in tRNA^{Asp} and protected in tRNA^{Phe}, whereas the situation is reversed for the N7 of G71; this occurs because of differences in stacking interactions. When a purine is stacked between two pyrimidines, the N7 is reactive, otherwise it is not; it is also unreactive when it is involved in tertiary interactions. Residues G18, G19, and G34 have N7 exposed and are reactive; they are located in loops. The solution structures largely agree with the crystal structures.

2.7 General principles of tRNA structure

The common feature of all these tRNAs is the overall L structure. Although there are some differences in the details of tertiary interactions, the RNA chain follows the same fold. The two helical arms of the L, built through stacking of the acceptor and T stems in one case and the anticodon and D stems in the other, are both based on the A-form RNA helix. Although their relative orientation with respect to each other may vary from one tRNA to another, it is very close to being orthogonal.

Levitt correctly predicted some interactions before any of the structures were determined (38), namely U8:A14, 9:12:23, G15:C48, 18:55, and 19:56. He also postulated some that were not found in the crystal structures; for example, he had A21 pairing with T54. While his prediction of the two limbs resulting from the stacking of the acceptor stem on the T stem and of the D stem on to the anticodon stem were correct, their relative orientation in the form of a sausage instead of an L was not.

This discussion has so far focused on cytoplasmic tRNAs. Plastids, i.e. mitochondria and chloroplasts, also possess their own translation machineries and pools of tRNAs (45). Mitochondrial tRNAs have some unique features (46). In some cases, they have truncated clover-leaf structures, i.e. a portion of the D or T arm may be absent. In principle, they can still fold in the manner of their cytoplasmic counterparts.

2.8 Nucleotide modifications in tRNA

In addition to the four standard ribonucleotides, tRNA, like many other RNAs, possesses modified nucleotides. It is the most extensively modified RNA species in the cell and possesses the greatest variety of such nucleotides. This subject has been recently reviewed in considerable detail (47–49). The patterns are similar in all phyla, which reflects common evolutionary origins. However, some modifications are specific to certain phylogenetic domains and/or species. Eukaryotic tRNAs are more extensively modified than prokaryotic and mitochondrial tRNAs.

There are more than 80 different types of modifications in all tRNAs. A tRNA species may possess a number of modified bases, all of which, with the exception of the Q base, are introduced post-transcriptionally by a variety of enzymes. There are at least 45 different modification enzymes in a bacterial cell, whose genes represent approximately 1% of the genome. By comparison, about 0.25% of the genome is used to encode the tRNA substrates. A pathway of several of these enzymes may be needed to produce one modified nucleoside. To date, 17 out of about 45 modification enzyme genes have been identified in *E. coli* (47). The structure of one of these enzymes, tRNA–guanine transglycosylase from *Zymomonas mobilis* (50), which changes the guanine in the wobble position of tRNA^{Asn}, tRNA^{Asp}, tRNA^{His}, and tRNA^{Tyr} to a hypermodified base, queuine, is based on an eight-stranded β barrel core; the parallel β strands are connected by simple helices, multiple helices, or even more elaborate combinations of helices and strands. The enzyme contains a zinc (Zn)-binding motif that is implicated in tRNA binding.

Modifications are introduced in a stepwise fashion at different stages during and after processing of tRNA precursors, in an intricate interplay of pathways; the timing depends on the processing stage, substrate concentration, and the activity of a given processing enzyme. As modified nucleotides inhibit RNAase P activity, the 5′ cleavage occurs early. Methylation of ribose moieties occurs in almost mature tRNA. In eukaryotes, some reactions take place in the nucleus while others occur in the cytoplasm (47).

Chemically, any single modification can add or enhance certain properties of a nucleotide base or sugar, which may include the introduction of transient or permanent charges, alteration or restriction of nucleoside or phosphodiester conformation, hindrance of canonical or non-canonical base pairing, facilitation of metal ion coordination, rearrangement of water structure, and formation of new interactions leading to new conformations and chemistries. Modifications thus extend the pool of functional groups in a nucleic acid beyond the four standard bases. They may be relatively simple, such as methylation (as in rT), thiolation (as in 4-thioU) or glycosidic bond substitution (as in pseudouridine, Ψ), or more complex, involving additions of amino acids or heterocyclic functional groups. However, even a simple methylation may alter hydrophobicity, inhibit Watson–Crick base pairing, or introduce a charge when added on to a heterocyclic nitrogen (e.g. N7) (49).

Structurally, modified and unmodified tRNAs are similar, either in solution (51,52) or in a complex with a protein (53). However, unmodified tRNAs are not as stable, as indicated by their lower melting temperatures (53–56) or chemical and enzymatic accessibility (55,57). Modifications thus enhance the stability of tRNA structure. Uridine modifications are very widespread, representing a large proportion of all modifications. The most frequently encountered are the D, Ψ, and thioU nucleotides (49). Ψ appears to stabilize the structure by reordering neighbouring water molecules (53). D (dihydrouridine) is a non-aromatic (saturated) version of U and is found in the D loop and sometimes in the variable loop. It alters the sugar pucker to C2′-*endo* and restricts backbone conformation (58,59). Thiouridines, such as 2-thio- and 4-thio-U, restrict nucleotide conformation (60). Methylations are also involved in structural stabilization through enhancement of metal binding and base stacking, restriction of conformational flexibility, and reordering of water (61,62).

Most modifications are not essential for aminoacylation, which has been demonstrated by a number of biochemical studies performed with unmodified tRNAs obtained by transcription *in vitro*. In *E. coli*, most tRNAs accept cognate amino acids. Examples include tRNA^{Val} (52), tRNA^{His} (63), tRNA^{Gln} (64), and tRNA^{Phe} (65). There are three notable exceptions, tRNA^{Ile}, tRNA^{Glu}, and tRNA^{Lys}. The mNm^{5}s^{2}U34 is a key determinant of tRNA^{Glu} identity (66,67). The absence of the same modification in tRNA^{Lys} reduces the rate of aminoacylation by two orders of magnitude (68). Aminoacylation of tRNA^{Ile} is similarly reduced when the lysidine modification of C at the wobble position 34, k^{2}C34, is replaced by a C (69). The kinetic parameters of most aminoacylations differ slightly when unmodified tRNAs are used, compared with modified tRNA; modifications may modulate interactions with aaRS (aminoacyl–tRNA synthetase). A notable exception is tRNA^{Asp} from *E. coli*, where the unmodified species can also be charged by ArgRS (70). Thus, modifications can constitute antideterminants, but not in all cases.

Modifications also play an important role in the way tRNAs interact with the ribosome and associated translation (initiation and elongation) factors. For example, 2′-O-ribosyladenosin[phosphate] at position 64 of eukaryotic initiator tRNA$^{Met}_i$ is likely to be a negative determinant for acceptance by the elongation factor eEF-1α (71–73). Moreover, modified nucleotides may strengthen tRNA–ribosome association (49,74). Furthermore, modified nucleotides at the wobble position of the anticodon (residue 34) modulate codon reading by enhancing the conformational flexibility or rigidity of the nucleotide; this extends or restricts the wobble read-out of the corresponding codon nucleotide (48).

3. tRNA in aminoacylation

The common structural fold shared by tRNAs enables them to interact with tRNA-processing enzymes and the protein synthesis apparatus. However, they show certain distinguishing features that are recognized by a cognate aminoacyl–tRNA synthetase (aaRS) and rejected by a non-cognate aaRS; these features, named identity determinants, were first identified in tRNA^{Ser} (75). They are distributed in differential patterns in different sets of tRNAs and comprise the necessary and sufficient elements for recognition by the cognate aaRS and rejection by non-cognate aaRS, i.e. the identity of a given set of isoacceptor tRNAs. They are located primarily in the anticodon loop, the acceptor arm, and a few base pairs in the T and D stems (76). Biochemical analyses using *in vivo* and *in vitro* techniques have led to the elucidation of the identity determinants for a number of tRNAs (77) by using two approaches: identity swapping and transplantation of identity elements. In the former, minimal changes are introduced into a tRNA such that it becomes recognized by the new aaRS. The experiment must also prove that the introduced elements constitute the identity of the new system (78). In the latter method, variants of a particular tRNA are synthesized and analysed for their capacity as substrates for the aaRS involved (79). Since efficient aminoacylation depends on the overall conformation of the tRNA as well as on the presence of the elements, tRNAs obtained in such a way are not optimized for the new amino acid acceptance (80).

Aminoacyl–tRNA synthetases (aaRSs) catalyse the esterification of the amino acid to one of the hydroxyl groups of the 3′-terminal adenosine of the tRNA via an aminoacyl–adenylate intermediate. The energy for the reaction is supplied by the hydrolysis of ATP (81). Each amino acid may be specified by several isoacceptor tRNA species, while, in general, there is one aaRS for each amino acid (81,82). Several reviews have been published on the subject of aaRSs (83–86).

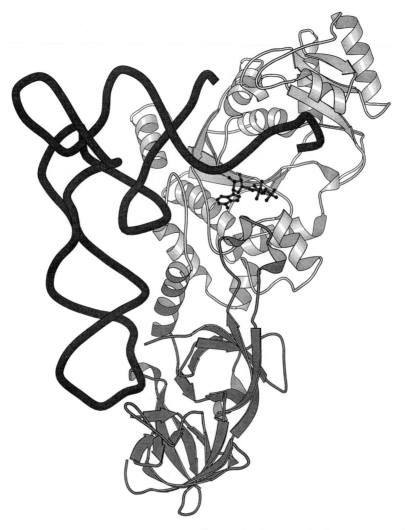

Fig. 19.12. Glutaminyl–tRNA synthetase: tRNA^{Gln} complex from *E. coli*. The acceptor arm-binding domain is in light grey and the anticodon-binding module is in dark grey. The tRNA is drawn with its phosphate chain traced as a thick line. (From ref. 85, with the permission of Cold Spring Harbor Laboratory Press.)

Although they catalyse what is essentially the same reaction, aaRSs are a diverse family of enzymes, whose quaternary structures can be monomeric, dimeric, and even tetrameric. Yet these enzymes can be grouped into two classes of ten members each (13,20), which are correlated to two structural and functional solutions to the organization of the active site domain. The active site domains of class I aaRSs contain the Rossmann fold nucleotide-binding motif, an alternating α–β structure with a central parallel β sheet and show signature amino acid sequences HIGH and KMSKS. These aaRSs esterify the amino acid to the 2′-OH of the 3′-terminal ribose. The active site modules of class II aaRSs are based on an antiparallel β sheet and have three concatenated homologous sequence motifs, 1, 2, and 3 (87,88); the latter two motifs form the catalytic site, while motif 1 is involved in the dimer interface, as these aaRSs are obligate dimers (88). These enzymes esterify the amino acid to the 3'-OH, with the exception of phenylalanyl–tRNA synthetase (PheRS), which acylates the 2'-OH (11–13,87). To the active site core domains that define the class, which typically consist of about 300 to 400 residues, are attached polypeptide modules that lead to different sizes and tRNA specificities of aaRS (88,89).

3.1 tRNA^{Gln} complexed with glutaminyl–tRNA synthetase

Glutaminyl–tRNA synthetase (GlnRS) is a class I aaRS. The enzyme from *Escherichia coli* is a monomer of 553 amino acid residues and has a molecular weight of 63 kDa (90). It is an elongated protein consisting of two major modules: the active site module consists of the parallel β sheet nucleotide-binding fold (the Rossmann fold) into which is inserted the acceptor-binding subdomain, and the anticodon-binding module comprises two β barrels (19) (Fig. 19.12). In the active site, the two motifs characteristic of class I aaRS, HIGH and MSK, interact with each other, forming a surface that binds the ATP molecule in an extended conformation. The 2'-OH of tRNA^{Gln} and the α-phosphate of ATP are within hydrogen bonding distance (91). GlnRS binds its cognate tRNA^{Gln} in what is considered a class I-characteristic mode: the acceptor arm of the tRNA interacts with the active site domain on the minor groove side, and the variable loop faces the solvent. The interface between the two extends over 2700 Å^2 (92).

The clover-leaf secondary structure of tRNA^{Gln} (Fig. 19.13) shows all the constant features and relatively typical variable features. It possesses a three-base pair D stem and a nine-nucleotide D loop. Its variable loop consists of five residues. All stems are GC-rich. Its structure was solved in complex with GlnRS and it is assumed that its uncomplexed structure resembles that of tRNA^{Phe}. Its overall folding is the same as that of tRNA^{Phe}, giving rise to the classical L-shaped structure that is 20 Å thick. However, both limbs of tRNA^{Gln} have undergone dramatic conformational change as they are induced to fit the enzyme. The terminal base pair of the acceptor stem is unravelled to facilitate the bending of the 3'-terminal CCA into the active site. The anticodon is spread out so as to maximize interactions with the protein.

The core of tRNA^{Gln} is very similar to that of tRNA^{Phe} and possesses most tertiary interactions found in the former (Fig. 19.14). The 4-thioU:A14 pairing is enhanced through a base–base contact with residue A21; in addition to the contact made by the base of A21 to the ribose of U8 also seen in tRNA^{Phe}, a similar contact exists between A14 and A21. The 12:23:9 triple is similar, although the nature of the bases, C12:G23:C9, is different. Unlike in tRNA^{Phe}, base pair G10:C25 forms no tertiary contact with A45; the latter, however, forms a twofold symmetrical purine–purine pair with A13, which also interacts with A22. As in tRNA^{Phe} the G15:C48 is a reverse Watson–Crick base pair. There is no residue 17, which makes the D arm shorter than that of tRNA^{Phe}. The G18: Ψ55 and G19:C56 interactions between the D and T loops are the same as in tRNA^{Phe}, as is the internal T loop pair T54:A58. The base of C20 contacts that of G19 and the ribose of G57. The base of G57 is stacked between those of G18 and G19. The mismatched purine–pyrimidine pair, A26:C44, at the

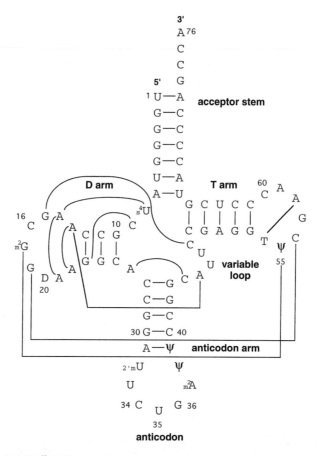

Fig. 19.13. *E. coli* tRNA^{Gln}. Clover-leaf representation. The tertiary interactions are shown by connecting lines.

bottom of the augmanted D helix replaces a purine–purine pair G26:A44 of tRNA^{Phe}. The bases of C16 and U46 project into the solvent.

There are three main regions in tRNA^{Gln} that interact with GlnRS: the acceptor arm, part of the D arm, and the anticodon loop (92) (Fig. 19.15). Biochemical analyses performed *in vitro* (64,93) and *in vivo* (94–96), in conjunction with analysis of the three-dimensional structure of the complex, have localized the identity of tRNA^{Gln} to the acceptor stem and the anticodon, with one element in the D stem, G10. In addition to the residues that are directly involved in protein–RNA interactions, tRNA^{Gln} possesses nucleotides that enable it to adopt the conformation that facilitates its binding to GlnRS. These residues are in the acceptor stem (G73 and base pair U1:A72) and in the anticodon loop (2'mU32, U33, m^{2}A37, and Ψ38) (19,92).

The three terminal base pairs in the acceptor arm of tRNA^{Gln} are the principal recognition elements for GlnRS, and the enzyme uses two loops and an α helix to interact directly with them. The first loop, tipped with Leu-136, denatures base pair

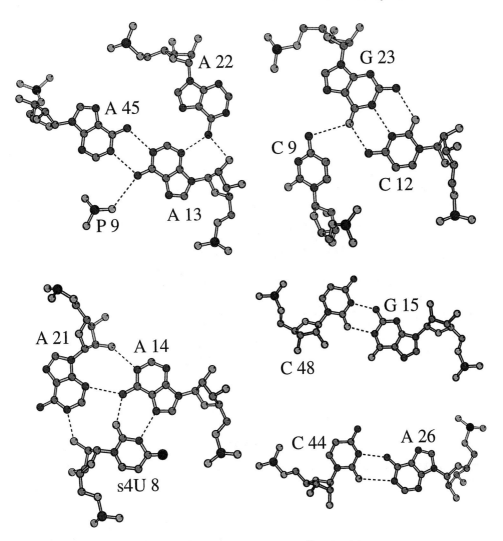

Fig. 19.14. Some representative tertiary interactions in tRNA^Gln. The following interactions are not shown, since they are very similar to the ones in tRNA^Phe: m²G18:Ψ55, G19:C56, and T54:A58. G10:C25 is a standard Watson–Crick base pair and does not participate in a base triple; therefore, it is not shown.

U1:A72, which facilitates the bending of the 3′-terminal CCA into the active site. This bend is stabilized by an intramolecular interaction within the tRNA: the exocyclic amino group of G73 hydrogen bonds with the phosphate moiety of residue 72 (Fig. 19.16a). The second loop (residues 179–184) lines up the backbone so that the peptide oxygen of Pro-181 hydrogen bonds with the exocyclic amino group of G2 and the peptide nitrogen of Ile-183 forms a water-mediated contact with C72 (Fig. 19.16b). Residue Asp-235 of the α helix interacts directly with G3 and contacts C70 through a water molecule (Fig. 19.16c). The helix extends into the active site (19).

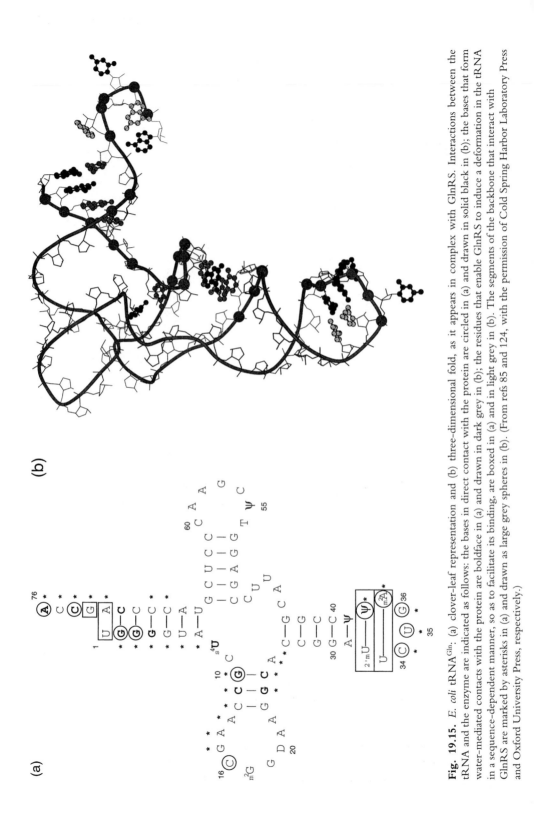

Fig. 19.15. *E. coli* tRNA^Gln. (a) clover-leaf representation and (b) three-dimensional fold, as it appears in complex with GlnRS. Interactions between the tRNA and the enzyme are indicated as follows: the bases in direct contact with the protein are circled in (a) and drawn in solid black in (b); the bases that form water-mediated contacts with the protein are boldface in (a) and drawn in dark grey in (b); the residues that enable GlnRS to induce a deformation in the tRNA in a sequence-dependent manner, so as to facilitate its binding, are boxed in (a) and in light grey in (b). The segments of the backbone that interact with GlnRS are marked by asterisks in (a) and drawn as large grey spheres in (b). (From refs 85 and 124, with the permission of Cold Spring Harbor Laboratory Press and Oxford University Press, respectively.)

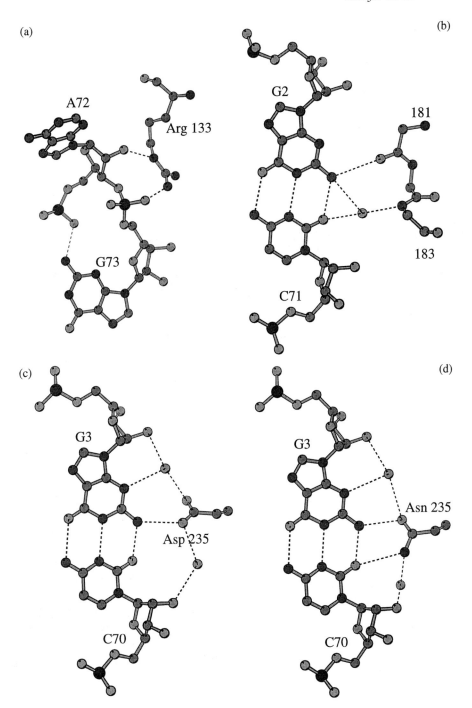

Fig. 19.16. Sequence-specific interactions between GlnRS and the acceptor arm of tRNA^Gln: (a) intramolecular interaction between G73 and the phosphate of A72; (b) interactions with base pair 2:71; interactions with base pair 3:70 by the (c) wild type and (d) mutant D235N GlnRS. (From ref. 164, with the permission of Cambridge University Press.)

Mutating residue 235 to Asn (Fig. 19.16d) or Gly results in changed interactions with base pair G3:C70, i.e. two direct hydrogen bonds or altered water structure, respectively (97). The GlnRS enzymes harbouring these mutations, which were isolated using an *in vivo* suppression screen (98,99), exhibit a slightly altered ability to glutaminylate wild type tRNA^{Gln}, while their ability to discriminate against a non-cogante U3:A70 base pair is lowered, which manifests itself in incorrect acylation of the amber suppressor derived from tRNA^{Tyr} (*supF*) with glutamine.

The anticodon bases of tRNA^{Gln} are essential recognition elements for GlnRS, as was shown very early by Seno *et al.* (100) and is seen in the crystal structure (92)

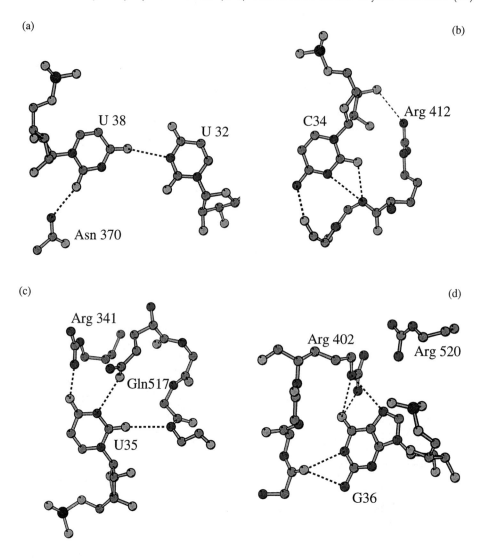

Fig. 19.17. (a) an additional non–Watson–Crick base pair in the anticodon loop of tRNA^{Gln}. Sequence-specific interactions between GlnRS and the anticodon loop of tRNA^{Gln}: bases (b) 34, (c) 35, and (d) 36. (From ref. 164, with the permission of Cambridge University Press.)

(Fig. 19.17). The anticodon loop undergoes a dramatic conformational change whereby the anticodon stem is extended by two non-Watson–Crick-type base pairs, which are not present in free tRNA^{Phe} (Fig. 19.17a). The three anticodon bases are splayed out so that they bind to complementary pockets in the C-terminal domain of GlnRS (Fig. 19.17b–d). The C34 binding cleft can accommodate both the C34 of tRNA$^{Gln}_2$ and the 2-thio-U34 of tRNA$^{Gln}_1$, the two isoacceptors. However, the U35- and G36-binding pockets are highly specific for these two bases. The three pockets share very similar structural arrangements. A polypeptide segment of 5 or 6 residues contains at least one positively charged residue that makes a salt bridge with the adjacent phosphate, while the aliphatic part of its side chain packs against either the base or the ribose. Each base is recognized through direct hydrogen bonding with the side chains or backbone of the peptide (92).

3.2 tRNA^{Asp} complexed with aspartyl–tRNA synthetase

Aspartyl–tRNA synthetase (AspRS) from yeast is a class II aaRS. The yeast enzyme is an α_2 dimer of two 557 residue, 63 kDa monomers (101). It is a compact, diamond-shaped dimer of two elongated monomers. Each AspRS subunit consists of two

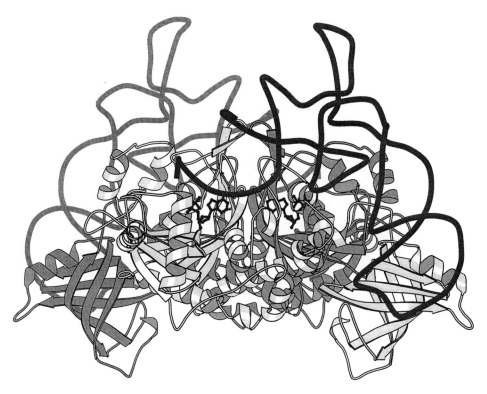

Fig. 19.18. Aspartyl:tRNA synthetase:tRNA^{Asp} complex from yeast. One monomer is in light grey and the other is in dark grey. The tRNA is drawn as a phosphate chain trace in a thick black and grey line. (From ref. 85, with the permission of Cold Spring Harbor Laboratory Press.)

modules connected by a hinge (Fig. 19.18). The N-terminal domain is a five-stranded β barrel (20) that has a topology similar to such unrelated proteins as staphyloccocal nuclease, verotoxin, and ribosomal protein S17. The motif is called the OB fold and is implicated in the binding of either oligonucleotides or oligosaccharides (102). The C-terminal module is the largest of the two domains and contains the catalytic site, which is composed of an antiparallel β sheet flanked by α helices, a topology charac-teristic of class II aaRSs. The N-terminal domain of one subunit interacts primarily with the C-terminal domain of the other. Most of the dimer interface is between the C-terminal core modules. Motif 1 and part of motif 2 form the dimer interface. Motifs 2 and 3 interact with the 3'-terminal CCA of tRNA^{Asp}, the amino acid, and ATP; the ATP adopts a bent conformation and binds in a manner characteristic of class II aaRSs (20,103). The ribose of the 3'-terminal adenosine is positioned in such a way that the 3'-OH can accept Asp from aspartyl-adenylate (104). The AspRS dimer binds tRNAs in a symmetrical fashion. Each monomer is complexed to a molecule of tRNA^{Asp} in what is considered a class II-characteristic mode. The acceptor arm of the tRNA interacts with the protein on the major groove side, and the variable loop side faces the protein. The buried surface has an area of 2500 Å^2, which represents 20% of the solvent-accessible surface of tRNA^{Asp} (103).

Since the structure of 'free', i.e. uncomplexed, tRNA^{Asp} is also known, a direct com-parison of tRNA^{Asp} in the two states is possible (Figs 19.6 and 19.19). Both limbs have undergone a protein-induced fit via a substantial conformational change; however, the change is most dramatic in the anticodon arm. The core region is virtually unchanged; all the interactions observed in the uncomplexed tRNA^{Asp} are maintained.

There are three regions in tRNA^{Asp} that form contacts with AspRS, of which each contains at least one putative identity element (103) (Fig. 19.19). They are located in the acceptor stem, the D stem, and the anticodon, while the bases that interact directly with the protein are in the acceptor stem and the anticodon loop (20,103,105). The three anticodon bases and residue G73 of the acceptor stem were found to be the main identity determinants, and base pair G10:C25 of the D stem is an accessory element. Yeast AspRS ignores the nature of the terminal base pair in the acceptor stem of tRNA^{Asp} (106), whereas the second base pair is a minor identity element in *E. coli* (107). Some residues enable the tRNA to adopt the conformation that facilitates its binding to AspRS but are not directly involved in protein–RNA interactions. They are G37 in the anticodon loop and base pair G10:C25 in the D stem; the latter stabilizes the conformation of the D stem near an important AspRS contact (20).

The acceptor stem of tRNA^{Asp} is positioned by motifs 1 and 2. The backbone of the motif 2 loop interacts with the base of G73 and the first base pair of the tRNA, which is undisrupted. The 3'-terminal GCCA of the tRNA is in a helical conforma-tion and interacts directly with the helices and loops of the protein that form part of the active site pocket (20). Two other loops contact C75 and A76. Most direct con-tacts involve the same subunit; only the phosphate of U1 interacts with Lys-293 of the other subunit (103).

The anticodon bases of tRNA^{Asp} are essential recognition elements for AspRS. The arm interacts with the N-terminal module on the major groove side and undergoes a protein-induced conformational change. This results in the bulging out

(a)

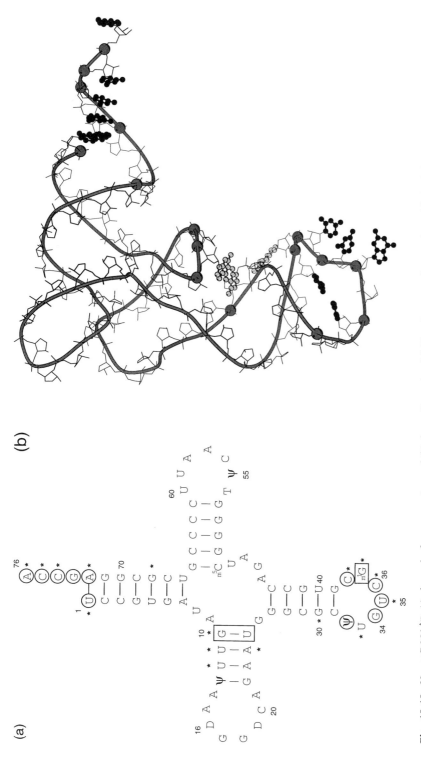

(b)

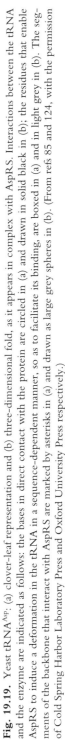

Fig. 19.19. Yeast tRNA^Asp: (a) clover-leaf representation and (b) three–dimensional fold, as it appears in complex with AspRS. Interactions between the tRNA and the enzyme are indicated as follows: the bases in direct contact with the protein are circled in (a) and drawn in solid black in (b); the residues that enable AspRS to induce a deformation in the tRNA in a sequence–dependent manner, so as to facilitate its binding, are boxed in (a) and in light grey in (b). The segments of the backbone that interact with AspRS are marked by asterisks in (a) and drawn as large grey spheres in (b). (From refs 85 and 124, with the permission of Cold Spring Harbor Laboratory Press and Oxford University Press respectively.)

of residue mG37, which shortens and bends the anticodon stem–loop; the residue forms an intramolecular hydrogen bond with the phosphate of residue 25 via its exocyclic amino group and thus stabilizes the conformation (Fig. 19.20a). The three anticodon bases are unstacked and spread out to maximize contacts with the protein; they are recognized by direct hydrogen bonding between the side chains or back-bone segments of the enzyme and the hydrogen bonding groups of the bases (20,103) (Fig. 19.20b–d).

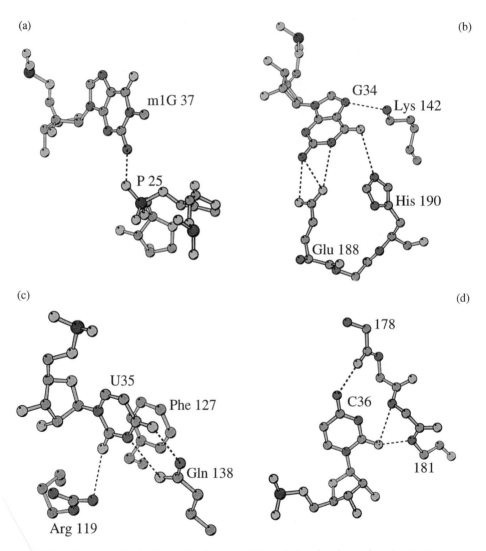

(a)

m1G 37

P 25

(b)

G34

Lys 142

His 190

Glu 188

(c)

U35

Phe 127

Gln 138

Arg 119

(d)

178

C36

181

19.20. (a) Intramolecular interaction between G37 and the phosphate of residue 25. Sequence-   interactions between AspRS and the anticodon loop of tRNA[Asp]: bases (b) 34, (c) 35, and (d) 36.   f. 164, with the permission of Cambridge University Press.)

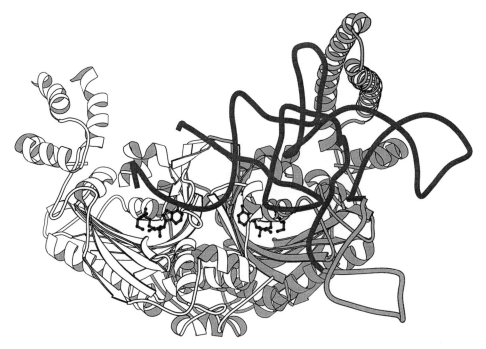

Fig. 19.21. Seryl:tRNA synthetase:tRNA^{Ser} complex from *T. thermophilus*. One monomer is in light grey and the other is in dark grey. The tRNA is drawn as a phosphate chain trace in thick black. The portion of the tRNA that was not seen in the electron density map and was modelled is shown as a light grey trace. (From ref. 85, with the permission of Cold Spring Harbor Laboratory Press.)

3.3 tRNA^{Ser} complexed with seryl–tRNA synthetase

Seryl–tRNA synthetase (SerRS) is a class II aaRS. The enzyme from *E. coli* is an α_2 dimer of 48 kDa subunits (108,109). Its counterpart from *T. thermophilus* is very similar (110). SerRS is a compact dimer with two helical appendages. Each monomer consists of two modules. The first 100 N-terminal residues form a 60 Å antiparallel coiled coil of two α helices. The core active site domain is made of a seven-stranded, mostly antiparallel, β sheet surrounded by α helices, a topology characteristic of class II aaRSs. All of the dimer interface is between the core modules; motif 1 and a portion of motif 2 constitute an important part of it (109). Motifs 2 and 3 form part of the active site platform, which interacts with ATP, seryl–adenylate (111), and with the acceptor end of tRNA^{Ser} (112) in a characteristic class II fashion. The tRNA binds across both subunits of the dimer; the major groove of the acceptor arm faces the active site domain of one subunit, whereas the variable arm and core of tRNA^{Ser} interact with the N-terminal appendage of the other subunit (21,112) (Fig. 19.21).

The clover-leaf secondary structure of tRNA^{Ser} from *T. thermophilus* appears to be very similar to that from *E. coli* described above (Fig. 19.22). In the core, many tertiary interactions are altered owing to the presence of the long variable arm, which removes the variable loop bases that are available for base triple formation in the augmented D helix of tRNA^{Phe}, tRNA^{Asp}, and tRNA^{Gln} (Fig. 19.23). As a result, the D stem base pairs, C10:G25 and C12:G23, do not participate in tertiary base-mediated interactions. The U8:A14:A21 interaction is analogous to the one observed in tRNA^{Asp}. Residue G9 interacts with a different pair, the mismatched G13:A22. The Levitt pair, G15:C48, is buttressed by the intra-D loop contact between G15 and D20A. The D loop lacks residue 17, but it possesses two additional residues between C20 and A21. The interactions between the D and T loops seen in other tRNAs so far, namely G18:Y55, G19:C56, and the base of G57 intercalating between G18 and G19, are preserved, as is the internal T loop strut T54:A58. The bases of U16 and C20 project into the solvent. Since tRNA^{Ser} comprises a large variable arm, it has introduced a feature that buttresses the arm and anchors it to the body of the tRNA. The base of G20B stacks upon the first base pair of the variable arm, A45:U47Q, and engages in van der Waals interactions with the edges of the bases of C48 and A21, while its sugar moiety interacts with C48. The usual mismatched base pair 26:44 is a twisted Watson–Crick A26:U44 pair; the base of residue 26 can also conceivably

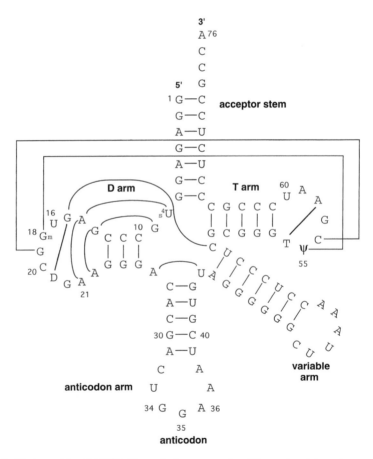

Fig. 19.22. *T. thermophilus* tRNA^{Ser}. Clover-leaf representation. The tertiary interactions are shown by connecting lines.

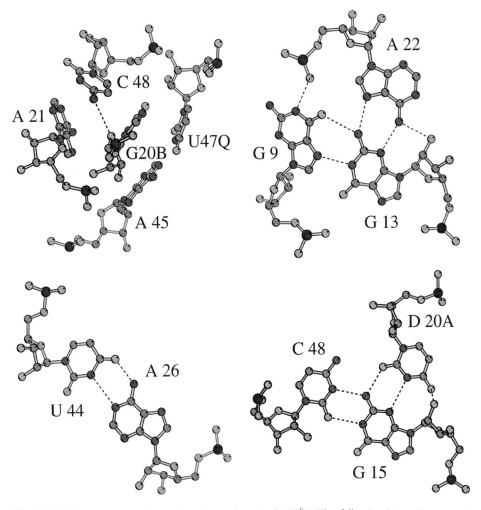

Fig. 19.23. Some representative tertiary interactions in tRNA^Ser. The following interactions are not shown, since they are very similar to the ones in tRNA^Phe: G18:Ψ55, G19:C56, and T54:A58; U8:A14:Q21 is the same as in tRNA^Asp. G10:C25 is a standard Watson–Crick base pair and does not participate in a base triple; therefore, it is not shown. Also shown is the stacking interaction of the variable arm on to G20B and the edges of bases A21 and C48.

interact with G43. The long variable loop inserts into the body at an angle, such that the entire molecule is not entirely flat.

The most striking feature of the seryl system is that SerRS does not interact with the anticodon of its cognate tRNA at all (21,113,114), since the tRNAs aminoacylated by the enzyme, five tRNA^Ser isoacceptors, and the tRNA^SeCys, possess a variety of anticodon sequences (109). There are four areas on the tRNA that it recognizes (Fig. 19.24): the 3′-end of the acceptor stem, the part of the anticodon stem at the base of the variable loop, part of the TΨC loop, and the base-paired portion of the long variable arm, as has been shown by chemical footprinting and enzymatic probes

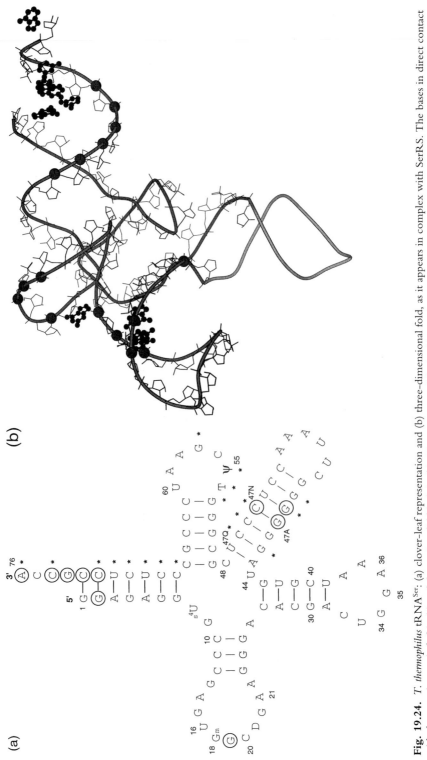

Fig. 19.24. *T. thermophilus* tRNA^Ser: (a) clover-leaf representation and (b) three-dimensional fold, as it appears in complex with SerRS. The bases in direct contact with the protein are circled in (a) and drawn in solid black in (b). The segments of the backbone that interact with SerRS are marked by asterisks in (a) and drawn as large grey spheres in (b). (From ref. 85, with the permission of Cold Spring Harbor Laboratory Press.)

of tRNA^{Ser} (113,114) and confirmed by X-ray crystallographic analysis of SerRS complexed with tRNA^{Ser} from *T. thermophilus* (21,112). Eight bases that are located in the acceptor and D arms were found to constitute the identity of tRNA^{Ser} (75,115), including the discriminator base, G73, and the first three base pairs of the acceptor stem. In addition, the length of the variable arm is an important factor (116).

The acceptor stem is recognized primarily by the motif 2 loop, which, in SerRS, is the longest in all the known class II aaRSs, such that it extends further down the major groove of the acceptor stem. It changes its conformation upon tRNA binding. Phe-262 forms van der Waals contacts with the hydrophobic edges of bases U68 and C69 and thus favours pyrimidines at those positions (Fig. 19.25a). Ser-261 interacts directly with G2 and possibly with C71; the backbone carbonyl oxygen of Phe-262 interacts with C71 as well (Fig. 19.25b). This is the most significant base-specific interaction. The discriminator base G73 is selected by Glu-258, which hydrogen bonds to the exocyclic 2-amino group. The protein interacts with the backbone from residue 66 to 71 (112) (Fig. 19.25a).

An important recognition feature of tRNA^{Ser} is the long variable arm, which interacts with the long, coiled-coil, N-terminal domain of the other monomer of SerRS. This protein module undergoes an induced change in its orientation and is stabilized upon tRNA binding (21); it also interacts with the T loop. There are very few contacts between the protein and nucleotide bases. One involves the tertiary base pair G19:C56; the peptide oxygen of Ala-555 hydrogen bonds to the exocyclic 2-amino group of G19 (Fig. 19.25c). The base pair stacks upon Pro-59 and Val-58. There is one notable interaction between the coiled-coil of SerRS and the minor groove of the variable arm of tRNA^{Ser}: Gln-545 interacts with both G47A and C47N (112) (Fig. 19.25d). SerRS makes many backbone interactions but few base-specific contacts with tRNA^{Ser}. It thus seems to recognize the unique shape rather than the sequence of its cognate tRNA (21,112,117).

3.4 Other aaRS systems and general principles

The modes of binding of tRNA to class I and class II aaRSs are mirror images of each other. The class I mode is characterized by the variable loop of the tRNA facing the solvent; the core domain of the enzyme interacts with the minor groove of the acceptor stem and the CCA terminus of the tRNA is distorted upon binding. The class II mode of binding is characterized by the variable loop of the tRNA facing the protein; the core domain of the enzyme interacts with the major groove of the acceptor helix. In addition, class I aaRSs are mostly monomeric, with the exception of TyrRS and TrpRS, while class II enzymes are mostly dimers.

The principles governing the acceptor arm binding can be extended to other aaRSs of the same class. In the case of class I aaRSs, the principles seen in the GlnRS:tRNA^{Gln}:ATP complex were shown to apply to two other aaRSs of known structure, MetRS (118,119) and GluRS (120). The active site domains of these two aaRSs are very similar to that of GlnRS, whereas the anticodon-binding domains are helical structures, unlike the double β barrel of GlnRS. TyrRS (121) and TrpRS (122) are both obligate dimers and are very similar to each other. Their active sites share the

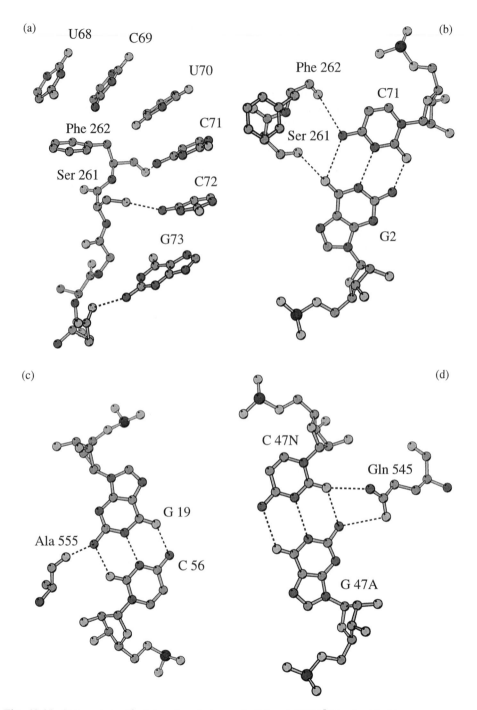

Fig. 19.25. Sequence-specific interactions between SerRS and tRNA^{Ser} in the (a), (b) acceptor stem, (c) D/T loop, and (d) variable loop.

Rossmann fold with the other three class I enzymes. A model has been proposed for tRNA^Tyr binding to TyrRS (123) that bears more resemblance to the class II mode of binding; however, the binding can conceivably occur in a class I fashion (124). In the case of class II aaRSs, the principles exemplified by AspRS and SerRS were shown to apply to other aaRSs of known structure. LysRS belongs to the same subgroup as AspRS and the structures of the two enzymes are very similar (125). Therefore, LysRS would be expected to bind its cognate tRNA in the same manner. This has been shown for the anticodon portion of the *in vitro* transcript of tRNA^Lys (126). GlyRS (127) and HisRS (128) share the active site fold with other class II aaRSs; they have a similar anticodon-binding C-terminal domain, which is different from that of AspRS. They were shown to bind their cognate tRNAs in a fashion similar to that of AspRS; in HisRS a simple superposition of the AspRS:tRNA^Asp complex brings the 3'-OH of the tRNA within 3 Å of the carbonyl carbon of histidyl–adenylate (128). PheRS is a dimer of class II dimers (129); in each of the two dimers one monomer is inactive. The PheRS tetramer thus binds two tRNA^Phe molecules.

Many aaRSs have been studied in complexes with amino acids, ATP, aminoacyl–adenylates, and analogues. Class I aaRSs bind ATP in an extended conformation, characteristic of other ATP-binding proteins, whereas class II aaRSs bind it in a new, bent conformation. These two distinct ATP conformations give rise to different angles of attack by the amino acid at the α-phosphate, which results in two distinct adenylate conformations. Furthermore, the tRNAs bind in different modes, positioning the 2'-OH of the terminal ribose in class I aaRSs and the 3'-OH of the terminal ribose in class II aaRSs in line to pick up the amino acid from the adenylate (86). In both classes, tRNA specificity results from idiosyncratic interaction with the cognate aaRS;

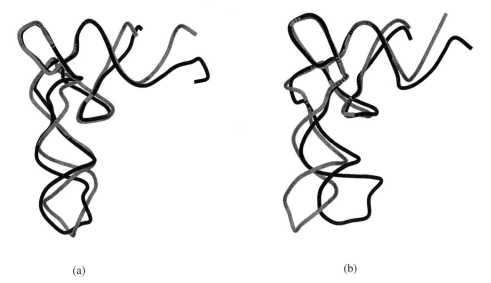

(a)   (b)

Fig. 19.26. Conformational changes in tRNA upon binding to its cognate aaRS. Superposition of (a) tRNA^Gln as bound to GlnRS (black) and free tRNA^Phe (light grey), and (b) tRNA^Asp as bound to AspRS (black) and uncomplexed (light grey). (From refs 85 and 124, with the permission of Cold Spring Harbor Laboratory Press and Oxford University Press, respectively.)

this includes direct base pair–protein contacts, backbone interactions, and sequence-dependent deformability.

Both tRNA^{Gln} and tRNA^{Asp} undergo dramatic conformational changes that are induced by their cognate aaRS to ensure complementary fit of their binding surfaces (Fig. 19.26). Both anticodon loops bend inwards, unstacking the anticodon bases so as to maximize their interactions with the protein. Other concomitant changes in the loop and stem aid in the process (19,20). In the acceptor stem of tRNA^{Gln}, the 3'-terminal CCA bends into the active site, which is facilitated by the melting of the U1:A72 base pair (19). In contrast, the acceptor arm and the CCA terminus of tRNA^{Asp} remains helical upon binding to the active site of AspRS (20,103). Conformational changes induced in tRNA^{Ser} by SerRS (21,112) are minimal, as the anticodon is not bound at all. The adjustments in the acceptor stem are probably of the same magnitude as seen in AspRS; these are difficult to ascertain since there is no reference structure of uncomplexed tRNA^{Ser}, which is different from tRNA^{Phe}.

In prokaryotes, such as *E. coli*, aminoacylated initiator tRNA$^{Met}_f$ (Met–tRNA$^{Met}_f$) is further modified before it enters the initiation stage of protein synthesis. This modification, the transfer of a formyl group from N-10 formyl-tetrahydrofolate to the amino group of the methionine esterified to the 3'-end of the tRNA is carried out by methionyl–tRNA$^{Met}_f$ formyltransferase. The enzyme is highly specific for initiator tRNA$^{Met}_f$ and discriminates against elongator tRNA$^{Met}_m$ (130). The key recognition element is the mismatched C1:A72 base pair in the acceptor stem (131). The protein has two domains, an N-teminal domain that contains a Rossmann fold and a β barrel C-terminal domain that resembles the anticodon-binding domain of AspRS. This domain and the flexible loop inserted in the N-terminal nucleotide-binding fold are implicated in tRNA binding. The N-terminal domain contains the active site. The modular organization of this enzyme is similar to that of aaRS (132).

4. tRNA in protein synthesis

4.1 Phe–tRNA^{Phe} bound to the elongation factor Tu

Once aminoacylated, a tRNA (aa–tRNA) is transported to the ribosome and positioned in the ribosomal A site by a protein known as the elongation factor (EF)-Tu in prokaryotes and eEF-1α in eukaryotes. This factor also ensures that the anticodon of the aa-tRNA recognizes the correct exposed codon of the messenger RNA. Its function is regulated by binding of GTP and GDP. It is active, i.e. capable of binding aa-tRNA, only when GTP is bound; once it positions the aa-tRNA in the A site of the elongating ribosome, the GTP is hydrolysed and the resulting EF-Tu:GDP is released from the ribosome. At this point its affinity for aa-tRNA is substantially reduced, and the factor needs to be recycled. Since GDP dissociates from EF-Tu at a very slow rate, another protein factor, EF-Ts, is needed for this recycling step. It accelerates the rate of exchange of GTP for GDP (133,134).

EF-Tu is a monomer of 405 residues with a molecular mass of 45 kDa. Its three-dimensional structure has been analysed in several functional states: as an inactive complex with GDP (135,136), as an active complex with the slowly hydrolysing GTP analogue GppNHp (137,138), as a ternary complex with Phe–tRNA^{Phe}, and as a GTP

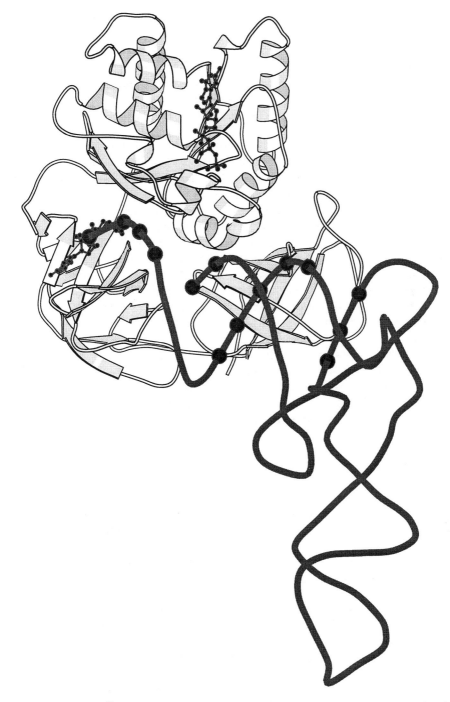

Fig. 19.27. Phe–tRNA^Phe complexed with the elongation factor Tu. The tRNA is drawn as a phosphate trace in solid black. The spheres indicate the portions of the backbone contacting the protein. There are no significant base interactions.

analogue EF-Tu (22) and a complex with the guanine nucleotide exchange factor EF-Ts (139).

EF-Tu consists of three domains (Fig. 19.27). Domain I is a β sheet of five parallel strands and one antiparallel strand surrounded on both sides by six major α helices. It contains a guanine nucleotide-binding site; hence it is also known as the G domain. The structure is similar to that of ras–p21 (135,137,138). Domains II and III are composed exclusively of antiparallel β sheets, each forming a β barrel. A large intramolecular movement occurs during the transition from the inactive GDP- to the active GTP-bound form (137,138). Domains II and III move as a rigid unit relative to domain I by a distance that exceeds one-third of the molecular diameter; the angle between the two units changes by about 90°. This results in a transition from a tight and mostly polar interface between domains I and II in the active form, to a substantial cavity separating the two domains in the inactive form.

The acceptor arm of aminoacylated tRNA binds to all three domains of EF-Tu, while the anticodon arm does not interact with the protein at all (Fig. 19.27). The aminoacylated CCA terminus is fixed in a narrow cleft between domains I and II (22), which is lined with several positively charged residues and is present only in the GTP-bound form (137,138). The amino acid-binding pocket can accommodate any one of the standard 20 amino acids. The protein interacts primarily with the sugar–phosphate backbone of the 5′-end of the acceptor helix also interacts with the junction of the three domains. The overall shape of the protein resembles that of the EF-G:GDP form. The tRNA itself changes its conformation only slightly upon binding to EF-Tu (22).

4.2 tRNA in the ribosome

The ultimate destination of aminoacylated tRNAs is the ribosome, where the amino acid is incorporated into a growing polypeptide according to the genetic message on the mRNA; the process occurs in three phases, initiation, elongation, and termination. The ribosome is a large RNA–protein complex that contains, in all species, a small and a large subunit (140). Each subunit is a complex between one or more large ribosomal RNA (rRNA) molecules and a number of relatively small, predominantly basic proteins. Ribosomes from prokaryotic organisms such as *E. coli* consist of 30S and 50S subunits, comprising 16S rRNA and 21 proteins, and 5S and 23S rRNA and 32 proteins, respectively (141). Eukaryotic ribosomes, such as those from yeast, are larger and are made of 40S and 60S subunits, which comprise 18S rRNA and about 30 proteins, and 5S, 5.8S, and 28S rRNA and about 40 proteins, respectively (142). Crystals of the particle and individual subunits have been available for some time; the determination of its three-dimensional structure by X-ray crystallography is a challenging long-term goal (143). Low resolution techniques such as electron microscopy, neutron scattering and diffraction, and chemical probing (144,145) have furnished much information on the structural organization of the ribosome and its subunits. Neutron scattering experiments have yielded a map of the relative locations of all ribosomal proteins in the *E. coli* ribosome (146,147). The structure of the ribosome and its interaction with its substrates, mRNA and tRNA, have been probed extensively by biochemical methods (134,148,149). Recently, the overall structure of the *E. coli* particle has been reconstructed from cryoelectron microscopic images at 23– 25 Å resolution. In the structure, the small subunit possesses a channel and the large subunit a bifurcating tunnel. The channel may accommodate the incoming mRNA, while the tunnel may serve as the exit pathway for the nascent peptide (150,151).

In addition to information on the overall structure of the ribosome, cryoelectron microscopy has pin-pointed three tRNA molecules bound to the A, P, and E sites of the ribosome (23) in what was an average structure and does not represent any physiological state of the ribosome, since only two tRNA sites are occupied at a time. However, the arrangement of tRNAs was determined at 20 Å resolution in two functional states of elongation, before and after translocation (24). Since there were no gross overall conformational changes between the two states at this resolution, which were also isomorphous to the vacant state, difference electron densities between the two states and relative to the vacant particles revealed the differential occupancies of the three sites and some other morphological changes. In the pre-translocational ribosomes, densities were observed corresponding to tRNAs in the A and P sites, while occupation by tRNA of the P and E sites was seen in the post-translocational state. As the P site is occupied in both states, it was not seen in a difference map between the two states. The A site was shown very clearly, while the density corresponding to the E site was more diffuse, probably reflecting the larger conformational heterogeneity of the site. In both the A and P site tRNAs, a thin line of density corresponding to the 3'-CCA terminus points towards the putative peptidyl transferase region of the 50 S subunit, while the regions corresponding to the anticodon arms lie in the neck of the 30 S subunit, the putative decoding region (24).

5. Perspectives

Transfer RNA is structurally and functionally a very versatile molecule. It can interact with many other molecules and serve as a substrate for many enzymes. The overall general features are used by enzymes such as tRNA precursor 5'- and 3'-processing nucleases, some modification enzymes, proteins such as translation factors, and ribonucleoprotein particles such as the ribosome. In addition, tRNAs posses certain distinguishing features that constitute their identity; these are recognized, within their common context, by specialized enzymes such as aminoacyl–tRNA synthetases, Met–tRNA$^{Met}_f$ formyltransferase, Glu–tRNA^{Gln} and Asp–tRNA^{Asn} amidotransferases (152) and many modification enzymes. All these general and specific encounters between tRNAs and associated molecules constitute an extensive structural and functional puzzle, only a few pieces of which we have begun to fathom, as we have seen in this chapter.

Many larger RNAs, such as those from some plant viruses and virusoids, are capable of structurally and functionally mimicking the versatility of tRNA (153). They do so at their 3'-termini, since these ends can be processed by RNAase P and tRNA nucleotidyl transferase, undergo aminoacylation, and interact with elongation factors. However, they do not participate in protein synthesis. Their primary role is to aid in viral replication. They may have co-evolved with tRNAs and associated molecules from common ancestors, as suggested by the genomic tag hypothesis of Weiner and Maizels (154).

Although the principal role of tRNAs in the cell is to take part in the message-directed protein synthesis, they are not confined to that purpose alone. They can participate in other cellular processes, such as priming reverse transcription (155) and regulation of gene expression (156), which reflect the roles played by the tRNA-like viral RNAs. In addition, they are involved in various other metabolic pathways, such as porphyrin biosynthesis (157).

The simple and sophisticated structure of tRNA, with its overall L-shape and two functional ends, one for mRNA codon reading and the other for amino acid attachment and transfer, makes it an adaptor molecule *par excellence*. It also makes it very adaptable to the many molecules it meets and associates with during its cellular career.

Acknowledgements

We thank S. Cusack for the latest atomic coordinates of the SerRS:tRNA[Ser] complex from *T. thermophilus*. All figures were made with program MOLSCRIPT (165).

References

1. Hoagland, M.B., Zamecnik, P.C. and Stephenson, M.L. (1957) *Biochim. Biophys. Acta* **24**, 215.
2. Holley, R.W., Apgar, J., Everett, G.A., Madison, J.T., Marquisse, M., Merrill, S.H., Penwick, J.R. and Zamir, R. (1965) *Science* **147**, 1462.
3. Sprinzl, M., Steegborn, C., Hübel, F. and Steinberg, S. (1996) *Nucl. Acids Res.* **24**, 68.
4. Crick, F.H.C. (1966) *J. Mol. Biol.* **19**, 548.
5. Dunn, D.B. (1959) *Biochim. Biophys. Acta* **34**, 286.
6. Smith, J.D. and Dunn, D.B. (1959) *Biochem. J.* **72**, 294.
7. Bernhardt, D. and Darnell, Jr, J.E., (1969) *J. Mol. Biol.* **42**, 43.
8. Altman, S. and Smith, J.D. (1971) *Nature New Biol.* 233, 35.
9. Altman, S., Kirsebom, L. and Talbot, S. (1995) in *tRNA: Structure, Biosynthesis, and Function*, (Söll, D. and RajBhandary, U., eds), p. 67. American Society for Microbiology, Washington, DC.
10. Deutscher, M.P. (1995) in *tRNA: Structure, Biosynthesis, and Function*, (Söll, D. and RajBhandary, U., eds), p. 51. American Society for Microbiology, Washington, DC.
11. Fraser, T.H. and Rich, A. (1975) *Proc. Natl. Acad. Sci. USA* **72**, 3044.
12. Sprinzl, M. and Cramer, M. (1975) *Proc. Natl. Acad. Sci. USA* **72**, 3049.
13. Eriani, G., Delarue, M., Poch, O., Gangloff, J. and Moras, D. (1990) *Nature* **347**, 203.
14. Kim, S.H., Suddath, F.L., Quigley, G.J., McPherson, A., Sussman, J.L., Wang, A.H.J., Seeman, N.C. and Rich, A. (1974) *Science* **185**, 435.
15. Robertus, J.D., Ladner, J.E., Finch, J.T., Rhodes, D., Brown, R.S., Clark, B.F.C. and Klug, A. (1974) *Nature* **250**, 546.
16. Moras, D., Comarmond, M.B., Fischer, J., Weiss, R., Thierry, J.C., Ebel, J.P. and Giegé, R. (1980) *Nature* **288**, 669.
17. Woo, N.H., Roe, B.A. and Rich, A. (1980) *Nature* **286**, 346.
18. Basavappa, R. and Sigler, P.B. (1991) *EMBO J.* **10**, 3105.
19. Rould, M.A., Perona, J.J., Söll, D. and Steitz, T.A. (1989) *Science* **246**, 1135.
20. Ruff, M., Krishnaswamy, S., Boeglin, M., Poterszman, A., Mitschler, A., Podjarny, A., Rees, B., Thierry, J.-C. and Moras, D. (1991) *Science* **252**, 1682.

21. Biou, V., Yaremchuk, A., Tukalo, M. and Cusack, S. (1994) *Science* **263**, 1404.

22. Nissen, P., Kjeldgaard, M., Thirup, S., Polekhina, G., Reshetnikova, L., Clark, B.F.C. and Nyborg, J. (1995) *Science* **270**, 1464.

23. Agrawal, R.K., Penczek, P., Grassucci, R.A., Li, Y., Leith, A., Nierhaus, K.H. and Frank, J. (1996) *Science* **271**, 1000.

24. Stark, H., Orlova, E.V., Rinke-Appel, J., Jünke, N., Mueller, F., Rodnina, M., Wintermeyer, W., Brimacombe, R. and van Heel, M. (1997) *Cell* **88**, 19.

25. Inokuchi, H. and Yamao, F. (1995) in *tRNA: Structure, Biosynthesis, and Function*, (Söll, D. and RajBhandary, U., eds), p. 17. American Society for Microbiology, Washington, DC.

26. Sprague, K.U. (1995) in *tRNA: Structure, Biosynthesis, and Function*, (Söll, D. and RajBhandary, U., eds), p. 31.   American Society for Microbiology, Washington, DC.

27. Sigler, P.B. (1975) *Annu. Rev. Biophys. Bioeng.* **4**, 477.

28. Dirheimer, G., Keith, G., Dumas, P. and Westhof, E. (1995) in *tRNA: Structure, Biosynthesis, and Function*, (Söll, D. and RajBhandary, U., eds), p. 93. American Society for Microbiology, Washington, DC.

29. Suddath, F.L., Quigley, G.J., McPherson, A., Sneden, D., Kim, J.J., Kim, S.H. and Rich, A. (1974) *Nature* **248**, 20.

30. Quigley, G.J., Wang, A., Seeman, N.C., Suddath, F.L., Rich, A., Sussman, J.L. and Kim, S.H. (1975) *Proc. Natl. Acad. Sci. USA* **72**, 4866.

31. Quigley, G.J. and Rich, A. (1976) *Science* **194**, 796.

32. Ladner, J.E., Jack, A., Robertus, J.D., Brown, R.S., Rhodes, D., Clark, B.F.C. and Klug, A. (1975) *Proc. Natl. Acad. Sci. USA* **72**, 4414.

33. Jack, A., Ladner, J.E. and Klug, A. (1976) *J. Mol. Biol.* **108**, 619.

34. Sussmann, J.L., Holbrook, S.R., Warrant, R.W., Church, G.M. and Kim, S.H. (1978) *J. Mol. Biol.* **123**, 607.

35. Rich, A. and RajBhandary, U.L. (1976) *Annu. Rev. Biochem.* **45**, 805.

36. Rich, A. (1977) *Acc. Chem. Res.* **10**, 388.

37. Kim, S.-H. (1978) *Adv. Enzymol.* **46**, 279.

38. Levitt, M. (1969) *Nature* **224**, 759.

39. Westhof, E., Dumas, P. and Moras, D. (1985) *J. Mol. Biol.* **184**, 119.

40. Moras, D., Dock, A.C., Dumas, P., Westhof, E., Romby, P., Ebel, J.P. and Giegé, R. (1986) *Proc. Natl. Acad. Sci. USA* **83**, 932.

41. Schevitz, R., Podjarny, A.D., Krishnanmachari, N., Hughes, J.J., Sigler, P.B. and Sussman, J.L. (1979) *Nature* **278**, 188.

42. Seong, B.L. and RajBhandary, U.L. (1987) *Proc. Natl. Acad. Sci. USA* **84**, 334.

43. Dock-Bregeon, A.C., Westhof, E., Giegé, R. and Moras, D. (1989) *J. Mol. Biol.* **206**, 707.

44. Romby, P., Moras, D., Dumas, P., Ebel, J.P. and Giegé, R. (1987) *J. Mol. Biol.* **195**, 193.

45. Martin, N.C. (1995) in *tRNA: Structure, Biosynthesis, and Function*, (Söll, D. and RajBhandary, U., eds), p. 127. American Society for Microbiology, Washington, DC.

46. Watanabe, K. and Osawa, S. (1995) in *tRNA: Structure, Biosynthesis, and Function*, (Söll, D. and RajBhandary, U., eds), p. 225. American Society for Microbiology, Washington, DC.

47. Björk, G.R. (1995) in *tRNA: Structure, Biosynthesis, and Function*, (Söll, D. and RajBhandary, U., eds), p. 165. American Society for Microbiology, Washington, DC.

48. Yokoyama, S. and Nishimura, S. (1995) in *tRNA: Structure, Biosynthesis, and Function*, (Söll, D. and RajBhandary, U., eds), p. 207. American Society for Microbiology, Washington, DC.

49. Agris, P.F. (1996) *Progr. Nucl. Acid Res. Mol. Biol.* **53**, 79.

50. Romier, C., Reuter, K., Suck, D. and Ficner, R. (1996) *EMBO J.* **15**, 2850.
51. Hall, K.B., Sampson, J.R., Uhlenbeck, O.C. and Redfield, A.G. (1989) *Biochemistry* **28**, 5794.
52. Chu, W.C. and Horowitz, J. (1989) *Nucl. Acids Res.* **17**, 7241.
53. Arnez, J.G. and Steitz, T.A. (1994) *Biochemistry* **33**, 7560.
54. Sampson, J.R. and Uhlenbeck, O.C. (1988) *Proc. Natl. Acad. Sci. USA* **85**, 1033.
55. Perret, V., Garcia, A., Puglisi, J., Grosjean, H., Ebel, J.P., Florentz, C. and Giegé, R. (1990) *Biochimie* **72**, 735.
56. Derrick, W.B. and Horowitz, J. (1993) *Nucl. Acids Res.* **21**, 4948.
57. Beresten, S., Jahn, M. and Söll, D. (1992) *Nucl. Acids Res.* **20**, 1523.
58. Emerson, J. and Sundaralingam, M. (1980) *Acta Cryst.* **B36**, 537.
59. Cadet, J., Ducolumb, R. and Hruska, F.E. (1980) *Biochim. Biophys. Acta* **563**, 206.
60. Agris, P.F., Sierzputowska-Gracz, H., Smith, W., Malkiewicz, A., Sochacka, E. and Nawrot, B. (1992) *J. Am. Chem. Soc.* **114**, 2652.
61. Chen, Y., Sierzputowska-Gracz, H., Guenther, R., Everett, K. and Agris, P.F. (1993) *Biochemistry* **32**, 10249.
62. Agris, P.F., Malkiewicz, A., Brown, S., Kraszewski, A., Nawrot, B., Sochacka, E., Everett, K. and Guenther, G. (1995) *Biochimie* **77**, 125.
63. Himeno, H., Hasegawa, T., Ueda, T., Watanabe, K., Miura, K. and Shimizu, M. (1989) *Nucl. Acids Res.* **17**, 7855.
64. Jahn, M., Rogers, M.J. and Söll, D. (1991) *Nature* **352**, 258.
65. Sampson, J.R., Behlen, L.S., DiRenzo, A.B. and Uhlenbeck, O.C. (1992) *Biochemistry* **31**, 4164.
66. Sylvers, L.A., Rogers, K.C., Shimizu, M., Ohtsuka, E. and Söll, D. (1993) *Biochemistry* **32**, 3836.
67. Rogers, K.C., Crescenzo, A.T. and Söll, D. (1995) *Biochimie* **77**, 66.
68. Tamura, K., Himeno, H., Asahara, H., Hasegawa, T. and Shimizu, M. (1992) *Nucl. Acids Res.* **20**, 2335.
69. Muramatsu, T., Nishikawa, K., Nemoto, F., Kuchino, Y., Nishimura, S., Miyazawa, T. and Yokoyama, S. (1988) *Nature* **336**, 179.
70. Perret, V., Garcia, A., Grosjean, H., Ebel, J.-P., Florentz, C. and Giegé, R. (1990) *Nature* **344**, 787.
71. Desgrès, J., Keith, G., Kuo, K.C. and Gehrke, C. (1989) *Nucl. Acids Res.* **17**, 868.
72. Kiesewetter, S., Ott, G. and Sprinzl, M. (1990) *Nucl. Acids Res.* **18**, 4677.
73. Förster, C., Chakraburtty, K. and Sprinzl, M. (1993) *Nucl. Acids Res.* **21**, 5679.
74. Koval'chuke, O.V., Potapov, A.P., El'skaya, A.V., Potapov, V.K., Krinetskaya, N.F., Dolinnaya, N.G. and Shabarova, Z.A. (1991) *Nucl. Acids Res.* **19**, 4199.
75. Normanly, J., Ogden, R.C., Horvath, S.J. and Abelson, J. (1986) *Nature* **321**, 213.
76. McClain, W.H. and Nicholas, H.B.J. (1987) *J. Mol. Biol.* **194**, 635.
77. Schulman, L.H. (1991) *Progr. Nucl. Acid Res. Mol. Biol.* **41**, 23.
78. Schulman, L.H. and Pelka, H. (1988) *Science* **242**, 765.
79. Normanly, J. and Abelson, J. (1989) *Annu. Rev. Biochem.* **58**, 1029.
80. Perret, V., Florentz, C., Puglisi, J.D. and Giegé, R. (1992) *J. Mol. Biol.* **226**, 323.
81. Schimmel, P. and Söll, D. (1979) *Annu. Rev. Biochem.* **48**, 601.
82. Yarus, M. (1972) *Nature New Biol.* **239**, 106.
83. Carter, Jr, C.W. (1993) *Annu. Rev. Biochem.* **62**, 715.
84. Meinnel, T., Mechulam, Y. and Blanquet, S. (1995) in *tRNA: Structure, Biosynthesis, and Function*, (Söll, D. and RajBhandary, U., eds), p. 251. American Society for Microbiology, Washington, DC..

85. Arnez, J.G. and Moras, D. (1998) *RNA Structure and Function*, (Grunberg-Manago, M. and Symons, R.W., eds), p. 465 Cold Spring Harbor Laboratory Press, Cold Spring Harbor.

86. Arnez, J.G. and Moras, D. (1997) *TIBS* **22**, 211.

87. Moras, D. (1992) *TIBS* **17**, 159.

88. Delarue, M. and Moras, D. (1993) *BioEssays* **15**, 1.

89. Jasin, M., Regan, L. and Schimmel, P. (1983) *Nature* **306**, 441.

90. Hoben, P., Royal, N., Cheung, A., Yamao, F., Biemann, K. and Söll, D. (1982) *J. Biol. Chem.* **257**, 11644.

91. Perona, J.J., Rould, M.A. and Steitz, T.A. (1993) *Biochemistry* **32**, 8758.

92. Rould, M.A., Perona, J.J. and Steitz, T.A. (1991) *Nature* **352**, 213.

93. Hayase, Y., Jahn, M., Rogers, M.J., Sylvers, L.A., Koizumi, M., Inoue, H., Ohtsuka, E. and Söll, D. (1992) *EMBO J.* **11**, 4159.

94. Ghysen, A. and Celis, J.E. (1974) *J. Mol. Biol.* **83**, 333.

95. Knowlton, R.G., Söll, L. and Yarus, M. (1980) *J. Mol. Biol.* **139**, 705.

96. Rogers, M.J. and Söll, D. (1988) *Proc. Natl. Acad. Sci. USA* **85**, 6627.

97. Arnez, J.G. and Steitz, T.A. (1996) *Biochemistry* **35**, 14725.

98. Inokuchi, H., Hoben, P., Yamao, F., Ozeki, H. and Söll, D. (1984) *Proc. Natl. Acad. Sci. USA* **81**, 5076.

99. Perona, J.J., Swanson, R.N., Rould, M.A., Steitz, T.A. and Söll, D. (1989) *Science* **246**, 1152.

100. Seno, T., Agris, P.F. and Söll, D. (1974) *Biochim. Biophys. Acta* **349**, 328.

101. Amiri, I., Mejdoub, H., Hounwanou, N., Boulanger, Y. and Reinbolt, J. (1985) *Biochimie* **67**, 607.

102. Murzin, A.G. (1993) *EMBO J.* **12**, 861.

103. Cavarelli, J., Rees, B., Ruff, M., Thierry, J.C. and Moras, D. (1993) *Nature* **362**, 181.

104. Cavarelli, J., Eriani, G., Rees, B., Ruff, M., Boeglin, M., Mitschler, A., Martin, F., Gangloff, J., Thierry, J.C. and Moras, D. (1994) *EMBO J.* **13**, 327.

105. Rudinger, J., Puglisi, J.D., Pütz, J., Schatz, D., Eckstein, F., Florentz, C. and Giegé, R. (1992) *Proc. Natl. Acad. Sci. USA* **89**, 5882.

106. Pütz, J., Puglisi, J.D., Florentz, C. and Giegé, R. (1991) *Science* **252**, 1696.

107. Nameki, N., Tamura, K., Himeno, H., Asahara, H., Hasegawa, T. and Shimizu, M. (1992) *Biochem. Biophys. Res. Commun.* **189**, 856.

108. Härtlein, M., Madern, D. and Leberman, R. (1987) *Nucl. Acids Res.* **15**, 1005.

109. Cusack, S., Berthet-Colominas, C., Härtlein, M., Nassar, N. and Leberman, R. (1990) *Nature* **347**, 249.

110. Fujinaga, M., Berthet, C.C., Yaremchuk, A.D., Tukalo, M.A. and Cusack, S. (1993) *J. Mol. Biol.* **234**, 222.

111. Belrhali, H., Yaremchuk, A., Tukalo, M., Berthet-Colominas, C., Rasmussen, B., Bösecke, P., Diat, O. and Cusack, S. (1995) *Structure* **3**, 341.

112. Cusack, S., Yaremchuk, A. and Tukalo, M. (1996) *EMBO J.* **15**, 2834.

113. Dock-Bregeon, A.C., Garcia, A., Giegé, R. and Moras, D. (1990) *Eur. J. Biochem.* **188**, 283.

114. Schatz, D., Leberman, R. and Eckstein, F. (1991) *Proc. Natl. Acad. Sci. USA* **88**, 6132.

115. Normanly, J., Ollick, T. and Abelson, J. (1992) *Proc. Natl. Acad. Sci. USA* **89**, 5680.

116. Himeno, H., Hasegawa, T., Ueda, T., Watanabe, K. and Shimizu, M. (1990) *Nucl. Acids Res.* **18**, 6815.

117. Asahara, H., Himeno, H., Tamura, K., Nameki, N., Hasegawa, T. and Shimizu, M. (1994) *J. Mol. Biol.* **236**, 738.

118. Brunie, S., Zelwer, C. and Risler, J.L. (1990) *J. Mol. Biol.* **216**, 411.

119. Perona, J.J., Rould, M.A., Steitz, T.A., Risler, J.L., Zelwer, C. and Brunie, S. (1991) *Proc. Natl. Acad. Sci. USA* **88**, 2903.
120. Nureki, O., Vassylyev, D.G., Katayanagi, K., Shimizu, T., Sekine, S., Kigawa, T., Miyazawa, T., Yokoyama, S. and Morikawa, K. (1995) *Science* **267**, 1958.
121. Brick, P., Bhat, T.N. and Blow, D.M. (1989) *J. Mol. Biol.* **208**, 83.
122. Doublié, S., Bricogne, G., Gilmore, C. and Carter, C.W. (1995) *Structure* **3**, 17.
123. Bedouelle, H. and Winter, G. (1986) *Nature* **320**, 371.
124. Arnez, J.G. and Moras, D. (1994) *RNA–Protein Interactions*, (Nagai, K. and Mattaj, I., eds), p. 52. Oxford University Press, Oxford.
125. Onesti, S., Miller, A.D. and Brick, P. (1995) *Structure* **3**, 163.
126. Cusack, S., Yaremchuk, A. and Tukalo, M. (1996) *EMBO J.* **15**, 6321.
127. Logan, D.T., Mazauric, M.H., Kern, D. and Moras, D. (1995) *EMBO J.* **14**, 4156.
128. Arnez, J.G., Harris, D.C., Mitschler, A., Rees, B., Francklyn, C.S. and Moras, D. (1995) *EMBO J.* **14**, 4143.
129. Mosyak, L., Reshetnikova, L., Goldgur, Y., Delarue, M. and Safro, M.G. (1995) *Nature Struct. Biol.* **2**, 537.
130. Mangroo, D. and RajBhandary, U.L. (1995) *J. Biol. Chem.* **270**, 12203.
131. Guillon, J.M., Meinnel, T., Mechulam, Y., Lazennec, C., Blanquet, S. and Fayat, S. (1992) *J. Mol. Biol.* **224**, 359.
132. Schmitt, E., Blanquet, S. and Mechulam, Y. (1996) *EMBO J.* **15**, 4749.
133. Miller, D.L. and Weissbach, H. (1977) *Molecular Mechanisms of Protein Biosynthesis*, (Weissbach, H. and Petska, S., eds), p. 323. Academic Press, New York.
134. Moazed, D. and Noller, H.F. (1989) *Nature* **342**, 142.
135. Jurnak, F. (1985) *Science* **230**, 32.
136. Kjeldgaard, M. and Nyborg, J. (1992) *J. Mol. Biol.* **223**, 721.
137. Berchtold, H., Reshetnikova, L., Reiser, C.O.A., Schirmer, N.K., Sprinzl, M. and Hilgenfeld, R. (1993) *Nature* **365**, 126.
138. Kjeldgaard, M., Nissen, P., Thirup, S. and Nyborg, J. (1993) *Structure* **1**, 35.
139. Kawashima, T., Berthet-Colominas, C., Wulff, M., Cusack, S. and Leberman, R. (1996) *Nature* **379**, 511.
140. Lake, J.A. (1981) *Sci. Am.* **245**, 84.
141. Wittmann, H.G. (1982) *Annu. Rev. Biochem.* **51**, 155.
142. Kozak, M. (1983) *Microbiol. Rev.* **47**, 1.
143. Yonath, A. and Wittmann, H.G. (1989) *TIBS* **14**, 329.
144. Moore, P.B. (1988) *Nature* **331**, 223.
145. Lake, J.A. (1985) *Annu. Rev. Biochem.* **54**, 507.
146. Capel, M.S., Engelman, D.M., Freeborn, B.R., Kjeldgaard, M., Langer, J.A., Ramakrishnan, V., Schindler, D.G., Schneider, D.K., Schoenborn, B.P., Sillers, I.-Y., Yabuki, S. and Moore, P. (1987) *Science* **238**, 1403.
147. Walleczek, J., Schüler, D., Stöffler-Meilicke, M., Brimacombe, R. and Stöffler, G. (1988) *EMBO J.* **7**, 3571.
148. Vonahsen, U. and Noller, H.F. (1995) *Science* **267**, 234.
149. Samaha, R.R., Green, R. and Noller, H.F. (1995) *Nature* **377**, 309.
150. Frank, J., Zhu, J., Penczek, P., Li, Y., Srivastava, S., Verschoor, A., Radermacher, M., Grassucci, R., Lata, R.K. and Agrawal, R.K. (1995) *Nature* **376**, 441.
151. Stark, H., Mueller, F., Orlova, E.V., Schatz, M., Dube, P., Erdemir, T., Zemlin, F., Brimacombe, R. and van Heel, M. (1995) *Structure* **3**, 815.
152. Ibba, M., Curnow, A.W. and Söll, D. (1997) *TIBS* **22**, 39.

153. Florentz, C. and Giegé, R. (1995) in *tRNA: Structure, Biosynthesis, and Function*, (Söll, D. and RajBhandary, U., eds), p. 2141. American Society for Microbiology, Washington, DC.

154. Weiner, A.M. and Maizels, N. (1987) *Proc. Natl. Acad. Sci. USA* **84**, 7383.

155. Wilson, S.H. and Abbotts, J. (1992) *Transfer RNA in Protein Synthesis*, (Hatfield, D.L., Lee, B.J. and Pirtle, R.M., eds), p. 1. CRC Press, Boca Raton.

156. Graffe, M., Dondon, J., Caillet, J., Romby, P., Ehresmann, C., Ehresmann, B. and Springer, M. (1992) *Science* **255**, 994.

157. Schön, A., Krupp, G., Gough, S., Berry-Lowe, S., Kannangara, C.G. and Söll, D. (1986) *Nature* **322**, 281.

158. Poterszman, A., Delarue, M., Thierry, J.-C. and Moras, D. (1994) *J. Mol. Biol.* **244**, 158.

159. Arnez, J.G., Augustine, J.G., Moras, D. and Francklyn, C.S. (1997) *Proc. Natl. Acad. Sci. USA* **94**, 7144.

160. Åberg, A., Yaremchuk, A., Tukalo, M., Rasmussen, B. and Cusack, S. (1997) *Biochemistry* **36**, 3084.

161. Czworkowski, J., Wang, J., Steitz, T.A. and Moore, P.B. (1994) *EMBO J.* **13**, 3661.

162. Ævarsson, A., Brazhnikov, E., Garber, M., Zheltonosova, J., Chirgadze, Y., Al, K.S., Svensson, L.A. and Liljas, A. (1994) *EMBO J.* **13**, 3669.

163. Moras, D. (1989) *Nucleic Acids:Crystallographic and Structural Data II*, (Saenger, W., ed.), p. 1. Springer-Verlag, Berlin, Heidelberg, New York.

164. Arnez, J.G. and Cavarelli, J. (1997) *Q. Rev. Biophys.* **30**, 195.

165. Kranlis, P.J. (1991) *J. Appl. Crystallogr.* **24**, 946.

Index